AF802818

Pediatric and Adolescent Musculoskeletal MRI

Pediatric and Adolescent Musculoskeletal MRI

A Case-Based Approach

J. Herman Kan, MD
Assistant Professor of Radiology and Radiological Sciences and Pediatrics, Vanderbilt University School of Medicine, Nashville, Tennessee, USA
Chief, Pediatric Musculoskeletal Section, Diagnostic Imaging, Monroe Carell Jr. Children's Hospital at Vanderbilt, Nashville, Tennessee, USA

Paul K. Kleinman, MD
Professor of Radiology, Harvard Medical School, Boston, Massachusetts, USA
Director, Division of Musculoskeletal Imaging, Department of Radiology, Children's Hospital Boston, Boston, Massachusetts, USA

Orthopedic Consultants

Mark C. Gebhardt, MD
Frederick W. and Jane M. Ilfeld Professor of Orthopedic Surgery, Harvard Medical School, Boston, Massachusetts, USA
Chief, Department of Orthopedic Surgery, Beth Israel Deaconess Medical Center, Boston, Massachusetts, USA
Associate in Orthopaedic Surgery, Division of Orthopaedic Surgery, Department of Orthopaedics, Children's Hospital Boston, Boston Massachusetts, USA

Mininder S. Kocher, MD, MPH
Associate Professor of Orthopedic Surgery, Harvard Medical School, Boston, Massachusetts, USA
Associate Director, Division of Sports Medicine, Department of Orthopaedics, Children's Hospital Boston, Boston, Massachusetts, USA

Springer

J. Herman Kan, MD
Assistant Professor of Radiology and Radiological Sciences and Pediatrics
Vanderbilt University School of Medicine
Nashville, TN, USA
and
Chief, Pediatric Musculoskeletal Section
Diagnostic Imaging
Monroe Carell Jr. Children's Hospital at Vanderbilt
Nashville, TN
USA

Paul K. Kleinman, MD
Professor of Radiology
Harvard Medical School
Boston, MA
USA
and
Director, Division of Musculoskeletal Imaging
Department of Radiology
Children's Hospital Boston
Boston, MA
USA

ISBN 978-1-4419-2218-2
eISBN: 10: 0-387-38325-5
eISBN: 13: 978-0-387-38325-5

9 8 7 6 5 4 3 2 1

springer.com

Foreword

The number of new radiology texts that appears each year continues to grow, and each must compete with other available works to be successful. With this in mind, Drs. Herman Kan and Paul Kleinman have authored a book, *Pediatric and Adolescent Musculoskeletal MRI: A Case-Based Approach,* that will clearly prove a very useful addition to the literature. What sets it apart from other texts is its organization, its clinical utility, and, above all else, its readability.

The case-based organization is very user friendly. More than 100 cases dealing with the most important musculoskeletal conditions that affect children and adolescents are presented. Osseous, articular, and soft tissue disorders are covered, including neoplastic, traumatic, infectious, dysplastic, and vascular conditions. In fact, a survey of the cases indicates that virtually all of the important musculoskeletal disorders affecting the immature skeleton are included, such that upon reviewing the entire text, the reader is exposed to the imaging findings of all disorders that he or she probably will encounter in clinical practice.

Each case stands alone as a pragmatic review of the condition being covered. Initially, a history along with one or more appropriate images, including MRIs, is presented. The reader is able to survey these images and arrive at a diagnosis, thus testing his or her diagnostic acumen. Subsequently, the imaging findings are summarized and the correct diagnosis is given. This section is then followed by one or more questions about the entity being presented, followed by a focused discussion (appropriate in length) of this condition and any others that might have been considered. Finally, a section dealing with an orthopedic perspective of the entity, a list of what the clinician needs to know, and answers to the aforementioned questions follow. Completing the case are images of additional examples of the disorder and related conditions, a discussion of the findings and of pitfalls and pearls, and appropriate references. In this fashion, the organization of each case is superb and without fault, and the illustrations are of excellent quality with well-placed arrows.

Currently available texts dealing with musculoskeletal MR imaging are confined to a discussion of disorders affecting the adult skeleton. Thus, a book dealing with advanced imaging of those disorders that affect the immature skeleton, some of which are confined to children and adolescents, is welcome indeed. This text clearly fills a void and, further, presents material in such a clear and concise way that it is easy to remember. It presents practical material and is fun to read. There are not many books that actually lead to enjoyable reading!

I commend the authors for fulfilling their goals, and I encourage radiologists, orthopedists, and others involved in diagnosing and treating musculoskeletal disorders in children and adolescents to purchase this book, to consult it often, and even to read it in its entirety! I am honored, indeed, to have an opportunity to write this Foreword.

Donald Resnick, MD
Professor of Radiology
University of California, San Diego

Preface

MRI has transformed the field of pediatric and adolescent musculoskeletal imaging. When the more senior (and gray haired) of the two authors completed his pediatric radiology training, orthopedic radiology was a primarily plain film based discipline, occasionally supplemented by arthrography. Although much could be gleaned from the humble radiograph regarding the nature of orthopedic disorders, MRI has provided elegant depictions and insights of classic pediatric entities that would surely amaze the likes of John Caffey and Edward Neuhauser. With this technique, new challenges have arisen to comprehend the imaging findings in these classic disorders, and a wide array of newly appreciated entities has emerged with the wide utilization of MRI by pediatric, orthopedic and sports medicine specialists.

Despite these developments, a textbook devoted to MRI of pediatric and adolescent musculoskeletal diseases has been unavailable. Those with an interest in this area have had to rely upon published articles, as well as orthopedic, musculoskeletal and pediatric radiology texts. The goal of this work is to bring the literature of musculoskeletal MRI in children and adolescents together in an authoritative, but user friendly format. Cases are presented as "unknowns" in an effort to provide a dynamic learning process. The reader is given a brief history and initial images. A description of findings with appropriate annotated images and supplementary images follows. The diagnosis is then revealed and a discussion ensues. The discussion attempts to cover the salient features of the entity with related cases where appropriate. A differential diagnosis is given and, where appropriate, additional examples are illustrated. The result is a text that contains 315 pediatric and adolescent musculoskeletal MRI cases presented within the context of 102 unknowns.

To place this material more squarely in a clinical context, the authors invited two clinicians, Mininder Kocher, MD, MPH, a pediatric sports medicine orthopedic surgeon, and Mark Gebhardt, MD, a pediatric orthopedic oncologist, to join the effort. Sections entitled "Orthopedist's Perspective" and "What the Clinician Needs to Know" are provided to inform the radiologist about the important clinical issues and what information is required to plan a management strategy. A modest bibliography for each case guides the reader to further discussions in original articles, reviews and other texts. The authors hope that this unique combination of both the radiologic and orthopedic points of view will enrich the readers' learning experience and provide useful relevant information for the referring clinician.

Although the authors have sought to provide a solid and current presentation of both common and, where appropriate, unusual entities, space considerations have required exclusion of other entities. Like most first efforts, it is likely that this book will grow in scope and will undergo refinements in future editions, but for the present, we hope that this will be a useful instructional tool and reference source for radiologists and clinicians interested in pediatric and adolescent musculoskeletal disorders.

J. Herman Kan, MD
Paul K. Kleinman, MD

Acknowledgments

The authors would like to thank Susan A. Connolly, MD, staff radiologist, and Ilse Castro-Aragon, MD, pediatric radiology fellow at Children's Hospital Boston for their useful comments and criticisms during the preparation of this text. In addition, we would like to thank all the radiologists and clinicians of Children's Hospital Boston for referring us these interesting cases. Not least of all, we would like to thank Susan Ivey and Marllely Dewitz for their assistance in preparation of the manuscript. Rob Albano of Springer Science+Business Media has been behind this effort from its inception, and Merry Post has done a thorough and conscientious job of providing editorial support and guidance.

We would also like to graciously acknowledge the following radiology departments and hospitals for allowing us to publish useful images from their institutions:

Ann Jaques Hospital, Newburyport, MA
Children's Hospital Boston, Boston, MA
Brigham and Woman's Hospital, Boston, MA
Boston Medical Center, Boston, MA
Down East Community Hospital, Machias, ME
Falmouth Hospital, Falmouth, MA
Hospital Episcopal San Lucas, Ponce, Puerto Rico
Portsmouth Regional Hospital, Portsmouth, NE
Metrowest Medical Center, Framingham, MA
Morton Hospital and Medical Center, Tautum, MA
Massachusetts General Hospital, Boston, MA
Newton-Wellesley Hospital, Newton, MA
Rhode Island Hospital, Providence, RI
Shields Medical Center, Cape Cod, MA
Southern New Hampshire Medical Center, Nashua, NH
Vanderbilt Children's Hospital, Nashville, TN
West Suburban Imaging Center, Boston, MA

Dr. Kan would like to thank his wife, Shin-Mei, for her understanding and support. She provided unconditional encouragement and companionship as he worked on this text at various coffee shops all over Boston, evenings and weekends. He would also like to thank Paul Kleinman, MD, for his wisdom, patience, and mentorship during this incredible project.

Dr. Kleinman would like to thank his wife and colleague, Patricia, for her support and perseverance during the many evenings and weekends spent on this text. He offers special thanks to his coauthor, Herman Kan, MD, for his vision, industry, and willingness to accept guidance and criticism throughout the course of this challenging project.

Contents

Case 1

History

This is a 13-year-old boy who fell two days ago and has had persistent pain and swelling in his right knee. Radiographs of his right knee (not shown) demonstrated a large joint effusion only.

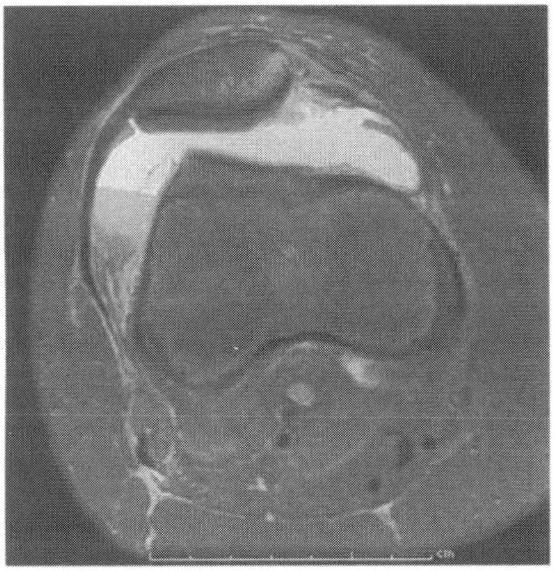

Figure 1A. Axial PD FS of the right knee.

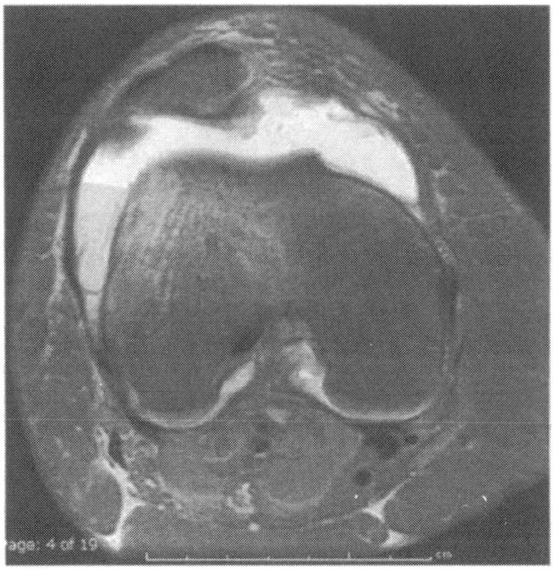

Figure 1B. Axial PD FS.

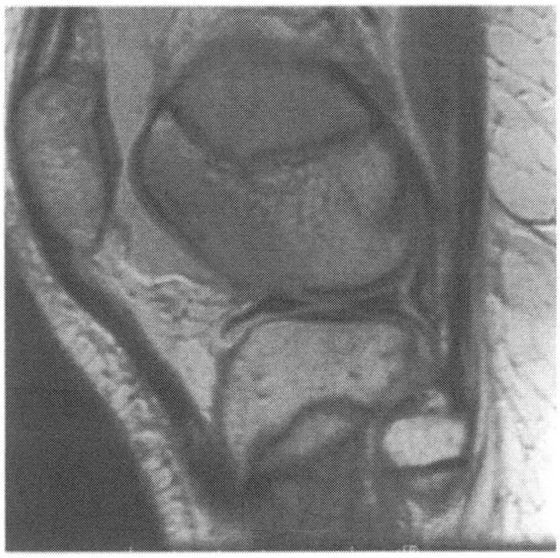

Figure 1C. Sagittal PD sections through the lateral femoral condyle.

Figures 1A (1A with annotations). There is a large joint effusion with fluid-fluid levels. Linear increased SI is present along the medial patella pole **(thin arrow)**. There is also edema within the medial retinaculum (*). A medial pole osteochondral fracture is present **(black arrowhead)**.

Figures 1B, 1C (1C with annotations). Edema is present within the anterior aspect of the lateral femoral condyle. An osteochondral fracture is better delineated on the sagittal view through the lateral femoral condyle **(thick arrows)**.

Figure 1D. A femoral condylar osteochondral loose fragment **(thick arrow)**, which originated from the donor site (Figure 1C with annotations), is identified. No internal derangement of the menisci, cruciate, or collateral ligaments was evident.

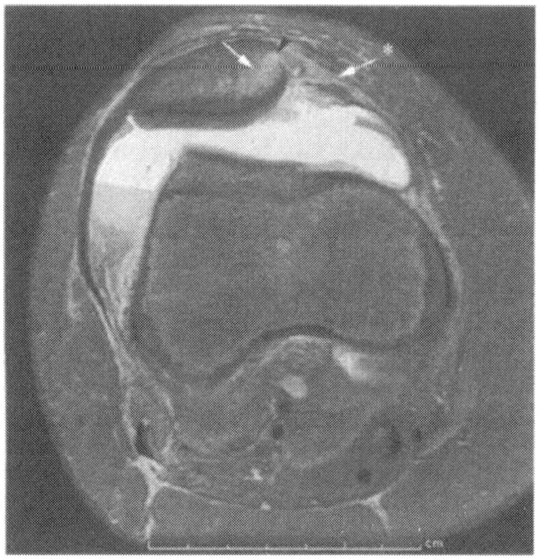

Figure 1A* Annotated.

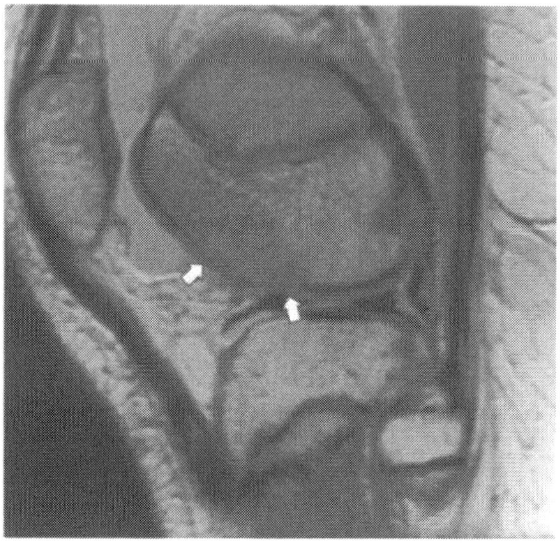

Figure 1C* Annotated.

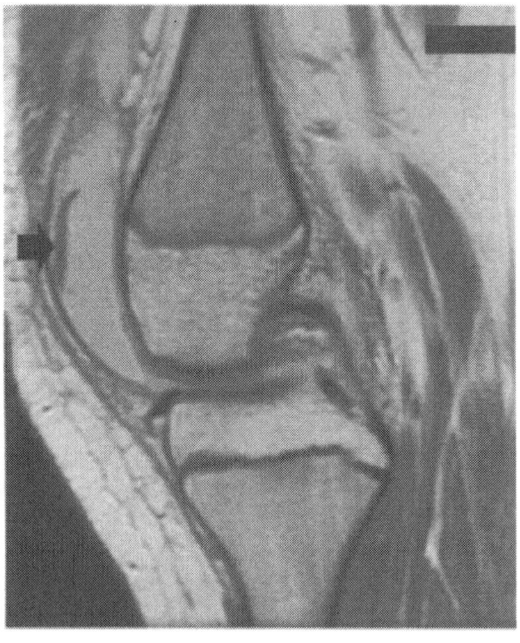

Figure 1D. Sagittal PD.

Diagnosis

Lateral patellar dislocation with displaced osteochondral fracture

Questions

1. What are two mechanisms leading to lateral patellar dislocation?
2. What pattern of bone bruising is seen with lateral patellar dislocation?

Discussion

Lateral patellar dislocation may occur in the setting of trauma with or without underlying patellofemoral dysplasia. The typical age of presentation is between 14 to 20 years with no gender predilection (1). A direct, medial blow or a valgus, twisting (external tibia rotation) injury while the knee is partly flexed are two traumatic mechanisms that may lead to lateral patella dislocation (Answer to Question 1). Anatomic causes that may predispose a patient to patellofemoral instability include patella alta (ratio of the patella tendon to the maximal diagonal length of the patella bone >1.2), a shallow femoral sulcus with hypoplasia of the femoral condyles (normal sulcus angles range from 134 to 155 degrees on a Merchant view), increased lateral patellar tilt, excessive knee valgus, increased femoral anteversion, increased Q-angle, and an abnormal extensor mechanism (2). In severe cases of patellofemoral dysplasia, the trochlear sulcus can even be convex. Causes of congenital patellar dislocation include, but are not limited to: Down's syndrome, Larsen's syndrome, Nail-Patella syndrome, diastrophic dysplasia, and arthrogryposis (3).

Osteochondral fractures and medial patella sleeve avulsion fractures are common occurrences in the setting of acute lateral patellar dislocation in children. In a study of 72 children with acute lateral patellar dislocation, 39% had fractures that were equally divided between osteochondral and medial capsular avulsion fractures (4). In another study, osteochondral injuries proven by arthroscopy were seen in 72% of cases of acute lateral dislocation in patients 12 to 19 years old (5). Osteochondral fractures and bone contusions usually occur when the patella reduces after lateral dislocation. Kissing bone contusions and osteochondral fractures may occur along the anterolateral femoral condyle, medial facet of the patella, and patella eminence (Figures 1E, 1F) (Answer to Question 2).

Medial patella sleeve avulsion fractures from lateral patellar dislocation preferentially occur in children because the patella is not completely ossified. The medial chondro-osseous junction is considered a relatively weak unit of the ligamentous-medial patella retinacular unit in children. The avulsion fragment may be missed entirely or underestimated on plain radiographs because the cartilaginous component is radiolucent. The medial retinaculum may also partially tear in the setting of medial patella sleeve avulsion fractures but usually does not completely tear. A complete medial retinacular tear should only be considered when no hypointense fibers are present and there are indirect signs of disruption such as lateral patella subluxation and/or a redundant and wavy medial retinaculum.

Intra-articular patellar dislocation is an entity that uniquely occurs in children (6). Intra-articular dislocation is when the patella dislocates and rests within the intercondylar notch. Intra-articular dislocations usually result from a quadriceps tendon tear at the chondro-osseous junction of the superior pole of the patella. The inferior pole of the patella is less commonly involved.

The objective of MRI in the assessment of acute lateral patellar dislocation is to suggest the diagnosis when it is unsuspected, to assess for surgical indications such as osteochondral fragments, and to assess for internal derangement. Additional features that should be assessed in patients with recurrent lateral patellar dislocation include the degree of patellofemoral dysplasia and the presence or absence of premature degenerative changes.

Surgical intervention was performed in this patient because of the presence of an osteochondral fracture with loose body.

Orthopedic Perspective

Traumatic patellar dislocation is a common knee injury in the pediatric athlete. The diagnosis can usually be made by history and clinical examination. However, in the context of an acute traumatic hemarthrosis after a twisting injury, physical examination can be difficult and differentiation must be made from ACL injury. MRI is often ordered after traumatic patellar dislocation to evaluate for an osteochondral injury and loose body, and the findings are often used to guide treatment. Patellar dislocations with osteochondral loose bodies are usually treated surgically. Small loose bodies are excised. However, some osteochondral fragments can be very large involving nearly the entire medial patellar facet or lateral femoral condyle. Lateral patellar dislocations without loose bodies are usually treated nonoperatively with bracing and rehabilitation.

What the Clinician Needs to Know

1. The diagnosis in cases where physical examination is difficult.
2. Presence and location of an associated osteochondral fracture/ loose body.

Answers

1. Direct, medial blow or a valgus, twisting (external tibia rotation) injury while the knee is partly flexed.
2. Anterolateral femoral condyle and medial patella facet or the patella eminence.

Additional Example

Acute Lateral Patellar Dislocation with Osteochondral Fragment

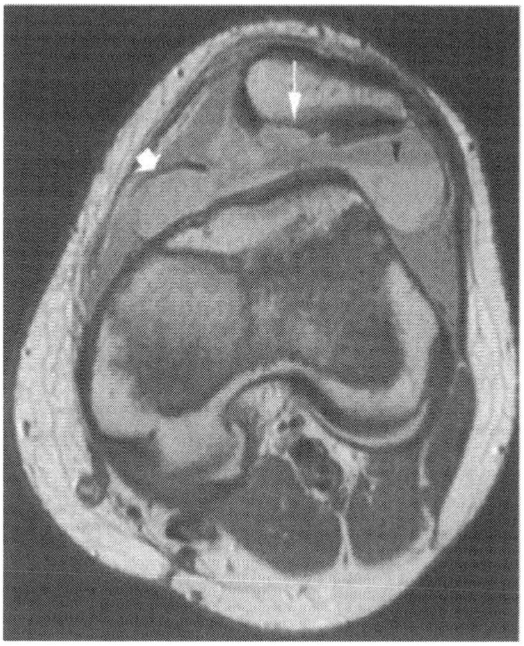

Figure 1E. Axial PD of the left knee.

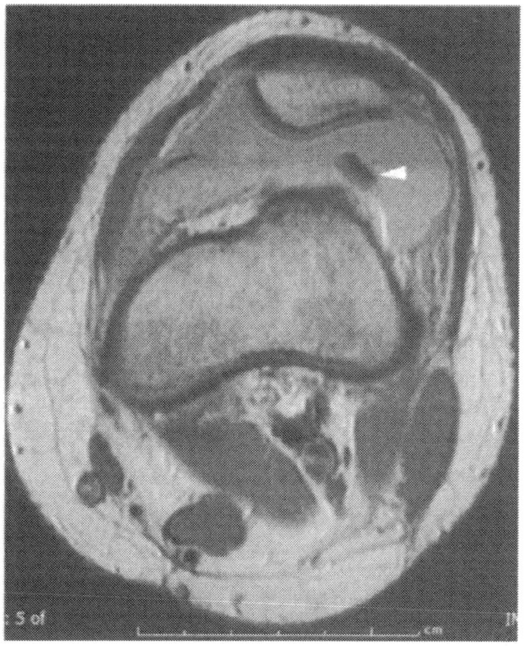

Figure 1F. Axial PD.

Findings

This is a 15-year-old girl with an acute left knee injury.

Figures 1E, 1F. A large hemarthrosis with a fluid-fluid level is present **(black arrowhead)**. There is a large osteochondral defect at the patella eminence **(white arrow)** and the daughter fragment is located laterally **(white arrowhead)**. The intermediate SI of the osteochondral fragment represents the patella articular cartilage and the hypointense SI represents the cortical bone fragment. Incidental note is made of a medial plica **(thick arrow)**.

Pitfalls and Pearls

1. Children and adolescents who dislocate their patella with trauma are often unaware that the dislocation has occurred. When the typical MRI findings are present, the diagnosis should be suggested—even in the absence of a history of dislocation.
2. Axial imaging at various stages of knee flexion may be able to demonstrate the dynamic nature of patella subluxation. The patella, when abnormal, tends to sublux with knee extension, and relocate with knee flexion.

References

1. Beasley LS, Vidal AF. Traumatic patellar dislocation in children and adolescents: Treatment update and literature review. *Curr Opin Pediatr* 2004; 16:29–36.

2. Hinton RY, Sharma KM. Acute and recurrent patellar instability in the young athlete. *Orthop Clin North Am* 2003; 34:385–396.
3. Koplewitz BZ, Babyn PS, Cole WG. Congenital dislocation of the patella. *AJR Am J Roentgenol* 2005; 184:1640–1646.
4. Nietosvaara Y, Aalto K, Kallio PE. Acute patellar dislocation in children: Incidence and associated osteochondral fractures. *J Pediatr Orthop* 1994; 14:513–515.
5. Nomura E, Inoue M, Kurimura M. Chondral and osteochondral injuries associated with acute patellar dislocation. *Arthroscopy* 2003; 19:717–721.
6. Ogden JA. Knee. In: *Skeletal Injury in the Child,* 3rd ed., Chapter 22. New York: Springer, 2000; 929–989.

Case 2

History

This is a 13-year-old girl with chronic right knee pain. She is otherwise well, without fever or systemic symptoms. CBC and ESR were normal.

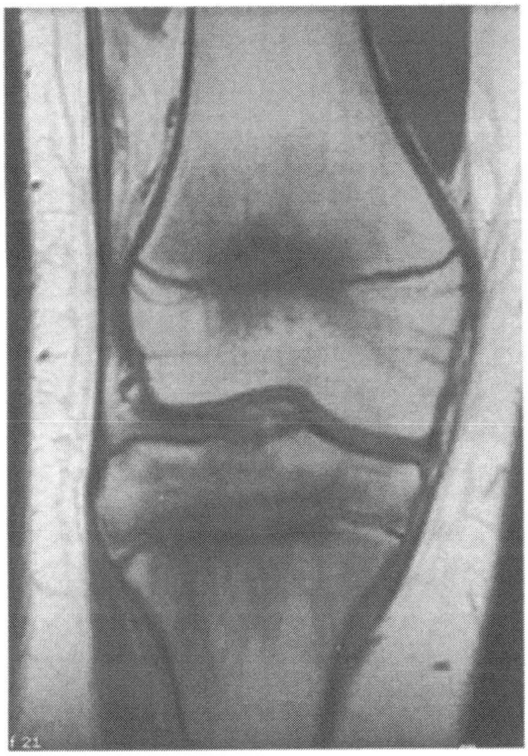

Figure 2A. Coronal T1 of the right knee.

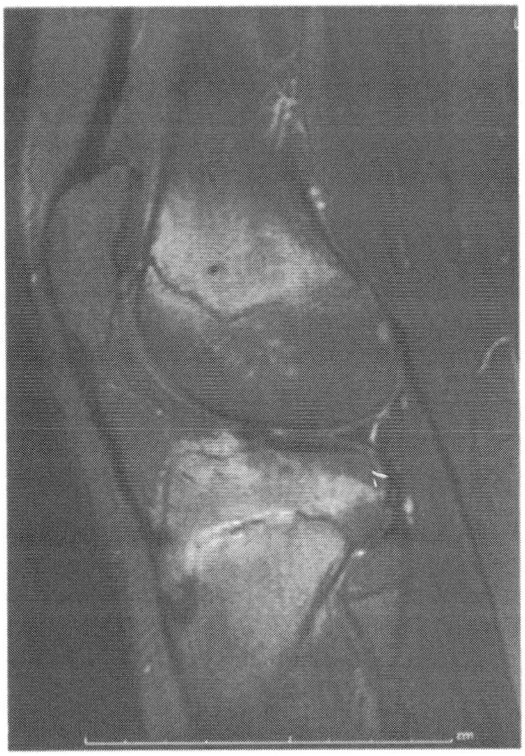

Figure 2B. Sagittal STIR.

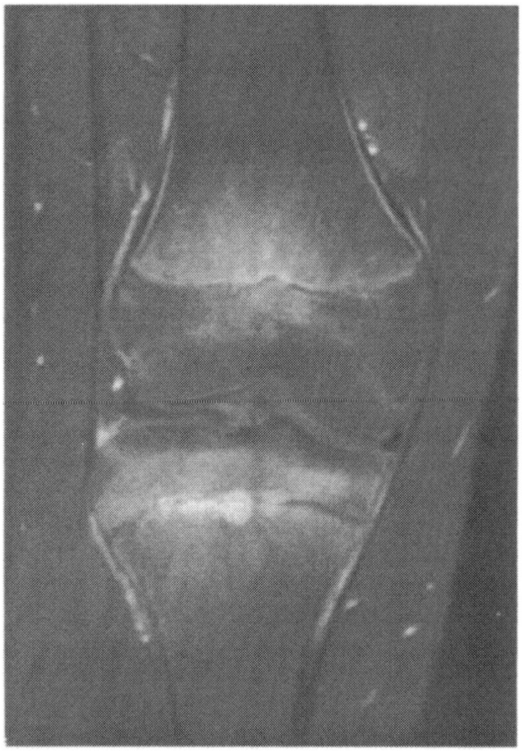

Figure 2C. Coronal T1 post-Gd FS.

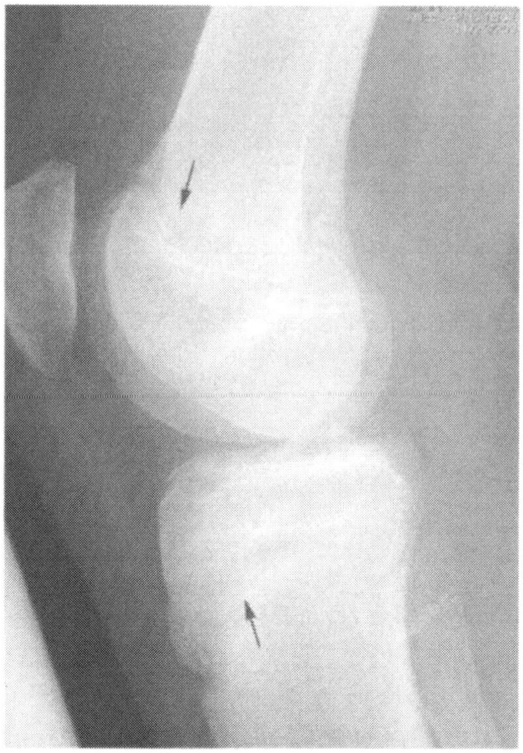

Figure 2D. Lateral plain radiograph.

Figures 2A, 2B, 2C. Focal ill-defined T1 hypointensity, STIR hyperintensity, and enhancement are present, centered at the physes of the distal femur and proximal tibia. There is minimal juxta-cortical edema in Hoffa's fat pad. There is also mild periosteal enhancement along the distal femoral and proximal tibial metaphyses. No intraosseous fluid collections or soft tissue abscess are seen.

Figure 2D. Osteolysis with marginal sclerosis involves the anterior aspects of the physeal margins of the metaphyses of the distal femur and proximal tibia **(arrows)**, corresponding to the signal abnormality on MRI.

Diagnosis

Chronic recurrent multifocal osteomyelitis (CRMO)

Questions

1. What MRI features are seen with pyogenic osteomyelitis and not CRMO?
2. What is the most common location for CRMO?

Discussion

Chronic recurrent multifocal osteomyelitis (CRMO) and pyogenic osteomyelitis share many features. Both may have osseous and adjacent soft tissue inflammation and they are typically located near the physis. Additional shared osseous findings include transphyseal spread, osteolysis or sclerosis, and varying degrees of periosteal reaction (1, 2). Pyogenic osteomyelitis is occasionally multifocal (Figures 2E, 2F), but unlike CRMO, there may be intraosseous and soft tissue abscesses, sequestra, and fistulous tracts (Answer to Question 1) (2). In the absence of these findings, CRMO and multifocal pyogenic osteomyelitis may be indistinguishable based on MRI features at individual sites.

CRMO is a diagnosis of exclusion after pyogenic osteomyelitis has been ruled out. By definition, an organism is not isolated by blood culture or biopsy. The term CRMO is a misnomer since it represents a non-pyogenic inflammatory disorder and is technically not considered osteomyelitis. CRMO is viewed as a seronegative arthropathy-like condition that occurs in children (3). Additional clinical features associated with CRMO include psoriasis, inflammatory bowel disease, recurrent arthritis, spondyloarthropathy, or sacroiliitis. CRMO is generally a self-limited disease and the majority of cases have no disability beyond childhood (4). SAPHO (synovitis, acne, pustulosis, hyperostosis, and osteitis) is considered the adult equivalent of CRMO.

CRMO has been observed most commonly in the tubular bones of the lower extremity (Answer to Question 2). The clavicles are next most common (see Case 11) (3). When the changes are restricted to the thorax and shoulder girdle, the term sternocostoclavicular hyperostosis is generally employed. Rarer locations include the spinal column and pelvic girdle (1, 5).

The principal differential diagnosis for CRMO is multifocal pyogenic osteomyelitis, and bone biopsy and culture are generally required for diagnosis. Less likely differential considerations include Langerhans cell histiocytosis, small round blue cell tumors, and trauma/stress reaction.

In this patient, the diagnosis of CRMO was invoked because of the multifocality and the absence of clinical or laboratory findings to suggest pyogenic osteomyelitis. A total body MRI showed no additional lesions. A biopsy showed changes of chronic osteomyelitis without bacterial organisms, and culture of the biopsy material showed no growth. Symptoms resolved with naproxen. She was not given antibiotics.

Orthopedic Perspective

Although patients with CRMO are usually not acutely ill, they may have erythema, low-grade fever, and abnormal laboratory values. The clinician relies on imaging to define the full extent of the process and identify a suitable site for open or percutaneous biopsy. Unlike pyogenic osteomyelitis, discrete fluid collections are not typical features of CRMO. Biopsy should target areas of granulation tissue and bone destruction, rather than bony sclerosis or nonspecific reactive edema to increase the diagnostic yield. Patients are typically treated with anti-inflammatory medications and followed by a rheumatologist.

What the Clinician Needs to Know

1. Is the lesion pyogenic osteomyelitis, CRMO, or tumor?
2. Are there other lesions?
3. Which lesion is best suited for percutaneous biopsy?
4. Is the process subsiding on follow-up studies? Active lesions demonstrate juxtacortical soft tissue edema, whereas the SI within inactive lesions is confined to bone (2).

Answers

1. Intraosseous or soft tissue abscesses, sequestra, and fistulous tracts.
2. Lower extremity tubular bones.

Additional Example

Multifocal Staphylococcus aureus Osteomyelitis

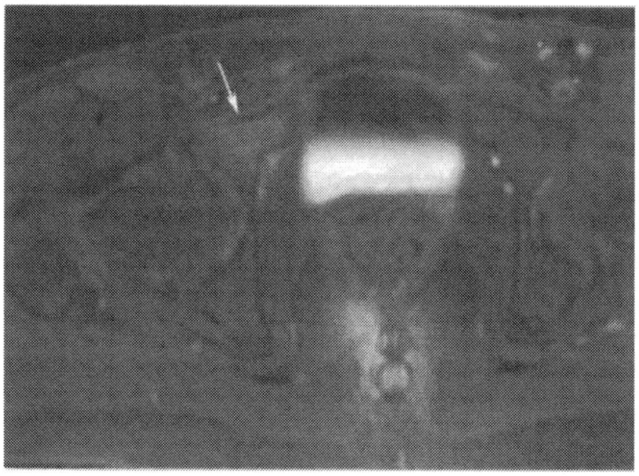

Figure 2E. Axial T1 post-Gd FS of the pelvis.

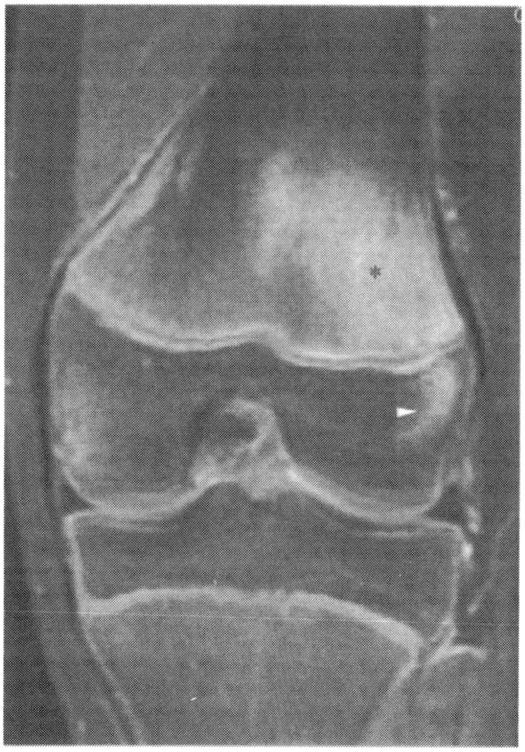

Figure 2F. Coronal T1 post-Gd FS of the left knee.

Findings

This is a 14-year-old boy with diffuse lower extremity pain and non-weight bearing. He had a severe upper respiratory tract infection 2 weeks prior to presentation. At the time of admission, he had a high fever and blood cultures were positive for *Staphylococcus aureus*.

Figures 2E, 2F. Two foci of abnormal enhancement are located in the right pubic ramus **(arrow)** and the left distal femoral metaphysis (*) and epiphysis **(arrowhead)**. Without the clinical history or positive blood cultures, the imaging findings do not allow differentiation of pyogenic osteomyelitis from CRMO, Langerhans cell histiocytosis, metastases, or stress reaction.

Pitfalls and Pearls

1. CRMO is a misnomer because it is not a bacterial infection of bone. CRMO represents a non-pyogenic inflammatory disorder that primarily affects bone.
2. The imaging features of pyogenic osteomyelitis and CRMO are similar in the majority of cases. Therefore, the diagnosis of CRMO should be made only if pyogenic osteomyelitis has been completely excluded.

3. If CRMO is a consideration, a bone scan is indicated to assess for other lesions, and to identify the optimal biopsy site.

References

1. Jurik AG, Egund N. MRI in chronic recurrent multifocal osteomyelitis. *Skeletal Radiol* 1997; 26:230–238.
2. Jurik AG. Chronic recurrent multifocal osteomyelitis. *Semin Musculoskelet Radiol* 2004; 8:243–253.
3. Earwaker JW, Cotten A. SAPHO: Syndrome or concept? Imaging findings. *Skeletal Radiol* 2003; 32:311–327.
4. Huber AM, Lam PY, Duffy CM, et al. Chronic recurrent multifocal osteomyelitis: Clinical outcomes after more than five years of follow-up. *J Pediatr* 2002; 141:198–203.
5. Anderson SE, Heini P, Sauvain MJ, et al. Imaging of chronic recurrent multifocal osteomyelitis of childhood first presenting with isolated primary spinal involvement. *Skeletal Radiol* 2003; 32:328–336.

Case 3

History

This is a 10-year-old boy with right knee swelling following remote minor knee trauma. Radiographs demonstrated only a large joint effusion (not shown).

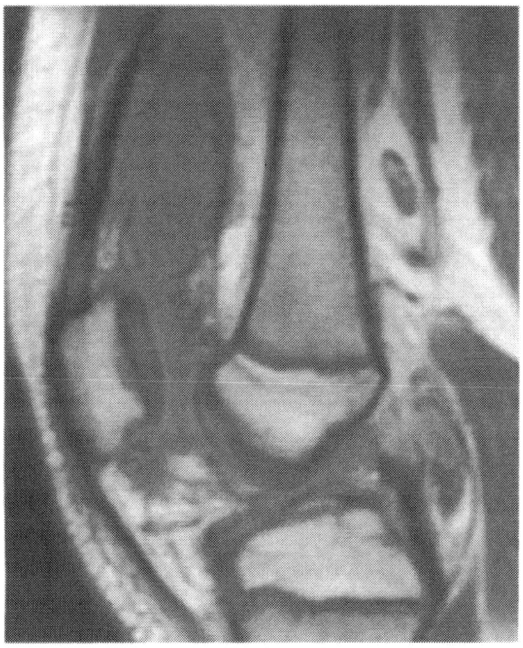

Figure 3A. Sagittal PD of the right knee.

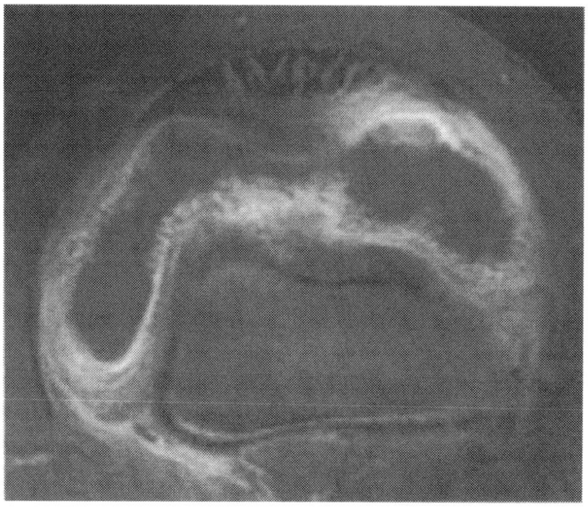

Figure 3B. Axial T1 post-Gd FS.

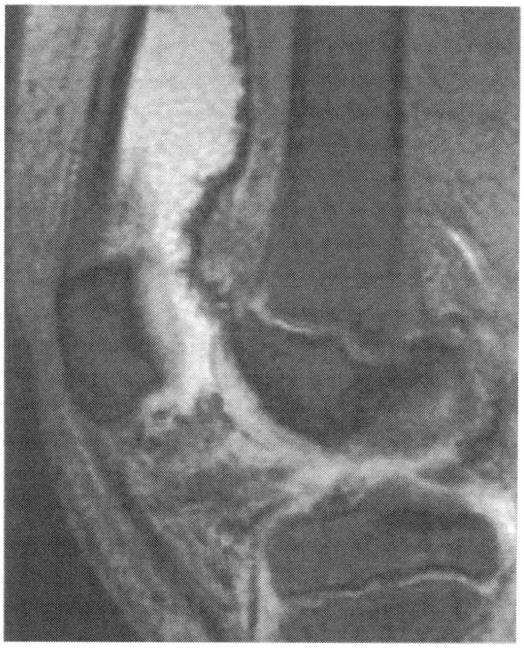

Figure 3C. Sagittal MPGR.

Figure 3A. A large joint effusion is present. The synovium is thickened with areas of hypointensity.

Figure 3B. Diffuse and nodular synovial enhancement is seen.

Figure 3C (3C with annotations). Marked susceptibility artifact **(arrows)** is present within thickened synovium and within Hoffa's fat pad **(arrows)**.

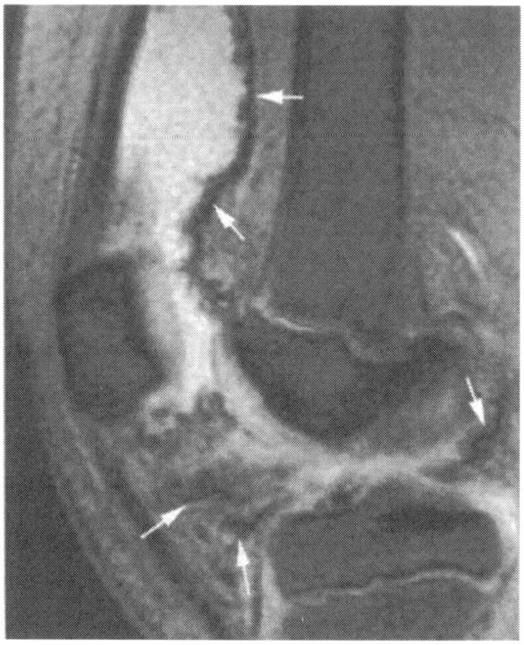

Figure 3C* Annotated.

Diagnosis

Pigmented villonodular synovitis (PVNS)

Questions

1. What are the MRI features of PVNS?
2. T/F: Post-arthroscopy blooming artifact on GRE sequences may be indistinguishable from that seen with recurrent PVNS.

Discussion

PVNS is a benign, synovial proliferation that may occur anywhere there is synovial tissue, including joints, bursae, and tendon sheaths (1). The two subtypes of PVNS are localized (most common) and diffuse. PVNS most commonly affects the knee followed by the hip. It usually is seen in the second and third decade of life and is less common in children (2). Giant cell tumor of the tendon sheath histologically resembles PVNS and may represent a localized extra-articular form of the disorder. It tends to occur in female patients in the third and fifth decade of life.

The histologic features of PVNS include synovial nodular hyperplasia associated with hemosiderin, variable lipid deposition, and fibrosis (1). On MRI, these lesions are generally hypointense on all imaging sequences and demonstrate blooming artifact on GRE sequences due to the presence of hemosiderin (Figure 3C) (3). Blooming artifact represents exaggerated signal loss artifact that may occur with hemosiderin or metal. Both the localized and diffuse forms of PVNS demonstrate variable enhancement (Answer to Question 1). Intense enhancement may be seen within these lesions related to active synovitis (3). The localized form of PVNS can be mistaken for a loose body or a mass (Figures 3D, 3E) (1). The diffuse form is often associated with a large joint effusion or hemarthrosis and may be a diagnostic challenge, since there is no normal synovial tissue on the image for comparison (Figures 3F, 3G). Although uncommon at the knee, juxta-articular bone erosions may occur in the less capacious joints, such as the hips and ankles (Figures 3H–3J).

The differential diagnosis for localized and diffuse PVNS includes: hematoma from acute or remote trauma, loose bodies, inflammatory arthritis with pannus, foreign body reaction, synovial venous malformations (AKA synovial hemangioma), hemophilic arthropathy, and synovial osteochondromatosis. All of these conditions may demonstrate hypointensity on T1 and T2W sequences due to the presence of blood products and, when focal, may suggest an intra-articular mass. Plain radiography or CT is often helpful to narrow the differential diagnosis, since PVNS rarely calcifies, whereas synovial osteochondromatosis, loose bodies, and synovial venous malformations may demonstrate calcifications (3).

The treatment for PVNS is synovectomy, but there is a high local recurrence rate. Postsurgical blooming artifact may be indistinguishable from recurrent PVNS on GRE sequences (Answer to Question 2) (2). Conventional SE post-Gd T1W sequences help distinguish PVNS recurrence from postsurgical changes (Figures 3K, 3L).

In this case, diffuse PVNS was pathologically confirmed. The patient underwent synovectomy and did well.

Orthopedic Perspective

There are very few tumors that affect the synovium, so the differential diagnosis from a tumor standpoint is short. Inflammatory arthropathies are the most common conditions in the differential diagnosis, but trauma, synovial venous malformations, and synovial chondromatosis should also be entertained. In developing countries, infection (tuberculosis) also enters the differential diagnosis. Of much more concern is a true intra-articular synovial sarcoma, which, although extremely rare, must be considered in the differential diagnosis. A biopsy is necessary to establish the correct diagnosis.

The nodular or localized form of PVNS is easier to treat, and complete excision is usually curative. Diffuse PVNS is an extremely frustrating disease for the patient and the treating surgeon. In the knee, complete synovectomy is difficult, often requiring anterior and posterior approaches, and in the hip it is almost impossible. Despite aggressive synovectomies, recurrence is common. Eventually, the destruction of the articular cartilage and adjacent bone will necessitate arthroplasty in unsuccessfully treated cases, and since these are generally young patients, this is often a poor alternative. External beam radiotherapy may control PVNS, but its use carries concerns, especially in children with open physes. The so-called radiation synovectomy, where radioactive agents are injected into the joint after synovectomy has some proponents, but its effectiveness has not been proven.

What the Clinician Needs to Know

1. Is this PVNS or an inflammatory arthritis?
2. Is this PVNS or an intra-articular sarcoma? This problem usually requires a tissue diagnosis, but the character and location of the lesion on MRI may affect the approach to biopsy.
3. Are the findings on a postoperative study due to recurrent/residual PVNS, or are they postoperative changes only?

Answers

1. Hypointensity on all imaging sequences, particularly with GRE sequences. Variable enhancement.
2. True.

Additional Examples

PVNS, Localized

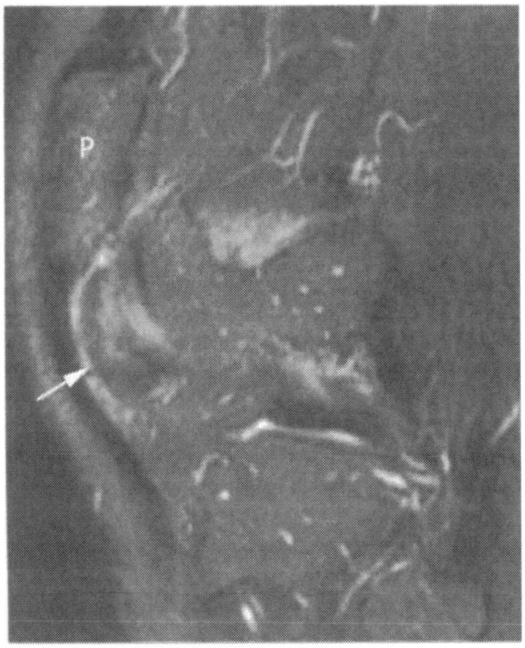

Figure 3D. Sagittal T2 FS.

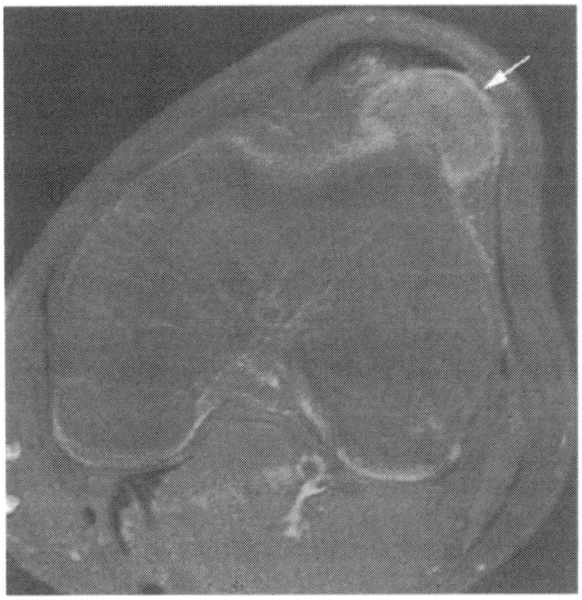

Figure 3E. Axial T1 post-Gd FS.

Findings

Figures 3D, 3E. This is a 13-year-old boy with a large, well-defined mass **(arrows)** that arises from the joint and extends inferiorly into Hoffa's fat pad. It is heterogeneously intermediate SI on T2 and demonstrates heterogeneous mild enhancement on post-Gd sequences. Blood products are inconspicuous. This lesion was pathologically confirmed localized PVNS.

Diffuse PVNS

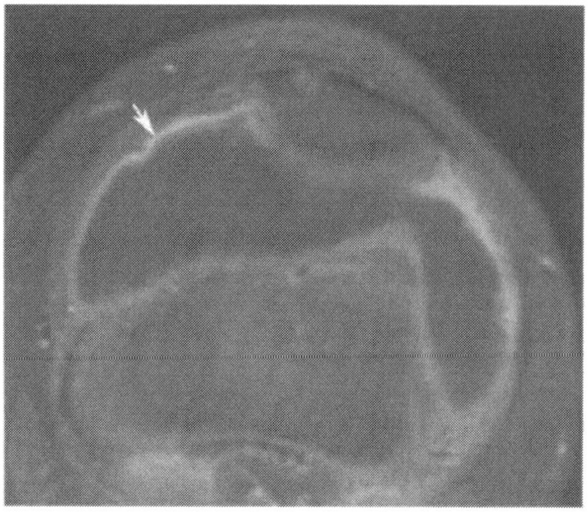

Figure 3F. Axial T1 post-Gd FS.

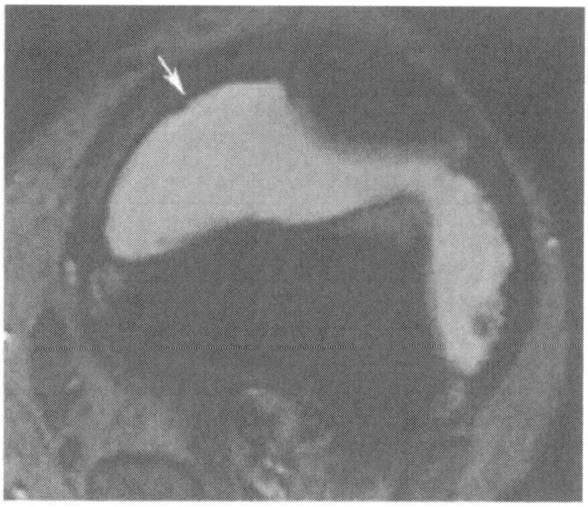

Figure 3G. Axial MPGR.

Findings

This is an 11-year-old boy who had a remote history of left knee trauma while skiing.
Figures 3F, 3G. There is a large joint effusion. There is diffuse synovial enhancement and hemosiderin deposition **(arrows)**. This was pathologically proven diffuse PVNS.

PVNS, Localized with Talar Neck Deformity

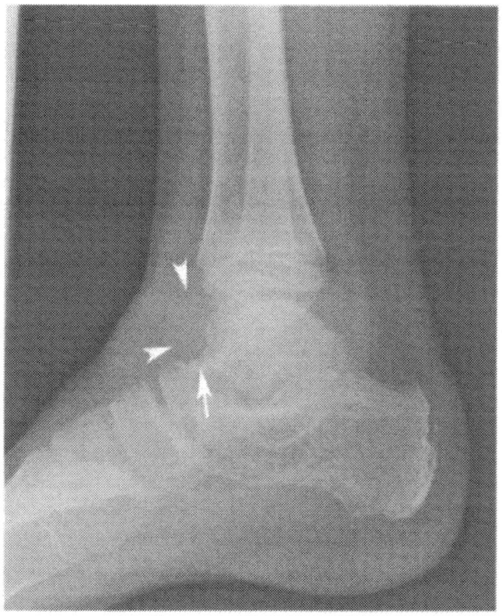

Figure 3H. Lateral radiograph of the left ankle.

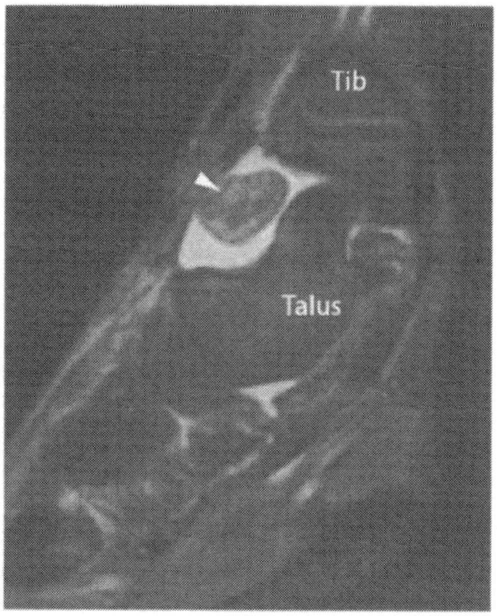

Figure 3I. Sagittal STIR.

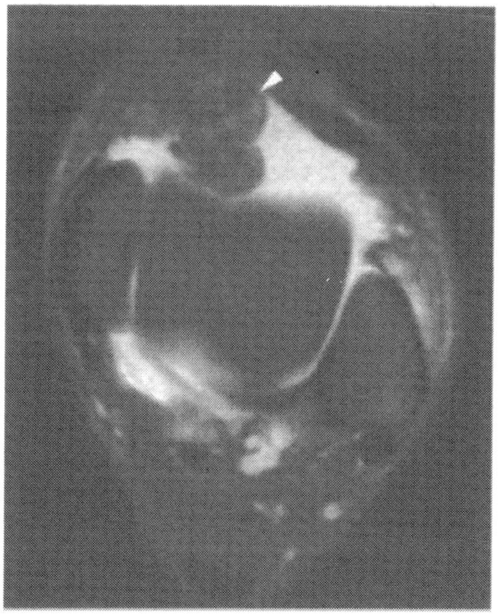

Figure 3J. Axial T2 FS.

Findings

This is a 9-year-old girl with chronic left ankle pain.

Figure 3H. There is an apparent intra-articular soft tissue mass **(arrowheads)** associated with concave deformity of the talar neck **(arrow)**.

Figures 3I, 3J. Fluid sensitive sequences demonstrate a moderate joint effusion and a hypointense lobulated mass with a hypointense rim **(arrowheads)** in the anterior joint space. This was pathologically confirmed localized PVNS. Tibia (Tib).

Recurrent Focal PVNS

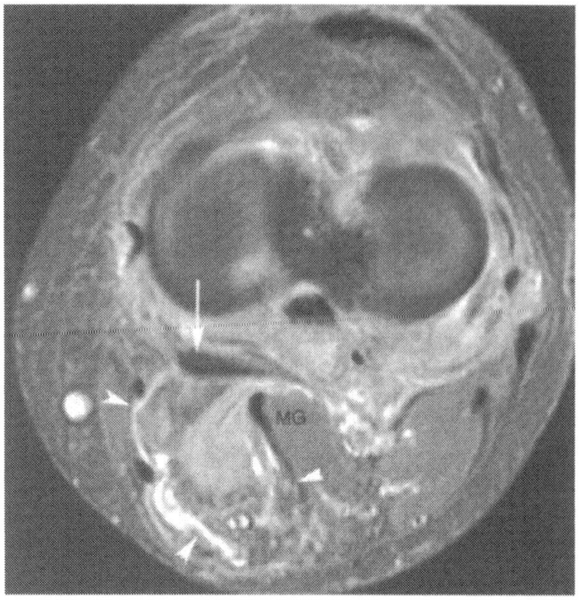

Figure 3K. Axial T1 post-Gd FS of the left knee.

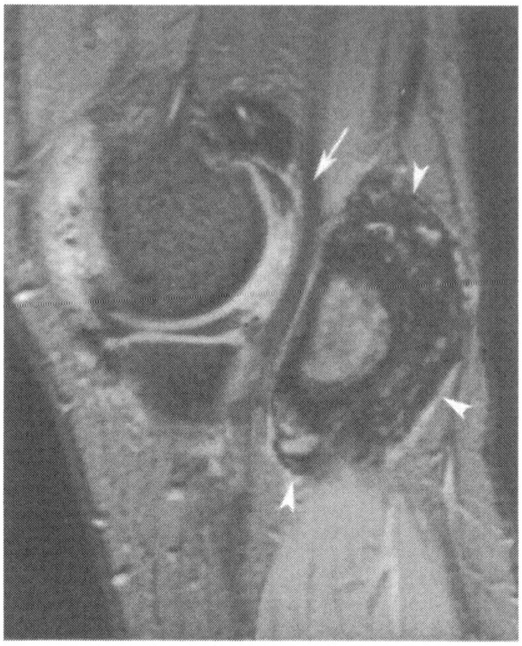

Figure 3L. Sagittal MPGR.

Findings

This 13-year-old girl with known left knee PVNS had undergone three prior synovectomies but continued to have recurrent knee pain.

Figure 3K. There is diffuse synovial enhancement with a more focal, heterogeneously enhancing mass **(arrowheads)** within the region of the semimembranosus bursa. Note location posterior to the semimembranosus tendon **(arrows)** and medial to the medial head of the gastrocnemius muscle (MG).

Figure 3L. This focal lesion shows central T2* hyperintensity and a thick peripheral rind of hemosiderin staining with blooming artifact **(arrowheads)** on this gradient echo sequence. The semimembranosus tendon is located anterior to this mass **(arrow)**. Recurrent PVNS was confirmed arthroscopically.

Pitfalls and Pearls

The absence of hemosiderin staining or blooming susceptibility artifact does not preclude the diagnosis of PVNS. Always include it in the differential diagnosis of a noncalcified intra-articular synovial-based mass.

References

1. Al-Nakshabandi NA, Ryan AG, Choudur H, et al. Pigmented villonodular synovitis. *Clin Radiol* 2004; 59:414–420.
2. Narvaez JA, Narvaez J, Aguilera C, De Lama E, Portabella F. MR imaging of synovial tumors and tumor-like lesions. *Eur Radiol* 2001; 11:2549–2560.
3. Masih S, Antebi A. Imaging of pigmented villonodular synovitis. *Semin Musculoskelet Radiol* 2003; 7:205–216.

Case 4

History

This is a 12-year-old boy with right hip pain and limp for 3 months after a minor fall.

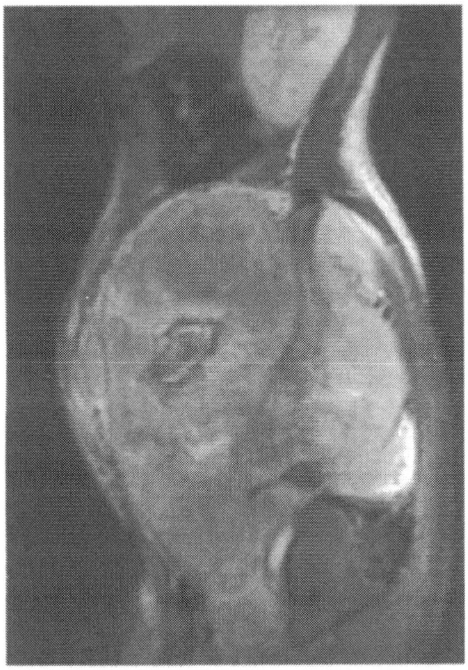

Figure 4A. Sagittal STIR through the right ilium.

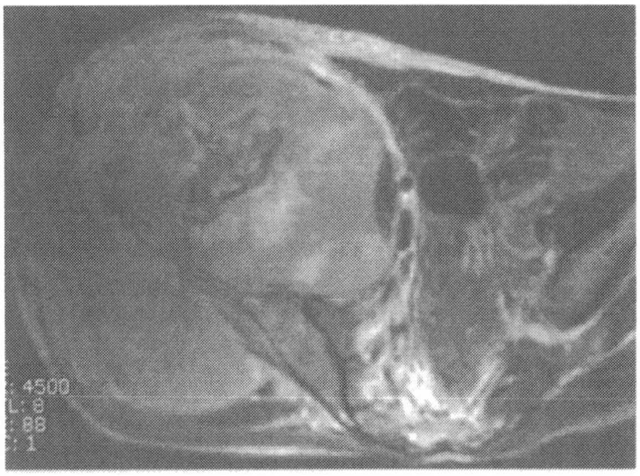

Figure 4B. Axial T2 FS.

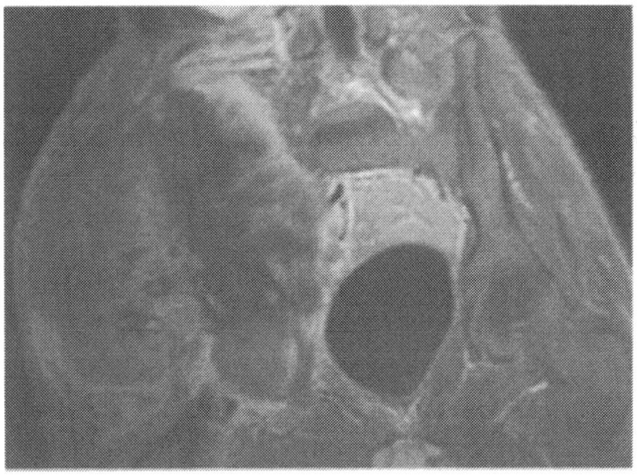

Figure 4C. Coronal T1 post-Gd FS.

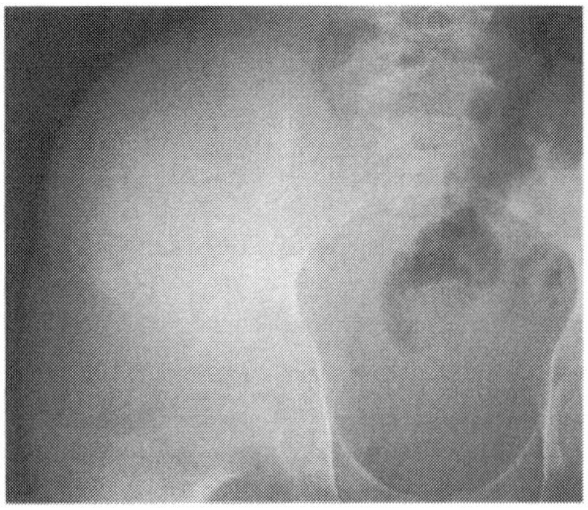

Figure 4D. AP radiograph.

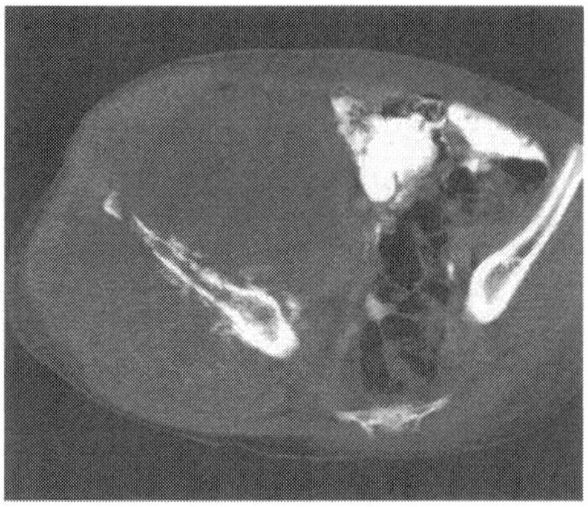

Figure 4E. CT.

Figures 4A, 4B (4A with annotations). A large heterogeneous mass is identified arising from the right ilium that is hypointense to intermediate SI on fluid-sensitive sequences. Centrally, there is a discrete area of heterogeneous SI that has a hypointense rim consistent with necrotic tumor **(arrowheads)**. Anteromedially, the mass extends into the pelvic space and displaces the iliac vessels medially. Posteriorly, the gluteal muscles are displaced and stretched over the mass.

Figure 4C. There is heterogeneous tumoral enhancement, with a sharply demarcated central zone of relatively decreased enhancement consistent with necrosis.

Figures 4D, 4E. Plain radiography and CT demonstrate moth-eaten destruction with a wide transition zone in the right ilium. Aggressive perpendicular periosteal new bone formation is seen on CT that correlates with hypointense regions on the MRI. There is some faint mineralization in the posterior extent of the lesion, but osseous tumor matrix is not evident.

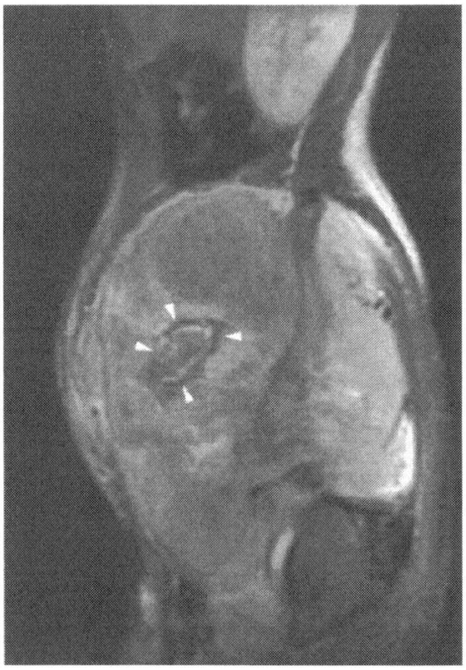

Figure 4A* Annotated.

Diagnosis

Ewing's sarcoma

Questions

1. T/F: Ewing's sarcoma rarely arises from the diaphysis of long bones.
2. Why is Ewing's sarcoma occasionally iso- to hypointense with respect to muscle on fluid-sensitive sequences?

Discussion

Ewing's sarcoma is second to osteosarcoma as the most common primary malignant tumor of bone in children (1). Ewing's sarcoma of bone belongs to the Ewing's family of tumors, which includes primitive neuroectodermal tumor (PNET), extraosseous Ewing's sarcoma, and Askin's tumor. The most common chromosomal translocation found in this family of tumors is at t(11;22) (2). Ewing's sarcoma of bone usually occurs in patients between 10 and 20 years of age, a younger population compared with osteosarcoma. Ewing's sarcoma generally arises from the medullary canal and much less often originates in the subperiosteal space and extraosseous soft tissues (3). Within the long bones, Ewing's sarcoma commonly occurs within the diaphysis (35%) or meta-diaphysis (59%) (Answer to Question 1) (4). Diaphyseal involvement is more frequently encountered with Ewing's sarcoma compared with conventional osteosarcoma. Ewing's sarcoma occurs in the axial and appendicular skeleton with about equal frequency, whereas the vast majority of osteosarcomas occur in the appendicular skeleton (1). Ewing's sarcoma of the pelvis is more common in older patients (late adolescence and young adults) than long bone lesions and usually carries a worse prognosis because of the higher incidence of metastasis and large tumor size at presentation (5).

Approximately 20–25% of all patients with Ewing's sarcoma have detectable metastasis, usually involving the lung, at the time of presentation (1). However, it is believed by some that micrometastases are present in all patients with Ewing's sarcoma at diagnosis (6). Features associated with a poor prognosis include axial location of the primary tumor, older age at presentation, and tumor size (7, 8).

On plain radiography, Ewing's sarcoma has a variable, but usually aggressive appearance. These changes include a wide zone of transition with permeative or moth- eaten osteolytic destruction. Codman's triangle, onion-skin and perpendicular (hair-on-end) aggressive periosteal reaction are common. Ewing's sarcoma often demonstrates a large extraosseous soft tissue component that may cause cortical saucerization. Cortical saucerization represents erosion and destruction of the outer table of the cortex due to subperiosteal tumor extension. These tumors may be osteolytic, osteoblastic, or mixed. Although they may provoke dramatic bone reaction, Ewing's sarcoma does not produce a mineralized osteoid matrix as seen with osteosarcoma.

In general, the extraosseous soft tissue component of Ewing's sarcoma tends to be larger compared with its intramedullary component, especially when the flat bones are involved (9). The soft tissue component is usually hypo- to isointense on T1, variable SI on fluid-sensitive sequences, and demonstrates heterogeneous enhancement (4). These tumors may be hypo- to isointense to muscle on fluid-sensitive sequences because of dense cellularity (Answer to Question 2). Tumoral signal heterogeneity on all imaging sequences may also result from tissue necrosis and hemorrhage. Approximately 4% of Ewing's sarcoma demonstrate skip metastasis at presentation (Figures 4F–4J) (10). Tumoral extension into the adjacent soft tissues may be distinguished from reactive soft tissue edema by its mass effect and solid tumoral enhancement.

The differential diagnosis of Ewing's sarcoma includes osteosarcoma, lymphoma/leukemia, metastasis (e.g., neuroblastoma), Langerhans cell histiocytosis (LCH), and osteomyelitis. The lack of a calcified matrix or an axial skeletal location makes osteosarcoma less likely. CT may be used for problem solving to assess for a mineralized osteoid matrix. Subacute and chronic osteomyelitis can mimic Ewing's sarcoma when there is mass-like granulation tissue present. In general, adjacent soft tissue inflammatory changes are often more extensive with osteomyelitis compared with Ewing's sarcoma. Distinguishing between Ewing's sarcoma and LCH may be difficult, but LCH tends to produce geographic rather than permeative bone destruction and to have a smaller extraosseous soft tissue mass. LCH is more likely to be encountered in the toddler and young child than Ewing's sarcoma.

The initial and postchemotherapy/presurgical MRI evaluation of Ewing's sarcoma is similar to osteosarcoma in regard to defining tumor extent (see Case 17). However, unlike osteosarcoma, decrease in tumor size after chemotherapy correlates with the degree of tumoral necrosis and a good response to chemotherapy (Figures 4K–4M) (11). Dynamic enhanced MRI (DEMRI) may distinguish between viable tumor and adjacent soft tissue edema and fibrosis after chemotherapy. Van der Woude et al. observed that viable tumor enhanced within 6 seconds of arterial enhancement, whereas peritumoral edema and fibrosis enhanced at 6 seconds or longer (12).

In this patient, Ewing's sarcoma was confirmed after right iliac bone biopsy and CT demonstrated pulmonary metastases. He was treated with three cycles of chemotherapy, but continued to show progression of metastatic disease during his most recent follow-up imaging evaluation.

Orthopedic Perspective

It is essential to make the diagnosis of Ewing's sarcoma/PNET and images are crucial in doing so. Unlike osteosarcoma, there is no matrix production by the tumor, and in sites such as the pelvis, large masses can go undetected because the radiographic changes are subtle. It is not uncommon for pelvic, sacral, spinal, or even lower-extremity tumors to present with symptoms that mimic a disc herniation and for the diagnosis to be missed for several weeks or months. At times, neurological findings such as a foot drop can reinforce this incorrect diagnosis. A high index of suspicion in a child with unexplained pain and a careful physical examination looking for a soft tissue mass are essential. Establishing the diagnosis always requires a biopsy, and a CT-directed needle biopsy is useful in this regard. It is important to get as many cores as possible because special processing of the tissue, including cytogenetics, is necessary to precisely differentiate among the various small round blue cell malignancies. Finding the characteristic translocation (chromosome t(11;22) being the most common) is helpful as are immunohistochemical stains. Since infection is in the differential diagnosis, a culture should always be obtained and the biopsy tract must be placed in a location that can be excised at the time of definitive resection. Staging of the patient to look for lung, other bone, bone marrow, and skip metastases is indicated, as in all sarcomas. The oncologist will include a bone marrow biopsy.

In the past, Ewing's sarcoma/PNET was treated primarily with chemotherapy and radiation of the primary tumor. Improvement of survival due to chemotherapy led to the recognition of secondary malignancies in a percentage of Ewing's sarcoma survivors (estimated 5–25%), usually in the site of the primary disease. Detection of a second malignancy, which is usually an osteosarcoma, is difficult, because following radiotherapy of the primary tumor, the bone remains abnormal indefinitely. Detecting the often subtle changes in this already abnormal-appearing bone can be difficult.

Survivors who develop pain in the primary site four or more years after completion of therapy should be carefully evaluated for this possibility. MRI is particularly useful in this regard.

There has been a gradual change in the treatment philosophy over the years, and currently the primary tumor is usually treated by resection and limb-sparing reconstruction whenever possible. As noted above, unlike osteosarcoma, Ewing's sarcoma/PNET often responds dramatically to neoadjuvant chemotherapy with marked reduction in the size of the soft tissue mass because of the lack of an osseous matrix as seen in osteosarcoma. Imaging postchemotherapy is essential to the surgeon to define the extent of the lesion within both bone and soft tissue and the relationship to the neurovascular structures. The planning and concepts are similar to osteosarcoma with the following caveat: Ewing's sarcoma/PNET frequently permeates into surrounding muscle with satellite nodules or fingers of tumor extending beyond what is apparent on the postchemotherapy MRI. Preoperative evaluation of high-quality images by the surgeon and radiologist is essential to avoid positive margins at the time of resection. Exclusion of skip metastases and defining the extent of the tumor within bone is important. In the pelvis, the sacroiliac joint and sacrum must be carefully evaluated for tumor extension.

What the Clinician Needs to Know

1. The clinician must be able to differentiate Ewing's sarcoma/PNET from other neoplasms and other conditions, such as disc herniation, that can mimic the clinical presentation.
2. Adequate tissue from a needle biopsy is essential to differentiate Ewing's sarcoma/PNET from other small round blue cell neoplasms and infection.
3. Survivors who have been treated by radiation to the primary must be monitored for a secondary malignancy.
4. Assessment of the local extent of the tumor following chemotherapy is essential to plan appropriate resections with the goal of tumor-free margins.

Answers

1. False.
2. Hypo- to isointense on fluid-sensitive sequences because of dense cellularity.

Additional Examples

Ewing's Sarcoma of the Femur

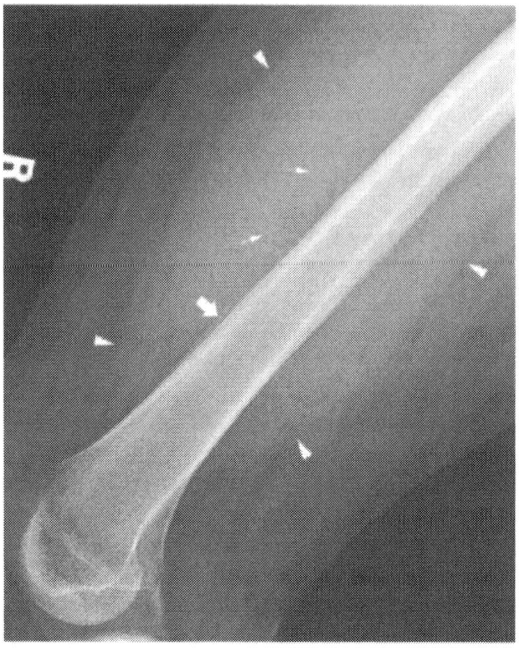

Figure 4F. Lateral radiograph of the right femur.

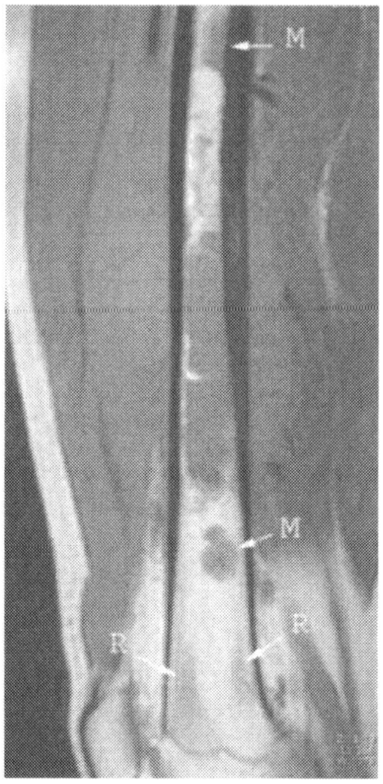

Figure 4G. Sagittal T1.

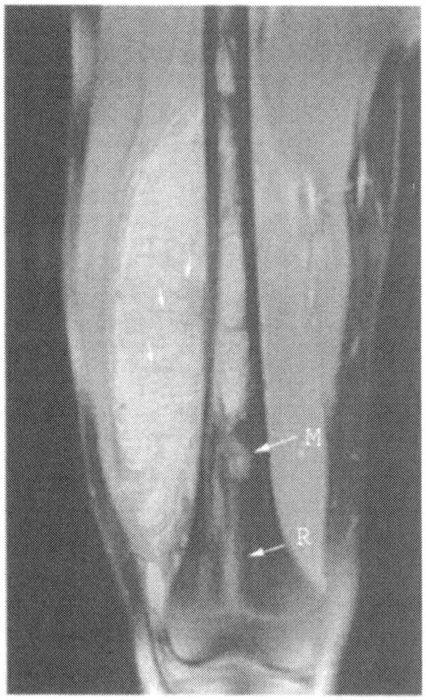

Figure 4H. Coronal PD FS.

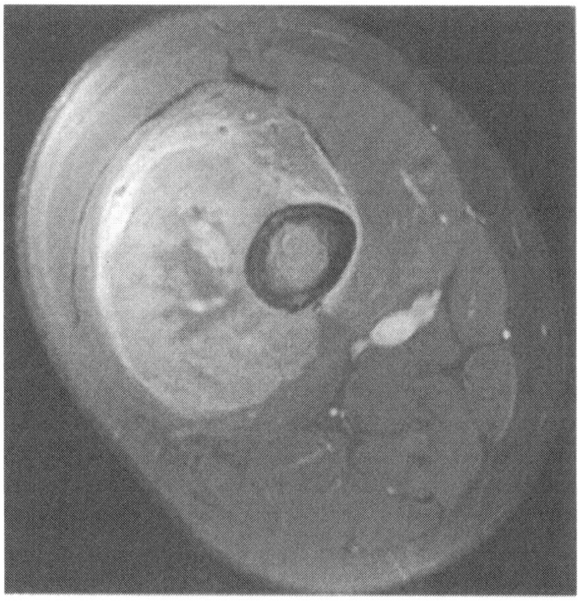

Figure 4I. Axial T1 post-Gd FS.

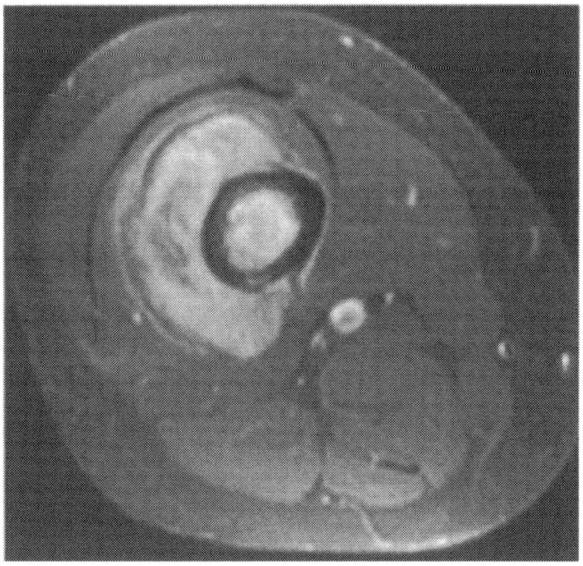

Figure 4J. Axial T1 post-Gd FS—3 months after chemotherapy.

Findings

This is a 17-year-old boy with right thigh fullness and pain.

Figure 4F. There is subtle permeative lesion of the distal diaphysis of the femur associated with soft tissue mass **(arrowheads)**, hair-on-end periostitis **(thin arrows)**, and a Codman's triangle **(thick arrow)**.

Figures 4G, 4H. MRI demonstrates an extensive soft tissue mass and diaphyseal marrow infiltration. Skip lesions are identified (M) that should be distinguished from normal, flame-shaped residual red marrow (R). On the PD FS sequence, note linear hypointense SI perpendicularly oriented striations within the mass **(arrows)** that correlate with the hair-on-end periosteal reaction seen on the plain radiograph (Figure 4F).

Figure 4I. There is diminished central enhancement, suggesting tumor necrosis. This lesion was a pathologically proven Ewing's sarcoma, and CT revealed pulmonary metastasis (not shown).

Figure 4J. A follow-up MRI after chemotherapy shows that the tumor is smaller. The patient received additional courses of chemotherapy and radiation therapy.

Ewing's Sarcoma of the Fibula

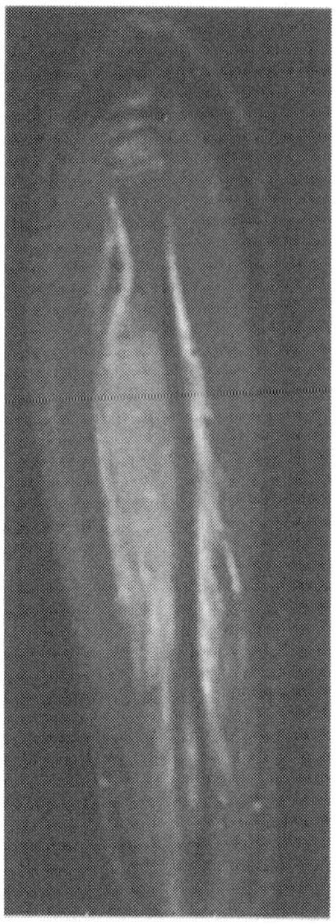

Figure 4K. Saggital STIR of the right calf.

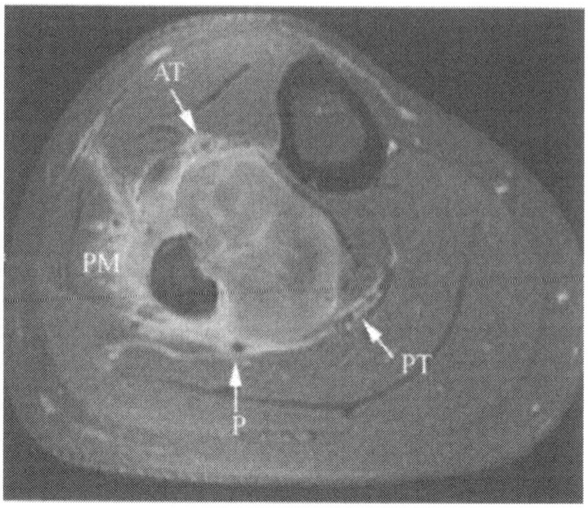

Figure 4L. Axial T1 post-Gd FS.

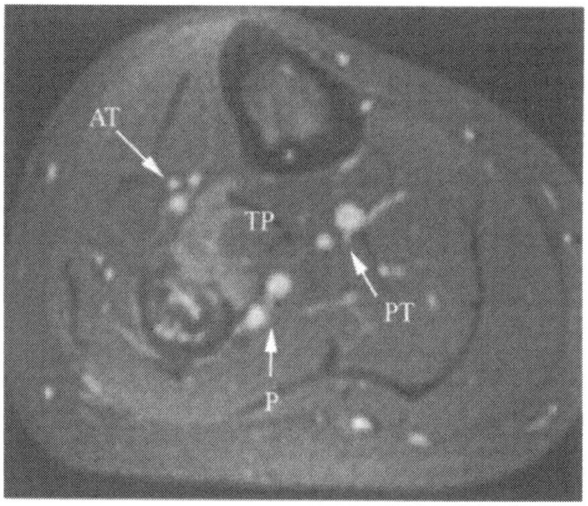

Figure 4M. Axial T1 post-Gd FS—10 weeks later.

Findings

This is a 13-year-old boy with an enlarging right calf mass.

Figures 4K, 4L. There is a proximal fibular diaphyseal mass with a substantial extraosseous component. The mass is heterogeneously hyperintense on STIR. The mass abuts and displaces the anterior tibial (AT), posterior tibial (PT), and peroneal (P) nerve-artery-vein complexes. The mass infiltrates the peroneus muscles (PM).

Figure 4M. After chemotherapy, the mass is significantly smaller with residual edema or tumor within the tibialis posterior muscle (TP). The mid-diaphysis of the fibula was resected along with portions of the tibialis posterior, flexor digitorum, and the interosseous membrane. Pathologic evaluation of the surgical specimen showed fibrosis without evidence of residual Ewing's sarcoma.

Pitfalls and Pearls

1. Ewing's sarcoma more commonly involves the diaphyses and flat bones, whereas osteosarcoma tends to involve the metaphyses.
2. Use an appropriate surface coil with a small field of view to delineate fine detail tumoral anatomy. Include additional large field of view T1W and fluid-sensitive sequences to assess the entire bone for skip metastases (7).

References

1. Arndt CA, Crist WM. Common musculoskeletal tumors of childhood and adolescence. *N Engl J Med* 1999; 341:342–352.
2. Carvajal R, Meyers P. Ewing's sarcoma and primitive neuroectodermal family of tumors. *Hematol Oncol Clin North Am* 2005; 19:501–525, vi–vii.
3. Saifuddin A, Whelan J, Pringle JA, Cannon SR. Malignant round cell tumours of bone: Atypical clinical and imaging features. *Skeletal Radiol* 2000; 29:646–651.
4. Kransdorf MJ, Smith SE. Lesions of unknown histogenesis: Langerhans cell histiocytosis and Ewing sarcoma. *Semin Musculoskelet Radiol* 2000; 4:113–125.
5. Cotterill SJ, Ahrens S, Paulussen M, et al. Prognostic factors in Ewing's tumor of bone: Analysis of 975 patients from the European Intergroup Cooperative Ewing's Sarcoma Study Group. *J Clin Oncol* 2000; 18:3108–3114.
6. Meyer JS, Mackenzie W. Malignant bone tumors and limb-salvage surgery in children. *Pediatr Radiol* 2004; 34:1030.
7. Fletcher BD. Imaging pediatric bone sarcomas: Diagnosis and treatment-related issues. *Radiol Clin North Am* 1997; 35:1477–1494.
8. Argon A, Basaran M, Yaman F, et al. Ewing's sarcoma of the axial system in patients older than 15 years: Dismal prognosis despite intensive multiagent chemotherapy and aggressive local treatment. *Jpn J Clin Oncol* 2004; 34:667–672.
9. Eggli KD, Quiogue T, Moser RP, Jr. Ewing's sarcoma. *Radiol Clin North Am* 1993; 31:325–337.
10. Brisse H, Ollivier L, Edeline V, et al. Imaging of malignant tumours of the long bones in children: Monitoring response to neoadjuvant chemotherapy and preoperative assessment. *Pediatr Radiol* 2004; 34:595–605.
11. van der Woude HJ, Bloem JL, Hogendoorn PC. Preoperative evaluation and monitoring chemotherapy in patients with high-grade osteogenic and Ewing's sarcoma: Review of current imaging modalities. *Skeletal Radiol* 1998; 27:57–71.
12. van der Woude HJ, Bloem JL, Verstraete KL, Taminiau AH, Nooy MA, Hogendoorn PC. Osteosarcoma and Ewing's sarcoma after neoadjuvant chemotherapy: Value of dynamic MR imaging in detecting viable tumor before surgery. *AJR Am J Roentgenol* 1995; 165:593–598.

Case 5

History

This is a 7-year-old girl who had a fracture of her right ankle 7 months ago.

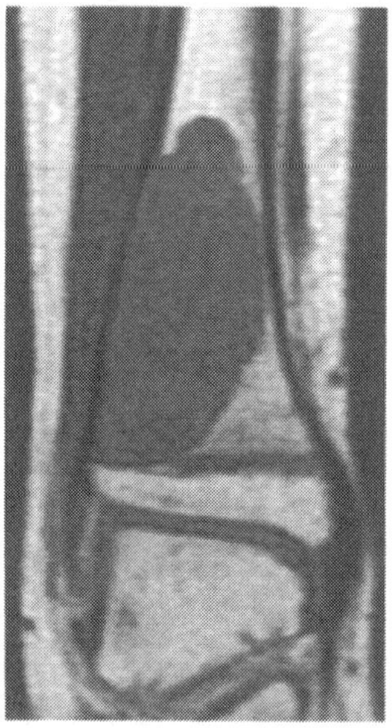

Figure 5A. Coronal T1 of the right distal tibia.

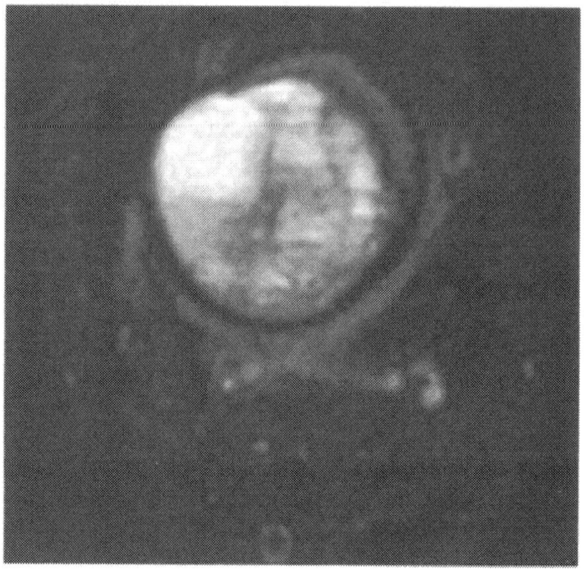

Figure 5B. Axial T2 FS.

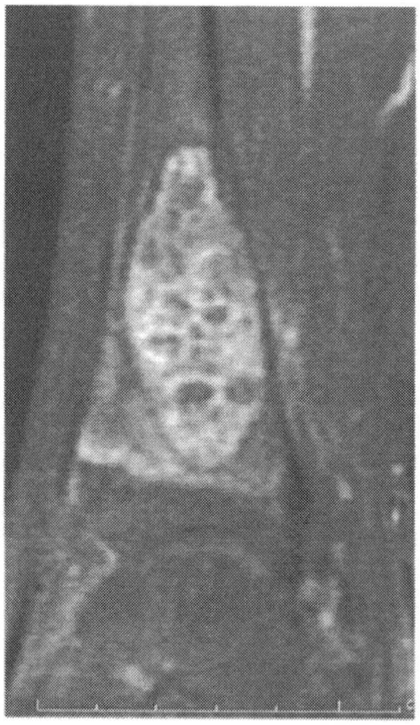

Figure 5C. Sagittal T1 post-Gd FS.

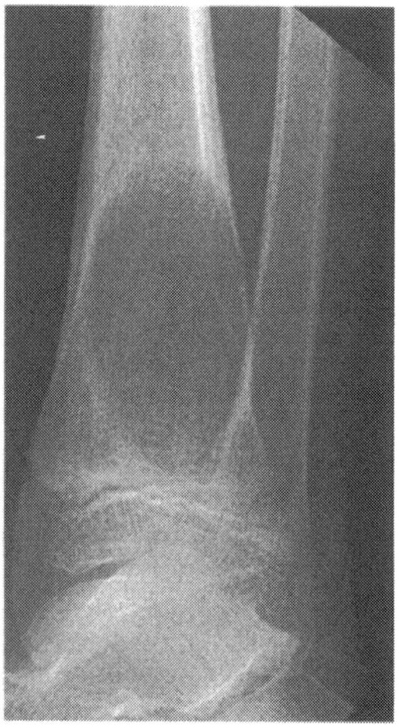

Figure 5D. Oblique radiograph.

Figures 5A, 5B, 5C (5B with annotations). A distal tibia metaphyseal lesion is identified that causes mild expansion of bone. This lesion is isointense on T1, hyperintense with fluid-fluid levels **(arrowhead)** on T2, and demonstrates diffuse septal and marginal enhancement. There is minor extraosseous soft tissue edema posterior and adjacent to this lesion. The cortex is thin **(thick arrow)** but intact. A well-defined sclerotic rim **(thin arrow)** separates the lesion from normal tibial marrow.

Figure 5D. The radiograph demonstrates a slightly expansile osteolytic lesion with a well-defined nonsclerotic transition zone. There is endosteal cortical thinning, but the cortex is intact. Mature appearing periosteal reaction is seen in the distal tibia metadiaphysis.

Figures 5E, 5F. Recurrent fluid filled lesional foci **(arrows)** are identified surrounded by hypointense allograft bone chips. Note multiple fluid-fluid levels **(arrowheads)**.

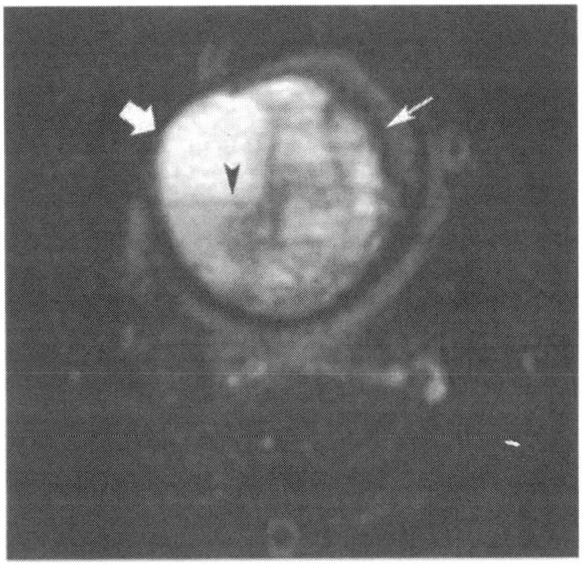

Figure 5B* Annotated.

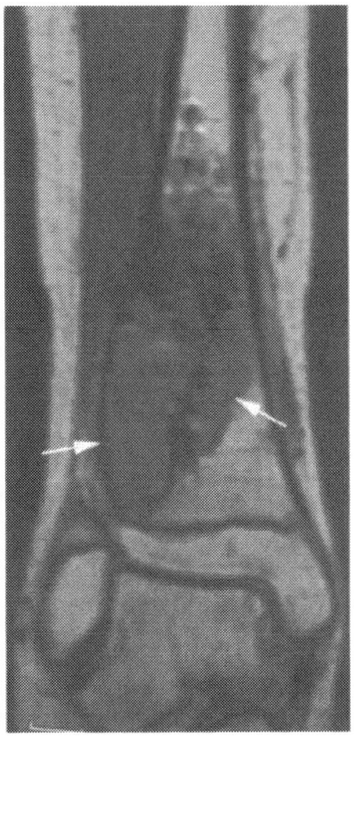

Figure 5E. Coronal T1—six months later and after treatment.

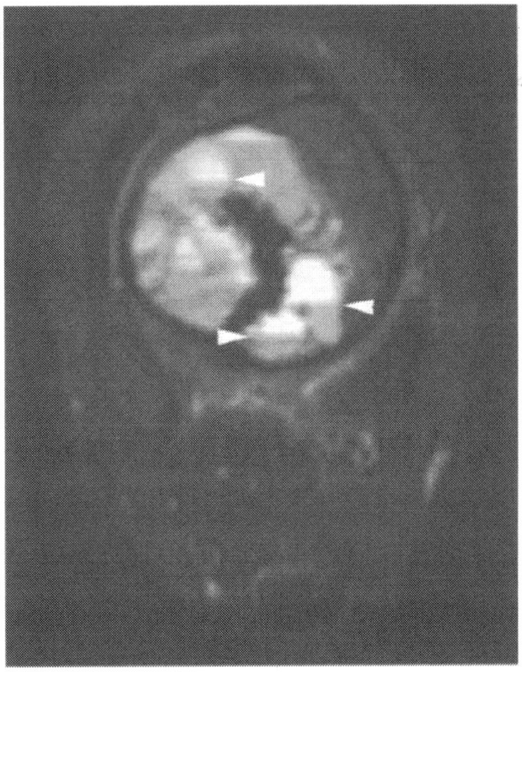

Figure 5F. Axial T2 FS.

Diagnosis

Aneurysmal bone cyst (ABC)

Questions

1. T/F: ABC may arise from the cortex.
2. How often does primary ABC demonstrate central enhancement?

Discussion

Aneurysmal bone cysts (ABCs) occur as primary lesions or arise secondary to preexisting bone tumors. The mean age of presentation of a primary ABC is 10.2 years (range 1.5 to 17 years) (1). In children, the five most common locations in order of frequency are the femur, tibia, spine, humerus, and pelvis (Figures 5G–5J). They are usually found in the metaphysis and are most often intramedullary, and less commonly cortical, surface (between the periosteum and cortex), or mixed (Answer to Question 1) (2). A large surface ABC may be indistinguishable from a cortical ABC when there is significant cortical destruction. Intramedullary lesions are usually slightly eccentric in the long tubular bones and central in the short tubular bones of the hands and feet. Secondary ABCs may arise from giant cell tumor, osteoblastoma, chondroblastoma, nonossifying fibroma, unicameral bone cyst, eosinophilic granuloma, fibrous dysplasia, and malignancies such as telangiectatic osteosarcoma or Ewing's sarcoma.

On plain radiography, ABCs often have an aggressive appearance. They are osteolytic, expansile, may show periosteal reaction, and contain multiple septations. The width of the transition zone is variable and the margins may show sclerosis, depending on aggressiveness of the lesion. As ABCs enlarge, they expand and thin cortical margins, although more rapidly growing lesions may obliterate the cortex and balloon the periosteal envelope (2).

On MRI, ABCs are well-defined, lobulated, and expansile lesions that are generally hypointense on T1 and are hyperintense on T2W sequences (3). Fluid levels related to layering blood products are often encountered and are best visualized on fluid-sensitive and GRE sequences. After gadolinium, multiple enhancing septa are often seen. Additional features include cortical disruption, periosteal reaction, and a hypointense sclerotic rim. In one study, as many as 1/3 of primary ABCs demonstrated perilesional soft tissue and marrow edema and approximately 12% demonstrated central enhancement, mimicking a primary neoplasm with secondary ABC (Answer to Question 2) (4). However, the presence of solid and/or enhancing components as well as cortical destruction in an ABC should always raise the consideration of an underlying primary lesion.

MRI, in conjunction with radiography, is used to assess for ABC recurrence after therapy. Recurrence shares similar features with the original primary ABC (Figures 5E, 5F). The reported incidence of primary ABC recurrence after surgical resection ranges from 10% to 59% in children (5). When there is recurrence following treatment of an ABC, an underlying primary bone lesion must always be ruled out. Imaging after treatment may be confusing because granulation tissue related to healing may mimic solid tumor.

In this patient, the distal tibial lesion was considered to be a primary ABC, since no solid enhancing components were identified, the cortex was intact, and the periosteal reaction had benign features. The lesion was pathologically proven to be a primary ABC and was curettaged and packed. The lesion was curettaged and packed again 8 months later, following recurrence (Figures 5E, 5F).

Orthopedic Perspective

The main concern for the orthopedist is to ensure that the lesion is in fact an ABC and not another tumor, such as telangiectatic osteosarcoma. ABC is a lesion that can mimic other tumors and can be secondary to other lesions. Although the presence of fluid-fluid levels is reassuring, they are not pathognomonic of ABC and are seen with osteosarcoma. Solid enhancing tumoral components should undergo percutaneous or open biopsy before a primary ABC is treated. Once the diagnosis has been assured, the extent of the lesion is important in planning treatment. At times both ABC and UBC can cross the growth plate; that is important to assess prior to treatment. Since ABC has a high local recurrence rate following curettage, monitoring for recurrence is important; MRI is very useful in this regard. Usual treatment is curettage and bone graft packing or, less commonly, sclerotherapy. Resection is occasionally done in expendable bones.

What the Clinician Needs to Know

1. Distinguish ABC from other lesions, especially osteosarcoma.
2. The relationship of the lesion to the growth plate is important in skeletally immature patients.
3. Since the recurrence rate following intralesional treatment is high, monitoring for recurrence post-treatment is important. MRI is particularly useful in this regard.

Answers

1. True.
2. Approximately 12% of cases.

Additional Example

Superior Pubic Ramus ABC, Recurrent

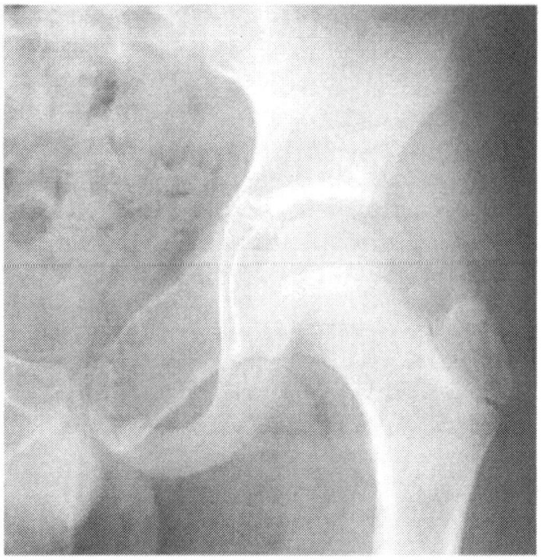

Figure 5G. AP radiograph.

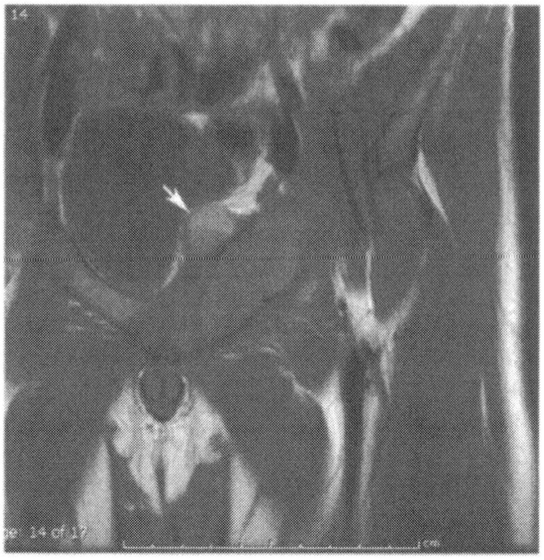

Figure 5H. Coronal T1.

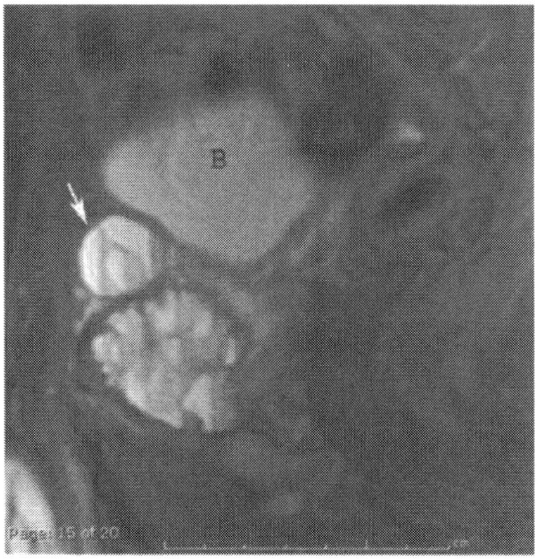

Figure 5I. Sagittal STIR of the left pelvis.

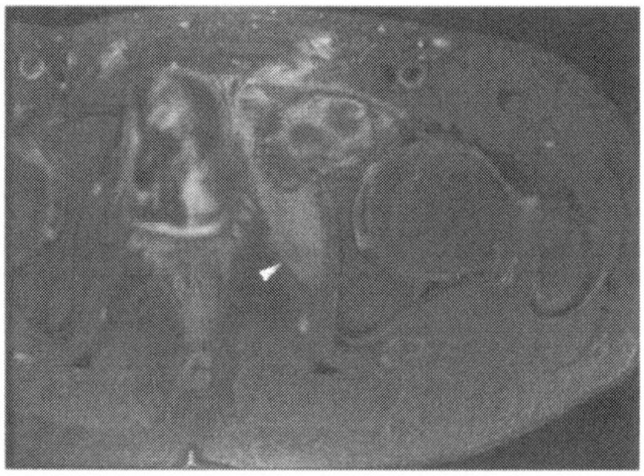

Figure 5J. Axial T1 post-Gd FS.

Findings

This 12-year-old boy was diagnosed with a primary ABC of the left superior pubic ramus 5 months earlier and was treated with sclerotherapy.

Figures 5G, 5H, 5I, 5J. Plain radiograph and MRI demonstrate recurrent ABC with an extraosseous component **(arrows)**. Fluid-fluid levels, septal and peripheral enhancement, and marginal sclerosis are seen. There is adjacent soft tissue enhancement, particularly within the obturator internus muscle **(arrowhead)**.

Pitfalls and Pearls

1. MRI is recommended in the work-up of suspected ABC to assess for underlying primary bone lesion and to exclude malignancy.
2. A unicameral bone cyst should be considered, rather than ABC, if there is only mild intramedullary expansion, metadiaphyseal location, and absent septations. Fluid-fluid levels may also be seen with a unicameral bone cyst, especially after minor trauma or pathologic fracture.

References

1. Cottalorda J, Kohler R, Sales de Gauzy J, et al. Epidemiology of aneurysmal bone cyst in children: A multicenter study and literature review. *J Pediatr Orthop B* 2004; 13:389–394.
2. Maiya S, Davies M, Evans N, Grimer J. Surface aneurysmal bone cysts: A pictorial review. *Eur Radiol* 2002; 12:99–108.
3. Sullivan RJ, Meyer JS, Dormans JP, Davidson RS. Diagnosing aneurysmal and unicameral bone cysts with magnetic resonance imaging. *Clin Orthop Relat Res* 1999:186–190.
4. Mahnken AH, Nolte-Ernsting CC, Wildberger JE, et al. Aneurysmal bone cyst: Value of MR imaging and conventional radiography. *Eur Radiol* 2003; 13:1118–1124.
5. Dormans JP, Hanna BG, Johnston DR, Khurana JS. Surgical treatment and recurrence rate of aneurysmal bone cysts in children. *Clin Orthop Relat Res* 2004:205–211.

Case 6

History

This is a 17-year-old boy with a several-year history of an enlarging painful left forearm mass.

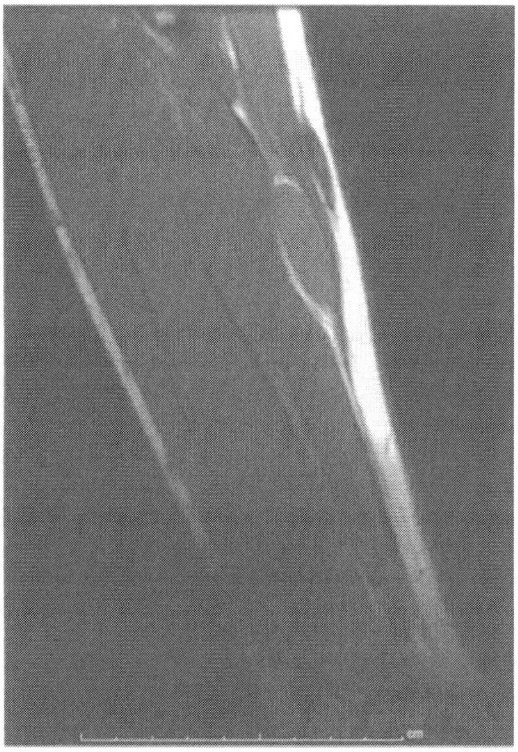

Figure 6A. Coronal T1 of the left forearm.

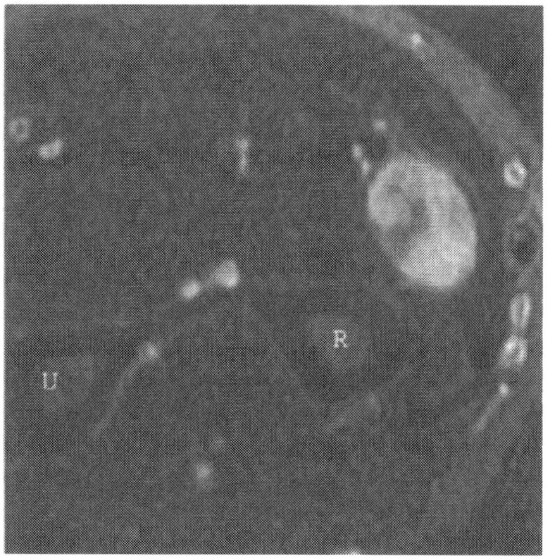

Figure 6B. Axial T2 FS.

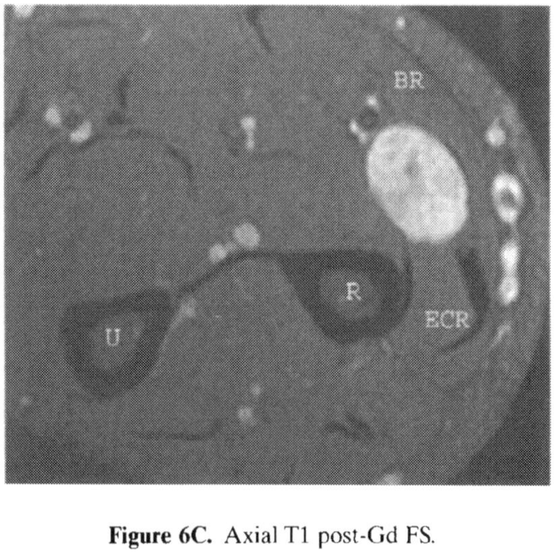

Figure 6C. Axial T1 post-Gd FS.

Figures 6A, 6B, 6C. There is a well-defined intermuscular forearm mass that is slightly hyperintense on T1, heterogeneously hyperintense on T2, and shows diffuse enhancement. A fat plane separates the mass from the normal extensor carpi radialis (ECR) and brachioradialis muscles (BR), best seen on the coronal T1W sequence. Radius (R), ulna (U).

Figures 6D, 6E. The mass is well-defined and hypervascular with small echogenic foci **(arrows)** suggesting calcifications.

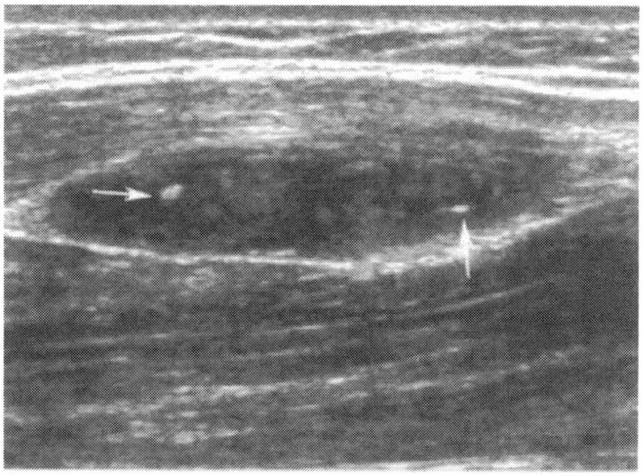

Figure 6D. Sagittal ultrasound.

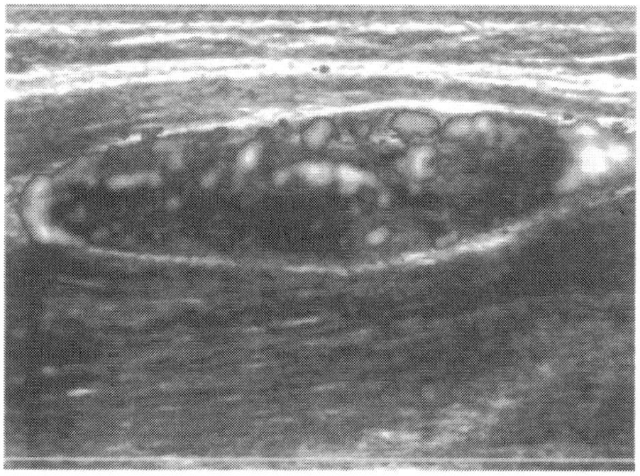

Figure 6E. Doppler ultrasound.

Diagnosis

Synovial sarcoma

Questions

1. T/F: Synovial sarcoma is more likely to contain internal calcification and/or cysts compared with extremity rhabdomyosarcomas.
2. Which imaging features of synovial sarcomas are associated with a poorer prognosis?

Discussion

Synovial sarcoma is the second most common soft tissue sarcoma (after rhabdomyosarcoma) in children. Approximately 30% of all patients with this tumor are under 20 years of age at presentation (1). The majority of lesions occur within 5 to 7 cm of a joint (1, 2). In a series of 34 patients, only 2 cases (6%) were intra-articular (2). The lower extremities are most commonly affected.

On MRI, synovial sarcomas may be sharply defined with respect to adjacent soft tissues due to the presence of a pseudocapsule. They may contain cystic components with fluid-fluid levels, hemorrhage, fibrous elements, and calcifications (3). Cystic change (approximately 38%) and calcifications (approximately 30%) may be found within synovial sarcomas, considerably greater than in other soft tissue sarcomas such as a rhabdomyosarcoma (Answer to Question 1) (3, 4). Calcifications tend to be punctate and centrally located, but a peripheral location may also occur. Jones et al. reported that 71% of cases abut or invade the adjacent bone (2). A triple signal pattern on T2W sequences (hypointense, isointense, and hyperintense) has been reported in 35% of synovial sarcomas (Figures 6F, 6G) (2). However, it may be seen in a variety of benign and malignant soft tissue tumors, especially if there has been minor trauma.

Factors associated with a poor prognosis in patients with synovial sarcoma include proximal location in a given extremity, size >5 cm, absence of calcification, presence of cysts, hemorrhage, and a triple signal pattern on T2W sequences (Answer to Question 2) (5).

Because synovial sarcomas have a variable appearance on MRI, a correct diagnosis by imaging prior to surgery is rarely made. Berquist et al. observed that synovial sarcoma was most often misclassified as benign by MRI compared with various soft tissue malignancies because its well-defined margins and cyst-like components mimicked hematomas (6). Synovial sarcoma may mimic other benign entities including ganglion and synovial cyst, lymphatic malformation, and venous malformation (Figures 6H, 6I) (3). Venous malformation or microcystic lymphatic malformation may be impossible to differentiate from a synovial sarcoma by MRI. Ultrasound may be more informative, particularly for characterizing a venous malformation, by documenting tubular confluent vessels and an absence of a discrete soft tissue mass. The clinical assessment is critical when a vascular malformation is a consideration prior to imaging. When the dominant feature is tumoral enhancement, alternative diagnoses include rhabdomyosarcoma and benign entities such as fibrous tumors and nodular fasciitis (Figures 6J–6L). Nodular fasciitis is a benign fibroblastic proliferation that tends to occur in older patients (20–40 years), has a female predominance, and may be subcutaneous, intermuscular, or intramuscular (7). It typically presents as a subcutaneous upper-extremity mass.

In this patient, the US and MRI could not provide a specific diagnosis and biopsy was advised. The diagnosis of synovial sarcoma was pathologically proven.

Orthopedic Perspective

Synovial sarcoma is a great mimic. Masses can be present for many years and ignored until some subtle change brings the patient to medical attention and a synovial sarcoma is then discovered. As shown here, synovial sarcoma can simulate benign lesions including peripheral nerve sheath tumors, vascular malformations, nodular fasciitis, and other sarcomas. Imaging may provide some clues to point the clinician in the right direction, but, as in all soft tissue lesions, general principles guiding the decision for biopsy should be applied. Solid masses that are large and deep to fascia should be viewed with suspicion, especially if they are painful. The imaging can guide the placement of open or needle biopsies to avoid contaminating neurovascular structures and allow excision of the biopsy tract with the definitive resection. As in other soft tissue sarcomas, the main goal of imaging is to define the mass relative to key neurovascular structures in planning the surgical approach. In adults, resection is frequently combined with radiotherapy, but its use in children is limited since the radiation fields may involve the physis, resulting in growth disturbance. The case of the synovial sarcoma adjacent to the knee joint described above is such an example. This patient underwent excision and radiation therapy to avoid a more debilitating operation but has a mild angular deformity due to the effects of radiation on the periphery of the distal femoral growth plate. As with all sarcomas, a metastatic screen with chest CT and regional lymph node assessment is necessary.

What the Clinician Needs to Know

1. Distinguish a synovial sarcoma from other soft tissue masses (benign and malignant). Synovial sarcoma is a great mimic.
2. Metastatic disease and lymph node involvement should be ruled out.
3. Imaging to demonstrate the extent of the lesion and its relationship to neurovascular structures and bone is needed to plan surgical treatment.
4. The relationship of the mass to the growth plates is important if adjuvant radiotherapy is planned.

Answers

1. True.
2. Proximal location, size >5 cm, absence of calcification, presence of cysts, presence of hemorrhage, and triple signal pattern on T2W sequences.

Additional Examples

Synovial Sarcoma

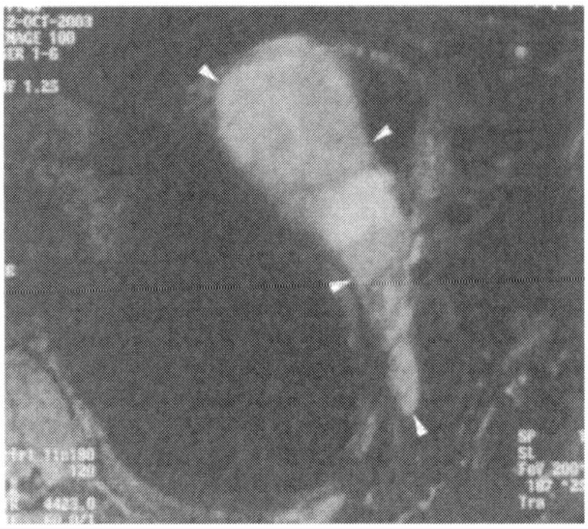

Figure 6F. Axial T2 FS of the left chest.

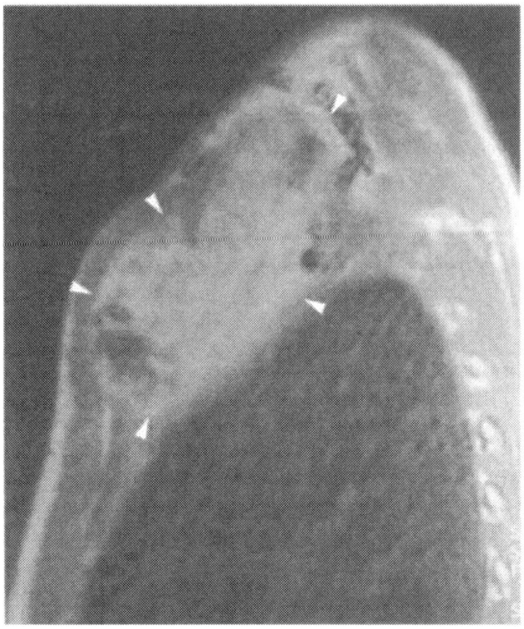

Figure 6G. Sagittal T1 post-Gd FS of the left chest.

Findings

This is a 13-year-old boy with an enlarging left chest wall/axillary mass.

Figures 6F, 6G. The mass is heterogeneous, demonstrating a triple signal pattern (hypointense, isointense, and hyperintense) on T2 and showing diffuse heterogeneous enhancement **(arrowheads)**. This was pathologically proven to be a synovial sarcoma.

Synovial Sarcoma Mimicking a Venous Malformation

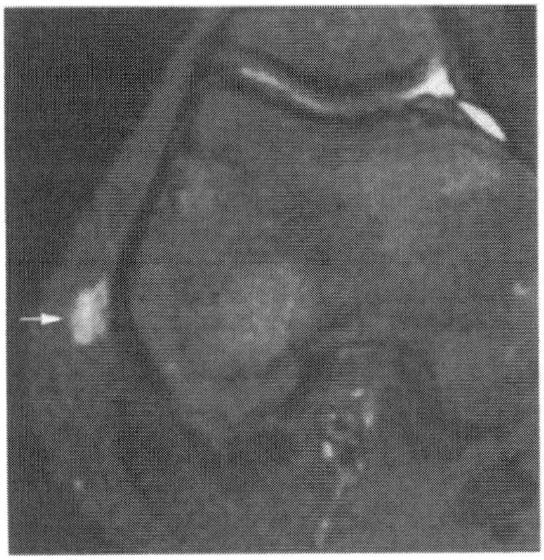

Figure 6H. Axial T2 FS of the right knee.

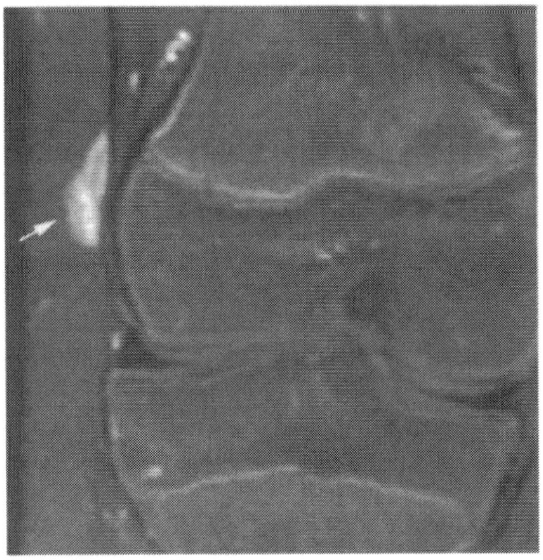

Figure 6I. Coronal T1 post-Gd FS.

Findings

This is a 9-year-old girl who noted a soft tissue bump along her lateral right knee. **Figures 6H, 6I.** MRI shows a small, well-defined, lobular soft tissue mass immediately superficial to the tensor fascia lata, which is hyperintense on T2 and homogeneously enhances **(arrows)**. Given its sharp lobulated margins, it was mistakenly diagnosed as a venous malformation. Because it continued to grow, the lesion was excised and was pathologically proven to represent a synovial sarcoma.

Nodular Fasciitis

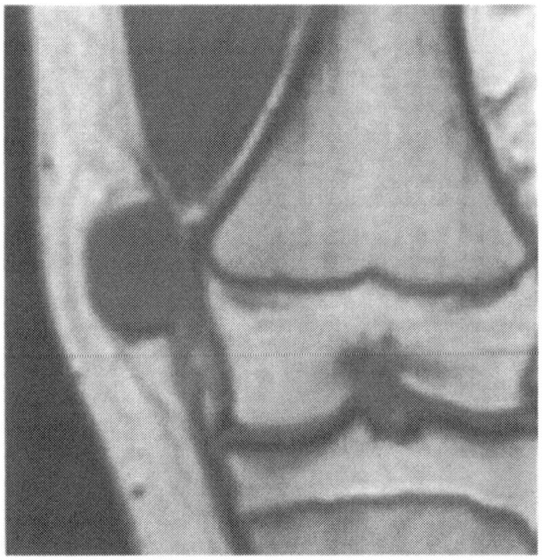

Figure 6J. Coronal T1 of the left knee.

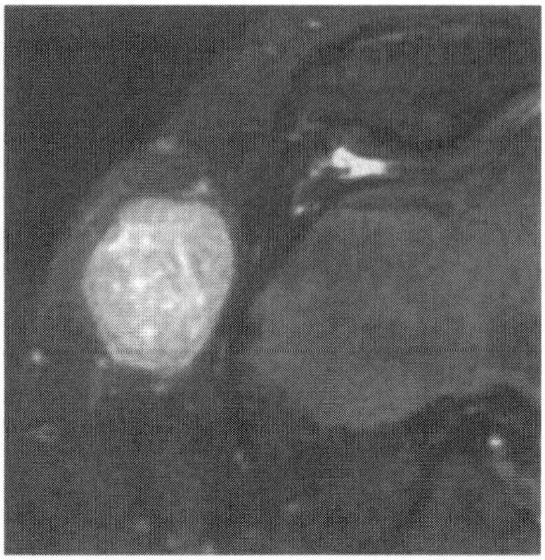

Figure 6K. Axial T2 FS.

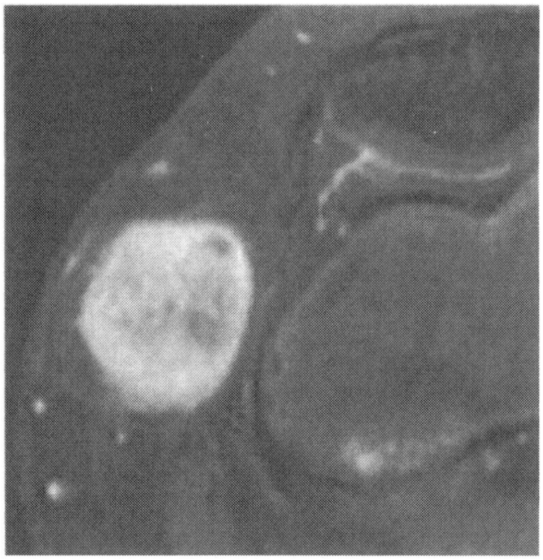

Figure 6L. Axial T1 post-Gd FS.

Findings

This 12-year-old girl noticed a painless mass along the medial aspect of her left knee.

Figures 6J, 6K, 6L. This mass is hypo- to isointense on T1, markedly hyperintense on T2, and shows diffuse enhancement, most intense in the periphery. This mass closely abuts but does not invade the medial retinaculum. This was surgically confirmed to represent nodular fasciitis with myxoid and hyaline stroma. As seen in this case, these benign tumors are indistinguishable from soft tissue sarcomas, and may have an even more ominous appearance when large.

Pitfalls and Pearls

1. Size for size, synovial sarcomas tend to be more heterogeneous in appearance and contain more calcifications, cysts, and hemorrhage compared with extremity rhabdomyosarcomas.
2. Sharp margins and the lack of peritumoral edema and enhancement do not rule out soft tissue malignancies. With a few exceptions, such as lipomas and certain vascular lesions, biopsy is usually indicated.

References

1. Valenzuela RF, Kim EE, Seo JG, Patel S, Yasko AW. A revisit of MRI analysis for synovial sarcoma. *Clin Imaging* 2000; 24:231–235.
2. Jones BC, Sundaram M, Kransdorf MJ. Synovial sarcoma: MR imaging findings in 34 patients. *AJR Am J Roentgenol* 1993; 161:827–830.
3. Nakanishi H, Araki N, Sawai Y, et al. Cystic synovial sarcomas: Imaging features with clinical and histopathologic correlation. *Skeletal Radiol* 2003; 32:701–707.
4. McCarville MB, Spunt SL, Skapek SX, Pappo AS. Synovial sarcoma in pediatric patients. *AJR Am J Roentgenol* 2002; 179:797–801.
5. Tateishi U, Hasegawa T, Beppu Y, Satake M, Moriyama N. Synovial sarcoma of the soft tissues: Prognostic significance of imaging features. *J Comput Assist Tomogr* 2004; 28:140–148.
6. Berquist TH, Ehman RL, King BF, Hodgman CG, Ilstrup DM. Value of MR imaging in differentiating benign from malignant soft-tissue masses: Study of 95 lesions. *AJR Am J Roentgenol* 1990; 155:1251–1255.
7. Wang XL, De Schepper AM, Vanhoenacker F, et al. Nodular fasciitis: Correlation of MRI findings and histopathology. *Skeletal Radiol* 2002; 31:155–161.

Case 7

History

This is a 5-year-old girl with a 2-week history of lower extremity weakness, especially when walking up stairs. Four days ago, she developed a rash on the chest, face, and dorsum of her fingers.

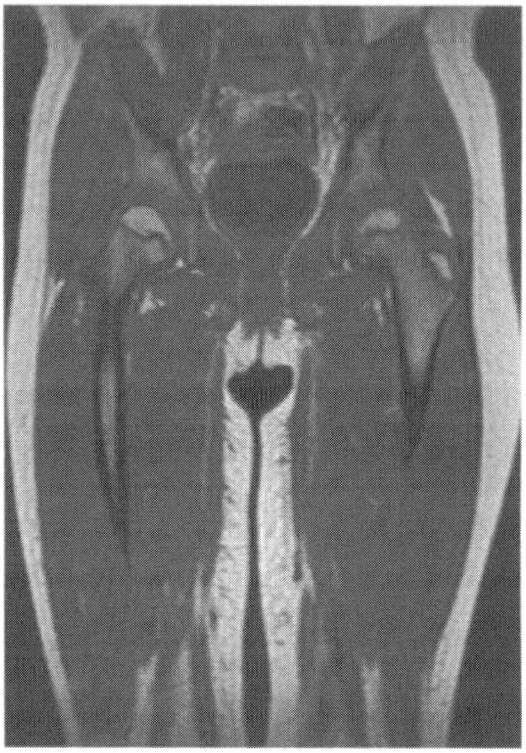

Figure 7A. Coronal T1.

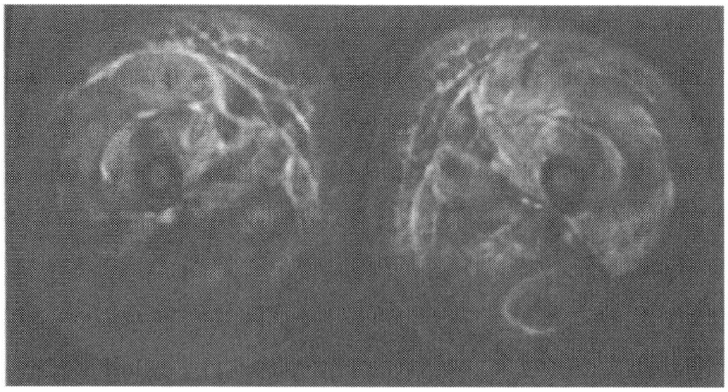

Figure 7B. Axial T2 FS.

Figure 7A. T1W images are normal.

Figure 7B (7B with annotations). There is symmetric increased quadriceps and adductor muscle, superficial fascial **(arrow)**, and subcutaneous **(arrowhead)** T2 hyperintensity present, consistent with edema. The hamstring muscles show mild involvement.

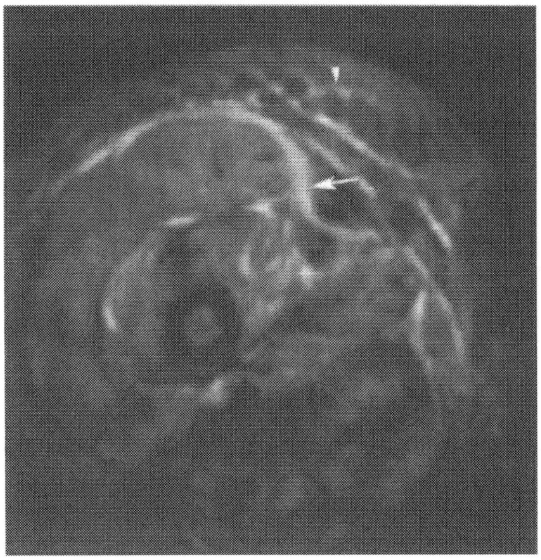

Figure 7B* Annotated, cropped to right side.

Diagnosis

Juvenile dermatomyositis

Question

1. Which clinical and radiologic features of dermatomyositis are more commonly seen in children compared with adult patients?

Discussion

The most common idiopathic inflammatory myopathy in children is dermatomyositis (1). Polymyositis and inclusion body myositis are rare in children. Dermatomyositis is a multisystem autoimmune process that primarily affects striated muscle and skin. The age range at presentation is 2 to 15 years and females are more commonly affected (1, 2). Patients with juvenile dermatomyositis are more likely to develop soft tissue calcifications, skin ulcerations, and arthritis, and suffer from vasculitis-like problems (e.g., skin and mucosal ulceration) compared with adult patients with dermatomyositis (Answer to Question 1) (1).

On MRI, diffuse, heterogeneous increased SI on water-sensitive sequences may be seen within muscle, superficial and deep fascia, skin, and subcutaneous fat. Enhancement after gadolinium administration usually parallels hyperintensity on fluid-sensitive sequences. Consequently, contrast administration is usually not necessary in the MRI evaluation of probable dermatomyositis. The proximal muscles of the appendicular skeleton are often symmetrically involved. The proximal lower extremity is usually affected first followed by the proximal upper-extremity muscles (3). Increased T2 relaxation times have correlated with muscle inflammation as determined clinically (4). For the large muscle groups of the thigh, the adductor muscles are often affected (89%) whereas the hamstring muscles are less commonly affected (56%) in patients with active disease (5).

The musculoskeletal complications of dermatomyositis include dystrophic calcification, muscle necrosis, and muscle atrophy with fatty replacement (Figures 7C, 7D). In one study, soft tissue calcifications were seen within 9 months once muscle edema was present on STIR sequences in 5 of 26 patients (6). Calcifications may be seen in 25% to 50% of patients with juvenile dermatomyositis (2). Muscle atrophy and fatty replacement are well delineated on T1W sequences (Figure 7C).

The differential diagnosis for the muscle and fascial inflammation is broad, and clinical correlation is generally required. The MRI features of dermatomyositis may mimic necrotizing fasciitis, strenuous exercise, pyomyositis, viral myositis, myositis ossificans, acute muscle denervation, metabolic myopathies, graft versus host disease following organ transplant (Figure 7E), medication-related myositis (e.g., diuretics), other rheumatologic conditions such as lupus, polyarteritis nodosa, and juvenile rheumatoid arthritis as well as undifferentiated inflammatory myopathy (Figures 7C, 7D) (3, 7–11). Inflammation of a single muscle or muscle group should not raise the diagnostic possibility of dermatomyositis, since this entity is only a consideration if there is a typical bilateral, symmetric, and proximal appendicular distribution of muscle inflammation.

The diagnosis of dermatomyositis in this case was made based on clinical history, rash, MRI features with a typical distribution of myositis, and abnormal muscle enzyme values. The patient was begun on methotrexate and corticosteroids. The MRI and clinical features of dermatomyositis made muscle biopsy unnecessary in this case.

Orthopedic Perspective

The clinical spectrum of juvenile dermatomyositis includes a characteristic skin rash; proximal muscle weakness; elevated laboratory values (including antinuclear antibodies [ANA], myositis-specific autoantibodies, and muscle enzyme values such as creatine kinase); and an abnormal MRI, electromyography (EMG), or muscle biopsy. EMG and muscle biopsy are invasive procedures that may be avoided when MRI shows characteristic features of dermatomyositis. However, muscle biopsy may still be required when the clinical presentation and MRI features are atypical. In these situations, the MRI findings may be used to map which muscle should be biopsied by identifying areas of myositis.

What the Clinician Needs to Know

1. Is there both muscle atrophy and myositis?
2. How does the degree and distribution of myositis compare with prior exams?
3. Where is the precise location of the muscle edema? Is it localized at the myotendinous junction? If so, are the abnormalities actually due to myotendinous tear rather than an idiopathic myositis?

Answer

1. Children are more likely to develop soft tissue calcifications, skin ulcerations, arthritis, and to suffer from vasculitis-like problems.

Additional Examples

Diffuse Myositis and Atrophy

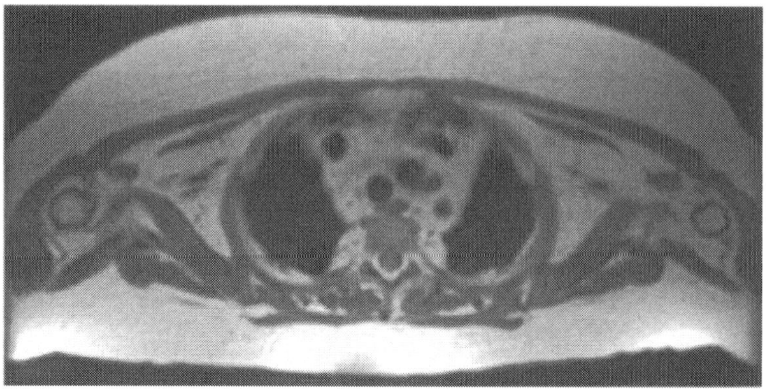

Figure 7C. Axial T1 through the shoulders.

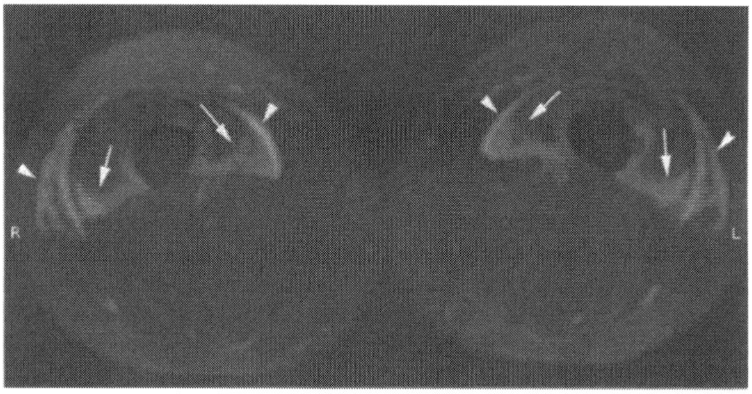

Figure 7D. Axial STIR through the distal thighs.

Findings

This is a 10-year-old boy with an undifferentiated inflammatory myopathy.

Figure 7C. There is global muscle atrophy with fatty change present in the shoulder girdle. There is also abundant subcutaneous, mediastinal, and subpleural fat, most likely a consequence of prior steroid therapy.

Figure 7D. Muscle atrophy is also present in the distal thigh muscles. Hyperintensity present within the distal quadricep muscles **(arrows)** and superficial fascia **(arrowheads)** are appreciated and compatible with active inflammatory changes.

Post-transplant Myositis

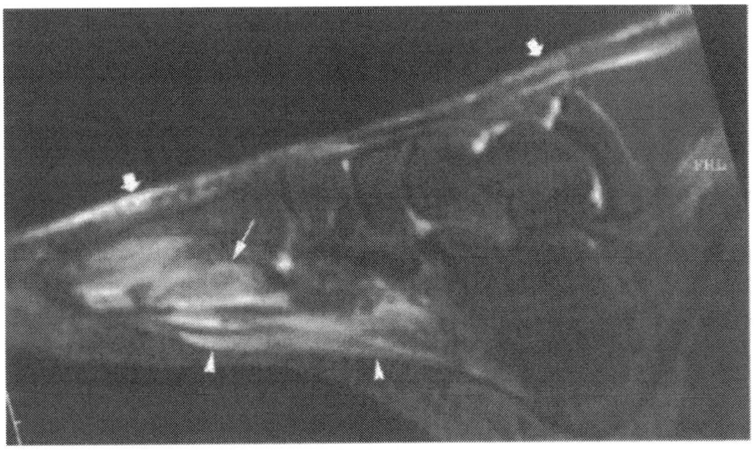

Figure 7E. Sagittal STIR of the left foot.

Findings

This is a 13-year-old girl who had a lung transplant six months ago because of end-stage lung disease related to cystic fibrosis (CF). She presented with severe left foot pain and the study was ordered to rule out osteomyelitis.

Figure 7E. There is diffuse increased STIR SI present along the plantar fascia **(arrowheads)** and muscles **(thin arrow)** including the flexor hallucis longus (FHL). There is also skin thickening and subcutaneous soft tissue inflammation **(thick arrows)**. There is normal marrow SI. The presumed etiology was post-transplant myositis versus drug-induced myositis. Less likely, post-transplant lymphoproliferative disorder (PTLD) would also be considered in this patient (follow-up 9 months later showed no evidence of PTLD).

Pitfalls and Pearls

1. Do not confuse muscle edema related to vigorous exercise with active disease in patients with known dermatomyositis (12).
2. Dermatomyositis involves mainly the proximal muscles of the appendicular skeleton with a symmetric bilateral distribution.
3. Make sure that similar sequence parameters are being used for comparison for follow-up exams. For example, a T2 FS sequence should be compared with a prior T2 FS sequence, not with a STIR.
4. Vasoocclusive crisis in sickle cell disease can produce MRI findings simulating myositis.

References

1. Wedderburn LR, Li CK. Paediatric idiopathic inflammatory muscle disease. *Best Pract Res Clin Rheumatol* 2004; 18:345–358.
2. Hanlon R, King S. Overview of the radiology of connective tissue disorders in children. *Eur J Radiol* 2000; 33:74–84.
3. Chan WP, Liu GC. MR imaging of primary skeletal muscle diseases in children. *AJR Am J Roentgenol* 2002; 179:989–997.
4. Maillard SM, Jones R, Owens C, et al. Quantitative assessment of MRI T2 relaxation time of thigh muscles in juvenile dermatomyositis. *Rheumatology (Oxford)* 2004; 43:603–608.
5. Hernandez RJ, Sullivan DB, Chenevert TL, Keim DR. MR imaging in children with dermatomyositis: Musculoskeletal findings and correlation with clinical and laboratory findings. *AJR Am J Roentgenol* 1993; 161:359–366.
6. Kimball AB, Summers RM, Turner M, et al. Magnetic resonance imaging detection of occult skin and subcutaneous abnormalities in juvenile dermatomyositis: Implications for diagnosis and therapy. *Arthritis Rheum* 2000; 43:1866–1873.
7. May DA, Disler DG, Jones EA, Balkissoon AA, Manaster BJ. Abnormal signal intensity in skeletal muscle at MR imaging: Patterns, pearls, and pitfalls. *Radiographics* 2000; 20 Spec No:S295–315.
8. Courtney AE, Doherty, CC, Herron B, McCarron MO, Connolly JK, Jefferson JA. Acute polymyositis following renal transplantation. *Am J Transplant* 2004; 4:1204–1207.
9. Miller ML, Levinson L, Pachman LM, Poznanski A. Abnormal muscle MRI in a patient with systemic juvenile arthritis. *Pediatr Radiol* 1995; 25 Suppl 1:S107–108.
10. Garton MJ, Isenberg DA. Clinical features of lupus myositis versus idiopathic myositis: A review of 30 cases. *Br J Rheumatol* 1997; 36:1067–1074.
11. Hom C, Ilowite NT. Other rheumatic diseases in adolescence: Dermatomyositis, scleroderma, overlap syndromes, systemic vasculitis, and panniculitis. *Adolesc Med* 1998; 9:69–83, vi.
12. Summers RM, Brune AM, Choyke PL, et al. Juvenile idiopathic inflammatory myopathy: Exercise-induced changes in muscle at short inversion time inversion-recovery MR imaging. *Radiology* 1998; 209:191–196.

Case 8

History

This is a 14-year-old female gymnast with a 4-month history of medial left foot pain.

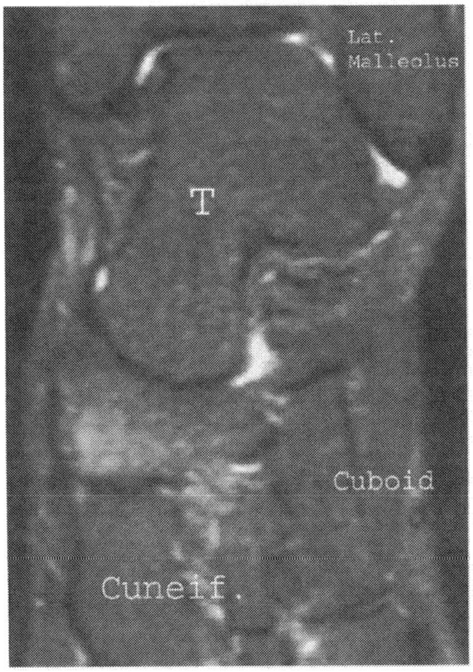

Figure 8A. Axial (footprint) STIR of the left foot.

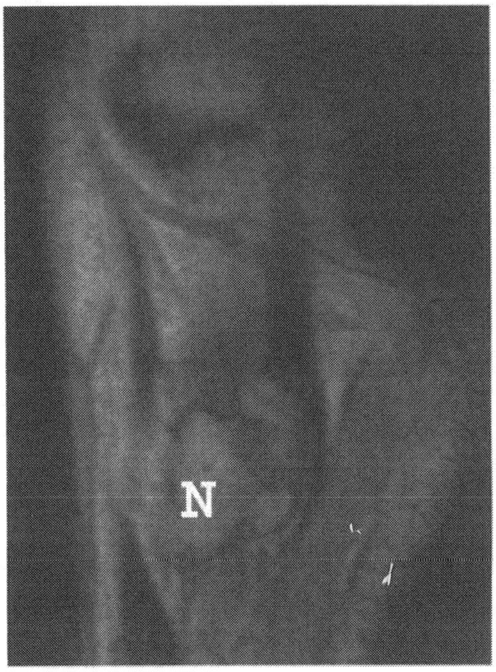

Figure 8B. Sagittal T1 of the medial aspect of the left foot.

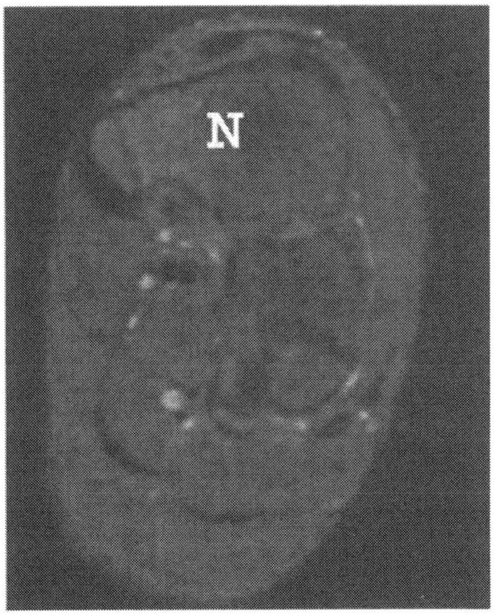

Figure 8C. Coronal STIR.

Figure 8A. There is diffuse increased SI throughout the navicular. Talus (T), cuneiforms (Cuneif), and lateral malleolus (lat. malleolus).

Figures 8B, 8C (8B with annotations). There is an accessory ossification center **(arrow)** medial to the navicular. There is mild increased STIR SI within the accessory ossification center and body of the navicular. The junction of the body and accessory ossification center of the navicular is serpiginous and is low SI on both T1 and STIR sequences. The normal appearing posterior tibial tendon (P) is seen inserting on the accessory ossification center. Navicular (N).

Figure 8D. The accessory navicular is dense. The margins of the synchondrosis are irregular and sclerotic with a somewhat curvilinear configuration **(arrows)**. Navicular (N), talus (T).

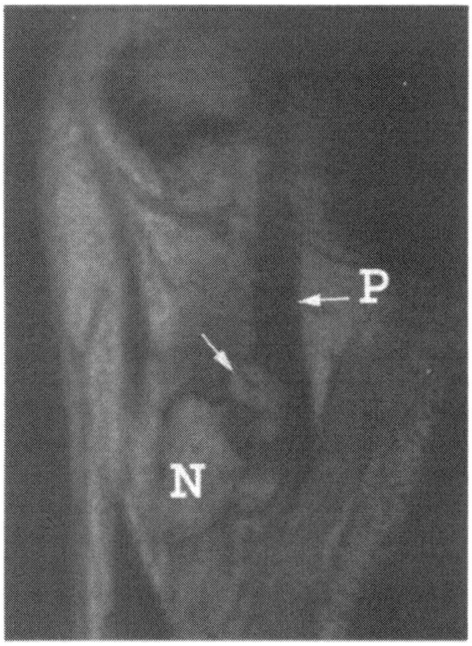

Figure 8B* Annotated.

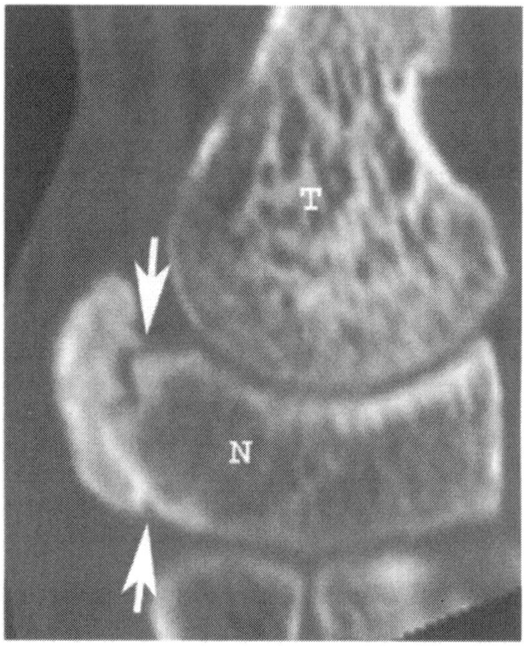

Figure 8D. Coronal-oblique CT reformat.

Diagnosis

Type 2 accessory navicular with stress change

Question

1. What are the three types of accessory navicular? Which type is most likely to be symptomatic?

Discussion

The ossification center of the navicular appears between 2 and 4 years of age (1). The medial aspect of the navicular is the last portion to ossify and may develop a secondary ossification center. The accessory ossification center is usually seen at 9 to 11 years in boys and approximately 7 to 9 years in girls (2). When the secondary ossification center fails to fuse with the body, an accessory navicular (AKA os tibiale externum) develops. An accessory navicular may be seen in 4–14% of the general population (2).

The accessory navicular may be classified into three types (Answer to Question 1) (2):

1. Separate ossicle imbedded within the posterior tibial tendon (PTT). This represents a sesamoid equivalent of the PTT.
2. Accessory ossification center that may be united to the body of the navicular by a cartilaginous or fibrous bar. The type 2 accessory ossification center is usually larger than a type 1. This is the most common accessory navicular to become symptomatic.
3. Cornuate navicular. This is a navicular that extends posteromedially and may represent a fused type 2 navicular.

Symptomatic type 2 accessory navicular bones may demonstrate edema of the body of the navicular, accessory navicular, and the insertion of the PTT (3). Medial avulsion injuries of a type 3 navicular (cornuate) may also occur (Figures 8E–8H). With frank avulsion injuries, displacement of the avulsed fragment and edema are present within the adjacent soft tissues and within the synchondrosis. Long-term changes related to a type 2 accessory navicular include PTT dysfunction and rupture as well as degenerative changes of the synchondrosis (4). PTT dysfunction may subsequently lead to a painful flatfoot; however, the association between an accessory navicular and a flatfoot deformity is unclear (5, 6).

The differential diagnosis for signal alterations of the navicular includes: stress fracture (Figures 8I, 8J), navicular osteonecrosis (Mueller Weiss syndrome), and Koehler's disease (see Case 99). In one study evaluating 55 navicular stress fractures, all fractures occurred along the central 1/3 of the proximal dorsal margin of the navicular bone (7). Mueller Weiss syndrome affects the adult population, may involve the entire navicular or may be confined to the lateral navicular, and may be accompanied by fragmentation and loss of navicular sagittal diameter along the lateral margin (8). Koehler's disease is an osteochondrosis of the navicular bone that tends to occur in children 3–7 years of age with little if any long-term sequelae (9, 10). A discussion of the osteochondroses may be found in Case 99.

An accessory navicular synchondrosis should not be mistaken for an avulsion or stress fracture. Fracture lines and fibrous synchondrosis are both low SI on all imaging sequences and may have associated marrow edema. However, fracture lines generally have straight, sharp margins, whereas fibrous synchondroses tend to have more irregular margins with a roughly curvilinear configuration. Ultimately, CT may be required to distinguish between a fracture line and a synchondrosis.

Based on the clinical and radiologic findings, the edema pattern in the navicular was probably due to chronic stress reaction related to the PTT dysfunction. This patient was asked to limit her activities and was given a foot boot and orthotic arch supports. She was symptom-free at follow-up.

Orthopedic Perspective

The accessory navicular may cause foot pain when associated with overuse in the setting of pes planus, pronation, and posterior tibial tendon dysfunction. The diagnosis is typically made by physical examination and plain radiographs. These patients may be referred for MRI when associated posterior tibial tendinopathy or alternative causes of a painful flatfoot are being considered. Nonoperative treatment of a symptomatic accessory navicular involves casting or bracing acutely, followed by physical therapy and orthotics. For recalcitrant cases, surgical excision of the accessory navicular bone may be required.

What the Clinician Needs to Know

1. Is there edema within the accessory navicular bone and PTT?
2. Classification of the accessory navicular (types 1–3).
3. Is there another cause for pes planus such as tarsal coalition?
4. Is there associated posterior tibial tendinopathy?

Answer

1. See discussion for classification. Of the three types, type 2 accessory navicular is most commonly symptomatic.

Additional Examples

Avulsion Fracture of a Type 3 Navicular (Cornuate)

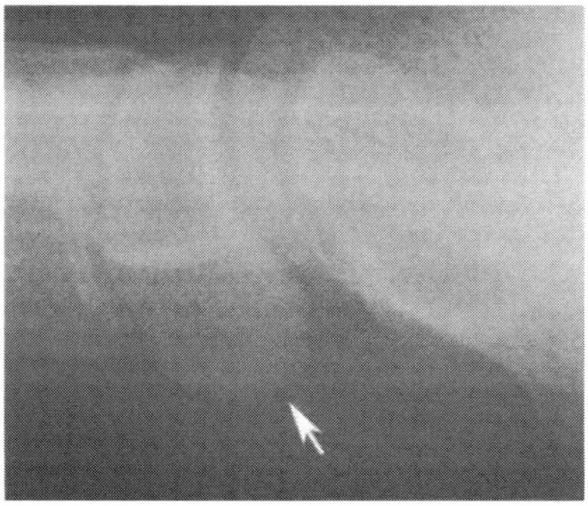

Figure 8E. Oblique radiograph of the left foot.

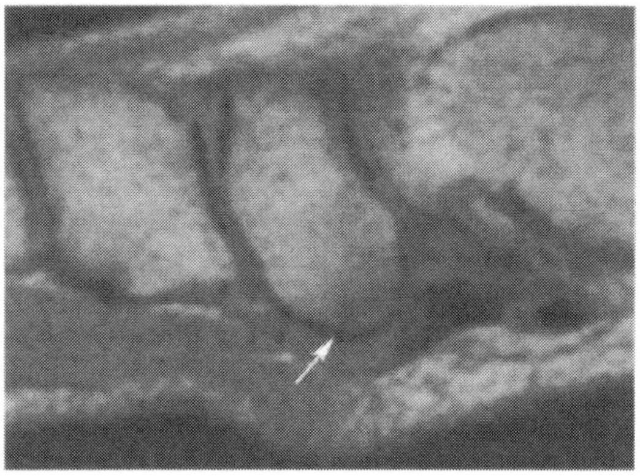

Figure 8F. Oblique-sagittal T1.

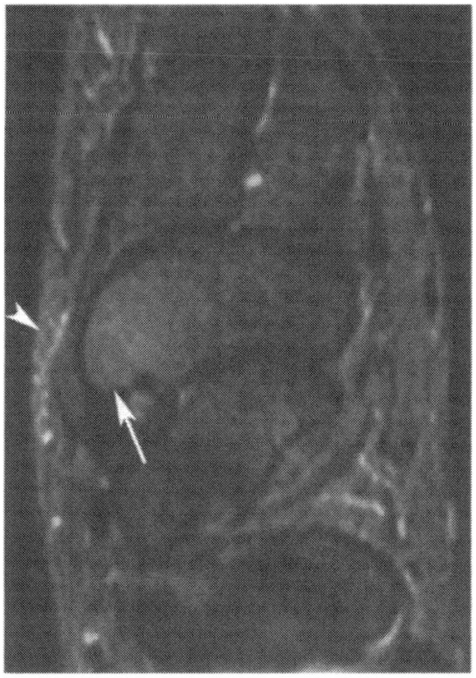

Figure 8G. Axial (footprint) STIR.

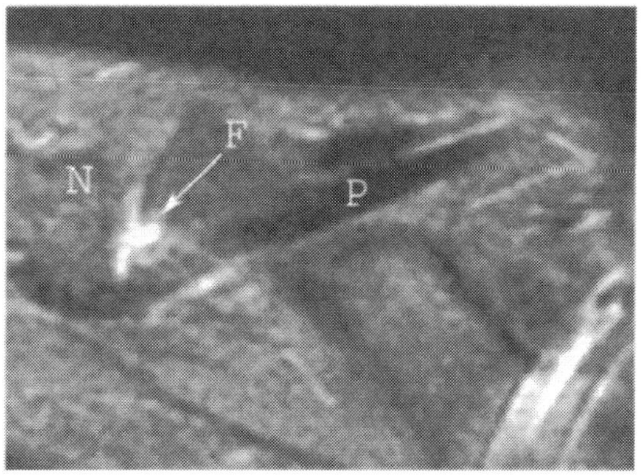

Figure 8H. Sagittal STIR through the PTT.

Findings

This 13-year-old girl developed acute medial foot pain while dancing.

Figure 8E. There is a subtle radiolucency of the medial navicular **(arrow)**. Note that the navicular extends too far medially, representing a type 3 navicular (cornuate).

Figures 8F, 8G, 8H. A sharply defined fracture line is evident **(arrows)**. Fluid signal is noted within and surrounding the posterior tibial tendon **(arrow, F)**, and there is edema within the medial navicular and adjacent soft tissues **(arrowhead)**. Navicular (N), Posterior tibial tendon (P).

Stress Fracture of the Navicular

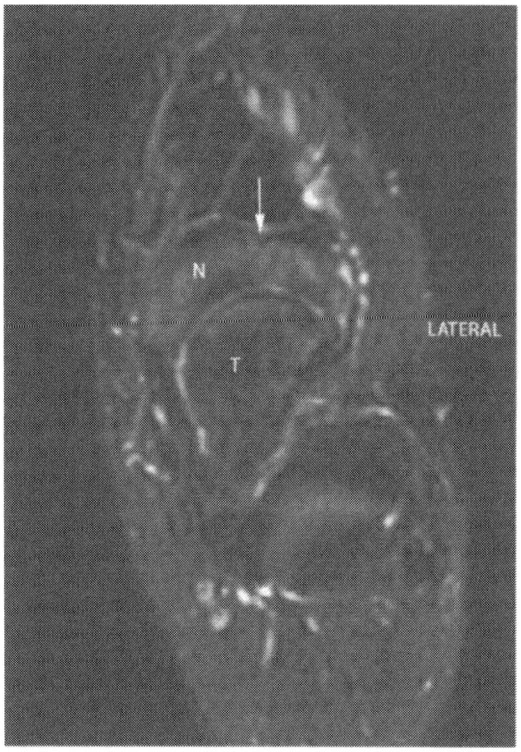

Figure 8I. Axial (footprint) STIR of the left foot.

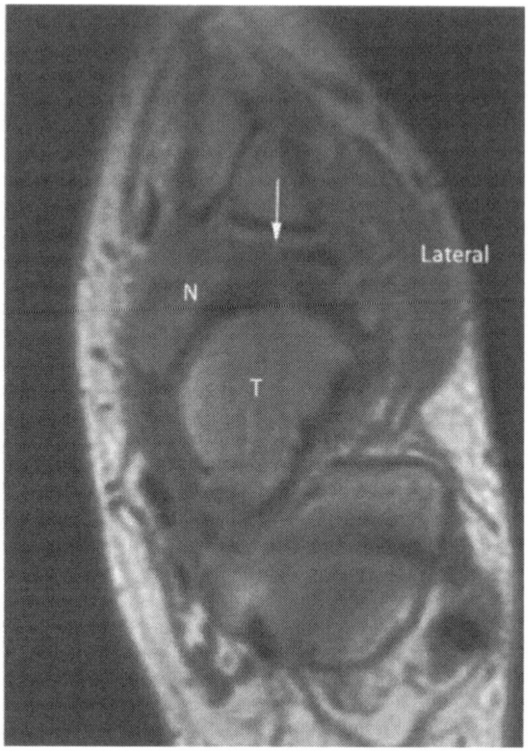

Figure 8J. Axial (footprint) T1.

Findings

This is an 18-year-old male with chronic foot pain.

Figures 8I, 8J. There is diffuse increased STIR SI within the navicular. A fracture line is identified along the dorsal aspect of the middle 1/3 of the navicular bone **(arrow)**. Navicular (N), talus (T).

Pitfalls and Pearls

1. Eighty percent of asymptomatic type 2 accessory navicular bones will fuse. In contrast, only 20% of symptomatic type 2 accessory navicular bones will fuse (2).

References

1. Vallejo JM, Jaramillo D. Normal MR imaging anatomy of the ankle and foot in the pediatric population. *Magn Reson Imaging Clin N Am* 2001; 9:435–446, ix.
2. Lawson JP. International Skeletal Society Lecture in honor of Howard D. Dorfman: Clinically significant radiologic anatomic variants of the skeleton. *AJR Am J Roentgenol* 1994; 163:249–255.
3. Miller TT, Staron RB, Feldman F, Parisien M, Glucksman WJ, Gandolfo LH. The symptomatic accessory tarsal navicular bone: Assessment with MR imaging. *Radiology* 1995; 195:849–853.

4. Chen YJ, Hsu RW, Liang SC. Degeneration of the accessory navicular synchondrosis presenting as rupture of the posterior tibial tendon. *J Bone Joint Surg Am* 1997; 79:1791–1798.
5. Kiter E, Erdag N, Karatosun V, Gunal I. Tibialis posterior tendon abnormalities in feet with accessory navicular bone and flatfoot. *Acta Orthop Scand* 1999; 70:618–621.
6. Sullivan JA, Miller WA. The relationship of the accessory navicular to the development of the flat foot. *Clin Orthop Relat Res* 1979:233–237.
7. Kiss ZS, Khan KM, Fuller PJ. Stress fractures of the tarsal navicular bone: CT findings in 55 cases. *AJR Am J Roentgenol* 1993; 160:111–115.
8. Haller J, Sartoris DJ, Resnick D, et al. Spontaneous osteonecrosis of the tarsal navicular in adults: Imaging findings. *AJR Am J Roentgenol* 1988; 151:355–358.
9. Gips S, Ruchman RB, Groshar D. Bone imaging in Kohler's disease. *Clin Nucl Med* 1997; 22:636–637.
10. Borges JL, Guille JT, Bowen JR. Kohler's bone disease of the tarsal navicular. *J Pediatr Orthop* 1995; 15:596–598.

Case 9

History

This is a 12-year-old girl with chronic, right-sided wrist pain.

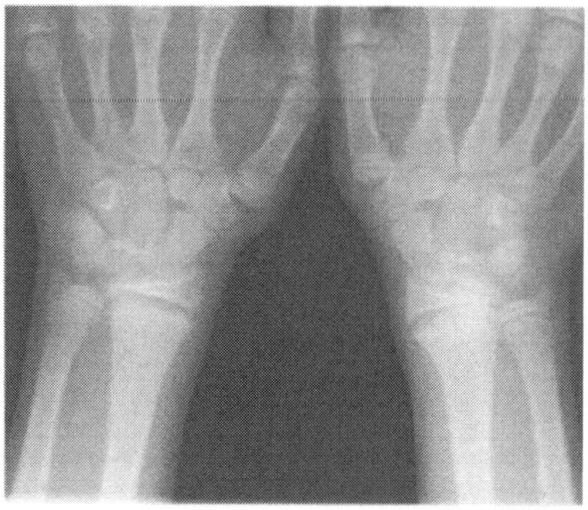

Figure 9A. PA radiograph.

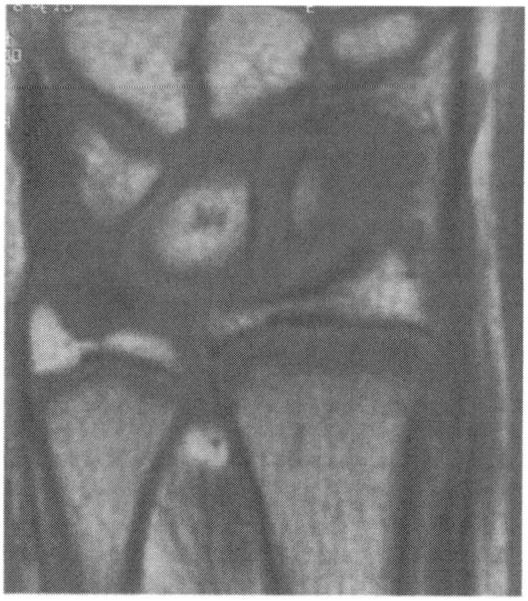

Figure 9B. Coronal T1 of the right wrist.

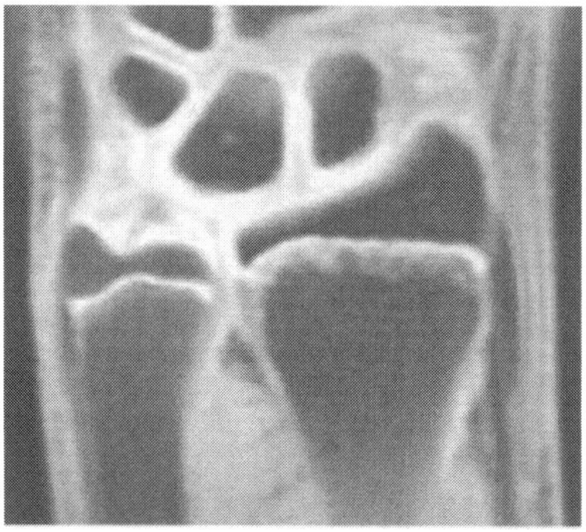

Figure 9C. Coronal 3D SPGR FS.

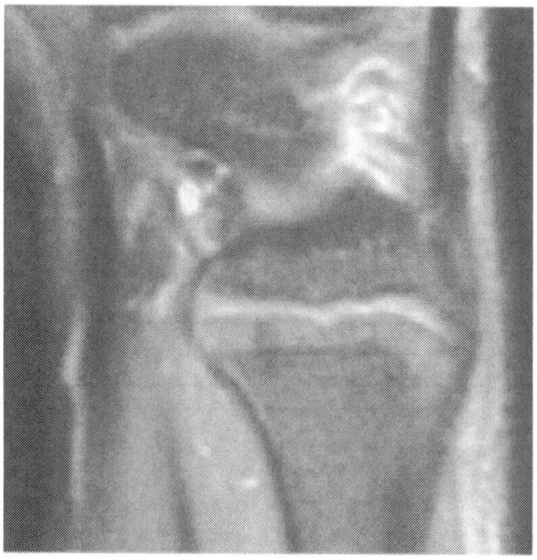

Figure 9D. Sagittal PD FS.

Figure 9A. The distal radial physes appear widened and irregular, particularly laterally. The distal ulnas are normal.

Figure 9B. On the right, the distal radial physis appears widened and the adjacent metaphyseal bone marrow shows relative T1 hypointensity. Note loss of the sharp zone of provisional calcification. Compare this with the normal distal ulna.

Figures 9C, 9D (9C with annotations). On cartilage sensitive sequences, the physis (P) is better defined and shares similar characteristics with the normal distal ulna physis. On the SPGR FS sequence, there are numerous foci of hyperintensity extending from the physis into the adjacent metaphyseal primary spongiosa **(arrowheads)**. They follow the SI of physeal cartilage and reflect disordered metaphyseal endochondral bone formation. No evidence of periphyseal marrow or soft tissue edema is seen on the PD FS sequence.

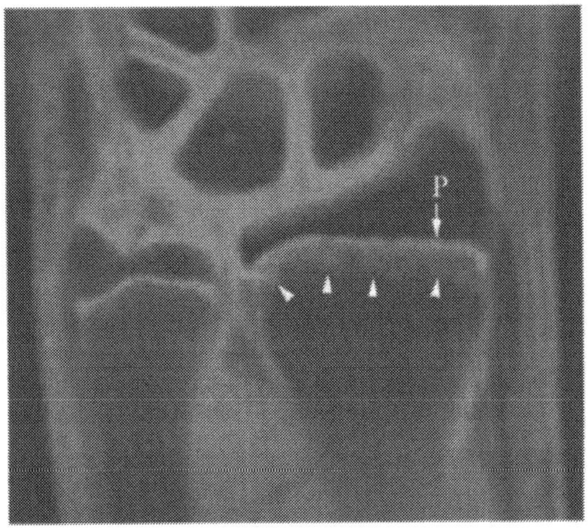

Figure 9C* Annotated.

Diagnosis

Gymnast's wrist

Questions

1. T/F: Physeal widening seen with the gymnast's wrist is a result of an acute Salter-Harris type 1 injury.
2. T/F: Apparent physeal widening is usually seen in both distal radius and ulna in the gymnast's wrist.

Discussion

The gymnast's wrist represents chronic physeal stress reaction to the distal forearm related to upper-extremity weight bearing (Answer to Question 1) (1). This leads to apparent physeal widening and metaphyseal irregularity. These changes are most pronounced in the distal radius. Apparent physeal widening is less commonly seen in the distal ulna since axial loading forces are borne mainly by the radius (Answer to Question 2). Ligamentous and tendinous injuries are less common since these structures are mechanically stronger than the physis.

Shih et al. evaluated the wrists in 47 gymnasts by MRI and found a high incidence of horizontal fractures, physeal widening, and metaphyseal cartilaginous rests within the radius (2). They concluded that apparent physeal widening seen on plain radiographs was usually not due to an acute physeal injury with separation but presumably due to disrupted metaphyseal endochondral ossification related to metaphyseal injury. Based on the studies by Jaramillo et al., metaphyseal injury may result in ischemia to the zone of hypertrophy, disruption of the zone of provisional calcification, and subsequent arrest of metaphyseal endochondral ossification (3). Physeal widening may normalize once the injurious activity is discontinued and normal endochondral bone formation resumes.

On MRI, cartilage-sensitive sequences such as a PD FS or 3D SPGR FS may be used to assess the physis and demonstrate metaphyseal cartilaginous extensions (Figures 9C, 9D). A Salter-Harris type 1 fracture (Figures 9E, 9F) may be distinguished from the gymnast's wrist by the presence of conspicuous osseous and juxta-cortical soft tissue edema, as well as subtle epiphyseal separation from the metaphysis.

A similar pattern of disrupted metaphyseal endochondral ossification may occur in patients with rickets, certain metaphyseal chondrodysplasias, and with the classic metaphyseal lesion of child abuse (4–6). These clinical entities and chronic trauma may all lead to apparent physeal widening that may be mistaken for a Salter-Harris injury on plain radiography.

In addition to evaluating the physis by MRI, other wrist injuries related to gymnastic activities should be assessed. These include stress reaction or stress fractures of the distal radius and scaphoid bone, osteonecrosis of the capitate, lunate impaction syndrome, and ligamentous injuries (7). Premature closure of the radial physis may occur especially in elite gymnasts. This may lead to radial shortening, secondary ulnar overgrowth, and the development of ulnar-impaction syndrome as well as an acquired Madelung's deformity (Figure 9G) (8).

This gymnast's symptoms resolved after 5 months of resting. Plain radiographs at 5 months returned to normal (not shown).

Orthopedic Perspective

The gymnast's wrist, a relatively common disorder, is usually diagnosed with a history of impact loading and upper-extremity weight bearing associated with pain over the distal radial physis. Plain radiographs are usually sufficient for confirmation. Cessation of upper-extremity weight bearing to allow for physeal healing is the mainstay of treatment for the gymnast's wrist. MRI is only ordered when there is diagnostic uncertainty, response to therapy is atypical, or complications such as ulna-carpal impaction or physeal arrest are of concern.

What the Clinician Needs to Know

1. Distinguish a true acute Salter-Harris physeal injury from the classic gymnast's wrist.
2. Assess healing of the physis to allow the athlete to return to gymnastics.
3. Rule out other injuries seen in gymnasts such as ligamentous injury and stress fractures of the carpal bones and distal radius.

Answers

1. False.
2. False.

Additional Examples

Acute Salter-Harris Injury to the Wrist

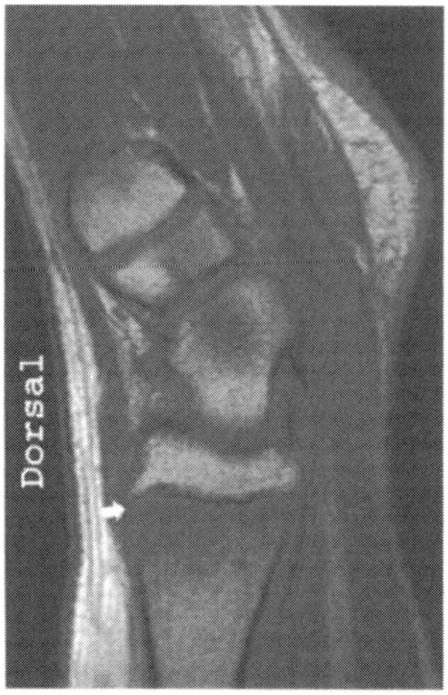

Figure 9E. Sagittal T1 of the left wrist.

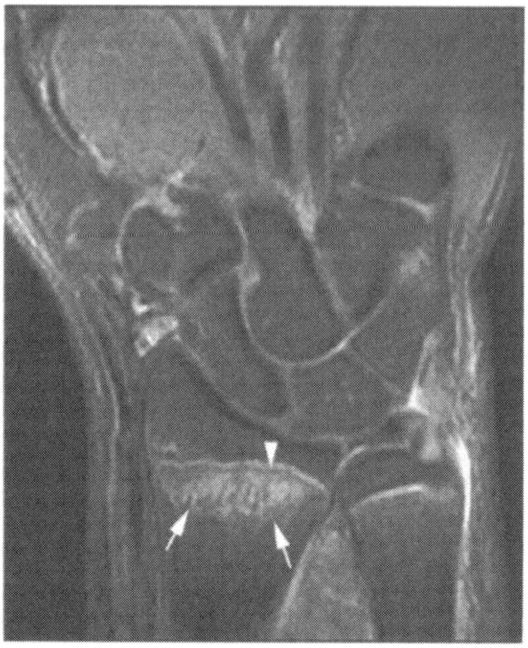

Figure 9F. Coronal STIR.

Findings

This is a 13-year-old boy with an acute injury to the left wrist.

Figures 9E, 9F. The dorsal physis is focally widened **(thick arrow)**, and there is significant metaphyseal marrow edema **(arrows)**—features consistent with an acute physeal injury. Also note soft tissue edema and signal abnormality in the region of the ulnar styloid. No metaphyseal cartilaginous rests are evident. Note that the hypointense zone of provisional calcification is intact **(arrowhead)**.

Madelung's Deformity with Physeal Bridge

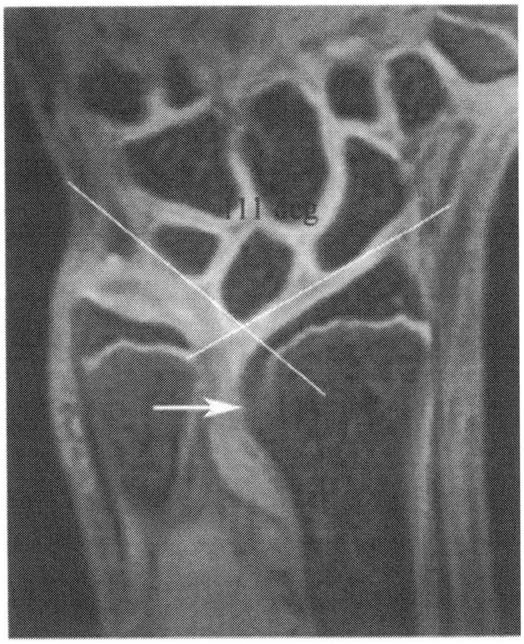

Figure 9G. Coronal 3D SPGR FS of the right wrist.

Findings

This 13-year-old girl gymnast had chronic right wrist pain.

Figure 9G. This coronal sequence through the volar aspect of the wrist shows a bony bar **(arrow)** bridging the anteromedial physis, causing an ulnar tilt of the distal radius and a Madelung's deformity. Note the decreased carpal angle, measuring 111 degrees, with resultant triangular orientation of the proximal carpal row. The carpal angle is measured by drawing one line along the proximal margin of the scaphoid and navicular and a second line along the proximal margin of the triquetrum and navicular. The normal carpal angle measurement is 131.5 degrees with one standard deviation of 7.2 degrees (9).

Pitfalls and Pearls

1. The gymnast's wrist and Salter-Harris type 1 distal radial fractures may both occur in gymnasts. Use fluid and cartilage sensitive sequences to help distinguish between these two conditions.
2. If there is widening of the distal ulna as well as the distal radial physis, consider a metabolic disorder.

References

1. Carter SR, Aldridge MJ, Fitzgerald R, Davies AM. Stress changes of the wrist in adolescent gymnasts. *Br J Radiol* 1988; 61:109–112.
2. Shih C, Chang CY, Penn IW, Tiu CM, Chang T, Wu JJ. Chronically stressed wrists in adolescent gymnasts: MR imaging appearance. *Radiology* 1995; 195:855–859.
3. Jaramillo D, Laor T, Zaleske DJ. Indirect trauma to the growth plate: Results of MR imaging after epiphyseal and metaphyseal injury in rabbits. *Radiology* 1993; 187:171–178.
4. Kleinman PK, Marks SC, Jr., Spevak MR, Belanger PL, Richmond JM. Extension of growth-plate cartilage into the metaphysis: A sign of healing fracture in abused infants. *AJR Am J Roentgenol* 1991; 156:775–779.
5. Camera A, Camera G. Distinctive metaphyseal chondrodysplasia with severe distal radius and ulna involvement (upper extremity mesomelia) and normal height. *Am J Med Genet A* 2003; 122:159–163.
6. Ecklund K, Doria AS, Jaramillo D. Rickets on MR images. *Pediatr Radiol* 1999; 29:673–675.
7. Gabel GT. Gymnastic wrist injuries. *Clin Sports Med* 1998; 17:611–621.
8. De Smet L, Claessens A, Lefevre J, Beunen G. Gymnast wrist: An epidemiologic survey of ulnar variance and stress changes of the radial physis in elite female gymnasts. *Am J Sports Med* 1994; 22:846–850.
9. Kosowicz J. The carpal sign in gonadal dysgenesis. *J Clin Endocrinol Metab* 1962; 22:949–952.

Case 10

History

This is a 9-year-old boy with bilateral ankle pain. There is no history of trauma, and he is relatively short for his age.

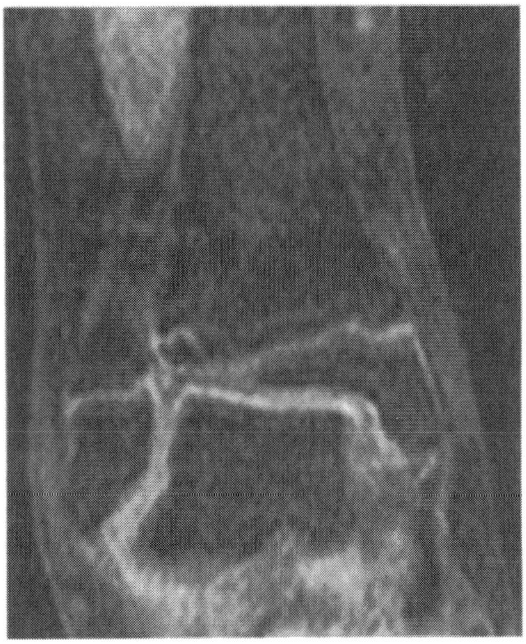

Figure 10A. Coronal PD FS of the right ankle.

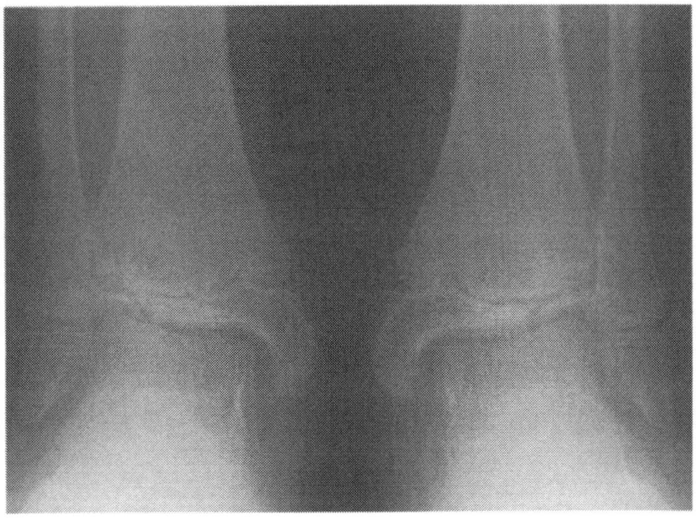

Figure 10B. Standing AP radiograph of both ankles.

Figures 10A (10A with annotations). Ankle valgus deformity is present. The tibial plafond and talar dome both have an upward lateral slant. The lateral aspect of the tibial epiphysis is thin and irregular compared with the medial side. The tibia physis has an abnormal undulation laterally **(arrow)**. Note that the distal fibula epiphysis is normal. Similar changes were present in the left ankle (not shown).

Figure 10B. Ankle valgus deformities are present bilaterally. The lateral aspects of the tibia epiphyses are small and fragmented with symmetric vertical clefts.

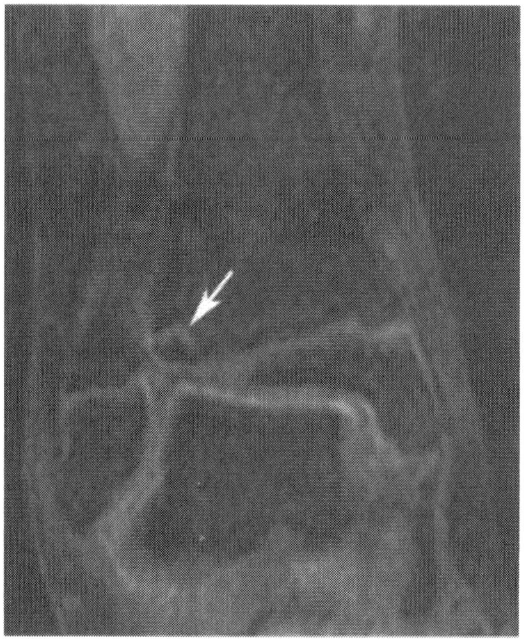

Figure 10A* Annotated.

Diagnosis

Multiple epiphyseal dysplasia (MED) with valgus tibiotalar slant

Questions

1. T/F: Deformity of the talar dome is usually present with valgus tibiotalar slant.
2. T/F: Patients with MED are predisposed to premature osteoarthritis.

Discussion

Tibiotalar slant generally describes an upward, lateral slant of the tibial plafond leading to an ankle valgus deformity. Abnormal tibiotalar slant may result from growth restriction of the lateral aspect of the distal tibia epiphysis and physis (1). This may be due to an ischemic, inflammatory, congenital, metabolic, or traumatic etiology. The talar dome is generally not affected and does not contribute to ankle valgus deformity (Answer to Question 1). Specific conditions that may lead to an acquired tibiotalar slant include: distal tibial physeal fractures, osteomyelitis/septic arthritis (Figure 10C), juvenile rheumatoid arthritis, hemophilia, sickle cell disease, rickets, epiphyseal dysplasia, pseudoachondroplasia, metaphyseal dysplasia, neurofibromatosis, fibrous dysplasia, and multiple hereditary exostosis (1). Overcorrected clubfoot, fibular shortening, and a hindfoot valgus deformity may also lead to an ankle valgus deformity (2, 3). Although tibiotalar slants are usually associated with an ankle valgus deformity, a varus tilt may rarely be noted (Figure 10D).

Multiple epiphyseal dysplasia (MED) is a disorder that leads to disrupted endochondral bone formation of the epiphyseal ossification center (4). On plain radiography, this leads to delayed appearance of the secondary ossification center and epiphyseal fragmentation. A similar appearance may be seen with spondyloepiphyseal dysplasia (SED). Radiographs of the spine are useful in distinguishing these entities. MED may have mild platyspondyly, but SED usually has marked platyspondyly. Early degenerative changes and epiphyseal flattening often occur because the secondary ossification centers do not develop normally and fail to provide adequate mechanical support for the epiphyseal and articular cartilage (Answer to Question 2) (4).

Physeal imaging with a cartilage sensitive sequence such as a 3D SPGR FS or PD FS is important for evaluating physeal bars resulting from physeal growth arrest in the setting of a tibiotalar slant. The physis is hyperintense on these sequences and loss of the physeal hyperintensity suggests either a fibrous or osseous bar. T1W sequences demonstrating hyperintense fatty marrow continuity between the metaphysis and epiphysis are compatible with an osseous bar. If the bar is hypointense on T1, the bar may be either osseous or fibrous. Documenting an open physis may be helpful should medial epiphyseodesis with a transphyseal screw be considered to arrest further medial overgrowth (3). The morphologic alterations of the articular cartilage are particularly important in patients with MED or SED, as well as inflammatory arthropathies, since these patients are predisposed to early degenerative changes. Assessment for bone marrow edema and ligamentous pathology is important as well, since the tibiotalar deformity results in altered ankle mechanics.

This patient with known MED was referred for MRI because of persistent bilateral ankle pain. The MRI was done preoperatively prior to bilateral medial epiphysiodesis.

Orthopedic Perspective

The MRI evaluation of valgus tibiotalar slant is usually performed as a preoperative examination since the diagnosis is usually evident based on plain radiography and history. The objective of early correction of tibiotalar slant is to allow normal gait and to prevent early degenerative changes related to altered weightbearing. Corrective surgical options include performing a medial epiphyseodesis to arrest progressive medial overgrowth or resection of a lateral distal tibial physeal bar to allow resumed lateral physeal growth. Epiphyseodesis is surgical fusion of the growth plate. Prior to surgical correction, documenting bone age to predict future skeletal growth and quantifying leg-length discrepancies with a scanogram are important to determine timing of an epiphyseodesis or physeal bar resection. Epiphyseodesis is preferred over physeal bar resection when there is >50% physeal bar and there is less than 2 years' growth remaining. Epiphyseodesis has a shorter convalescence (approximately 2 weeks) compared with physeal bar resection (6–8 weeks). If tibiotalar slant is unilateral, imaging the normal contralateral side is helpful to determine the normal ankle angle and shape. Physeal surgical intervention is a less invasive procedure than a corrective tibial wedge osteotomy after the physis is fused.

What the Clinician Needs to Know

1. Is there surgically correctable cause of the tibiotalar slant, such as a fibrous or osseous physeal bar?
2. Are there early articular cartilage degenerative changes related to altered weight bearing?

Answers

1. False.
2. True.

Additional Examples

Valgus Tibiotalar Slant from Prior Septic Arthritis

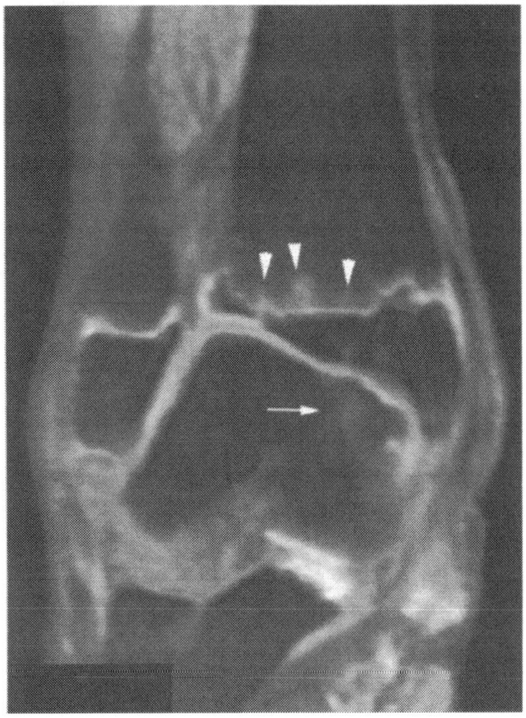

Figure 10C. Coronal SPGR FS of the right ankle.

Findings

This is an 11-year-old boy who had septic arthritis of the right ankle as an infant.

Figure 10C. There is a valgus tibiotalar slant with relative undergrowth of the lateral tibia epiphysis, but there is no evidence of a physeal bar. There are degenerative changes involving the medial articular surfaces of the tibiotalar joint including articular cartilage thinning, subchondral bone irregularity, and increased SI within the talar bone marrow **(arrow)**. Note multiple tongues of metaphyseal cartilaginous rests **(arrowheads)**, related to disturbance of metaphyseal endochondral ossification.

Varus Tibiotalar Slant from Prior Osteomyelitis

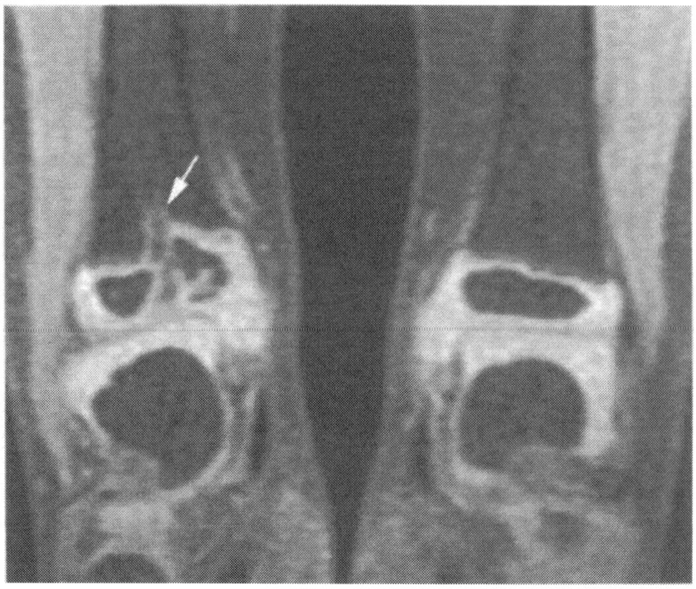

Figure 10D. Coronal SPGR FS of both ankles.

Findings

This 2-year-old girl was evaluated for right ankle varus deformity. She had distal tibial osteomyelitis one year ago.

Figure 10D. The medial physis of the right tibial metaphysis is abnormal with cartilaginous extensions within the medial metaphysis **(arrow)**, indicating disrupted metaphyseal endochondral ossification. The secondary ossification center of the tibia epiphysis is fragmented medially. Note that the height of the medial half of the epiphysis is increased, resulting in a "coned" appearance of the ossification center, with corresponding "cupping" of the adjacent metaphysis. Compare these changes with the normal left ankle.

Pitfalls and Pearls

1. When bilateral tibiotalar slants are encountered in an otherwise normal patient, consider MED, and request views of the knees and hips.

References

1. Griffiths H, Wandtke J. Tibiotalar tilt: A new slant. *Skeletal Radiol* 1981; 6:193–197.
2. Stevens PM, Belle RM. Screw epiphysiodesis for ankle valgus. *J Pediatr Orthop* 1997; 17:9–12.
3. Davids JR, Valadie AL, Ferguson RL, Bray EW, 3rd, Allen BL, Jr. Surgical management of ankle valgus in children: Use of a transphyseal medial malleolar screw. *J Pediatr Orthop* 1997; 17:3–8.
4. Bassett GS. Orthopaedic aspects of skeletal dysplasias. *Instr Course Lect* 1990; 39:381–387.

Case 11

History

This is a 9-year-old girl who noticed progressive swelling and pain over her right clavicle. She has a history of psoriatic arthritis that is well controlled with medications.

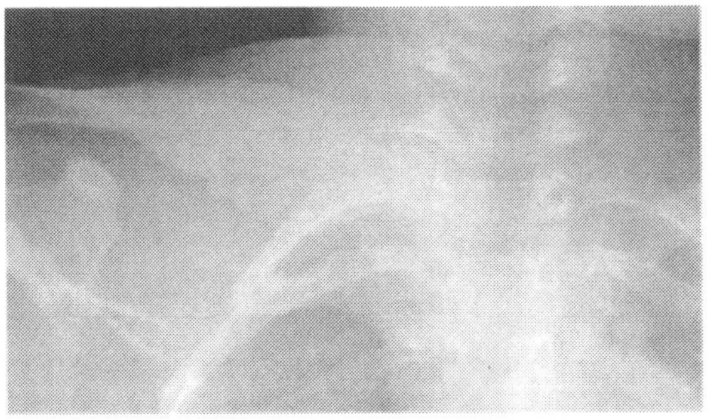

Figure 11A. Lordotic AP radiograph of the right clavicle.

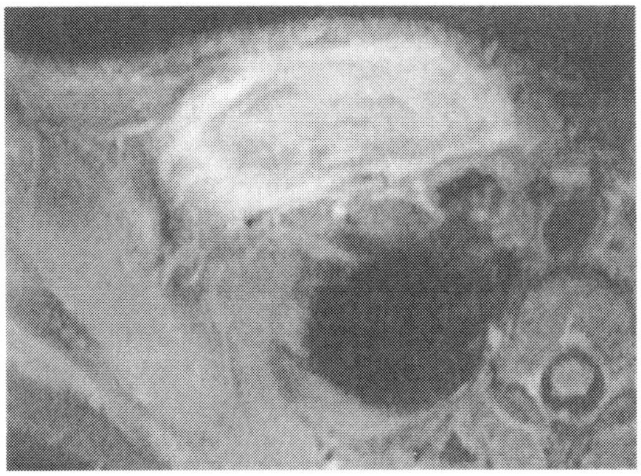

Figure 11B. Axial STIR of the right clavicle.

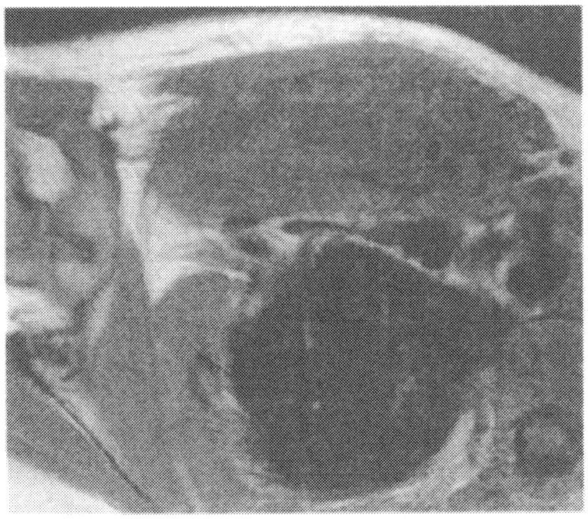

Figure 11C. Axial T1.

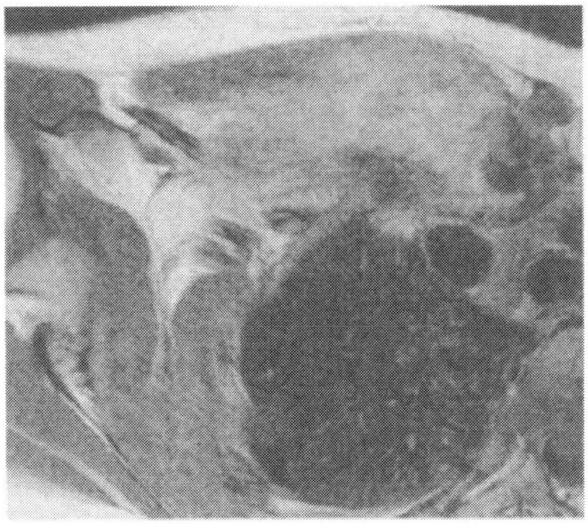

Figure 11D. Axial T1 post-Gd.

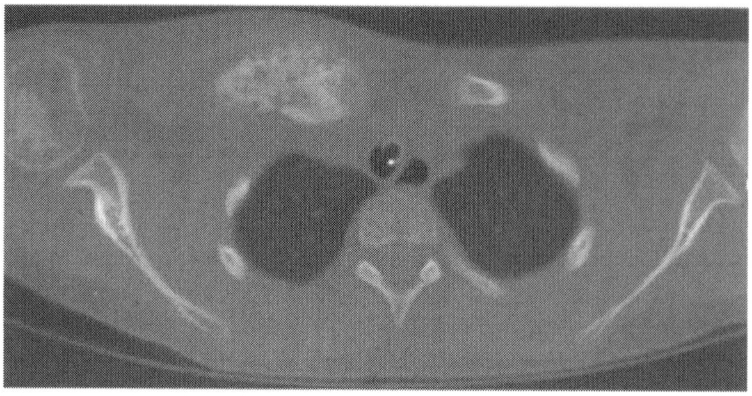

Figure 11E. Axial CT.

Figure 11A. There is marked mature periosteal new bone formation involving the medial aspect of the right clavicle.

Figures 11B, 11C, 11D. The clavicle is enlarged and is isointense on T1, hyperintense on STIR, and demonstrates heterogeneous enhancement. There is mild edema in the juxtacortical soft tissues.

Figure 11E. On CT, multiple lucencies are evident within the enlarged, sclerotic clavicle.

Diagnosis

Sternocostoclavicular hyperostosis

Questions

1. T/F: An organism is frequently isolated by culture or bone biopsy in cases of sternocostoclavicular hyperostosis.
2. T/F: Both sternocostoclavicular hyperostosis and pyogenic osteomyelitis of the clavicle preferentially affect the medial end of the clavicle.

Discussion

The etiology for sternocostoclavicular hyperostosis is controversial. It is a chronic sclerosing disorder of the anterior chest wall that may be related to chronic recurrent multifocal osteomyelitis (CRMO), seronegative spondyloarthritis, or both (1). Garre's osteomyelitis is another term that has also been used to describe sclerosing, inflammatory disorders of bone. However, terms such as Garre's sclerosing osteomyelitis and CRMO may be misleading since an organism is rarely isolated by blood culture and bone biopsy (Answer to Question 1).

The plain radiographic features of sternocostoclavicular hyperostosis include diffuse osseous enlargement from marked periosteal new bone deposition. Lytic changes are usually minimal. The medial end of the clavicle is often affected, and the lateral end is usually unaffected, even with severe cases (1).

On MRI, diffuse osteitis (bone inflammation), osseous proliferation, and adjacent soft tissue edema may be identified within the shoulder girdle (hypointense on T1, hyperintense on fluid-sensitive sequences with enhancement after Gd administration). Juxtacortical soft tissue edema may be seen in the setting of active osteitis (2). Areas of hypointensity on all imaging sequences may be identified, reflecting bony sclerosis.

The clinical and radiographic features of sternocostoclavicular hyperostosis vary between children and adults. In contrast to adults, where diffuse involvement of the anterior chest wall, ankylosis, and ligamentous ossification are characteristic, isolated clavicular hyperostosis is typical in childhood (3). Long tubular bone osteitis may be seen in both children and adults in the setting of CRMO and SAPHO syndrome (synovitis, acne, pustulosis, hyperostosis, osteitis), respectively.

Sternocostoclavicular hyperostosis should be differentiated from acute pyogenic osteomyelitis of the clavicle (Figures 11F, 11G). Radiographs in acute pyogenic clavicular osteomyelitis may be normal or may demonstrate only modest periosteal reaction. On MRI, extensive marrow and extraosseous soft tissue edema, as well as areas of frank lytic destruction may be seen. Sternoclavicular septic arthritis, reactive pleural effusions, septic thrombophlebitis, sequestra, cloaca, sinus tracts, and intra- and extraosseous soft tissue abscesses may also occur in the setting of acute pyogenic clavicular osteomyelitis. These are not seen with sternocostoclavicular hyperostosis. As in sternocostoclavicular hyperostosis, the medial head of the clavicle is more commonly affected with osteomyelitis (Answer to Question 2) (4).

Ewing's sarcoma (Figures 11H, 11I) and fibrous dysplasia may mimic sternocostoclavicular hyperostosis. Ewing's sarcoma usually shows an aggressive appearance with permeative or moth-eaten osteolysis, rather than bone expansion. Furthermore, a substantial extraosseous soft tissue mass is typically present. Fibrous dysplasia produces medullary expansion with well-defined cortical margins with little if any periosteal reaction, unless a pathologic fracture is present.

This patient had a history of psoriatic arthritis with polyarticular involvement. A biopsy of the right clavicle showed changes of chronic osteomyelitis. An organism was not isolated, and there was no evidence of neoplasm. Given the history of psoriatic arthritis, a diagnosis of CRMO with sternocostoclavicular hyperostosis was made. She was given naproxen and no antibiotics. At 1-month follow-up, her clavicular pain had improved but the clavicular swelling persisted. For a complete discussion on CRMO, see Case 2.

Orthopedic Perspective

Differentiation of sternocostoclavicular hyperostosis from pyogenic osteomyelitis is important since sternocostoclavicular hyperostosis related to CRMO is typically treated with anti-inflammatory medications and pyogenic osteomyelitis is treated with antibiotics. Biopsy is often required when blood cultures are negative or when clavicular changes mimic a primary bone sarcoma. Biopsy of juxtaclavicular soft tissue disease, as well as bony cortex and medulla, will optimize diagnostic yield. If an abscess is present related to pyogenic osteomyelitis, the size and location should be characterized to facilitate aspiration.

What the Clinician Needs to Know

1. Which components of the clavicular lesion are amenable to biopsy?
2. Does the inflammatory process extend to the sternoclavicular joint?

Answers

1. False.
2. True.

Additional Examples

Acute Clavicular Osteomyelitis from Staphylococcus Aureus

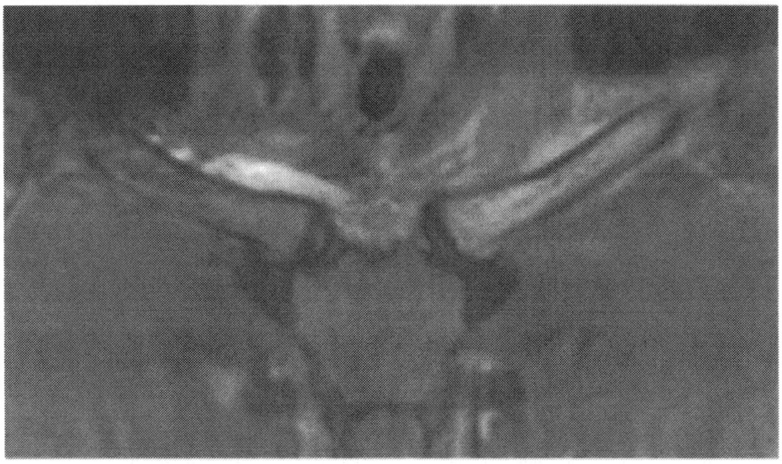

Figure 11F. Oblique coronal STIR.

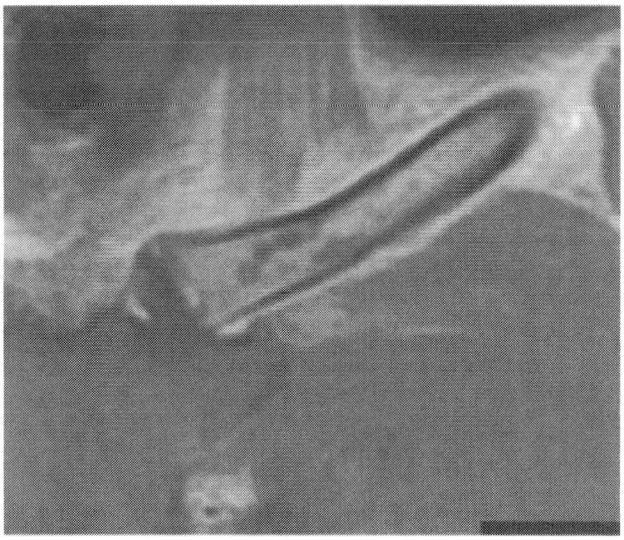

Figure 11G. Coronal T1 post-Gd FS.

Findings

This is an 18-year-old boy with acute left shoulder pain and fever. Radiographs (not shown) were normal.

Figures 11F, 11G. There is diffuse, increased STIR SI and heterogeneous enhancement of the left clavicle and periosteum. There is no cortical thickening or medullary expansion. Nonspecific juxtaclavicular STIR hyperintensity is noted above the right clavicle. Blood cultures were positive for *Staphylococcus aureus*.

Ewing's Sarcoma

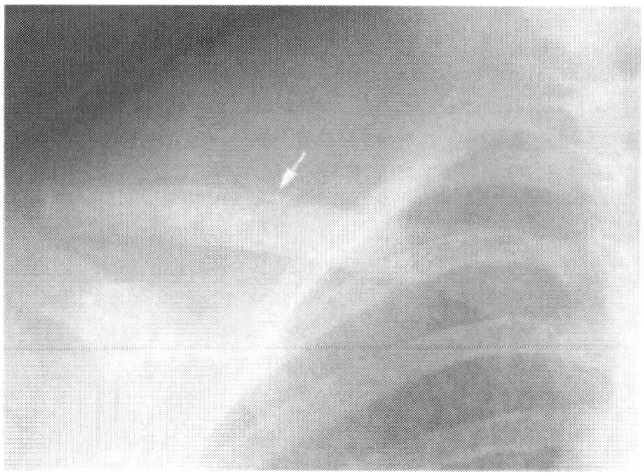

Figure 11H. AP radiograph of the right clavicle.

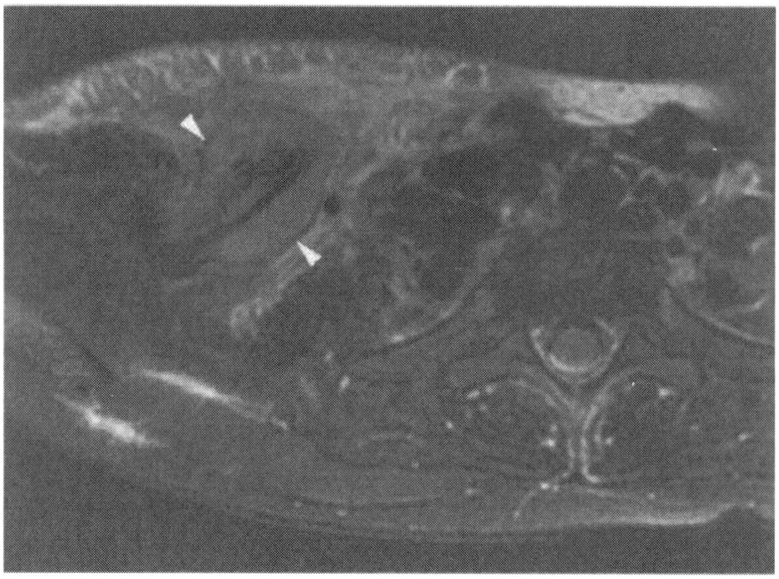

Figure 11I. Axial T2 FS.

Findings

This patient presented with a 2-month history of right jaw pain.

Figure 11H. There is a mixed permeative-sclerotic lesion involving the middle to lateral third of the clavicle with a Codman's triangle **(arrow)**.

Figure 11I. An intermediate SI soft tissue mass is evident arising from the clavicle and confined by the periosteal envelope **(arrowheads)**. This was a pathologically proven Ewing's sarcoma.

Pitfalls and Pearls

1. Beware—the plain radiographic findings of clavicular hyperostosis may be dramatic and alarming, raising the ominous consideration of Ewing's sarcoma. The MRI findings may reinforce this diagnostic concern.
2. Patients with sternocostoclavicular hyperostosis may manifest other sites of bony disease, so do a bone scan to assess for other evidence of CRMO.
3. MRI may be helpful to follow disease activity. Active sternocostoclavicular hyperostosis shows increased enhancement and T2 SI compared with normal bone. Osseous structures remain hypointense on T1W and T2W sequences with inactive disease (3).

References

1. Azouz EM, Jurik AG, Bernard C. Sternocostoclavicular hyperostosis in children: a report of eight cases. *AJR Am J Roentgenol* 1998; 171:461–466.
2. Jurik AG. Chronic recurrent multifocal osteomyelitis. *Semin Musculoskelet Radiol* 2004; 8:243–253.
3. Earwaker JW, Cotten A. SAPHO: Syndrome or concept? Imaging findings. *Skeletal Radiol* 2003; 32:311–327.
4. Jurik AG, Egund N. MRI in chronic recurrent multifocal osteomyelitis. *Skeletal Radiol* 1997; 26:230–238.

Case 12

History

This is an 11-year-old girl with a 1-year history of knee pain.

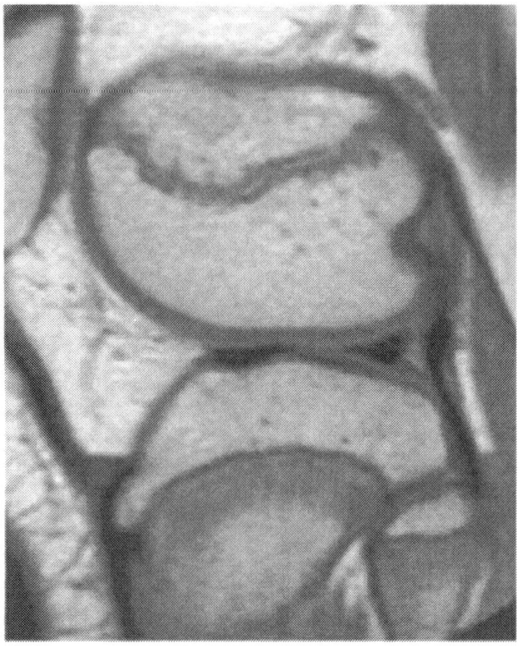

Figure 12A. Sagittal PD, right knee.

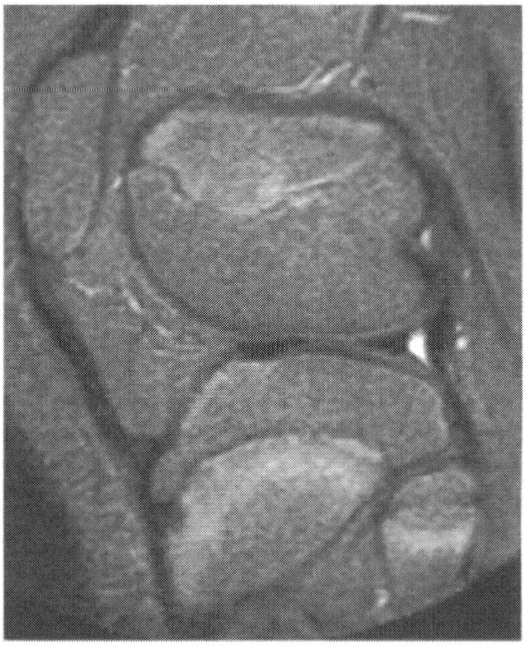

Figure 12B. Sagittal STIR, right knee.

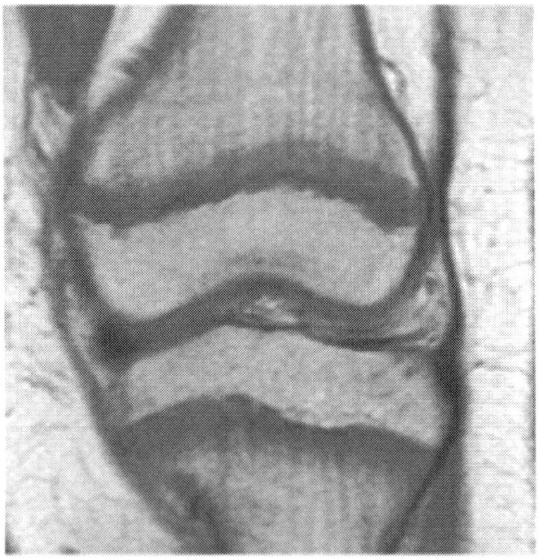

Figure 12C. Coronal T1, left knee.

Figures 12A (12A with annotations). On the right, there is physeal widening of the proximal tibia and fibula. Note that the epiphyseal margins of the proximal tibial and fibular physes are sharp **(arrows)** and the metaphyseal margins are ill-defined (*), blending with the adjacent metaphyses.

Figure 12B. There is mild increased SI within the proximal physes of the tibia and fibula.

Figure 12C. On the left, there is also physeal widening of the distal femur and proximal tibia.

Figure 12D. Physeal widening is seen in the proximal tibia and fibula **(arrowheads)**. The distal femoral physis is ill-defined.

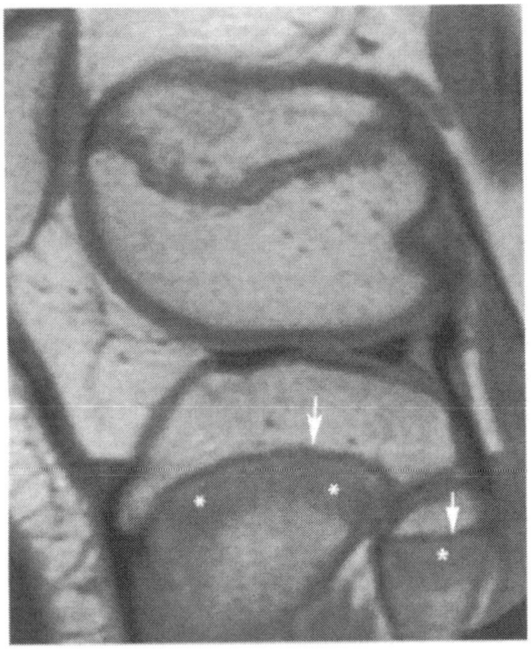

Figure 12A* Annotated.

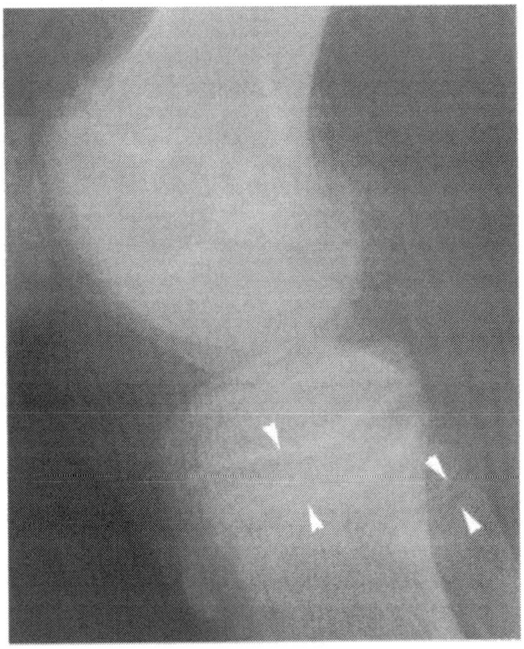

Figure 12D. Lateral radiograph of the right knee.

Diagnosis

Rickets (Vitamin D deficiency)

Questions

1. Which zone of the physis widens as a result of rickets?
2. What are the causes of a widened physis?

Discussion

The physis is divided into several components: germinal matrix (epiphyseal side), proliferative zone, and hypertrophic zone (metaphyseal side) (Diagram 12A). The portion of the hypertrophic zone which undergoes progressive cartilaginous mineralization is referred to as the zone of provisional calcification (ZPC) (1). In the setting of rickets, the regulated chondrocyte apoptosis in the hypertrophic zone is disrupted, leading to poor metaphyseal endochondral ossification (2). Failure of chondrocyte apoptosis causes widening of the hypertrophic zone with poor delineation of the zone of provisional calcification (Answer to Question 1).

The radiographic features of rickets include physeal widening, irregularity, and metaphyseal cupping. Osteopenia, bowing deformities, and insufficiency fractures may also be seen. With treatment, the physeal thickness normalizes and the ZPC mineralizes, resulting in a dense metaphyseal band (3).

MRI better delineates the rachitic changes in the physis. On fluid and cartilage sensitive sequences, the normal and rachitic physes are both hyperintense. Fluid sensitive sequences distinguish the relative hypointense epiphyseal cartilage from the hyperintense physeal cartilage (2). In the setting of rickets, physeal widening and blurred metaphyseal margins due to arrested endochondral ossification are evident. The ZPC and primary spongiosa of the metaphysis, which are normally hypointense on all imaging sequences, are poorly defined. Metaphyseal cartilaginous rests may occur when there is focal arrested endochondral ossification. Similar findings may be seen in the epiphyseal ossification center with blurring of the cortical margin due to arrested endochondral ossification of the spherical growth plate. MRI may detect insufficiency fractures and healing rickets before they are radiographically apparent (2). Referral for MRI is of little value when the clinical and plain radiographic features of rickets are obvious.

The differential diagnosis of a widened physis includes chronic trauma, rickets, chronic systemic illness, skeletal dysplasias such as metaphyseal chondrodysplasia and leukemia (Answer to Question 2). The various causes of hepatic rickets include: dietary, renal, (e.g., biliary atresia), and metabolic conditions such as X-linked hypophosphatemia (vitamin D resistant rickets) (3). Rickets may also be seen as a paraneoplastic process (oncogenic osteomalacia) due to benign and malignant tumors such as a hemangiopericytoma, non-ossifying fibroma, giant cell tumors, and osteosarcomas (4). Leukemic infiltration may produce apparent physeal widening and periosteal reaction.

In general, the rapidly growing physes (knees, wrists) will be abnormal in the setting of significant rickets or systemic illness, whereas chronic physeal stress injury will usually show changes isolated to the weight bearing physis (Figures 12E, 12F). However, as seen in the primary case (Figures 12A–12D), physeal changes may be nonuniform. The diagnosis of chronic physeal stress injury should be considered once systemic etiologies have been excluded. Metaphyseal cartilaginous rests may occur with any process that disrupts endochondral bone formation, including chronic stress injury (Figures 12E, 12F).

In this patient, the key to the diagnosis of rickets is the presence of multiple abnormal physes. Further questioning revealed that the patient had not consumed milk products for the past 6 years. Family history and clinical work-up revealed no underlying disorder. Vitamin D serologic level was under 7 ng/ml (normal: 9.0 to 74.0 ng/ml). She was treated with vitamin D and calcium supplements and did well.

Orthopedic Perspective

Dietary rickets is relatively rare in older children. Risk factors for dietary rickets include African-American descent, being breast-fed, and living in geographic locations where there is limited sunlight exposure (5). A metabolic work-up is advised in a child of any age when there are globally widened physes, constitutional symptoms such as general malaise or poor school performance, and growth delay. The initial diagnosis of dietary rickets should be made well before patients are referred to MRI for knee pain, as seen in this patient. The atypical features of this case that led to a delay in diagnosis were the nonuniform physeal widening and older age of presentation. Differentiation between physeal stress injury and dietary rickets can usually be determined by clinical context (e.g., the athlete with overuse and lack of deformity). Patients with dietary rickets are usually referred to an endocrinologist for further work-up. From an orthopedic perspective, patients with dietary rickets are treated with activity limitations only if there is pain.

What the Clinician Needs to Know

1. Are multiple physes involved, suggesting a metabolic disorder such as rickets?
2. Are there Looser zone fractures and bowing deformities present?

Answers

1. Hypertrophic zone.
2. Chronic trauma, rickets, chronic systemic illness, skeletal dysplasias such as metaphyseal chondrodysplasia, oncogenic osteomalacia, and leukemia.

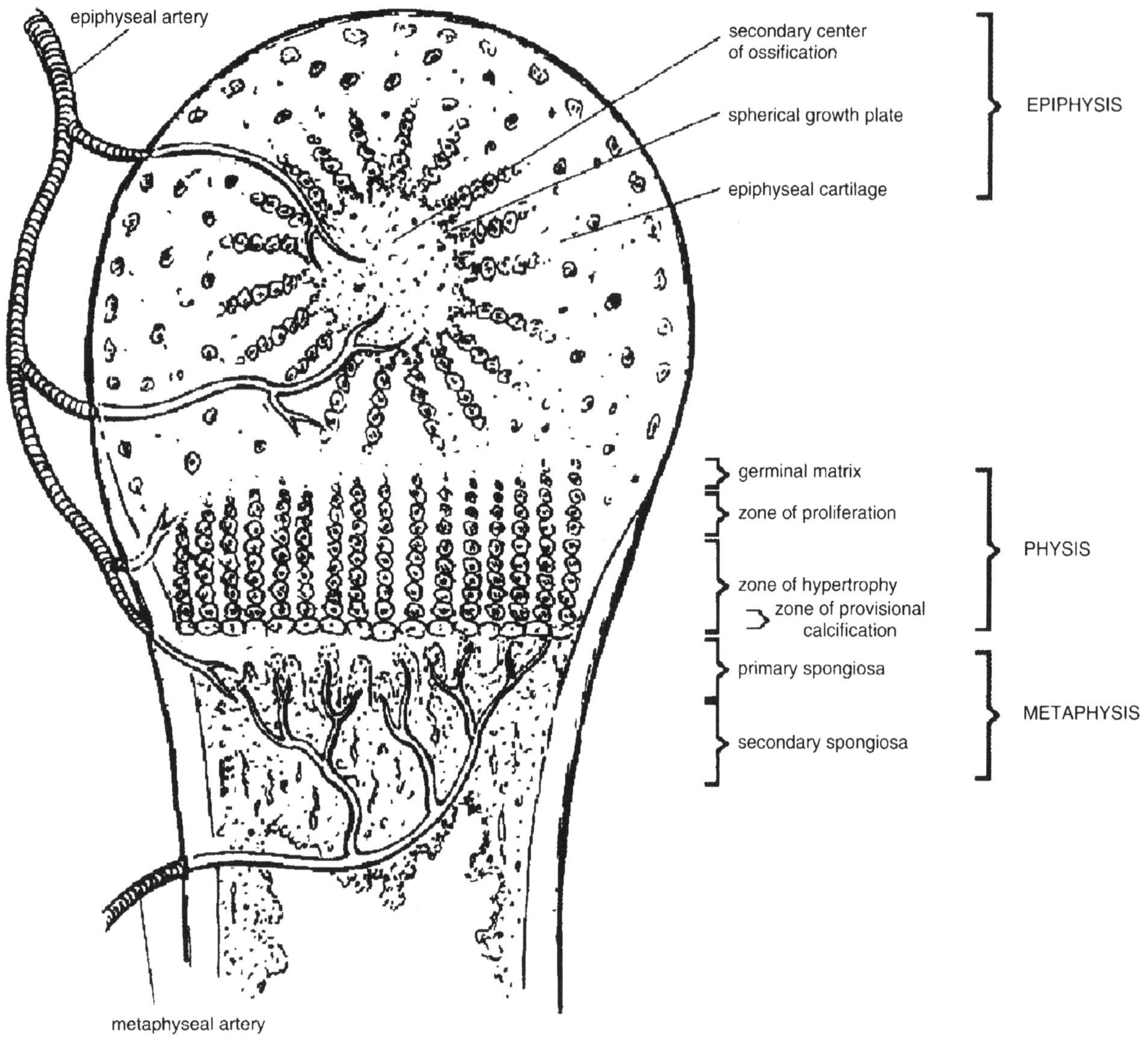

Diagram 12A. Anatomy of the metaphysis, physis, and epiphysis.

Additional Example

Physeal Stress Reaction

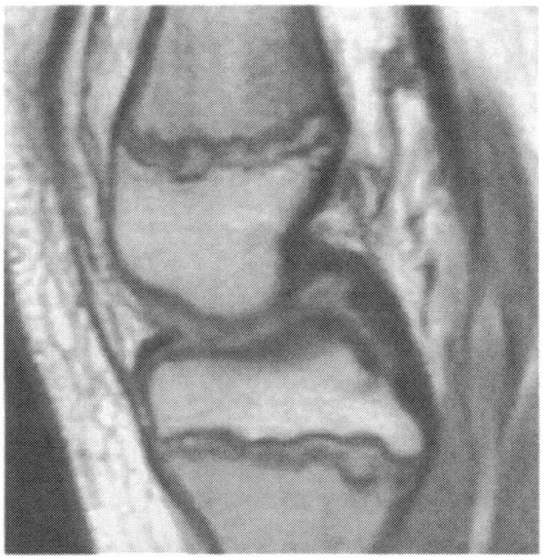

Figure 12E. Sagittal PD of the right knee.

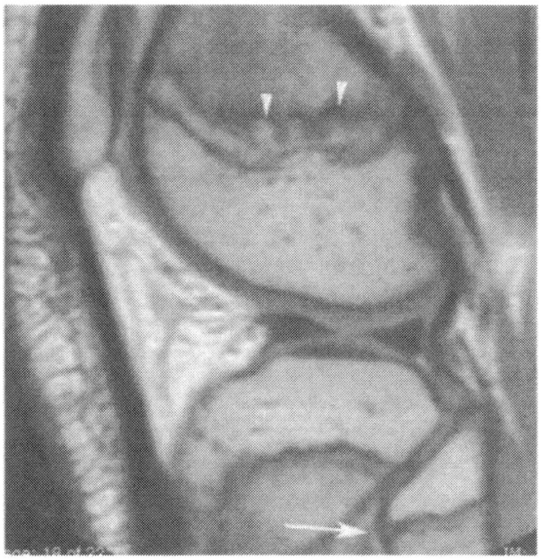

Figure 12F. Sagittal PD, lateral to Figure 12E.

Findings

This active 10-year-old girl had chronic knee pain, right greater than left.

Figures 12E, 12F. There is abnormal physeal widening and irregularity in the distal femur and proximal tibia. Note multiple metaphyseal cartilaginous rests **(arrowheads)**. The physis of the fibula is normal **(arrow)**. Metabolic work-up was normal and these findings were attributed to chronic physeal stress reaction related to repetitive trauma.

Pitfalls and Pearls

1. Dietary rickets does not appear in normal term infants until 6 months of age due to maternal stores of vitamin D (3). Therefore, consider alternative diagnoses, including underlying metabolic disorders, such as hypophosphatasia, metaphyseal chondrodysplasia, or child abuse, when metaphyseal widening and irregularity are seen in young infants.
2. There has been a recent resurgence of dietary rickets, and the radiologist may be the first to make the diagnosis based on the distinctive plain radiograph and MRI findings (5).

References

1. Ecklund K, Jaramillo D. Imaging of growth disturbance in children. *Radiol Clin North Am* 2001; 39:823–841.
2. Ecklund K, Doria AS, Jaramillo D. Rickets on MR images. *Pediatr Radiol* 1999; 29:673–675.
3. States LJ. Imaging of metabolic bone disease and marrow disorders in children. *Radiol Clin North Am* 2001; 39:749–772.
4. Sundaram M, McCarthy EF. Oncogenic osteomalacia. *Skeletal Radiol* 2000; 29:117–124.
5. Eugster EA, Sane KS, Brown DM. Minnesota rickets: Need for a policy change to support vitamin D supplementation. *Minn Med* 1996; 79:29–32.

Case 13

History

This is a 15-year-old girl with persistent ankle pain. Her history is significant for a twisting ankle injury while jumping on a trampoline 1.5 years ago.

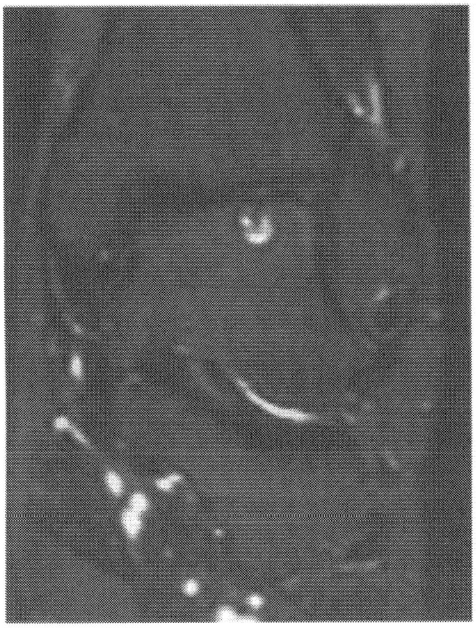

Figure 13A. Coronal STIR of the left ankle.

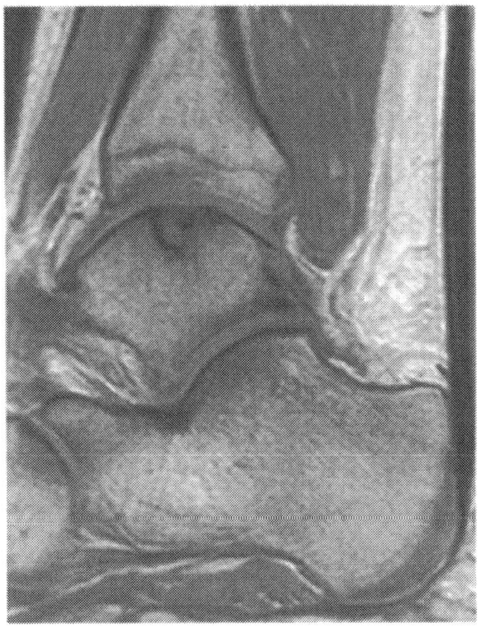

Figure 13B. Sagittal T2.

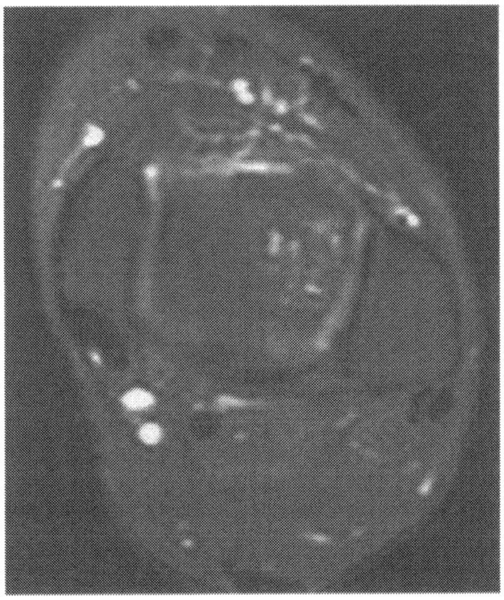

Figure 13C. Axial T2 FS.

Figure 13A, 13B, 13C (13B, 13C with annotations). There is an osteochondral lesion in the lateral talar dome. The overlying articular cartilage is slightly thinned and flattened **(white arrowheads)**. There are cyst-like changes noted beneath the osteochondral lesion **(arrows)**. There is no surrounding edema, displaced osteochondral fragment, or tracking of fluid into the lesion to indicate instability. There is no evidence of loose body.

Figure 13D. A lateral talar dome osteochondral lesion is present **(black arrowheads)** with slight fragmentation of the subchondral cortex **(arrow)**. The lesion has a slightly sclerotic rim, consistent with early healing.

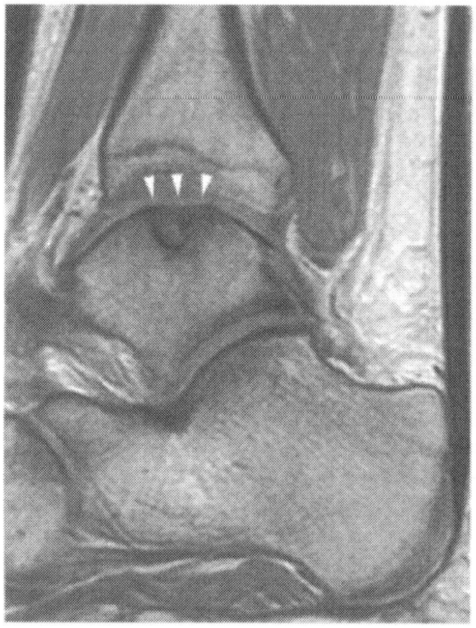

Figure 13B* Annotated.

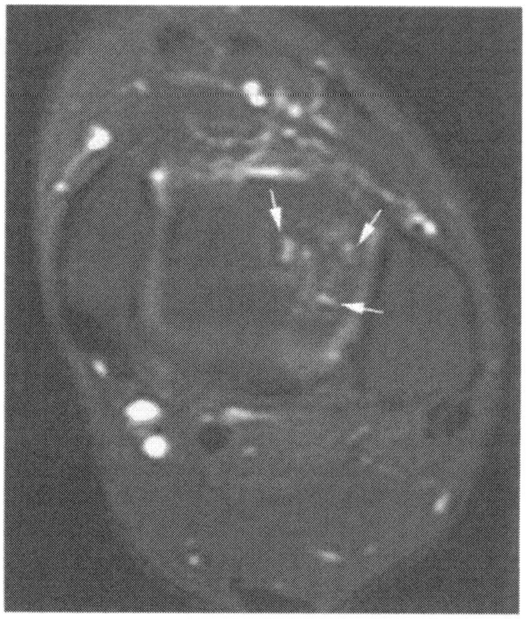

Figure 13C* Annotated.

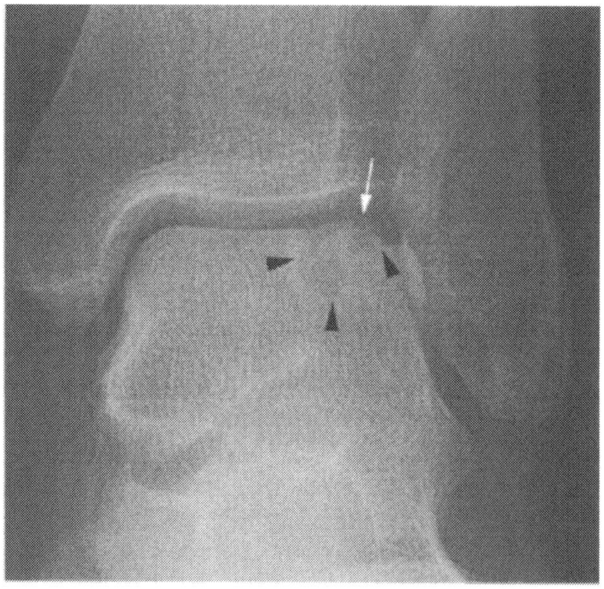

Figure 13D AP radiograph of the ankle.

Diagnosis

Osteochondritis dissecans (AKA osteochondral lesion) of the medial talar dome

Questions

1. What is the mechanism of injury for talar dome osteochondritis dissecans?
2. What is the most common location for talar stress injury?

Discussion

The talus is separated into three components: the talar head, neck, and body. Three important disorders that affect the talus include: osteochondritis dissecans (OCD), which usually involves the posteromedial and anterolateral corners of the talar dome, osteonecrosis, which tends to involve the talar body, and stress reaction or fractures, which most commonly involve the talar neck.

Talar dome OCD is felt to arise secondary to chronic microtrauma (Answer to Question 1). In one study evaluating 24 children with 26 lesions, the medial corner (Figure 13E) (19/26), the lateral corner (5/26), and central talar dome (3/26) were all affected (1). Like OCD elsewhere, it is important to assess for instability utilizing both fluid-sensitive and cartilage-sensitive sequences. Features suggesting instability include: loose bodies, overlying cartilaginous thinning, subchondral cysts at the interface of the lesion and parent bone, and fluid SI tracking between the lesion and parent bone. OCD should be distinguished from an acute osteochondral fracture of the talar dome. An acute osteochondral fracture of the lateral or medial talar dome will often have significant edema of both the talus and opposing malleolus (kissing contusions) and may also have soft tissue edema and ligamentous tears of the opposite side of the ankle. There is a continuum between acute osteochondral fracture and OCD, and an unrecognized flake fracture of the lateral dome may progress to a typical talar OCD. It should be noted that OCD may be asymptomatic, particularly with medial lesions, and that the finding may be encountered incidentally when a study is obtained for an acute injury. Medial talar OCD may be bilateral, and one or both lesions may be asymptomatic.

Talar osteonecrosis (Figures 13F–13H) most commonly occurs after trauma. It may be due to drugs such as corticosteroids, vasculitis, inherited disorders (e.g., sickle cell and Gaucher's disease), or it may be iatrogenic (e.g., clubfoot treatment). Trauma is an important etiology for osteonecrosis of the talar body. The blood supply to the talar body is dependent on branches of the anterior tibial artery and deltoid artery that enter the talar neck (2). Therefore, talar neck fractures may lead to talar body osteonecrosis if these vessels are injured. The complication of talar dome osteonecrosis is early degenerative changes with chondrolysis and talar dome collapse. Osteonecrosis tends to be geographically defined with serpiginous margins and variable SI depending on the age of the insult. A "double-line sign" may be seen surrounding the infarcted region on T1 and T2W sequences, with an outer low SI sclerotic rim and an inner hyperintense rim (3). Uncomplicated osteonecrosis tends to follow fat SI on all imaging sequences. Osteonecrosis complicated by collapse or insufficiency fracture may follow blood SI (hyperintense T1, hyperintense T2), fluid (hypointense T1, hyperintense T2), or fibrous tissue (hypointense T1, hypointense T2).

Stress injuries most commonly involve the talar neck but may occur elsewhere, including the talar body and posteromedial talus (Figure 13I) (Answer to Question 2) (4). Stress injury of the talar neck may occur secondary to repetitive dorsiflexion with anterior tibial impingement. Ballet dancers may develop chronic talar stress changes and posterior impingement due to prolonged plantar flexion (*en pointe*) positioning. Stress changes range from nonspecific marrow edema representing stress reaction to a true fracture with a T1 hypointense and variable T2 hyperintense line that disrupts normal trabecular bone (5). Fracture lines should be distinguished from hypointense bony trabeculae and vascular grooves. Since talar neck stress injuries may occur from repetitive dorsiflexion or plantar flexion, one should also evaluate for coexistent edema or osteophytes of the anterior or posterior margins of the tibia plafond that may reflect tibiotalar impingement. It should be kept in mind that patchy nonspecific marrow edema may be seen in the tarsal bones in athletes without a history of acute trauma and that striking talar edema may be observed in ballet dancers without significant complaints referable to the ankle (Figure 13J) (6).

This patient was treated with a non-weightbearing cast for two months because the OCD was considered stable. After cast removal, she was asked to protect her ankle and to refrain from sports for 6 weeks. Because of persistent symptoms, she was referred to physical therapy. Eventually, she underwent arthroscopic debridement and drilling of the base of the OCD.

Orthopedic Perspective

Some orthopedists may use the Berndt and Hardy plain radiographic classification of talar dome OCD and request that the lesion be staged (I—small subchondral trabecular compression fracture not seen radiographically, II—incomplete avulsion or separation of the fragment, III—complete avulsion without displacement, and IV—detached and rotated fragment, or fragment is free within the joint) (7).

For medial and lateral talar dome OCD, the treatment approach is the same, although prognosis is better with medial lesions. Talar dome OCD is managed nonoperatively if it is an early lesion with an intact articular surface in a patient with open growth plates. Lesions that are considered unstable with a fissured articular surface, or lesions that have failed approximately 6 months of nonoperative treatment, are treated arthroscopically. The procedure depends on the stage of the lesion. Drilling is performed for stable lesions, fixation for unstable lesions, and excision for loose bodies. If the ankle is symptomatic or develops joint space narrowing after excision, then mosaicoplasty and microfracture technique may be employed.

The treatment for talar stress fracture often requires casting and crutches, whereas talar stress reaction may be treated with activity restriction. Patients with talar dome osteonecrosis are treated symptomatically and with activity restriction to prevent subchondral collapse.

What the Clinician Needs to Know

1. OCD stability versus instability.
2. Distinguish stress reaction from stress fracture.
3. What is the status of the talar dome in the setting of an acute talar neck fracture? The talar dome may develop secondary osteonecrosis after its vascular supply is disrupted in the setting of a talar neck fracture.

Answers

1. Chronic microtrauma.
2. Talar neck.

Additional Examples

Stable Osteochondritis Dissecans of the Medial Talar Dome

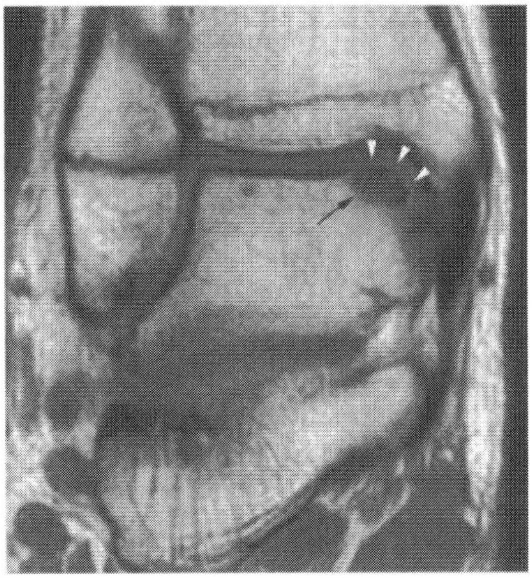

Figure 13E. Coronal T1 of the right ankle.

Findings

This is a 14-year-old boy with persistent right ankle pain.

Figure 13E. There is a small OCD lesion **(arrow)** with no evidence of instability. Note that the overlying articular cartilage is intact and of normal thickness **(arrowheads)**.

Osteonecrosis of the Anterolateral Talar Dome

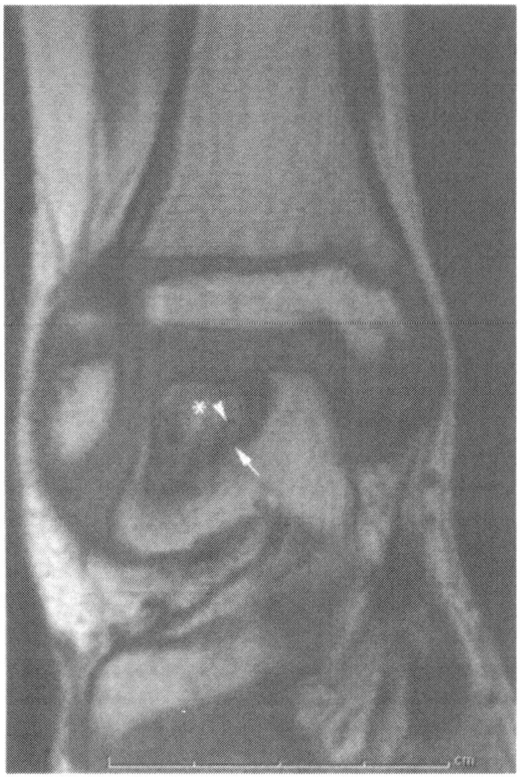

Figure 13F. Coronal T1 of the right ankle.

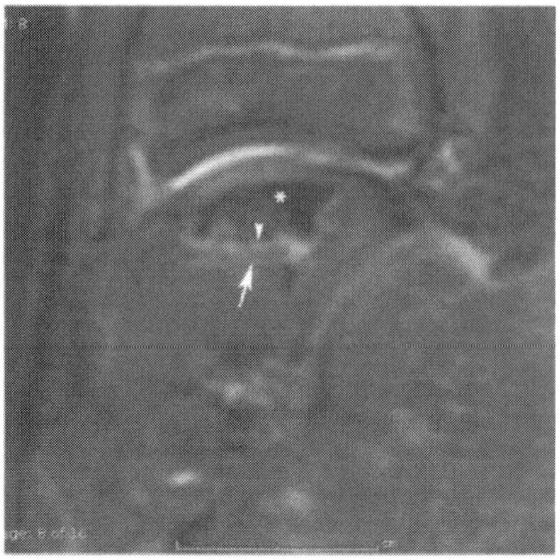

Figure 13G. Sagittal STIR.

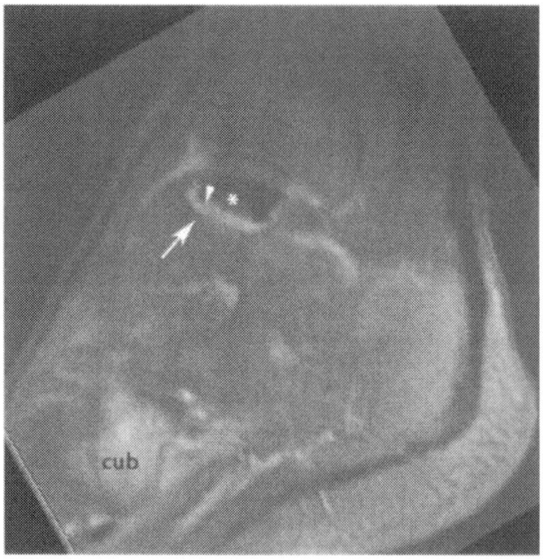

Figure 13H. Sagittal T1 post-Gd FS.

Findings

This 6-year-old boy with acute lymphoblastic leukemia had acute right ankle pain.

Figures 13F, 13G, 13H. There is articular collapse present along the anterolateral dome of the talus. A double-line sign is present with an outer hypointense sclerotic rim **(arrows)** and inner hyperintense and enhancing rim **(arrowheads)** (AKA penumbra sign). The infarcted region follows fat SI on all imaging sequences (*). The cuboid (cub) shows diffuse enhancement, compatible with stress reaction.

Stress Reaction of the Talus

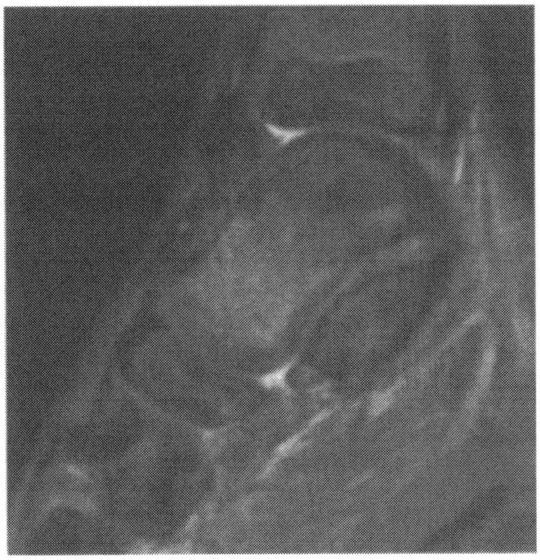

Figure 13I. Sagittal STIR of the right medial ankle.

Findings

This is an 8-year-old boy with a 3-week history of chronic right ankle pain.

Figure 13I. There is increased SI throughout the medial talar neck. No discrete fracture line is seen on the STIR sequence or the T1W sequence (not shown). Therefore, stress reaction, not stress fracture, was diagnosed.

Stress Reaction of the Talus

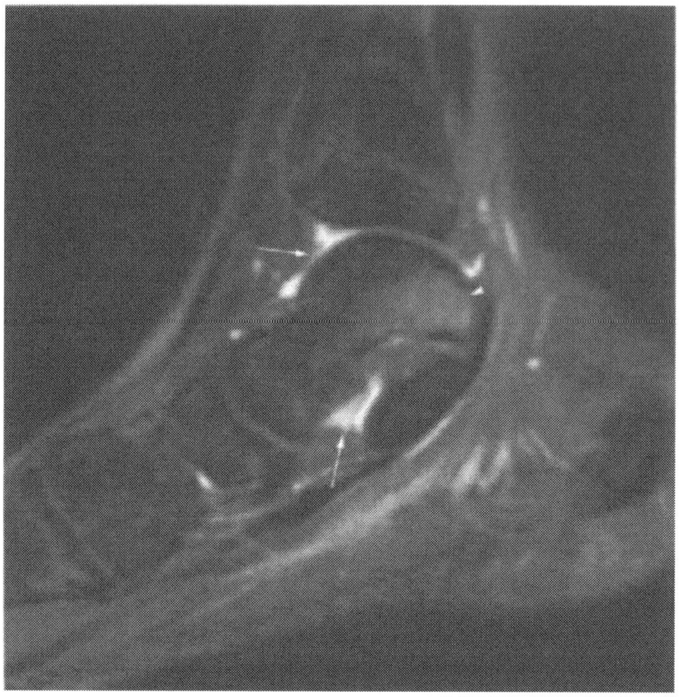

Figure 13J. Sagittal STIR of the left foot.

Findings

This is a 15-year-old ballet dancer with focal pain confined to the midshaft of the second metatarsal.

Figure 13J. There is diffuse increased SI present in the posterior talar body **(arrowhead)** without a discrete fracture line. There are small joint effusions in the ankle and subtalar joints **(arrows)**. Since the patient had intermittent ankle pain in the past, but had none at the time of this examination, these findings were considered to be nonspecific stress changes commonly seen in ballet dancers.

Pitfalls and Pearls

1. Direct or indirect MR arthrography or CT arthrography may be used to better assess stability of OCD.
2. If the abbreviation OCD is used, make sure that the treating physician understands that the term refers to osteochondritis dissecans, not an acute osteochondral fracture.
3. Talar edema is a nonspecific finding and may be asymptomatic. This finding must be correlated with the history and physical examination.

References

1. Letts M, Davidson D, Ahmer A. Osteochondritis dissecans of the talus in children. *J Pediatr Orthop* 2003; 23:617–625.
2. Adelaar RS, Madrian JR. Avascular necrosis of the talus. *Orthop Clin North Am* 2004; 35:383–395, xi.
3. Mitchell DG, Rao VM, Dalinka MK, et al. Femoral head avascular necrosis: Correlation of MR imaging, radiographic staging, radionuclide imaging, and clinical findings. *Radiology* 1987; 162:709–715.
4. Umans H, Pavlov H. Insufficiency fracture of the talus: Diagnosis with MR imaging. *Radiology* 1995; 197:439–442.
5. Rosenberg ZS, Beltran J, Bencardino JT. From the RSNA Refresher Courses, Radiological Society of North America: MR imaging of the ankle and foot. *Radiographics* 2000; 20 Spec No:S153–179.
6. Gigena LM, Chung CB, Lektrakul N, Pfirrmann CW, Sung MS, Resnick D. Transient bone marrow edema of the talus: MR imaging findings in five patients. *Skeletal Radiol* 2002; 31:202–207.
7. Herring JA. Lower extremity injuries. In: *Tachdjian's Pediatric Orthopaedics*, 3 ed., Chapter 42. Philadelphia: W.B. Saunders Company, 2002; 2251–2438.

Case 14

History

This is an 11-year-old boy who had a twisting injury 6 days ago. Radiographs were normal (not shown).

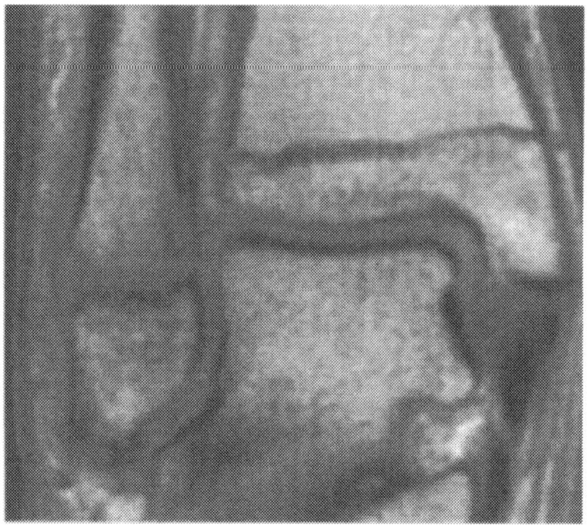

Figure 14A. Coronal T1 of the right ankle.

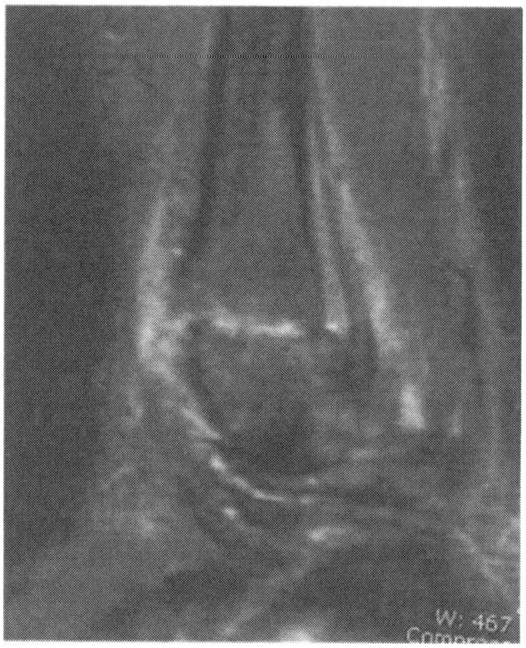

Figure 14B. Sagittal STIR.

Figures 14A, 14B (14B with annotations). The fibular physis is widened. There is mildly decreased SI on T1 and increased SI on STIR in the distal fibular epiphysis, physis, metaphysis, and juxtacortical soft tissues. The tibial physis is normal. On the sagittal view, the distal fibular epiphysis is minimally displaced posteriorly. Note stripping of posterior periosteum from the adjacent metaphysis (**arrow**) with subperiosteal fluid signal.

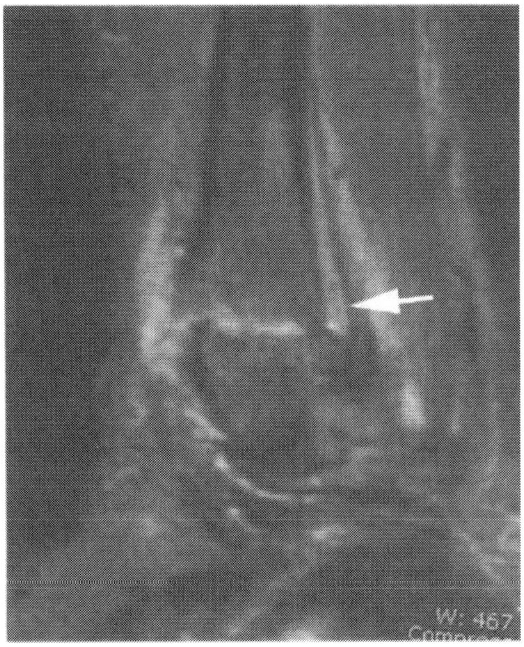

Figure 14B* Annotated.

Diagnosis

Salter-Harris Type 1 fracture of the distal fibula

Questions

1. T/F: Ligamentous disruptions are more common than physeal injuries in young children.
2. What are some normal variations of the metaphyseal-epiphyseal complex that should not be mistaken for fractures on MRI?

Discussion

In the adult patient, the weakest component of the musculoskeletal unit is the myotendinous junction and ligaments. In the skeletally immature patient, however, the weakest component is the physis (Answer to Question 1) (1). The initial task of MRI interpretation of the pediatric patient with trauma and normal radiographs is to assess the status of the physis and rule out an occult fracture. If there is significant soft tissue swelling near an otherwise normal physis on plain radiography, a Salter-Harris type fracture should be considered.

The Salter-Harris classification of physeal fractures provides a practical approach to effectively communicate the extent of fracture and its consequences (Diagram 14A). The type 2 Salter-Harris fracture is the most common and represents a metaphyseal fracture that extends to the physis. The extent of physeal injury is generally less severe with lower grade Salter-Harris fractures. Similarly, the potential for physeal growth disturbance and physeal bar formation is also less likely with lower grade Salter-Harris fractures.

Failure to recognize and treat a physeal injury may lead to growth disturbance and physeal bars. In the proper clinical context, MRI may explain persistent pain in an extremity, despite normal radiographs. Lohman et al. showed that MRI changed the diagnosis or Salter-Harris classification of 21% of 60 consecutive ankle injuries in children that were initially evaluated by plain radiography (2). However, they did not show a significant change in management, even when MRI helped with the detection of radiographically occult fractures and delineated bone bruising. In an another study that had a more selected population, Stuart et al. found 5 out of 10 Salter-Harris injuries with MRI that were not initially detected on radiographs (3).

A practical approach to MRI evaluation of trauma without plain radiographic abnormality is to use fluid-sensitive sequences in conjunction with T1 or high spatial resolution PD sequences. The high signal-to-noise ratio afforded by T1 and PD sequences is well suited to the identification of subtle hypointense fracture lines, as well as associated tendinous or ligamentous injury. T1 or PD sequences obtained in the same imaging plane as the fluid-sensitive sequences are especially useful (Figures 14C, 14D). Stress reaction, bone contusion, and fracture represent a spectrum of bone injuries. Stress reaction demonstrates nonspecific marrow edema, a bone contusion may show subtle trabecular discontinuity, and a fracture demonstrates a discrete, linear hypointensity reflecting trabecular and/or cortical disruption.

There is a spectrum of anatomic variations of the metaphyseal/epiphyseal complex that should not be misinterpreted as a fracture on MRI. Therefore, correlation with radiographs is essential in the MRI evaluation of trauma. Anatomic variations that should not be mistaken for a fracture include: normal physeal undulation, a fibrous synchondrosis, and accessory ossicles (Answer to Question 2). Physeal undulation may be distinguished from a true fracture by the absence of edema on fluid-sensitive sequences. It should be noted that developmental variants, particularly in the foot and ankle, may become symptomatic in the context of chronic trauma (Figures 14E–14G). Although comparison views of the opposite extremity are generally discouraged in the evaluation of pediatric trauma, they may be useful in differentiating a symptomatic developmental variant from a fracture.

The MRI in this patient was ordered because a Salter-Harris 1 injury of the fibula was suspected on clinical grounds. No other fractures or ligamentous injuries were seen and the patient was treated with casting. Without appropriate history and the presence of slight epiphyseal displacement on MRI, the marrow edema with a subperiosteal fluid collection would be indistinguishable from osteomyelitis.

Orthopedic Perspective

Salter-Harris 1 distal fibular physeal ankle injuries are extremely common and represent the "pediatric ankle sprain." Usually, the diagnosis can be made clinically in the patient with an inversion ankle injury with negative radiographs and tenderness over the distal fibular physis. Treatment is typically immobilization with a cast, brace, or boot for 3 to 4 weeks. If MRI shows a normal physis and the injury is confined to the soft tissues, then the patient may be splinted (for comfort) and not treated with a cast. If a Salter-Harris injury is present, then the patient is typically immobilized with casting for 4 to 6 weeks.

What the Clinician Needs to Know

Is there a Salter-Harris Type 1 fracture and other associated occult injuries?

Answers

1. False.
2. Normal physeal undulation, fibrous synchondroses, and accessory ossicles.

Additional Examples

Salter-Harris type 2 fracture of the proximal tibia

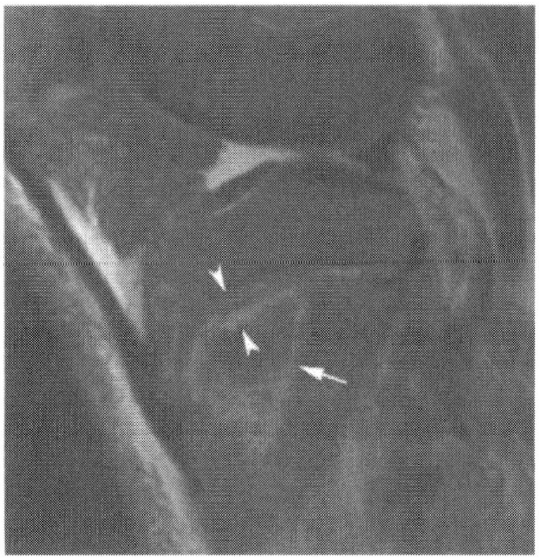

Figure 14C. Sagittal STIR of the right knee.

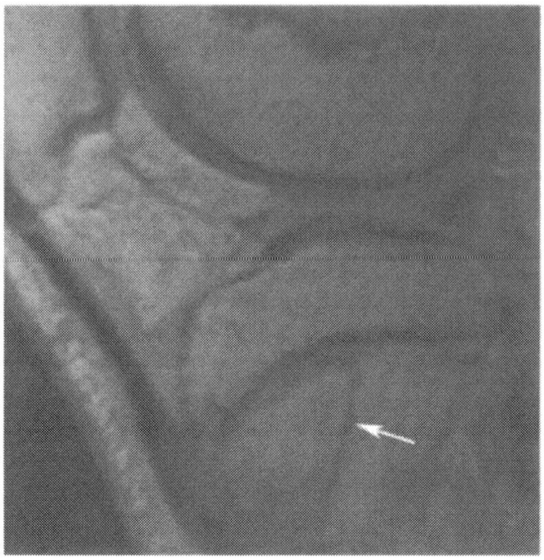

Figure 14D. Sagittal PD.

Findings

This 12-year-old boy fell on his knees and had persistent pain despite negative radiographs (not shown).

Figure 14C. There is linear increased SI within the tibia metaphysis (**arrow**) extending to the widened physis (**arrowheads**).

Figure 14D. The fracture line (**arrow**) and physeal widening is better delineated with a high-resolution PD sequence.

Stress Reaction in an Os Subfibulare

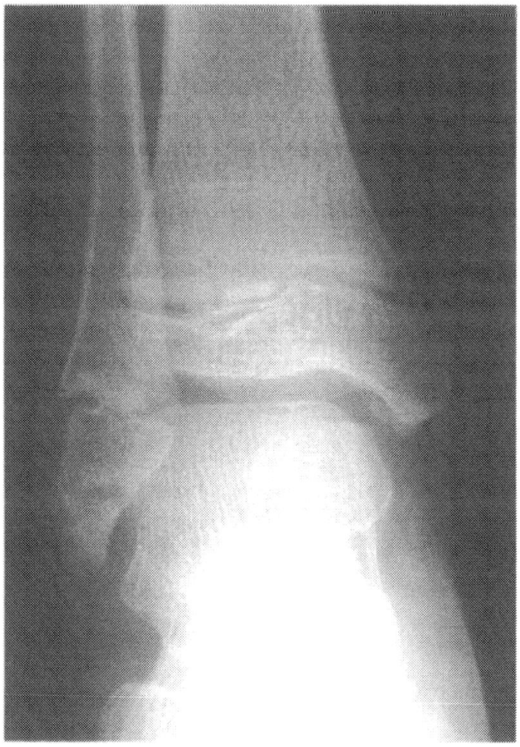

Figure 14E. AP radiograph of the right ankle.

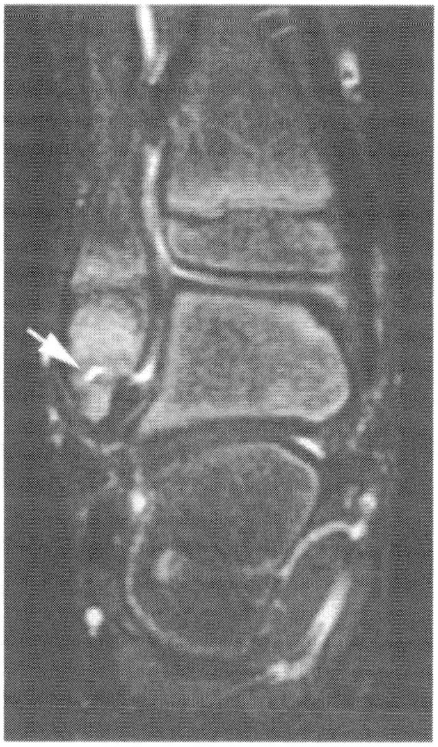

Figure 14F. Coronal STIR.

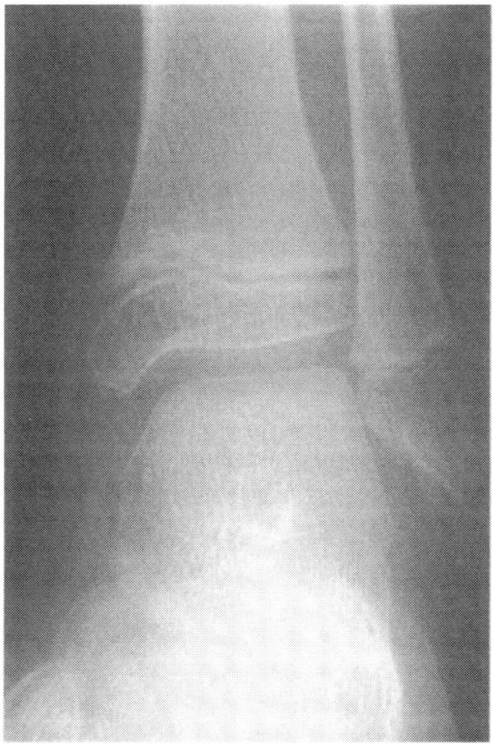

Figure 14G. AP radiograph of the left ankle.

Findings

This 8-year-old girl had multiple right ankle inversion injuries and chronic pain at the lateral malleolus.

Figure 14E. The radiograph shows a well corticated ossicle below the tip of the lateral malleolus.

Figure 14F. There is hyperintense SI within the distal fibular epiphysis and ossicle with an intervening fluid collection **(arrow)**. It is not possible to differentiate a symptomatic os subfibulare from an ununited fracture.

Figure 14G. The comparison radiograph of the left ankle shows a similar ossicle, providing support that the changes on the right represent a symptomatic normal variant.

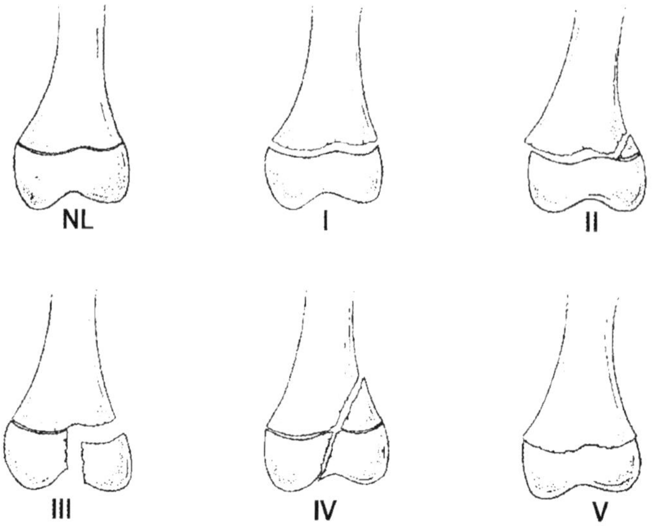

Type 1. Epiphyseal separation and dislocation without osseous extension.
Type 2. Metaphyseal fracture line extends to the physis. This is the most common type of Salter-Harris fracture.
Type 3. Epiphyseal fracture line extends to the physis.
Type 4. Fracture line crosses the metaphysis, physis, and epiphysis.
Type 5. Physeal compression fracture.
Diagram 14A. The Salter-Harris classification (NL: normal). (Diagram reprinted from Kleinman PK, with permission from Elsevier) (4).

Pitfalls and Pearls

1. It may be difficult to differentiate a SH 1 fracture from osteomyelitis by MRI in the patient with both a history of trauma and inflammatory signs and symptoms. In this context, contrast enhanced imaging is critical, to assess for other MRI features of infection.
2. Many suspected SH fractures are treated with immobilization, despite negative radiographs, and thus MRI should be performed only in selected cases where management is likely to be affected by the study.

References

1. Ohashi K, Brandser EA, el-Khoury GY. Role of MR imaging in acute injuries to the appendicular skeleton. *Radiol Clin North Am* 1997; 35:591–613.
2. Lohman M, Kivisaari A, Kallio P, Puntila J, Vehmas T, Kivisaari L. Acute paediatric ankle trauma: MRI versus plain radiography. *Skeletal Radiology* 2001; 30:504–511.
3. Stuart J, Boyd R, Derbyshire S, Wilson B, Phillips B. Magnetic resonance assessment of inversion ankle injuries in children. *Injury* 1998; 29:29–30.
4. Kleinman P. *Diagnostic Imaging of Child Abuse*. 2nd ed. St. Louis: C.V. Mosby, 1998.

Case 15

History

This is a 10-year-old girl with a 1-week history of left sided hip pain and decreased range of motion.

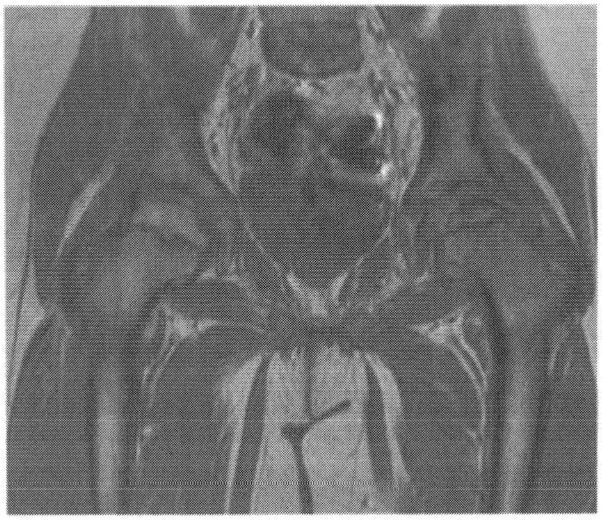

Figure 15A. Coronal T1.

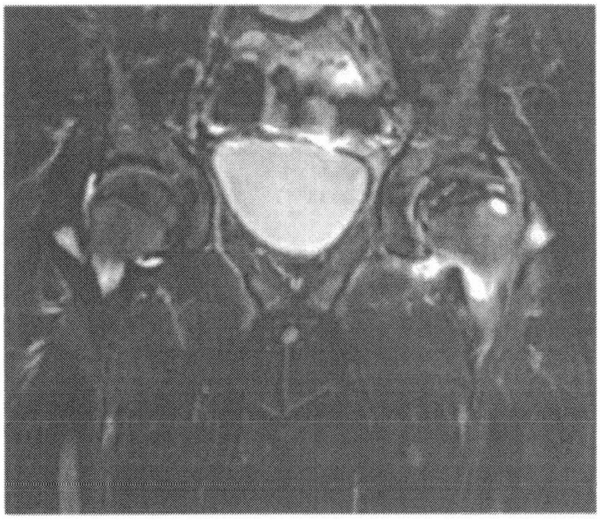

Figure 15B. Coronal T2 FS.

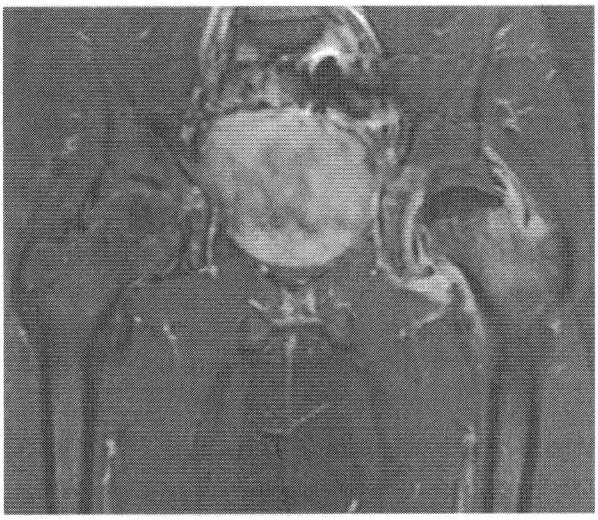

Figure 15C. Coronal T1 post-Gd FS.

Figures 15A, 15B, 15C (15B with annotations). The left capital femoral epiphysis is small, aspherical, and hypointense on all imaging sequences. Metaphyseal **(arrowhead)** and epiphyseal **(thin arrow)** T2 hyperintense cysts are present. Fluid signal defines a crescentic subchondral fracture **(thick arrow)**. There is significant metaphyseal marrow edema, joint effusion, and synovial enhancement. With the exception of a small joint effusion, the right hip is normal.

Figure 15D. The left capital femoral epiphysis is flattened, sclerotic, and there is a crescent sign present **(arrow)**. The right hip is normal.

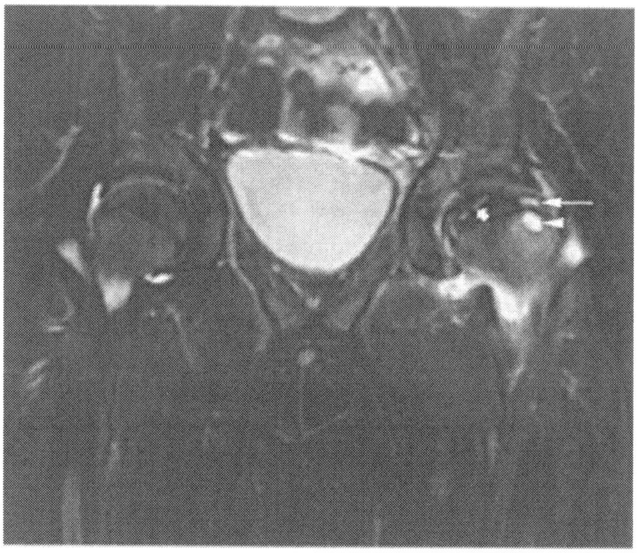

Figure 15B* Annotated.

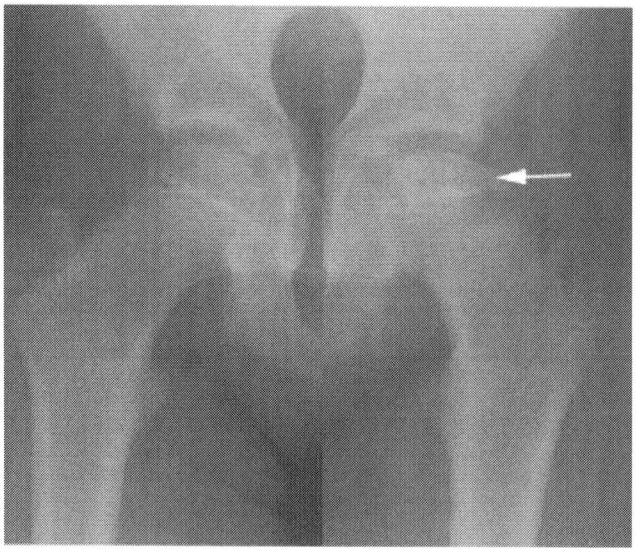

Figure 15D. AP radiograph.

Diagnosis

Legg-Calve-Perthes (LCP) disease (Catterall grade 3 or Herring lateral column class B)

Questions

1. What causes lateral hip subluxation in patients with chronic changes of LCP?
2. T/F: LCP may lead to osteochondritis dissicans and loose bodies.

Discussion

Legg-Calve-Perthes (LCP) disease is idiopathic ischemic necrosis (AKA avascular necrosis or epiphyseal osteonecrosis) of the capital femoral epiphysis. It presents in children between 3 and 12 years of age and is most commonly seen between 5 and 7 years (1). There is a male predominance and it occurs bilaterally in 10% to 20% of patients. Ischemic necrosis occurs in the young child, when the capital femoral epiphyseal blood supply is most tenuous. The capital femoral epiphysis receives the majority of its blood supply from the deep branch of the medial femoral circumflex artery (2). Ischemic necrosis is believed to occur less often in the young infant or the older adolescent because of additional epiphyseal vascular supply from transphyseal vessels present in these patients.

The early plain radiographic findings of LCP include: subchondral lucency (crescent sign) (Figure 15D), increased epiphyseal density or osteopenia, small epiphysis, lateral subluxation, and cystic changes within the epiphysis and metaphysis (Figure 15B). Advanced changes of LCP include: epiphyseal asphericity including flattening and fragmentation, femoral neck shortening, and lateral subluxation. Herring and others have created classification systems to grade and predict the natural history of LCP (3). Herring's lateral pillar classification system uses the AP radiograph of the hips and compares the height of the lateral 15% to 30% of the affected capital femoral epiphysis with the normal contralateral side. Group A lesions demonstrate fragmentation without loss of height. Group B lesions demonstrate fragmentation with up to 50% loss of height. Group C lesions demonstrate fragmentation with greater than 50% loss of height.

The MRI features of early LCP include hypointensity on T1 and variable increased SI on fluid-sensitive sequences in the epiphysis, physis, and metaphysis, as well as synovitis (Figure 15C) (4, 5). On post-Gd sequences, peripheral enhancement of the epiphysis may be seen during early LCP (Figures 15E–15I); central enhancement may be seen later (4). Dynamic post-Gd and diffusion weighted sequences have also been proposed for detection of early LCP (6, 7). During the early arterial phase, absent enhancement may be seen in the ischemic femoral head. With diffusion weighted sequences, restricted diffusion of the ischemic regions may occur (Figure 15I).

MRI features that may predict future femoral head deformity in the setting of LCP include: greater than 2/3 epiphyseal involvement by low SI on both T1 and T2W sequences, metaphyseal edema, physeal edema, physeal bars, metaphyseal cartilaginous rests, and metaphyseal cysts (8, 9). Metaphyseal cartilaginous rests are tongue-like projections that extend from the physis and are related to disrupted endochondral bone formation of similar SI of cartilage elsewhere. Metaphyseal cysts have more well-defined margins compared with cartilaginous rests and follow fluid SI on all imaging sequences. Reconstitution of high T1 SI in the lateral quarter of the epiphysis may predict less femoral head deformity and uncovering in the future (8).

MRI features of advanced LCP include: medial and lateral epiphyseal cartilaginous overgrowth, coxa plana, and premature femoral and triradiate cartilage physeal arrest. Joint effusion, medial epiphyseal cartilage overgrowth, synovial hypertrophy, and lateral growth arrest seen by MRI may explain lateral subluxation and acetabular-femoral incongruity seen with LCP (Answer to Question 1). Late complications of LCP include CAM-type femoral-acetabular impingement (usually anterior) with labral degeneration (see Case 71), secondary degenerative changes with articular cartilage thinning and collapse, and osteochondritis dissicans with loose bodies (Answer to Question 2) (10, 11). Synovitis may be seen in both early and late stages of LCP (5).

LCP should not be confused with normal variations of capital femoral epiphyseal ossification (Figures 15J–15L). Although patients with this pattern of ill-defined subchondral lucency and sclerosis (Figure 15J) are occasionally referred for MRI, the diagnosis is usually readily apparent on radiographs and further imaging is rarely necessary.

Meyer's dysplasia is an ill-defined and not universally accepted entity that is felt by some authorities to represent a self-limited osteochondrosis of the capital femoral epiphysis. In one study, the average age of presentation was 2 years, 8 months (range: 16 months to 5 years, 6 months) (12). No long-term femoral head deformity or disability is seen with Meyer's dysplasia, although some cases may go on to develop typical LCP. In contrast to LCP, a normal femoral head is never documented prior to the appearance of the typical radiographic alterations. Fragmentation of the capital femoral epiphysis may be seen. The changes seen with Meyer's dysplasia in some cases may relate to normal, but asynchronous, ossification of the capital femoral epiphysis. In other instances, cases labeled as Meyer's dysplasia may reflect mild early onset LCP that resolves without residual radiologic changes. It is a diagnosis that should be made with caution.

This patient was newly diagnosed with LCP disease. The acuity of her symptoms was probably related to subchondral collapse of the capital femoral epiphysis. Septic arthritis/osteomyelitis was excluded based on clinical exam in conjunction with the radiographic and MRI features of typical LCP. She was given crutches to limit weightbearing and a hyperabduction brace to help contain the capital femoral epiphysis.

Orthopedic Perspective

LCP disease is a vexing condition to treat because the etiology is not established and the indications for treatment are unclear. LCP is usually classified and followed by radiographs; however, MRI provides a better representation of epiphyseal cartilaginous anatomy, the extent of ischemic necrosis, and long-term complications such as femoral head asphericity, CAM-type femoroacetabular impingement, and loss of femoral head and acetabular congruity. MRI is typically not needed for staging. However, MRI may be utilized for problem solving when symptoms are atypical or used to help explain the patient's symptoms (e.g., is there active synovitis related to LCP and is there new epiphyseal collapse?). The prognosis of LCP disease depends on the sphericity and anatomy of the femoral head after progression through the stages of fragmentation, healing, ossification, and remodeling.

The treatment for LCP is rest until the patient is asymptomatic, as well as restriction of high-impact activities through the healing phase. The healing phase is determined when fragmentation and collapse are stabilized and re-ossification is evident. Free fibular bone grafts, which may be used for ischemic necrosis in older adolescents, are not performed on young patients because of growth plate issues.

The loss of normal acetabular-femoral congruity that occurs as a long-term complication of LCP may require pelvic or femoral osteotomies to preserve congruity.

What the Clinician Needs to Know

1. LCP staging including percentage of femoral head containment by the acetabulum.
2. Internal derangement such as an osteochondral flap or labral tear.

Answers

1. Lateral subluxation may occur secondary to joint effusion, medial epiphyseal overgrowth, synovial hypertrophy, and lateral growth arrest.
2. True.

Additional Examples

LCP, Atypical Acute Presentation

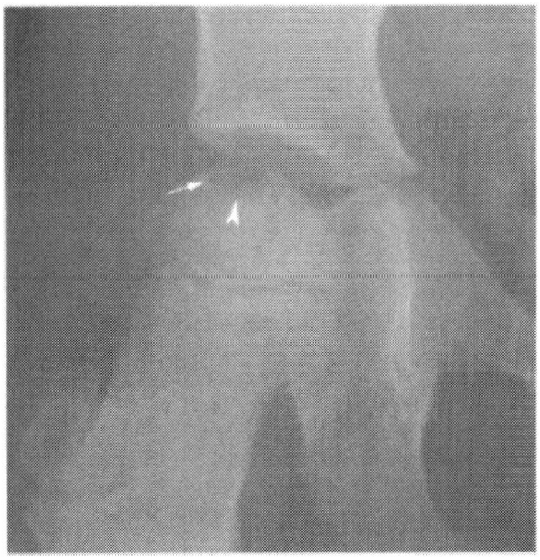

Figure 15E. AP radiograph of the right hip.

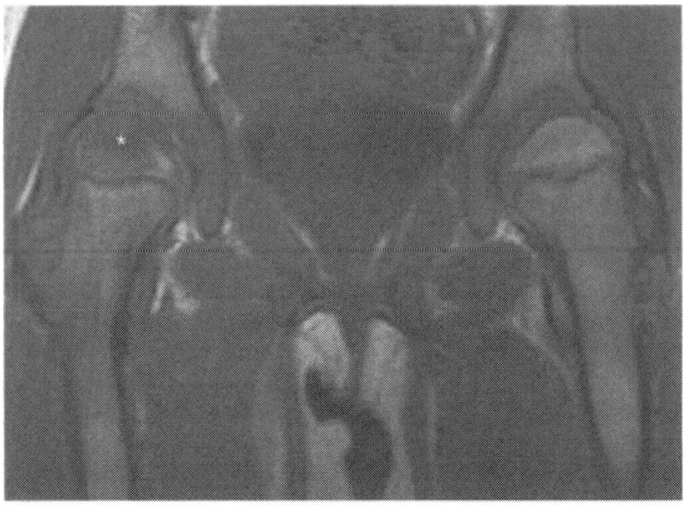

Figure 15F. Coronal T1.

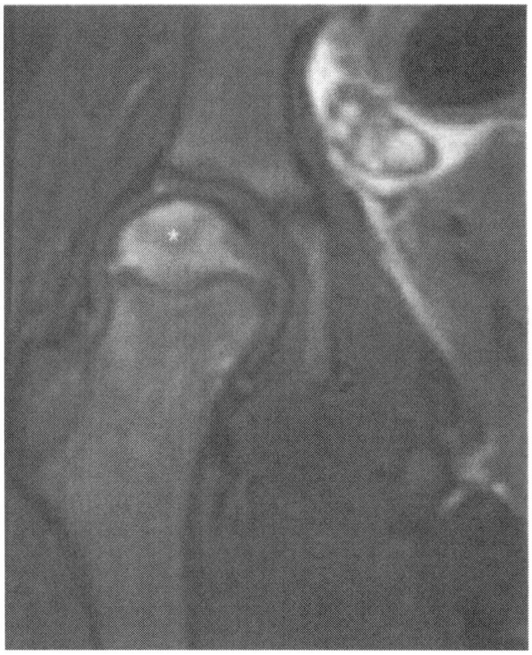

Figure 15G. Coronal STIR.

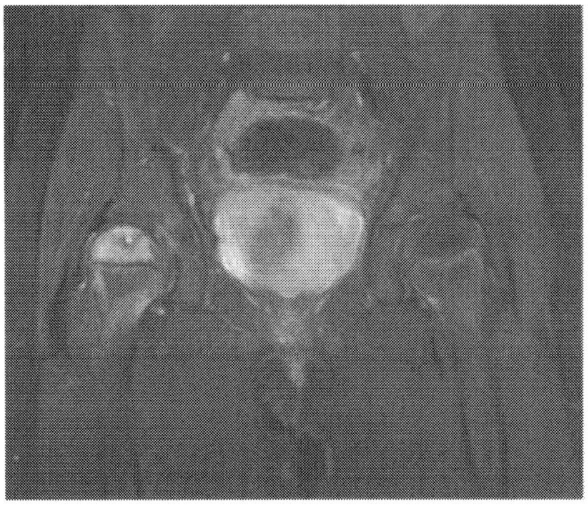

Figure 15H. Coronal T1 post-Gd FS.

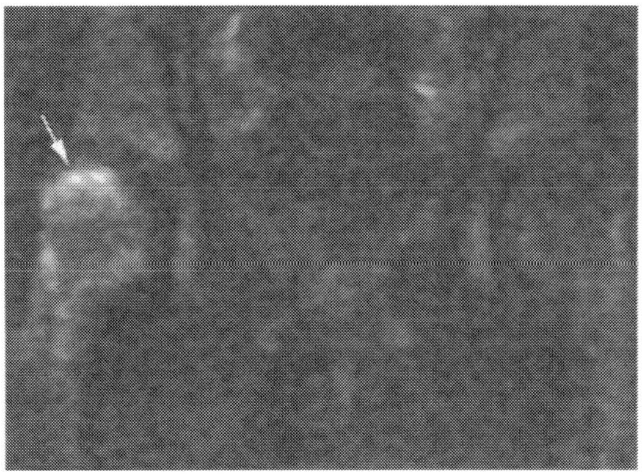

Figure 15I. Coronal diffusion (DWI).

Findings

This 4-year-old girl had acute right hip pain.

Figure 15E. The radiograph shows a subchondral crescentic lucency **(arrow)** within a larger lucent region **(arrowhead)**.

Figure 15F. There is significant central epiphyseal hypointensity (*).

Figures 15G, 15H. There is subchondral STIR hyperintensity and enhancement surrounding a central epiphyseal zone of hypointensity (*).

Figure 15I. Subchondral restricted diffusion is seen **(arrow)**. These findings, in conjunction with the clinical history and follow-up, are compatible with LCP.

Normal Capital Femoral Epiphyseal Developmental Irregularity

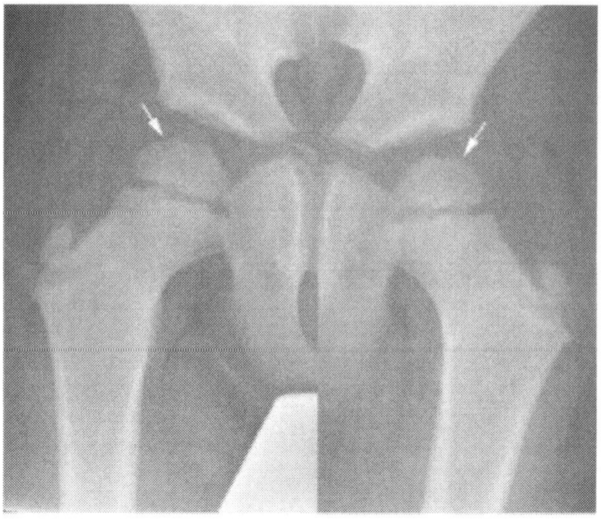

Figure 15J. AP radiograph.

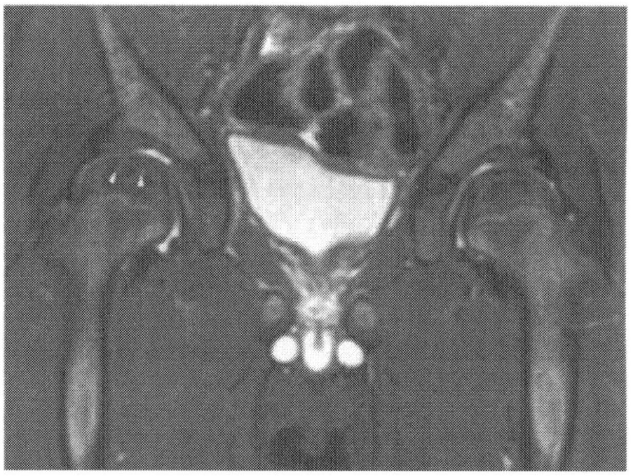

Figure 15K. Coronal STIR.

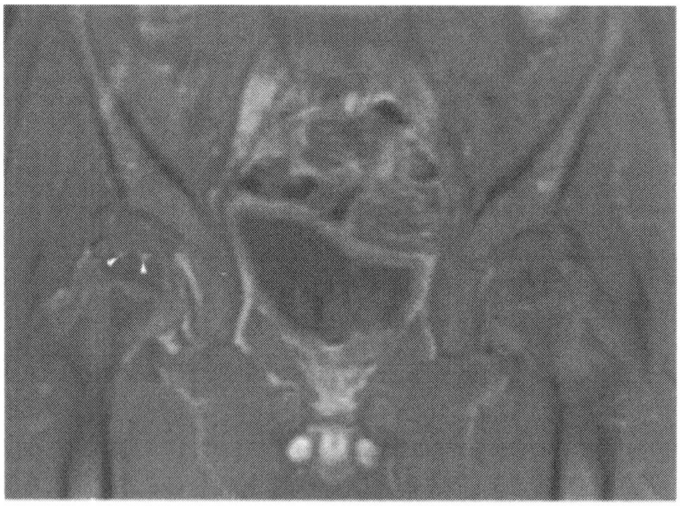

Figure 15L. Coronal T1 post-Gd FS.

Findings

This is a 4-year-old boy with acute onset right hip pain. MRI was ordered based on the concerns of LCP on radiographs.

Figure 15J. Ill-defined subchondral lucency and sclerosis are noted in the capital femoral epiphyses **(arrows)**.

Figures 15K, 15L. STIR and post-Gd T1 sequences demonstrate a normal appearance of the capital femoral epiphyses. On the right, there are focal areas of subchondral STIR hyperintensity and enhancement, related to the normal pre-ossification zone beneath the spherical growth plate **(arrowheads)**.

Pitfalls and Pearls

1. If advanced LCP-like changes are present bilaterally, consider congenital etiologies, such as multiple epiphyseal dysplasia and pseudoachondroplasia.
2. Since developmental variants of the femoral head may be mistaken for LCP, be sure to evaluate good quality radiographs prior to MRI to avoid an unnecessary study.

References

1. Wenger DR, Ward WT, Herring JA. Legg-Calve-Perthes disease. *J Bone Joint Surg Am* 1991; 73:778–788.
2. Gautier E, Ganz K, Krugel N, Gill T, Ganz R. Anatomy of the medial femoral circumflex artery and its surgical implications. *J Bone Joint Surg Br* 2000; 82:679–683.
3. Herring JA, Kim HT, Browne R. Legg-Calve-Perthes disease, Part I: Classification of radiographs with use of the modified lateral pillar and Stulberg classifications. *J Bone Joint Surg Am* 2004; 86-A:2103–2120.
4. Hochbergs P, Eckervall G, Wingstrand H, Egund N, Jonsson K. Epiphyseal bone-marrow abnormalities and restitution in Legg-Calve-Perthes disease: Evaluation by MR imaging in 86 cases. *Acta Radiol* 1997; 38:855–862.
5. Hochbergs P, Eckerwall G, Egund N, Jonsson K, Wingstrand H. Synovitis in Legg-Calve-Perthes disease: Evaluation with MR imaging in 84 hips. *Acta Radiol* 1998; 39:532–537.
6. Sebag G, Ducou Le Pointe H, Klein I, et al. Dynamic gadolinium-enhanced subtraction MR imaging—a simple technique for the early diagnosis of Legg-Calve-Perthes disease: Preliminary results. *Pediatr Radiol* 1997; 27:216–220.
7. Jaramillo D, Connolly SA, Vajapeyam S, et al. Normal and ischemic epiphysis of the femur: Diffusion MR imaging study in piglets. *Radiology* 2003 June; 227(3):825–832.
8. Lahdes-Vasama T, Lamminen A, Merikanto J, Marttinen E. The value of MRI in early Perthes' disease: An MRI study with a 2-year follow-up. *Pediatr Radiol* 1997; 27:517–522.
9. Jaramillo D, Hoffer FA. Cartilaginous epiphysis and growth plate: Normal and abnormal MR imaging findings. *AJR Am J Roentgenol* 1992; 158:1105–1110.
10. DeAngelis NA, Busconi BD. Hip arthroscopy in the pediatric population. [Review] [11 refs]. *Clinical Orthopaedics & Related Research* 2003 Jan (406):60–63.
11. Linden B, Jonsson K, Redlund-Johnell I. Osteochondritis dissecans of the hip. *Acta Radiol* 2003; 44:67–71.
12. Khermosh O, Wientroub S. Dysplasia epiphysealis capitis femoris: Meyer's dysplasia. *J Bone Joint Surg Br* 1991; 73:621–625.

Case 16

History

This is a 2-year-old boy with swollen hands.

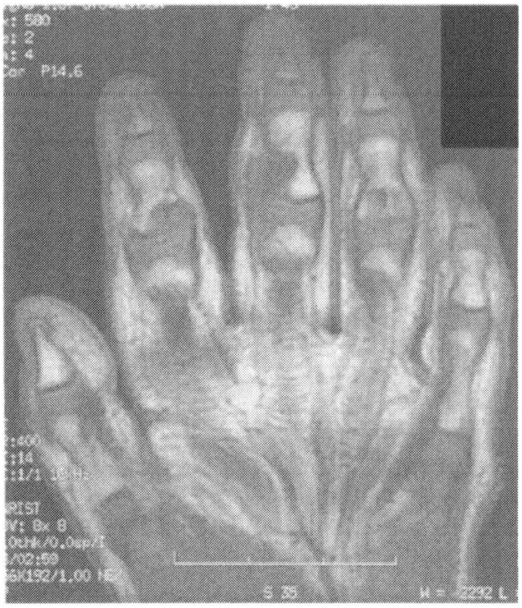

Figure 16A. Coronal T1 of the left hand.

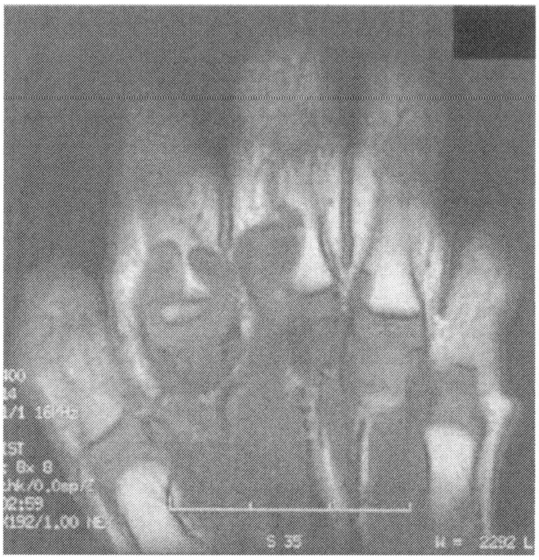

Figure 16B. Coronal T1 more dorsal to Figure 16A.

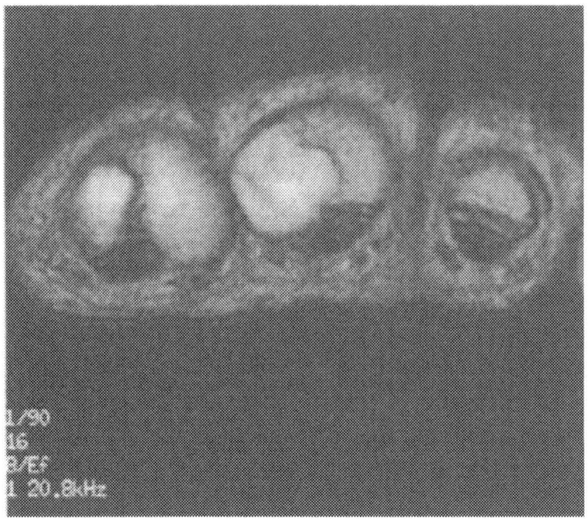

Figure 16C. Axial T2 FS digits 2–4.

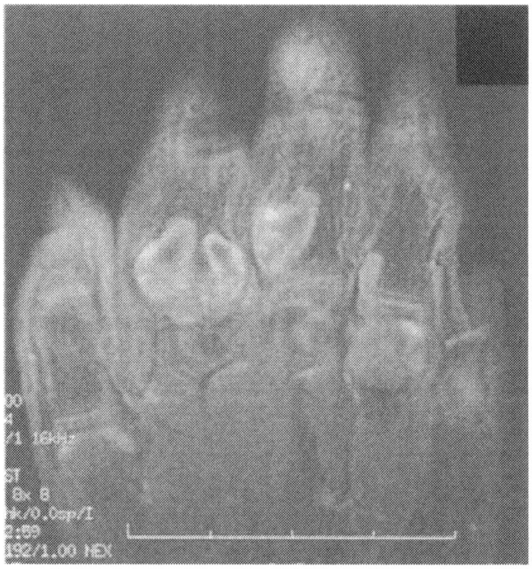

Figure 16D. Coronal T1 post-Gd FS.

Figures 16A, 16B, 16C, 16D. There are multiple, intramedullary, sharply defined exophytic metaphyseal masses arising from the phalanges of the left hand that are isointense on T1, hyperintense on T2, and demonstrate peripheral enhancement. They follow the SI of cartilage on all sequences.

Figure 16E. The lesions present on MRI correlate radiographically with well-defined, eccentrically located osteolytic lesions. Note additional lesions in the 2nd and 3rd metacarpals.

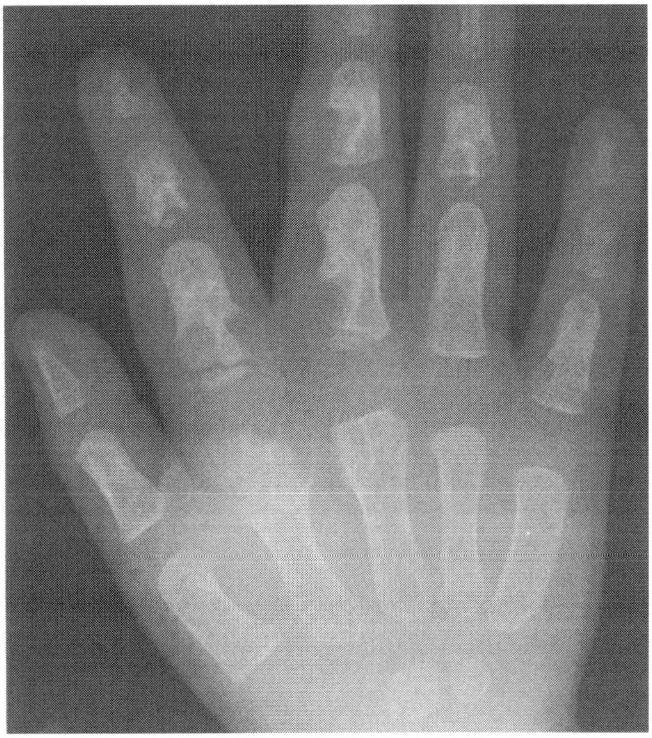

Figure 16E. PA radiograph of the left hand.

Diagnosis

Enchondromatosis (AKA Ollier's disease)

Questions

1. What is the most common location for an enchondroma?
2. Patients with Ollier's disease and Maffucci's syndrome are at risk for what neoplasms?

Discussion

Enchondromas are benign cartilaginous tumors arising from the medullary cavity. They are most often found in the small bones of the hand (most frequently in the proximal phalanx), followed by the femur, humerus, and tibia (Answer to Question 1) (1). Within the long tubular bones, they most often occur in the diaphysis. Isolated enchondromas are usually found the during third and fourth decade of life, with a mean age of presentation of 38 years (2). In contrast to patients with isolated enchondromas, Ollier's disease, as well as the related disorder, Maffucci's syndrome, often presents in infancy or in early childhood (3, 4).

Ollier's disease and Maffucci's syndrome are both nonhereditary enchondromatoses characterized by multiple enchondromas associated with significant orthopedic deformity. Maffucci's syndrome resembles Ollier's disease but also manifests multiple soft tissue spindle cell hemangiomas (5). Based on the Mulliken and Glowacki classification of vascular anomalies, these spindle cell hemangiomas are probably a combination of venous malformations (since they contain phleboliths) and hemangioendotheliomas (6). Patients with enchondromatosis, particularly those with Maffucci's syndrome, have a high incidence of chondrosarcomatous transformation of enchondromas (15% to 30%) and also have a higher risk of developing gliomas, gastrointestinal adenocarcinomas, ovarian tumors, and pancreatic carcinoma (Answer to Question 2) (3). A third enchondromatosis syndrome is metachondromatosis, a very rare condition characterized by multiple enchondromas and osteochondromas (7). Unlike conventional osteochondromas, osteochondromas associated with metachondromatosis point toward, rather than away from the joint.

On plain radiography, enchondromas are osteolytic lesions with well-defined, sclerotic margins, reflecting their slow growing benign character. They often demonstrate a ring-and-arc pattern of calcification that is usually easily identified when the long tubular bones of the extremities are involved (2). This pattern of calcification is less well appreciated when the short tubular bones of the hands and feet are affected. The exophytic appearance of these lesions may incorrectly suggest multiple exostoses. Enchondromas may sometimes demonstrate channel-like metaphyseal lucencies with an "organ pipe" appearance in the long tubular bones (Figures 16F–16H) (8). This finding is more commonly seen in patients with Ollier's disease.

On MRI, enchondromas usually follow similar SI with respect to hyaline cartilage on all imaging sequences. The lesions have a multilobular appearance and are sharply defined with respect to adjacent normal bone. These lesions are usually hypointense to isointense on T1, markedly hyperintense on fluid and cartilage sensitive sequences, and may demonstrate peripheral enhancement and central ring-and-arc enhancement (1, 9, 10). Small hyperintense foci may be seen within enchondromas (11). Low SI septa and mineralized matrix may be identified between and within cartilaginous lobules.

Asymptomatic enchondromas are not generally referred for MRI, since the radiographic features are sufficient to suggest a benign entity and are often diagnostic. MRI may be requested in the setting of a pathologic fracture when lesional characterization is necessary prior to treatment. When enchondromas involve the long tubular bones of the appendicular skeleton, chondrosarcoma may be considered, particularly in the context of pain. However, sarcomatous transformation of enchondromas is rare in children. In addition to pain, features that suggest chondrosarcoma in adult patients include size greater than 5–6cm, greater than 2/3 endosteal scalloping of the cortex, cortical break, extraosseous soft tissue mass, marked uptake on bone scintigraphy, and periosteal reaction (11).

This patient was diagnosed with Ollier's disease after biopsy confirmed that these lesions were enchondromas.

Orthopedic Perspective

The main concern for the orthopedist is in making the diagnosis, which is usually not difficult given the characteristic appearance of enchondroma described and illustrated above. When multiple lesions are present, the differential diagnosis is relatively short and includes multicentric conditions such as neurofibromatosis and Langerhans cell histiocytosis. Since malignant degeneration is extremely uncommon in childhood, the orthopedist is primarily concerned with limb length inequalities (Ollier's disease is usually more prominent on one side of the body), angular deformities, and potential fracture risk. Occasionally solitary enchondromas will be a concern for fracture and require curettage and bone graft, but most often enchondromas are asymptomatic and can be observed.

What the Clinician Needs to Know

1. Enchondroma needs to be distinguished from other radiolucent lesions of bone.
2. The presence of multicentric disease should be assessed if clinically suspected.
3. Malignant degeneration in childhood is extremely rare.
4. Angular deformities and limb length inequality should be assessed clinically and at times with standing radiographs and scanograms in patients with Ollier's disease.
5. Fracture risk should be assessed and prophylactic treatment offered to patients with large enchondromas suspected of being at risk.

Answers

1. Small bones of the hand.
2. Chondrosarcoma, gliomas, gastrointestinal adenocarcinomas, ovarian tumors, and pancreatic carcinoma.

Additional Example

Ollier's Disease Variant

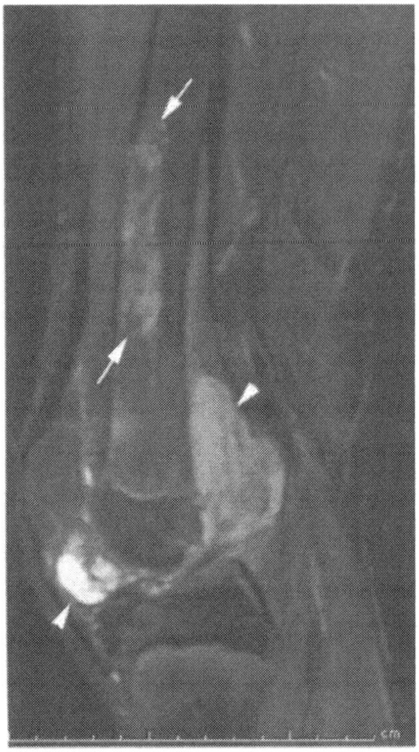

Figure 16F. Sagittal STIR of the left distal femur and knee.

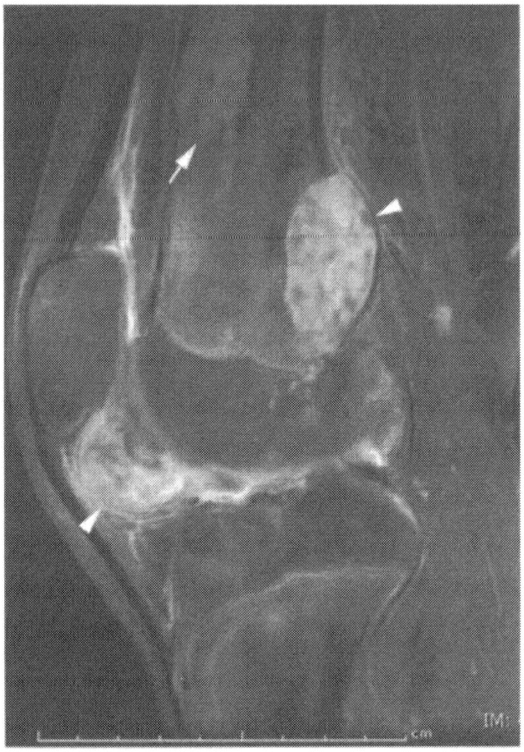

Figure 16G. Sagittal T1 post-Gd FS.

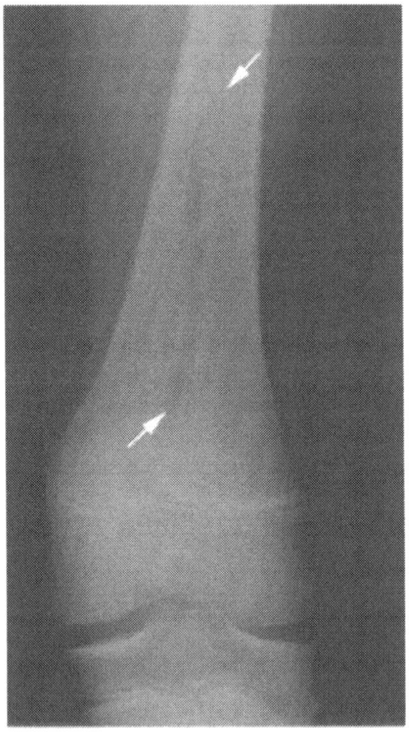

Figure 16H. AP radiograph.

Findings

This is a 13-year-old boy who presented with a 1-year history of difficulty bending his knee.

Figures 16F, 16G. There are multiple hyperintense juxtacortical and intra-articular cartilaginous masses **(arrowheads)** identified that are hyperintense on STIR and demonstrate heterogeneous enhancement. There are hyperintense vertical intramedullary columns **(arrows)** that follow the SI of the cartilaginous masses.

Figure 16H. The MRI findings correlate with the channel-like lucencies on the radiograph characteristic of Ollier's disease **(arrows)**. Complete work-up, including biopsy, led to the diagnosis of an unusual variant of Ollier's disease. Despite its size, the posterior juxtacortical distal femoral mass showed no pathologic evidence of malignancy.

Pitfalls and Pearls

The ring-and-arc calcifications (plain radiograph and CT) and the central ring-and-arc enhancement patterns (MRI) are not specific for enchondromas and also can be seen with osteochondromas, periosteal chondromas, and chondrosarcomas.

References

1. Flemming DJ, Murphey MD. Enchondroma and chondrosarcoma. *Semin Musculoskelet Radiol* 2000; 4:59–71.
2. Brien EW, Mirra JM, Kerr R. Benign and malignant cartilage tumors of bone and joint: Their anatomic and theoretical basis with an emphasis on radiology, pathology and clinical biology. I. The intramedullary cartilage tumors. *Skeletal Radiol* 1997; 26:325–353.
3. Zwenneke Flach H, Ginai AZ, Wolter Oosterhuis J. Best cases from the AFIP. Maffucci syndrome: Radiologic and pathologic findings. Armed Forces Institutes of Pathology. *Radiographics* 2001; 21:1311–1316.
4. Kosaki N, Yabe H, Anazawa U, Morioka H, Mukai M, Toyama Y. Bilateral multiple malignant transformation of Ollier's disease. *Skeletal Radiol* 2005; 34:477–484.
5. Fanburg JC, Meis-Kindblom JM, Rosenberg AE. Multiple enchondromas associated with spindle-cell hemangioendotheliomas: An overlooked variant of Maffucci's syndrome. *Am J Surg Pathol* 1995; 19:1029–1038.
6. Mulliken JB, Glowacki J. Hemangiomas and vascular malformations in infants and children: A classification based on endothelial characteristics. *Plast Reconstr Surg* 1982; 69:412–422.
7. Herman TE, Chines A, McAlister WH, Gottesman GS, Eddy MC, Whyte MP. Metachondromatosis: Report of a family with facial features mildly resembling trichorhinophalangeal syndrome. *Pediatr Radiol* 1997; 27:436–441.
8. Resnick D, Kyriakos M, Greenway GD. Tumors and tumor-like lesions of bone: Imaging and pathology of specific lesions. In: Resnick D, ed., *Diagnosis of Bone and Joint Disorders*, 4th ed., Section 29, Chapter 76. Philadelphia: W.B. Saunders, 2002; 3763–4128.
9. Aoki J, Sone S, Fujioka F, et al. MR of enchondroma and chondrosarcoma: Rings and arcs of Gd-DTPA enhancement. *J Comput Assist Tomogr* 1991; 15:1011–1016.
10. De Beuckeleer LH, De Schepper AM, Ramon F, Somville J. Magnetic resonance imaging of cartilaginous tumors: A retrospective study of 79 patients. *Eur J Radiol* 1995; 21:34–40.
11. Murphey MD, Flemming DJ, Boyea SR, Bojescul JA, Sweet DE, Temple HT. Enchondroma versus chondrosarcoma in the appendicular skeleton: Differentiating features. *Radiographics* 1998; 18:1213–1237; quiz 1244–1215.

Case 17

History

This is a 17-year-old girl who sustained a right tibial fracture 2 months ago. An MRI was subsequently ordered because the fracture was not healing properly.

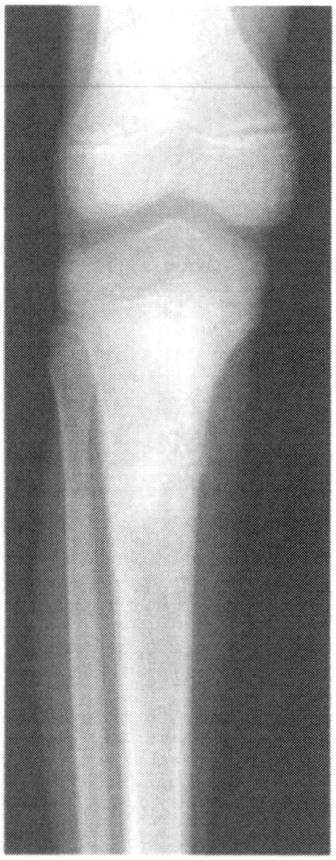

Figure 17A. AP radiograph of the right tibia 2 months earlier.

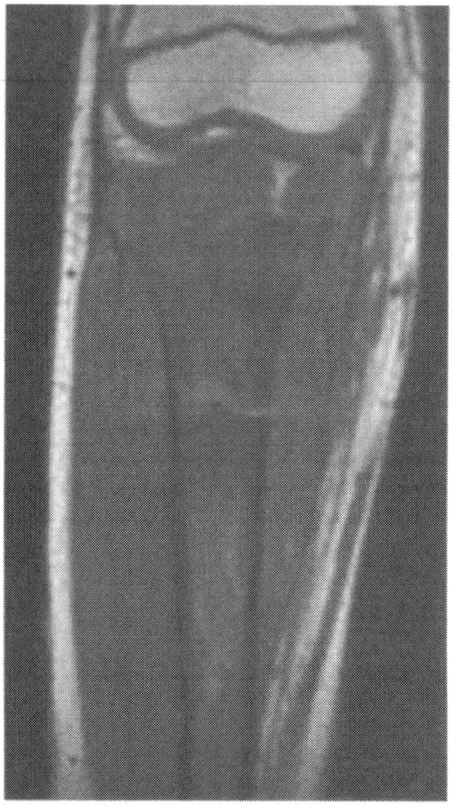

Figure 17B. Coronal T1 at presentation.

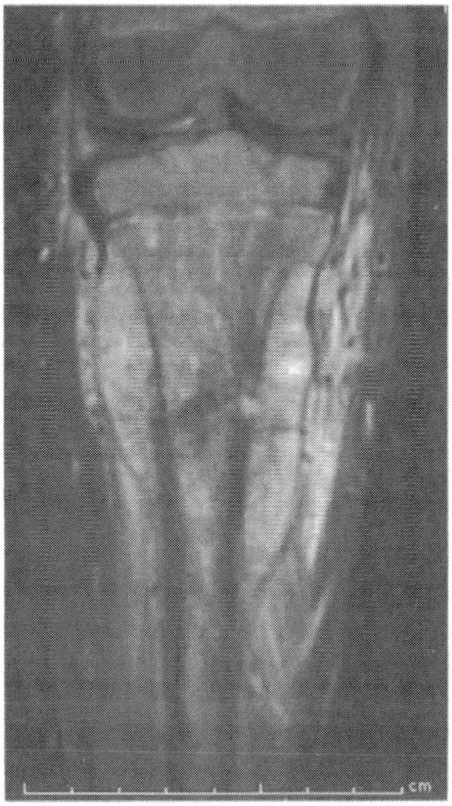

Figure 17C. Coronal STIR.

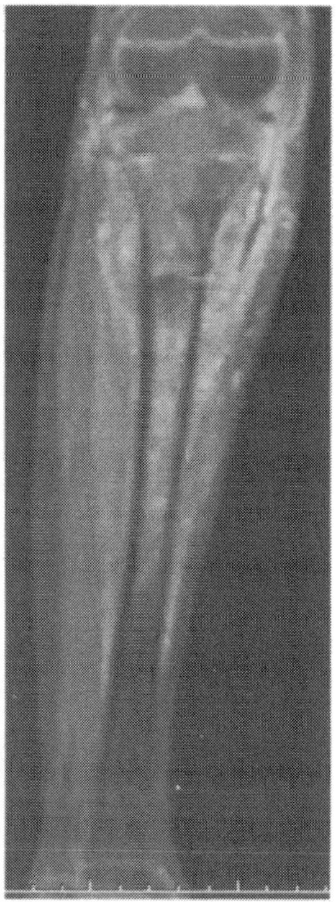

Figure 17D. Coronal T1 post-Gd FS.

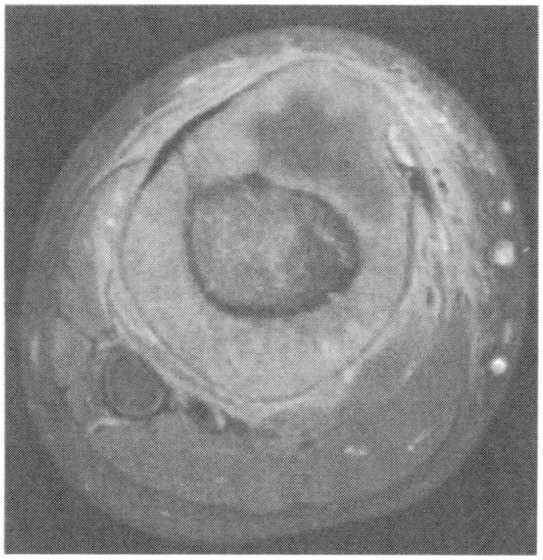

Figure 17E. Axial T1 post-Gd FS.

Figure 17A. There is a faint transverse fracture with medial cortical disruption of the proximal tibial metadiaphysis. There is associated ill-defined radiolucency and sclerosis extending to the physis. Note lack of expected callus formation.

Figures 17B, 17C, (17C with annotations). There is a large mass of heterogeneous SI centered in the proximal tibial metaphysis, extending into the diaphysis and across the physis into the epiphysis. There is diffuse marrow replacement with low SI on T1W and heterogeneous hyperintensity on STIR. The tumor demonstrates transcortical extension with the largest soft tissue component lying within the periosteal envelope **(arrowheads)**. A pathologic fracture **(arrow)** is evident, corresponding to the finding on the initial radiograph.

Figures 17D, 17E. There is tumoral enhancement with a juxtacortical nonenhancing component located anteriorly.

Figure 17F. A pathologic fracture through a densely sclerotic proximal tibia lesion is identified associated with hair-on-end periosteal reaction and cloud-like calcifications anteriorly. This soft tissue mineralization correlates with the anterior juxtacortical region of nonenhancement present in Figure 17E.

Figure 17G. Increasing heterogeneous SI (now predominantly hypointense) and decreased juxta-cortical soft tissue mass are present within the proximal tibia. There is now mild increased STIR SI within the distal femoral epiphysis and metaphysis, which may reflect a response to granulocyte colonizing-stimulating factor (G-CSF).

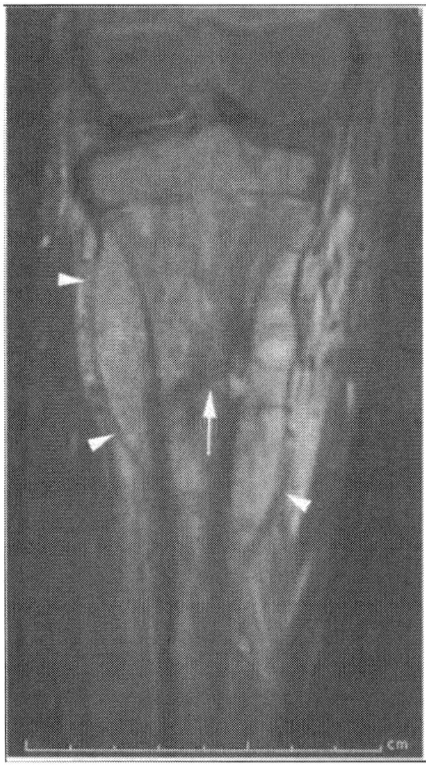

Figure 17C* Annotated.

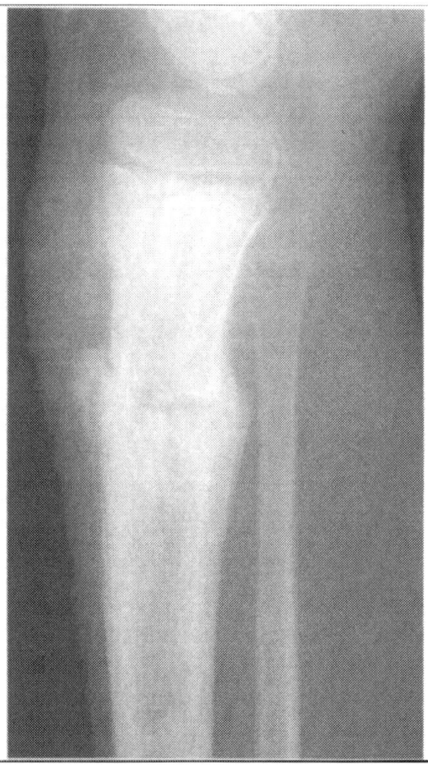

Figure 17F. Lateral radiograph at time of MRI.

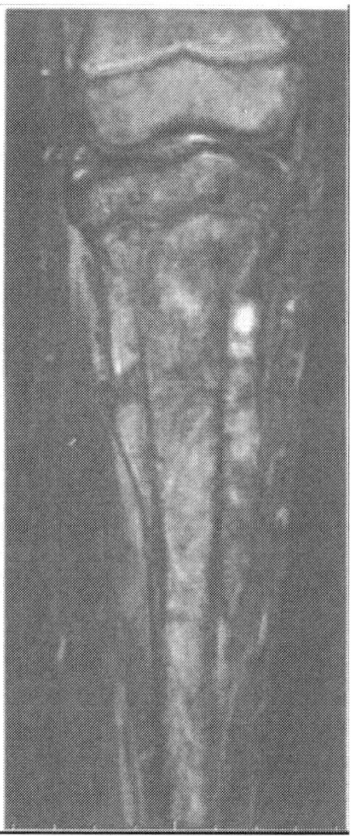

Figure 17G. Coronal STIR 3 months after initiation of chemotherapy.

Diagnosis

Conventional osteosarcoma

Questions

1. T/F: A minority of metaphyseal osteosarcomas violate the physis and extend into the epiphysis.
2. What percentage of conventional osteosarcomas demonstrates skip metastasis?

Discussion

Osteosarcoma is the most common primary malignant bone tumor in children and adolescence. The majority of patients present between 15 and 25 years of age and males are more commonly affected (1). Osteosarcomas may be divided based on location: medullary, surface, or extraskeletal. The high-grade intramedullary osteosarcoma (AKA conventional osteosarcoma) is most common, accounting for approximately 75% of all osteosarcomas. The other intramedullary subtypes of osteosarcoma include: telangiectatic, low-grade, small cell, osteosarcomatosis, and gnathic tumors. Patients with hereditary retinoblastoma and Li Fraumeni syndrome have a higher risk of developing osteosarcoma (2).

The most common location for conventional osteosarcoma is the knee (distal femur followed by the proximal tibia) (2). Tumors most commonly arise from the metaphysis with only 2% to 11% arising from the diaphysis (1). At pathologic examination, the majority of conventional osteosarcomas arising from the metaphysis cross the physis and extend into the epiphysis (75–88%) (Answer to Question 1) (1). Unlike Ewing's sarcoma, osteosarcoma rarely affects the axial skeleton. In a study of 1603 patients with osteosarcoma, 13% of patients with high-grade intramedullary tumors had metastasis at presentation (3). The most common location for metastasis is the lung.

On plain radiography, osteosarcoma often demonstrates juxtacortical soft tissue and intramedullary osteoid matrix calcification that may be characterized as fluffy or cloud-like (Figures 17H–17L). Sclerotic, osteolytic, or mixed marrow replacement may be seen with a wide or narrow zone of transition. Cortical destruction, rather than expansion, points to the rapid growth and osteoclastic resorption by the tumor. Aggressive periosteal reaction with onion skin, sunburst, and Codman triangle patterns are associated with these tumors, as opposed to the wavy, dense, or thick periosteal reaction typically seen with benign processes. However, aggressive periosteal reaction can be seen with nonsarcomatous entities such as Langerhans cell histiocytosis (LCH), stress reaction, and osteomyelitis.

In general, osteosarcoma is isointense on T1, variably hyperintense on fluid sensitive sequences, and demonstrates heterogeneous enhancement. It may be difficult to distinguish osteosarcoma from benign and other malignant lesions based on the MRI features alone; plain radiographs and CT are often necessary. Osteoblastic osteosarcomas containing a significant amount of osteoid calcification may be hypointense on all imaging sequences. An alternative consideration for hypointense zones within an osteosarcoma is blood products related to intratumoral hemorrhage. Fluid-fluid levels within cystic components may be seen with conventional and telangiectatic osteosarcoma and may develop after the initiation of chemotherapy (Figures 17M, 17N). However, fluid-fluid levels are much less commonly seen compared with aneurysmal bone cyst. Telangiectatic osteosarcoma is distinguished from conventional osteosarcoma by the presence of greater than 90% hemorrhagic, cystic, or necrotic components that are lined by viable sarcomatous cells prior to the initiation of treatment (4).

The differential diagnosis for osteosarcoma includes Ewing's sarcoma, lymphoma, LCH, osteomyelitis, osteoid osteoma, stress fracture, and unicameral and aneurysmal bone cyst. Although benign bone tumors tend to lack a juxtacortical soft tissue mass, the presence of a soft tissue mass is by no means diagnostic of a malignant lesion. Osteomyelitis, LCH, osteoid osteoma, and bone cysts complicated by fracture may all have significant soft tissue signal abnormalities suggesting malignancy. However, when osteoid matrix calcification is evident on plain radiography or CT, osteosarcoma is the diagnosis until proven otherwise.

The role of MRI at initial evaluation is to define extent of tumor including the degree of marrow replacement, cortical destruction, skip lesions, regional metastasis (Figures 17M, 17N), extraosseous soft tissue extent (Figures 17O–17Q), intra-articular spread, and involvement of neurovascular structures. Since cortical bone is relatively porous, high-grade intramedullary osteosarcomas may extend into the juxta-cortical soft tissues without gross cortical destruction. Skip lesions are seen in approximately 3% of all osteosarcomas at the time of presentation (Answer to Question 2) (5).

After chemotherapy, the role of MRI is to determine tumor response and define the tumor margins prior to resection. Patients with a greater than 90% tumor necrosis at pathologic examination have a higher disease-free survival rate (1). For osteosarcoma, an increase in tumor volume on MRI may correlate with a poor response to therapy, but stability or decrease in tumor volume is indeterminate (5). Determining response to chemotherapy is challenging because the osteoid matrix of osteosarcoma may regress slowly despite extensive tumoral necrosis. In addition, intra-tumoral hemorrhage, cystic change, and juxta-tumoral edema may exaggerate the true size of viable tumor. Dynamic-enhanced MRI (DEMRI) may be used to identify residual tumor (Figures 17H–17L). Viable tumor enhances more readily compared with adjacent tumoral fibrosis and edema. After chemotherapy, enhancement may be considered pathologic if it is seen within 6 seconds after arterial enhancement (6). For surgical considerations, a 1 cm tumor-free soft tissue resection margin and a 2 to 3 cm bony margin is desired (Figures 17O–17Q) (7). T1W sequences tend to be most accurate for determining true tumor margins, whereas STIR sequences often exaggerate tumor extent.

MRI of osteosarcoma and Ewing's sarcoma before and after neoadjuvant chemotherapy should include a large FOV to image the entire extremity with T1W and fluid sensitive sequences to rule out skip lesions and define intraosseous tumor extent. Suitable measurements to guide the orthopedic surgeon should be obtained and annotated on the images. Images with a smaller field of view, centered on the lesion, should then be acquired with an appropriate surface coil to assess for transphyseal, intra-articular, and extracompartmental soft tissue extension. Axial images are particularly useful to assess the integrity of adjacent neurovascular structures, as well as tumor seeding of the needle tract or surgical biopsy site.

In this patient, conventional osteosarcoma (chondroblastic predominant) was proven pathologically. She had been receiving G-CSF for neutropenia related to chemotherapy. The signal changes in the distal femur were presumably related to both G-CSF therapy and altered mechanics (Figure 17G). After chemotherapy, a wide resection of the right proximal tibial osteosarcoma was performed with an expandable prosthesis. At the time of surgery, the joint and distal femur were inspected and there was no evidence of tumor invasion. Please see Case 50 for a discussion on marrow response to G-CSF therapy.

Orthopedic Perspective

When presented with a new extremity osteosarcoma patient, the orthopedic oncologist is interested in knowing the local extent of the primary tumor. It is also important to evaluate the patient for distant disease by performing a chest CT to exclude pulmonary metastases and a radionuclide bone scan to look for skip metastases and lesions in other parts of the skeleton (synchronous bony metastases).

Most patients with extremity osteosarcoma will be considered for a limb-sparing procedure, although amputation remains in the armamentarium of treatment options. To be a suitable limb salvage candidate, it must be possible to widely resect the tumor with pathologically negative margins, ideally 5 to 10mm of soft tissue margin and 2 to 3cm of bone margin. The MRI is extremely helpful in determining bony margins and coronal images of the entire involved bone, and measurements of disease extent from palpable landmarks (such as the joint line) are essential. Careful evaluation of the joint to determine rare instances where the tumor is intra-articular is also important. Axial images are the most helpful to assess soft tissue margins and to give the orthopedist information about (1) the extent of soft tissue to resect, (2) the relationship of the soft tissue mass to the major nerves and blood vessels, and (3) indications about whether soft tissue flaps will be necessary for closure. Imaging also helps with the selection of reconstruction type (prosthesis, allograft, rotationplasty, etc.) based on predictions of how much bone and soft tissue will remain after resection of the tumor and whether the articular surfaces can be spared. Final determination of a surgical plan most often awaits repeat imaging following the administration of neoadjuvant chemotherapy. In most instances the edema surrounding the tumor is less evident and the true extent of the tumor is easier to discern after chemotherapy. In osteosarcoma, shrinking of the tumor is seldom evident because of the bony matrix, but having some idea of the response to chemotherapy and the precise extent of the lesion after chemotherapy is helpful.

In extremity osteosarcoma the choice of the reconstruction involves a complex decision-making process involving the treatment team, the patient, and family. Age is a key factor, because skeletally immature patients have the additional considerations of limb length inequality if growth plates are resected. For diaphyseal tumors, it is sometimes possible to spare the growth plates and reconstruct with an allograft bone segment or a vascularized fibular graft. When epiphyses are resected, the surgeon must either choose an implant that has the capability of expansion over time or plan standard limb equalization procedures in the future (epiphyseodeses, limb lengthening or contralateral limb shortening procedures).

For patients at or near the end of growth, reconstructions are similar to adults and involve the use of osteoarticular allografts, modular endoprosthesis, or a combination of allografts and standard joint replacements. Each of these reconstruction options has differences in functional outcome and complication risks, and each has relative advantages and disadvantages. It is up to the surgeon and patient to match the most appropriate reconstruction for a given lesion. It should be noted that none of these options should be considered if the tumor cannot be reliably resected with negative margins. An amputation remains a very functional option, especially in the lower extremity, and should not be considered an admission of defeat.

What the Clinician Needs to Know

1. Is there intra-articular extension of the tumor?
2. Is there neurovascular involvement?
3. How far is the tumor from the physis or the articular surface?
4. How far does the tumor extend from a palpable anatomic landmark?
5. What is the response to neoadjuvant chemotherapy?

Answers

1. False.
2. 3%.

Additional Examples

Conventional Osteosarcoma with Cloud-like Osteoid Matrix

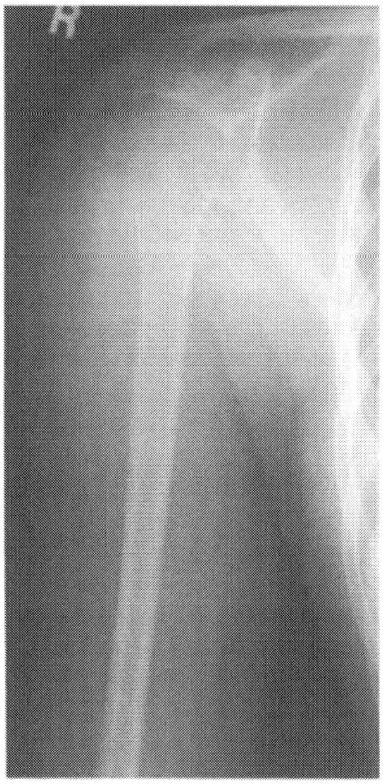

Figure 17H. AP radiograph of the right humerus.

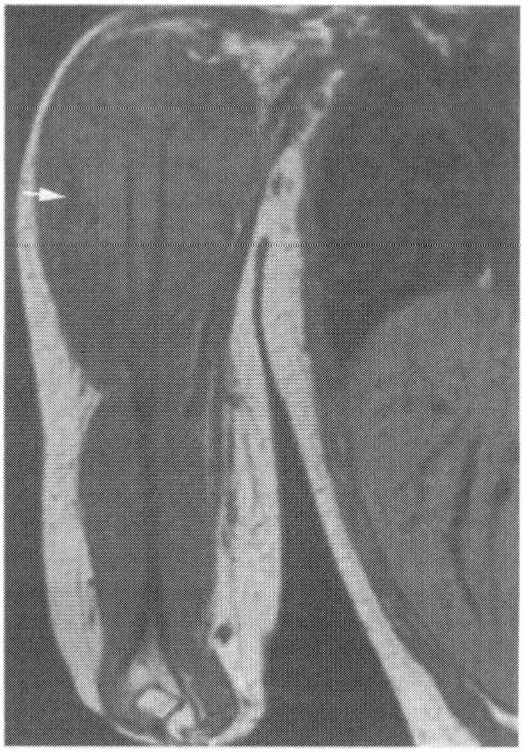

Figure 17I. Coronal T1.

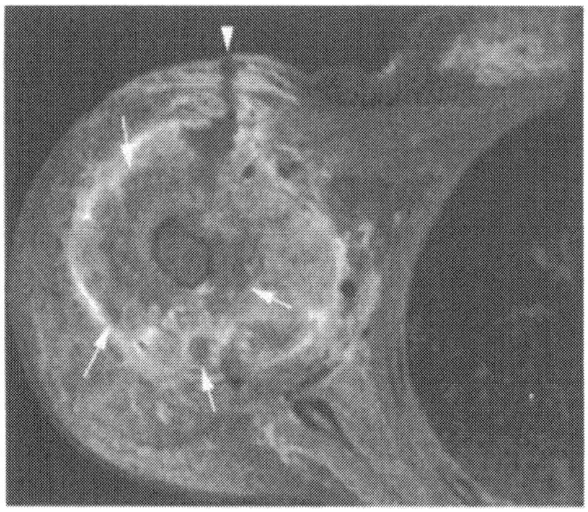

Figure 17J. Axial T1 post-Gd FS.

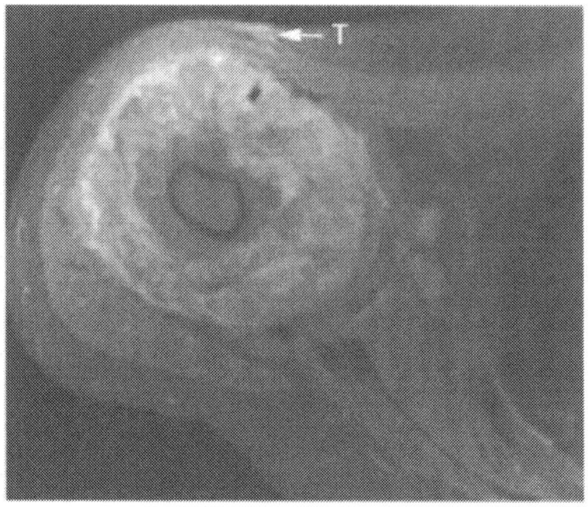

Figure 17K. Axial T1 post-Gd FS—3 months after therapy.

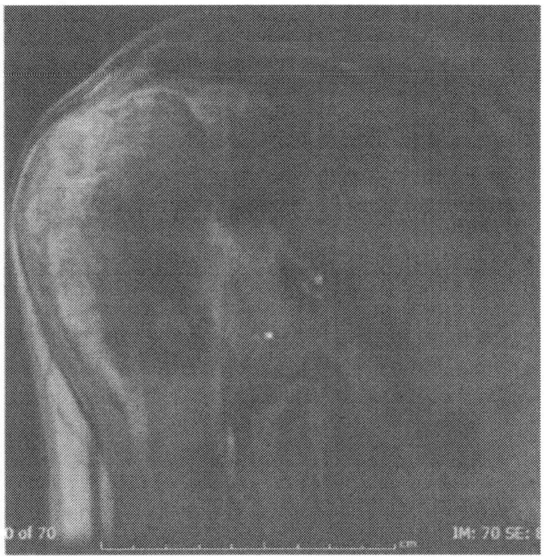

Figure 17L. Coronal, dynamic post-Gd SPGR (DEMRI) (2 seconds after initial arterial enhancement).

Findings

This is a 15-year-old girl with an enlarging arm mass that has become painful in the preceding 3 weeks.

Figure 17H. Permeative osteolysis and surrounding cloud-like osteoid matrix calcification are seen involving the proximal humerus.

Figure 17I. There is diffuse marrow replacement with a large soft tissue component **(arrow)**.

Figure 17J. The lesion demonstrates heterogeneous peripheral enhancement. Serpiginous and geographic regions of low SI in the soft tissue mass correlate with osteoid matrix **(arrows)**. A biopsy tract is identified anteriorly **(arrowhead)**. This was pathologically proven conventional osteosarcoma.

Figure 17K. After three months of chemotherapy, the lesion size is essentially unchanged, there is heterogeneous peripheral enhancement of the tumor, and enhancing soft tissue along the needle tract is evident **(arrow, T)**.

Figure 17L. Although peripheral enhancement persists, there is diminished central enhancement on early arterial phase dynamic SPGR imaging. At pathologic resection, there was 95% tumoral necrosis despite no change in tumor size on postchemotherapy MRI.

Fibular Conventional Osteosarcoma with Regional Metastases and Fluid-Fluid Levels

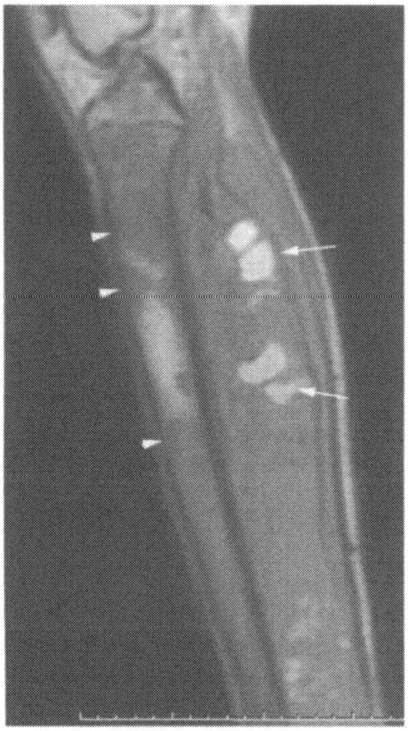

Figure 17M. Sagittal PD of the right calf.

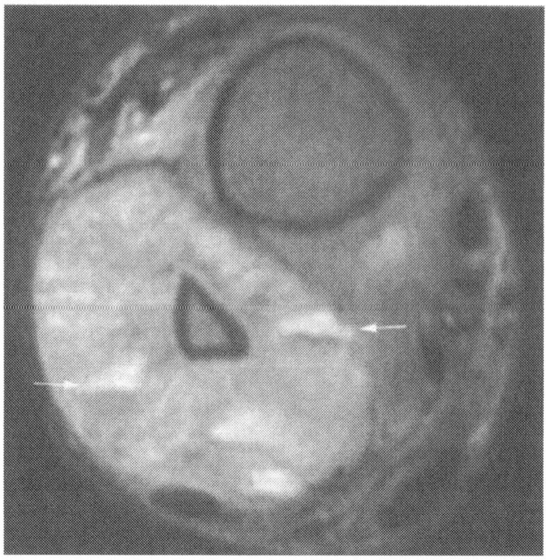

Figure 17N. Axial T2 FS.

Findings

This is a 16-year-old boy who initially presented with chronic calf swelling and pain. The diagnosis of osteosarcoma was made, and he received one cycle of chemotherapy prior to the MRI.

Figures 17M, 17N. This conventional fibular osteosarcoma shows a large soft tissue component. Multiple "cysts" are present with fluid-fluid levels **(arrows)**. Multiple regional metastases are present within the tibia **(arrowheads)**.

Conventional Osteosarcoma Confined to the Metadiaphysis

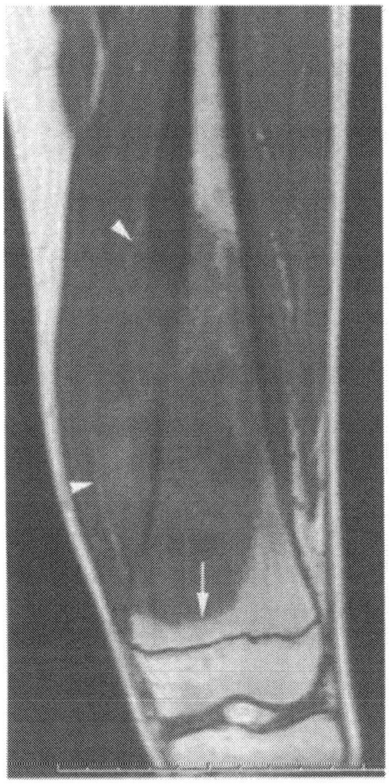

Figure 17O. Coronal T1 of the distal left femur (after chemotherapy).

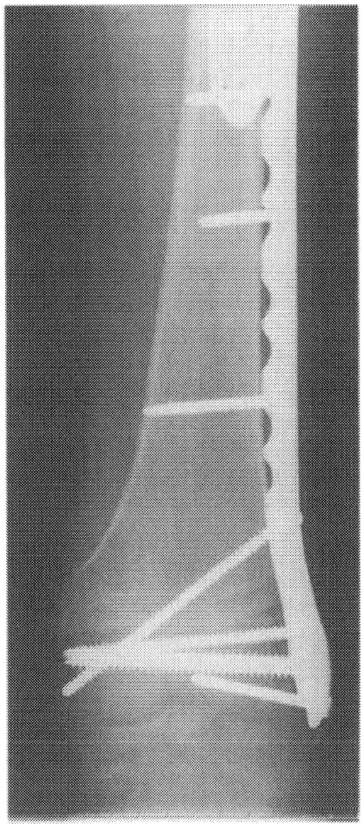

Figure 17Q. AP radiograph after resection.

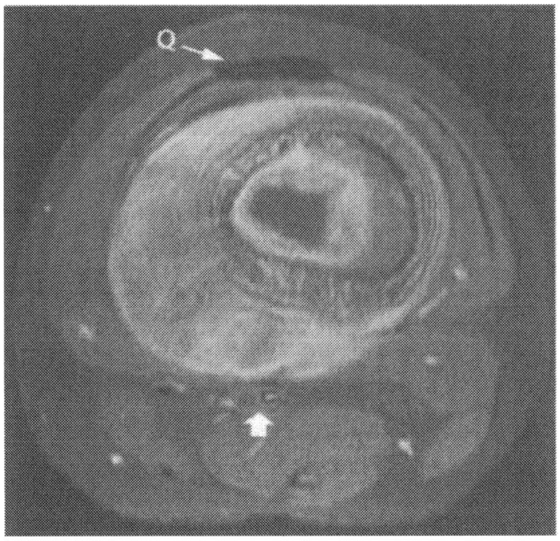

Figure 17P. Axial T1 post-Gd FS.

Findings

This is a 14-year-old girl with a conventional osteosarcoma of the distal femur. The MRI was ordered for preoperative planning after chemotherapy.

Figures 17O, 17P. A significant extraosseous soft tissue component is present **(arrowheads)**. The tumor does not involve the popliteal vessels/tibial nerve complex **(thick arrow)** or the quadriceps femoris tendon **(arrow, Q)**. The tumor margin was approximately 1 cm away from the physis **(thin arrow)**. Therefore, the native epiphysis was preserved during tumor resection.

Figure 17Q. A distal femoral diaphyseal and metaphyseal resection was performed, and an intercalary allograft was fixed with a condylar plate to the native epiphysis. Preoperative MRI examination informed the surgeon that the epiphysis, quadriceps femoris tendon, and popliteal vessel/tibial nerve complex could be preserved at the time of tumor resection.

Pitfalls and Pearls

Use a surface coil with an appropriate field of view to characterize the primary lesion. In addition, use a larger field of view (and coil change if necessary) to acquire sagittal or coronal T1/STIR images in order to rule out skip lesions.

References

1. Murphey MD, Robbin MR, McRae GA, Flemming DJ, Temple HT, Kransdorf MJ. The many faces of osteosarcoma. *Radiographics* 1997; 17:1205–1231.
2. Arndt CA, Crist WM. Common musculoskeletal tumors of childhood and adolescence. *N Engl J Med* 1999; 341:342–352.
3. Bacci G, Longhi A, Bertoni F, et al. Primary high-grade osteosarcoma: Comparison between preadolescent and older patients. *J Pediatr Hematol Oncol* 2005; 27:129–134.
4. Murphey MD, wan Jaovisidha S, Temple HT, Gannon FH, Jelinek JS, Malawer MM. Telangiectatic osteosarcoma: Radiologic-pathologic comparison. *Radiology* 2003; 229:545–553.
5. Brisse H, Ollivier L, Edeline V, et al. Imaging of malignant tumours of the long bones in children: Monitoring response to neoadjuvant chemotherapy and preoperative assessment. *Pediatr Radiol* 2004; 34:595–605.
6. van der Woude HJ, Bloem JL, Verstraete KL, Taminiau AH, Nooy MA, Hogendoorn PC. Osteosarcoma and Ewing's sarcoma after neoadjuvant chemotherapy: Value of dynamic MR imaging in detecting viable tumor before surgery. *AJR Am J Roentgenol* 1995; 165:593–598.
7. Meyer JS, Mackenzie W. Malignant bone tumors and limb-salvage surgery in children. *Pediatr Radiol* 2004; 34:1030.

Case 18

History

This is a 10-year-old girl with acute left hip pain for one week. She is unable to bear weight, has a temperature of 38.1 C, a chronic reactive protein (CRP) of 11.4 mg/dl (normal <0.50), erythrocyte sedimentation rate (ESR) of 95 mm/hr (normal 0–20), and a WBC of 9.5 k/uL.

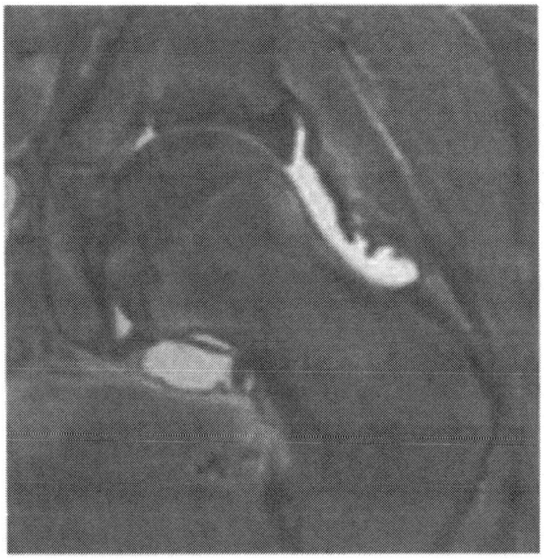

Figure 18A. Coronal STIR of the left hip.

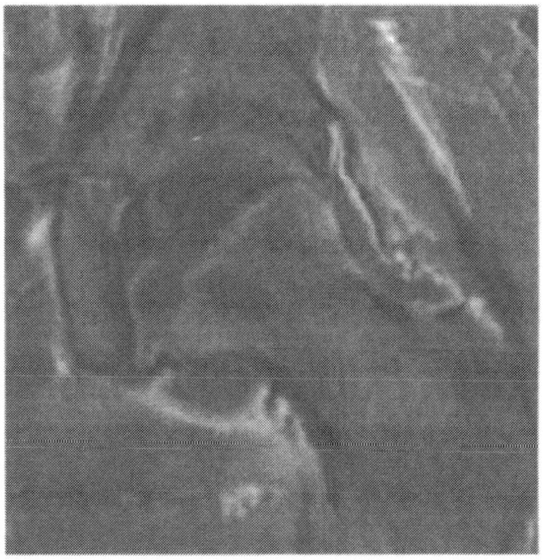

Figure 18B. Coronal T1 post-Gd FS.

Figures 18A, 18B, (18A with annotations). There is a moderate size joint effusion and synovial enhancement. There is mildly increased STIR SI and enhancement within the obturator externus muscle **(arrows)** and gluteus minimi muscle **(arrowhead)**. There is also mildly increased STIR SI within the metaphysis **(thick arrows)**.

Figure 18C. Pre-contrast images show flame-shaped areas of mildly decreased SI within the metaphysis **(thick arrows)** that correspond to the regions of increased STIR SI (Figure 18A), compatible with residual red marrow.

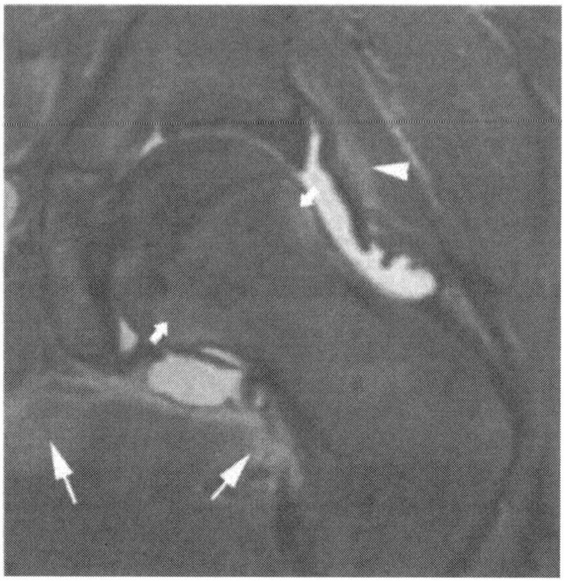

Figure 18A* Annotated.

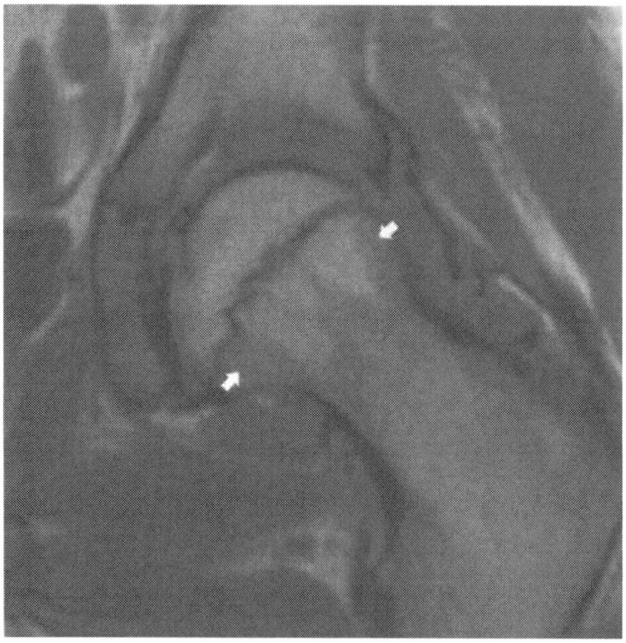

Figure 18C. Coronal T1.

Diagnosis

Septic arthritis

Questions

1. T/F: Marrow edema on MRI is a classic feature of transient synovitis.
2. What are some clinical features of septic arthritis?

Discussion

Bacterial entry into the joint may occur from direct extension with osteomyelitis, hematogeneous seeding, iatrogenic or a penetrating injury, and direct inoculation from adjacent soft tissue infection. *Staphylococcus aureus* is the most common organism isolated in both osteomyelitis and pyogenic septic arthritis in children (1, 2). *Neisseria gonorrhoeae* may be isolated in sexually active adolescents. Like osteomyelitis, septic arthritis most commonly occurs in the lower extremity (knee 35%, hip 35%, and ankle 10%), and there is often a preceding history of trauma (3). Less commonly, septic arthritis involves the upper extremity (Figures 18D, 18E).

The MRI features of septic arthritis include: synovial thickening and enhancement, juxta-articular marrow edema, adjacent osteomyelitis, and inflammatory changes within the surrounding muscles, tendons, and subcutaneous tissues (4, 5). In a study by Graif et al., MRI evaluation of 11 nonseptic and 19 septic joints in adult patients showed that erosions and coexistent marrow edema were more often seen in patients with septic arthritis; however, other features such as synovial thickening or enhancement, effusion, cartilage loss, and juxta-articular soft tissue edema were nonspecific (6). Residual red marrow should not be confused with edema on fluid sensitive sequences. Residual red marrow is typically flame shaped, hypointense with respect to muscle on T1, and may be iso- to mildly hyperintense on fluid-sensitive sequences (Figures 18A–18C).

The MRI features of transient synovitis (Figures 18F–18G) and septic arthritis of the hip are generally indistinguishable on MRI. When marrow edema is evident on fluid sensitive sequences, the diagnosis of a septic arthritis is favored. In one study, marrow edema was present in 8 of 9 patients with septic arthritis and in 0 of 14 patients with transient synovitis (Answer to Question 1) (4). However, septic arthritis should not be excluded when marrow edema is absent, without supportive clinical features. Kocher et al. retrospectively evaluated 282 cases of hip pain in children and found four independent clinical predictors of septic arthritis: presence of a fever (oral temperature >38.5 C), non-weightbearing, ESR >40 mm per hour, and a WBC >12,000 (Answer to Question 2) (7). The predictability of septic arthritis with none of the four predictors was less than 0.2 %, whereas it was 99.6% when all four predictors were present.

Other diagnostic considerations for septic arthritis include adjacent osteomyelitis, acute Legg-Calve-Perthes disease, trauma, and inflammatory arthropathies. All of these entities may cause a joint effusion, synovial inflammation, adjacent soft tissue, and marrow edema. Therefore, the MRI features of septic arthritis are nonspecific and require correlation with clinical and laboratory data.

MRI is used to evaluate complications of septic arthritis, including epiphyseal ischemia, growth arrest, secondary osteomyelitis, and premature degenerative changes. Epiphyseal ischemia may develop due to increased intracapsular pressure generated from the infected joint fluid (8). This can be irreversible and lead to ischemic necrosis (Figures 18H–18J). The epiphyseal manifestations of ischemic necrosis range from a delayed appearance of the secondary center of ossification to complete osteolysis. The spectrum of premature degenerative changes related to septic arthritis includes chondrolysis with joint space narrowing, subchondral erosions (Figures 18K, 18L), joint contractures, and complete ankylosis.

In this case, a left hip aspiration was subsequently performed with sonographic guidance and the white cell count was 80,000/mm^3. The patient subsequently underwent surgical drainage, and *Staphylococcus aureus* was cultured from the joint fluid. After 7 days of IV cefazolin, the patient was discharged home with a 4-week course of oral cephalexin and did well.

Orthopedic Perspective

The child presenting with an irritable hip poses a diagnostic dilemma in differentiating between septic arthritis and transient synovitis. Septic arthritis is a surgical emergency treated with surgical drainage and IV antibiotics, with the potential for serious sequelae including growth disturbance and ischemic necrosis. The diagnosis of septic arthritis is made when there is a positive culture from joint aspiration or a white blood cell count in the joint fluid >50,000 cells/mm^3 with a positive blood culture. Transient synovitis, on the other hand, is treated supportively with a benign prognosis. The differentiation is based on clinical exam and laboratory findings as noted above. A joint aspirate should be obtained to determine the diagnosis in all cases in which the diagnosis is not clear. MRI can be useful in septic arthritis to identify associated proximal femoral and acetabular osteomyelitis in cases that are not responding to conventional treatment. MRI is rarely used in the initial diagnostic work-up of suspected septic arthritis since the clinical exam, hip ultrasound, and joint aspiration are usually sufficient.

What the Clinician Needs to Know

1. Is there associated osteomyelitis in cases that are not responding to conventional treatment?
2. On follow-up studies, are there complications of septic arthritis such as growth disturbance and ischemic necrosis?

Answers

1. False.
2. Fever, nonweightbearing, ESR >40mm per hour, and a WBC >12,000.

Additional Examples

Septic Arthritis of the Elbow, Staphylococcus aureus

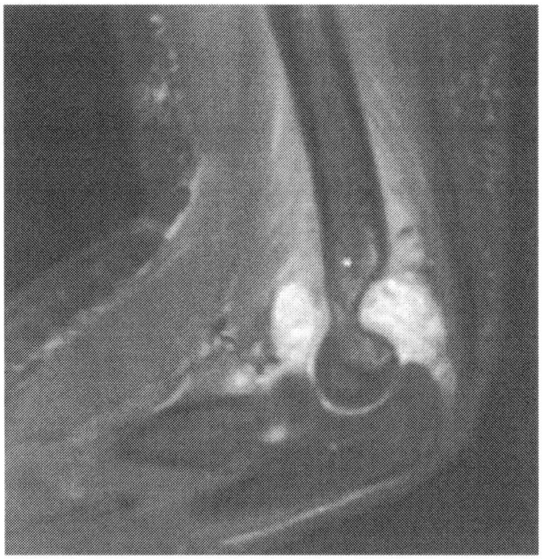

Figure 18D. Sagittal STIR of the left elbow.

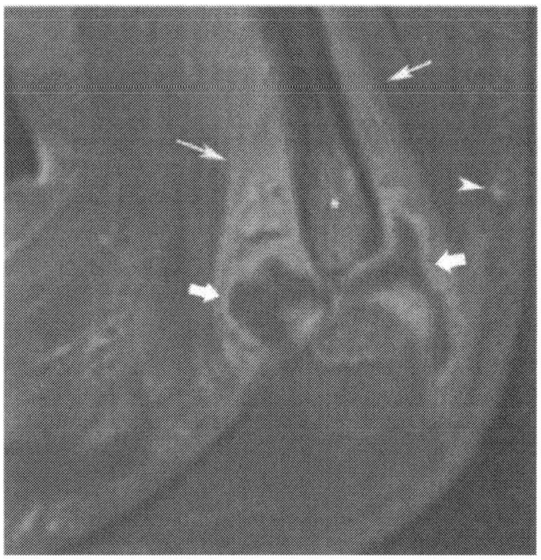

Figure 18E. Sagittal T1 post-Gd FS.

Findings

This is a 9-year-old boy with a 4-day history of fever and limited range of motion of the left elbow.

Figure 18D. There is a moderate-size joint effusion and reactive marrow edema in the distal humerus (*).

Figure 18E. The synovium is thickened and shows increased enhancement **(thick arrows)**. Enhancement is also present within the distal humeral marrow (*) and the adjacent muscles **(thin arrows)**. Reticular increased SI is also seen within the anterior and posterior **(arrowhead)** subcutaneous tissues of the elbow.

Transient Synovitis

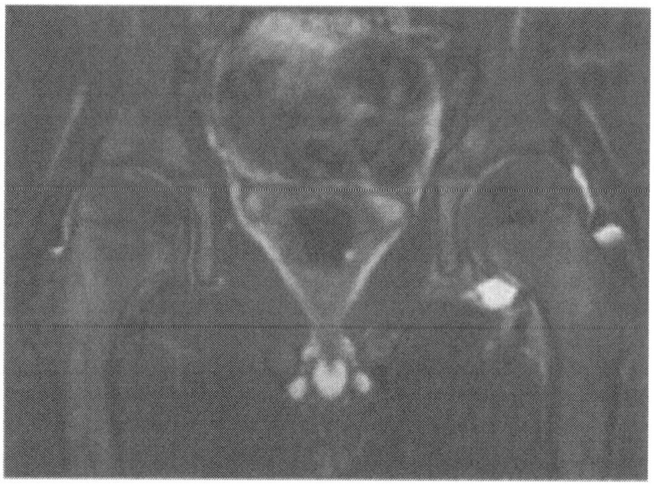

Figure 18F. Coronal T2 FS.

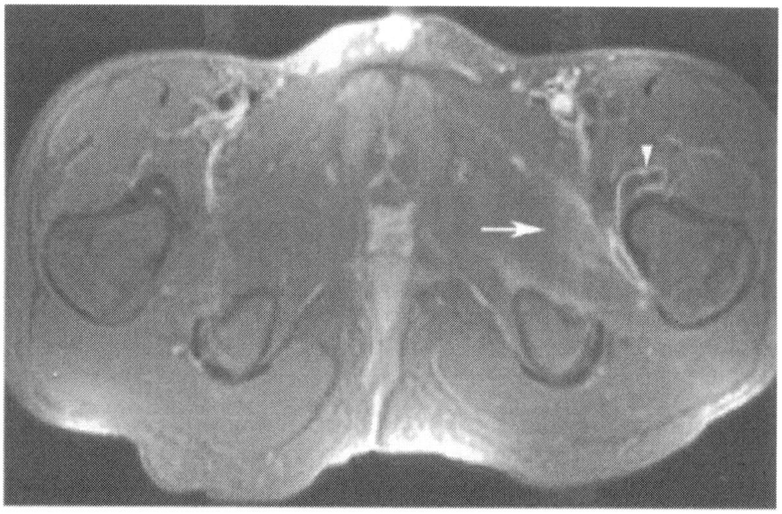

Figure 18G. Axial T1 post-Gd FS.

Findings

This is an 8-year-old boy with acute left hip pain.

Figures 18F, 18G. The right hip is normal. On the left, there is a small joint effusion, mild synovial thickening and enhancement **(arrowhead)**, and obturator externus muscle enhancement present **(arrow)**. The marrow signal is normal. These findings are indistinguishable from septic arthritis (Figures 18A–18C). However, this patient was afebrile, was weightbearing on the left hip, had an ESR value of 14 mm per hour (<20 normal), and a WBC of 9000. Blood cultures were negative. Therefore, this patient was diagnosed with transient synovitis. A hip aspiration was not performed.

Septic Arthritis Complicated by Ischemic Necrosis

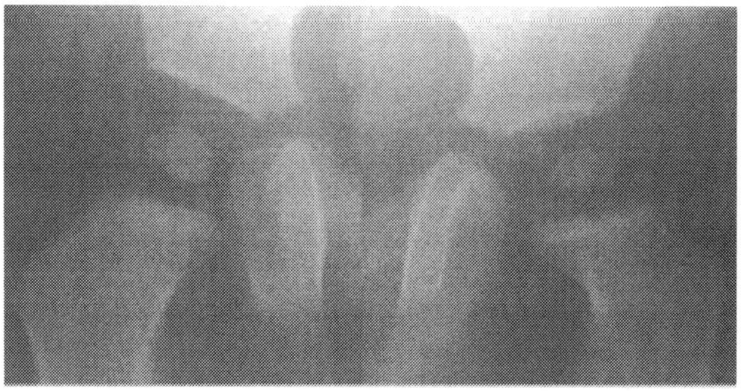

Figure 18H. AP radiograph of the hips at presentation.

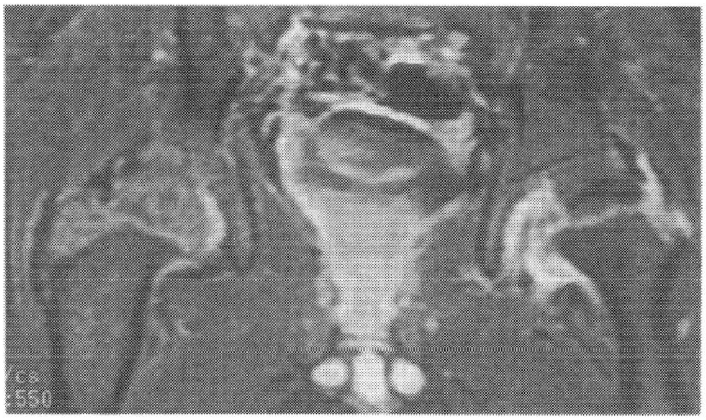

Figure 18I. Coronal T1 post-Gd FS 6 months later.

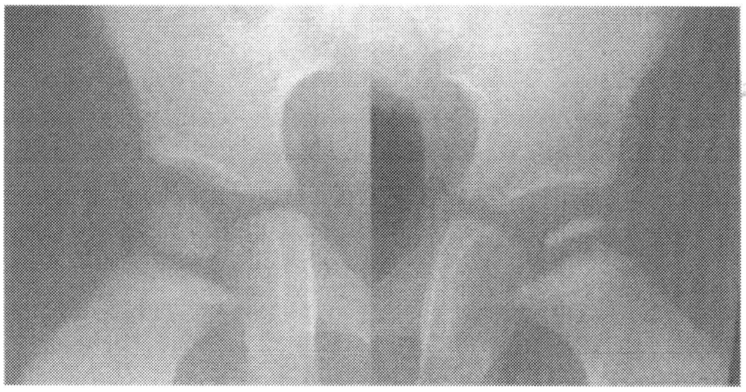

Figure 18J. AP radiograph of the hips 8 months later.

Findings

This is a 10-month-old boy with *Staphylococcus aureus* septic arthritis of the left hip.

Figure 18H. Initial radiograph show a slightly smaller left capital femoral epiphysis.

Figures 18I, 18J. Subsequent MRI and AP radiograph demonstrate collapse of the left capital femoral epiphysis. Abnormal synovial enhancement is present. The findings of ischemic necrosis complicating septic arthritis are indistinguishable from Legg-Calve-Perthes disease.

Prior Septic Arthritis with Severe Degenerative Changes

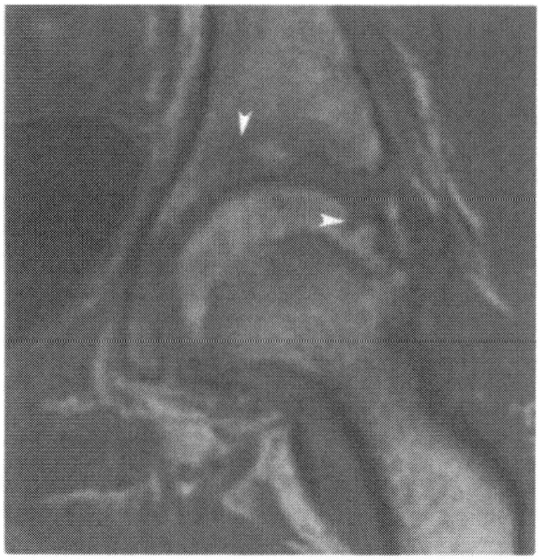

Figure 18K. Coronal T1 of the left hip.

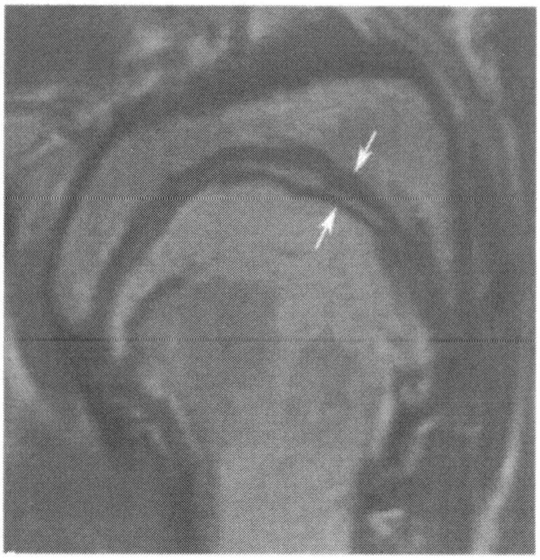

Figure 18L. Sagittal PD.

Findings

This is a 12-year-old girl who had left hip septic arthritis 4 years ago and now has worsening joint contractures and hip pain. This was a preoperative MRI for surgical planning.

Figure 18K. Irregularity of the sourcil (acetabular roof) and erosions of the femoral head are present **(arrowheads)**.

Figure 18L. The femoral head is aspherical, and there is severe joint space narrowing. The articular cartilage is diffusely attenuated (intermediate SI between the irregular hypointense cortices of the femoral head and acetabulum [**arrows**]).

Pitfalls and Pearls

1. Do not confuse normal residual red marrow for edema with fluid sensitive sequences. Residual red marrow is typically flame shaped and is hypointense on T1 and mildly hyperintense on fluid sensitive sequences.
2. Image both hips, since transient synovitis is often associated with bilateral effusions, but septic arthritis is usually unilateral.

References

1. Howard AW, Viskontas D, Sabbagh C. Reduction in osteomyelitis and septic arthritis related to *Haemophilus influenzae* type B vaccination. *J Pediatr Orthop* 1999; 19:705–709.
2. Luhmann JD, Luhmann SJ. Etiology of septic arthritis in children: An update for the 1990s. *Pediatr Emerg Care* 1999; 15:40–42.
3. Waagner DC. Musculoskeletal infections in adolescents. *Adolesc Med* 2000; 11:375–400.
4. Lee SK, Suh KJ, Kim YW, et al. Septic arthritis versus transient synovitis at MR imaging: Preliminary assessment with signal intensity alterations in bone marrow. *Radiology* 1999; 211:459–465.

5. Ecklund K, Vargas S, Zurakowski D, Sundel RP. MRI features of lyme arthritis in children. *AJR Am J Roentgenol* 2005; 184:1904–1909.
6. Graif M, Schweitzer ME, Deely D, Matteucci T. The septic versus nonseptic inflamed joint: MRI characteristics. *Skeletal Radiol* 1999; 28:616–620.
7. Kocher MS, Zurakowski D, Kasser JR. Differentiating between septic arthritis and transient synovitis of the hip in children: An evidence-based clinical prediction algorithm. *J Bone Joint Surg Am* 1999; 81:1662–1670.
8. Jaramillo D, Treves ST, Kasser JR, Harper M, Sundel R, Laor T. Osteomyelitis and septic arthritis in children: Appropriate use of imaging to guide treatment. *AJR Am J Roentgenol* 1995; 165:399–403.

Case 19

History

This is a 9-year-old girl with chronic left foot pain.

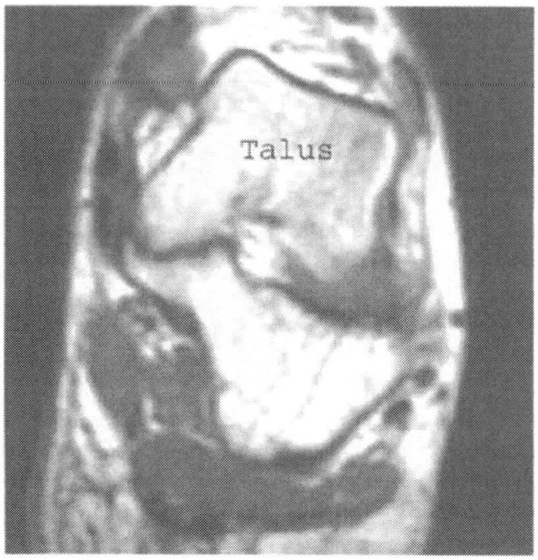

Figure 19A. Coronal T1 of the left ankle.

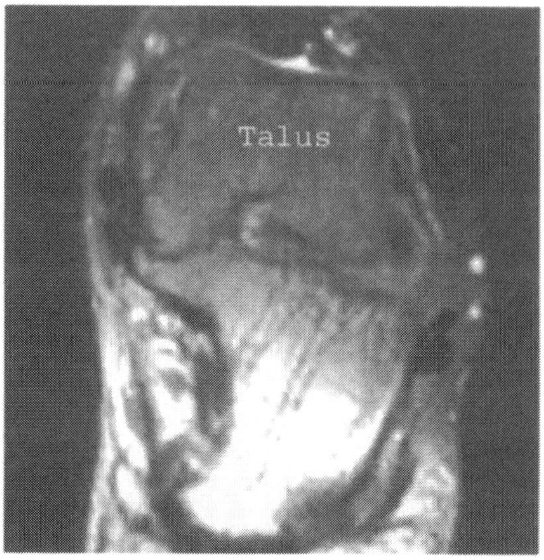

Figure 19B. Coronal T2 FS.

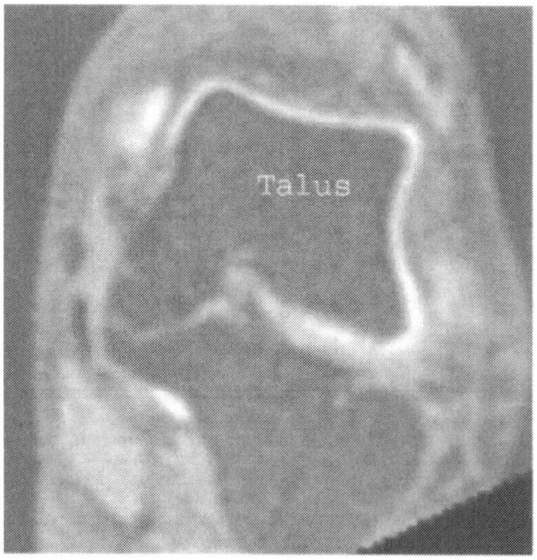

Figure 19C. Coronal 3D SPGR FS.

Figures 19A, 19B, 19C. There is marked narrowing and irregularity of the middle facet of the talocalcaneal joint. There is a plantar tilt of the joint line and the apposing bony margins have a bulky, broadened configuration. On T1 and T2W sequences, the middle facet joint is hypointense; on the SPGR sequence, it follows the SI of articular cartilage.

Figure 19D. On the left, the middle facet of the talocalcaneal joint is fused. On the right, the corresponding joint space is normal.

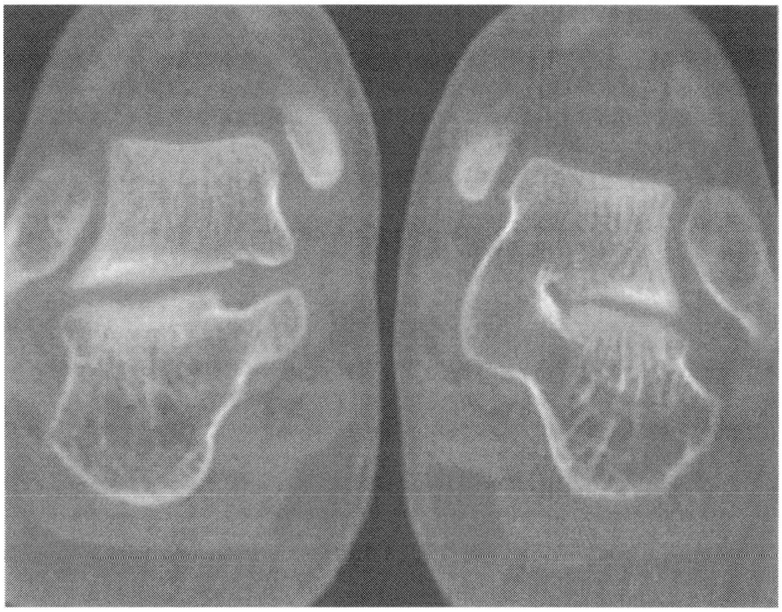

Figure 19D. Coronal CT, 5 years later.

Diagnosis

Talocalcaneal coalition, middle facet

Questions

1. What percentage of tarsal coalitions is bilateral?
2. T/F: Osseous and fibrous coalitions may be indistinguishable by MRI.

Discussion

Tarsal coalition is a cartilaginous, fibrous, or osseous fusion anomaly that may affect the talocalcaneal, calcaneonavicular, talonavicular, or calcaneocuboid articulations (1). Rarely, tarso-metatarsal coalitions may be seen (Figures 19E, 19F). Most cases are congenital but may not develop symptoms until osseous fusion begins around 8 to 16 years of age. Although tarsal coalitions usually occur sporadically, the trait may be inherited. In one study, Leonard et al. evaluated 31 patients with tarsal coalition and found that 39% of first-degree relatives also had some type of tarsal coalition (2). Coalitions are occasionally noted in association with dysmorphic syndromes (Figures 19E, 19F). Acquired coalitions may result from prior trauma, infection, nonpyogenic arthropathies, prior surgery, and clubfoot deformity (3).

Tarsal coalitions affect 1% of the population; approximately 50% are bilateral (Answer to Question 1) (3). The most common locations for tarsal coalitions are the calcaneonavicular and talocalcaneal (subtalar) joints; in the latter, the middle facet is typically involved. Subtalar coalitions have also been observed immediately posterior to the sustentaculum tali (between the middle and posterior subtalar joint) (Figures 19G–19I) (4).

Radiographs of the feet in cases of tarsal coalition may show hindfoot valgus and a flatfoot deformity (1). Several other radiographic features have been described; these are usually found in adolescents and adult patients. Radiographic findings are usually diagnostic of calcaneonavicular coalitions, but detection of subtalar coalitions on plain radiographs may be problematic since the subtalar joint is often poorly seen. Radiographic features suggesting talocalcaneal coalitions include the C-sign (C-shaped talus on lateral views), a dysmorphic-appearing sustentaculum, anterior talar beaking, "ball-and-socket" ankle joint (concave tibia articular surface and rounded talar dome), and an indistinct middle facet joint (5, 6). Radiographic features suggesting a calcaneonavicular coalition include: anterior talar beak, an "anteater's nose" deformity (elongated anterior process of the calcaneus), a short talar neck, and a broad and laterally tapered navicular (5).

Multiplanar CT is generally used to delineate osseous and fibrous subtalar, multiple, and complex coalitions. Osseous coalitions show cortical continuity between the affected tarsal bones. Fibrous coalitions may have joint narrowing with subchondral sclerosis, facet irregularity, and occasionally subchondral cysts (6).

Although CT is the modality of choice to evaluate tarsal coalitions, patients may be referred for MRI of the foot/ankle for other clinical problems, and the study may reveal an incidental asymptomatic coalition. Patients with symptomatic coalitions may also be referred for MRI if the condition is not considered clinically, and there are concerns of a traumatic or neoplastic process. Lastly, a coalition may be strongly suspected on clinical grounds, and, despite a normal or equivocal CT, MRI will be requested. If normal joint fluid SI is absent and the joint is indistinct, the diagnosis of coalition should be raised. Fibrous coalitions may show hypointensity on all imaging sequences, cartilaginous coalitions may show similar SI to cartilage elsewhere (Figures 19J–19L), and osseous coalitions may show variable SI (6). The diagnosis of an osseous coalition on MRI is made when marrow SI is continuous between the affected tarsal bones, particularly on T1W sequences. When osseous coalitions are sclerotic, they may appear hypointense on all imaging sequences and are indistinguishable from fibrous coalitions (Answer to Question 2). Soft tissue and osseous edema in the hindfoot and midfoot may be seen (Figure 19H) due to restricted foot motion leading to abnormal stress. Since patients with tarsal coalitions may present with a peroneal spastic flatfoot, the peroneal tendons should be inspected for signs of tendinopathy and tenosynovitis.

This patient was treated with casting and physical therapy after the initial MRI was obtained. The CT was requested 5 years later when she was having recurrent foot pain. Although there was complete osseous fusion of the middle talocalcaneal joint, there was also posterior talocalcaneal joint space narrowing (not shown) and spasticity. These were considered negative factors for a successful resection. Therefore, the patient was asked to continue performing strengthening exercises and surgery was not considered at that time.

Orthopedic Perspective

Clinically, patients with tarsal coalition present with rigid flatfeet. There is often pain along the medial midfoot or posterolateral ankle pain associated with peroneal muscle spasm (peroneal spastic flatfoot). Patients may have a history of frequent ankle sprains. Therefore, tarsal coalition may mimic a tendinopathy and the diagnosis by MRI may be an unexpected surprise for both the radiologist and orthopedist. Initial treatment is supportive with immobilization, physical therapy, and orthotics. Recalcitrant cases are often treated with resection of the coalition.

What the Clinician Needs to Know

1. Is the coalition fibrous, cartilaginous, or osseous? Is CT necessary to further characterize and quantify the coalition?
2. Is there more than one coalition?
3. Are there stress reactions or occult fractures evident that may have been caused by altered mechanics due to the coalition?

Answers

1. Approximately 50%.
2. True.

Additional Examples

Fifth Metatarsal-Calcaneal Cartilaginous Coalition

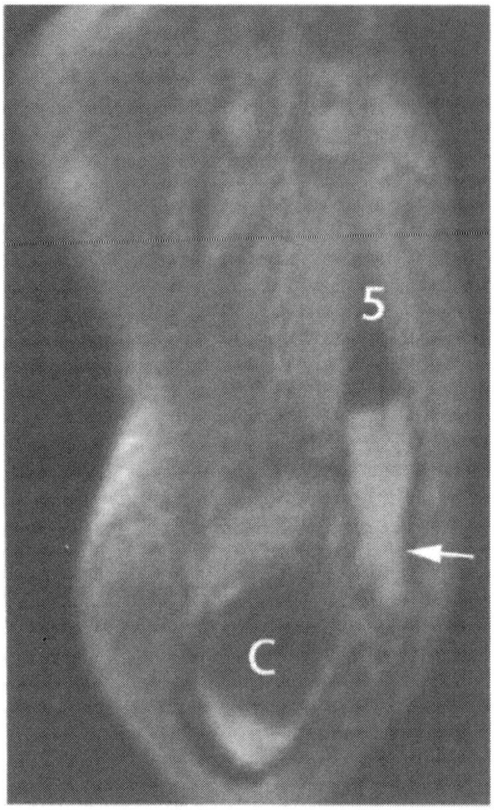

Figure 19E. Axial (footprint) PD FS of the left foot.

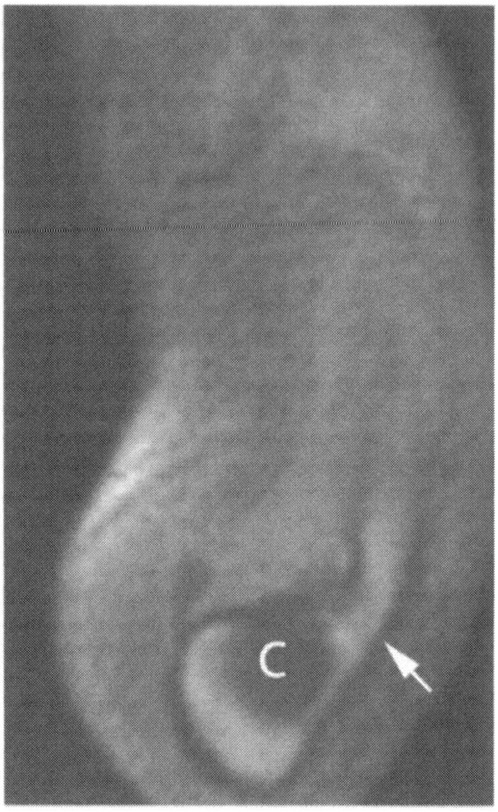

Figure 19F. Axial (footprint) PD FS.

Findings

This is a 2-year-old girl with Townes-Brocks syndrome (Anus-Hand-Ear syndrome) who had known clubfoot deformities.

Figures 19E, 19F. There is a cartilaginous coalition **(arrows)** between the base of the fifth metatarsal (5) and the calcaneus (C). A similar coalition was present on the right (not shown).

Subtalar Coalition Located Immediately Posterior to the Sustentaculum Tali

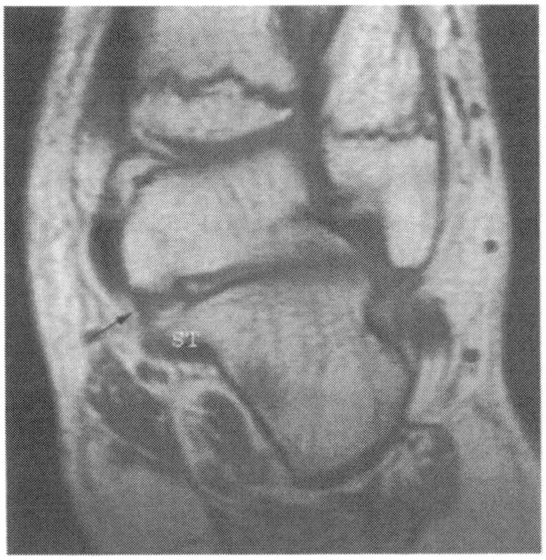

Figure 19G. Coronal T1 of the left foot.

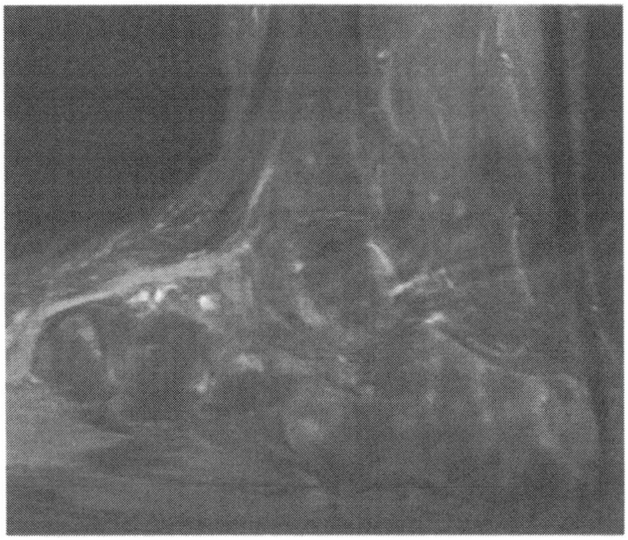

Figure 19H. Sagittal STIR.

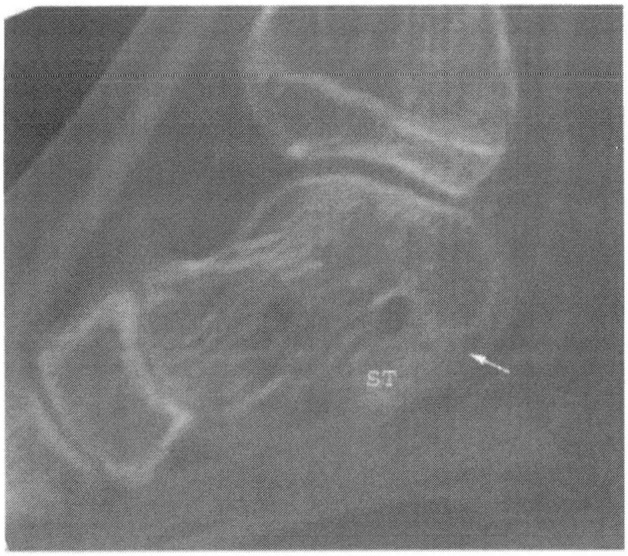

Figure 19I. Sagittal reformat CT, 1 year later.

Findings

This is an 18-year-old boy with a painful left flat foot.

Figures 19G, 19H. Immediately posterior to the sustentaculum tali (ST), a fibrous or fibrocartilaginous coalition is identified **(arrow)**. Patchy marrow edema is seen in the hindfoot possibly related to stresses from altered weightbearing.

Figure 19I. Subsequent CT shows an osseous coalition in the same region **(arrow)**.

Calcaneonavicular Fibrocartilaginous Coalition

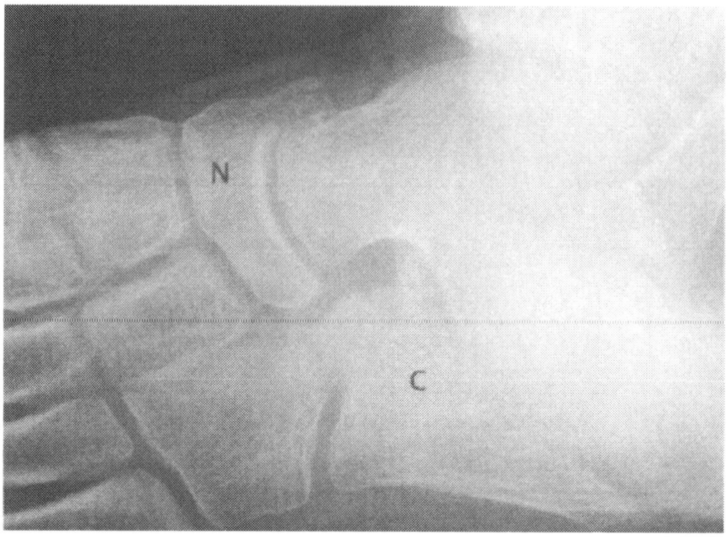

Figure 19J. Oblique radiograph of the left foot.

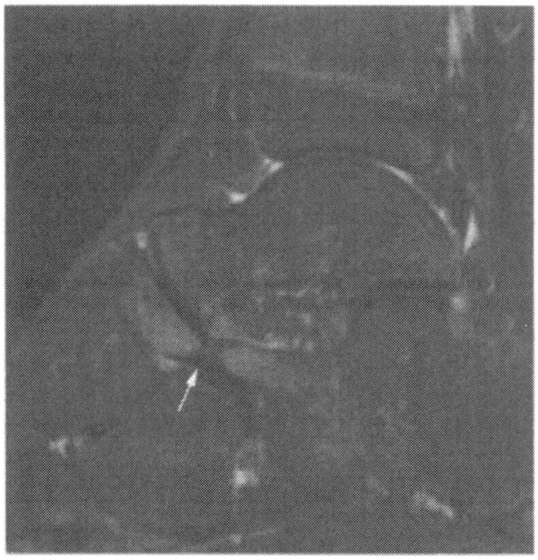

Figure 19K. Sagittal STIR.

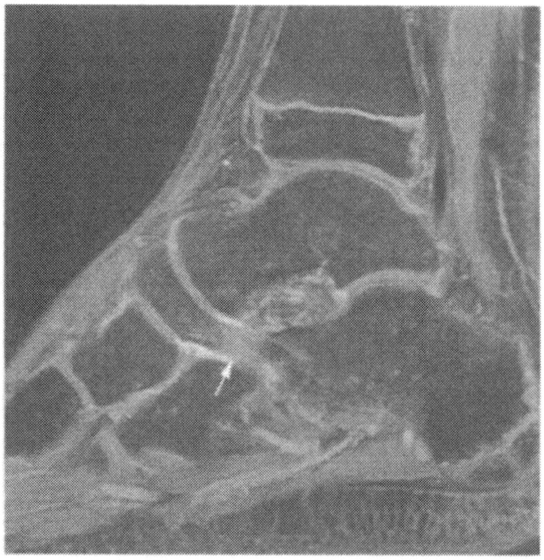

Figure 19L. Sagittal 3D SPGR FS.

Findings

This is 13-year-old boy with a spastic flatfoot.

Figure 19J. The calcaneonavicular joint is slightly irregular. Calcaneus (C), navicular (N).

Figure 19K. Juxta-articular marrow edema in the navicular, calcaneus, and talus is present. No joint fluid is seen within the calcaneonavicular joint **(arrow)**.

Figure 19L. SPGR sequence demonstrates a cartilaginous calcaneonavicular coalition **(arrow)**.

Pitfalls and Pearls

1. Patients with unsuspected tarsal coalition may be referred for MRI because of foot pain or peroneal symptoms. Therefore, the intertarsal joints should always be carefully evaluated in these patients.
2. Peroneal contracture, tendinopathy, tenosynovitis, and spastic flatfoot may occur due to restricted subtalar motion in the setting of talocalcaneal coalitions (3).

References

1. Kulik SA, Jr., Clanton TO. Tarsal coalition. *Foot Ankle Int* 1996; 17:286–296.
2. Leonard MA. The inheritance of tarsal coalition and its relationship to spastic flat foot. *J Bone Joint Surg Br* 1974; 56B:520–526.
3. Bohne WH. Tarsal coalition. *Curr Opin Pediatr* 2001; 13:29–35.
4. Lee MS, Harcke HT, Kumar SJ, Bassett GS. Subtalar joint coalition in children: new observations. *Radiology* 1989; 172:635–639.
5. Crim JR, Kjeldsberg KM. Radiographic diagnosis of tarsal coalition. *AJR Am J Roentgenol* 2004; 182:323–328.
6. Newman JS, Newberg AH. Congenital tarsal coalition: Multimodality evaluation with emphasis on CT and MR imaging. *Radiographics* 2000; 20:321–332; quiz 526–327, 532.

Case 20

History

This is a 5-year-old girl who has a longstanding bowleg deformity, left greater than right. She also has left in-toeing.

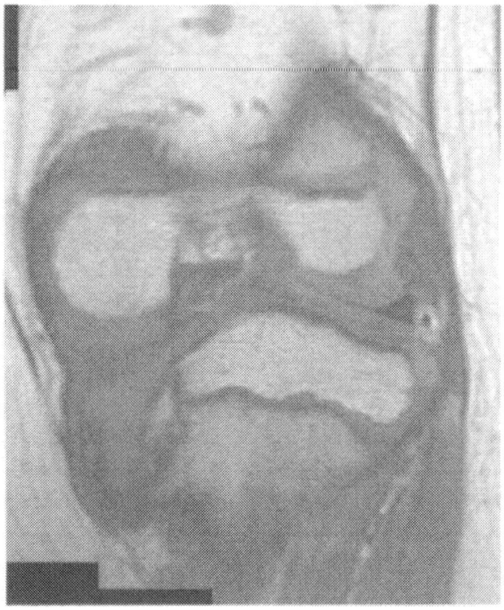

Figure 20A. Coronal T1 of the left knee.

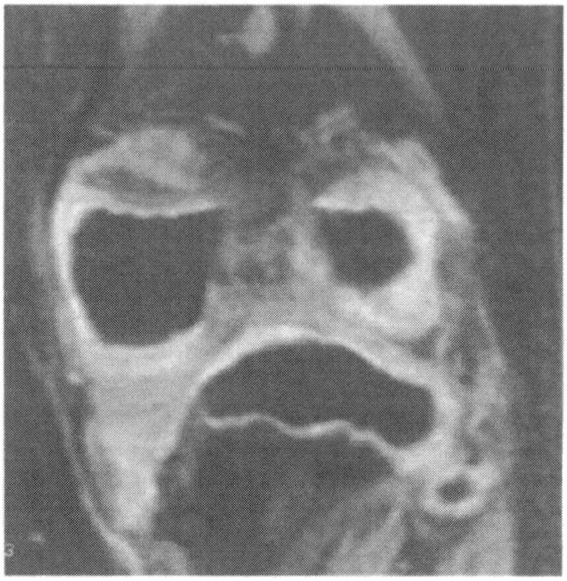

Figure 20B. Coronal-oblique 3D SPGR FS through the posterior knee.

Figure 20A. Deficient ossification of the medial tibial epiphysis is apparent. The cartilaginous medial epiphysis is normal in vertical dimension but is displaced distally, resulting in an abnormal slope of the medial tibial plateau.

Figures 20B, (20B with annotations). The SPGR sequence shows a physeal bar spanning the posteromedial aspect of the proximal tibial physis **(arrow)**.

Figure 20C. There is tibia vara. The epiphysis of the proximal tibia is deficient and fragmented medially. There is a sharp downward sloping of the medial physis and metaphysis. The medial metaphysis is beaked and sclerotic.

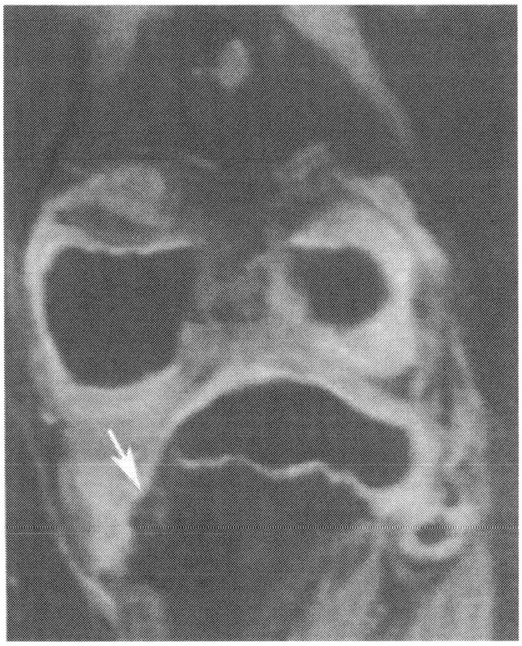

Figure 20B* Annotated.

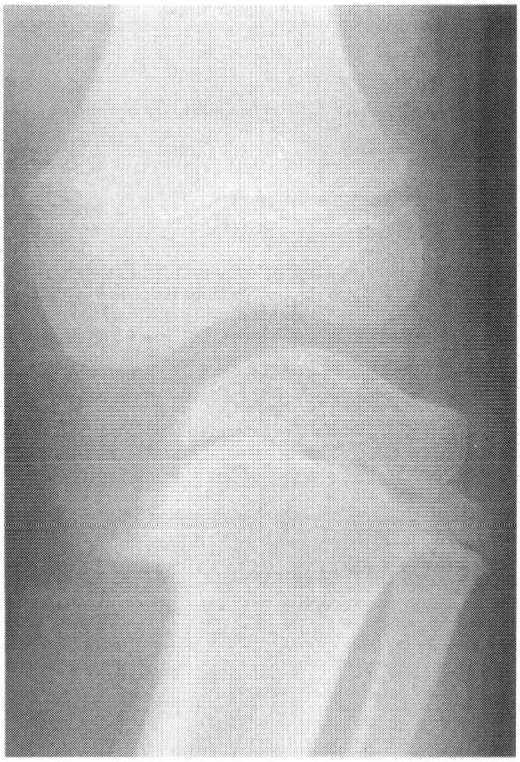

Figure 20C. AP radiograph of the left knee.

Diagnosis

Infantile Blount's disease

Questions

1. What are the two types of Blount's disease?
2. At what age does physiologic genu varus convert to a normal mild genu valgus?

Discussion

Blount's disease is generally considered an osteochondrosis of the medial aspect of the proximal tibia that leads to a progressive varus deformity. Infantile Blount's disease is usually seen in patients under 3 years of age, and adolescent Blount's disease is seen after 8 years of age (Answer to Question 1) (1). Infantile Blount's disease tends be progressive, and up to 60% of cases may be bilateral. Adolescent Blount's disease tends to be mild and self-limited, although the deformity may be exaggerated during growth spurts (Figures 20D, 20E). The proposed etiologies for Blount's disease resemble those for other osteochondroses and include: repetitive microtrauma (most likely), genetic causes, endocrine disorders, infection, and/or local ischemia. Unlike most osteochondrosis, the insult related to Blount's disease may affect the epiphysis, physis, and metaphysis (2). Additional risk factors for Blount's disease include early walking, obesity, and African descent.

The radiographic findings of Blount's disease include: proximal metaphyseal beaking with an angular downsloping of the medial tibia metaphysis, medial physeal irregularity, fragmentation and sclerosis of the medial tibia epiphysis, decreased medial epiphyseal height, increased tibia-femoral angle for age, and a metaphyseal-diaphyseal angle >11 degrees (1). The normal tibia-femoral angles depend on age. In newborns, an average of 17 degrees of genu varus is seen, whereas an average of 2 degrees of genu valgus is seen at 2 years of age (Answer to Question 2) (3).

Additional features of Blount's disease that are usually better visualized by MRI include: the presence of a medial osseous or fibrous physeal bar, large medial meniscus with degeneration, partial to complete epiphyseal cartilage compensation for decreased height of the osseous component of the medial epiphysis, and the cartilaginous component of the metaphyseal beak (Figure 20F) (4). Epiphyseal, physeal, and metaphyseal abnormalities tend to affect the medial or posteromedial proximal tibia. In a study evaluating five patients with adolescent Blount's disease by MRI, Synder et al. noted proximal tibial and distal femoral physeal widening and cartilaginous invaginations into the medial tibial metaphysis in all subjects (5).

Plain radiographic and MRI evaluation of bowlegs in the infant is challenging since it may be difficult to separate physiologic bowing from infantile Blount's disease. The apex of angulation of physiologic bowing is at the knee joint, whereas the apex of angulation of tibia varus is at the proximal tibial physis. Both conditions may have medial tibial metaphyseal beaking. However, physiologic bowing tends to have a gentler medial downward slope of the physis and metaphysis compared with the sharp angular deformity, sclerosis, and irregularity seen with infantile Blount's disease. In addition, the medial epiphyseal height of the proximal tibia is usually normal in patients with physiologic bowing (Figures 20G, 20H). Several authors have observed MRI findings, which may suggest future development of Blount's disease. Iwasawa et al. observed that toddlers with physiologic bowleg deformities that had posteromedial physeal undulation, altered SI within the medial tibial metaphysis, and increased T2 SI within the posteromedial tibial epiphyseal cartilage were less likely to resolve and were at increased risk of developing Blount's disease (6). Mukai et al. observed that toddlers evaluated for physiologic bowing that had abnormal increased T2 SI along the medial perichondrial region of the epiphysis, irregular width of the physis, and depression of the medial physis had a higher chance of developing infantile Blount's disease (7).

The patient subsequently underwent left tibial physeal bar resection and external rotation osteotomies of the tibia and fibula and did well.

Orthopedic Perspective

The diagnosis of tibia varus is typically established on plain radiographs by measuring the metaphyseal-diaphyseal angle and assessing the medial aspect of the proximal tibial physis. A metaphyseal-diaphyseal angle greater than 11 degrees is usually treated by bracing or surgical correction. Surgical correction of infantile Blount's disease is considered when the child is approximately 4–5 years old and the deformity has not corrected with bracing. The role of MRI is limited in the initial diagnosis and is reserved for preoperative planning. Documenting a physeal bar and the status of the medial joint (early degenerative changes, meniscal tear) are important for surgical planning and prognosis. A scanogram and bone age are routinely obtained on all patients prior to surgical intervention. Epiphyseodesis is preferred over physeal bar resection in older children (>12 years) or when there is less than 2 years of growth remaining. Epiphyseodesis is a less invasive procedure and requires a shorter convalescence period compared with physeal bar resection (approximately 2 weeks versus 6–8 weeks). If an epiphyseodesis is performed, the contralateral knee should be evaluated and a bone age performed to predict future leg-length discrepancies. If there is future growth potential, performing a contralateral epiphyseodesis may be indicated. These physeal surgical interventions have considerably less morbidity and shorter convalescence compared with a wedge tibial osteotomy typically performed after skeletal maturity.

What the Clinician Needs to Know

1. Distinguish between infantile Blount's disease and physiologic genu varus.
2. Identify surgically correctable causes of Blount's disease such as a fibrous or osseous medial physeal bar of the tibia.
3. Are there secondary degenerative changes in the knee joint related to altered mechanics due to Blount's disease?

Answers

1. Infantile (1–3 years) and adolescent (after 8 years).
2. Approximately 2 years of age.

Additional Examples

Adolescent Blount's Disease of the Right Knee

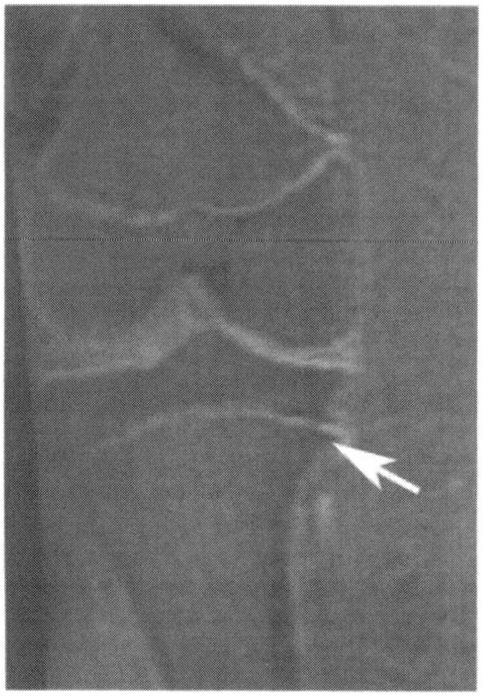

Figure 20D. Coronal 3D SPGR FS of the right knee.

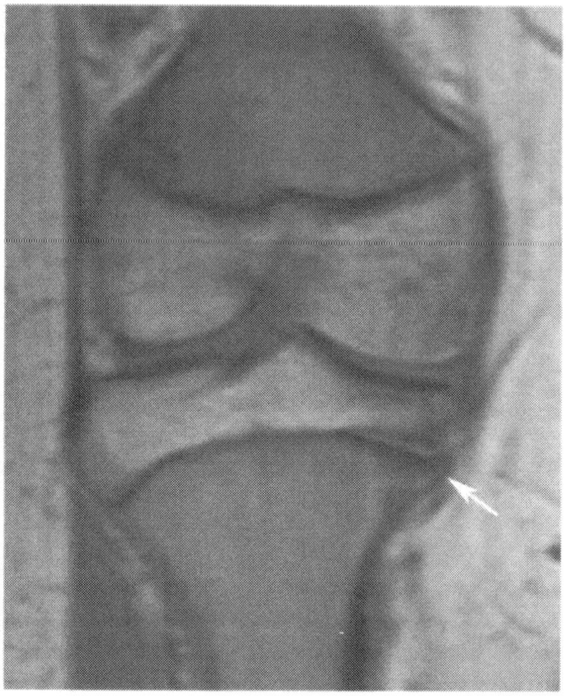

Figure 20E. Coronal T1.

Findings

This 13-year-old girl had mild right tibia vara.

Figures 20D, 20E. MRI shows diminished height of the medial tibia epiphysis and mild metaphyseal beaking **(arrows)**. The medial tibia physis is open and slightly widened compared with the lateral side. This patient subsequently underwent lateral epiphysiodesis of the right proximal tibia.

Infantile Blount's Disease of the Left Knee

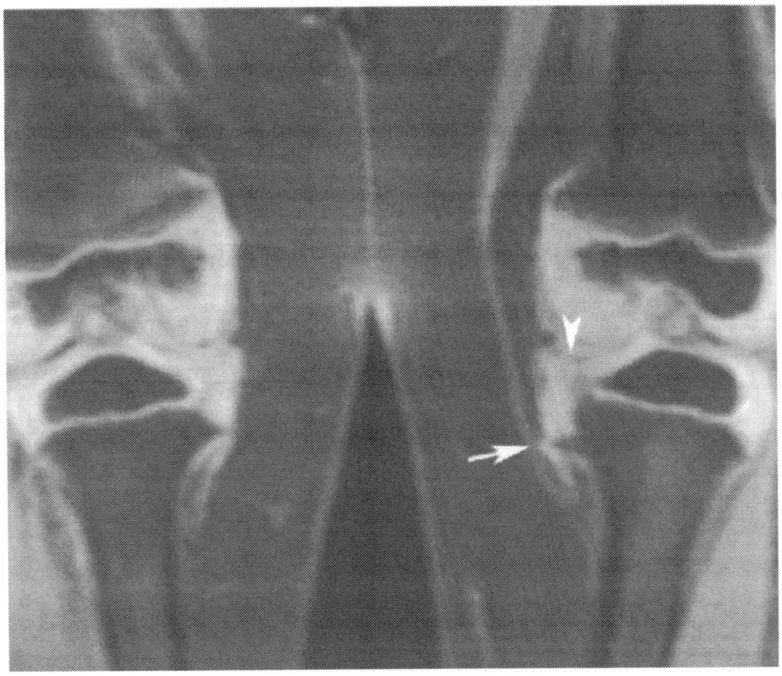

Figure 20F. Coronal 3D SPGR FS of both knees.

Findings

This 3-year-old girl had bilateral tibia vara that was followed since 20 months of age. The right genu varus deformity was improving over time but the left genu varus deformity was progressing.

Figure 20F. Note the gentle medial downsloping of the normal right tibia metaphysis. Contrast this with the abnormal left side, which has characteristic features of infantile Blount's disease. These findings include: metaphyseal beaking **(arrow)**, diminished height of the osseous component of the medial epiphysis, and medial tibia plateau depression **(arrowhead)**. This patient subsequently underwent left tibial valgus osteotomy.

Bilateral Physiologic Genu Vara

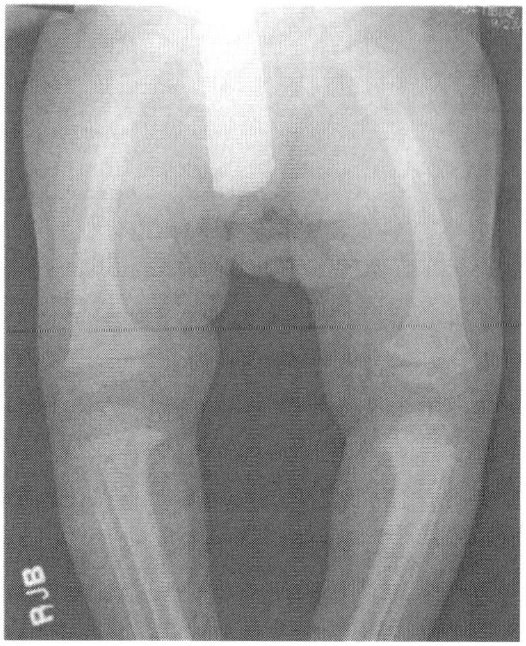

Figure 20G. Standing AP radiograph of both legs.

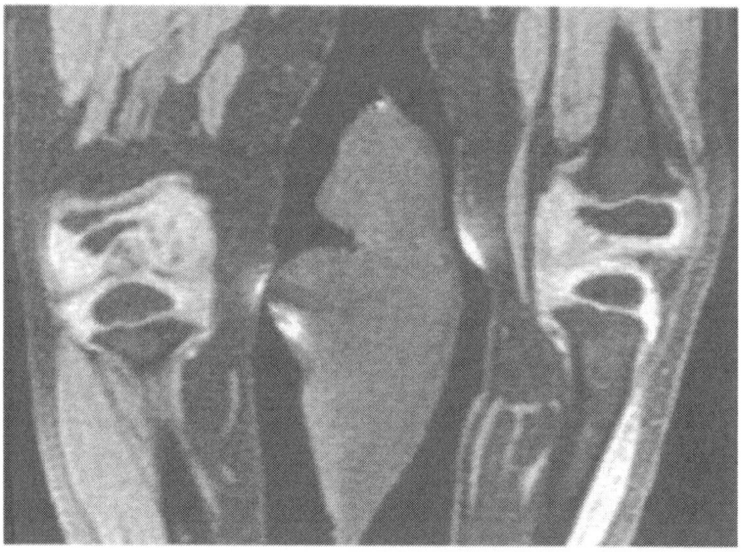

Figure 20H. Coronal 3D SPGR FS.

Findings

This is a 2-year-old boy who was noted to be bowlegged, with an early onset of walking at 8 months of age.

Figure 20G. There is bilateral genu vara with prominence of the medial tibial metaphyses, left greater than right.

Figure 20H. The medial aspects of the proximal tibia epiphyses are normal in height with a gentle medial physeal sloping compatible with physiologic bowing. There is no evidence of infantile Blount's disease.

Pitfalls and Pearls

1. Patients with Blount's disease usually have associated internal tibial torsion, leading to in-toeing.
2. The transition from physiologic genu varus to genu valgus occurs at approximately 2 years of age.
3. Rule out metabolic conditions and bone dysplasias associated with bowing (e.g., rickets, metaphyseal dysplasias, and osteogenesis imperfecta).
4. Genu varus refers to varus bowing centered at the knee. Tibia vara (Blount's disease) refers to varus bowing centered at the proximal tibia.

References

1. Do TT. Clinical and radiographic evaluation of bowlegs. *Curr Opin Pediatr* 2001; 13:42–46.
2. de Sanctis N, Della Corte S, Pempinello C, Di Gennaro G, Gambardella A. Infantile type of Blount's disease: Considerations concerning etiopathogenesis and treatment. *J Pediatr Orthop B* 1995; 4:200–203.
3. Ozonoff MB. The lower extremity. In: *Pediatric Orthopedic Radiology,* 2 ed., Chapter 4. Philadelphia: W.B. Saunders, 1992; 304–396.
4. Ducou le Pointe H, Mousselard H, Rudelli A, Montagne JP, Filipe G. Blount's disease: Magnetic resonance imaging. *Pediatr Radiol* 1995; 25:12–14.
5. Synder M, Vera J, Harcke HT, Bowen JR. Magnetic resonance imaging of the growth plate in late-onset tibia vara. *Int Orthop* 2003; 27:217–222.
6. Iwasawa T, Inaba Y, Nishimura G, Aida N, Kameshita K, Matsubara S. MR findings of bowlegs in toddlers. *Pediatr Radiol* 1999; 29:826–834.
7. Mukai S, Suzuki S, Seto Y, Kashiwagi N, Hwang ES. Early characteristic findings in bowleg deformities: Evaluation using magnetic resonance imaging. *J Pediatr Orthop* 2000; 20:611–615.

Case 21

History

This is a 4-year-old boy with acute lymphoblastic leukemia (ALL) on cyclic chemotherapy. He presented with episodic right hip pain. The initial radiographs of the right hip were normal (not shown).

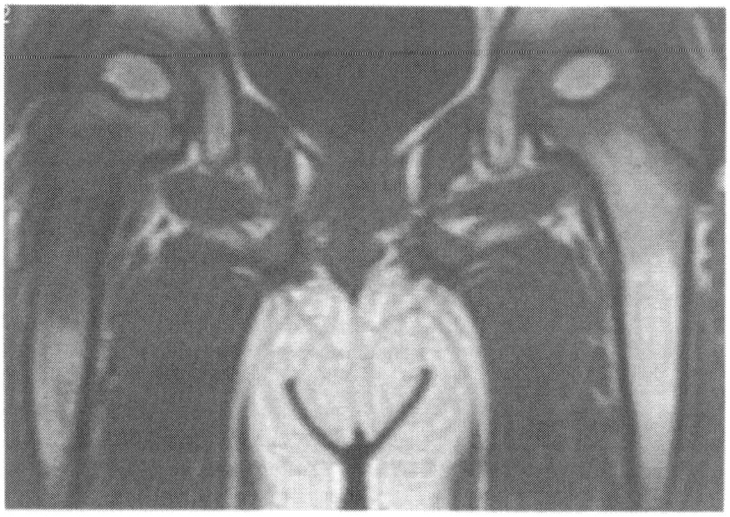

Figure 21A. Coronal T1.

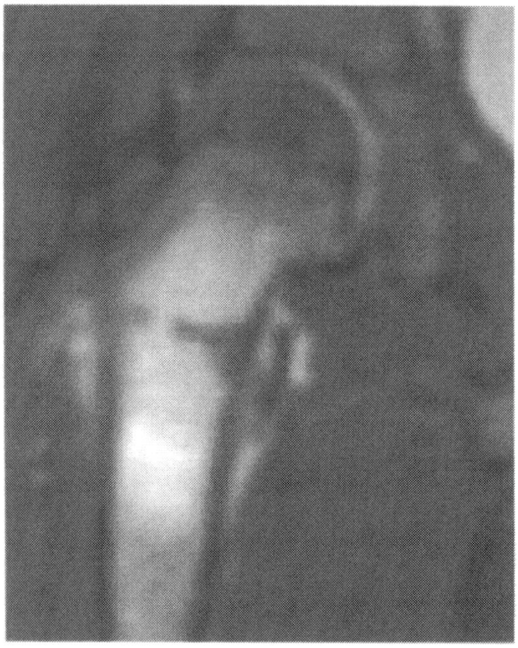

Figure 21B. Coronal STIR.

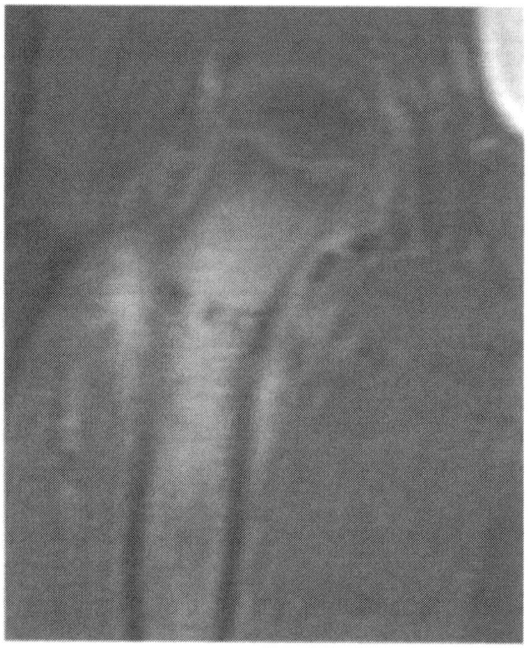

Figure 21C. Coronal T1 post-Gd FS.

Figures 21A, 21B, 21C, (21C with annotations). There is diffuse decreased T1 SI, increased STIR SI, and diffuse enhancement within the right femoral neck and intertrochanteric region. A hypointense linear fracture line is evident **(arrows)**. In addition, there is mild periosteal and juxtacortical soft tissue enhancement **(arrowheads)**. A discrete intraosseous or juxtacortical soft tissue mass is not identified.

Figure 21D. There is focal sclerosis and faint periosteal reaction seen within the intertrochanteric region **(arrow)**.

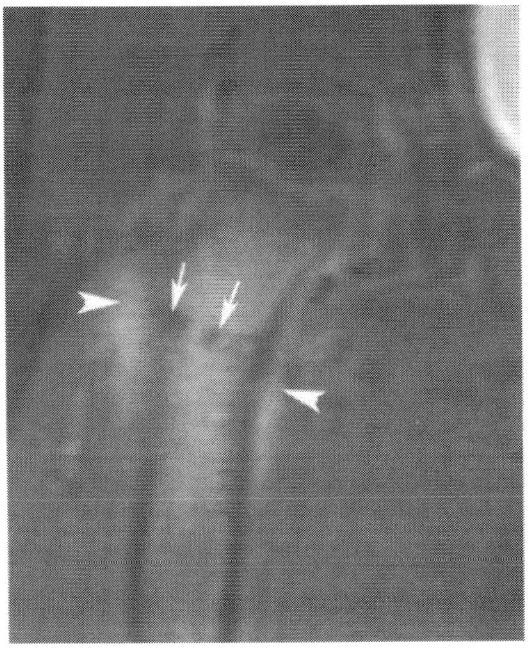

Figure 21C* Annotated.

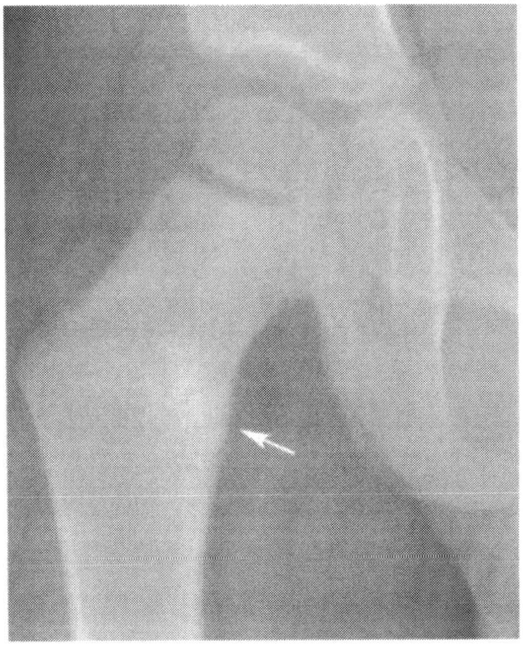

Figure 21D. AP radiograph of the right hip–6 weeks later.

Diagnosis

Insufficiency fracture of the proximal femur

Questions

1. What is the difference between a fatigue and an insufficiency fracture?
2. What are some causes of marrow alterations on MRI in patients with ALL?

Discussion

Fatigue and insufficiency fractures are both overuse or stress injuries. Fatigue fractures occur in normal bone, whereas insufficiency fractures occur in osteopenic bone that is architecturally normal (Answer to Question 1). Insufficiency fractures are different from pathologic fractures. Pathologic fractures occur where tumor, infection, or a marrow replacement process compromises the underlying bone architecture.

Children with ALL are predisposed to insufficiency fractures since their bones may become osteopenic due to the primary disease or its treatment (1). During and shortly after therapy, fractures in children with ALL have been observed to occur six times more often than in normal controls (1).

The MRI diagnosis of an insufficiency fracture is straightforward in this case because a distinct fracture line is observed. With acute or subacute fractures, the fracture line is hypointense on all imaging sequences (Figures 21E, 21F) (2). With healing, there may be increased fluid SI and enhancement along the fracture line and the periosteum due to granulation tissue and other processes related to healing. A discrete fracture line may not be identified in the subacute setting, which may result in nonspecific MRI findings. Alternative causes for alterations in marrow SI in cases of ALL include osteonecrosis (Figures 21G–21I), diffuse tumor infiltrate, red marrow reconversion, and osteomyelitis (Answer to Question 2). In the absence of a discrete fracture line, the latter three considerations may be impossible to distinguish from stress reaction.

This patient did well with conservative therapy. Both clinical and radiologic follow-up showed a healing insufficiency fracture (Figure 21D). There was no evidence of an underlying pathologic fracture on follow-up.

Orthopedic Perspective

MRI may be preferred over bone scintigraphy in differentiating a simple stress reaction or fracture from a pathologic fracture in oncology patients. The anatomic detail and the ability to distinguish pathology near the physis (where physeal physiologic uptake on bone scintigraphy may obscure adjacent pathology) make MRI well suited in these circumstances.

A nondisplaced stress fracture is usually treated with resting, although high risk areas such as the femoral neck or intertrochanteric region may require earlier intervention with screw fixation. Both insufficiency and stress fractures are managed similarly, although healing is expected to take longer in patients with metabolic bone disease.

What the Clinician Needs to Know

1. Is the fracture complete or incomplete?
2. Degree of angulation deformity, if present.
3. Has a pathologic fracture been definitively excluded?

Answers

1. Stress fractures occur in normal bone whereas insufficiency fractures occur in osteopenic bone that is architecturally normal.
2. Osteonecrosis, diffuse tumor infiltrate, red marrow reconversion, and osteomyelitis.

Additional Examples

Insufficiency Fracture of the Talus and Calcaneal Osteosarcoma

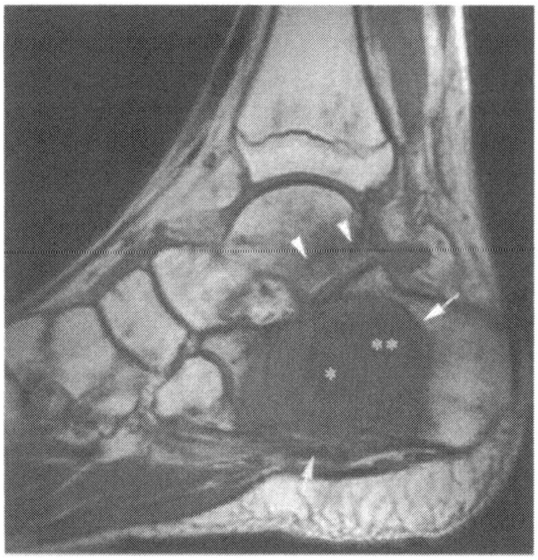

Figure 21E. Sagittal T1 of the left foot.

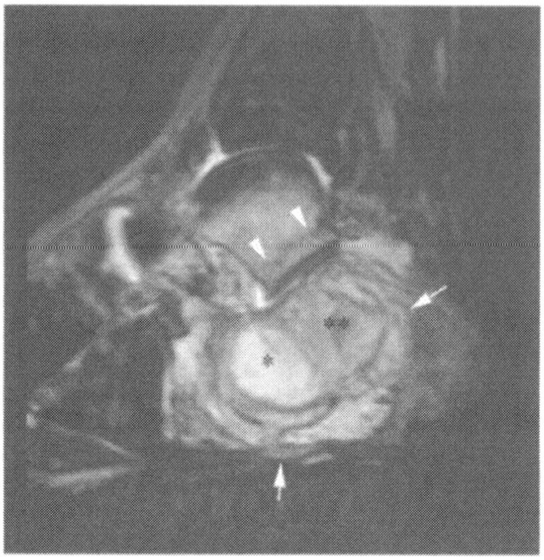

Figure 21F. Sagittal STIR.

Findings

Figures 21E, 21F. This is a 15-year-old boy with biopsy proven calcaneal osteosarcoma **(arrows)**, containing both solid (**) and cystic elements (*). He was undergoing chemotherapy and had recently increased his level of activity. Along the posterior talus, just above the posterior subtalar joint, there is a hypointense line on both T1 and STIR sequences **(arrowheads)**. There is marrow edema within the talus. This is compatible with an insufficiency fracture of the talus, presumably related to altered weightbearing and osteopenia secondary to chemotherapy. Direct extension or metastasis was considered unlikely since a discrete mass was not identified within the talus.

Old Bone Infarction with Acute Stress Reaction

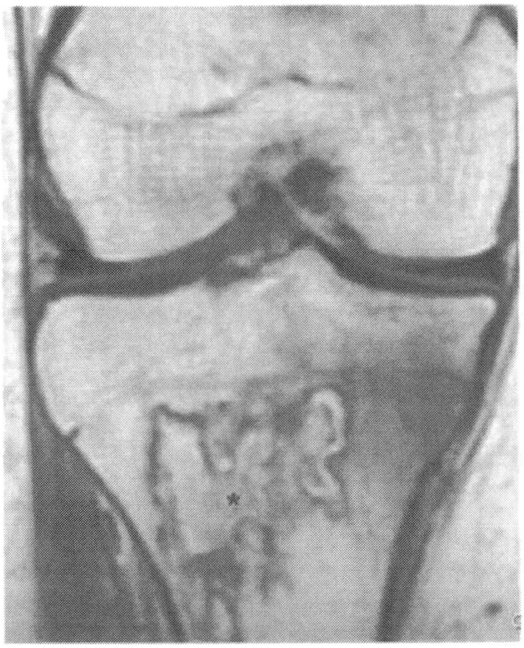

Figure 21G. Coronal T1.

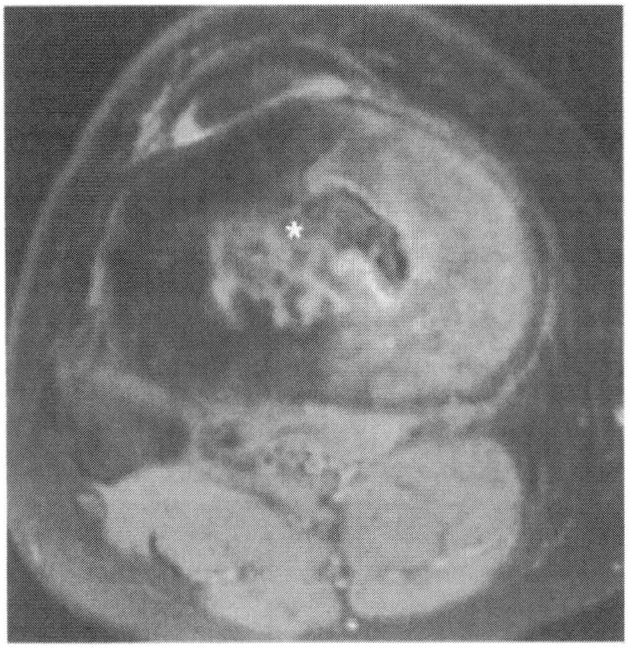

Figure 21H. Axial PD FS.

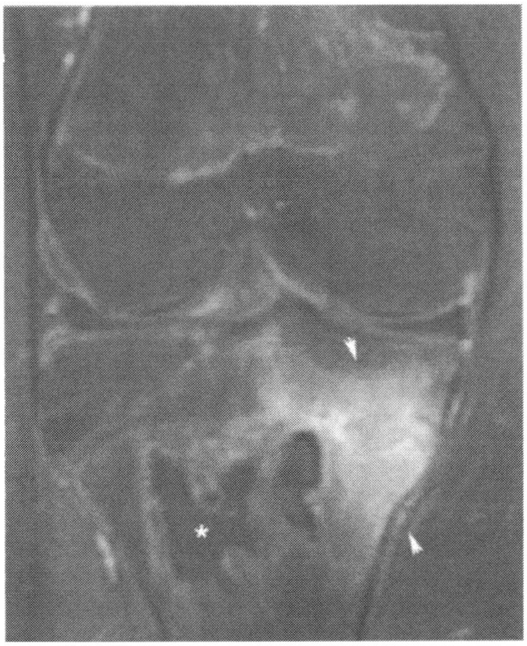

Figure 21I. Coronal T1 post-Gd FS.

Findings

This is a 16-year-old girl with a history of ALL. She recently completed her last round of chemotherapy. While running, she felt acute knee pain.

Figures 21G, 21H, 21I. A subacute to old proximal tibia bone infarction is present containing fat SI centrally (*). There is diffuse increased fluid SI and enhancement seen in the medial aspect of the proximal tibial metaphysis, epiphysis, and adjacent soft tissues **(arrowheads)**. Although enhancement suggested osteomyelitis, the clinical work-up suggested stress injury. Symptoms resolved without antibiotics.

Pitfalls and Pearls

1. The most common radiographic finding in ALL is diffuse osteopenia. Other findings include radiolucent or radiodense metaphyseal bands (AKA leukemic lines; see Case 62), periostitis, focal osteolytic lesions, osteosclerosis, and synovitis (3).
2. In the setting of trauma, the term *stress fracture* should be applied only if a discrete fracture line is visible. If there is bone and adjacent soft tissue edema without a fracture line, the term *stress reaction* is more appropriate.

References

1. van der Sluis IM, van den Heuvel-Eibrink MM, Hahlen K, Krenning EP, de Muinck Keizer-Schrama SM. Altered bone mineral density and body composition, and increased fracture risk in childhood acute lymphoblastic leukemia. *J Pediatr* 2002; 141:204–210.
2. Spitz DJ, Newberg AH. Imaging of stress fractures in the athlete. *Radiol Clin North Am* 2002; 40:313–331.
3. Resnick D, Haghighi P. Lymphoproliferative and myeloproliferative disorders. In: *Diagnosis of Bone & Joint Disorders*, 4 ed. Philadelphia: W.B. Saunders, 2002; 2291–2345.

Case 22

History

This is a 12-year-old girl with left calf swelling for 1 week. She has no history of trauma. Radiographs of the knee were normal (not shown).

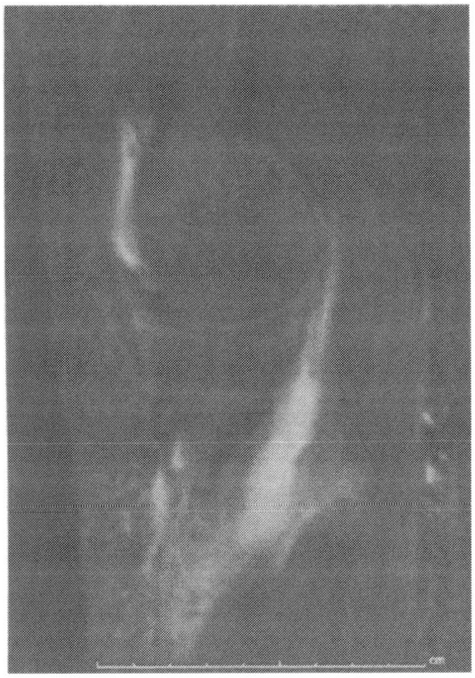

Figure 22A. Sagittal STIR of the medial compartment of the left knee.

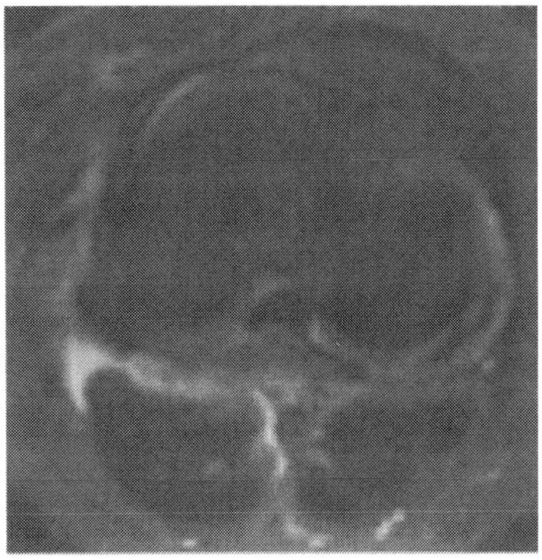

Figure 22B. Axial T2 FS.

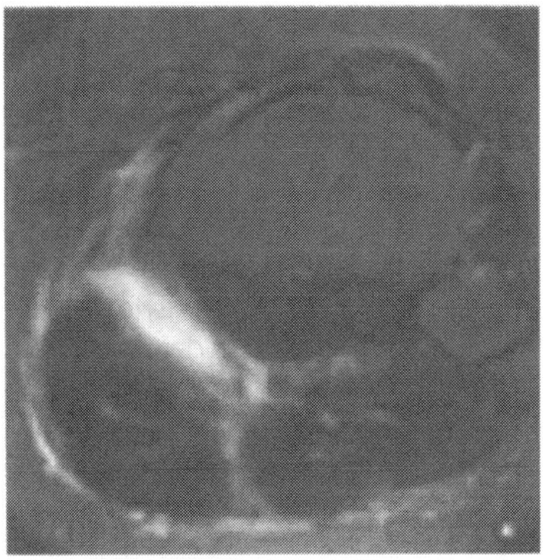

Figure 22C. Axial T2 FS.

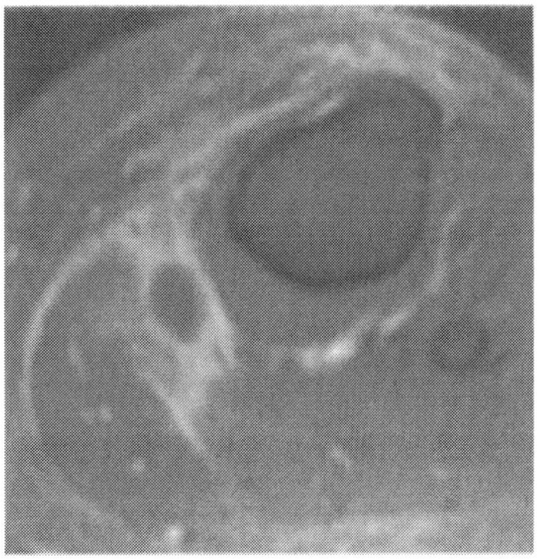

Figure 22D. Axial T1 post-Gd FS.

Figures 22A, 22B, 22C (22A, 22B with annotations). There is an ill-defined fluid collection in the posterior soft tissues of the knee that insinuates between the semimembranosus tendon and the medial head of the gastrocnemius muscle. There is a small joint effusion. Medial head of the gastrocnemius muscle (MG), semimembranosus tendon (SM).

Figures 22D, (22D with annotations). The fluid collection (*) shows rim enhancement. The anterior fascia of the medial calf shows diffuse enhancement **(arrows)**.

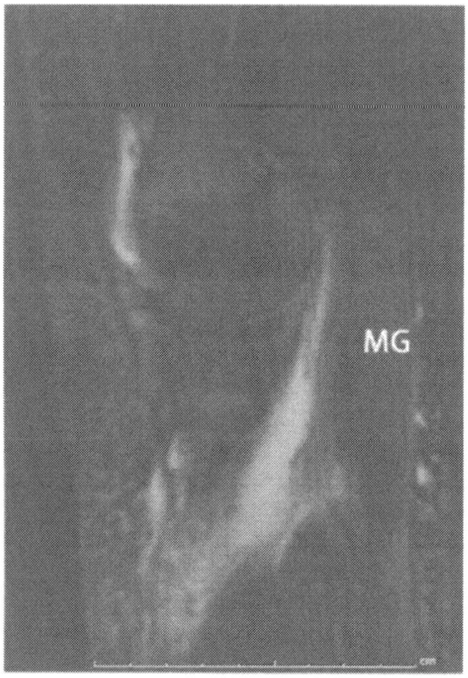

Figure 22A* Annotated.

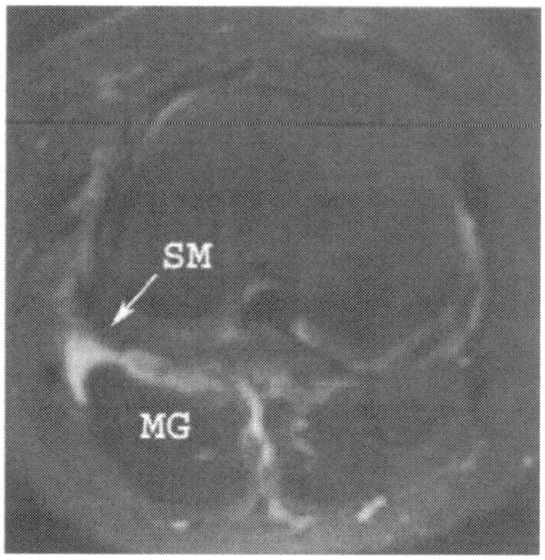

Figure 22B* Annotated.

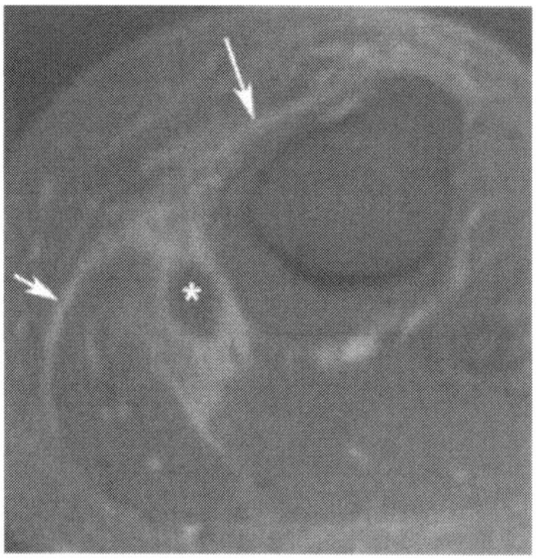

Figure 22D* Annotated.

Diagnosis

Ruptured Baker's cyst

Questions

1. Are children, or adults, with Baker's cysts less likely to have associated internal derangements of the knee?
2. What are some secondary causes of a Baker's cysts?

Discussion

Baker's cysts (AKA Popliteal cyst) are found in the posteromedial aspect of the popliteal fossa. They often communicate with the posteromedial joint capsule and are frequently associated with a joint effusion. Fluid accumulates within the gastrocnemius-semimembranosus bursa, both superficial and deep with respect to the medial head of the gastrocnemius muscle. The superficial component tends to be larger than the deep component, and these two bursae communicate with each other (Figure 22E) (1).

In children, Baker's cysts often are asymptomatic and present because of a palpable mass. In one study of asymptomatic children under 15 years of age, the prevalence was 2.4% (2). Unlike Baker's cysts in the adult population, they are infrequently associated with internal derangements of the knee (Answer to Question 1) (1, 3). The common occurrence of Baker's cysts in asymptomatic young children, without associated joint effusion, suggests a primary/developmental etiology that may be familial in some cases (4). Trauma appears to be the most common cause of secondary Baker's cysts and blood products may be found within the lesions (Figures 22F, 22G). Baker's cysts may also develop secondary to pyogenic and degenerative processes, Lyme, and juvenile rheumatoid arthritis, as well as pigmented villonodular synovitis (PVNS) (Answer to Question 2) (1).

The MRI findings of a ruptured Baker's cyst may be misleading. Rim enhancing fluid collections as well as subcutaneous and fascial enhancement may suggest an infectious etiology such as cellulitis, abscess, pyomyositis, pyogenic arthritis with soft tissue extension, or an infected Baker's cyst. The characteristic soft tissue changes centered over the gastrocnemius-semimembranosus bursa should first suggest a ruptured Baker's cyst. Ultimately, clinical correlation is required to rule out infection.

This patient had no history of trauma, and she was initially evaluated for deep venous thrombosis with Doppler ultrasound. This study was negative, and she was referred to MRI. She had no clinical or laboratory signs of infection and was treated with compression stockings and ibuprofen, with resolution of symptoms.

Orthopedic Perspective

Popliteal cysts usually present in children as an asymptomatic posterior knee mass. Unlike adults, popliteal cysts in children are rarely associated with an internal derangement. The diagnosis is usually made clinically but may be confirmed by ultrasound or MRI. The majority of popliteal cysts are treated with observation and resolve spontaneously. The symptomatic and/or ruptured popliteal cyst may be treated with ice, nonsteroidal anti-inflammatory agents, and physical therapy until asymptomatic. The natural history of a popliteal cyst unassociated with internal derangement is spontaneous resolution in approximately one year.

What the Clinician Needs to Know

1. Is the Baker's cyst intact or ruptured?
2. Is there an underlying etiology for the Baker's cyst such as an internal derangement (e.g., torn meniscus)?

Answers

1. Children with Baker's cysts are less likely to have internal derangements of the knee as compared with adults.
2. Trauma, arthritis (pyogenic, degenerative, Lyme, and juvenile rheumatoid arthritis), and PVNS.

Additional Examples

Large Baker's Cyst

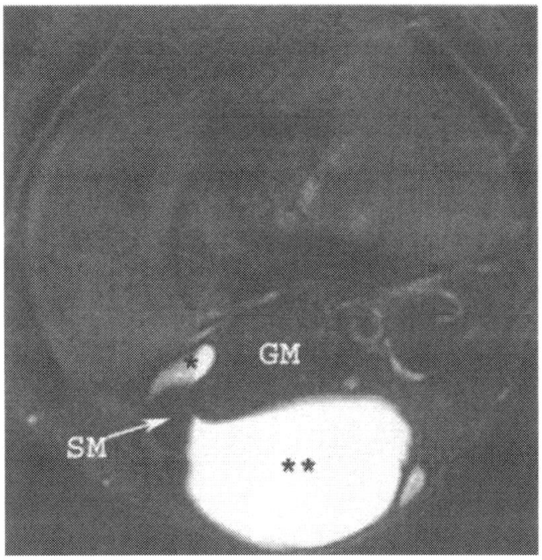

Figure 22E. Axial T2 FS of the left knee.

Findings

This is a 5-year-old boy who presented with a large posteromedial knee mass.

Figure 22E. There is a large Baker's cyst located in the superficial bursa (**) communicating with a small cyst located in the deep bursa (*). There was no internal derangement of the knee. Medial head of the gastrocnemius muscle (GM), semimembranosus tendon (SM).

Ruptured Baker's Cyst with Internal Hemorrhage

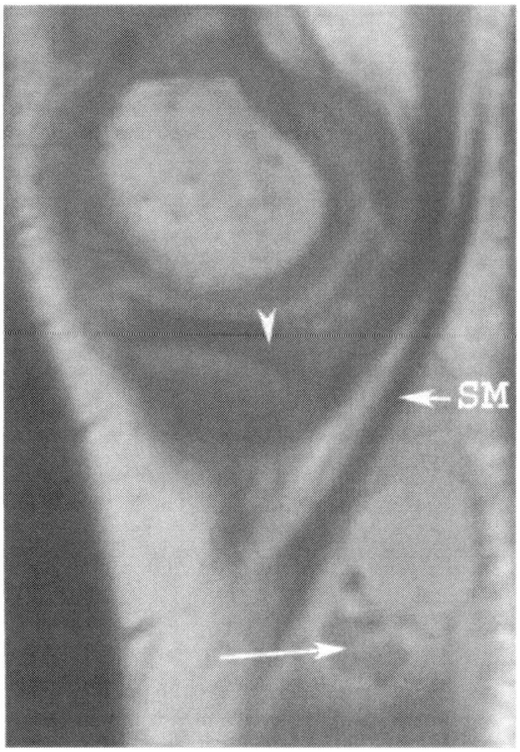

Figure 22F. Sagittal PD of the right knee.

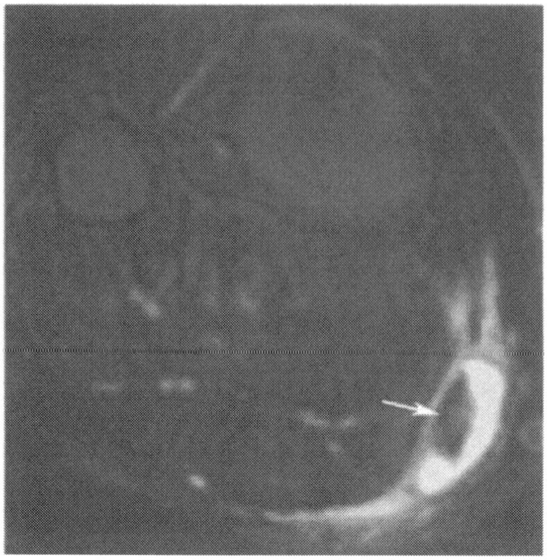

Figure 22G. Axial T2 FS.

Findings

This is an 8-year-old girl who presented with acute knee pain.

Figures 22F, 22G. Within the ruptured Baker's cyst, there is a globular region that is hypointense on both PD and T2W sequences **(arrows)**, compatible with hemorrhage. Note the peripheral, linear hyperintense signal within the posterior horn of the medial meniscus **(arrowhead)** related to normal meniscal vascularity in this child. This should not be mistaken for a degenerative tear. Semimembranosus tendon (SM).

Pitfalls and Pearls

1. Unlike adult patients, Baker's cysts in children are rarely associated with internal derangements of the knee.
2. Not all rim enhancing soft tissue fluid collections are caused by infection. Post-traumatic soft tissue seromas and hematomas may all show rim enhancement.
3. When a child or adolescent presents with a poorly defined, painful fluid collection in the calf, consider a ruptured Baker's cyst and look for the typical fluid signal between the medial head of the gastronemius and semimembranosis tendon.

References

1. De Maeseneer M, Debaere C, Desprechins B, Osteaux M. Popliteal cysts in children: Prevalence, appearance and associated findings at MR imaging. *Pediatr Radiol* 1999; 29:605–609.
2. Seil R, Rupp S, Jochum P, Schofer O, Mischo B, Kohn D. Prevalence of popliteal cysts in children: A sonographic study and review of the literature. *Arch Orthop Trauma Surg* 1999; 119:73–75.
3. Miller TT, Staron RB, Koenigsberg T, Levin TL, Feldman F. MR imaging of Baker cysts: Association with internal derangement, effusion, and degenerative arthropathy. *Radiology* 1996; 201:247–250.
4. Toyama WM. Familial popliteal cysts in children. *Am J Dis Child* 1972; 124:586–587.

Case 23

History

This is a 5-year-old boy undergoing treatment for acute lymphoblastic leukemia (ALL), now with knee pain.

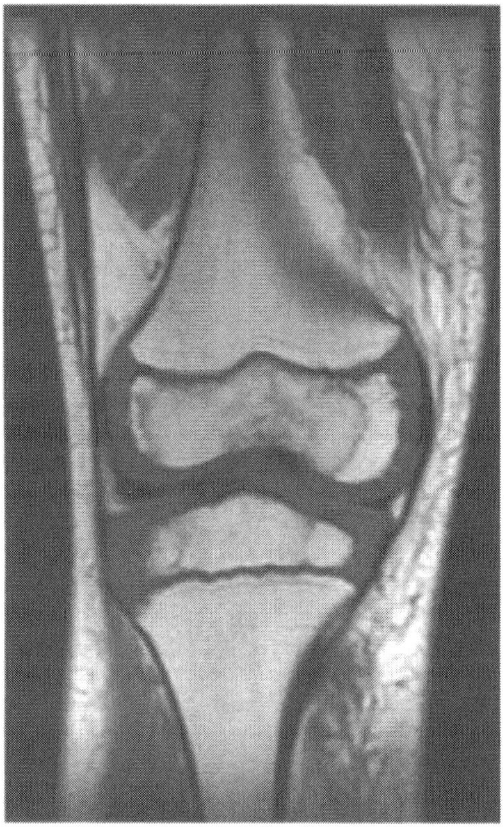

Figure 23A. Coronal T1 of the right knee.

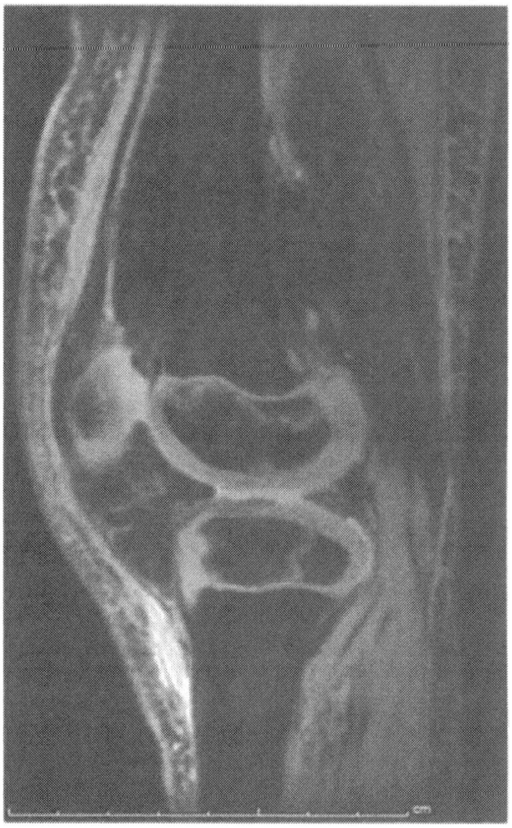

Figure 23B. Sagittal PD FS.

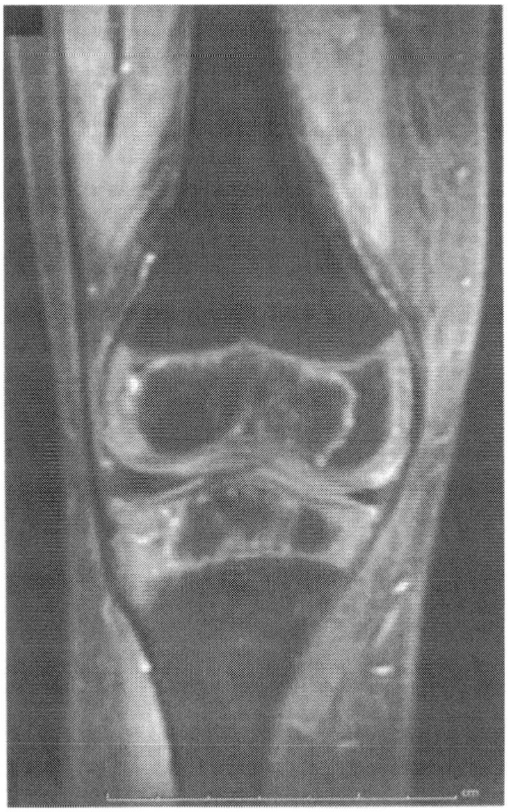

Figure 23C. Coronal T1 post-Gd FS.

Figures 23A, 23B, 23C. Well-defined lesions are noted to involve all but the periphery of the distal femoral and proximal tibial epiphyses. The serpiginous margins are hypointense on T1, hyperintense on PD FS, and demonstrate enhancement. The centers of these lesions are heterogeneously hypointense on T1 with respect to yellow marrow, with patchy hyperintensity and enhancement on PD FS and post-Gd images, respectively. Subcutaneous soft tissue edema and enhancement are also present. There is also more focal edema present over the tibial tubercle.

Diagnosis

Osteonecrosis

Questions

1. T/F: Osteonecrosis may occur as a complication of Salter-Harris physeal injuries.
2. What is the explanation for the MRI double-line sign seen with osteonecrosis?

Discussion

Osteonecrosis is the broad term used to describe bone death related to ischemia. When osteonecrosis affects an epiphysis or subarticular bone, it may be referred to as ischemic necrosis, whereas osteonecrosis of the metaphysis or diaphysis is referred as bone infarction (1). Alternative terms used to describe epiphyseal or subarticular osteonecrosis include avascular necrosis (AVN) and aseptic necrosis. Entities associated with osteonecrosis include sickle cell disease and other hemoglobinopathies, medications such as corticosteroids, radiation, malignancy (e.g., acute lymphoblastic leukemia [ALL] and its treatment), metabolic conditions including Gaucher's disease, chronic renal failure, vasculitidies such as systemic lupus erythematosus, juvenile rheumatoid arthritis, and idiopathic causes including Legg-Calve-Perthes disease, Freidberg's infraction, Koehler's disease and Kienböck's disease. An element of osteonecrosis may occur following major Salter-Harris physeal injuries, particularly at the knees (Answer to Question 1). The findings are often inconspicuous on plain radiographs but striking on MRI (Figures 23D, 23E).

The MRI features of epiphyseal and metadiaphyseal osteonecrosis are similar. In the acute setting, significant marrow and soft tissue edema are nonspecific findings (Figures 23F–23I) (2, 3). Osteonecrosis complicated by insufficiency fracture may also be associated with significant marrow and soft tissue edema. With epiphyseal osteonecrosis, a reactive joint effusion may also be seen. The MRI findings of acute osteonecrosis may be indistinguishable from osteomyelitis, neoplasms such as lymphoma, and stress reaction. However, Umans et al. reported that contrast-enhanced MRI may be helpful for distinguishing these entities. They observed that osteonecrosis demonstrates thin peripheral enhancement, whereas osteomyelitis shows geographic or thick and irregular peripheral enhancement around a nonenhancing center (4).

In the subacute or chronic setting, the margins of osteonecrosis may have a characteristic serpiginous contour and a "double-line" sign may be seen (Figures 23J, 23K) (5). With this sign, the outer rim represents bony sclerosis and is hypointense on all imaging sequences. The inner T2 hyperintense zone reflects granulation tissue (Answer to Question 2). Gadolinium is not required to make the diagnosis of osteonecrosis once a double-line sign is evident, but may be helpful to differentiate granulation tissue and necrotic bone. The center of osteonecrosis may follow similar SI to fat, blood, fluid, and fibrous tissue (5). Usually a mixture of these various components is evident with chronic osteonecrosis (Figures 23L, 23M) (1). The explanation for alterations in the normal marrow SI after osteonecrosis includes microfracture, irreversible ischemia, and mineralization. Transphyseal osteonecrosis can occur despite the traditional concept that the epiphysis and metadiaphysis have relatively separate vascular anatomy after the age of 18 months (Figures 23J, 23K).

The complications of osteonecrosis include fracture and subchondral bone collapse, premature degenerative disease, superimposed osteomyelitis, and rarely sarcomatous transformation in adults.

This patient underwent bone marrow transplantation 1 month earlier for the treatment of ALL. He subsequently developed graft versus host disease and knee pain. The imaging features were characteristic of subacute epiphyseal osteonecrosis (ischemic necrosis) and correlated with his clinical presentation.

Orthopedic Perspective

Osteonecrosis per se is of little clinical significance to the patient or the orthopedist unless it is acute and painful (as in sickle cell disease) or unless it causes subchondral collapse and joint deterioration. Differentiation of acute osteonecrosis from osteomyelitis is often difficult on imaging grounds, and bone aspiration/biopsy with appropriate cultures may be required. Treatment of uncomplicated osteonecrosis is symptomatic until the pain resolves. Conservative treatment of the acute phase with limited weightbearing (crutches or braces) and anti-inflammatory drugs for pain are usually sufficient, and long term sequellae are unusual. Patients with leukemia or bone marrow transplantation often have multifocal osteonecrosis, making it difficult to limit the forces across the joints while healing takes place. Fortunately for leukemic patients, epiphyseal osteonecrosis does not seem to lead to as much joint collapse and osteoarthritis as might be imagined.

There are no proven surgical approaches to prevent collapse of the involved joints, although core decompression (to attempt to restore blood supply) or osteotomies (to unload the affected portion of the joints) are occasionally used. Some surgeons have advocated vascularized fibular grafts to restore blood supply to the femoral head, but this practice is not widely accepted. The worst complication of osteonecrosis is a sarcoma, which is fortunately rare and occurs primarily in adults.

What the Clinician Needs to Know

1. Osteonecrosis must be differentiated from osteomyelitis, a particular problem in patients with sickle cell disease, where the two conditions may coexist.
2. For epiphyseal osteonecrosis, the extent of involvement of the epiphysis and the integrity of the subchondral bone is important to assess. The clinician is interested in maintaining the articular surface and avoiding collapse of the subchondral bone during the healing phase of osteonecrosis.

Answers

1. True.
2. The outer rim represents bone reaction and is hypointense on all imaging sequences. The inner T2 hyperintense zone represents granulation tissue.

Additional Examples

Bone Infarction After Trauma

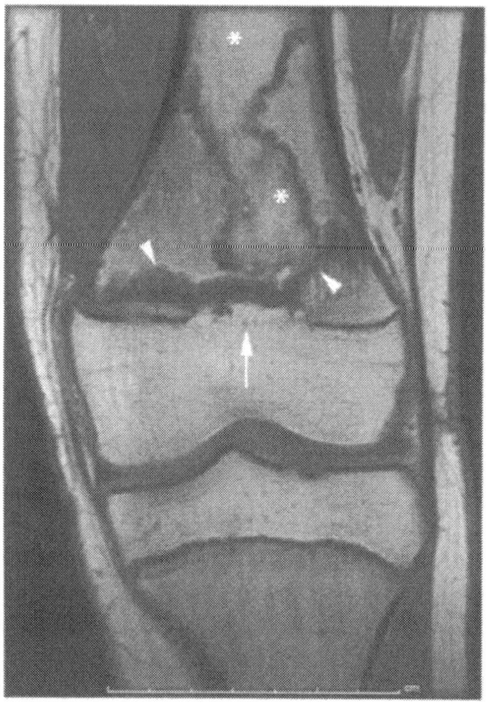

Figure 23D. Coronal T1 of the left ankle.

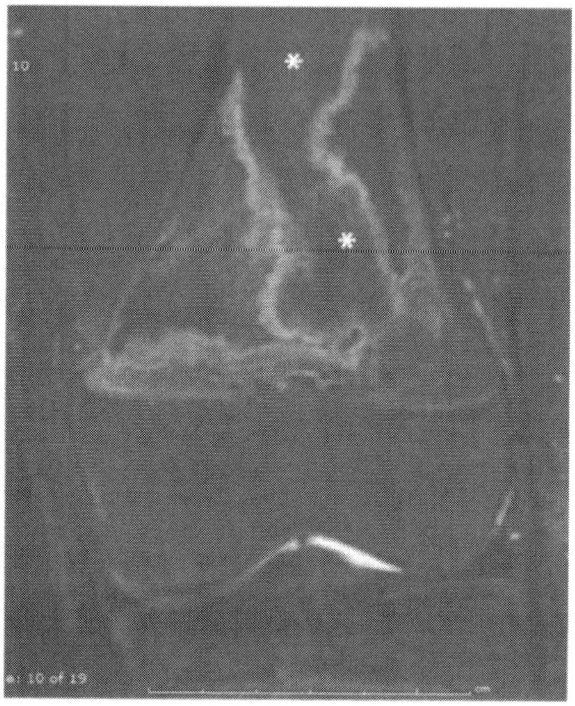

Figure 23E. Coronal STIR.

Findings

This is a 13-year-old boy who sustained a distal femoral Salter-Harris Type 2 injury 5 months ago while playing basketball.

Figures 23D, 23E. A large central geographic area is evident (*) that follows the SI of fat. The serpiginous margins manifest the double-line sign. There is also a large central physeal bar **(arrow)** that is in continuity with the fracture line **(arrowheads)**.

Osteonecrosis of the Talus with Atypical Features

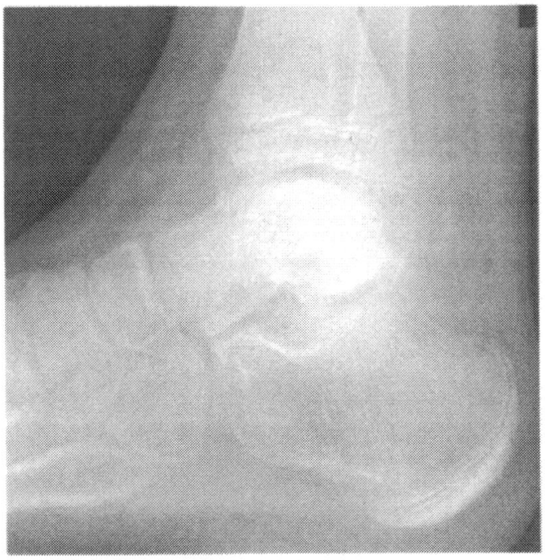

Figure 23F. Lateral radiograph of the left ankle.

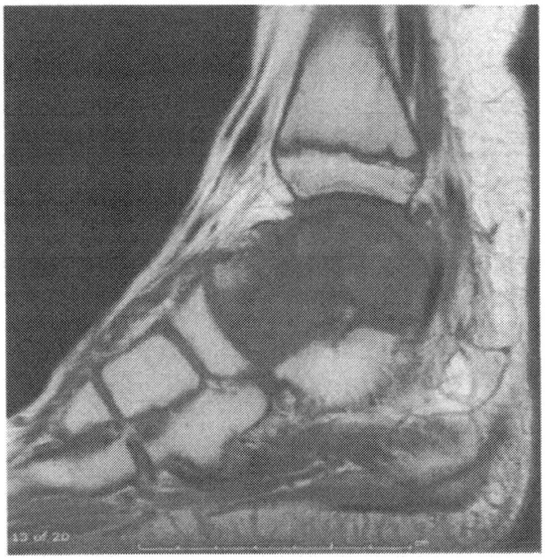

Figure 23G. Sagittal T1.

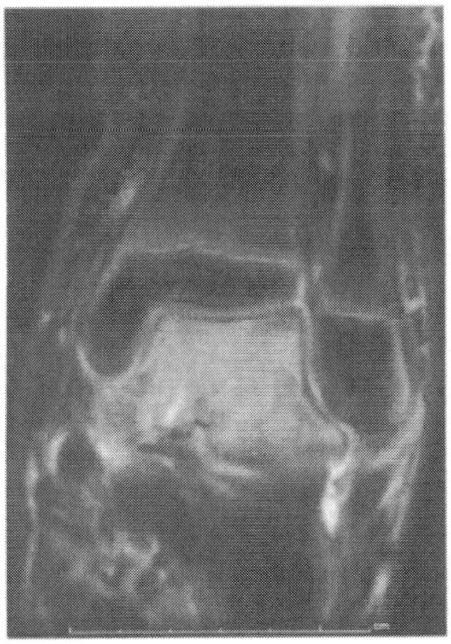

Figure 23H. Coronal PD FS.

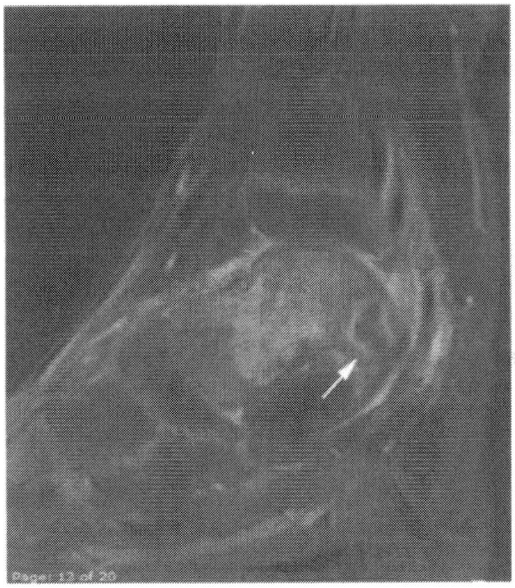

Figure 23I. Sagittal T1 post-Gd FS.

Findings

This is an 11-year-old boy with a 4-month history of left ankle pain, no fever, and no history of trauma or underlying disease.

Figure 23F. The talus is densely sclerotic without collapse.

Figure 23G. There is diffuse loss of T1 fat SI within the talus.

Figures 23H, 23I. There is diffuse PD hyperintensity within the talus with homogeneous enhancement, except for a rim-enhancing region posteriorly **(arrow)**. There is also extraosseous soft tissue edema and enhancement. Because of the atypical imaging features and absence of clinical clues to the etiology, biopsy was performed and osteonecrosis was pathologically proven.

Osteonecrosis, Crossing Physis

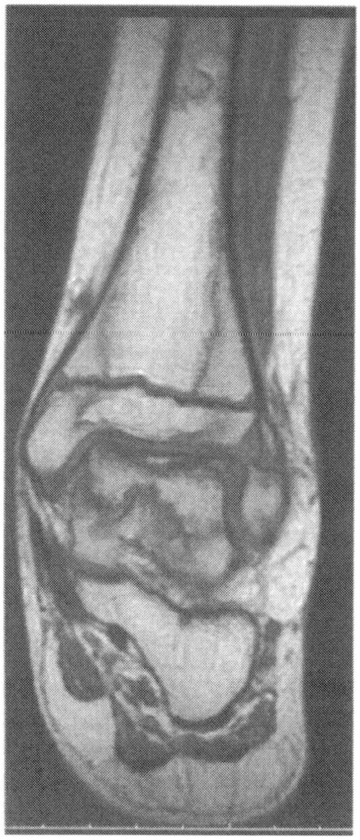

Figure 23J. Coronal T1 of the left ankle.

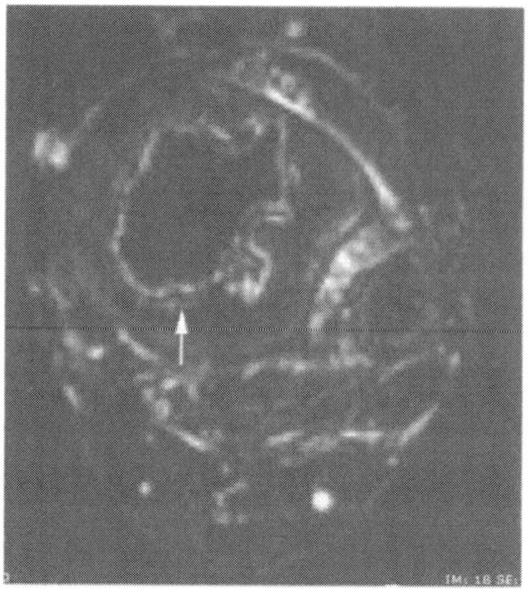

Figure 23K. Axial T2 FS through the tibia metaphysis.

Findings

This is an 11-year-old boy who had a bone marrow transplant 1.5 years earlier for acute myeloid leukemia.

Figures 23J, 23K. There is osteonecrosis of the tibial metadiaphysis, epiphysis, and talar dome. A double-line sign is evident along the margin of the process **(arrow)**.

Chronic Osteonecrosis, Sickle Cell Disease

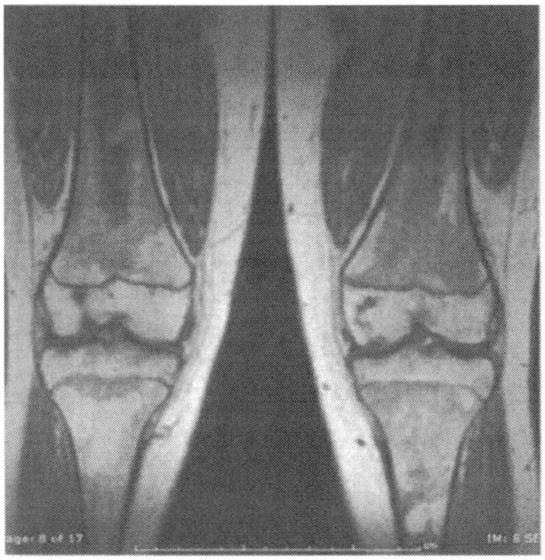

Figure 23L. Coronal T1.

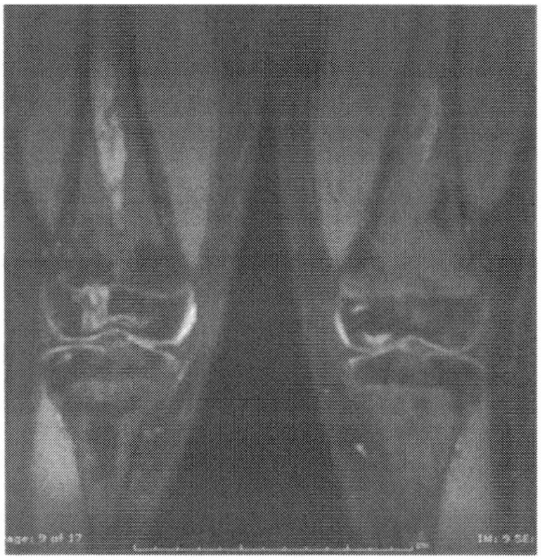

Figure 23M. Coronal PD FS.

Findings

This is a 15-year-old girl with bilateral knee pain and known sickle cell disease.

Figures 23L, 23M. Well-defined serpiginous zones of abnormal SI in the condyles and distal diaphyses of both femurs are present compatible with osteonecrosis. There is significant residual red marrow in the distal femur and proximal tibial metadiaphyses as well as the proximal tibia epiphyses, compatible with chronic anemia related to sickle cell disease.

Pitfalls and Pearls

1. MRI features characteristic of osteonecrosis include the double-line sign and sharply defined serpiginous margins.
2. Osteonecrosis may simulate the MRI features of infection and tumor. Include it in the differential for focal marrow processes without cortical destruction.

References

1. Saini A, Saifuddin A. MRI of osteonecrosis. *Clin Radiol* 2004; 59:1079–1093.
2. Imhof H, Breitenseher M, Trattnig S, et al. Imaging of avascular necrosis of bone. *Eur Radiol* 1997; 7:180–186.
3. Frush DP, Heyneman LE, Ware RE, Bissett GS, 3rd. MR features of soft-tissue abnormalities due to acute marrow infarction in five children with sickle cell disease. *AJR Am J Roentgenol* 1999; 173:989–993.
4. Umans H, Haramati N, Flusser G. The diagnostic role of gadolinium enhanced MRI in distinguishing between acute medullary bone infarct and osteomyelitis. *Magn Reson Imaging* 2000; 18:255–262.
5. Mitchell DG, Rao VM, Dalinka MK, et al. Femoral head avascular necrosis: Correlation of MR imaging, radiographic staging, radionuclide imaging, and clinical findings. *Radiology* 1987; 162:709–715.

Case 24

History

This is a 13-year-old boy with vague right knee pain for the last 6 months.

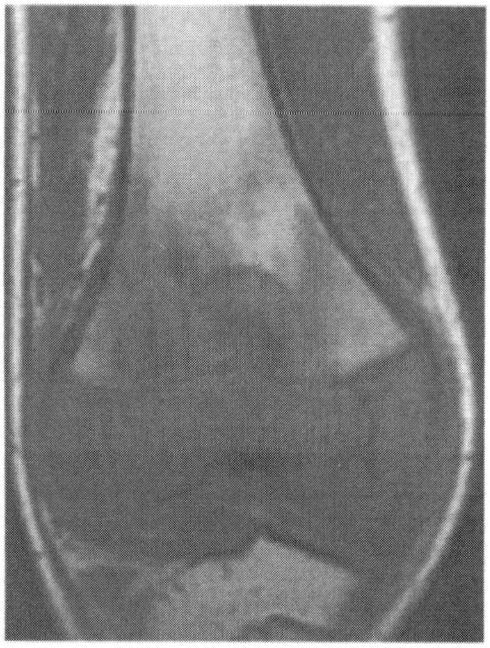

Figure 24A. Coronal T1 of the right knee.

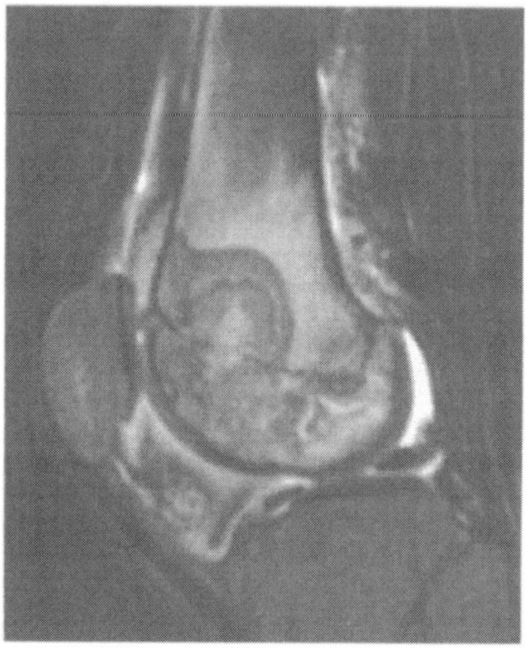

Figure 24B. Sagittal T2 FS.

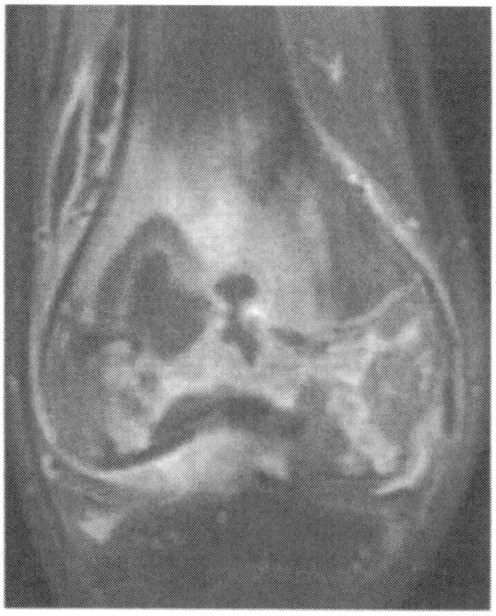

Figure 24C. Coronal T1 post-Gd FS.

Figures 24A, 24B, 24C, (24C with annotations). There is complex signal abnormality involving both the epiphysis and metaphysis, spreading across the physis of the distal femur **(arrow)**. The lesion has sharp geographic margins and is heterogeneously isointense on T1 and isointense to mildly hyperintense on T2. The process shows a multilayered enhancement pattern, with non-enhancing centers (*). There is associated marrow edema with enhancement. Significant synovial enhancement is also evident **(arrowhead)**.

Figure 24D. There is a mixed osteolytic/sclerotic lesion involving the distal femoral epiphysis.

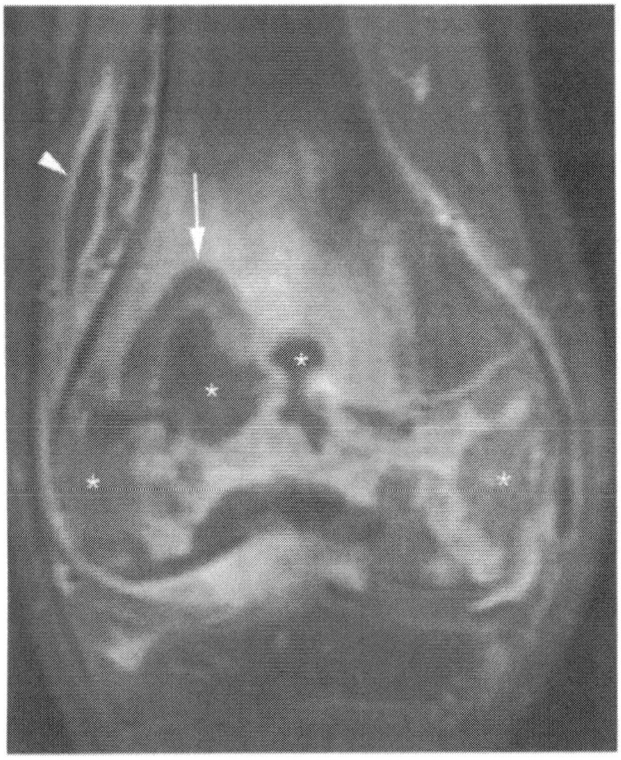

Figure 24C* Annotated.

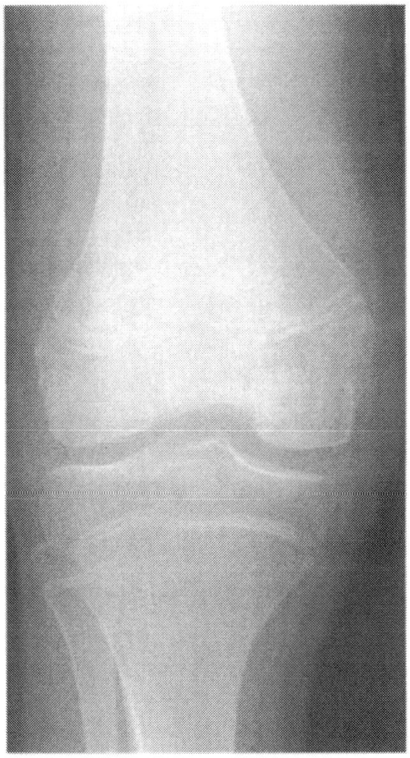

Figure 24D. AP radiograph of the right knee.

Diagnosis

Primary bone lymphoma

Question

1. What are the primary small round blue cell tumors of bone?

Discussion

The diagnostic criteria for primary bone lymphoma require a pathologic diagnosis of lymphoma without evidence of nodal or distant disease within 6 months of presentation (1). Primary non-Hodgkin's lymphoma (NHL) of bone may occur at any age, although it is rare in children under 10 years and most commonly occurs in patients between 50 and 60 years of age (1). In a series of 31 pediatric patients with primary NHL of bone, 68% were histologically classified as large-cell lymphoma followed by lymphoblastic lymphoma and small, noncleaved cell lymphoma (Burkitt's) (2). In another series, the incidence of primary bone involvement in NHL at initial presentation in children was 6.8% (3). In this same series, the lower extremities were most often involved and tumor usually involved the metadiaphysis. Epiphyseal involvement, as noted in this patient, was reported in 39% of cases (3). Acute lymphoblastic leukemia (Figures 24E, 24F) should be distinguished from disseminated lymphoblastic lymphoma when more than 25% of the bone marrow is replaced by tumor cells (4).

Plain radiography of primary bone lymphoma may be normal, demonstrate permeative or moth-eaten osteolysis, or show mixed osteolytic/blastic destruction (Figure 24D). In Mulligan et al.'s series, 70% (166/237) showed pure lytic bone destruction (5). Aggressive periosteal reaction, cortical destruction, juxtacortical soft tissue mass, sequestra, pathologic fracture, and transphyseal extension may also be seen.

On MRI, primary lymphoma of bone appears as a relatively discrete lesion with marrow replacement. Adjacent soft tissue mass with or without cortical disruption may also be seen. In Rosenthal et al.'s series, 23% of primary bone NHL in children showed extraosseous soft tissue abnormality and 27% demonstrated periosteal reaction by MRI (Figures 24G–24J) (3). There is a wide spectrum of MRI signal alterations of primary bone lymphoma. Tumors have been described as hypointense to hyperintense on T1, hypointense to hyperintense on T2, and demonstrate heterogeneous tumoral enhancement (6–8). Hermann et al. observed that hypointensity on T1 and T2W sequences may be related pathologically to extensive fibrosis in primary bone lymphoma (8). In their series, this was not a feature of other primary round cell tumors of bone. However, White et al. observed no correlation between T2 hypointensity and degree of fibrosis (6).

Primary lymphoma of bone may mimic benign and other malignant lesions on both plain radiography and MRI. The appearance may simulate osteonecrosis before and after therapy (7). With cortical destruction, these lesions may simulate other primary small round blue cell tumors of bone such as Ewing's sarcoma, small cell osteosarcoma, and primitive neuroectodermal tumor (PNET) (Answer to Question 1) (9). When primary lymphoma extends to or originates in the epiphysis, it may mimic osteomyelitis or chondroblastoma. The rim enhancing transphyseal foci evident in this case suggest osteomyelitis with intraosseous fluid collections.

Other oncologic manifestations of NHL that may be seen in the musculoskeletal system include extranodal involvement of the subcutaneous soft tissues (Figure 24K–24L) and disseminated NHL. Primary bone lymphoma should be distinguished from disseminated NHL with secondary bone deposits since this may affect therapy.

In this patient, the differential diagnosis included a primary bone tumor and therefore open biopsy was performed. The lesion was pathologically proven primary large B cell lymphoma of bone. The patient was subsequently referred to a pediatric oncologist close to his home.

Orthopedic Perspective

The treatment of NHL of bone is usually nonsurgical, so the main concern of the orthopedist is in making the diagnosis. The differential is extensive as noted above, and NHL can mimic many other lesions. Biopsy can be difficult because tumor necrosis is often present or cells demonstrate "squeeze artifact" on histologic sections from bone, making the cytology difficult to interpret. Sufficient tissue must be available for the pathologist to process the tissue for immunohistochemical stains to differentiate NHL from the other small round blue cell tumors. It is reasonable to start with a needle biopsy, especially if there is a soft tissue mass, but an open biopsy may be necessary to get sufficient tissue. In childhood, bone lymphoma is treated by chemotherapy alone in most cases. The orthopedist may become involved later in the treatment if osteonecrosis (presumably from corticosteriod administration) becomes an issue. Assessment of infarcts near weightbearing joints is important because subchondral collapse can lead to joint destruction and arthritis.

Patients with leukemia may present with bone involvement as the first manifestation. They may also present with a joint effusion or pain around a joint suggestive of infection. A high index of suspicion is necessary to avoid missing the diagnosis of leukemia if the first presentation is a bone lesion. The blood smear may be negative and a biopsy is needed.

What the Clinician Needs to Know

1. The clinician needs to establish the diagnosis since bone lymphoma can mimic tumors and non-neoplastic processes.
2. A needle biopsy is often sufficient to diagnose lymphoma, but open biopsy may be necessary if more tissue is needed. A culture should always be performed since infection is in the differential diagnosis.
3. MRI of joints affected by bone infarcts from lymphoma therapy is useful to assess potential articular damage.

Answer

1. Small cell osteosarcoma, Ewing's sarcoma, PNET, and primary lymphoma of bone.

Additional Examples

Acute Lymphoblastic Leukemia

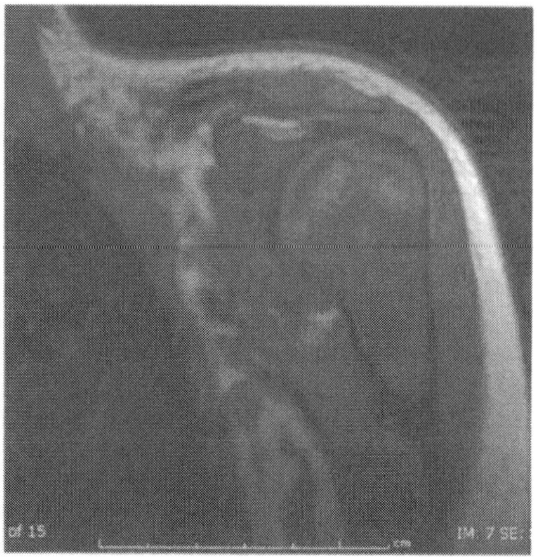

Figure 24E. Coronal oblique T1 of the left shoulder.

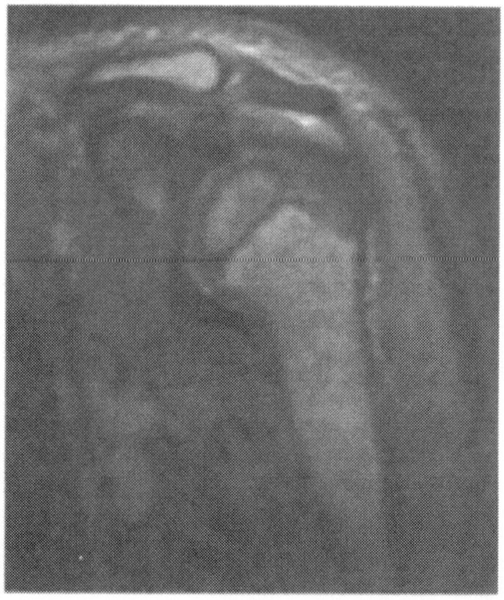

Figure 24F. Coronal oblique STIR.

Findings

This is a 4-year-old girl who presented with left shoulder pain and decreased range of motion.

Figures 24E, 24F. There is diffuse marrow replacement, including the epiphysis of the proximal humerus. Laboratory work-up demonstrated severe neutropenia. The diagnosis of acute lymphoblastic leukemia was made shortly after this examination.

Multifocal Primary Lymphoma of Bone

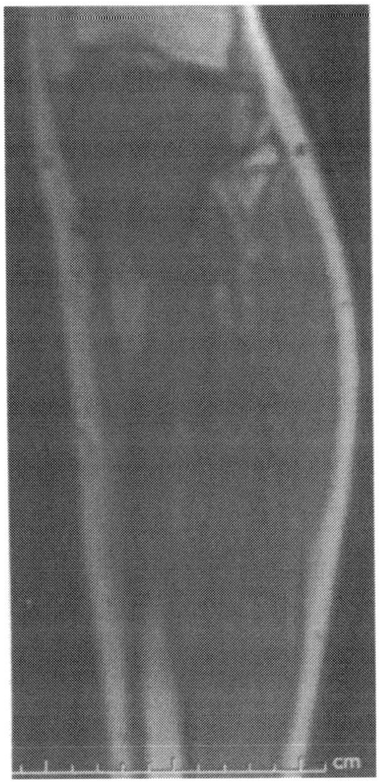

Figure 24G. Coronal T1 of the right tibia.

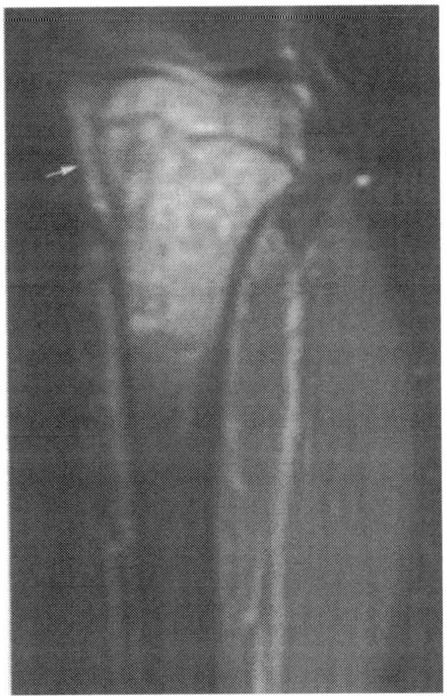

Figure 24H. Coronal STIR.

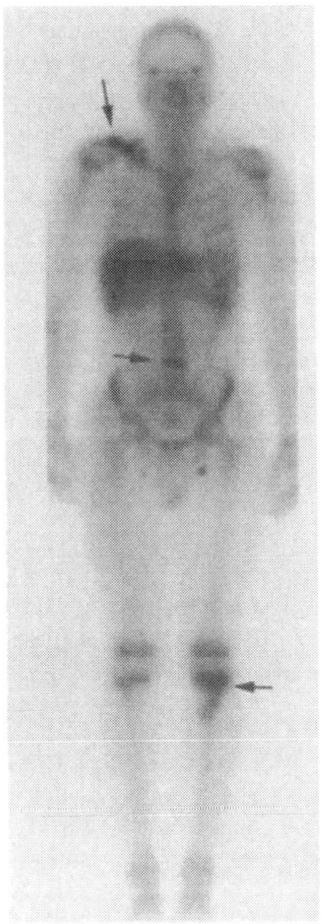

Figure 24J. Gallium-67 (Ga-67) whole body scan.

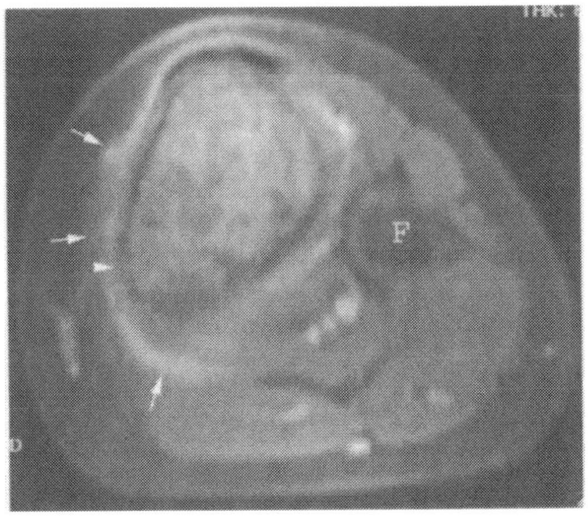

Figure 24I. Axial T1 post-Gd FS.

Findings

This is a 16-year-old boy who presented with a 1-month history of right knee pain and a palpable mass with no history of trauma.

Figures 24G, 24H, 24I. There is a discrete proximal tibial mass that is centered in the metaphysis and crosses the physis to involve the epiphysis. There is also extraosseous soft tissue extension and/or inflammation **(arrows)**. The cortex of the proximal tibia is transgressed **(arrowhead)**. Fibula (F).

Figure 24J. There is focal Ga-67 uptake within the left proximal tibia as well as additional areas of increased uptake present in the L4 vertebral body and the right clavicle **(arrows)**. The patient underwent open biopsy of the tibial lesion, and this was pathologically proven to be multifocal large B-cell lymphoma. No nodal or soft tissue masses were present; therefore the diagnosis was made of primary bone lymphoma with multifocal involvement. He was treated with CHOP (cyclophosphamide, doxorubicin, vincristine, and prednisone) therapy.

Subcutaneous Lymphoblastic Lymphoma

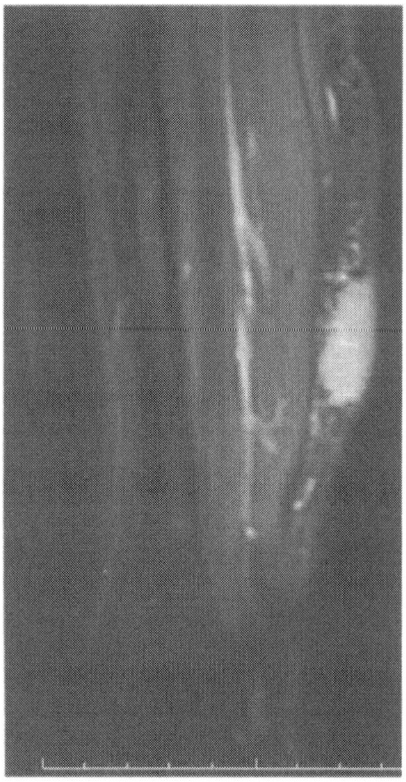

Figure 24K. Sagittal STIR.

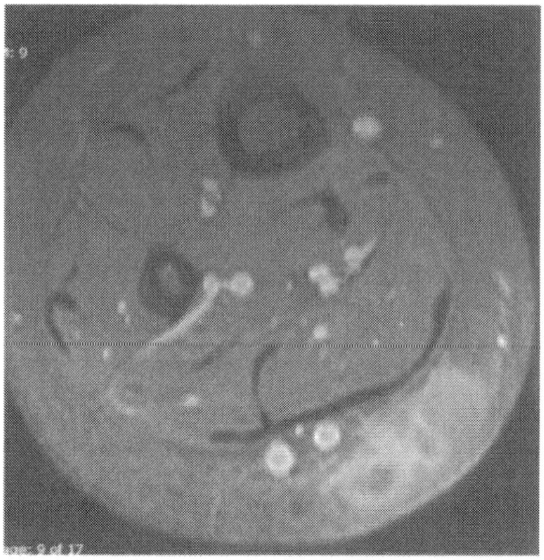

Figure 24L. Axial T1 post-Gd FS.

Findings

This is a 7-year-old girl who noticed a painless enlarging posterior calf mass over the course of several months.

Figures 24K, 24L. There is an infiltrative subcutaneous solid mass that is hyperintense on STIR and demonstrates heterogeneous peripheral and central enhancement. This was pathologically proven to be lymphoblastic lymphoma.

Pitfalls and Pearls

1. Lymphoma of bone has a variable appearance and may mimic benign and other malignant tumors ranging from osteomyelitis to Ewing's sarcoma. Keep it in mind for medullary lesions with nonspecific MRI features.
2. Make sure to include lymphoma in the differential diagnosis of atypical lytic epiphyseal lesions in childhood.

References

1. Krishnan A, Shirkhoda A, Tehranzadeh J, Armin AR, Irwin R, Les K. Primary bone lymphoma: Radiographic-MR imaging correlation. *Radiographics* 2003; 23:1371–1383; discussion 1384–1377.
2. Suryanarayan K, Shuster JJ, Donaldson SS, Hutchison RE, Murphy SB, Link MP. Treatment of localized primary non-Hodgkin's lymphoma of bone in children: A Pediatric Oncology Group study. *J Clin Oncol* 1999; 17:456–459.
3. Rosenthal H, Kolb R, Gratz KF, Reiter A, Galanski M. Bone manifestations in non-Hodgkin's lymphoma in childhood and adolescence. *Radiologe* 2000; 40:737–744.
4. Sandlund JT, Downing JR, Crist WM. Non-Hodgkin's lymphoma in childhood. *N Engl J Med* 1996; 334:1238–1248.
5. Mulligan ME, McRae GA, Murphey MD. Imaging features of primary lymphoma of bone. *AJR Am J Roentgenol* 1999; 173:1691–1697.
6. White LM, Schweitzer ME, Khalili K, Howarth DJ, Wunder JS, Bell RS. MR imaging of primary lymphoma of bone: Variability of T2-weighted signal intensity. *AJR Am J Roentgenol* 1998; 170:1243–1247.
7. Mengiardi B, Honegger H, Hodler J, Exner UG, Csherhati MD, Bruhlmann W. Primary lymphoma of bone: MRI and CT characteristics during and after successful treatment. *AJR Am J Roentgenol* 2005; 184:185–192.
8. Hermann G, Klein MJ, Abdelwahab IF, Kenan S. MRI appearance of primary non-Hodgkin's lymphoma of bone. *Skeletal Radiol* 1997; 26:629–632.
9. Saifuddin A, Whelan J, Pringle JA, Cannon SR. Malignant round cell tumours of bone: Atypical clinical and imaging features. *Skeletal Radiol* 2000; 29:646–651.

Case 25

History

This is an 8-year-old boy with right knee pain for 2 years.

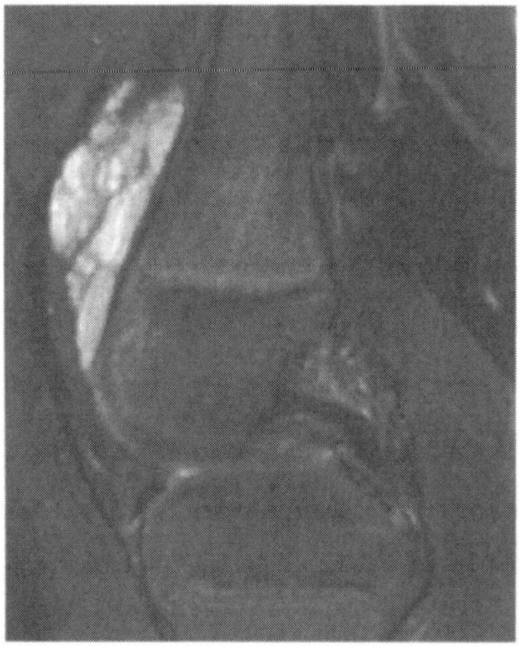

Figure 25A. Sagittal T2 FS through the medial aspect of the right knee.

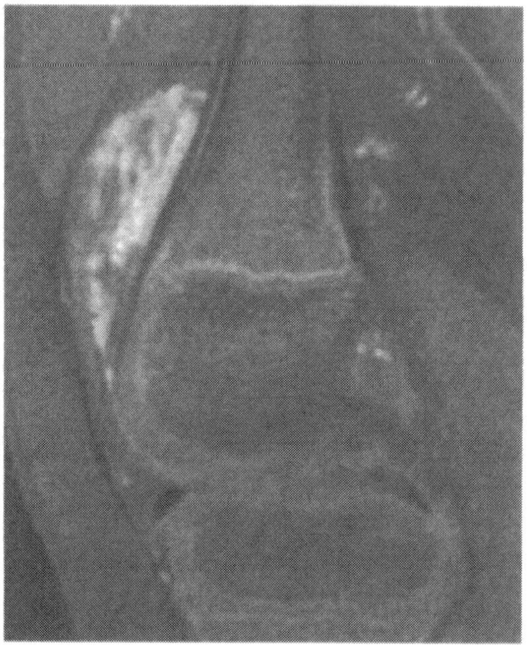

Figure 25B. Sagittal T1 post-Gd FS.

Figures 25A, (25A with annotations). Within the suprapatellar bursa, there is a lobulated, confluent tubular mass that is moderately T2 hyperintense. Fluid-fluid levels are present **(arrowheads)**.

Figure 25B. Some of the tubular structures enhance, while others do not.

Figure 25C. Fluid-fluid levels are again seen **(arrowhead)**, and a small hypointense phlebolith is also identified **(arrow)**.

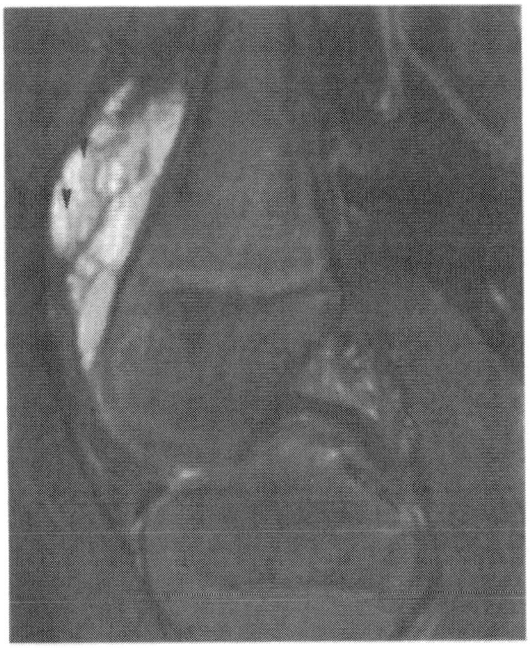

Figure 25A* Annotated.

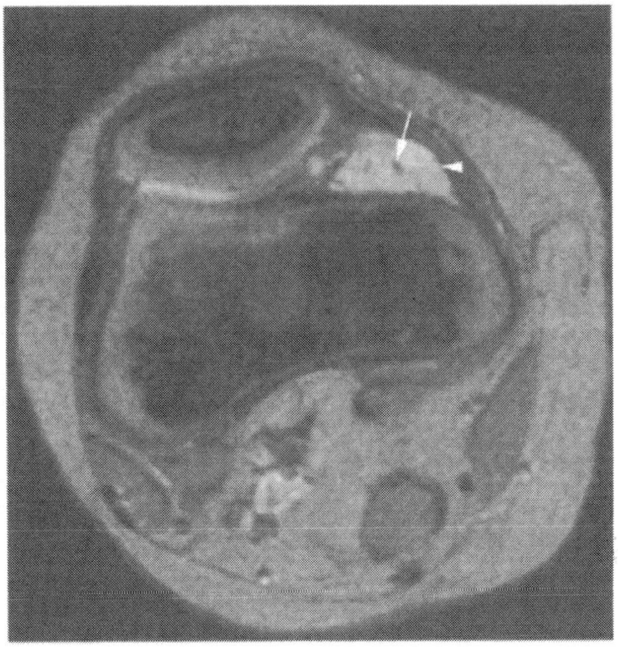

Figure 25C. Axial GRE.

Diagnosis

Synovial venous malformation

Question

1. T/F: Untreated, synovial vascular malformations may cause early degenerative changes that may be indistinguishable from hemophilic arthropathy.

Discussion

The most common location for a synovial venous malformation (AKA synovial hemangioma) is the knee (1, 2). Most cases present during childhood and adolescence, and there is often an extra-articular superficial or deep soft tissue component. Synovial inflammation related to recurrent hemarthrosis may lead to synovial hyperplasia, early physeal closure, advanced epiphyseal maturation, epiphyseal overgrowth, erosions, and degenerative changes (3, 4). These changes may resemble hemophilic arthropathy (Answer to Question 1) or an inflammatory arthritis such as juvenile rheumatoid arthritis. Chronic synovial inflammation may have a trophic effect on the epiphysis, leading to epiphyseal overgrowth or ballooning.

The MRI features of synovial venous malformations are similar to venous malformations elsewhere (see Case 67). Synovial venous malformations are isointense on T1 and hyperintense on T2 with respect to muscle. A serpiginous architecture with lobular margins is characteristic. They demonstrate patchy tubular enhancement related to slow flow and/or thrombus within dilated venous structures. Fluid-fluid levels are also features of venous malformations and are best seen on T2W sequences (5). A solid soft tissue component and tumoral enhancement are absent. Gradient echo (GRE) images may demonstrate both phleboliths and intra-articular hemosiderin staining. Phleboliths are more often observed when synovial venous malformations show extra-articular extension (Figures 25D–25G) (1). Since venous malformations have a tendency to bleed, there may be localized consumptive coagulopathy with low platelet and clotting factors (2).

The presence of blood products may help narrow the differential diagnosis to pigmented villonodular synovitis (PVNS) and hemophilic arthropathy, but it is the identification of abnormal vascular structures within and around the joint that permits a specific diagnosis of a venous malformation. Intra-articular blood products are less commonly observed with pyogenic and nonpyogenic arthropathies. GRE sequences are most sensitive for detecting intra-articular blood breakdown products (Figure 25H). Synovial venous malformations should not be confused with juxta-articular neoplasms such as a synovial cell sarcoma (see Case 6).

Because of the high risk of hemarthrosis and early degeneration, this patient's venous malformation was resected. Prior to resection, sclerotherapy was performed to decrease the size of the lesion.

What the Clinician Needs to Know

1. Accurately diagnose an intra-articular venous malformation and distinguish it from other tumor-like intra-articular processes such as PVNS, post-traumatic hemarthrosis, and septic arthritis.

2. Describe the extent of intra- and extra-articular components.
3. Is there secondary articular destruction and synovial hypertrophy related to recurrent hemarthrosis?

Answer

1. True.

Additional Examples

Diffuse Venous Malformation with Synovial Component

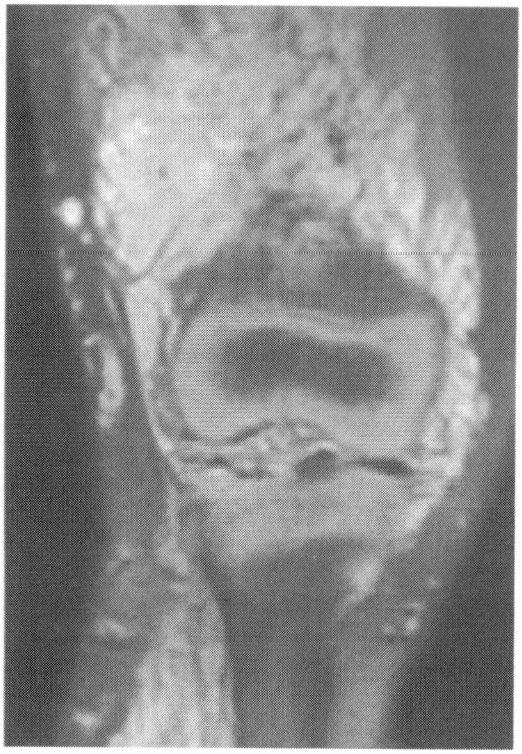

Figure 25D. Coronal PD FS of the right knee.

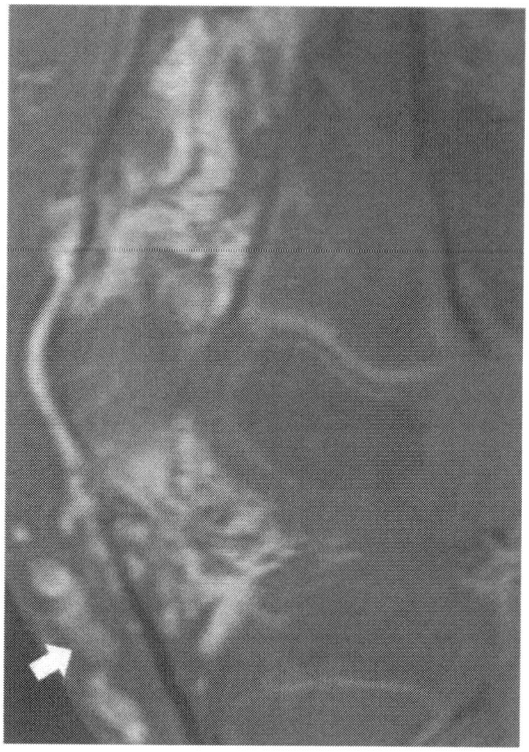

Figure 25E. Sagittal T1 post-Gd FS.

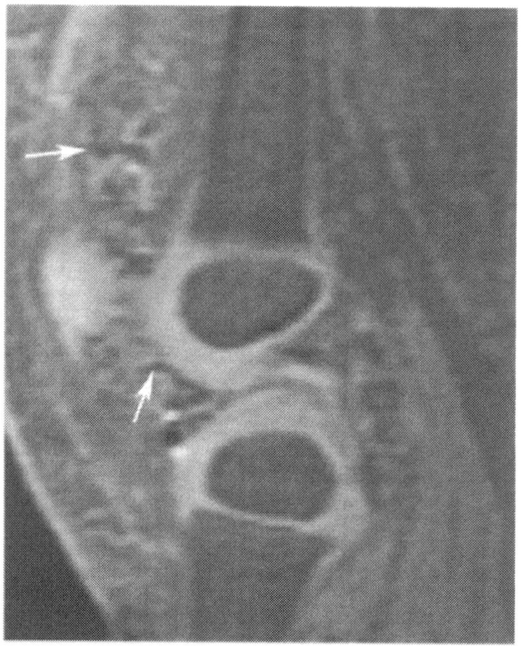

Figure 25F. Sagittal 3D SPGR FS.

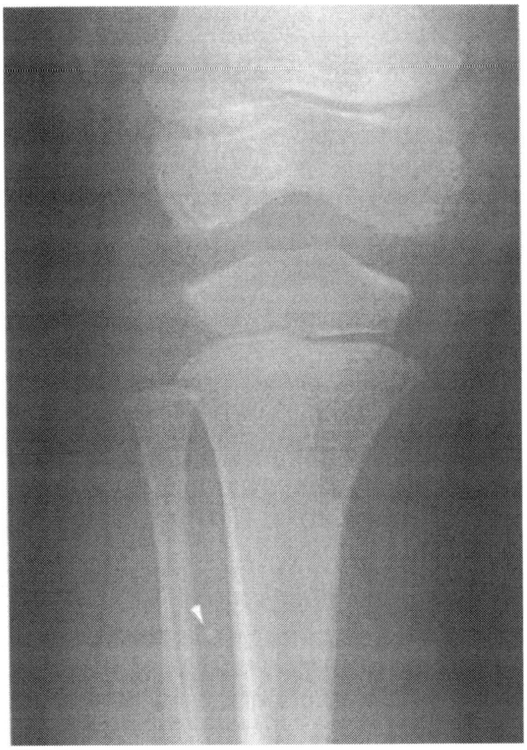

Figure 25G. AP radiograph.

Findings

This is a 2½-year-old girl who had progressive bluish discoloration of her lower leg beginning at 8 months of age.

Figures 25D, 25E. There are multiple confluent tubular structures identified within the suprapatellar bursa, Hoffa's fat pad, and muscles. Note a large superficial varicose vein anteriorly **(thick arrow)**.

Figure 25F. There are multiple tubular areas of susceptibility artifact **(arrows)**, compatible with thrombi.

Figure 25G. An extra-articular phlebolith is present **(arrowhead)**. Multiple growth recovery lines are incidentally noted in the distal femur and proximal tibial metaphysis.

Synovial Venous Malformation with Blood Products

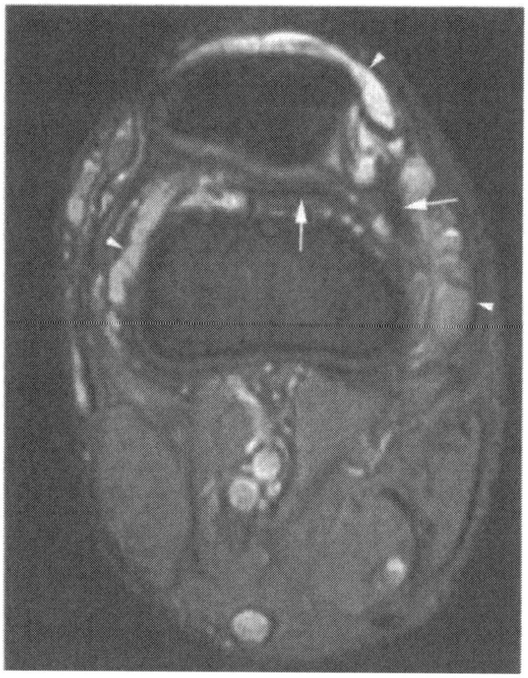

Figure 25H. Axial FFE T2.

Findings

This is a 16-year-old boy who was referred to a rheumatologist because of chronic right knee pain.

Figure 25H. There is extensive intra- and extra-articular venous malformation present **(arrowheads)**. Intra-articular susceptibility artifact is present **(thin arrows)** related to chronic hemarthrosis.

Pitfalls and Pearls

The long term radiographic and MRI features of synovial venous malformations resemble an inflammatory arthritis and are related to recurrent hemarthrosis.

References

1. Greenspan A, Azouz EM, Matthews J, 2nd, Decarie JC. Synovial hemangioma: Imaging features in eight histologically proven cases, review of the literature, and differential diagnosis. *Skeletal Radiol* 1995; 24:583–590.
2. Enjolras O, Ciabrini D, Mazoyer E, Laurian C, Herbreteau D. Extensive pure venous malformations in the upper or lower limb: A review of 27 cases. *J Am Acad Dermatol* 1997; 36:219–225.
3. Ramseier LE, Exner GU. Arthropathy of the knee joint caused by synovial hemangioma. *J Pediatr Orthop* 2004; 24:83–86.
4. Narvaez JA, Narvaez J, Aguilera C, De Lama E, Portabella F. MR imaging of synovial tumors and tumor-like lesions. *Eur Radiol* 2001; 11:2549–2560.
5. Burrows PE, Laor T, Paltiel H, Robertson RL. Diagnostic imaging in the evaluation of vascular birthmarks. *Dermatol Clin* 1998; 16:455–488.

Case 26

History

This is a 7-year-old boy with no history of trauma. He presents for imaging because his left leg is shorter than his right.

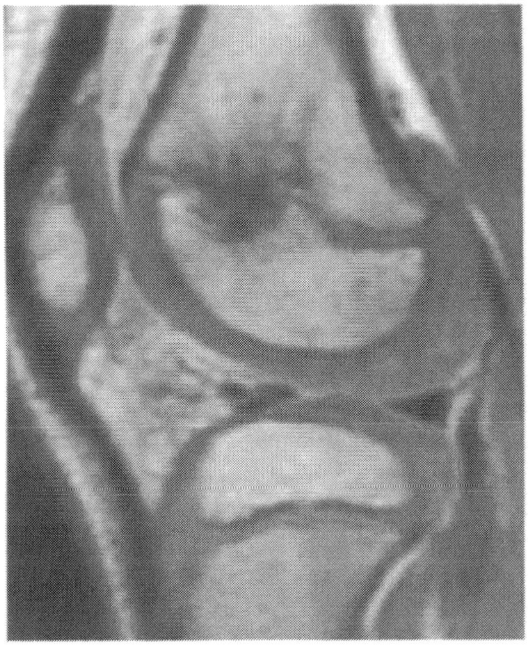

Figure 26A. Sagittal PD through the lateral aspect of the left knee.

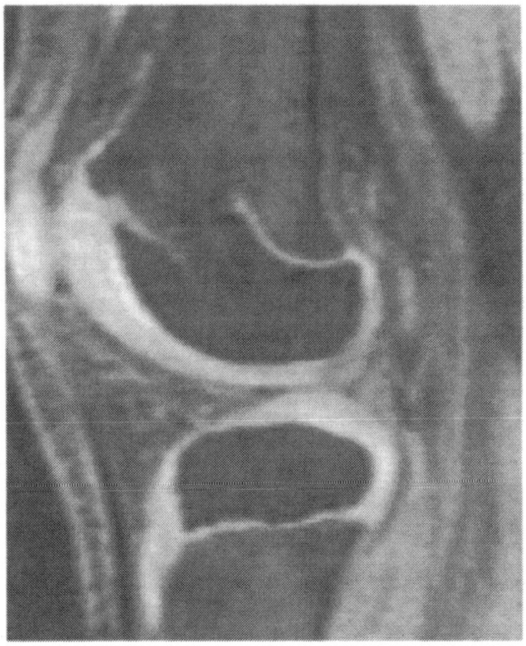

Figure 26B. Sagittal 3D SPGR FS.

Figures 26A, 26B. A sharply demarcated physeal bridge is identified in the lateral aspect of the distal femoral physis. On the PD image, it is hypointense to marrow and it is isointense to marrow on the SPGR sequence.
Figure 26C. A sclerotic, physeal bony bridge is identified.

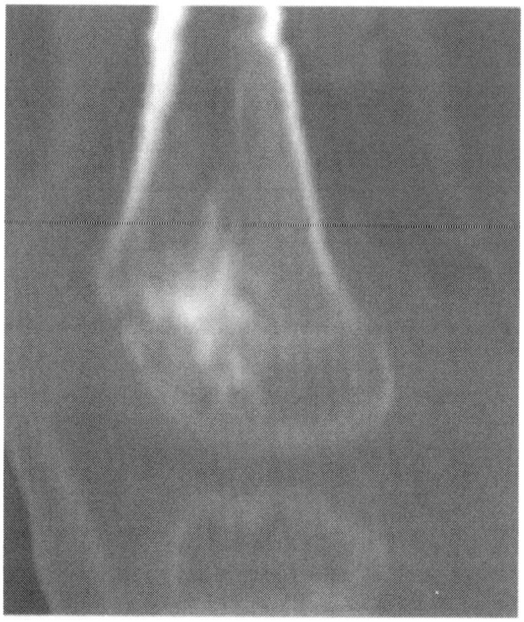

Figure 26C. Sagittal reformatted CT.

Diagnosis

Bony physeal bar

Questions

1. What is the most common cause of a physeal bar?
2. Are physeal bars more likely to occur with epiphyseal or metaphyseal arterial disruption?

Discussion

A wide array of conditions is associated with the development of physeal bars (Figures 26D–26H) (Table 26A). A traumatic etiology is most common (Answer to Question 1) (1). On occasion, as in this case, no explanation for a physeal bar is apparent. Post-traumatic physeal bars most commonly occur in the lower extremities and involve the distal physes of the femur and tibia (Figures 26E–26H) (2). Physeal bars involving the proximal portions of the tubular bones are more frequently due to nontraumatic etiologies.

Regardless of etiology, physeal bars tend to involve the central physis of the distal femur and the anteromedial physis of the distal tibia (2). The predilection for this latter location has been attributed to Kump's bump, a normal focal undulation of the anteromedial physis, which is the earliest site of physiologic distal tibial physeal fusion (3).

Experimental studies suggest that the development of physeal bars may be due to ischemia to the germinal matrix of the physis. The vascular supply to the germinal matrix is epiphyseal, whereas the supply to the zone of provisional calcification is metaphyseal. Jaramillo et al. showed that physeal bar formation occurred in rabbits when the epiphyseal arteries were ligated (Answer to Question 2) (4). When metaphyseal arteries were ligated, apparent physeal widening occurred due to stunted endochondral bone formation, but physeal bars did not develop. In addition, focally disrupted endochondral bone formation may occur after metaphyseal ischemia, resulting in focal cartilaginous tongues within the metaphysis rather than diffuse physeal widening. Physeal bar formation may occur after indirect or direct physeal injury (3). Indirect injury includes any mechanism that disrupts the germinal matrix or its epiphyseal arterial supply. Direct injury includes physeal compression fracture or a vertical transphyseal fracture. This promotes transphyseal vascularity with the potential for physeal bar formation.

Table 26A. Conditions associated with physeal bar formation.

Trauma (most common)
Infection
Ischemia
Radiation
Corticosteroids
Hypervitaminosis A
Blount's disease
Legg-Calve-Perthes disease
Developmental dysplasia of the hip
Madelung deformity
Bone dysplasias

MRI evaluation of physeal bars is superior to radiography because of its ability to delineate the physis and more accurately determine the size of a physeal bar. CT may also be used to delineate physeal bars (5). Factors favoring MRI rather than CT in this context include the possibility that a fibrous bar may be missed, physeal cartilage is not well delineated, and CT entails ionizing radiation. The MRI appearance of physeal bars depends on the size and composition. Physeal bars may be fibrous, osseous, or contain granulation tissue. On 3D SPGR FS sequences, all tend to appear similar in SI to adjacent marrow. Bony bars containing fatty marrow are isointense with respect to adjacent marrow (Figure 26I), whereas fibrous bars or dense bony bars may be hypointense on all imaging sequences (6). Smaller bars tend to be hypointense compared with larger bars (5). CT may be used to distinguish between a fibrous and dense bony bar (Figures 26E–26H). On CT, transphyseal bone elements are present with a dense bony bar and absent with a fibrous bar. In the acute or subacute setting, transphyseal granulation tissue may be present related to healing, which may appear hypointense on T1 and hyperintense on fluid-sensitive sequences.

This patient had no known history of trauma, acute or remote infection, or metabolic bone condition. He was referred for imaging because of a leg length discrepancy and a left genu valgus deformity. He underwent physeal bar resection and did well.

Orthopedic Perspective

The treatment for physeal arrest depends on the amount of growth remaining and the extent of the physeal bar. In younger patients with substantial growth remaining and physeal bars <50% of the area of the physis, excision of the physeal bar may be considered. In patients with less growth remaining (<2 years) or large physeal bars, ipsilateral epiphyseodesis, contralateral epiphyseodesis, or ipsilateral leg lengthening are performed depending on the severity of the leg-length discrepancy and/or angular deformity. In general, a leg-length discrepancy of 1 to 2 cm is well tolerated by patients as well as minor joint angulation deformities. MRI data are used in conjunction with the bone age and scanogram to determine the correct time and surgical approach for physeal bar treatment. Epiphyseodesis is a less invasive procedure with a shorter convalescence (2 weeks) compared with physeal bar resection (6 to 8 weeks).

What the Clinician Needs to Know

1. The size (percentage cross-sectional area) and location of the physeal arrest.
2. Are there angulation deformities present secondary to physeal arrest?

Answers

1. Trauma.
2. Physeal bars are more likely to occur after epiphyseal arterial disruption.

Additional Examples

Meningococcemia with Physeal Bar

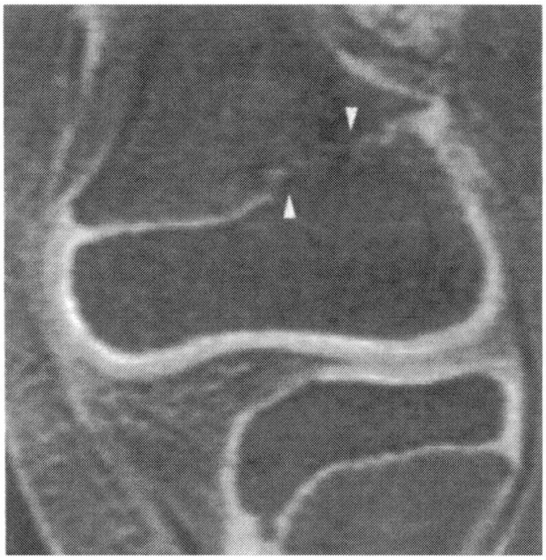

Figure 26D. Coronal 3D SPGR FS of the right knee.

Findings

This 9-year-old boy with a history of neonatal meningococcemia had a varus deformity of the distal femur evident on plain radiographs (not shown).

Figure 26D. MRI reveals a varus tilt of the physis and a medial physeal bar **(arrow heads)**. This medial location is somewhat unusual since distal femoral physeal bars tend to develop centrally, regardless of etiology (2).

Distal Tibia Physeal Bar

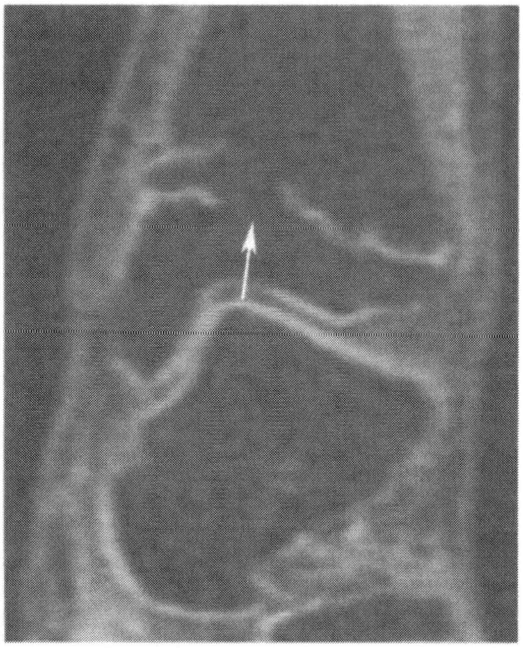

Figure 26E. Coronal 3D SPGR FS of the left ankle.

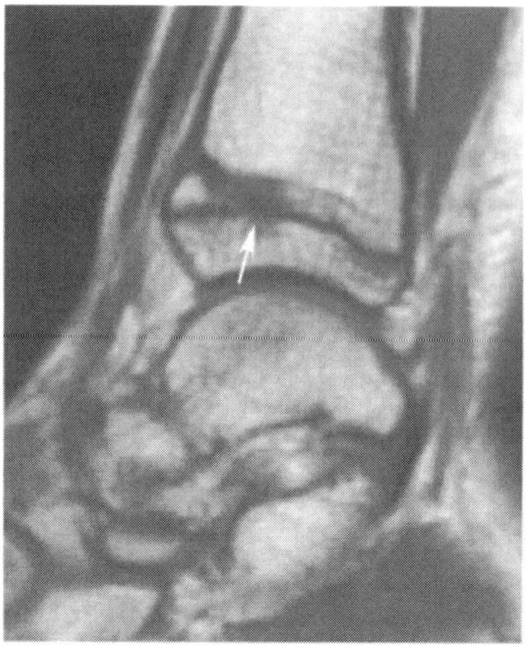

Figure 26F. Sagittal T1 of the medial left ankle.

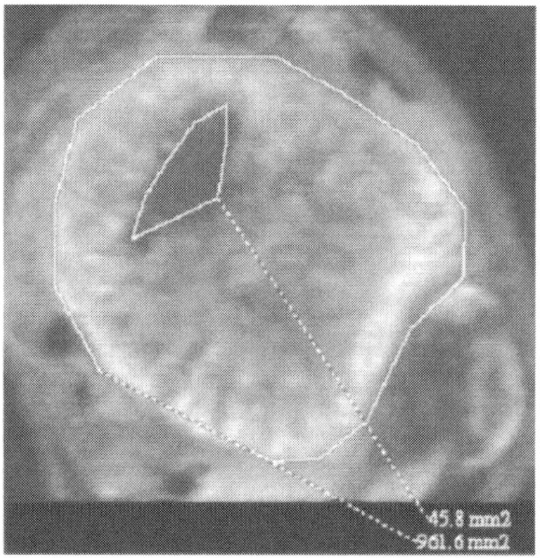

Figure 26G. Axial 3D SPGR FS physeal map (physeal bar quantification).

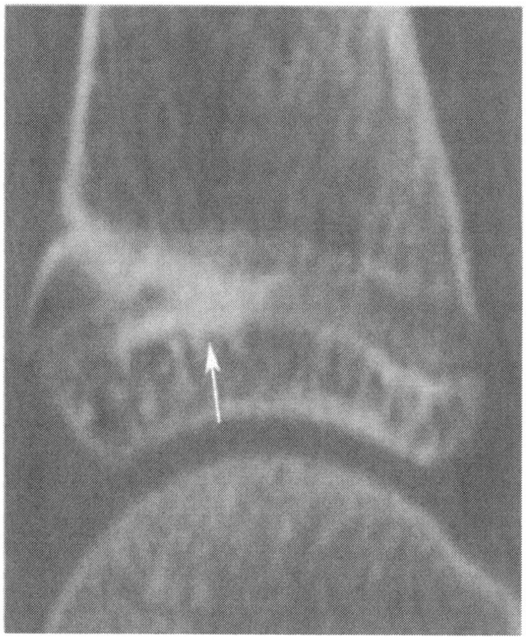

Figure 26H. Sagittal reformatted CT of the medial left ankle.

Findings

This 11-year-old girl had a Salter-Harris (SH) type 4 fracture of her distal tibia several months earlier and now has a varus ankle deformity.

Figures 26E, 26F, 26G, 26H. An anteromedial bony physeal bar is present related to her prior SH fracture **(arrows)**. Based on the 3D SPGR map, the bar occupies only 5% of the growth plate and is suitable for surgical resection. This patient had a good outcome after physeal bar resection with fat graft interposition.

Legg-Calve-Perthes Disease with Bony Physeal Bar

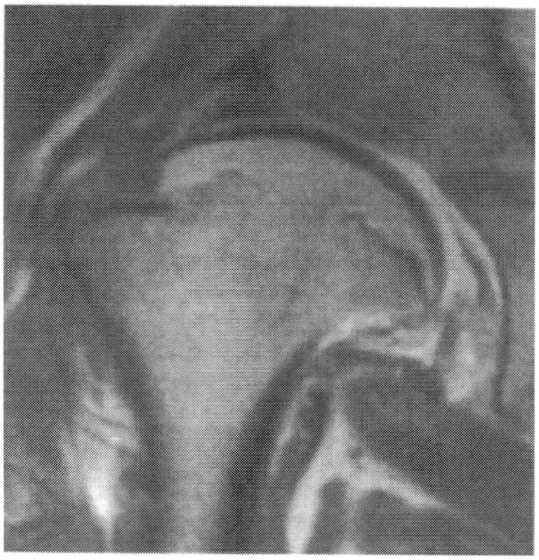

Figure 26I. Coronal PD of the right hip.

Findings

This 15-year-old boy had a history of Catteral grade 4 Legg-Calve-Perthes disease.
Figure 26I. Severe changes of lateral growth arrest with shortening of the femoral neck are evident. There is continuity of marrow fat signal across a large bony physeal bar.

Pitfalls and Pearls

1. Peripheral physeal bars may lead to angular deformities, whereas central physeal bars lead to generalized growth restriction.
2. Be sure to include T1W images when evaluating for bony bars to identify continuity of fatty marrow signal across the physis.

References

1. Borsa JJ, Peterson HA, Ehman RL. MR imaging of physeal bars. *Radiology* 1996; 199:683–687.
2. Ecklund K, Jaramillo D. Patterns of premature physeal arrest: MR imaging of 111 children. *AJR Am J Roentgenol* 2002; 178:967–972.
3. Ecklund K, Jaramillo D. Imaging of growth disturbance in children. *Radiol Clin North Am* 2001; 39:823–841.
4. Jaramillo D, Laor T, Zaleske DJ. Indirect trauma to the growth plate: Results of MR imaging after epiphyseal and metaphyseal injury in rabbits. *Radiology* 1993; 187:171–178.
5. Loder RT, Swinford AE, Kuhns LR. The use of helical computed tomographic scan to assess bony physeal bridges. *J Pediatr Orthop* 1997; 17:356–359.
6. Jaramillo D, Hoffer FA, Shapiro F, Rand F. MR imaging of fractures of the growth plate. *AJR Am J Roentgenol* 1990; 155:1261–1265.

Case 27

History

This 10-year-old boy had a comminuted supracondylar fracture of the humerus 2 months ago. At the time, he had closed reduction with percutaneous pinning. The pins were routinely removed one month after the reduction. There was persistent drainage from the cutaneous pin tracts and restricted elbow range of motion at his 2-month follow-up visit. An MRI was subsequently ordered.

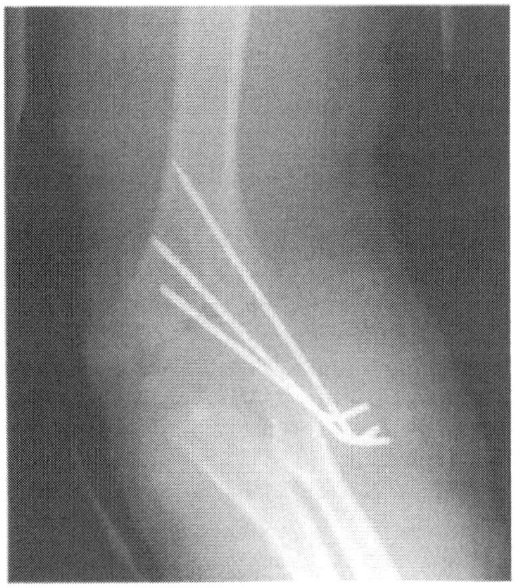

Figure 27A. AP radiograph of the left elbow 1 week after closed reduction.

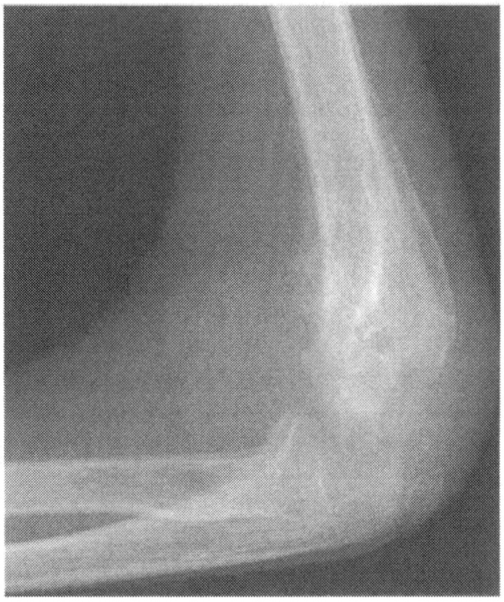

Figure 27B. Lateral radiograph of the left elbow just prior to MRI (2 months later).

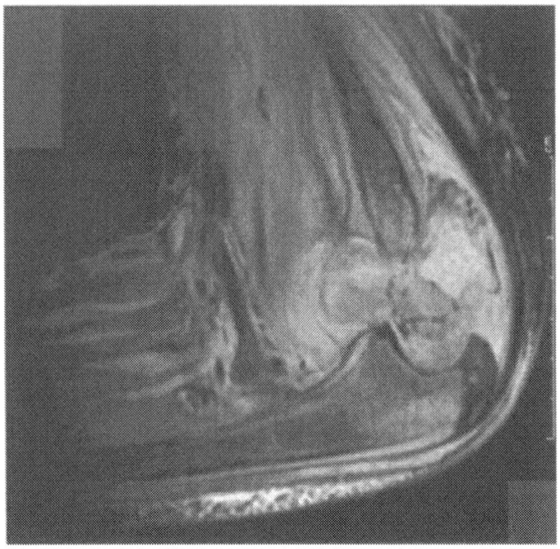

Figure 27C. Sagittal STIR.

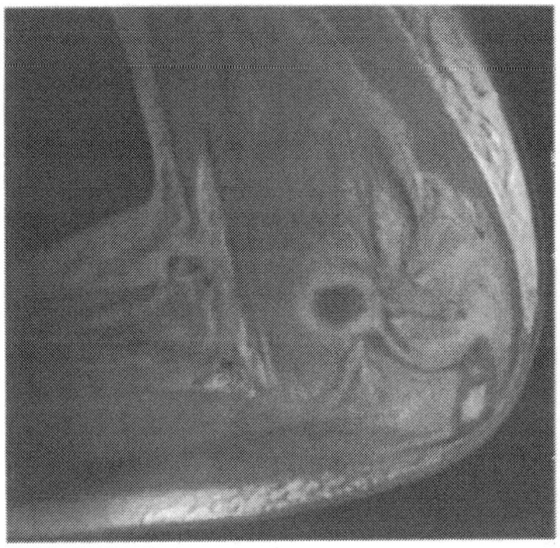

Figure 27D. Sagittal T1 post-Gd FS.

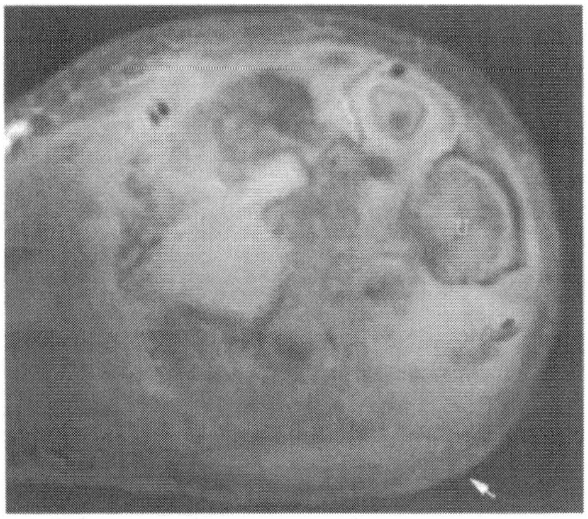

Figure 27E. Axial T1 post-Gd FS (through the proximal ulna region).

Figure 27A. Initial radiograph shows lateral-entry K-wire fixation of a supracondylar fracture of the humerus.

Figure 27B. There are expected postsurgical changes with periosteal reaction and a dense joint effusion after pin removal.

Figures 27C, 27D, (27C with annotations). There is a large joint effusion with thick synovial enhancement. There is also diffuse subcutaneous soft tissue, muscle, and marrow edema. Periosteal reaction is again seen **(arrowheads)**.

Figure 27E. There is synovial and soft tissue enhancement with extension to the subcutaneous soft tissues associated with skin thickening **(arrow)**. U = Ulna.

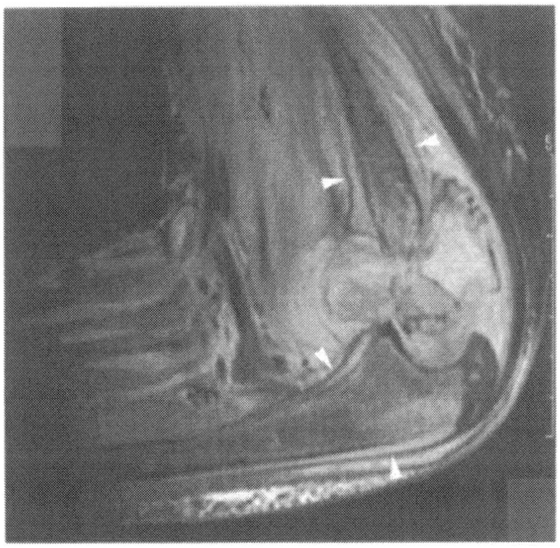

Figure 27C*. Annotated.

Diagnosis

Septic arthritis after fracture and internal fixation

Questions

1. What is the incidence of osteomyelitis complicating closed fractures with internal fixation?
2. T/F: The marrow enhancement pattern after hardware removal is helpful for distinguishing postoperative changes from osteomyelitis.

Discussion

The incidence of osteomyelitis with open fractures ranges from 2% to 16%, but with closed fractures with internal fixation, the frequency is much lower, ranging from 0.5 to 1.5% (Answer to Question 1) (1, 2). *Staphylococcus* is the most frequently isolated organism related to orthopedic hardware, accounting for approximately 50% of cases (3). MRI is occasionally performed to rule out osteomyelitis after removal of hardware placed for fracture fixation. However, distinguishing acute or chronic postoperative changes from infection in this setting is difficult, if not impossible by MRI (4). The acute and chronic soft tissue, periosteal, and marrow changes after hardware removal are nonspecific. Granulation tissue that forms during normal fracture healing and as a response to infection is often indistinguishable by MRI. Furthermore, juxtacortical soft tissue edema and hematomas/seromas related to recent trauma may also be difficult to distinguish from pyomyositis or a developing abscess. Therefore, clinical correlation is essential.

Early after hardware removal, the marrow may normally show diffuse edema and enhancement that can be mistaken for osteomyelitis. Additionally, altered bone mechanics and stresses may cause persistent marrow enhancement and increased SI on fluid-sensitive sequences. Marrow enhancement may persist up to a year after hardware removal (4). Therefore, contrast-enhanced imaging may not help distinguish osteomyelitis from normal postoperative marrow (Answer to Question 2). To make matters more confusing, surgically created cortical tracts (e.g., pin tracts) may mimic a cloaca (bone to soft tissue) or sinus tract (bone to skin), and osteonecrosis related to trauma may mimic a sequestrum or abscess.

Other complications after hardware removal include fracture nonunion and synostosis (5). Acute and late complications of supracondylar fractures include neurovascular injury, cubitus varus deformity, and loss of motion (6). Ulnar nerve injury may also occur with percutaneous medial-entry pin fixation.

Operative findings in this case showed no evidence of osteomyelitis. However, there was significant synovial hypertrophy with articular cartilage thinning and cultures from the joint grew *Pseudomonas*. The elbow joint was debrided and the patient was placed on IV tobramycin and ceftazidime. The patient was discharged home and given antibiotics through a PICC.

Orthopedic Perspective

The standard treatment for displaced supracondylar humerus fractures is closed reduction and percutaneous pinning. The percutaneous pins protrude out the skin, are under the cast, and are typically removed in clinic 3 to 4 weeks after surgery. Infection can occur along the pin sites, but this is unusual. Pin tract infections can be treated with local wound care and oral antibiotics. Osteomyelitis related to percutaneous pin reduction is rare and is treated more aggressively with surgical drainage and IV antibiotics. Despite nonspecificity of fluid collections and marrow edema in the setting of trauma or hardware removal, these areas of abnormality should be described. When the concern is infection, the clinician must know where to place a needle for drainage or culture, even if a post-traumatic hematoma or seroma is likely.

What the Clinician Needs to Know

The location and size of intraosseous and soft tissue fluid collections. Is the collection amenable to percutaneous aspiration and will the needle track endanger neurovascular structures?

Answers

1. 0.5%–1.5%.
2. False.

Pitfalls and Pearls

Do not confuse the normal postoperative marrow edema pattern after hardware removal with osteomyelitis.

References

1. Kaim AH, Gross T, von Schulthess GK. Imaging of chronic posttraumatic osteomyelitis. *Eur Radiol* 2002; 12:1193–1202.
2. Coles CP, Gross M. Closed tibial shaft fractures: Management and treatment complications. A review of the prospective literature. *Can J Surg* 2000; 43:256–262.
3. Widmer AF. New developments in diagnosis and treatment of infection in orthopedic implants. *Clin Infect Dis* 2001; 33 Suppl 2:S94–S106.
4. Ledermann HP, Kaim A, Bongartz G, Steinbrich W. Pitfalls and limitations of magnetic resonance imaging in chronic posttraumatic osteomyelitis. *Eur Radiol* 2000; 10:1815–1823.
5. Stern PJ, Drury WJ. Complications of plate fixation of forearm fractures. *Clin Orthop* 1983:25–29.
6. Skaggs DL, Cluck MW, Mostofi A, Flynn JM, Kay RM. Lateral-entry pin fixation in the management of supracondylar fractures in children. *Journal of Bone & Joint Surgery—American Volume* 2004 Apr; 86-A(4):702–707.

Case 28

History

This is a 4-year-old girl with a right shoulder deformity.

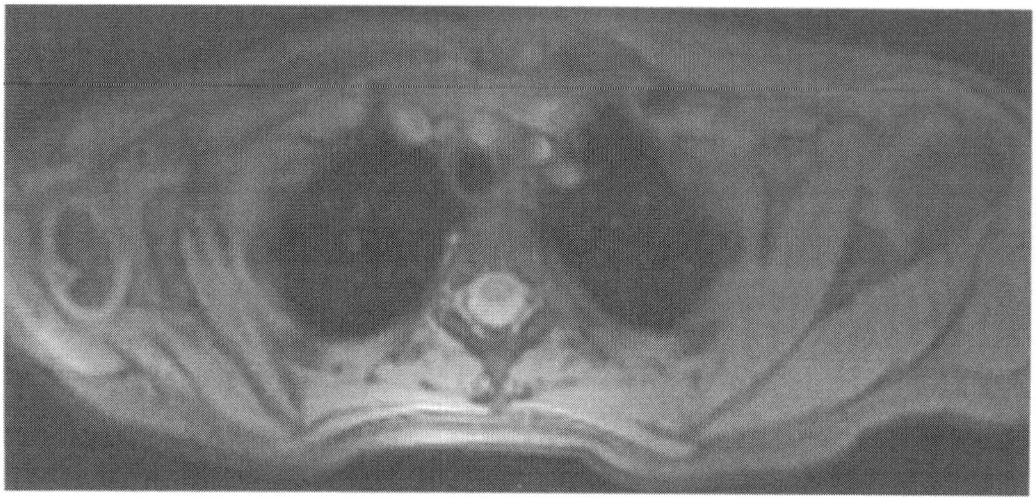

Figure 28A. Axial MPGR of both shoulders.

Figures 28A, (28A with annotations). The left shoulder is normal. On the right, a pseudoglenoid is present and the shoulder is abnormally retroverted. The right humeral head is internally rotated, smaller than the left humeral head, and has a dysplastic shape. The right shoulder girdle muscles are globally smaller compared with the left, particularly the subscapularis and infraspinatus muscles. The right glenoid fossa is directed posteriorly with a retroversion angle of 63 degrees (see calculation method below). This compares with a retroversion angle of 3 degrees on the normal left side. The percentage of the humeral head anterior to the scapular spine line is 0% on the right and 52% on the left (see calculation method below).

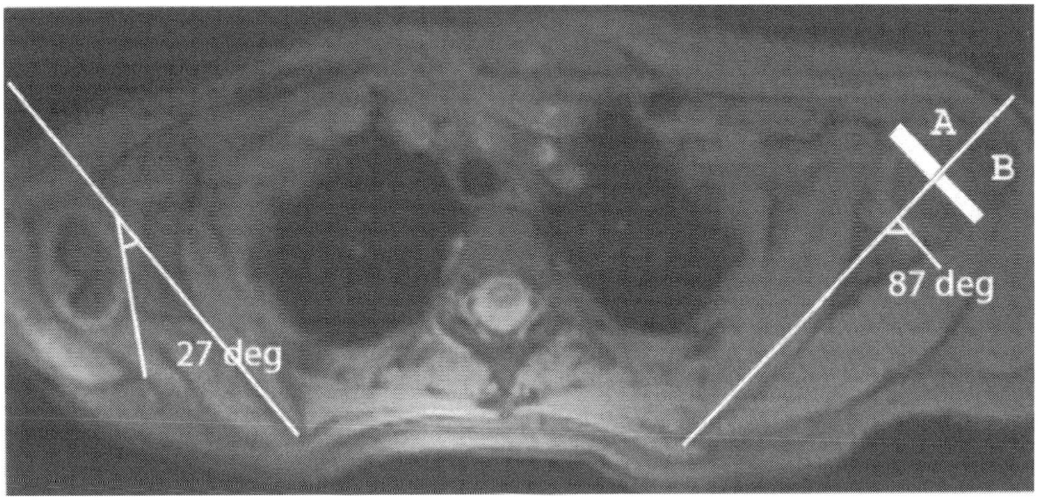

Figure 28A*. Annotated.

Diagnosis

Neonatal brachial plexopathy with glenohumeral dysplasia

Questions

1. T/F: Patients with brachial plexopathy have a higher incidence of posterior shoulder subluxation and dislocation.
2. T/F: The normal glenoid is mildly retroverted.

Discussion

Glenohumeral dysplasia, resulting from brachial plexus birth trauma, may develop as early as the first year of life (1). In some of these patients, the deformity may be progressive and occur because of unopposed humeral head internal rotation and loss of normal shoulder proprioception (2). The persistent internal rotational deformity may produce abnormal loading forces on the humeral head and the posterior glenoid epiphysis. This leads to progressive posterior glenoid deformity with increased glenoid retroversion and posterior humeral head subluxation (Answer to Question 1). The humeral head may also be decreased in size and dysplastic, and there may be a delay in appearance of the secondary ossification center. The muscles of the shoulder girdle atrophy as a result of brachial plexus injury, particularly the subscapular and infraspinous muscles (3).

Radiography, ultrasound, CT, MRI, and arthrography may be utilized for evaluation of glenohumeral dysplasia. The benefit of MRI is that epiphyseal and articular cartilage is better delineated, the images are reproducible, and there is no radiation exposure. Glenohumeral joint orientation and measurements should be made using the articular cartilage of the glenoid or pseudoglenoid as the reference point (2). The unaffected shoulder should be imaged and used for comparison. The axial images should also be used to assess for brachial plexus pseudomeningocele, an additional finding in cases of brachial plexus injury (4).

The normal glenoid fossa is mildly concave and slightly retroverted. The normal concave shape is augmented by the fibrocartilaginous labrum. In two separate studies, the average degree of glenoid retroversion in the unaffected shoulder in patients with brachial plexus palsy was 5.5 to 8 degrees (Answer to Question 2) (2, 5). The degree of retroversion of normal shoulders decreases with age, with an average adult value of 1.7 degrees (5). The degree of glenoid retroversion is calculated by first drawing a line that connects the medial scapular tip to the mid-point of the glenoid (Figure 28A with annotations). A second line is drawn that connects the anterior and posterior glenoid. Note that the hyperintense glenoid hyaline articular cartilage is used as a reference point for the glenoid measurement, not the hypointense fibrocartilaginous labrum. The angle generated from the posteromedial quadrant is subtracted by 90 degrees to give the glenoid version index. If the value is negative, then the glenoid is retroverted. If the value is positive, then the glenoid is anteverted. For children of any age with brachial plexus birth trauma, the average glenoid retroversion was 24 to 26 degrees in two separate studies (2, 5). These two indices are particularly useful in characterizing the anatomic alterations in mild cases of neonatal brachial plexopathy (Figure 28B).

Calculating glenoid version may be difficult when there is significant glenohumeral dysplasia. The normal glenohumeral joint is concentric when the humeral head is round and centered within a concave glenoid. With brachial plexopathy and progressive glenohumeral deformity, the glenoid fossa may also become flat, biconcave (two concavities within the same imaging plane) (Figure 28C), or develop a pseudoarticulation (pseudoglenoid—two concavities in different imaging planes) (6). The posterior concavity may be used for glenoid version calculation when the biconcave or pseudoglenoid configuration is present (5).

The percentage of humeral head anterior to the scapular spine (subluxation index) may be calculated by extending the longitudinal scapular line used in the glenoid version calculation through the humeral head (Figure 28A with annotations). The percentage of humeral head lying anterior to this bisecting line is then calculated [A/(A + B)]. In one study, the subluxation index in patients with glenohumeral dysplasia was 28% (range: 0% to 51%), as apposed to approximately 45% (range: 34% to 54%) in the normal contralateral shoulder (5). Poyhia et al. observed that increased rotator cuff muscle atrophy, particularly the subscapular and infraspinous muscle, correlated with increased posterior subluxation of the humeral head and glenoid retroversion in the setting of brachial plexus birth injury (7).

This patient subsequently had an open reduction. A derotational osteotomy of the humerus with internal fixation was performed with improvement on follow-up.

Orthopedic Perspective

MRI quantifies the degree of glenohumeral dysplasia related to brachial plexopathy and helps guide management. Open reduction may be considered when there is significant subluxation and glenoid retroversion. A glenoid osteotomy may be necessary to correct glenoid retroversion. MRI may also be used to determine humeral version. If substantial internal torsion exists, humeral osteotomy may be necessary. Finally, MRI can help assess muscle bulk and quality. This is important to know prior to performing external rotator shoulder tendon transfers during open glenohumeral reduction.

What the Clinician Needs to Know

1. Glenoid version angle, subluxation index, and glenohumeral congruity.
2. Are there any barriers to concentric reduction of the humeral head?
3. Degree of shoulder girdle muscle atrophy.

Answers

1. True.
2. True.

Additional Examples

Mild Brachial Plexopathy

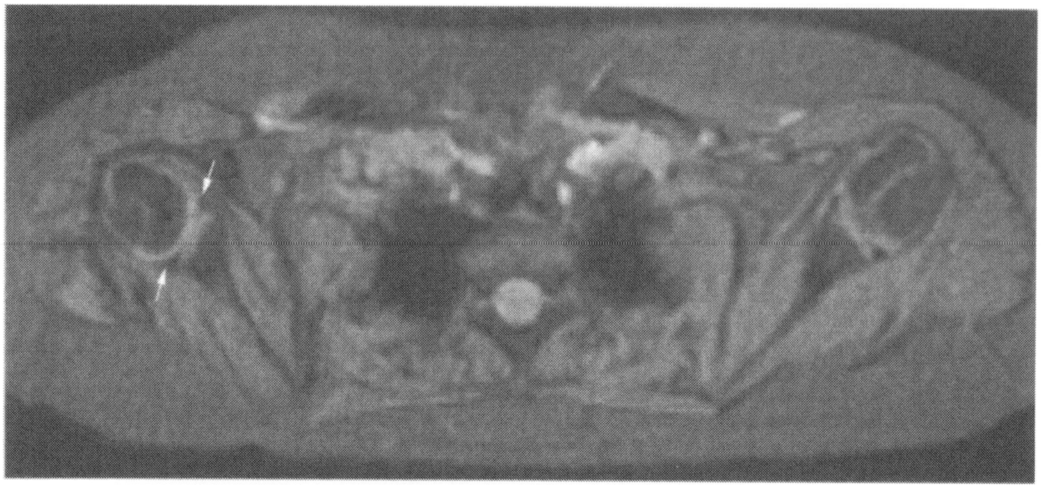

Figure 28B. Axial MPGR.

Findings

This is a 2-year-old girl with a right neonatal brachial palsy with mild glenohumeral dysplasia.

Figure 28B. On the right, the glenoid is abnormally retroverted, the humeral head is mildly subluxed posteriorly, and the shoulder girdle muscles are also globally diminished in bulk as compared to the left shoulder. Note that the anterior and posterior labra impart a concave contour to the relatively flat glenoid articular surface. The left shoulder is normal. The right glenoid is retroverted by 32 degrees. The left glenoid is retroverted by 9 degrees. The right glenoid subluxation index is 39%, and the left glenoid subluxation index is 48%.

Brachial Plexopathy with a Biconcave Glenoid

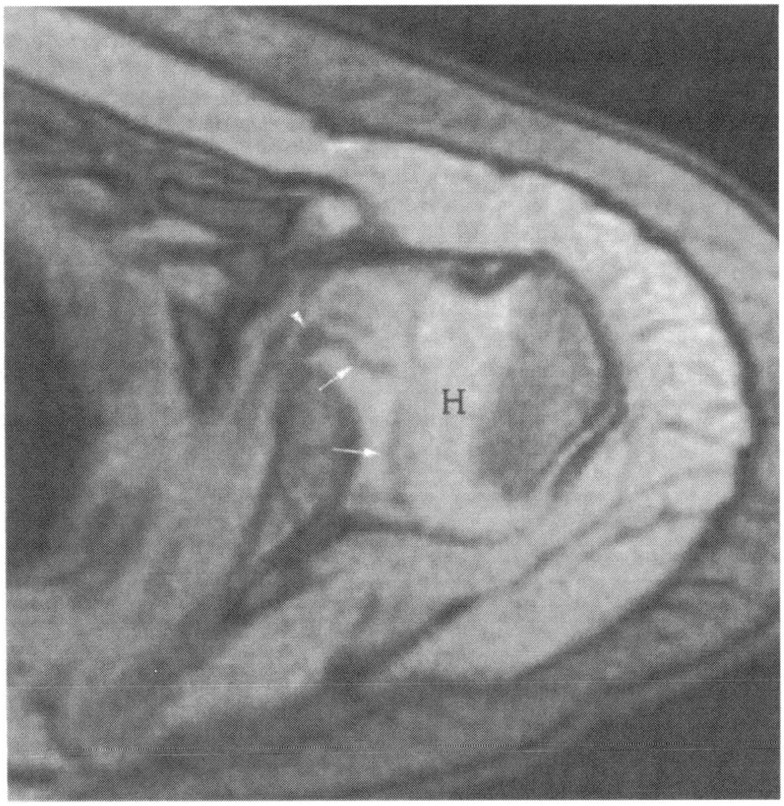

Figure 28C. Axial MPGR, left shoulder.

Findings

This is a 17-month-old girl with a left brachial palsy with moderate glenohumeral dysplasia.

Figure 28C. The glenoid contour **(arrows)** is biconcave, and the labrum is hypoplastic anteriorly **(arrowhead)** and not identified posteriorly. The dysplastic humeral head (H) articulates with the posterior glenoid concavity. This deformity is considered a biconcave glenoid, rather than a pseudoglenoid, since both glenoid articular surfaces are in the same imaging plane.

Pitfalls and Pearls

The MRI findings associated with brachial plexopathy may be quite subtle. The MRI should always include the opposite shoulder, and high quality images are essential. If diminished shoulder musculature is evident, the glenoid version and subluxation indices should be calculated to confirm the disorder.

References

1. Moukoko D, Ezaki M, Wilkes D, Carter P. Posterior shoulder dislocation in infants with neonatal brachial plexus palsy. *J Bone Joint Surg Am* 2004; 86-A:787–793.
2. Waters PM, Smith GR, Jaramillo D. Glenohumeral deformity secondary to brachial plexus birth palsy. *J Bone Joint Surg Am* 1998; 80:668–677.
3. Poyhia TH, Nietosvaara YA, Remes VM, Kirjavainen MO, Peltonen JI, Lamminen AE. MRI of rotator cuff muscle atrophy in relation to glenohumeral joint incongruence in brachial plexus birth injury. *Pediatr Radiol* 2005.
4. Miller SF, Glasier CM, Griebel ML, Boop FA. Brachial plexopathy in infants after traumatic delivery: Evaluation with MR imaging. *Radiology* 1993; 189:481–484.
5. Kozin SH. Correlation between external rotation of the glenohumeral joint and deformity after brachial plexus birth palsy. *J Pediatr Orthop* 2004; 24:189–193.
6. Kon DS, Darakjian AB, Pearl ML, Kosco AE. Glenohumeral deformity in children with internal rotation contractures secondary to brachial plexus birth palsy: Intraoperative arthrographic classification. *Radiology* 2004; 231:791–795.
7. Poyhia TH, Nietosvaara YA, Remes VM, Kirjavainen MO, Peltonen JI, Lamminen AE. MRI of rotator cuff muscle atrophy in relation to glenohumeral joint incongruence in brachial plexus birth injury. *Pediatr Radiol* 2005; 35:402–409.

Case 29

History

This is a 2-year-old girl with an abnormal gait and a short right leg. She is status post partial amputation of the right foot. She also has left developmental dysplasia of the hip.

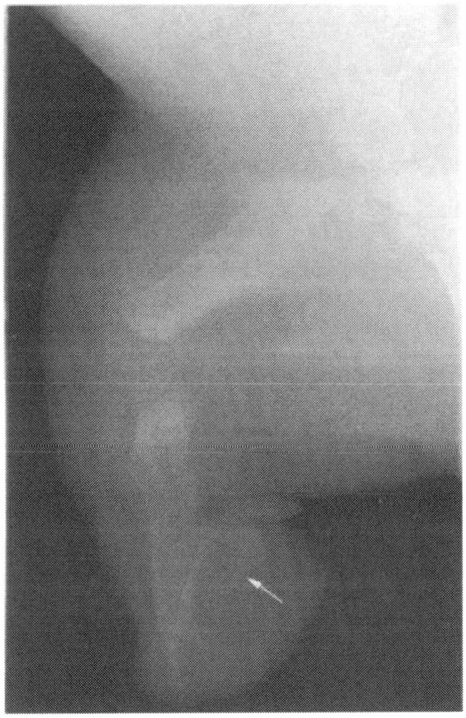

Figure 29A. Frog lateral radiograph of the right lower extremity.

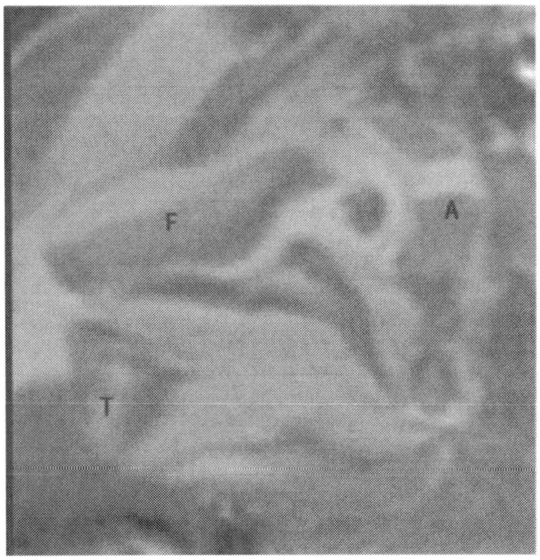

Figure 29B. Coronal 3D SPGR FS of the right hip.

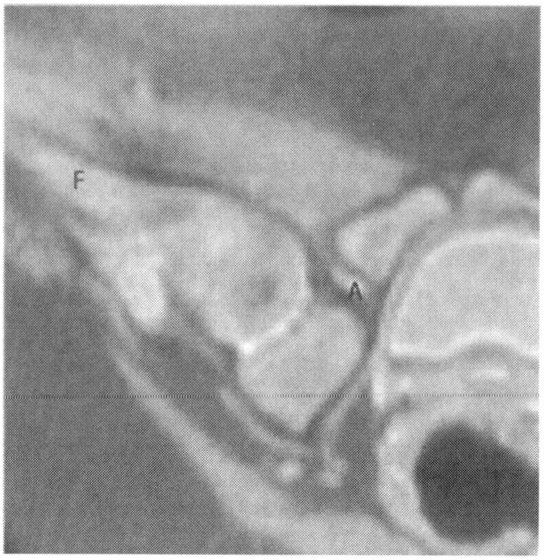

Figure 29C. Axial PD FS.

Figure 29A. The right femur is short. A fibula is not identified. The tibia is short and dysplastic, and the calcaneus overrides the tibial shaft **(arrow)**. The capital femoral epiphysis is well seated within a shallow, dysplastic acetabulum. There is a coxa varus deformity. The relationship of the femoral head and neck is unclear.

Figures 29B, 29C. The capital femoral ossification center and femoral diaphysis are bridged by cartilage. The capital femoral epiphysis is posteriorly situated within a shallow acetabulum. The right knee is dysplastic, and the tibia is dislocated posteriorly. Femoral diaphysis (F), tibia (T), and acetabulum (A).

Diagnosis

Proximal focal femoral deficiency (PFFD), Aitken class B

Questions

1. What drug is associated with the development of PFFD?
2. T/F: Fibular hemimelia may been seen with PFFD.

Discussion

Proximal focal femoral deficiency (PFFD) is a rare congenital anomaly affecting the proximal femur and acetabulum. It usually occurs sporadically and has also been described in association with maternal thalidomide ingestion (Answer to Question 1) as well as maternal diabetes (1, 2). Various other musculoskeletal anomalies have been associated with PFFD including, but not limited to: ipsilateral fibular hemimelia, tibia shortening, clubfoot, talocalcaneal coalition, and joint deformities of the ankle and knee (Answer to Question 2) (3). The muscles around the hip may be smaller compared with the contralateral side, but the sartorius muscle may be hypertrophied and contribute to flexion deformities of the hip and knee (4). Bilateral PFFD has been observed in up to 15% of cases. It is notable that this infant with right-sided PFFD also had left developmental dysplasia of the hip (images not shown).

There is a spectrum of femoral and acetabular deformities seen with PFFD that have been classified by Aitken and others (Diagram 29A) (3). Imaging features used to classify PFFD include: the location of the capital femoral epiphysis with respect to the acetabulum, the presence of acetabular deformity, the degree of femoral shaft shortening, and the presence of a subtrochanteric pseudoarthrosis between the capital femoral epiphysis and femoral shaft. The presence of foot, tibia, and fibula deformities is not included in the classification scheme of PFFD. Infantile coxa vara is a separate entity, although there may be superficial overlap of radiographic findings with Aitken class A PFFD. Infantile coxa vara tends to present later at 1 to 2 years of age; the deformity is mild with minimal femoral shortening and is more often bilateral than PFFD (5).

PFFD classification has traditionally been based on plain radiography and is modified as the relationship of the joint and ossification centers become radiographically apparent with maturation. Since MRI provides a multiplanar depiction of the morphology of the process, including the cartilaginous component, it has advantages over plain radiographic PFFD staging. The objective of MRI evaluation in patients with PFFD is to define osseous, cartilaginous, joint, and muscular anatomy at an earlier age to help guide appropriate management. In one study, MRI, US, and CT arthrography were able to distinguish deformities with a fixed acetabulum and capital femoral epiphysis relationship from a relatively mobile hip (6). These findings were not readily apparent on plain radiography.

Fibular hemimelia is a congenital hypoplasia or absence of the fibula (Figures 29D–29F). It may be an isolated deformity or may be associated with PFFD (7). Congenital anomalies of the knee that may be seen with fibular hemimelia include: lateral femoral condyle hypoplasia, absent or hypoplastic cruciate ligaments, menisci, and a small high-riding patella (8). Various other musculoskeletal anomalies that have been reported include short tibia with anterior bowing, and ankle, foot, and upper-extremity anomalies. MRI is useful in these patients to assess the status of the ligaments, tendons, and cartilaginous structures of the knee.

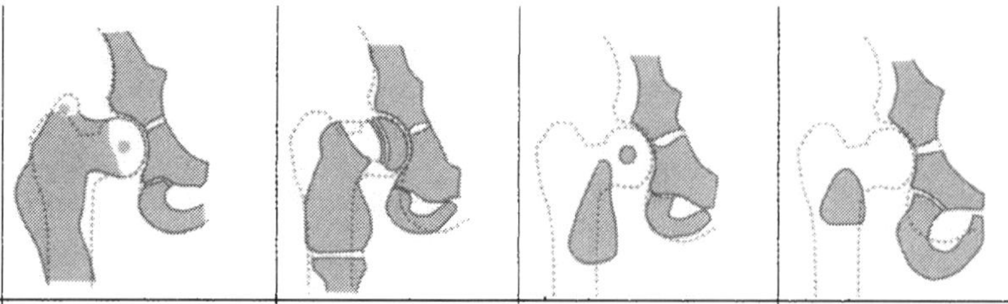

class	A	B	C	D
femoral head	present	present	absent/ small	absent
acetabulum	normal	mild to moderate dysplasia	severe dysplasia	absent
other	mildest form, may present with coxa varus deformity	no osseous connection between the femoral head and shaft	tapered proximal femur, femoral shaft appears to articulate with the acetabulum	

Diagram 29A. Aitken classification of proximal focal femoral deficiency. (Adapted from Herring JA. Limb deficiencies. In: Tachdjian's Pediatric Orthopaedics, Herring JA et al., eds, chapter 32. Philadelphia: WB Saunders Company, 2002; 1745–1810, with permission from Elsevier.)

Popliteal pterygium syndrome is another rare disorder of the lower extremity that shows autosomal dominant inheritance and is associated with facial and urogenital anomalies (9). The characteristic feature of this disorder is a flexion deformity of the knee due to a popliteal web containing a fibrous band that extends from the ischium to the calcaneal region (Figures 29G–29J). MRI may help accurately define the contents of the popliteal web and its relationship with muscles (particularly hamstring and calf muscles) and neurovascular structures. The sciatic nerve is often foreshortened and may be intimately associated with the fibrous band. Other extremity deformities seen with popliteal pterygium syndrome include digital anomalies, contracted heel cord, hypoplastic patella, scoliosis, and hypoplastic tibia (9).

This patient was considered to have an Aitken class B deformity based on plain radiographs. MRI better delineated the degree of acetabular deformity and the cartilaginous continuity of the femoral shaft and femoral head. At 4 years of age, the patient is mobile with the aid of a prosthesis and is receiving physical therapy. Surgical correction of the right hip is not planned at this time.

Orthopedic Perspective

Traditionally, treatment of PFFD has been based on plain radiographs according to the Aitken classification. The problem with this classification is that grading may be modified with age and surgical and physical therapy management goals may change with progressive ossification on radiographs. MRI may provide earlier information regarding the degree of hip deformity and stability as well as associated deficiencies of the knee joint that are not well delineated by plain radiographs. Long-term prognosis and treatment plans may be determined at an earlier age by employing MRI in the diagnostic algorithm. The surgical treatment considerations include joint stabilization that may entail an arthrodesis and correction of leg-length discrepancies by either ipsilateral leg-lengthening osteotomies and/or contralateral epiphysiodesis. Since an

MRI staging system of PFFD has not been established, the MRI report should provide a detailed description of the relevant findings.

What the Clinician Needs to Know

1. Congruity and stability of the hip joint.
2. Associated congenital deformities of the knee, fibula, and foot.

Answers

1. Thalidomide.
2. True.

Additional Example

Fibular Hemimelia

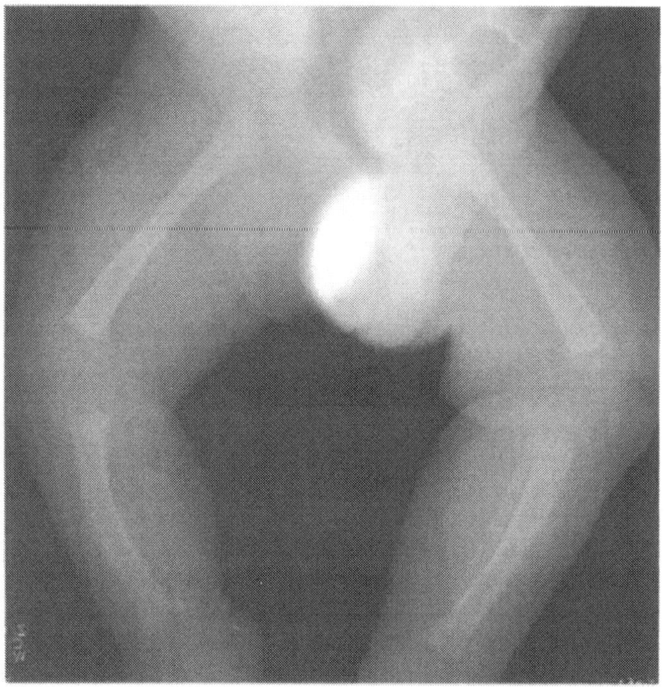

Figure 29D. AP radiograph of the legs.

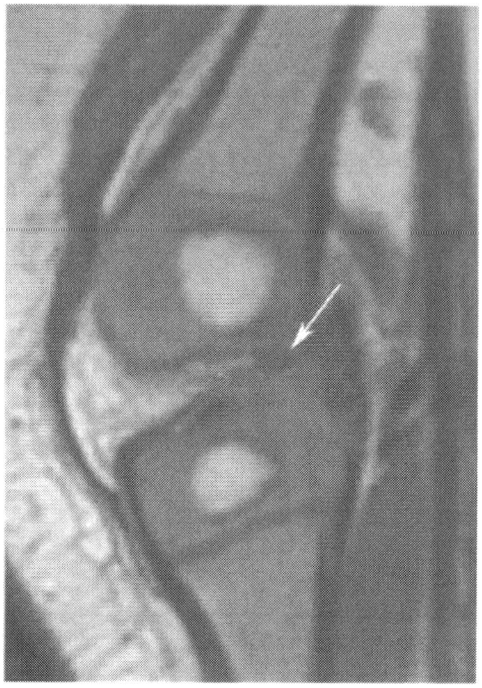

Figure 29E. Sagittal PD right knee.

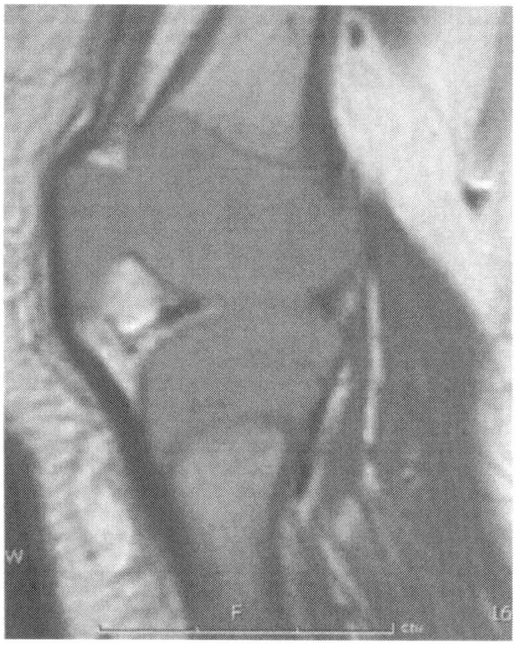

Figure 29F. Sagittal PD.

Findings

This is a 13-month-old boy with known right fibular hemimelia.

Figure 29D. There is absence of a right fibula. The right tibia is shorter than the left and there is mild anterior bowing deformity. There is a relatively normal articulation of the femur and tibia.

Figures 29E, 29F. A hypoplastic PCL is identified **(arrow)**. The menisci are hypoplastic, and the ACL is absent. A normal appearing cartilaginous patella and patella tendon are present.

Popliteal Pterygium Syndrome

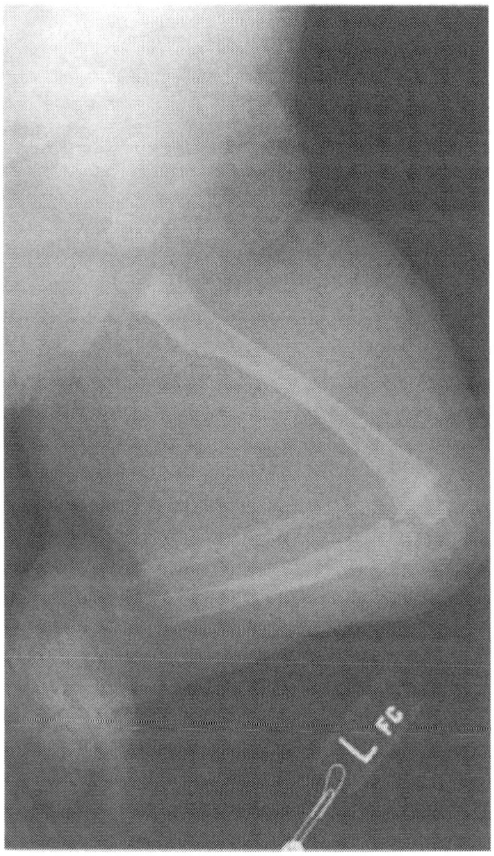

Figure 29G. AP radiograph of the left lower extremity.

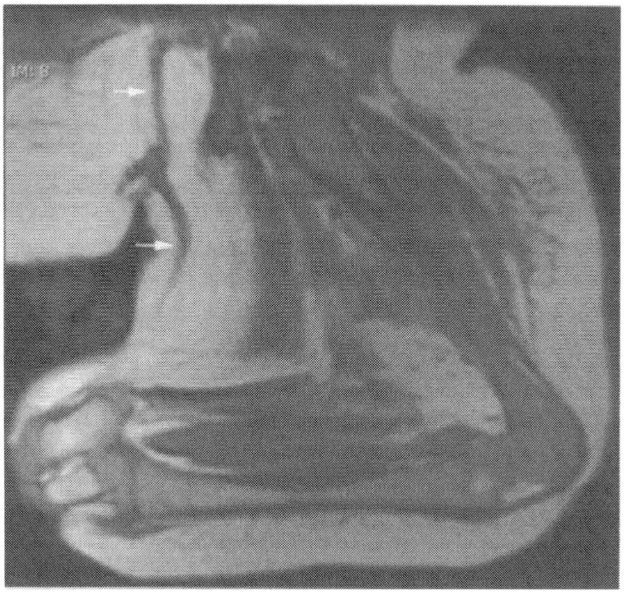

Figure 29H. Coronal PD.

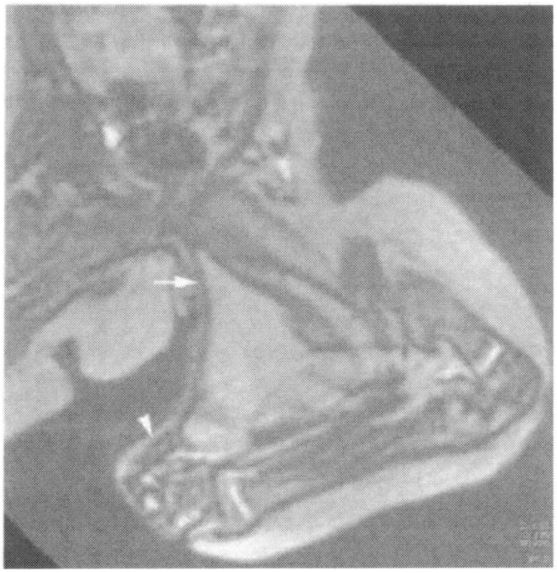

Figure 29I. Coronal GRE.

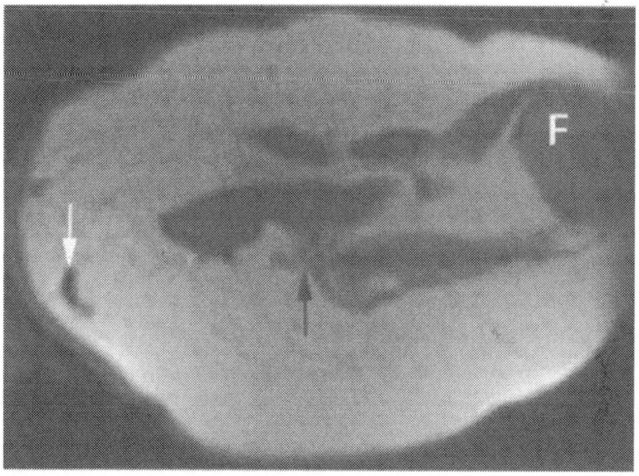

Figure 29J. Axial T1 (through the distal left thigh).

Findings

This is a 2-year-old boy with known popliteal pterygium syndrome.

Figure 29G. The left knee is held in flexion and there is a broad soft tissue band bridging the popliteal fossa. There is also a clubfoot deformity.

Figures 29H, 29I. MRI demonstrates a predominantly fat containing pterygium containing a long fibrous band **(white arrows)** extending from the ischium to the plantar muscles of the hindfoot **(arrowhead)**.

Figure 29J. An axial view through the distal thigh shows that the hamstring muscles and neurovascular bundle (at the sciatic nerve/tibial nerve junction) **(black arrow)** are contained within the fatty component of the pterygium and are separate from the fibrous band **(white arrow)**. Due to the nature of the deformity, a full reconstruction was not performed. Operatively, a hamstring muscle release, stretching of the neurovascular bundle, and a below-knee amputation were performed.

Pitfalls and Pearls

1. When the acetabulum and proximal femur are predominantly cartilaginous, MRI may provide more accurate and early characterization of PFFD than plain radiographs.
2. MRI evaluation of fibular hemimelia is not performed to evaluate the fibular deformity. Rather, it is used to delineate articular anatomy including potential tendinous, ligamentous, and cartilaginous anomalies of the knee.

References

1. Hamanishi C. Congenital short femur: Clinical, genetic and epidemiological comparison of the naturally occurring condition with that caused by thalidomide. *J Bone Joint Surg Br* 1980; 62:307–320.
2. Hitti IF, Glasberg SS, Huggins-Jones D, Sabet R. Bilateral femoral hypoplasia and maternal diabetes mellitus: Case report and review of the literature. *Pediatr Pathol* 1994; 14:567–574.
3. Anton CG, Applegate KE, Kuivila TE, Wilkes DC. Proximal femoral focal deficiency (PFFD): More than an abnormal hip. *Semin Musculoskelet Radiol* 1999; 3:215–226.
4. Laor T, Burrows PE. Congenital anomalies and vascular birthmarks of the lower extremities. *Magn Reson Imaging Clin N Am* 1998; 6:497–519.
5. Pavlov H, Goldman AB, Freiberger RH. Infantile coxa vara. *Radiology* 1980; 135:631–640.
6. Court C, Carlioz H. Radiological study of severe proximal femoral focal deficiency. *J Pediatr Orthop* 1997; 17:520–524.
7. Fordham LA, Applegate KE, Wilkes DC, Chung CJ. Fibular hemimelia: More than just an absent bone. *Semin Musculoskelet Radiol* 1999; 3:227–238.
8. Laor T, Jaramillo D, Hoffer FA, Kasser JR. MR imaging in congenital lower limb deformities. *Pediatr Radiol* 1996; 26:381–387.
9. Oppenheim WL, Larson KR, McNabb MB, Smith CF, Setoguchi Y. Popliteal pterygium syndrome: An orthopaedic perspective. *J Pediatr Orthop* 1990; 10:58–64.
10. Herring JA. Limb deficiencies. In: *Tachdjian's Pediatric Orthopaedics,* Herring JA, ed., Chapter 32. Philadelphia: W.B. Saunders Company, 2002; 1745–1810.

Case 30

History

This is a 9-year-old boy with a 4-week history of right knee pain and swelling. The patient had minor knee trauma while skiing six weeks ago.

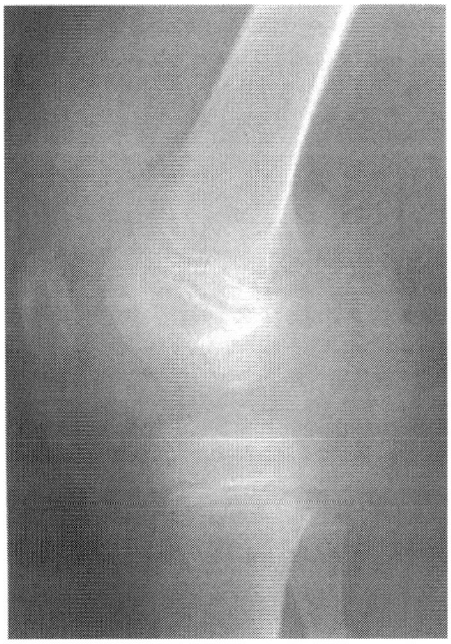

Figure 30A. Lateral radiograph of the right knee.

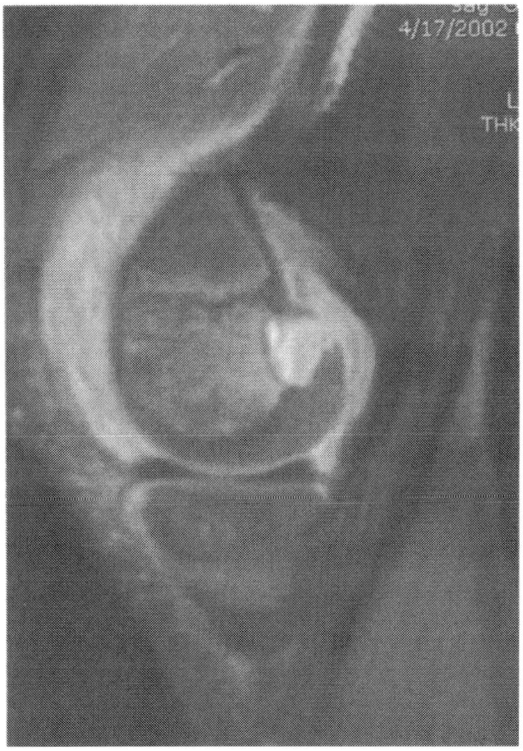

Figure 30B. Sagittal STIR.

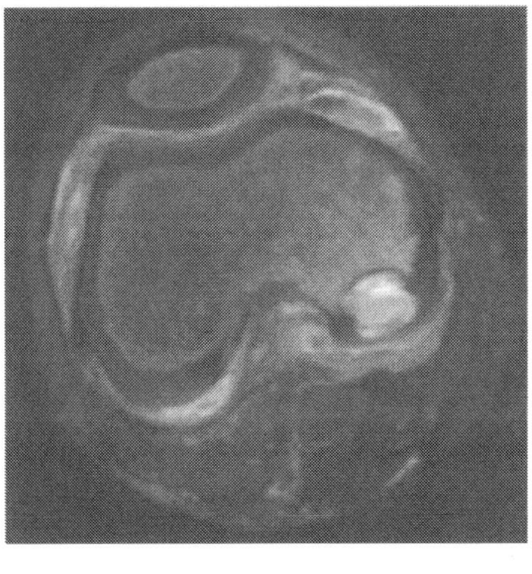

Figure 30C. Axial T2 FS.

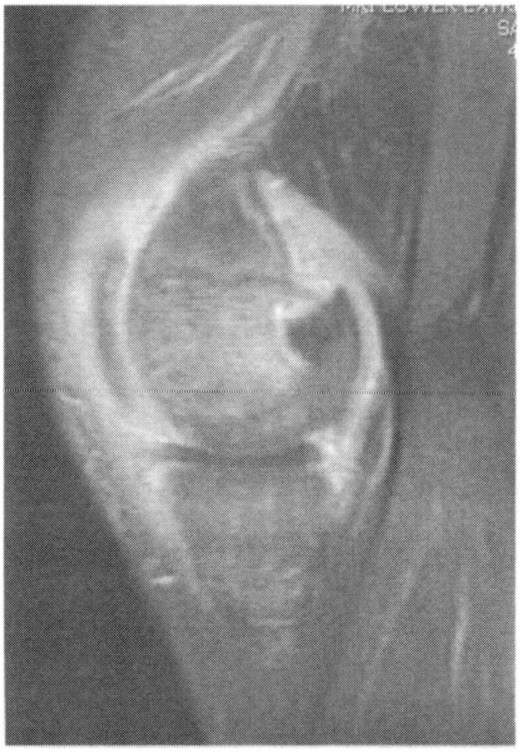

Figure 30D. Sagittal T1 post-Gd FS.

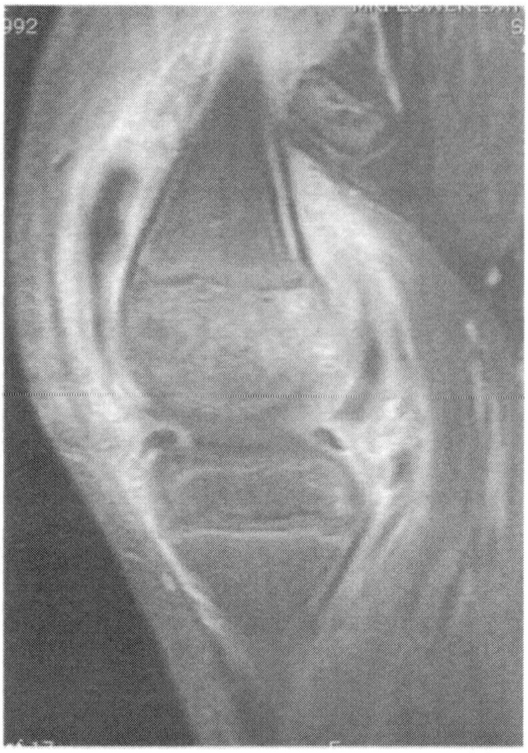

Figure 30E. Sagittal T1 post-Gd FS.

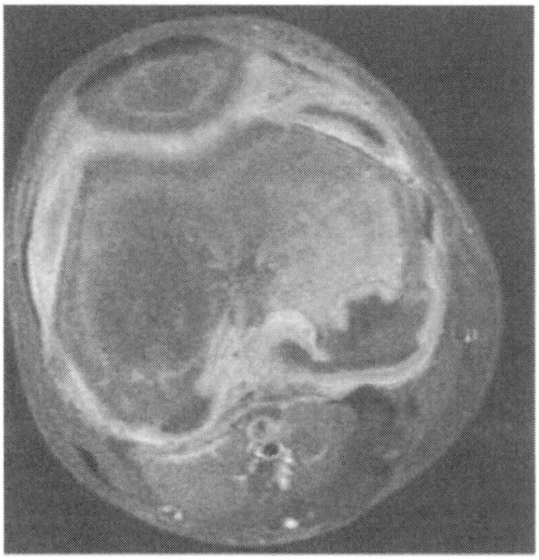

Figure 30F. Axial T1 post-Gd FS.

Figure 30A. There is an osteolytic defect in the posterior aspect of the medial femoral condyle. There is generalized soft tissue swelling and a large suprapatellar joint effusion.

Figures 30B, 30C. There is a hyperintense fluid collection at the epiphyseal chondro-osseous junction that corresponds to the osteolytic defect seen on the radiograph.

Figures 30D, 30E, 30F, (30F with annotations). The collection extends beyond the articular cartilage and is surrounded by thick peripheral rim enhancement **(arrow)**. In addition, there is a moderate size joint effusion with thick synovial enhancement **(arrowheads)**.

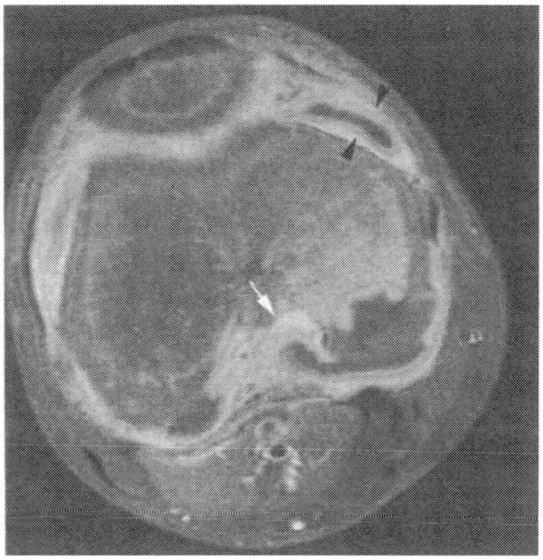

Figure 30F*. Annotated.

Diagnosis

Epiphyseal osteomyelitis and abscess

Questions

1. What is the most common site for epiphyseal osteomyelitis?
2. What is the differential diagnosis for an epiphyseal T2 hyperintense lesion?

Discussion

The most common location for epiphyseal osteomyelitis is the distal femur (Answer to Question 1) (1). Epiphyseal osteomyelitis can occur by hematogenous seeding, or less commonly, with direct inoculation from a penetrating injury (Figures 30G, 30H). Hematogenous epiphyseal osteomyelitis can begin in the epiphysis or extend from the metaphysis by transphyseal spread. Primary epiphyseal osteomyelitis is the cause in this patient (Figures 30A–30F) because the growth plate and the metaphyseal marrow are normal.

The physes and the epiphyseal spherical growth plate share similar anatomic features (2). Both regions abut vascular sinusoids where organisms may lodge and proliferate due to slow flow and diminished phagocytic activity. These sinusoids can be damaged with minor trauma leading to venous stasis. Experimental studies have shown that hematogenous osteomyelitis is more likely to develop in the setting of trauma (3). A history of preceding minor trauma is often present in patients with osteomyelitis (4).

Differential diagnostic considerations for isolated epiphyseal lesions include neoplasms such as chondroblastoma, osteoid osteoma (intramedullary subtype), and Langerhans cell histiocytosis (LCH). Inflammatory synovial processes, such as juvenile rheumatoid arthritis (JRA), pigmented villonodular sinovitis (PVNS), and synovial osteochondromatosis can also produce focal erosive lesions simulating primary epiphyseal pathology (Answer to Question 2). These neoplastic and nonpyogenic inflammatory lesions may show nonspecific increased signal intensity on fluid sensitive sequences within the epiphysis and adjacent structures. Epiphyseal osteomyelitis with abscess demonstrates rim enhancement, whereas tumors such as LCH and chondroblastoma may demonstrate solid central enhancement. When epiphyseal osteomyelitis consists only of enhancing granulation tissue, differentiation from these noninfectious entities by MRI may be impossible. Thus, radiography and clinical and laboratory findings are often necessary to help differentiate these entities.

This patient underwent arthroscopic debridement of the knee joint and received antibiotics. Synovial fluid cytology yielded 123,000-cells/mm^3 with 93% neutrophils. No organism was recovered from arthroscopic drainage or from blood cultures. The patient did well except for $^1/_2$ inch overgrowth of the ipsilateral limb noted 2 years later.

Orthopedic Perspective

Decisions regarding the management of juxta-articular infections are heavily dependent on MRI findings. It is critical to differentiate edema and granulation tissue from purulent exudate. Substantial intraosseous, intra-articular, subperiosteal, and soft tissue abscesses usually require surgical drainage. Osteomyelitis without abscess can usually be treated with antibiotics alone. Long term complications of epiphyseal osteomyelitis

OSTEOMYELITIS CPG ALGORITHM

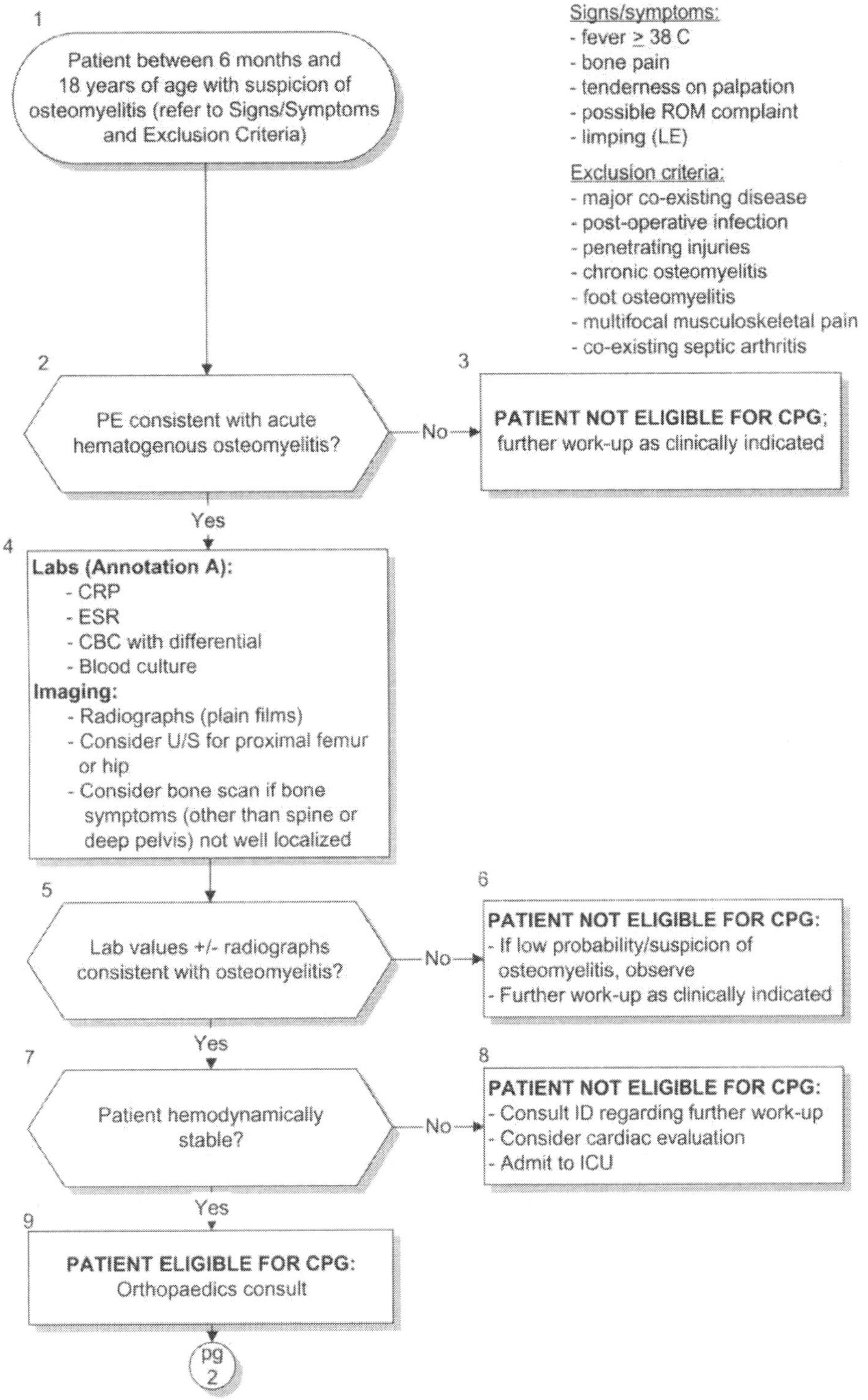

CLINICAL PRACTICE GUIDELINES DISCLAIMER STATEMENT

"This Clinical Practice Guideline is designed to provide clinicians an analytical framework for evaluation and treatment of a particular diagnosis or condition. This Clinical Practice Guideline is not intended to establish a protocol for all patients with a particular condition, nor is it intended to replace a clinician's clinical judgment. A clinician's adherence to this Clinical Practice Guideline is voluntary. It is understood that some patients will not fit the clinical conditions contemplated by this Clinical Practice Guideline and that the recommendations contained in this Clinical Practice Guideline should not be considered inclusive of all proper methods or exclusive of other methods of care reasonably directed to obtaining the same results. Decisions to adopt any specific recommendation of this Clinical Practice Guideline must be made by the clinician in light of available resources and the individual circumstances presented by the patient."

Diagram 30A. Osteomyelitis clinical, imaging, and treatment algorithm. (Courtesy of Dr. Kocher.)

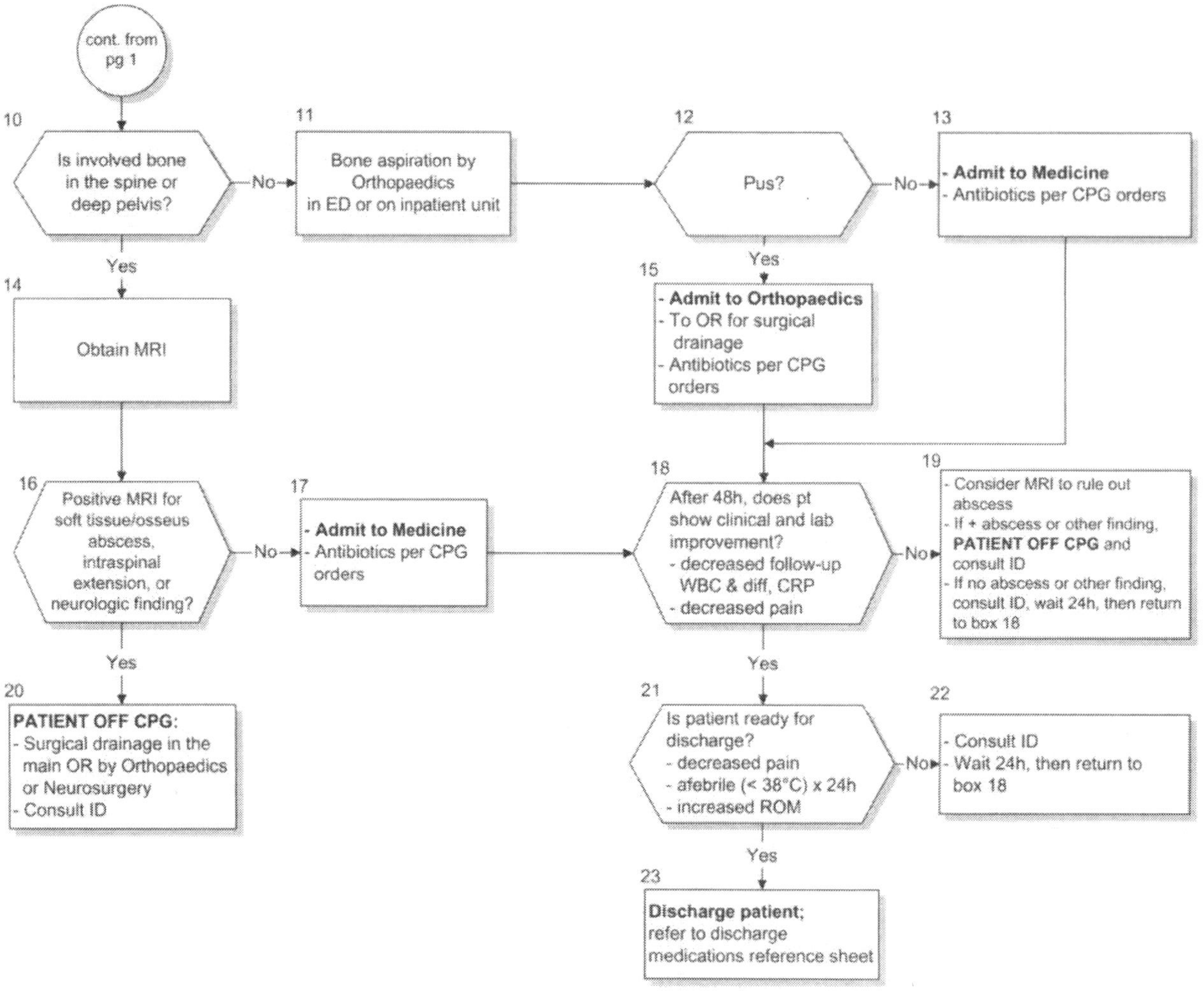

OSTEOMYELITIS CPG ALGORITHM ANNOTATIONS

Annotation A: Preliminary Work-up Laboratory Information

Test	Blood Specimen Amount	Type of Tube Used with Patient Label	Form to Complete:	For:	Send Specimen To:	Results returned:
CRP C-Reactive Protein	Minimum 1mL	Mint green top tube	Hematology/ Chemistry	Core Lab	Lab Control	1 day
Sed Rate (ESR) Erythrocyte Sedimentation Rate	2 mL	Lavender top tube	Hematology/ Chemistry	Hematology	Lab Control	< 4 hours
CBC Plt Diff Complete Blood Count, Platelets and Differential	Minimum 1mL * **Mix specimen by gentle inversion x10**	Lavender top tube	Hematology/ Chemistry	Hematology	Lab Control	Stat: 30 min Routine: 1 hour
Blood Culture	5 mL: • 1 mL minimum into Aerobic btl	Blood culture bottles	Bacteriology ***State clinical diagnosis**	Bacteriology	Lab Control	Prelim: 24 hrs and 48 hrs. Final negative: 6 days

Diagram 30A. (*Continued*)

include epiphyseal growth disturbance and deformity, since the spherical growth plate may be disrupted, as well as early degenerative arthritis due to destruction of articular cartilage. The clinical and imaging work-up for osteomyelitis at our institution is illustrated in Diagram 30A.

What the Clinician Needs to Know

1. Distinguish solid tumor (e.g., osteoid osteoma, chondroblastoma) versus a rim enhancing epiphyseal lesion (more likely osteomyelitis).
2. Is the physis preserved? Physeal destruction may lead to future growth disturbance.
3. Is the epiphyseal lesion secondary to a synovial process (e.g., erosions from PVNS) or is it a primary epiphyseal process?
4. Are there intraosseous, intra-articular, or subperiosteal abscesses associated with epiphyseal osteomyelitis?

Answers

1. Distal femur.
2. Epiphyseal osteomyelitis, LCH, chondroblastoma, intra-articular osteoid osteoma (intramedullary subtype), and primary synovial processes with secondary epiphyseal destruction including PVNS, JRA, and synovial osteochondromatosis.

Additional Example

Epiphyseal Osteomyelitis from Penetrating Trauma

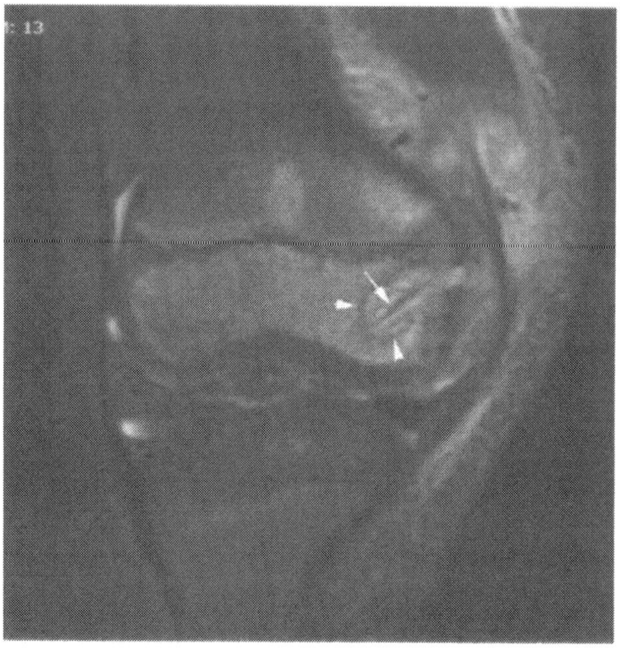

Figure 30G. Coronal STIR of the right knee.

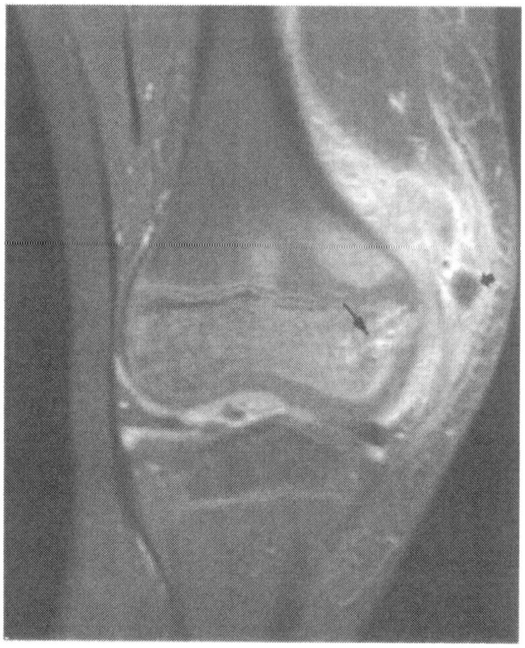

Figure 30H. Coronal T1 post-Gd FS.

Findings

This is a 7-year-old girl who had been impaled by a broken fishing pole one year ago. At the time of the injury, the knee was debrided, but she continued to have intermittent knee pain.

Figures 30G, 30H. There is a linear foreign body within the epiphysis that is hypointense on both STIR and the post-Gd sequence **(thin arrows)**. STIR hyperintense and enhancing granulation tissue surrounds the foreign body. Granulation tissue is surrounded by a hypointense rim of sclerosis **(arrowheads)** on the STIR image, pointing to the chronicity of the process. Significant edema and enhancement of the epiphysis and medial soft tissues are evident. There is also a loculated fluid collection in the medial soft tissues **(thick arrow)**. At surgery, the tip of a broken fishing rod was identified and removed from the medial femoral condyle.

Pitfalls and Pearls

Neoplastic epiphyseal lesions that may demonstrate exuberant marrow and soft tissue edema include osteoid osteoma and chondroblastoma.

References

1. Sorensen TS, Hedeboe J, Christensen ER. Primary epiphyseal osteomyelitis in children: Report of three cases and review of the literature. *J Bone Joint Surg Br* 1988; 70:818–820.
2. Jaramillo D, Hoffer FA. Cartilaginous epiphysis and growth plate: Normal and abnormal MR imaging findings. *AJR Am J Roentgenol* 1992; 158:1105–1110.
3. Morrissy RT, Haynes DW. Acute hematogenous osteomyelitis: A model with trauma as an etiology. *J Pediatr Orthop* 1989; 9:447–456.
4. Waagner DC. Musculoskeletal infections in adolescents. *Adolesc Med* 2000; 11:375–400.
5. Osteomyelitis CPG Algorithm and Annotations, Children's Hospital Boston Internal Web site, Clinical Practice Guidelines Program Reference Page. Boston, Massachusetts. September 2002.

Case 31

History

This is a 4-year-old boy with sickle cell disease with a new left limp, thigh pain, and fever. He also has a history of periorbital rhabdomyosarcoma and is on chemotherapy. The initial radiographs of the left femur obtained 2 days earlier were normal (not shown).

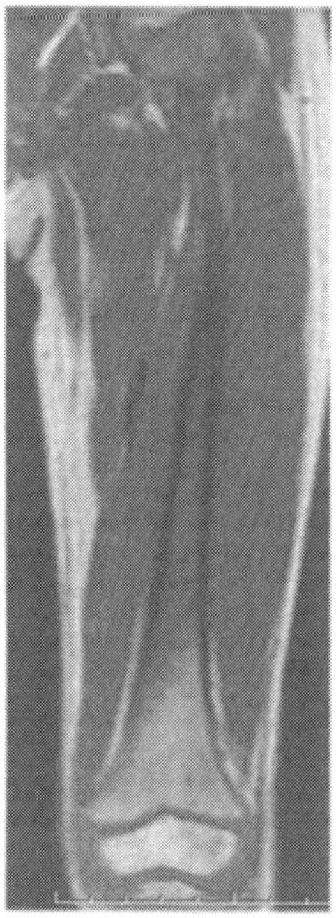

Figure 31A. Coronal T1 of the left femur.

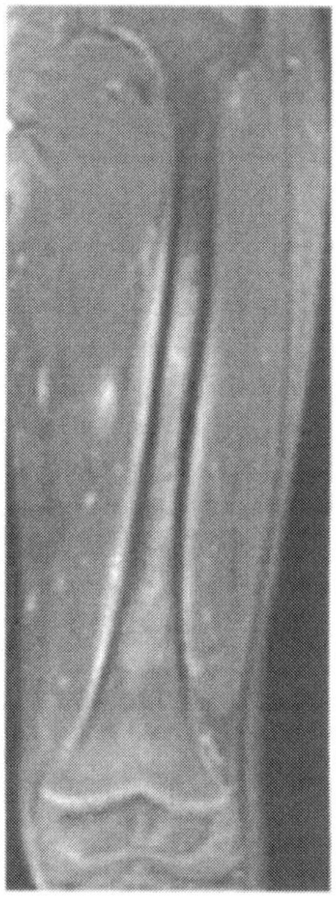

Figure 31C. Coronal T1 post-Gd FS.

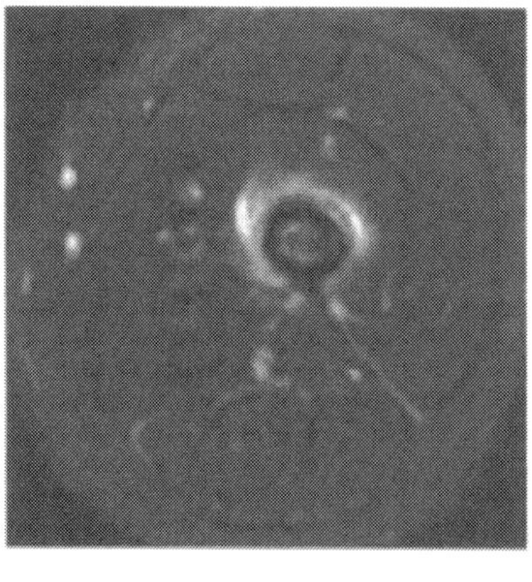

Figure 31B. Axial T2 FS.

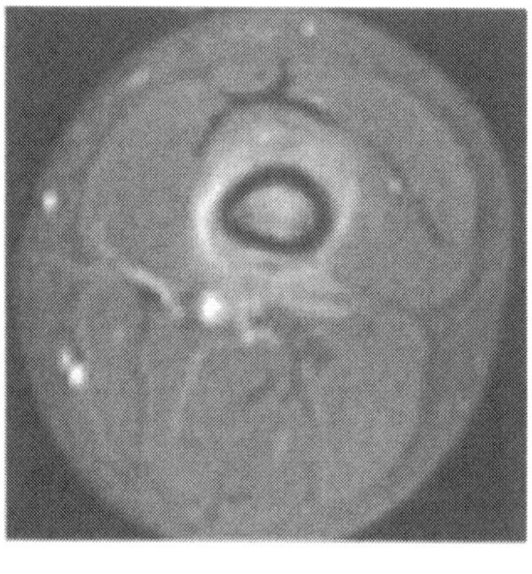

Figure 31D. Axial T1 post-Gd FS.

Figures 31A, 31B, 31C, 31D. There is abnormal T1 hypointensity within the diaphysis. There is increased T2 SI and enhancement within the diaphyseal medullary cavity and periosteum of the femur and adjacent soft tissues. The cortex is intact. There is no subperiosteal new bone formation, soft tissue mass, or fluid collection.

Figures 31E, 31F. Compared to the initial study there is persistent, but decreased, T2 SI and enhancement in the marrow of the femoral diaphysis and adjacent soft tissues. Juxta-cortical soft tissue enhancement is evident **(thick arrows)**. The axial image demonstrates a circumferential zone of enhancing tissue **(arrowhead)** located between the hypointense periosteal new bone and the cortex. No focal mass is seen. There are bilateral diaphyseal foci of enhancing bone marrow containing serpiginous hypointense bands typical of chronic bone infarction **(thin arrows)**.

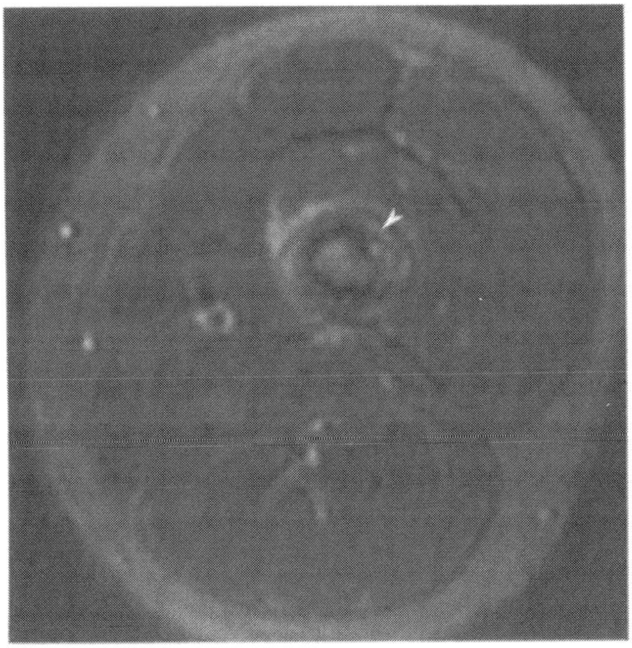

Figure 31E. Axial T2 FS through the left mid-femur 6 weeks later.

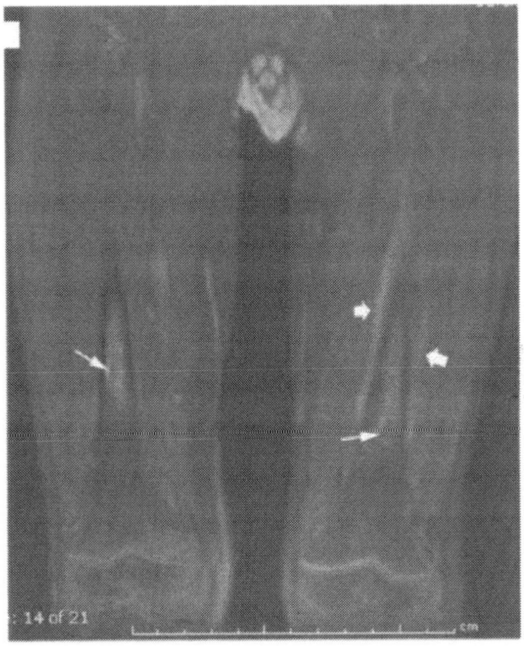

Figure 31F. Coronal T1 post-Gd FS.

Diagnosis

Acute bone marrow infarction of the left femur

Questions

1. What is the most common organism isolated in sickle cell osteomyelitis?
2. T/F: MRI easily distinguishes between early osteomyelitis and acute bone infarction in patients with sickle cell disease.

Discussion

Salmonella is the most common organism isolated in sickle cell patients with osteomyelitis (Answer to Question 1) (1). *Staphylococcus aureus* is the second most common. Although rare, sickle cell patients also have a higher incidence of osteomyelitis from other gram-negative bacilli. The higher incidence of salmonella and other gram-negative bacilli infection is speculated to result from GI mucosal infarction related to sickle cell disease. This leads to mucosal breakdown and intravascular inoculation of these gastrointestinal organisms (2). Although sickle cell osteomyelitis usually involves the long bone diaphyses, the axial skeleton may also be affected (Figures 31G, 31H).

Both osteomyelitis and osteonecrosis related to sickle cell disease may occur in the diaphysis (2). Some authors have suggested that the marrow enhancement pattern may be helpful to differentiate these two entities, especially when there are no clear signs of osteomyelitis, such as a fistulous tract, sequestra, or a juxta-cortical soft tissue abscess. With acute bone infarction, thin peripheral rim enhancement may be evident surrounding a nonenhancing center (3, 4). Peripheral thin hypointensity surrounding a central region of enhancement has also been described in the setting of acute bone infarction (3). With osteomyelitis, peripheral rim enhancement is irregular and thickened, surrounds a non-enhancing center, and is indicative of an intraosseous abscess. However, these authors and others have observed that distinguishing osteomyelitis and acute bone infarction is challenging, if not impossible by MRI (Figure 31I–31L) (Answer to Question 2) (2, 5). Both acute bone infarction and osteomyelitis may have abnormal marrow, cortical, and juxtacortical soft tissue edema (5). Therefore, clinical and microbiologic correlation is critical. Follow-up MRI may be of value in problematic cases.

Chronic bone infarction, on the other hand, has a more typical MRI appearance and is easier to distinguish from osteomyelitis. Juxtacortical soft tissue edema usually is absent, serpiginous low SI lines are present within the necrotic area on all imaging sequences, and the central zone may be low SI on all imaging sequences (due to sclerosis) or may follow fat SI (4).

The differential diagnosis for this case included metastatic disease (given the history of orbital rhabdomyosarcoma), osteomyelitis, and stress reaction. For this patient, metastatic disease was considered unlikely since the marrow signal abnormality was diffuse and a discrete mass was not identified. Stress reaction was considered less likely since a discrete fracture line was not seen and serpiginous hypointense lines were evident on the follow-up MRI exam (Figure 31F), suggesting bone infarction.

During this patient's hospital course, blood cultures were negative and the clinical examination was not consistent with osteomyelitis. He was treated only symptomatically for vaso-occlusive crisis and bone infarction. The diagnosis was confirmed on the follow-up MRI six weeks later (Figure 31E, 31F) that showed decreasing marrow and periosteal edema with typical changes of bone infarction (serpiginous low SI lines).

Orthopedic Perspective

The patient with sickle cell disease who presents with leg pain and limp often poses a diagnostic dilemma: acute bone crisis versus osteomyelitis. The differentiation is important since osteomyelitis is treated with antibiotics, whereas bone crisis is treated supportively. Clinical and laboratory findings are usually used to differentiate the two entities, as osteomyelitis has findings of an infectious process: fever, erythema, elevated serum white blood cell count, elevated CRP, and positive blood cultures. MRI can be useful in cases of clinical uncertainty.

What the Clinician Needs to Know

1. Is this vaso-occlusive crisis or osteomyelitis?
2. What is the location, extent, and age (if possible) of bone infarction?

Answers

1. Salmonella.
2. False.

Additional Examples

Salmonella Vertebral Osteomyelitis

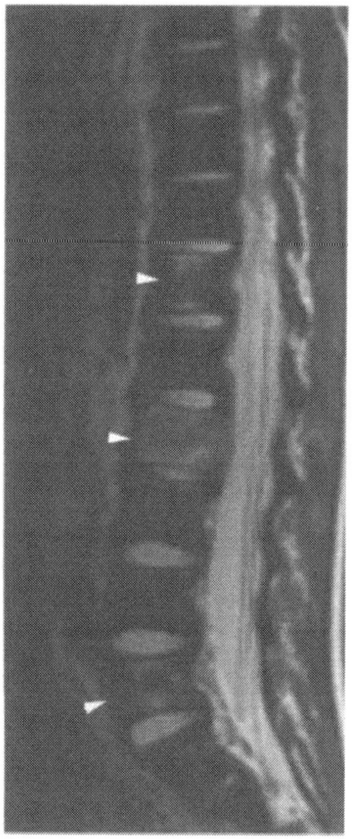

Figure 31G. Sagittal T2 FS.

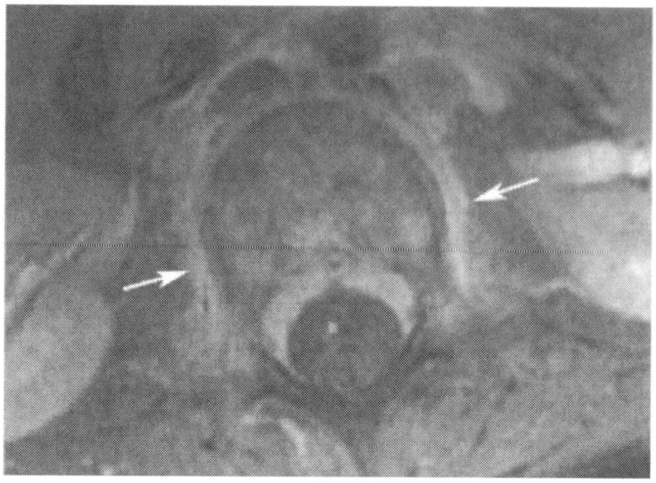

Figure 31H. Axial T1 post-Gd FS (through L2).

Findings

This 4-year-old boy with sickle cell disease presented with severe lower back pain and fever.

Figure 31G. There is increased T2 SI most prominent at L2, but also seen at T12 and L5 **(arrowheads)**. The abnormal SI is located just beneath the cartilaginous endplate, the metaphyseal equivalent zones of the vertebral bodies.

Figure 31H. An axial image through L2 shows mild vertebral body and juxtacortical soft tissue enhancement **(arrows)**. No soft tissue or intraosseous abscess is seen. Vertebral osteomyelitis and osteonecrosis could not be differentiated on imaging grounds alone. The diagnosis of osteomyelitis was made based on the clinical picture and on blood cultures positive for salmonella.

Salmonella Osteomyelitis Within a Region of Bone Infarction—a Diagnostic Dilemma

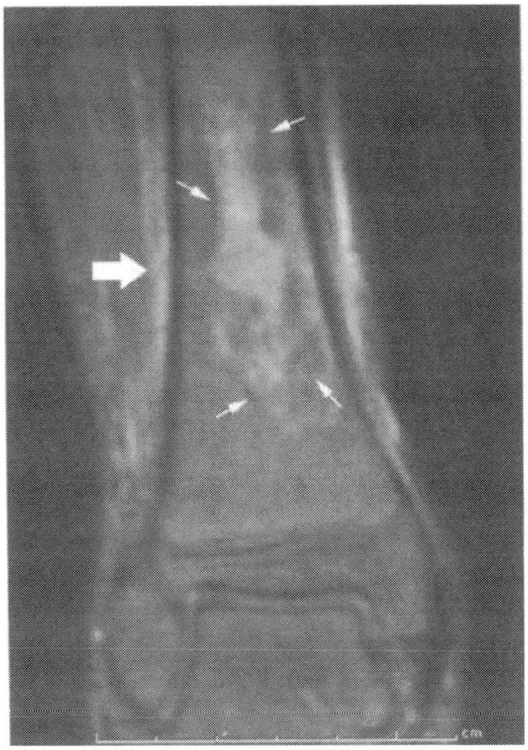

Figure 31I. Coronal STIR of the right tibia.

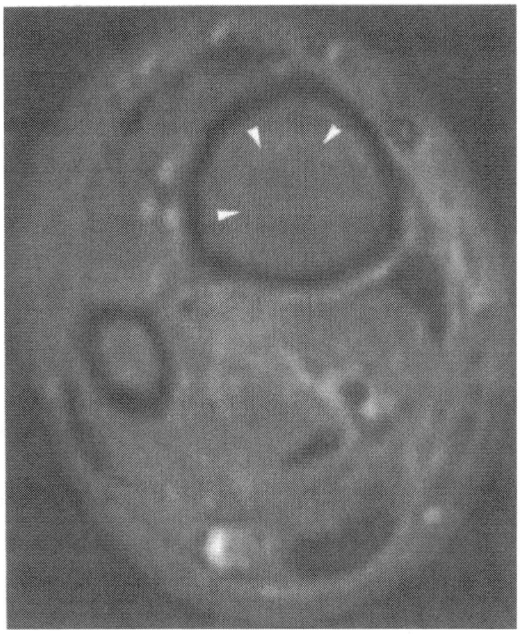

Figure 31J. Axial T1 post-Gd FS.

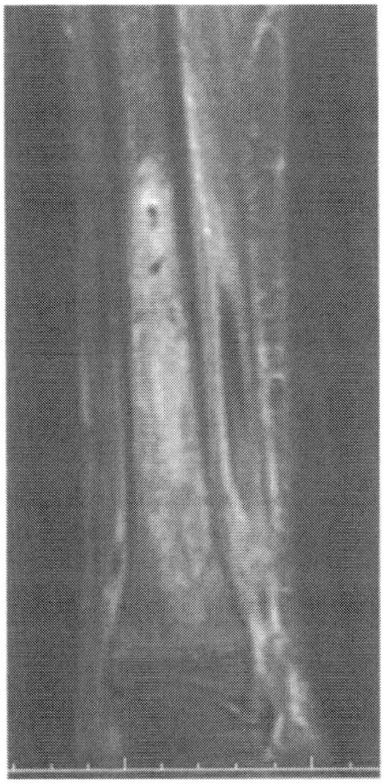

Figure 31K. Coronal STIR of the right tibia, 6 weeks later.

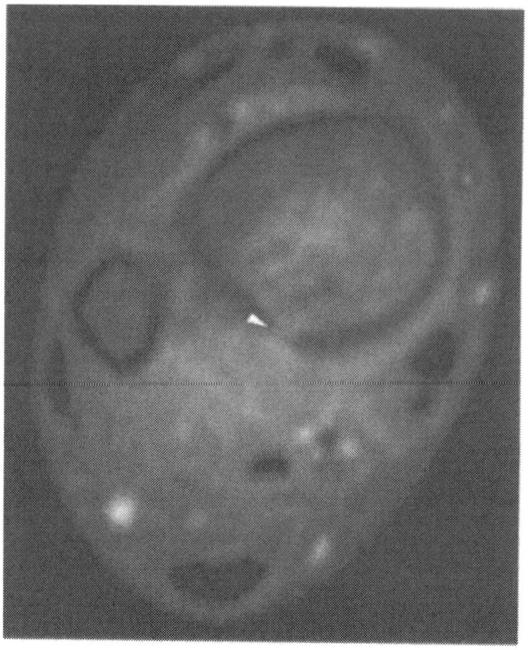

Figure 31L. Axial T1 post-Gd FS, 6 weeks later.

Findings

This is a 13-year-old boy with known sickle cell disease who presented with worsening right ankle pain.

Figures 31I, 31J. The initial study shows well-defined, map-like regions of STIR hyperintensity within the metadiaphysis of the distal tibia. Peripheral and central serpiginous hypointense lines are evident **(arrows)**. On the axial image, there is a thin rim of enhancement surrounding a nonenhancing center **(arrowheads)**. There is also juxtacortical soft tissue edema **(thick arrow)**. The constellation of MRI findings suggests acute or chronic bone infarction. However, blood cultures were positive for salmonella.

Figures 31K, 31L. A follow-up examination six weeks later shows increasing diffuse heterogeneous marrow and juxtacortical soft tissue edema and enhancement, associated with a cloaca (**arrowhead**). Wedge biopsy demonstrated multiple intraosseous microabscesses, compatible with osteomyelitis. This case demonstrates that osteomyelitis and bone infarction may coexist in sickle cell patients, and the exclusion of osteomyelitis in the context of bone infarction may be impossible on MRI.

Pitfalls and Pearls

1. The incidence of bone infarction in sickle cell patients is 50 times more common than osteomyelitis (2).
2. Both osteomyelitis and bone infarction may have adjacent soft tissue edema and contrast enhancement. Follow-up imaging is often helpful for problematic cases.

References

1. Burnett MW, Bass JW, Cook BA. Etiology of osteomyelitis complicating sickle cell disease. *Pediatrics* 1998; 101:296–297.
2. Lonergan GJ, Cline DB, Abbondanzo SL. Sickle cell anemia. *Radiographics* 2001; 21:971–994.
3. Umans H, Haramati N, Flusser G. The diagnostic role of gadolinium enhanced MRI in distinguishing between acute medullary bone infarct and osteomyelitis. *Magn Reson Imaging* 2000; 18:255–262.
4. Saini A, Saifuddin A. MRI of osteonecrosis. *Clin Radiol* 2004; 59:1079–1093.
5. Frush DP, Heyneman LE, Ware RE, Bissett GS, 3rd. MR features of soft-tissue abnormalities due to acute marrow infarction in five children with sickle cell disease. *AJR Am J Roentgenol* 1999; 173:989–993.

Case 32

History

This is a 3½-year-old girl with a 3-month history of back and right hip pain.

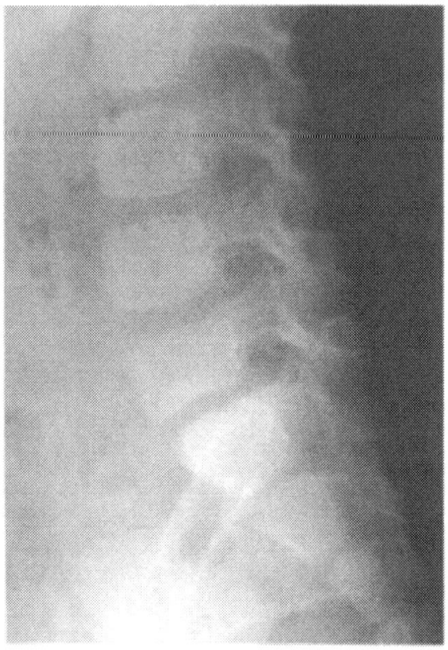

Figure 32A. Lateral lumbar radiograph.

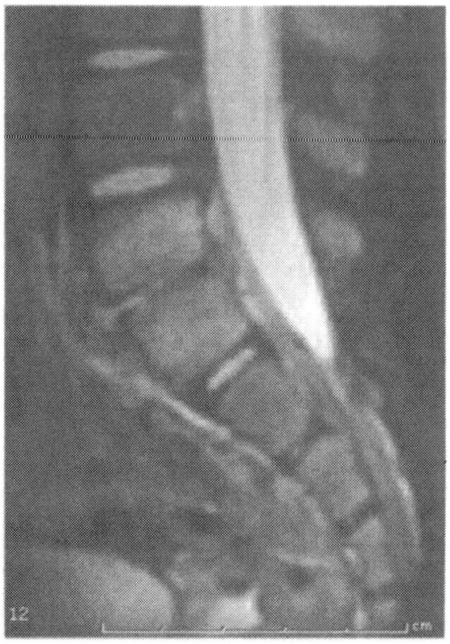

Figure 32B. Sagittal STIR.

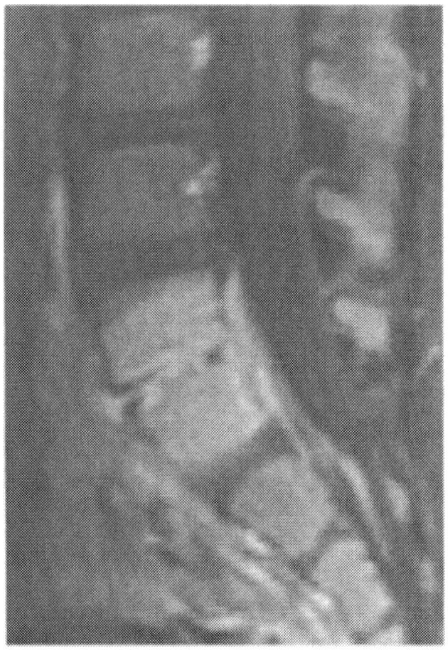

Figure 32C. Sagittal T1 post-Gd FS.

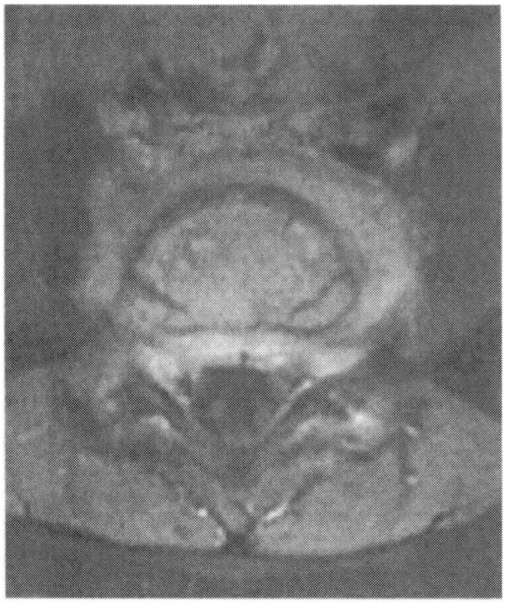

Figure 32D. Axial T1 post-Gd FS (through the S1 vertebral body).

Figure 32A. There is subtle focal disk space narrowing with mild endplate irregularity at the L5-S1 level.

Figures 32B, 32C, (32B with annotations). There is diffuse, increased STIR SI and enhancement present in the L5 and S1 vertebral bodies. The L5-S1 disk space is narrowed and shows enhancement. The L5-S1 endplates are disrupted **(thin arrows)**. Compare these findings with the normal L4–5 endplates **(thick arrows)** and intervertebral disk thickness and SI.

Figures 32D, (32D with annotations). Presacral and epidural soft tissue enhancement is better appreciated in the axial plane **(arrowheads)**.

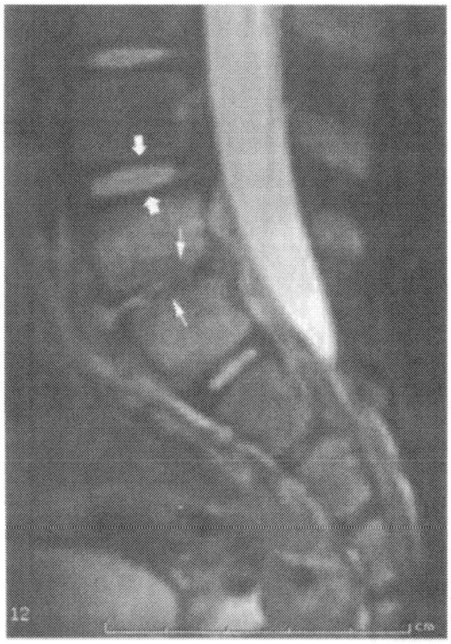

Figure 32B* Annotated.

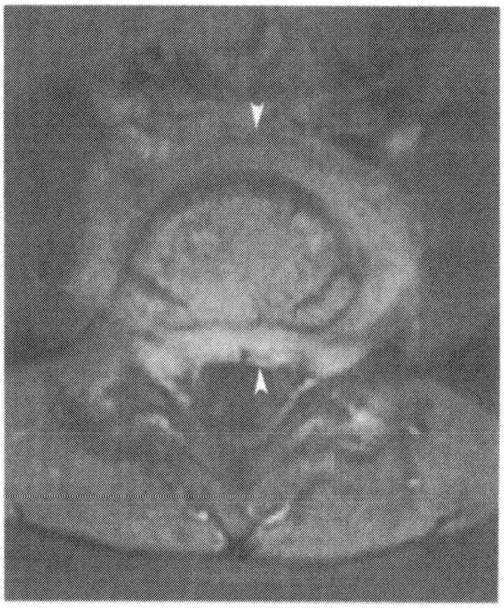

Figure 32D* Annotated.

Diagnosis

L5-S1 spondylodiscitis

Questions

1. T/F: An organism is isolated in the majority of cases of spondylodiscitis in infants and young children.
2. T/F: Isolated discitis is more common in early childhood than in adolescence.

Discussion

In infants and young children, the etiology of spondylodiscitis is controversial since an organism is usually not isolated (Answer to Question 1) (1). Blood and aspirate/biopsy cultures have been reported as negative 50% to 70% of the time. The presumed cause for childhood spondylodiscitis is a low-grade pyogenic or viral infection of the disk. As with adults, when cultures are positive, *Staphylococcus aureus* is most commonly isolated (2). The lumbar spine was most commonly affected (31 out of 42 patients, 74%) in one study evaluating spondylodiscitis in children ranging from 0 to 15 years of age (3).

One study that used both clinical and radiologic criteria showed that the mean age for discitis was 2.8 years in contrast to 7.5 years for vertebral osteomyelitis with or without discitis (Answer to Question 2) (1). The unique vascular supplies of the intervertebral disk of young and older children may explain this epidemiologic finding. In young children, crossing endplate vessels perfuse the disk and hematogenous bacterial seeding of the disk may occur. Infection centers at the disk, and secondary vertebral body osteomyelitis or reactive edema may be seen. For older children, these crossing endplate vessels regress; vertebral arterioles terminate in sinusoids located in the metaphyseal equivalent regions of the vertebral body (4). Therefore, spondylodiscitis in older children may begin at the endplate and secondarily involve the disk. Since there is often a delay between presentation and diagnosis, isolated discitis is uncommon and most patients present with clinical and radiologic features of spondylodiscitis.

The radiographic and MRI features that suggest spondylodiscitis include: intervertebral disk space narrowing, vertebral end-plate destruction with loss of the normal hypointense cortical margin, and presacral and paravertebral phlegmon or abscess. Inflammatory changes with T1 hypointensity, T2 hyperintensity, and enhancement, centered at either the intervertebral disk or the vertebral endplate are characteristic (5). Early osteomyelitis usually begins along the anterior subchondral vertebral endplate (5). Primary discitis may be suggested if rapid disk space narrowing is identified without evidence of endplate destruction.

The differential diagnosis for spondylodiscitis includes neoplasms and post-traumatic or degenerative changes. Unlike spondylodiscitis, neoplasms such as Langerhans cell histiocytosis, leukemia, and neuroblastoma metastases tend to center at the vertebral body or pedicle rather than the vertebral endplate or disk. In addition, these neoplasms may involve multiple noncontiguous levels at presentation. Both infection and neoplasm may demonstrate an extraosseous soft tissue component, which may be initially confusing without appropriate clinical history. When there is disc space narrowing and endplate irregularity, it may be impossible to differentiate sequelae of prior infection and disk degenerative processes such as Schmorl's nodes (Figures 32E, 32F). Schmorl's nodes are intraosseous endplate disk herniations that occur anteriorly, posteriorly, centrally, or laterally. They may be seen in isolation or in the setting of

Scheuermann's disease, limbus vertebra, disk degeneration, and disk herniation (Figure 32E) (6).

This patient was given a six-week course of IV antibiotics. An infectious agent was not found by blood cultures, and aspiration/biopsy was not performed. At one-year follow-up, she was doing well but was still having 1 to 2 episodes per week of minor back pain occurring in the morning.

Orthopedic Perspective

Radiographs are usually the first study requested in the radiologic work-up of back pain in children. Bone scintigraphy with SPECT and/or CT may be requested if radiographs are negative and there is a high clinical concern for spondylolysis. MRI evaluation is pursued if there are radiographic features of spondylodiscitis to assess for complications. Despite negative radiographs, in the face of a high clinical concern for discitis, spondylodiscitis, or other etiology for discogenic pain, MRI may still be indicated.

Since MRI may reliably distinguish between discitis and vertebral osteomyelitis, as well as assess the extent of inflammatory disease, it can have a substantial impact on management. Important imaging features affecting management include drainable fluid collections, presence and extent of epidural inflammation, and signs of cord compression. Aspiration/biopsy may be performed in cases that are not responding to medical management. Isolated disciitis may be treated with antibiotics for 3 to 4 weeks, whereas spondylodiscitis may require 6 to 8 weeks of antibiotic therapy.

What the Clinician Needs to Know

1. Differentiate between discitis and spondylodiscitis, if possible.
2. Identify the vertebral level that is affected and the presence of both intraspinal and paraspinal drainable collections.

Answers

1. False.
2. True.

Additional Examples

L5-S1 Disk Herniation and S1 Posterior Schmorl's Node

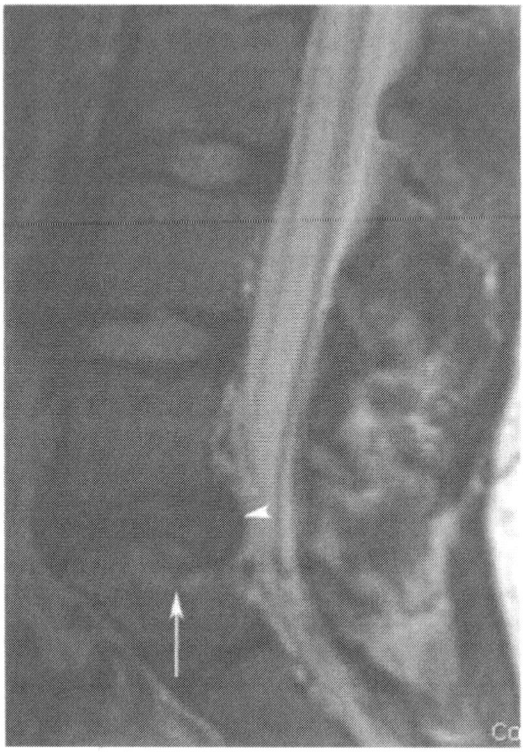

Figure 32E. Sagittal T2.

Findings

This is a 14-year-old girl hockey player with chronic back pain.

Figure 32E. Note L5-S1 disk dessication and narrowing, S1 posterior endplate Schmorl's node **(arrow)**, and disk herniation **(arrowhead)**.

L4-L5 Disk Herniation and L5 Anterior Schmorl's Node

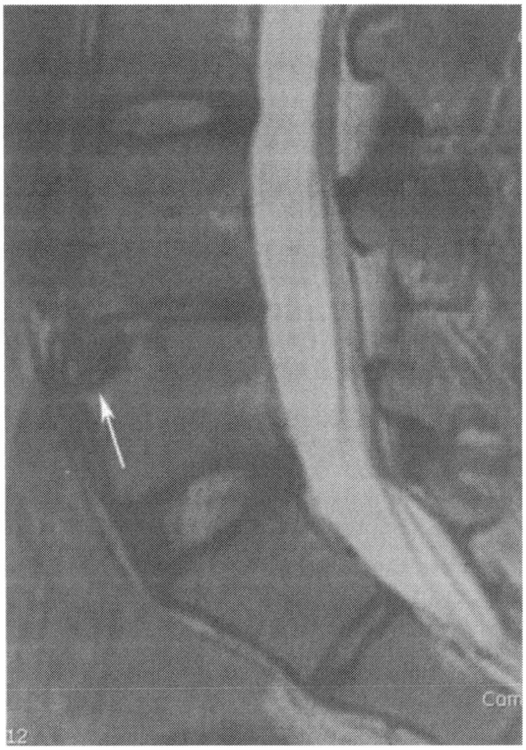

Figure 32F. Sagittal T2.

Findings

This is a 14-year-old girl ballet dancer with chronic back pain.

Figure 32F. Note L4–5 disk dessication, narrowing, and anterior L5 endplate Schmorl's node **(arrow)**. When this pattern interferes with fusion of the anterior ring apophysis, a limbus vertebrae may result.

Pitfalls and Pearls

1. Spondylodiscitis in children under 3 years of age often presents with nonspecific symptoms such as an inability to walk, with or without back pain (1). If osteomyelitis is suspected and lumbar imaging is negative, look elsewhere for infection such as the pelvic girdle and the lower extremity.
2. In the older child with subacute/chronic back pain, it may be impossible to clinically distinguish an infectious process from atypical Scheuermann's disease.

References

1. Fernandez M, Carrol CL, Baker CJ. Discitis and vertebral osteomyelitis in children: An 18-year review. *Pediatrics* 2000; 105:1299–1304.
2. Mendonca RA. Spinal infection and inflammatory disorders. In: *Magnetic Resonance Imaging of the Brain and Spine*, Atlas SW, ed. Philadelphia: Lippincott Williams and Wilkins, 2002; 1855–1972.

3. Garron E, Viehweger E, Launay F, Guillaume JM, Jouve JL, Bollini G. Nontuberculous spondylodiscitis in children. *J Pediatr Orthop* 2002; 22:321–328.
4. Song KS, Ogden JA, Ganey T, Guidera KJ. Contiguous discitis and osteomyelitis in children. *J Pediatr Orthop* 1997; 17:470–477.
5. Mahboubi S, Morris MC. Imaging of spinal infections in children. *Radiol Clin North Am* 2001; 39:215–222.
6. Swischuk LE, John SD, Allbery S. Disk degenerative disease in childhood: Scheuermann's disease, Schmorl's nodes, and the limbus vertebra: MRI findings in 12 patients. *Pediatr Radiol* 1998; 28:334–338.

Case 33

History

This is a 9-year-old girl with a 2-week history of left hip pain and inability to bear weight. She has a history of spastic diplegia but is very active. She normally walks with crutches.

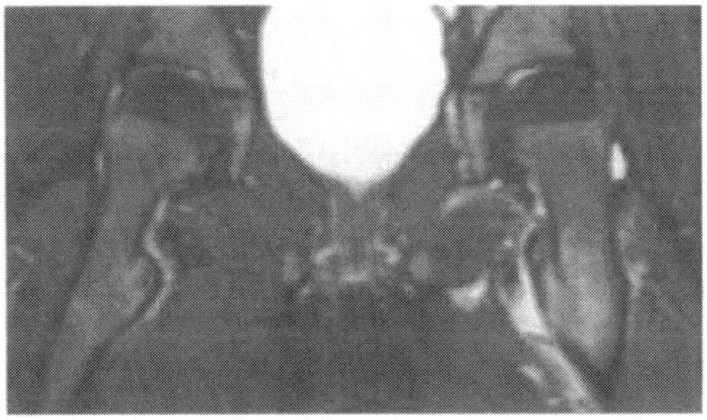

Figure 33A. Coronal STIR.

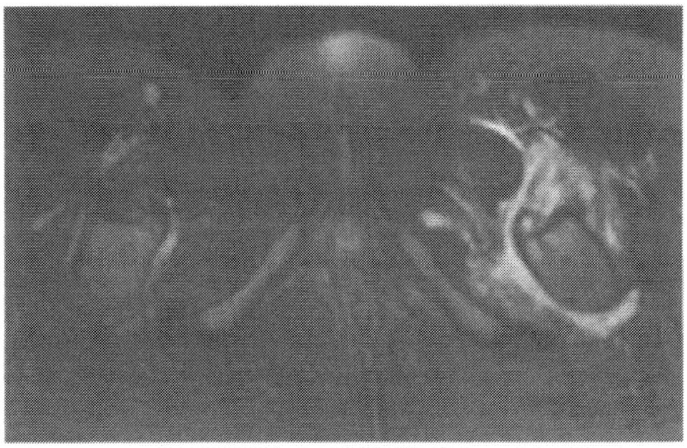

Figure 33B. Axial T2 FS.

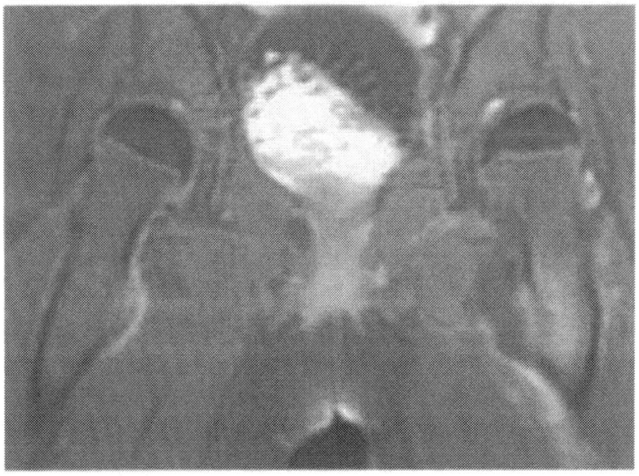

Figure 33C. Coronal T1 post-Gd FS.

Figures 33A, 33B, (33B with annotations). Bilateral coxa valga deformity is seen. On the left, there is increased fluid SI in the expected location of the apophysis of the lesser trochanter, at the site of insertion of the iliopsoas tendon. The apophysis of the lesser trochanter is displaced **(arrow)**. Compare with the normal location on the right **(arrow)**. There is also diffuse increased SI throughout the iliopsoas muscle and tendon **(arrowheads)**.

Figures 33C, (33C with annotations). The lesser trochanter apophysis **(arrow)** is displaced proximally. The base of the lesser trochanter is irregular with enhancement that extends into the intertrochanteric bone marrow **(arrowhead)**.

Figure 33D. The radiograph obtained at presentation was reported as negative. The right hip was not imaged at the time. In retrospect, the apophysis of the lesser trochanter is avulsed.

Figure 33E. A follow-up radiograph 9 months later shows healing of the left lesser trochanteric avulsion **(arrowhead)** and a normal right lesser trochanter **(arrow)**.

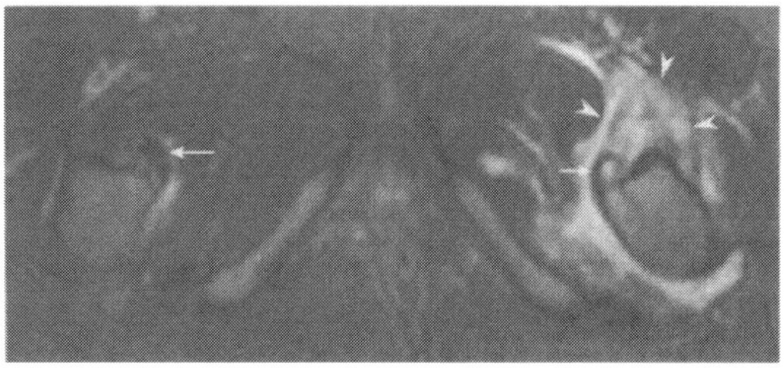

Figure 33B* Annotated.

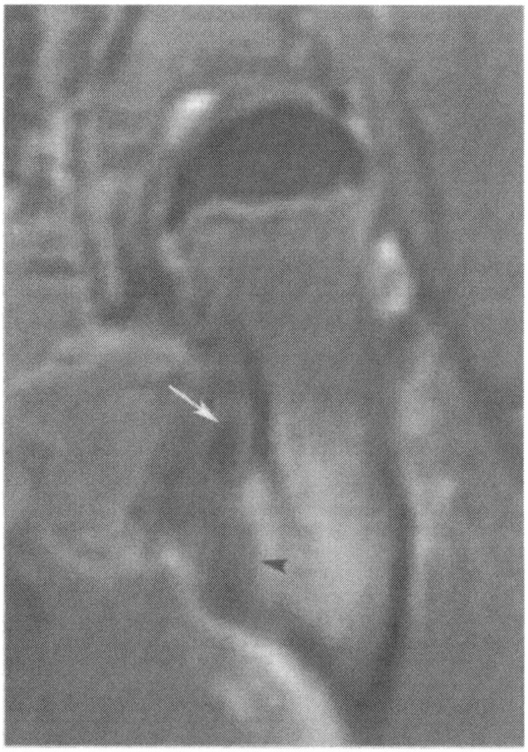

Figure 33C* Annotated detail.

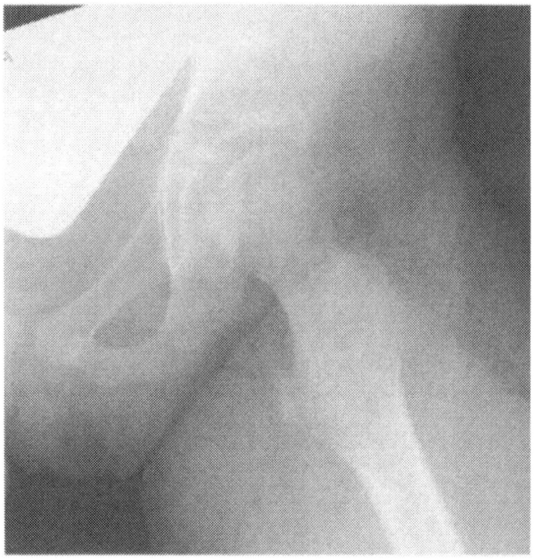

Figure 33D. Initial frog lateral radiograph of the left hip.

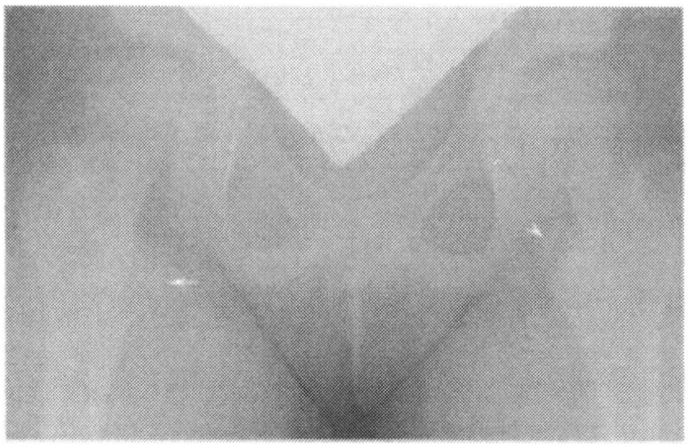

Figure 33E. AP radiograph of the pelvis 9 months later.

Diagnosis

Avulsion injury of the lesser trochanter

Questions

1. What is the most common avulsion injury of the pelvic girdle?
2. T/F: Avulsion injuries around the pelvis may be misdiagnosed as neoplasms.

Discussion

Avulsion injuries most commonly occur during periods of rapid growth when the physis is weakest. In one study evaluating 203 avulsion injuries in adolescent competitive athletes, the most commonly observed avulsion injury of the pelvic girdle was the ischial tuberosity, the point of origin of the hamstring muscles (Answer to Question 1). This site was followed by the anterior inferior iliac spine (AIIS) and the anterior superior iliac spine (ASIS), the points of origin of the rectus femoris and the sartorius, respectively (1). Lesser trochanter avulsion injuries due to tensile forces from the iliopsoas tendon are much less common. In adolescent and adult patients, myotendinous tears are more likely to occur compared with avulsion injuries once the physis closes (2). Lesser trochanter fractures are usually pathologic in adult patients and may be secondary to metabolic bone disease (e.g., renal osteodystrophy) or metastatic disease (3).

MRI evaluation of avulsion injuries in the acute or subacute setting may be challenging because there are usually significant and nonspecific soft tissue and bony signal abnormalities. These MRI findings can be misleading and suggest a neoplasm or infection. As with metaphyseal equivalent regions elsewhere, the lesser trochanter is also predisposed to infections and neoplasms (Figures 33F–33H). The most important imaging feature of an avulsion injury is properly identifying the displaced apophysis (Figure 33C) and the absence of a solid enhancing soft tissue mass. Without these findings, differentiating a traumatic injury from osteomyelitis or tumor may be difficult, since all may show hypointensity on T1, hyperintensity on fluid sensitive sequences, and abnormal contrast enhancement (Answer to Question 2). It is therefore essential to be aware of the MRI features of avulsive injuries to avoid confusion with more serious differential considerations. Subtle apophyseal avulsion injuries involving the pelvis and proximal femurs may be missed if the contralateral side is not imaged.

Another traumatic injury that may present for MR imaging is a myotendinous tear. The myotendinous junction is the weakest component of the myotendinous unit once the physis fuses. The clinical classification of myotendinous tears is first degree (stretch injury), second degree (partial tear) (Figures 33I–33L), and third degree (complete tear) (4). The MRI appearance of first-degree injuries includes edema and hemorrhagic products at the myotendinous junction without evidence of a discrete tear. T1W sequences are usually normal. Second-degree tears may demonstrate a visible partial disruption of the myotendinous unit with tendinous irregularity, thickening, or thinning. Third-degree tears demonstrate complete disruption of the myotendinous unit with muscle and tendon retraction. Alternative considerations for myotendinous tears include tendinopathy (edema and increased SI confined to a thickened tendon), delayed onset muscle soreness (DOMS), and direct muscle contusion. DOMS may be indistinguishable from first-degree myotendinous injury.

This patient initially presented with left hip pain after a fall. Because radiographs were normal, she was sent home. She was admitted a week later because of her severe hip pain and inability to bear weight. A normal hip US was performed prior to MRI.

She was treated conservatively and at 9-month follow-up, her only symptoms were minor left paralumbar pain related with sitting.

Orthopedic Perspective

Apophyseal avulsion fracture of the lesser trochanter is an acute traumatic injury caused by forceful contracture of the iliopsoas tendon. This injury often occurs around the time of rapid growth when the apophysis is relatively weak and the iliopsoas is often tight. Clinically, the patient presents with hip pain after an acute injury. The lesser trochanter is difficult to palpate; however, the patient will have weakness and pain with flexion. The diagnosis can usually be made by clinical examination and radiographs, including both hips for comparison. MRI is rarely needed for diagnosis but may be requested when the correct diagnosis is not made initially, as in this case. Treatment is almost always nonoperative.

What the Clinician Needs to Know

1. Degree of apophyseal displacement and alternative causes for hip or pelvis pain (e.g., avulsion injury versus stress fracture, infection, neoplasm).
2. When a myotendinous tear is present, is the tear incomplete or complete?

Answers

1. Ischial tuberosity.
2. True.

Additional Examples

Cortical Aneurysmal Bone Cyst (ABC) Adjacent to the Lesser Trochanter

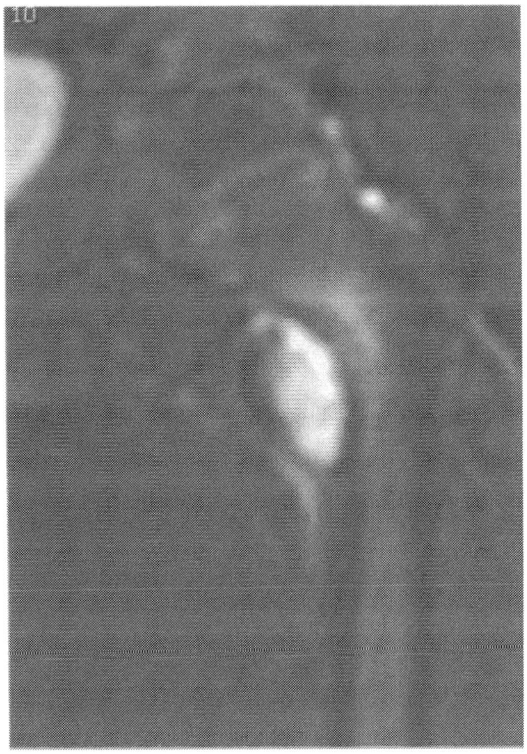

Figure 33F. Coronal STIR of the left proximal femur.

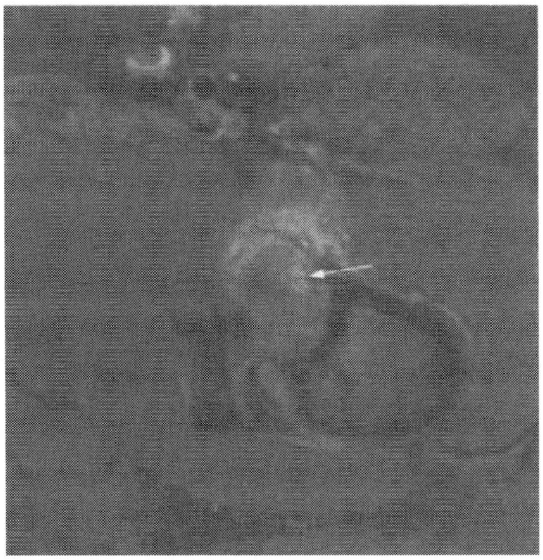

Figure 33G. Axial T1 post-Gd FS.

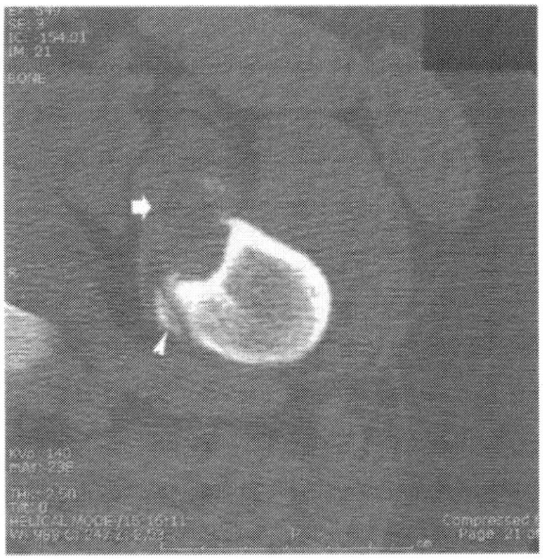

Figure 33H. CT.

Findings

This is a 13-year-old boy who had left hip pain for three weeks.

Figures 33F, 33G. There is a cortically based hyperintense mass in the region of the lesser trochanter on the STIR image. Mild peripheral and adjacent soft tissue edema and enhancement are seen. A central area of nodular enhancement is also noted **(thin arrow)**.

Figure 33H. CT reveals a mass with a thin expanded cortical rim **(thick arrow)** immediately anterior to the lesser trochanteric apophysis **(arrowhead)**. This was a pathologically confirmed ABC.

Grade 2 Myotendinous Tear of the Iliopsoas Muscle

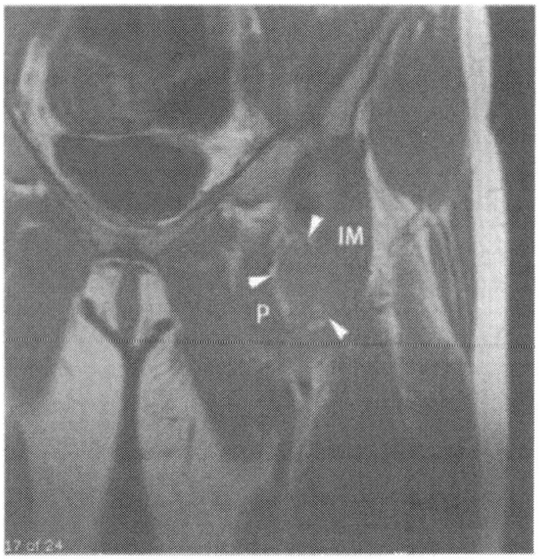

Figure 33I. Coronal T1 of the proximal left thigh.

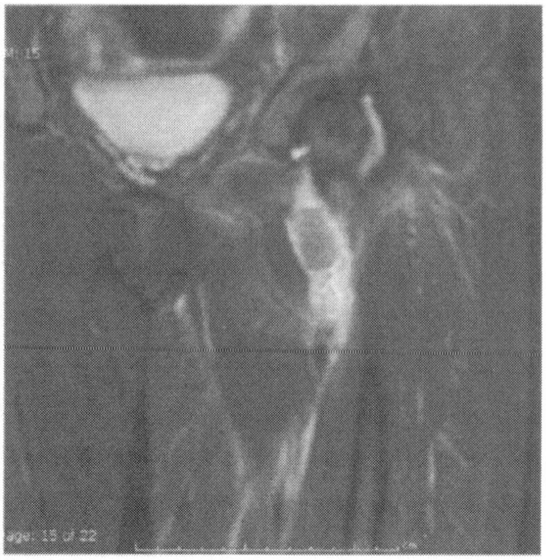

Figure 33J. Coronal STIR.

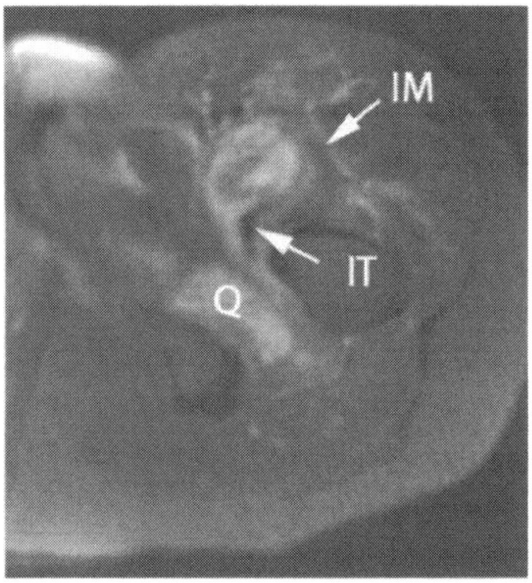

Figure 33K. Axial T2 FS.

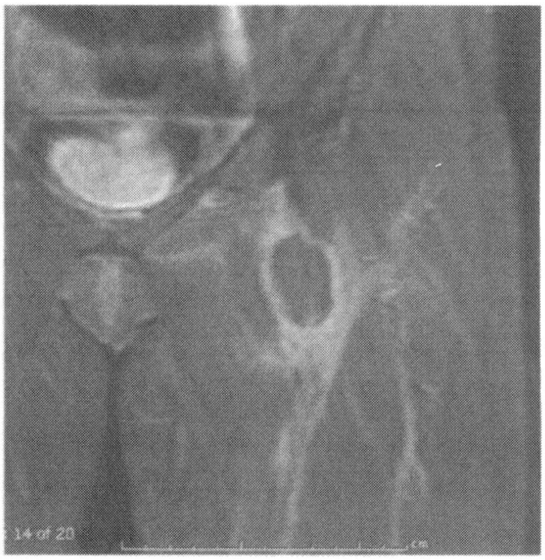

Figure 33L. Coronal T1 post-Gd FS.

Findings

This is an 18-year-old girl with left thigh and pelvic pain for one week. There was no specific traumatic event, but echymosis was evident in the left thigh.

Figures 33I, 33J. An ovoid mass is noted at the iliopsoas myotendinous (IM) junction that is isointense to muscle on T1 **(white arrowheads)** and heterogeneously mildly hyperintense on STIR. The pectineus muscle (P) is displaced medially by the mass.

Figure 33K. The mass lies between the iliopsoas tendon (IT) and iliopsoas muscle belly (IM) and is surrounded by T2 hyperintense fluid signal. There is feathery stranding of the iliopsoas and neighboring muscles (Q = quadratus femoris muscle). There are intact strands of muscle fibers that connect the iliopsoas muscle and iliopsoas tendon located posteriorly.

Figure 33L. The mass does not centrally enhance. The findings are compatible with a grade 2 myotendinous tear and hematoma of the iliopsoas muscle.

Pitfalls and Pearls

Avulsive injuries around the pelvis and hips may simulate more serious conditions, such as osteomyelitis and neoplasm. Correlation with history is often useful, and plain radiographs are essential.

References

1. Rossi F, Dragoni S. Acute avulsion fractures of the pelvis in adolescent competitive athletes: Prevalence, location and sports distribution of 203 cases collected. *Skeletal Radiol* 2001; 30:127–131.
2. Ecklund K. Magnetic resonance imaging of pediatric musculoskeletal trauma. *Top Magn Reson Imaging* 2002; 13:203–217.
3. Bencardino JT, Palmer WE. Imaging of hip disorders in athletes. *Radiol Clin North Am* 2002; 40:267–287, vi–vii.
4. Palmer WE, Kuong SJ, Elmadbouh HM. MR imaging of myotendinous strain. *AJR Am J Roentgenol* 1999; 173:703–709.

Case 34

History

This is a 2-year-old girl with a history of bilateral knee clicking.

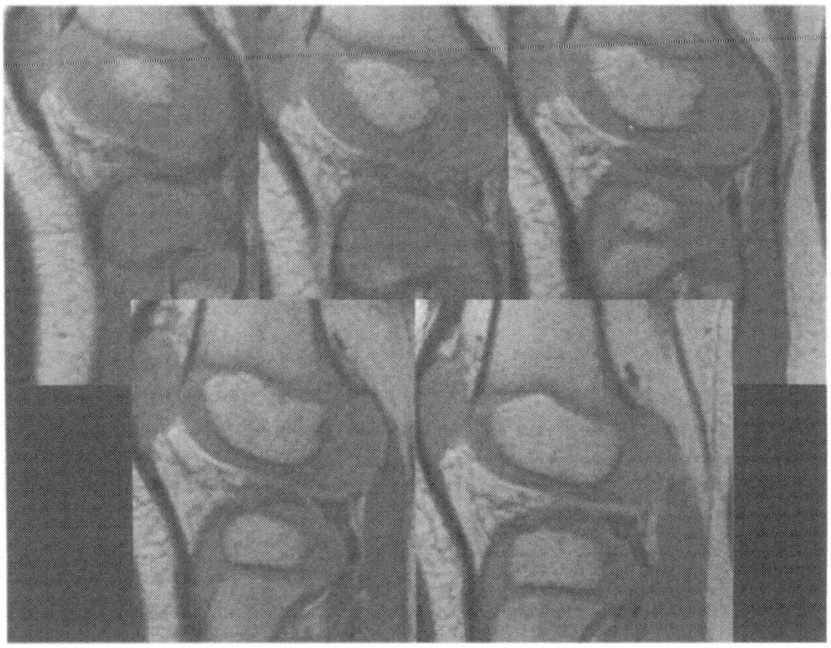

Figure 34A. Sagittal PD lateral to medial, 3 mm contiguous sections of the left knee.

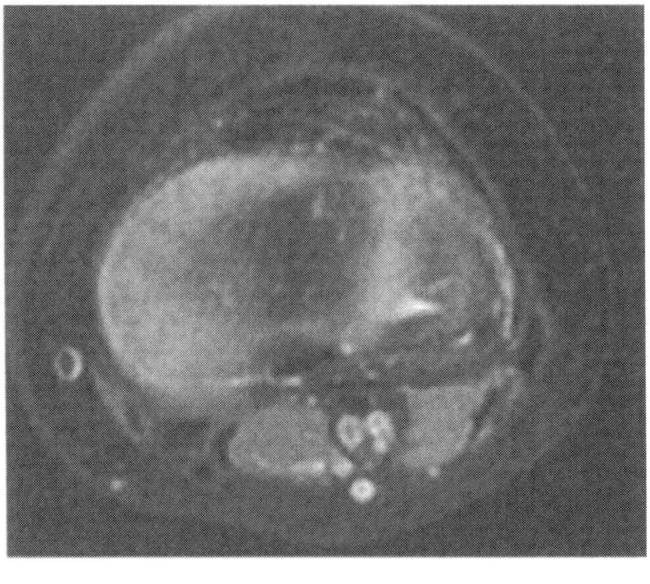

Figure 34B. Axial PD FS (through the meniscus).

Figures 34A, (34A with annotations). The transverse dimension of the body of the lateral meniscus is increased; meniscal bowties are visible on more than three contiguous slices. The central and peripheral height of the meniscus is similar. There is abnormal globular intrasubstance increased SI present. There is also a vertical-oblique tear at the junction of the body and posterior horn that extends from the superior to the inferior articular surface **(arrows)**.

Figures 34B, (34B with annotations). There is a radial tear involving the central zone at the junction of the body and posterior horn of the lateral meniscus **(arrow)**. The lateral meniscus has a semicircular configuration rather than a normal C-shape. Anterior horn (A), body (B), and posterior horn (P).

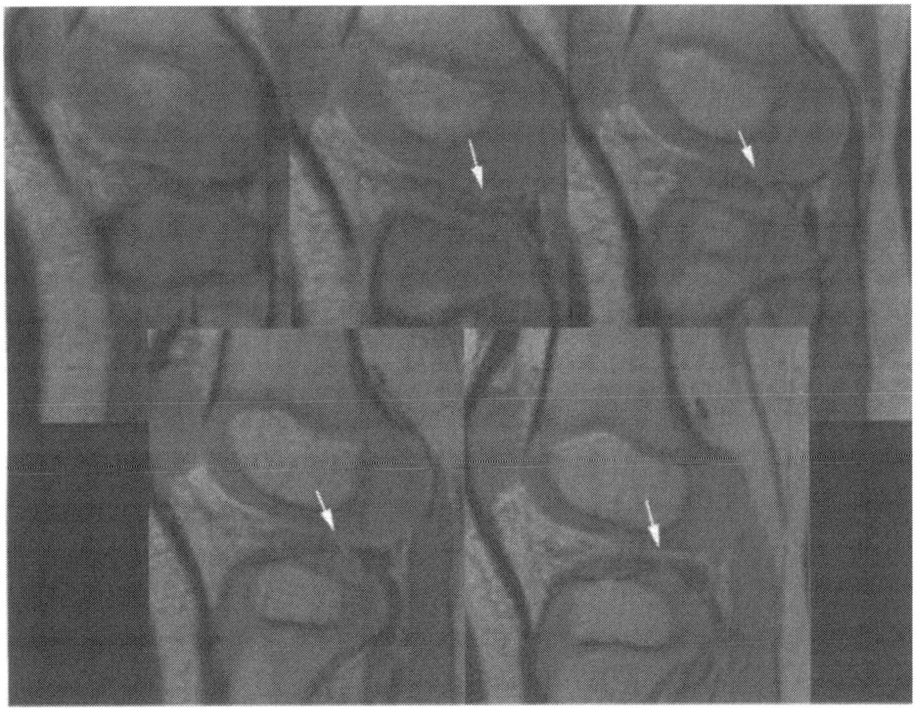

Figure 34A* Annotated.

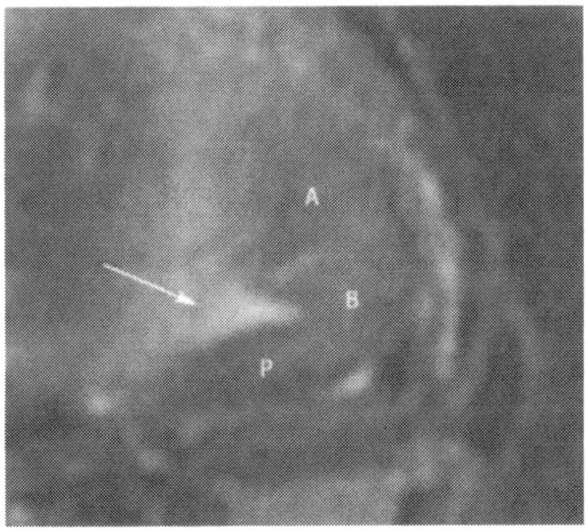

Figure 34B* Annotated detail.

Diagnosis

Discoid lateral meniscus (type 2) with a radial tear

Questions

1. What is the most common meniscal tear associated with a discoid lateral meniscus?
2. What is the Wrisberg variant of the discoid lateral meniscus?

Discussion

The reported incidence of a discoid lateral meniscus is 3% to 5% of the general population (1). The incidence of bilaterality is up to 20%. The majority of patients are asymptomatic. Discoid medial meniscus is less commonly found (0.6% of the general population in one study) (2). Because of their abnormal anatomy and mechanics, discoid menisci are predisposed to tears and degeneration (Figures 34C, 34D). The most common type of tear is a degenerative horizontal cleavage tear (Answer to Question 1). In one study, 69.5% of discoid meniscus demonstrated a meniscal tear at arthroscopy (3).

The arthroscopic classification of a discoid lateral meniscus is: type 1 (complete), which covers the entire tibial plateau; type 2 (incomplete), which covers less than 80% of the plateau; and type 3, which is the Wrisberg variant (1). The Wrisberg variant lacks a normal posterior attachment to the meniscofemoral ligaments (Answer to Question 2). Consequently, the posterior horn is free floating and can be symptomatic. A complementary classification system of discoid meniscus, initially based on conventional arthrography, distinguishes discoid meniscus into six morphologic types: slab (flat meniscus extending to the notch, with all portions having almost equal thickness), biconcave, wedge, asymmetric anterior, forme fruste (enlarged but normal meniscus shape), and grossly torn (not typeable due to deranged anatomy) (4).

The meniscus is divided into three parts: the anterior horn, body, and posterior horn. The normal lateral meniscus has a C-shape in the axial plane. The anterior and posterior horn of the lateral meniscus should have similar transverse and vertical dimensions. The bowtie sign represents the anterior and posterior horn connected by the meniscal body. With a discoid lateral meniscus, the C-shape becomes more semicircular because the thickened body extends medially toward the intercondylar notch. In adult patients, the normal transverse meniscal thickness is 9 to 12 mm (5). Therefore, on 4 mm sagittal images, the bowtie appearance of the meniscus should not be visible on more than three contiguous slices. This principle should be adjusted accordingly for the smaller dimensions of the menisci in the child and young adolescent. Although this bowtie pattern in excessive sagittal slices is a reliable indicator of a discoid lateral meniscus, absence of this sign does not rule out an incomplete discoid meniscus in the young child. Discoid lateral menisci may also demonstrate height asymmetry between the anterior and posterior horn as well as displacement due to unstable tears (Figures 34C, 34D) (6, 7).

Meniscal tears are classified based on shape (vertical, horizontal, or complex (Diagram 34A) and location (peripheral or central). Vertical tears are subdivided based on direction: radial, longitudinal, or parrot-beak (contains radial and longitudinal components) (Diagram 34B). A bucket-handle tear is a vertical, longitudinal tear with a displaced inner fragment (see Case 54).

The history of knee clicking (snapping knee), as noted in this patient, is common in children with an unstable discoid meniscus (1). A discoid meniscus with a parrot-beak tear was also noted in the opposite knee (not shown). At arthroscopy, this patient had

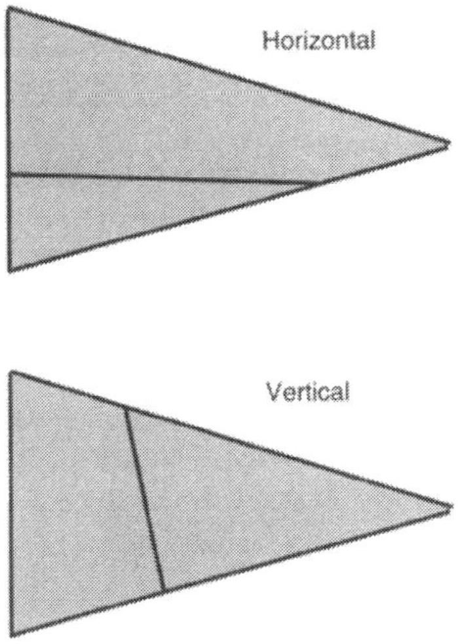

Diagram 34A. Meniscal tear direction.

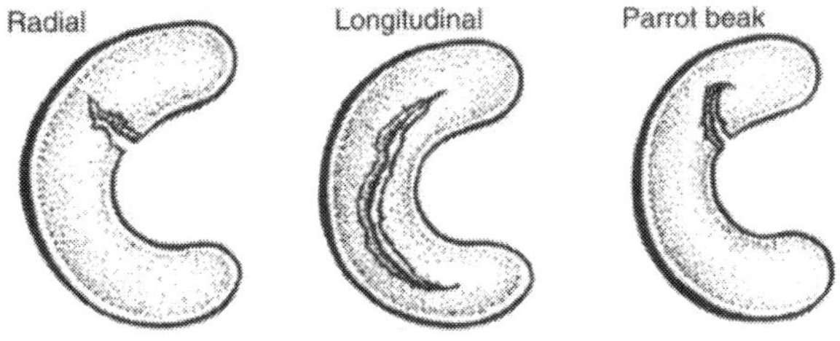

Diagram 34B. Subtypes of vertical meniscal tears. (Reprinted from Anderson M (8), with permission from Elsevier.)

an incomplete discoid menisci (type II) with a radial tear. The patient had saucerization and debridement of discoid menisci in both knees and did well.

Orthopedic Perspective

The presentation and treatment of discoid meniscus are highly variable and are related to the type of lesion present, peripheral stability of the meniscus, and the presence of an associated meniscal tear. Stable discoid menisci without associated tears may remain asymptomatic, identified only as incidental findings during MRI or arthroscopy. These cases are typically observed. Unstable discoid menisci more commonly present in younger children and often produce the so-called "snapping knee syndrome" with a painless and palpable, audible, or visible snap with knee motion, especially near terminal extension. If the popping is symptomatic and functionally limiting, surgery is recommended. In children with stable discoid lateral menisci, symptoms often present when the meniscus tears. Signs and symptoms of a meniscal tear include: pain, swelling, catching, locking, and limited motion. Unlike acute meniscal tears, such symptoms may present insidiously without previous trauma. For cases with meniscal tears, surgery is performed. Contemporary surgery for discoid lateral meniscus involves arthroscopic reshaping of the meniscus (saucerization) to a 6–8 mm rim with concomitant meniscal repair if unstable (Figures 34E, 34F). Total meniscectomy is avoided as it leads to early degenerative joint disease. Asymptomatic discoid menisci that are incidentally identified on MRI are observed and are not routinely referred to arthroscopy.

What the Clinician Needs to Know

1. Classification of the discoid meniscus: complete, incomplete, or Wrisberg.
2. Meniscal tearing (tear vs. no tear, type of tear, location of tear), and stability (peripheral detachment vs. stable).

Answers

1. Degenerative horizontal cleavage tear.
2. Absent posterior meniscal attachment to the meniscofemoral ligament.

Additional Examples

Discoid Lateral Meniscus (Type 2)

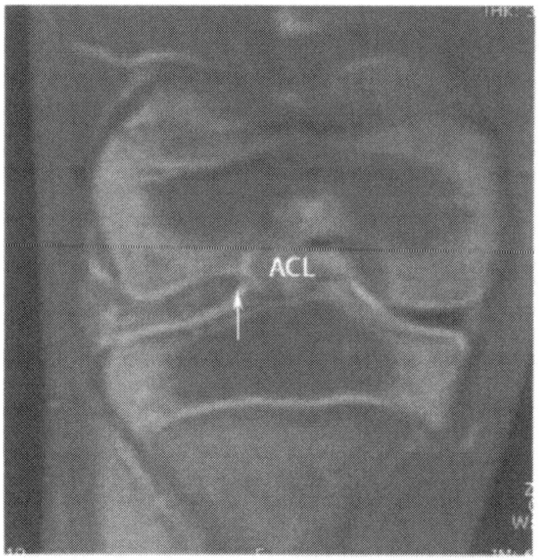

Figure 34C. Coronal PD FS of the right knee.

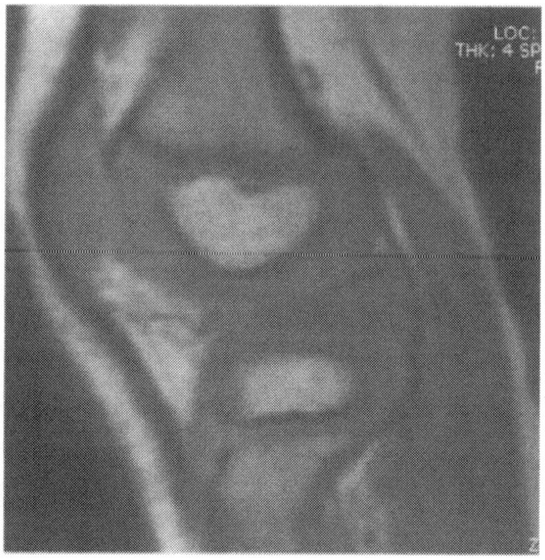

Figure 34D. Sagittal PD FS.

Findings

This 4-year-old boy presented with right knee snapping.

Figure 34C. The body of the meniscus extends far medially to the intercondylar notch (**arrow**). Anterior cruciate ligament (ACL).

Figure 34D. The entire lateral meniscus is mildly subluxed posteriorly. The posterior horn is taller than the anterior horn (wedge shape) and demonstrates diffuse increased intrasubstance signal. A degenerated and hypermobile lateral discoid meniscus was confirmed at arthroscopy.

Arthroscopic Appearance of Discoid Lateral Meniscus with Horizontal Cleavage Tear

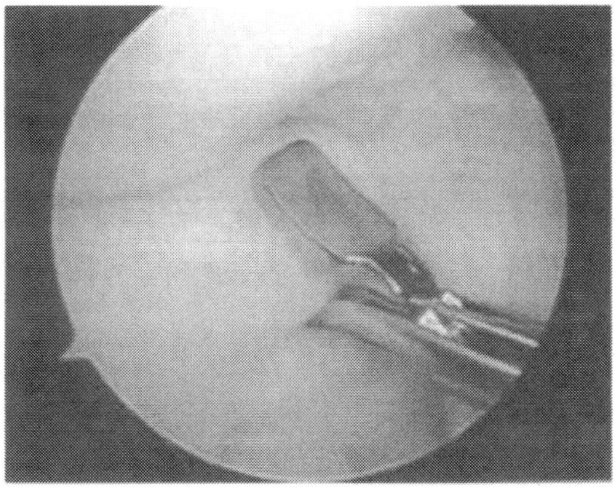

Figure 34E. Pre-saucerization appearance of a complete discoid lateral meniscus covering the lateral tibial plateau and extending into the intercondylar notch.

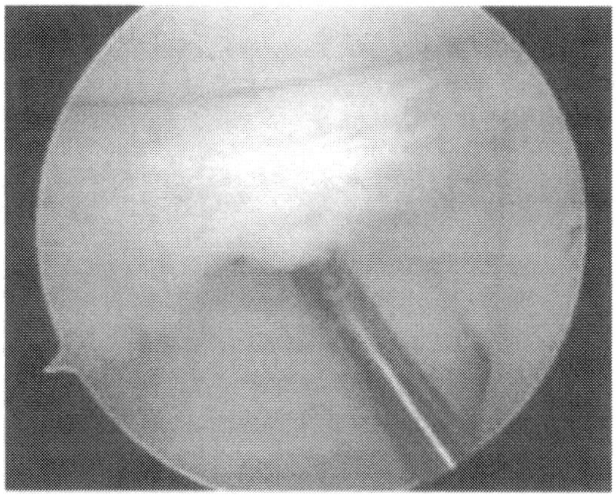

Figure 34F. Probing the unstable superior flap of the horizontal cleavage tear. (Reprinted from Kocher MS (1) with permissoin from Elsevier.)

Pitfalls and Pearls

1. The incomplete discoid lateral meniscus (type 2) may be subtle and easily overlooked on MRI. If there is a strong clinical suspicion of a discoid meniscus, the diagnosis should still be considered despite a negative MRI, and diagnostic arthroscopy may be necessary (9). A solid understanding and experience with the subtle MRI features of discoid lateral menisci are invaluable for proper diagnosis.
2. Don't forget to assess the menisci on the axial images.

References

1. Kocher MS, Klingele K, Rassman SO. Meniscal disorders: Normal, discoid, and cysts. *Orthop Clin North Am* 2003; 34:329–340.
2. Rohren EM, Kosarek FJ, Helms CA. Discoid lateral meniscus and the frequency of meniscal tears. *Skeletal Radiol* 2001; 30:316–320.
3. Klingele KE, Kocher MS, Hresko MT, Gerbino P, Micheli LJ. Discoid lateral meniscus: Prevalence of peripheral rim instability. *J Pediatr Orthop* 2004; 24:79–82.
4. Hall FM. Arthrography of the discoid lateral meniscus. *AJR Am J Roentgenol* 1977; 128:993–1002.
5. Helms CA. The meniscus: Recent advances in MR imaging of the knee. *AJR Am J Roentgenol* 2002; 179:1115–1122.
6. Strouse PJ, Koujok K. Magnetic resonance imaging of the pediatric knee. *Top Magn Reson Imaging* 2002; 13:277–294.
7. Ryu KN, Kim IS, Kim EJ, et al. MR imaging of tears of discoid lateral menisci. *AJR Am J Roentgenol* 1998; 171:963–967.
8. Anderson MW. MR imaging of the meniscus. *Radiol Clin North Am* 2002; 40:1081–1094.
9. Kocher MS, DiCanzio J, Zurakowski D, Micheli LJ. Diagnostic performance of clinical examination and selective magnetic resonance imaging in the evaluation of intraarticular knee disorders in children and adolescents. *Am J Sports Med* 2001; 29:292–296.

Case 35

History

This is a 1-year-old-boy with a 1-week history of a left leg limp and a temperature of 103 degrees. There was no history of trauma.

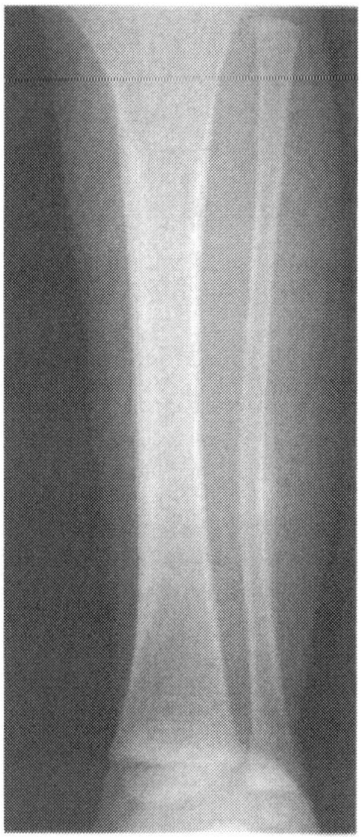

Figure 35A. AP radiograph of the left tibia and fibula.

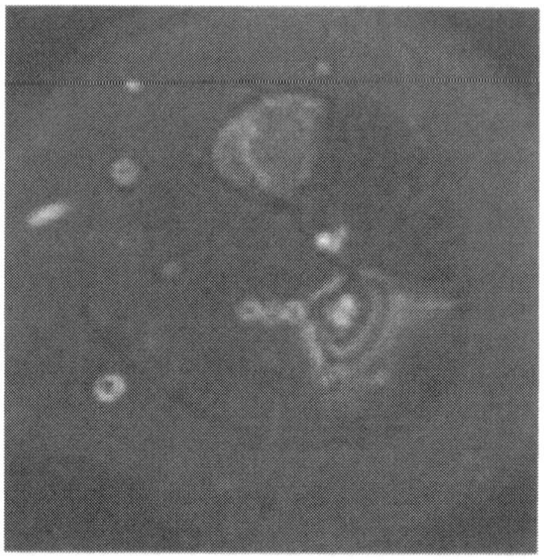

Figure 35B. Axial T2 FS.

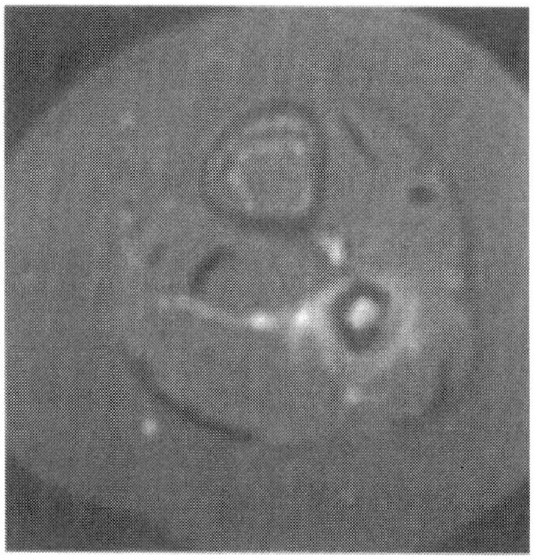

Figure 35C. Axial T1 post-Gd FS.

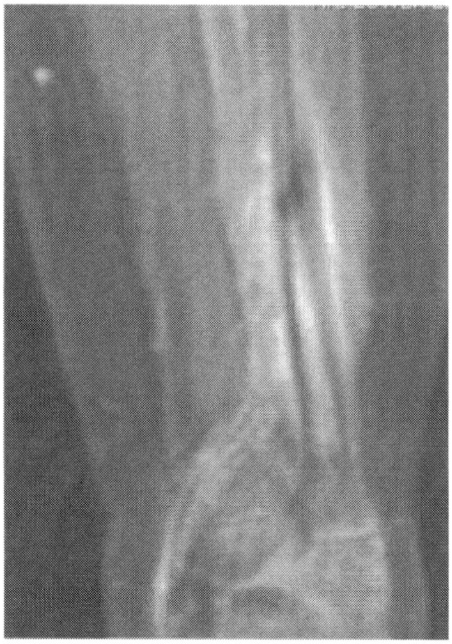

Figure 35D. Coronal T1 post-Gd FS.

Figure 35A. There is a mild valgus bowing deformity of the fibula. There is periosteal reaction present along the concavity of the bone.

Figures 35B, 35C, (35B with annotations). There is corresponding periosteal elevation and subperiosteal T2 hyperintensity **(arrow)**. Juxta-cortical soft tissue hyperintensity and enhancement are evident.

Figure 35D, (35D with annotations). There is an ill-defined focus of hypointensity in the distal 1/3 of the fibular diaphysis **(arrow)**, most likely representing a fracture line. There is also marrow and adjacent soft tissue enhancement.

Figure 35E. There is a focus of increased uptake in the distal 1/3 of the left fibula **(arrow)**.

Figure 35F. A subtle linear fracture line is present at the site of maximal periosteal reaction **(arrow)**.

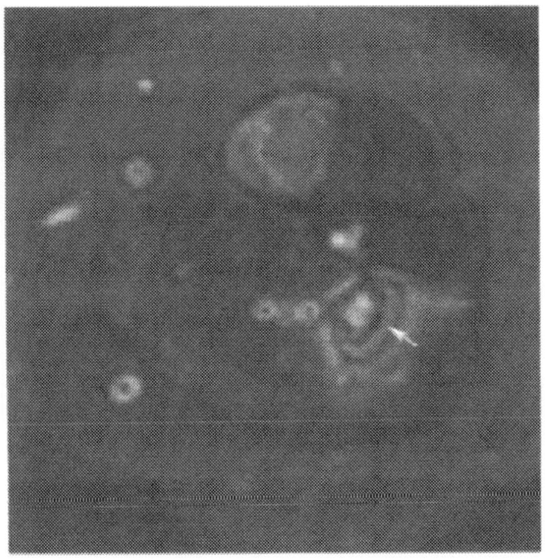

Figure 35B* Annotated.

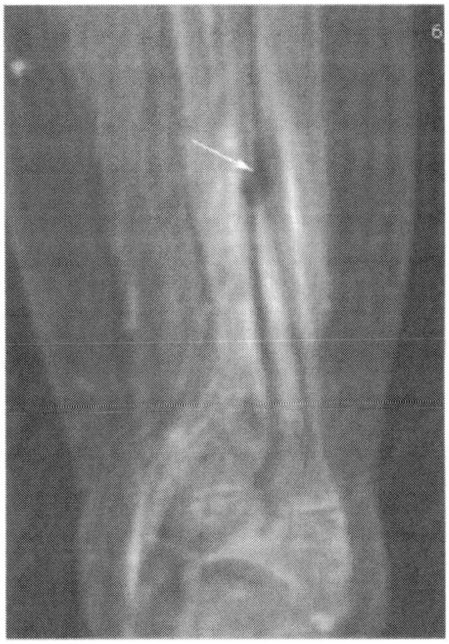

Figure 35D* Annotated.

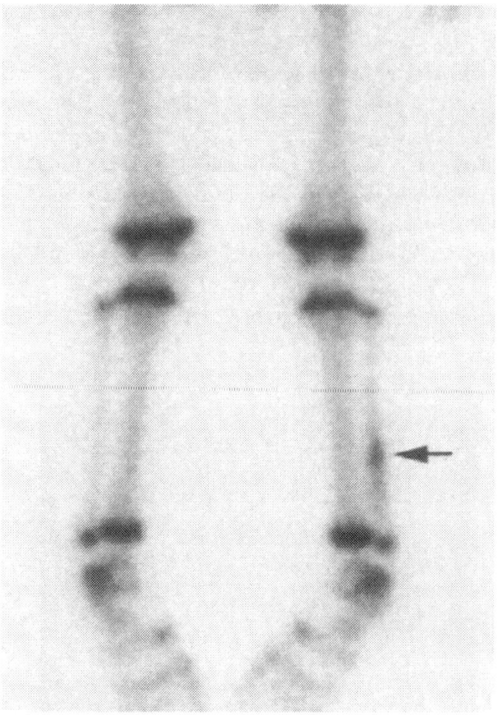

Figure 35E. Tc-99m MDP bone scintigraphy.

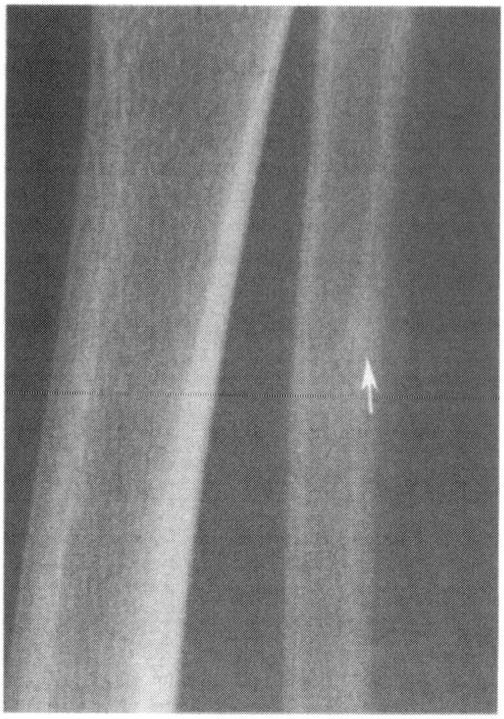

Figure 35F. High-detail oblique radiograph obtained after the MRI.

Diagnosis

Toddler's fracture of the fibula

Questions

1. T/F: Toddler's fractures may occur with or without a specific traumatic event.
2. What mechanism leads to a toddler's fracture of the fibula? How does this contrast with the mechanism of injury leading to a toddler's fracture of the tibia?

Discussion

The term *toddler's fracture* has historically been used to describe an oblique or spiral fracture of the distal tibia in children from 9 months to 3 years of age (1). The proposed mechanism is a twisting injury with early ambulation. A temporal history of trauma is occasionally obtained. Currently, the term toddler's fracture spans a clinical and radiologic spectrum ranging from an acute fracture occurring with a discrete traumatic event, to a subacute stress injury associated with pain and unwillingness to bear weight (Answer to Question 1). Aside from the tibia, other toddler's fracture sites include the fibula, metatarsals, and tarsal bones (Figures 35G, 35H) (2).

The fibula is a relatively uncommon site for a toddler's fracture, although fibular stress fractures are well-described in the older child and may even occur bilaterally. John et al. have noted that fibula bowing and stress fractures may occur while learning to walk and are due to vertical axial loading injuries (3). Ogden believes that fibular toddler's fractures, unlike tibial injuries, do not result from twisting forces (4). This is because the proximal and distal tibial-fibula joint mechanism protects the fibula during twisting (Answer to Question 2).

The most common presentation of a toddler's fracture is an inability to bear weight and negative plain radiographs at initial presentation (5). Other findings include warmth, point tenderness, and ankle pain with dorsiflexion. When clinical findings and plain radiographs are equivocal, MRI may be requested to evaluate for infection or neoplasm. MRI may support these concerns, since stress fractures may show variable soft tissue and osseous signal abnormality in the absence of a clear fracture line. Therefore, requests for MRI in this context should be carefully screened in light of clinical and radiographic findings to avoid an unnecessary study.

This patient was initially treated for osteomyelitis because of fever and the periosteal reaction evident on radiographs. Antibiotics were discontinued once the diagnosis of a healing fracture was confirmed on a subsequent high-detail radiograph (Figure 35F). Blood cultures during his hospitalization were negative.

Orthopedic Perspective

The diagnosis of a toddler's fracture is usually made based on both the clinical exam (refusal to walk and point tenderness to palpation) and radiographs. The child should be treated with a short leg walking cast for 3 to 4 weeks even if the radiographs are negative as long as there is a high clinical suspicion for a toddler's fracture. If the radiographs are initially negative, a follow-up study in approximately 2 to 3 weeks is recommended to evaluate for a fracture and periosteal reaction. As seen in this case, MRI evaluation may also be useful although the findings may be misleading particularly if a discrete fracture line is not seen.

What the Clinician Needs to Know

1. Location and extent of a toddler's fracture.
2. Are the changes related to stress reaction or a true toddler's fracture?

Answers

1. True.
2. A toddler's fracture of the fibula results from vertical axial loading. A toddler's fracture of the tibia is due to a twisting injury.

Additional Example

Toddler's Fracture of the Cuboid

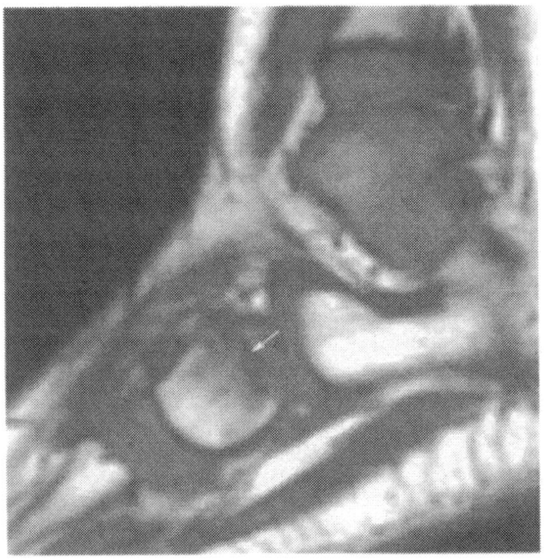

Figure 35G. Sagittal T1 of the left foot.

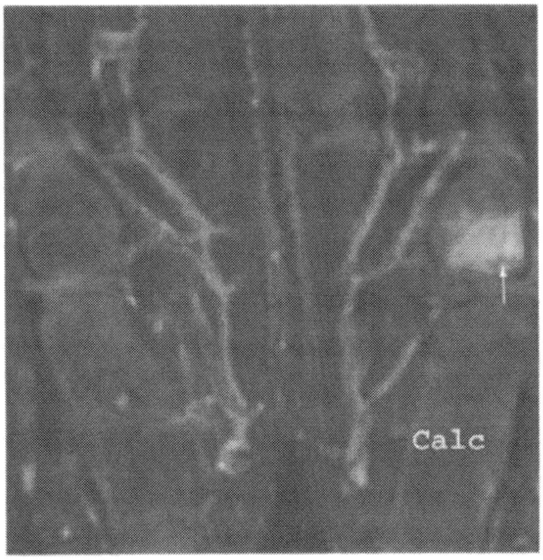

Figure 35H. Axial (footprint) STIR.

Findings

This 2-year-old girl presented with left foot pain.

Figures 35G, 35H. Within the cuboid, there is abnormal marrow SI surrounding a hypointense fracture line **(arrows)**. Calcaneus (Calc).

Pitfalls and Pearls

The toddler's fracture is a clinical and plain radiograph diagnosis. Oblique and/or high-detail radiographs may show a subtle fracture line, which effectively excludes infection or tumor and obviates the need for additional imaging.

References

1. Dunbar JS, Owen HF, Nogrady MB, McLeese R. Obscure tibial fracture of infants: The toddler's fracture. *J Can Assoc Radiol* 1964; 15:136–144.
2. Blumberg K, Patterson RJ. The toddler's cuboid fracture. *Radiology* 1991; 179:93–94.
3. John SD, Moorthy CS, Swischuk LE. Expanding the concept of the toddler's fracture. *Radiographics* 1997; 17:367–376.
4. Ogden JA. Tibia and fibula. In: *Skeletal Injury in the Child*, 3rd ed., Chapter 23. New York: Springer, 2000; 990–1090.
5. Halsey MF, Finzel KC, Carrion WV, Haralabatos SS, Gruber MA, Meinhard BP. Toddler's fracture: Presumptive diagnosis and treatment. *J Pediatr Orthop* 2001; 21:152–156.

Case 36

History

This is an 8-year-old boy with right knee pain.

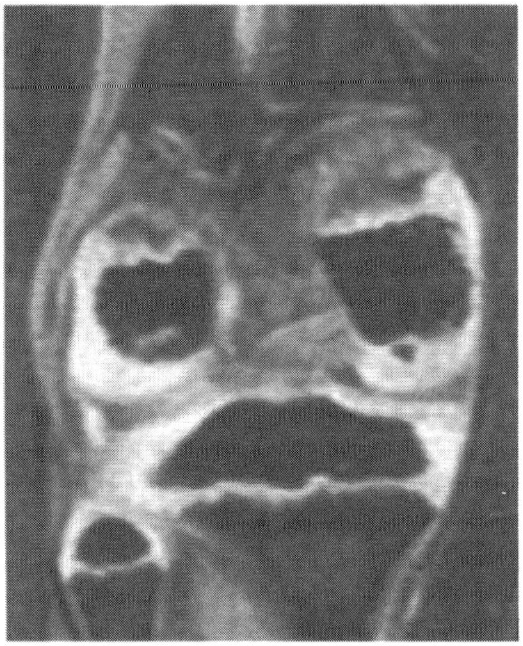

Figure 36A. Oblique coronal 3D SPGR FS of the right knee through the posterior femoral condyles.

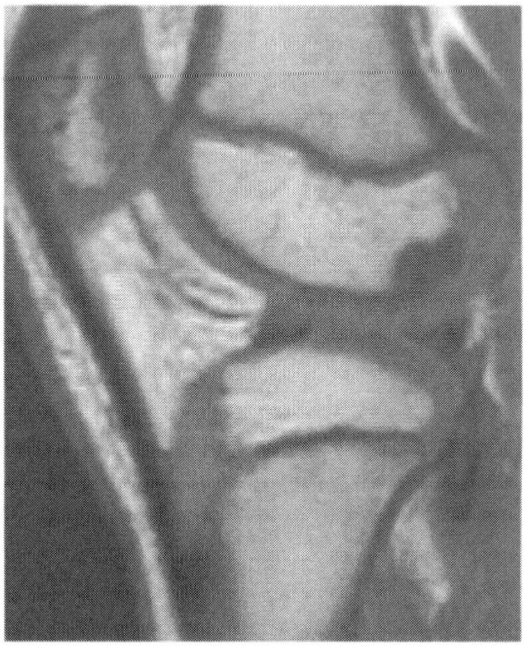

Figure 36B. Sagittal PD through the lateral femoral condyle.

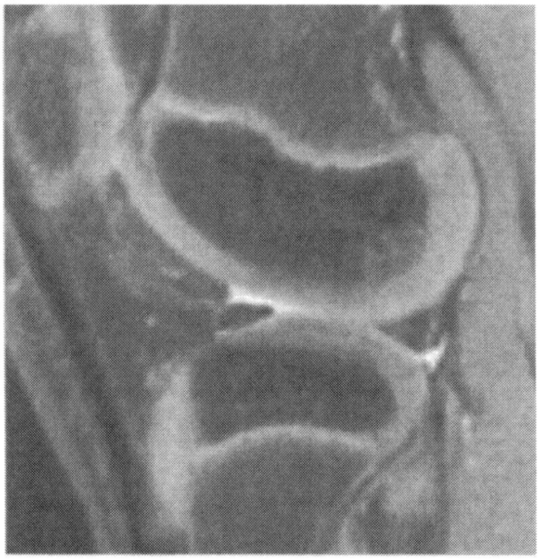

Figure 36C. Sagittal PD FS of the lateral femoral condyle.

Figure 36A. Subchondral bone irregularity and fragmentary ossification is present along the medial and lateral femoral condyles. The overlying cartilage is intact.

Figure 36B. A subchondral semicircular hypointensity is present in the posterior aspect of the lateral femoral condyle. It extends from medullary bone to the spherical growth plate of the epiphysis.

Figure 36C. With fat saturation, this same region shows a similar SI to the adjacent epiphyseal cartilage. No epiphyseal or articular cartilaginous defect is seen.

Figure 36D. There is subchondral sclerosis and radiolucency in the lateral femoral condyle **(arrow)**. There is subchondral fragmentary ossification of the medial femoral condyle **(arrowhead)**.

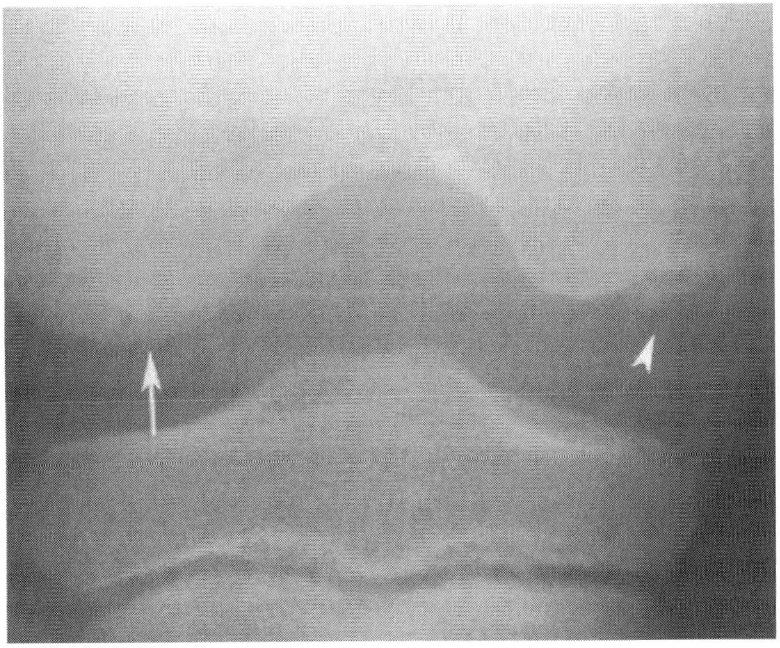

Figure 36D. Tunnel view of the right knee.

Diagnosis

Normal developmental irregularity of the femoral condyles

Questions

1. What is the principal differential diagnostic consideration for this normal developmental variant?
2. What is the most common location for developmental irregularities of the femoral condyle?

Discussion

The spectrum of findings for normal developmental irregularity of the femoral condyle ranges from subtle, marginal irregularities to discrete condylar bone fragments (1). These developmental condylar irregularities occur within the subchondral bone, located just beneath the epiphyseal spherical growth plate. They may show variable lucency, sclerosis, and fragmentation on radiography. These normal variations should not be confused with osteochondritis dissecans (OCD) (AKA osteochondral lesion) (Answer to Question 1).

Caffey et al. observed asymptomatic distal femoral condylar irregularity or discrete bone fragments in 66% of boys and 41% of girls aged 3–13 years (1). These condylar irregularities and fragments were frequently bilateral and the most common location was the posterior aspect of the lateral femoral condyle (Answer to Question 2). These condylar irregularities and fragments were most commonly seen in the youngest children and progressively decreased with age. Caffey concluded that these radiographic findings were normal variations of condylar ossification.

These developmental irregularities are located beneath the spherical growth plate, and they are generally considered stable. On MRI, the overlying epiphyseal and articular cartilage has normal thickness and SI. With maturation, the subchondral bone fragments and irregularity incorporate with the epiphyseal ossification center with no long term consequences. On occasion, impressive fluid signal may be encountered within and surrounding these femoral condylar irregularities (Figures 36E, 36F). This raises the interesting possibility that this anatomic variant in ossification of the femoral condyles may become symptomatic in the face of abnormal stresses. Clues that help distinguish this developmental variant from an OCD include location, stability of the lesion, bilaterality, and the patient's age.

OCD occurs in patients 10 to 20 years of age and has a male predominance (2). It may be related to acute or chronic traumatic injury conforming to a pattern classically described as an OCD. OCD is generally found more anterior than developmental irregularities and most commonly occurs along the lateral aspect of the medial femoral condyle (Figures 36G–36J) (2). OCD may be in situ with an intact articular cartilage covering or may be detached, or displaced. OCD with an intact articular cartilage covering is considered stable; when osteochondral detachment or displacement is present, OCD is considered unstable. When an OCD is in situ, MRI features that suggest instability on T2W sequences include: >5 mm hyperintense (fluid-bright) line separating the fragment and parent bone, a discrete >5 mm diameter cyst beneath the OCD, >5 mm articular surface defect, and a hyperintense SI line traversing articular cartilage and subchondral bone plate, extending into the OCD (3).

Direct MR arthrography may be used to evaluate in situ OCD if noncontrast knee MRI is equivocal. If contrast insinuates between the OCD and parent bone, the lesion is unstable (4). Knee MRI with IV gadolinium may also be used to assess OCD.

Enhancement between the lesion and parent bone may indicate granulation tissue and suggests instability, whereas nonenhancement suggests stability (5). The use of IV gadolinium may also help determine vascularity of the OCD fragment (6).

This patient was treated conservatively and was asked to limit his activities. He continued to have intermittent knee pain for 1.5 years. Follow-up MRI showed no signs of instability of the femoral condylar irregularities.

Orthopedic Perspective

Normal developmental irregularities of the distal femoral epiphysis are common and should not be confused with OCD. Normal developmental irregularities usually occur in younger children. OCD is generally seen in older children and adolescents, and there is usually a history of repetitive overuse from sports or other activities. The lesion is painful to palpation and joint loading. Normal developmental irregularities of the distal femoral epiphysis can usually be differentiated based on history, physical exam, and plain radiographs. MRI is usually not necessary unless the patient's symptoms are atypical and alternative considerations, such as internal derangements of the knee, are also being considered.

OCD has a variable clinical outcome depending on the stability of the lesion. Stable OCD has a good clinical outcome, whereas unstable OCD may lead to loose bodies and secondary degenerative changes. OCD that is stable is usually treated conservatively. Symptomatic but stable OCD that fails conservative therapy and unstable OCD may be treated arthroscopically with drilling to promote healing, in situ pinning of the bone fragment, or chondrocyte reimplantation (7).

What the Clinician Needs to Know

1. Distinguish a normal developmental irregularity from OCD.
2. Staging OCD (stable vs. unstable).
3. For OCD, are there secondary degenerative changes or loose bodies?

Answers

1. Osteochondritis dissecans.
2. Posterior aspect of the lateral femoral condyle.

Additional Examples

Normal Developmental Irregularity of the Femoral Condyle with Possible Stress Changes

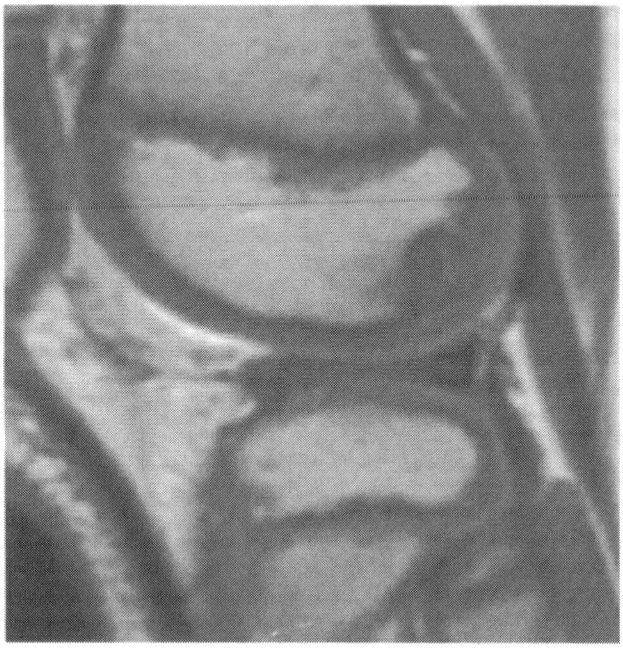

Figure 36E. Sagittal PD of the left knee.

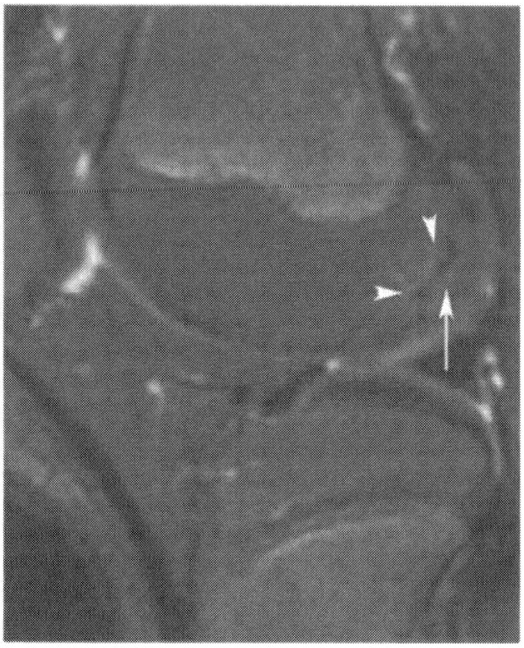

Figure 36F. Sagittal T2 FS.

Findings

This 11-year-old boy presented with left knee pain.

Figure 36E. There is a semicircular hypointense lesion present in the posterior aspect of the lateral femoral condyle.

Figure 36F. There is mild hyperintensity in the accessory ossification center **(arrow)** and parent bone **(arrowheads)**. The abnormal SI may be secondary to minor stress changes. Alternatively, this may represent increased T2 SI related to normal epiphyseal cartilage ossification. This lesion is not classic OCD.

Unstable Osteochondritis Dissecans

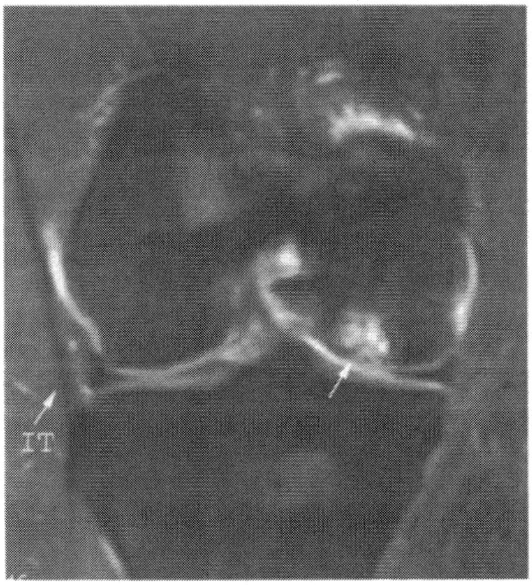

Figure 36G. Coronal PD FS of the right knee.

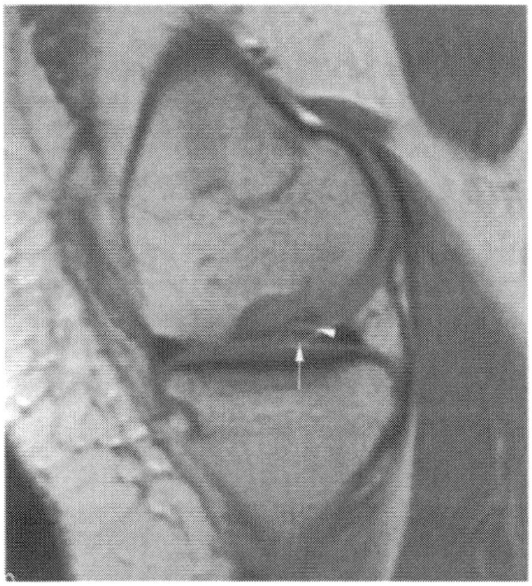

Figure 36H. Sagittal PD—medial compartment.

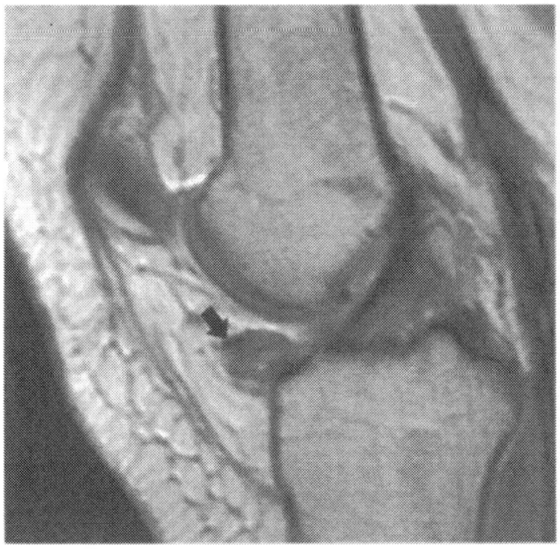

Figure 36I. Sagittal PD.

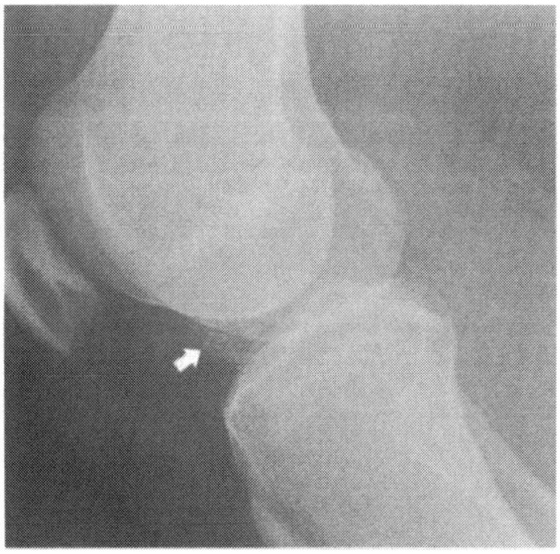

Figure 36J. Lateral radiograph.

Findings

This is a 17-year-old boy with right knee pain and locking.

Figures 36G, 36H. There is an osteochondral lesion present along the lateral aspect of the medial femoral condyle. There is articular cartilage thinning over the lesion **(thin arrows)** as well as fluid SI undercutting the cartilage defect **(arrowhead)**. Notice that the OCD location is more anterior compared with normal developmental irregularity of the femoral condyles (Figures 36B, 36C). Iliotibial Band (IT).

Figure 36I. An osteochondral loose body is present adjacent to Hoffa's fat pad **(thick arrow)**.

Figure 36J. This corresponds to a loose body on the lateral radiograph **(thick arrow)**.

Pitfalls and Pearls

1. The diagnosis of normal femoral condylar irregularities is made by radiography, not MRI. MRI should be used for problem solving or when alternative diagnoses are being considered.
2. Some orthopedists and radiologists prefer using the term *osteochondral lesion* in place of *OCD*.

References

1. Caffey J, Madell SH, Royer C, Morales P. Ossification of the distal femoral epiphysis. *J Bone Joint Surg Am* 1958; 40-A:647–654 passim.
2. Robertson W, Kelly BT, Green DW. Osteochondritis dissecans of the knee in children. *Curr Opin Pediatr* 2003; 15:38–44.
3. De Smet AA, Ilahi OA, Graf BK. Reassessment of the MR criteria for stability of osteochondritis dissecans in the knee and ankle. *Skeletal Radiol* 1996; 25:159–163.
4. Kramer J, Stiglbauer R, Engel A, Prayer L, Imhof H. MR contrast arthrography (MRA) in osteochondrosis dissecans. *J Comput Assist Tomogr* 1992; 16:254–260.
5. Bohndorf K. Osteochondritis (osteochondrosis) dissecans: A review and new MRI classification. *Eur Radiol* 1998; 8:103–112.
6. Peiss J, Adam G, Casser R, Urhahn R, Gunther RW. Gadopentetate-dimeglumine-enhanced MR imaging of osteonecrosis and osteochondritis dissecans of the elbow: Initial experience. *Skeletal Radiol* 1995; 24:17–20.
7. Herring JA. Disorders of the knee. In: *Tachdjian's Pediatric Orthopedics*, 3rd ed., Herring JA, ed., Chapter 20. Philadelphia: W.B. Saunders Company, 2002; 789–838.

Case 37

History

This is a 9-year-old girl with a history of developmental dysplasia of the left hip and prior inominate osteotomy.

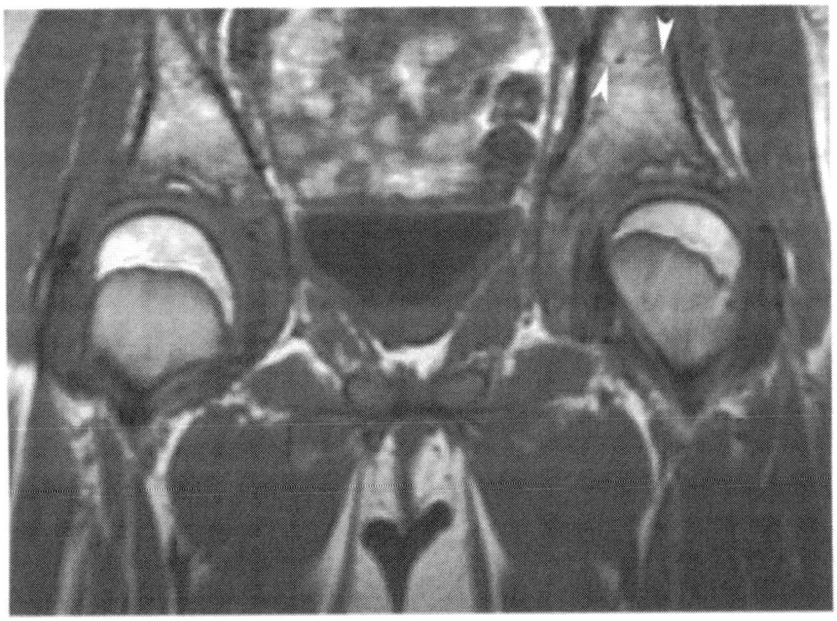

Figure 37A. Coronal T1.

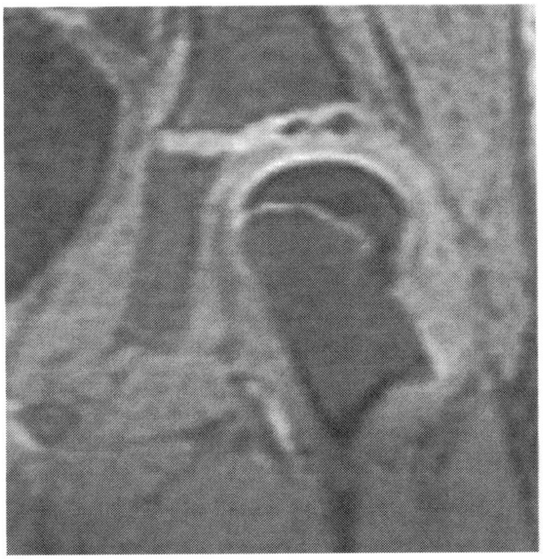

Figure 37B. Coronal 3D SPGR FS of the left hip.

Figure 37A. There has been a left inominate osteotomy, evidenced by a transverse hypointense line extending through the inferior ilium **(arrowheads)**. There is a lateral tilt of the left capital femoral epiphysis, and the physis has an excessive lateral downward slant. The height of the left femoral epiphysis is less than the normal right hip.

Figures 37B, (37B with annotations). A lateral physeal bar is present, best seen on the SPGR sequence **(thin arrow)**. The left femoral head is uncovered laterally. The secondary ossification center of the sourcil (acetabular roof) shows fragmentary ossification **(thick arrow)**.

Figure 37C. On the left, the femoral head deformity is again identified, but the lateral physeal bar is not seen. A healed inominate osteotomy is present **(arrowhead)**. The sourcil shows fragmentary ossification bilaterally, left **(thick arrow)** greater than right.

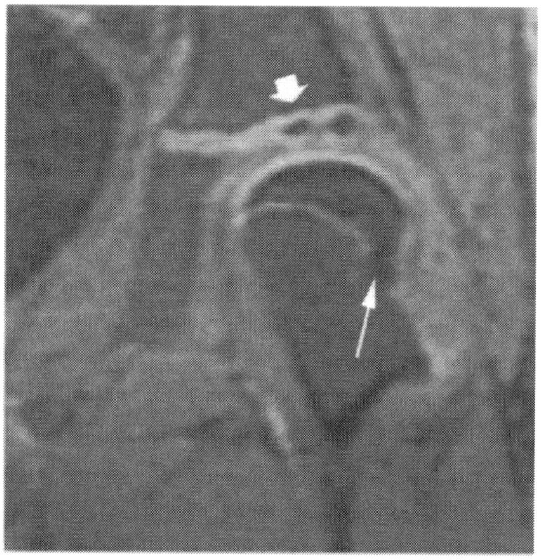

Figure 37B* Annotated.

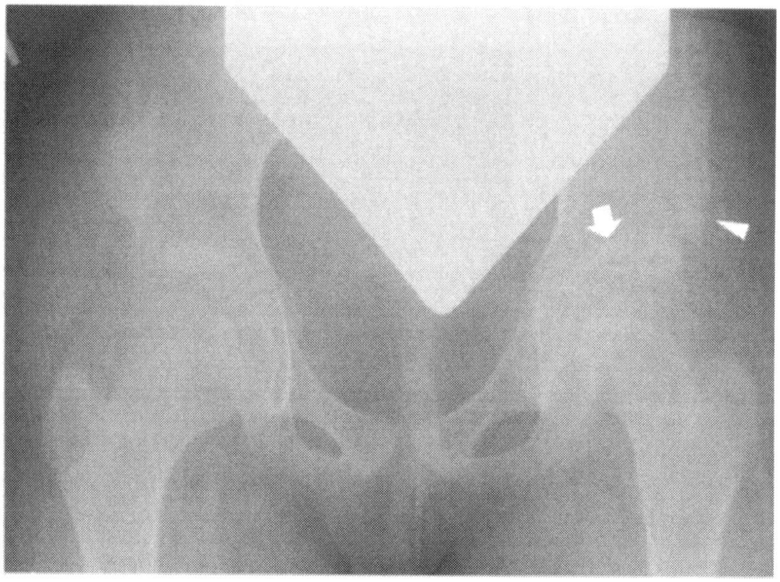

Figure 37C. AP radiograph.

Diagnosis

Treated developmental dysplasia of the hip (DDH) complicated by ischemic necrosis and physeal bar

Questions

1. T/F: Untreated DDH is a risk factor for femoral head ischemic necrosis.
2. T/F: The long term femoral head and acetabular deformity of treated developmental dysplasia of the hip complicated by ischemic necrosis may be indistinguishable from Legg-Calve-Perthes disease.

Discussion

Untreated developmental dysplasia of the hip (DDH) may lead to premature degenerative changes, but does not result in ischemic necrosis (Answer to Question 1). However, the therapy for DDH, which includes bracing, closed or open reduction, may lead to secondary femoral head and physeal ischemia. This may be explained by hip hyperabduction and kinking of the superior branches of the medial circumflex artery that supplies the femoral head and physis (1, 2). The greater trochanter apophysis and physis are unaffected since they have an extra-articular blood supply. If the femoral head and physeal ischemia is irreversible, a growth disturbance of both the capital femoral epiphysis and physis may occur. The growth disturbance preferentially occurs at the vertex of the proximal femoral physis, which is the most metabolically active area of growth (3). The vertex of the proximal femoral physis is at the junction of the femoral head and greater trochanteric physis.

The radiographic and MRI features of irreversible ischemic injury from treated DDH include: femoral neck shortening, coxa magna, coxa plana, greater trochanter overgrowth, epiphyseal and sourcil fragmentation, and physeal bars. Physeal bars tend to occur at the vertex of the proximal femur and inhibit lateral growth of the femoral head, while medial growth is unaffected. Coxa valga, lateral tilt of the femoral head, and lateral uncovering of the capital femoral epiphysis may subsequently occur.

The irreversible ischemic changes to the femoral head due to treated DDH are similar to the deformities seen with Legg-Calve-Perthes disease (LCP) Catterall types 3 and 4 (Answer to Question 2) (4). The key to distinguishing DDH with femoral head ischemic necrosis and LCP disease lies in the evaluation of the acetabulum, since the acetabular dysplasia seen with DDH is usually more severe than that noted with LCP.

Patients with untreated DDH have a radiologic spectrum of abnormalities different from treated DDH that reflects abnormal femoroacetabular mechanics. The imaging features of untreated DDH include early degenerative changes and hypertrophy of the anterosuperior labrum, periacetabular ganglia, a shallow acetabulum, and increased femoral anteversion (5, 6).

Indirect MR arthrography may be performed in patients with DDH for assessment of early degenerative changes of the articular cartilage, but is suboptimal for evaluation of labral tears (Figures 37D, 37E). Degenerated articular cartilage demonstrates increased enhancement compared with normal articular cartilage. Articular cartilage is normally composed of abundant negatively charged glycosaminoglycans (GAG) proteins, whereas degenerated cartilage has decreased concentrations of GAG proteins. Therefore, normal articular cartilage will repel the negatively charged gadolinium and degenerated cartilage will attract gadolinium and enhance (7).

This patient was initially treated with both closed and open reduction of bilateral DDH. After this MRI, she underwent a proximal femoral varus osteotomy and medial epiphysiodesis to improve her femoral head coverage.

Orthopedic Perspective

DDH represents a spectrum of abnormalities of the pediatric hip from a dislocated hip, to a dislocatable hip, to a located hip with a shallow acetabulum. The natural history of untreated DDH is often progressive degenerative joint disease. During closed reduction and casting, hyperabduction may lead to ischemic necrosis. Surgical correction of DDH via femoral and/or acetabular osteotomy may also lead to ischemic necrosis by disruption of the tenuous blood supply to the femoral head. The late diagnosis of DDH is often made on plain radiographs alone. Referral for MRI is done to determine whether the patient may benefit from corrective osteotomy and reduction or arthroscopic repair of labral and articular cartilage damage caused by abnormal mechanics.

What the Clinician Needs to Know

1. Presence of ischemic necrosis.
2. Presence of growth disturbance of the proximal femoral physis.
3. Extent of acetabular dysplasia.
4. Intra-articular derangements such as CAM-type femoroacetabular impingement (see Case 71), labral tears, and ganglion cysts.

Answers

1. False.
2. True.

Additional Example

Untreated DDH

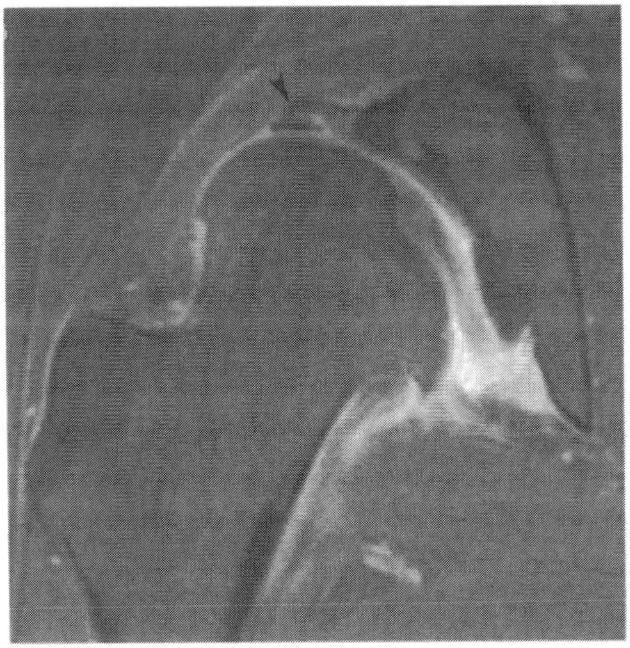

Figure 37D. Coronal T1 FS of the right hip 30 minutes after IV Gd.

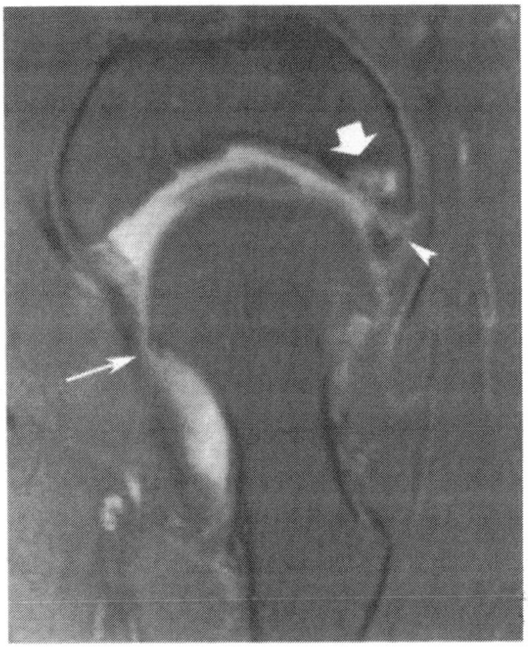

Figure 37E. Sagittal T1 FS of the right hip 30 minutes after IV Gd.

Findings

This 16-year-old boy with a history of congenital hypotonia presented with a 2-month history of right hip pain, worse while playing basketball.

Figures 37D, 37E. There is lateral uncovering of an aspherical femoral head, rim osteophytes **(thin arrow)**, and a subchondral cyst **(thick arrow)**. The superior **(black arrowhead)** and posterior **(white arrowhead)** labrum show increased SI, indicating degeneration. Although linear enhancement is present in the anterior labrum extending from the undersurface, there was no labral tear at surgery. This patient subsequently underwent a periacetabular osteotomy to improve femoral head coverage.

Pitfalls and Pearls

1. LCP and ischemic necrosis related to treated DDH may cause similar long term deformity of the femoral head. Look at the acetabulum—acetabular dysplasia is usually more severe with DDH compared with LCP.
2. Anterosuperior labral tears are most common in both DDH and femoroacetabular impingement (see Case 71) without a history of DDH (6).
3. Be cautious in diagnosing labral tears with indirect hip MR arthrography—direct MR arthrography is more sensitive and specific.

References

1. Gautier E, Ganz K, Krugel N, Gill T, Ganz R. Anatomy of the medial femoral circumflex artery and its surgical implications. *J Bone Joint Surg Br* 2000; 82:679–683.
2. Ogden JA. Changing patterns of proximal femoral vascularity. *J Bone Joint Surg Am* 1974; 56:941–950.
3. Ecklund K, Jaramillo D. Imaging of growth disturbance in children. *Radiol Clin North Am* 2001; 39:823–841.
4. Delaunay S, Dussault RG, Kaplan PA, Alford BA. Radiographic measurements of dysplastic adult hips. *Skeletal Radiol* 1997; 26:75–81.
5. Anda S, Terjesen T, Kvistad KA, Svenningsen S. Acetabular angles and femoral anteversion in dysplastic hips in adults: CT investigation. *J Comput Assist Tomogr* 1991; 15:115–120.
6. Leunig M, Podeszwa D, Beck M, Werlen S, Ganz R. Magnetic resonance arthrography of labral disorders in hips with dysplasia and impingement. *Clin Orthop* 2004:74–80.
7. Kim YJ, Jaramillo D, Millis MB, Gray ML, Burstein D. Assessment of early osteoarthritis in hip dysplasia with delayed gadolinium-enhanced magnetic resonance imaging of cartilage. *J Bone Joint Surg Am* 2003; 85-A:1987–1992.

Case 38

History

These are three patients with the same disorder.

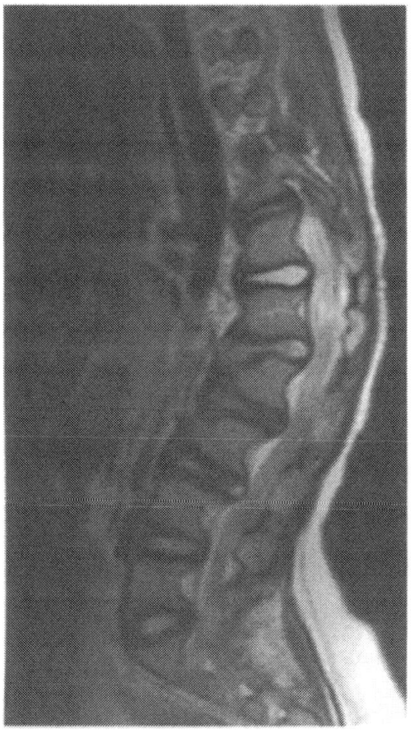

Figure 38A. 4 year old. Sagittal T2 of the thoracolumbar spine.

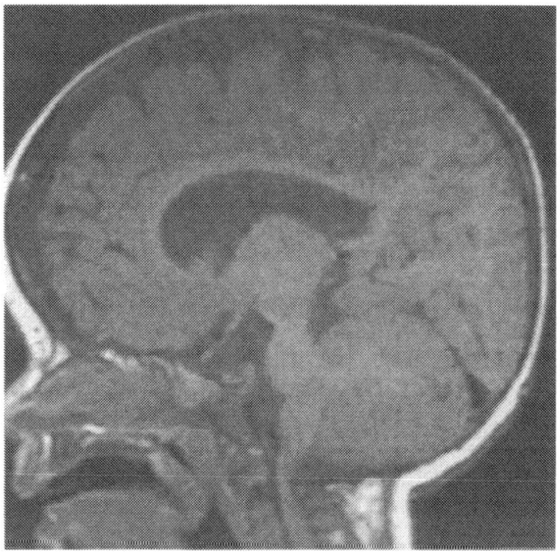

Figure 38B. 9 month old. Sagittal T1 of the brain.

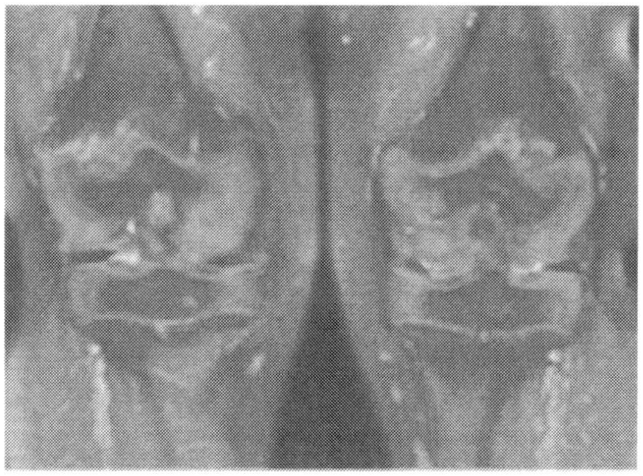

Figure 38C. 7 year old. Coronal PD FS of the knees.

Figures 38A, (38A with annotations). There is a moderate thoracolumbar kyphosis present (gibbus deformity). In addition, there is anterior beaking of multiple vertebral bodies **(arrowheads)** and posterior vertebral scalloping **(arrows)**, resulting in bullet-shaped vertebra. The anteroposterior diameter of the spinal canal is reduced.

Figures 38B, (38B with annotations). There is focal stenosis at the level of the occipitoatlantal junction and kinking of the cervical spinal cord at the level of C1 **(arrow)**. In addition, there is enlargement of the frontal extra-axial CSF space.

Figures 38C, (38C with annotations). The distal femoral metaphyses are flared and cupped **(arrowheads)**, and the epiphyses show a corresponding cone-shaped appearance.

Figures 38D, 38E. The fibula is overgrown, approaching the height of the tibia. The physis of the proximal fibula **(arrows)** is almost at the same level as the tibial physis. Metaphyseal cupping **(arrowheads)** and flaring as well as coned-shaped epiphyses are evident on the radiograph.

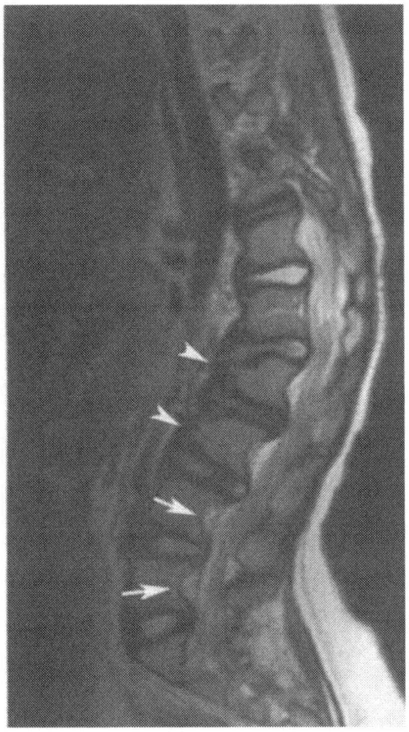

Figure 38A* Annotated.

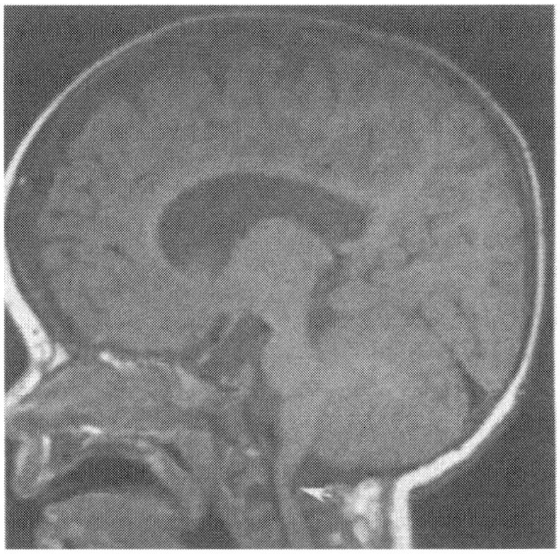

Figure 38B* Annotated.

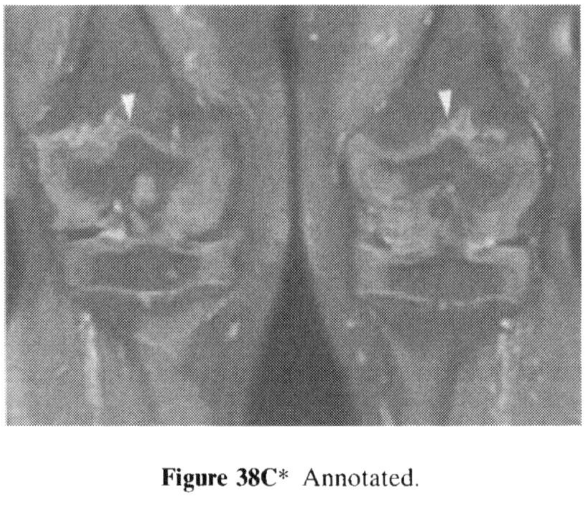

Figure 38C* Annotated.

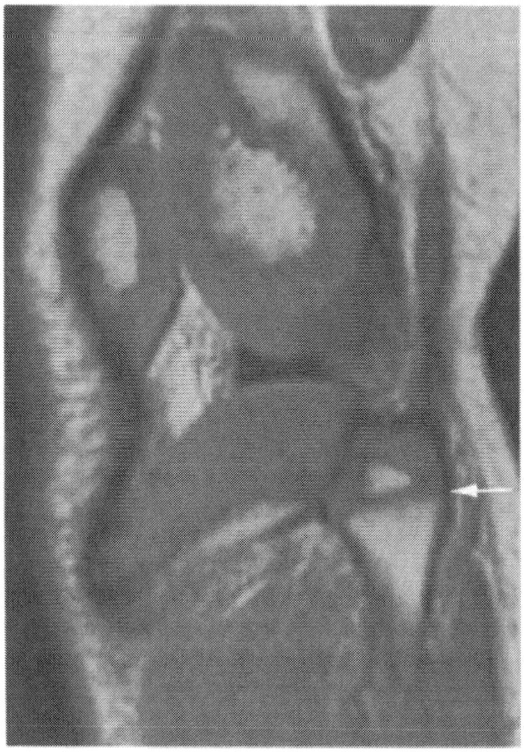

Figure 38D. Sagittal PD, lateral aspect of the right knee (same patient as in Figure 38C).

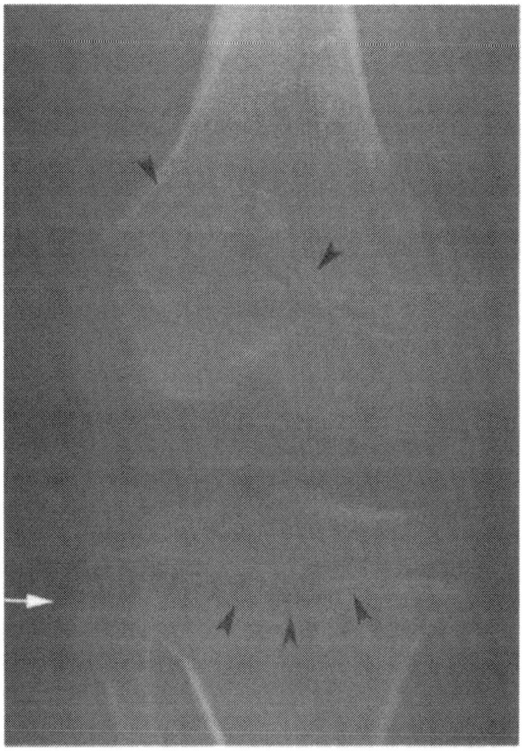

Figure 38E. AP radiograph of the right knee (same patient as in Figure 38C).

Normal Fibula-Tibia Relationship

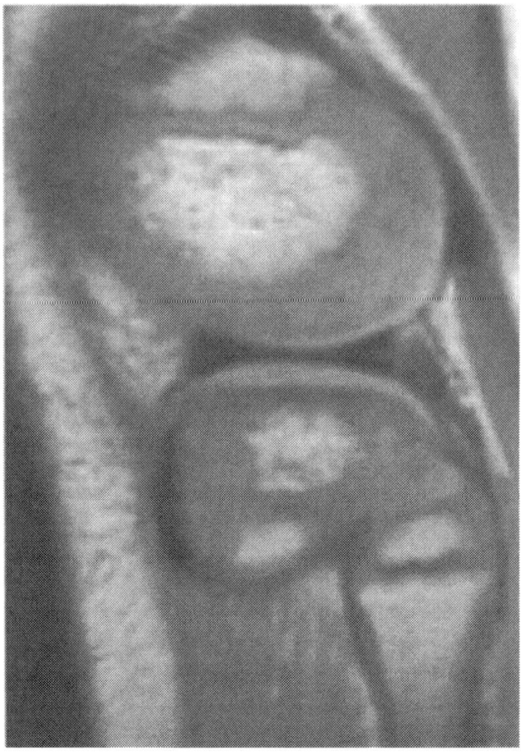

Figure 38F. Sagittal PD, lateral aspect of the knee in a normal 7 year old.

Findings

Figure 38F. This age-matched normal patient shows a normal fibula-tibia height relationship with the physis of the proximal fibula below that of the tibia.

Diagnosis

Achondroplasia

Questions

1. What are some conditions that display metaphyseal cupping and cone-shaped epiphyses?
2. What is the differential diagnosis of a gibbus deformity?

Discussion

Achondroplasia, hypochondroplasia, and thanatophoric dysplasia are related short-limb skeletal dysplasias that result from a defective growth factor receptor gene on chromosome 4p (1). Of the three, achondroplasia is the most common. The inheritance pattern is autosomal dominant, but the majority of cases are sporadic. It is a rhizomelic (proximal) dwarfism due to stunted endochondral but preserved membranous bone growth. Histologically, the thickness of the zone of proliferation of the growth plate is diminished (1).

Plain radiographic changes of achondroplasia are best seen at the ends of the tubular bones. Metaphyseal cupping (V-shaped physeal margin), metaphyseal flaring, and cone-shaped epiphyses are characteristic findings. The tubular bones are short and broad: the diaphyseal width of the tubular bones is unaffected because periosteal ossification is normal. Fibular overgrowth is a distinctive finding with achondroplasia (Figures 38D, 38E) (2). Other radiographic manifestations include: squared iliac wings, trident configuration of the metacarpals and phalanges, narrowed sciatic notch, flat acetabular roofs, coxa vara deformity, and kyphoscoliosis (1).

There are several conditions that result in physeal injury or growth disturbance that share similar metaphyseal/epiphyseal features of achondroplasia. These include a variety of bone dysplasias, physeal trauma, infarction, osteomyelitis/septic arthritis (Figure 38G), meningococcemia (Figure 38H), and metabolic conditions such as scurvy and hypervitaminosis A (Answer to Question 1) (3–5). Metaphyseal cupping and cone-shaped epiphyses reflect an insult to the central germinal matrix, resulting in focal endochondral growth arrest. Central physeal bar formation may subsequently occur (Figure 38H).

Patients with achondroplasia are at high risk for developing spinal stenosis due to three factors. They may develop hypoplasia of the spinal canal and vertebral column, kyphoscoliosis, and hyperplasia of the intervertebral discs that tend to bulge and herniate due to abnormal axial loading forces (6). They also are at high risk for developing cervicomedullary stenosis at the level of the foramen magnum, because the skull base arises from endochondral bone while the rest of the skull develops from membranous bone (Figure 38B) (7). Cervicomedullary stenosis may lead to obstructive hydrocephalus, myelomalacia, and a syrinx at the level of the cervicomedullary junction. Other vertebral manifestations of achondroplasia include: anterior vertebral body beaking, posterior vertebral scalloping, progressive narrowing of the lumbar interpediculate transverse dimensions, and kyphoscoliosis (1). They may also develop a thorocolumbar kyphosis (gibbus deformity) (Figure 38A). A gibbus deformity may also be seen in patients with mucopolysaccharidosis, infections such as tuberculosis, and myelodysplasia (Answer to Question 2) (8).

Orthopedic Perspective

The diagnosis of achondroplasia is made well before patients are referred for MRI evaluation. The utility of MR imaging is to assess for complications related to achondroplasia including central spinal canal and cerebello-pontine angle stenosis, presence and severity of kyphoscoliosis, premature degenerative changes related to abnormal weightbearing, as well as the status of the physis. Laminectomy and spinal fusion may be required. In the lower extremities, genu varus deformities may develop. The orthopedist is interested in defining physeal status including angulation deformities and physeal bars. Leg lengthening and joint angulation correction may require a combination of osteotomy, epiphysiodesis, and physeal bar resection. For example, halting fibular overgrowth with epiphysiodesis may help prevent ankle varus and genu varus deformities.

What the Clinician Needs to Know

1. Quantify the degree of central canal compromise, if present.
2. Describe features of physeal growth disturbance.

Answers

1. Achondroplasia and other bone dysplasias, physeal trauma, infarction, osteomyelitis/septic arthritis, meningococcemia, and metabolic conditions such as scurvy and hypervitaminosis A.
2. Achondroplasia, mucopolysaccharidosis, infections such as tuberculosis, and myelodysplasia.

Additional Examples

Neonatal Septic Arthritis

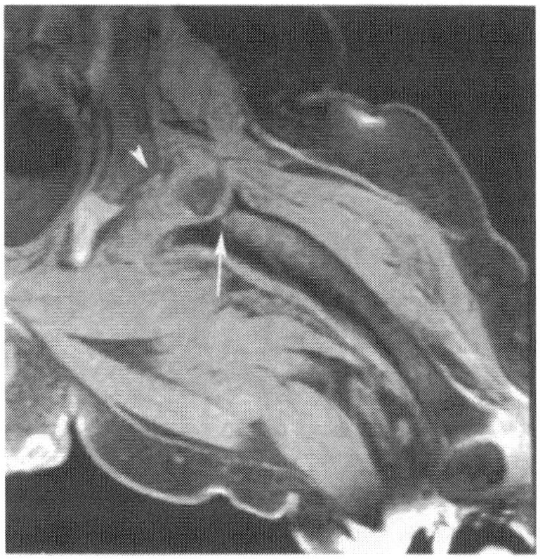

Figure 38G. Coronal 3D SPGR FS.

Findings

This 15-month-old girl had osteomyelitis with septic arthritis of her left hip as a neonate. The left hip was debrided at the time and cultures grew gram-positive cocci.

Figure 38G. The hip is laterally subluxed and the acetabulum is shallow **(arrowhead)**. The proximal femur physis is cupped shaped **(arrow)** and there is metaphyseal flaring.

Neonatal Meningococcemia

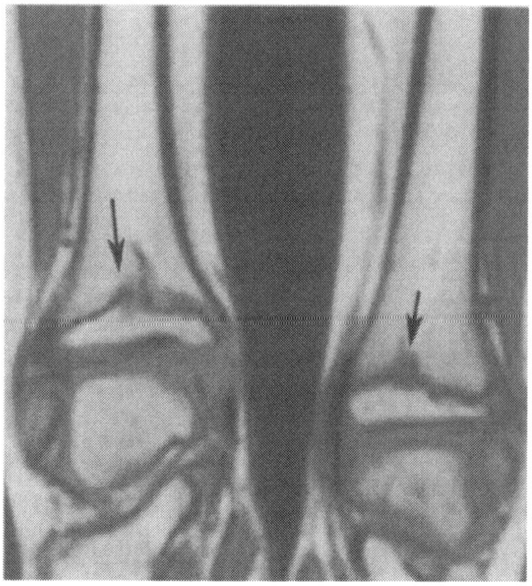

Figure 38H. Coronal T1.

Findings

This 6-year-old girl had a history of fulminant neonatal meningoencephalitis and sepsis.

Figure 38H. Mild metaphyseal cupping and early physeal bar formation **(arrows)** are evident. She subsequently underwent physeal bar resection and did well.

Pitfalls and Pearls

In addition to achondroplasia, posterior vertebral scalloping can be seen in patients with neurofibromatosis type 1, chronic hydrocephalus, intraspinal neoplasms, ankylosing spondylitis, and other conditions that may cause dural ectasia such as Marfan's syndrome (9–11).

References

1. Lemyre E, Azouz EM, Teebi AS, Glanc P, Chen MF. Bone dysplasia series. Achondroplasia, hypochondroplasia and thanatophoric dysplasia: Review and update. *Can Assoc Radiol J* 1999; 50:185–197.
2. Nehme AM, Riseborough EJ, Tredwell SJ. Skeletal growth and development of the achondroplastic dwarf. *Clin Orthop* 1976:8–23.
3. Patriquin HB, Trias A, Jecquier S, Marton D. Late sequelae of infantile meningococcemia in growing bones of children. *Radiology* 1981; 141:77–82.
4. Kumar SJ, Forlin E, Guille JT. Epiphyseometaphyseal cupping of the distal femur with knee-flexion contracture. *Orthop Rev* 1992; 21:67–70.
5. Hoeffel JC, Lascombes P, Mainard L, Durup de Baleine D. Cone epiphysis of the knee and scurvy. *Eur J Pediatr Surg* 1993; 3:186–189.
6. Morgan DF, Young RF. Spinal neurological complications of achondroplasia: Results of surgical treatment. *J Neurosurg* 1980; 52:463–472.

7. Kao SC, Waziri MH, Smith WL, Sato Y, Yuh WT, Franken EA, Jr. MR imaging of the craniovertebral junction, cranium, and brain in children with achondroplasia. *AJR Am J Roentgenol* 1989; 153:565–569.
8. Levin TL, Berdon WE, Lachman RS, Anyane-Yeboa K, Ruzal-Shapiro C, Roye DP, Jr. Lumbar gibbus in storage diseases and bone dysplasias. *Pediatr Radiol* 1997; 27:289–294.
9. Sponseller PD, Ahn NU, Ahn UM, et al. Osseous anatomy of the lumbosacral spine in Marfan syndrome. *Spine* 2000; 25:2797–2802.
10. Abello R, Rovira M, Sanz MP, et al. MRI and CT of ankylosing spondylitis with vertebral scalloping. *Neuroradiology* 1988; 30:272–275.
11. Tsirikos AI, Ramachandran M, Lee J, Saifuddin A. Assessment of vertebral scalloping in neurofibromatosis type 1 with plain radiography and MRI. *Clin Radiol* 2004; 59:1009–1017.

Case 39

History

This is a 6-year-old girl with left hand and wrist pain, swelling, and decreased range of motion for several months.

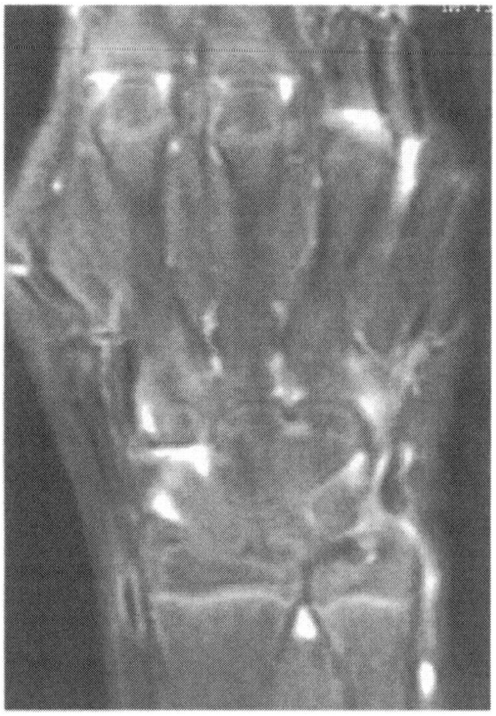

Figure 39A. Coronal T1 post-Gd FS of the left wrist.

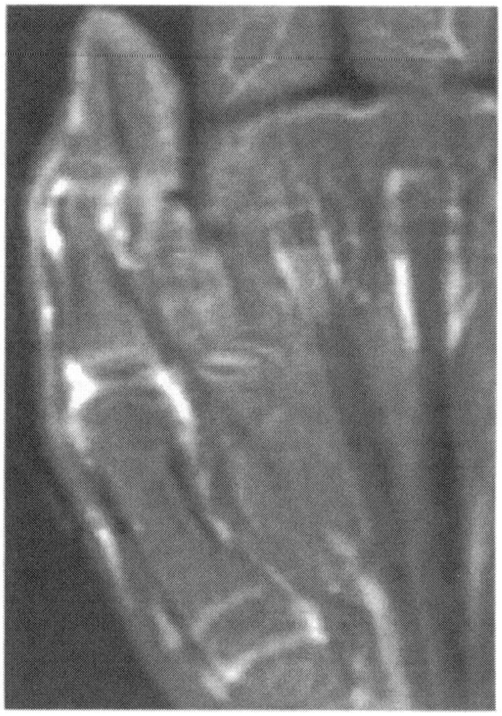

Figure 39B. Coronal T1 post-Gd FS of the first digit.

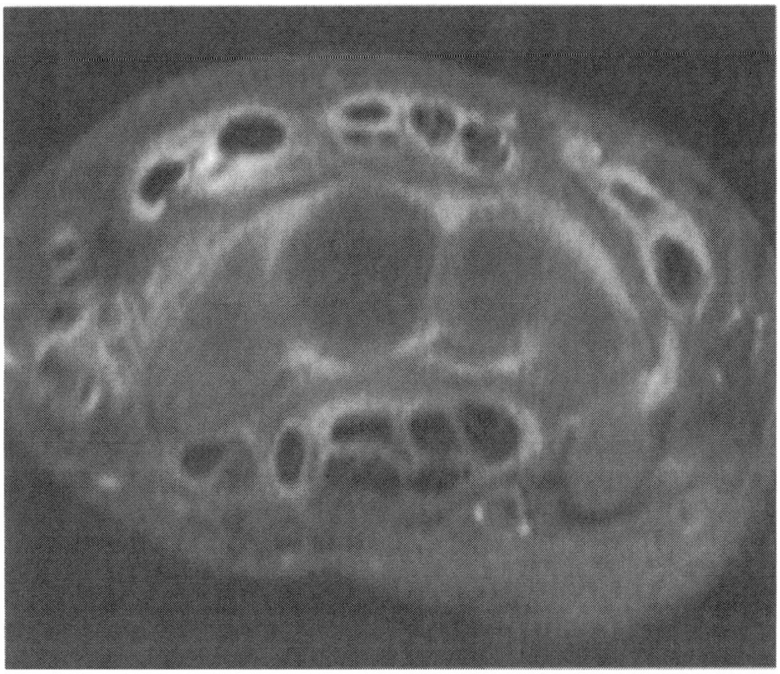

Figure 39C. Axial T1 post-Gd FS through the carpal bones.

Figures 39A, 39B. There is extensive synovial enhancement in the distal radioulnar, intercarpal, metacarpal phalangeal, and first digit IP joints.

Figure 39C. There is also diffuse tendon sheath enhancement involving the extensor tendons and, to a lesser degree, the flexor tendons of the wrist.

Diagnosis

Juvenile rheumatoid arthritis (JRA)

Questions

1. What MRI findings may predict future destructive arthropathy in patients with JRA?
2. T/F: In JRA, tenosynovitis of the wrist more commonly affects the extensor tendons than the flexor tendons.

Discussion

The diagnosis of juvenile rheumatoid arthritis (JRA) (AKA juvenile idiopathic arthritis) may be made when a patient presents under the age of 16 years with arthritic symptoms for a minimum of six weeks (1, 2). JRA is separated into four subtypes. The rheumatoid factor negative subtypes include: oligoarticular (4 joints or fewer are involved), polyarticular (greater than 4 joints involved), and systemic (AKA Still's disease—polyarticular disease with hepatosplenomegaly and fevers). The rheumatoid factor positive subtype is polyarticular, seen most commonly in females, usually occurs in older children, and manifests symptoms similar to its adult counterpart. The alternative nomenclature for JRA accepted by the International League of Associations of Rheumatologists (ILAR) is juvenile idiopathic arthritis (JIA) (3).

The plain radiographic features of JRA include diffuse soft tissue swelling, juxta-articular or diffuse osteopenia, periosteal reaction, joint space narrowing, marginal erosion, and joint malalignment (4). Within the hand and wrist, inflammatory changes are usually confined to the distal radial-ulnar joint, ulnar prestyloid recess, carpal bones, MCP, and proximal interphalangeal (PIP) joints. Erosions and joint space narrowing are late manifestations of disease since children have relatively thick articular and epiphyseal cartilage, which protects the subchondral bone from synovial inflammation. Periostitis is also a unique feature of JRA and is not seen in adult patients with RA unless there has been trauma or infection.

The MRI features that may be used to measure response to therapy and predict future articular destruction and erosions are: the degree of synovial hypertrophy and enhancement, subchondral bone edema seen on fluid-sensitive sequences, and subchondral bone enhancement (Answer to Question 1) (5–8). Modest synovial enhancement in the absence of synovial thickening or joint effusion may be a normal finding and can be difficult to distinguish from early or mild synovitis. In one study evaluating normal adult volunteers, 44% showed mild synovial enhancement in the hand (9). Subchondral edema may predict future sites of erosions and articular damage, whereas the extent of synovial hypertrophy may predict the future number of erosions in a given joint (10).

Inactive erosions, subchondral cysts, active erosions, and pre-erosive osteitis should be differentiated on MRI in combination with plain radiography. Inactive erosions and subchondral cysts do not enhance. Inactive erosions may contain fat SI, and may be hypointense on fluid-sensitive sequences. Active erosions and pre-erosive osteitis, however, show enhancement and increased SI on fluid-sensitive sequences (6). Pre-erosive osteitis (Figures 39D, 39E) is distinguished from active erosions (Figures 39F–39H) by the presence of normal bony cortex and cartilage overlying the area of subchondral edema. The diagnosis of a subchondral erosion should not be made by MRI unless the overlying cartilage and cortex is abnormal or there is plain radiographic evidence of an erosion.

Tenosynovitis and myositis may be primary manifestation of JRA or may be secondary to an adjacent arthritis. When tenosynovitis is primary, or is out of proportion to the degree of intra-articular inflammation, a seronegative spondyloarthropathy should also be considered (see Case 69) (11). In JRA, the extensor tendons of the wrist are more commonly affected than the flexor tendons (Answer to Question 2) (5, 12). The MRI features of tenosynovitis include tendon thickening, enhancement and edema, and fluid within the surrounding tendon sheath. Tenosynovitis may eventually lead to tendon rupture and care should be taken to determine tendon integrity on sagittal or coronal sequences.

This patient had a known diagnosis of polyarticular JRA, rheumatoid factor negative. The MRI was not ordered to confirm the clinical diagnosis, but to determine disease activity prior to methotrexate therapy.

Orthopedic Perspective

JRA is always in the differential diagnosis of atraumatic joint swelling in a child. Single or multiple joints may be involved. When a single joint is involved, the possibility of a septic arthritis must always be excluded. JRA is diagnosed based on clinical criteria, including morning stiffness and symptoms lasting for at least 6 consecutive weeks. The diagnosis of JRA is usually known well before these patients are evaluated by MRI, particularly those patients who suffer from polyarticular disease. Occasionally, the child with pauciarticular JRA may be discovered when an MRI is requested to evaluate unexplained joint pain or swelling. However, the principal role of MRI in the evaluation of JRA is to determine disease activity prior to and after changes in medical therapy.

What the Clinician Needs to Know

1. Disease activity, as determined by subchondral bone edema, synovial hypertrophy, and active erosions.
2. Do not state the obvious: the clinician usually knows that the patient has an inflammatory arthritis. The clinician is ordering an MRI to determine the level of disease activity prior to changes in medical therapy.

Answers

1. Synovial hypertrophy and enhancement, subchondral increased SI on fluid-sensitive sequences, and subchondral bone enhancement.
2. True.

Additional Examples

JRA, Pre-erosive Osteitis of the Scaphoid Bone

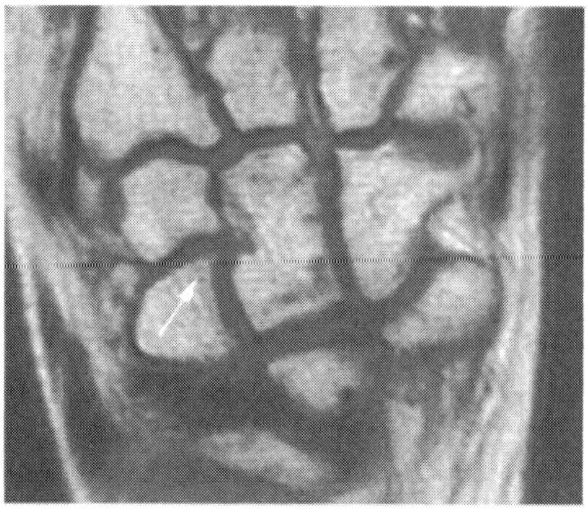

Figure 39D. Coronal T1 of the right wrist.

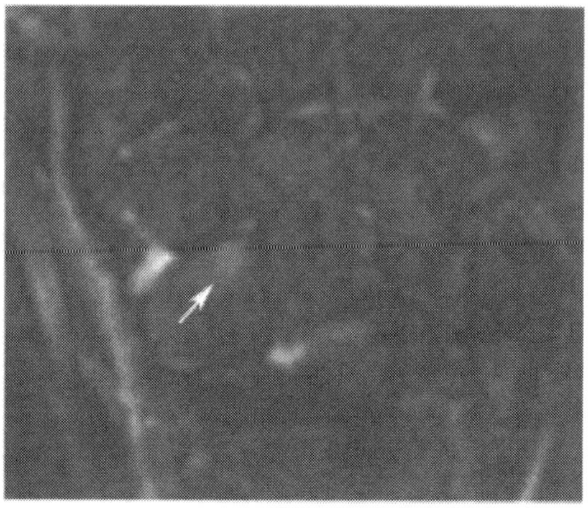

Figure 39E. Coronal STIR.

Findings

This is a 13-year-old girl with known JRA and increasing right wrist pain.

Figures 39D, 39E. There is increased STIR SI in the distal pole of the scaphoid **(arrow)**. On T1W sequences, the cortex and marrow SI are maintained **(arrow)**. Therefore, this lesion is compatible with osteitis without erosion.

JRA, Active Erosions

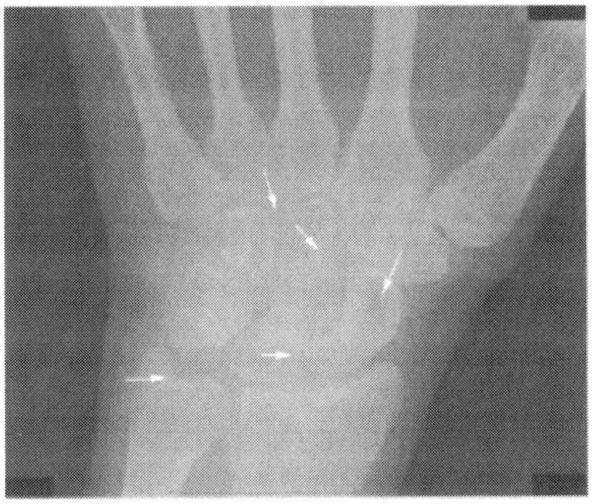

Figure 39F. PA radiograph of the left wrist.

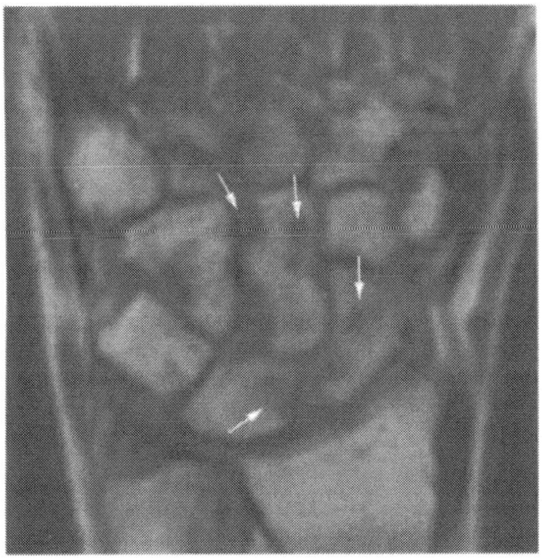

Figure 39G. Coronal T1.

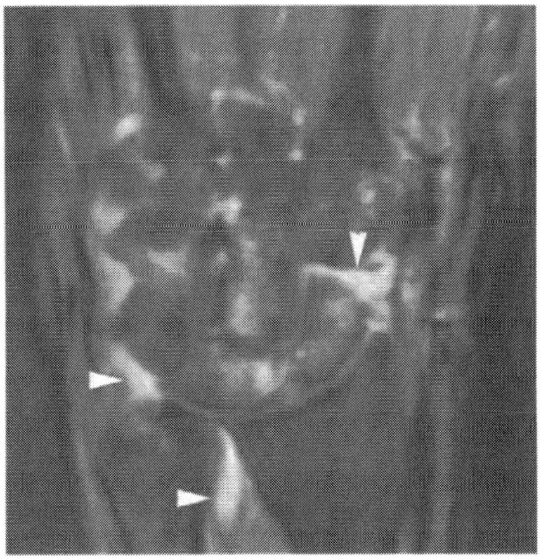

Figure 39H. Coronal T1 post-Gd FS.

Findings

This is an 18-year-old girl with polyarticular JRA, RF negative, with worsening left wrist pain.

Figure 39F. Plain radiograph demonstrates multiple carpal erosions **(arrows)**.

Figure 39G. These, and additional carpal erosions, are defined by cortical destruction and hypointense marrow on the T1W image **(arrows)**.

Figure 39H. Post-Gd image demonstrates enhancement of these active lesions. In addition, distal radioulnar joint, pre-styloid ulnar recess, and intercarpal joint synovial thickening and enhancement are evident **(arrowheads)**.

Pitfalls and Pearls

1. The diagnosis of oligoarticular JRA is often one of exclusion, once other considerations such as septic arthritis, Lyme arthritis, and trauma have been excluded. A period of observation may be required to confirm the diagnosis.
2. Active erosions, containing pannus, will show subchondral bone and cartilaginous disruption on plain radiography and MRI. Pre-erosive osteitis, however, will show intact overlying cortex and cartilage.
3. JRA is associated with Turner's and Down's syndrome (4).

References

1. Buchmann RF, Jaramillo D. Imaging of articular disorders in children. *Radiol Clin North Am* 2004; 42:151–168, vii.
2. Cohen PA, Job-Deslandre CH, Lalande G, Adamsbaum C. Overview of the radiology of juvenile idiopathic arthritis (JIA). *Eur J Radiol* 2000; 33:94–101.
3. Petty RE, Southwood TR, Manners P, et al. International League of Associations for Rheumatology classification of juvenile idiopathic arthritis: Second revision, Edmonton, 2001. *J Rheumatol* 2004; 31:390–392.
4. Wihlborg C, Babyn P, Ranson M, Laxer R. Radiologic mimics of juvenile rheumatoid arthritis. *Pediatr Radiol* 2001; 31:315–326.
5. Ostergaard M, Szkudlarek M. Magnetic resonance imaging of soft tissue changes in rheumatoid arthritis wrist joints. *Semin Musculoskelet Radiol* 2001; 5:257–274.
6. Guermazi A, Taouli B, Lynch JA, Peterfy CG. Imaging of bone erosion in rheumatoid arthritis. *Semin Musculoskelet Radiol* 2004; 8:269–285.
7. McQueen FM. Magnetic resonance imaging in early inflammatory arthritis: What is its role? *Rheumatology (Oxford)* 2000; 39:700–706.
8. Gylys-Morin VM, Graham TB, Blebea JS, et al. Knee in early juvenile rheumatoid arthritis: MR imaging findings. *Radiology* 2001; 220:696–706.
9. Partik B, Rand T, Pretterklieber ML, Voracek M, Hoermann M, Helbich TH. Patterns of gadopentetate-enhanced MR imaging of radiocarpal joints of healthy subjects. *AJR Am J Roentgenol* 2002; 179:193–197.
10. Savnik A, Malmskov H, Thomsen HS, et al. MRI of the wrist and finger joints in inflammatory joint diseases at 1-year interval: MRI features to predict bone erosions. *Eur Radiol* 2002; 12:1203–1210.
11. Jevtic V, Watt I, Rozman B, Kos-Golja M, Demsar F, Jarh O. Distinctive radiological features of small hand joints in rheumatoid arthritis and seronegative spondyloarthritis demonstrated by contrast-enhanced (Gd-DTPA) magnetic resonance imaging. *Skeletal Radiol* 1995; 24:351–355.
12. Narvaez JA, Narvaez J, Roca Y, Aguilera C. MR imaging assessment of clinical problems in rheumatoid arthritis. *Eur Radiol* 2002; 12:1819–1828.

Case 40

History

This is a 5-month-old girl with a 2-month history of an enlarging left supraclavicular mass.

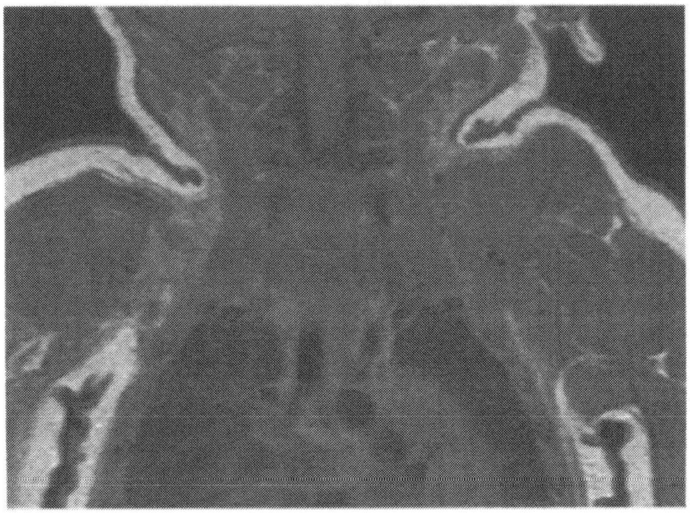

Figure 40A. Coronal T1.

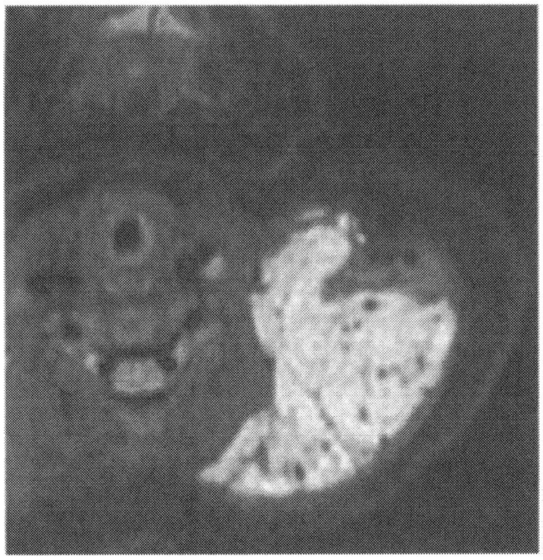

Figure 40B. Axial STIR.

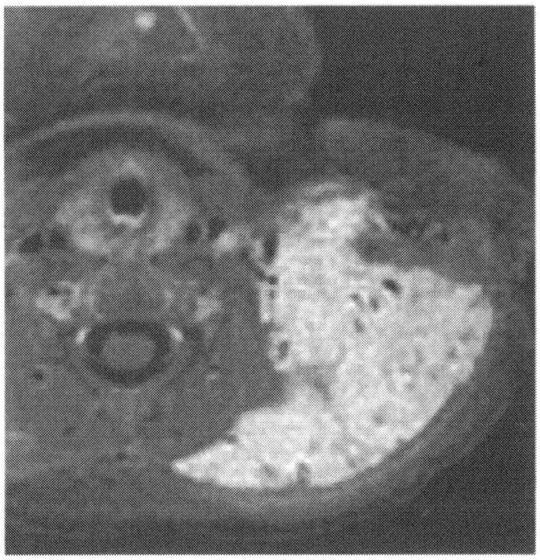

Figure 40C. Axial T1 post-Gd FS.

Figures 40A, 40B, 40C, (40C with annotations). There is a large, well-defined mass in the supraclavicular region. It is isointense on T1, markedly hyperintense on fluid-sensitive sequences, and shows intense enhancement with respect to muscle. Although there are flow voids present on both STIR and post-Gd sequences **(arrows)**, the mass is mainly comprised of solid enhancing tissue.

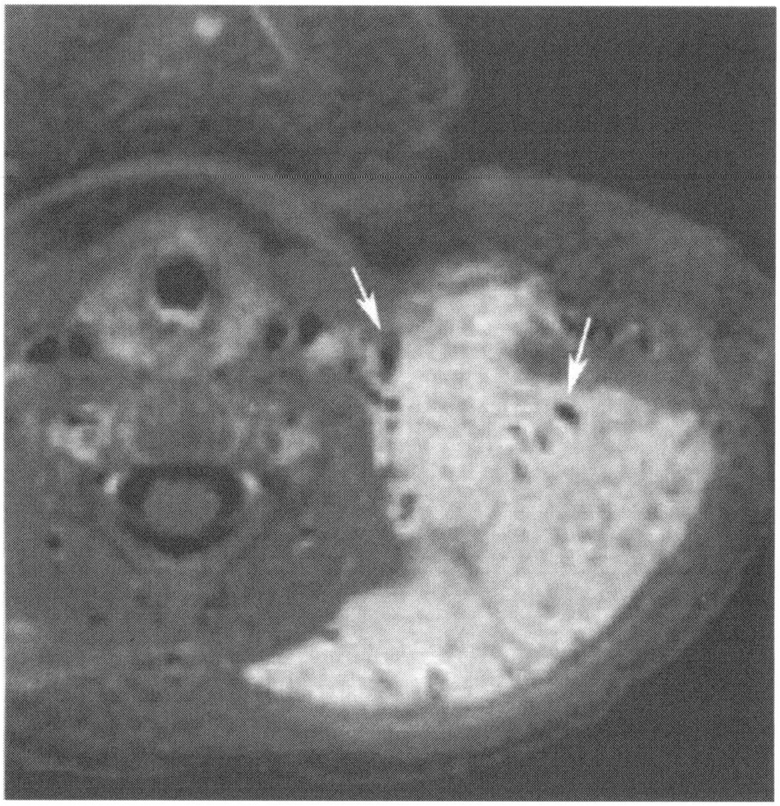

Figure 40C* Annotated.

Diagnosis

Infantile hemangioma

Questions

1. T/F: The most common location for infantile hemangioma is the head and neck region.
2. What are the three growth phases of an infantile hemangioma?

Discussion

Infantile hemangiomas are common benign vascular tumors that are usually identified during the first few weeks of life (1). By definition, they are not evident at birth. They are most commonly found in the head and neck region (60%), followed by the trunk (25%) (Figures 40D, 40E), and extremities (15%) (Answer to Question 1). They occur much more frequently in girls and premature infants. Multiple cutaneous hemangiomas are frequently associated with clinically silent hepatic hemangiomas (Figures 40D, 40E). In one study, the incidence of clinically silent hepatic hemangiomas in patients with 6 or more small cutaneous hemangiomas or a large cutaneous hemangioma was almost 25% (2). Infantile hemangiomas are true vascular neoplasms and should be distinguished from vascular malformations, such as capillary, venous (see Case 67), lymphatic (see Case 59), venolymphatic, and arteriovenous malformations (3).

There are three growth phases of an infantile hemangioma: proliferating, involuting, and involuted (Answer to Question 2) (4). Hemangiomas become evident during the proliferative phase, a period of rapid growth that typically extends through the first year of life. On MRI, hemangiomas are solid, well-defined masses with lobulated morphology and distinct flow voids. They are isointense on T1, markedly hyperintense on T2 ("light-bulb" bright), and show uniformly intense enhancement following gadolinium. During this phase, the presence of flow voids in a hemangioma may suggest an arteriovenous malformation (AVM); however, a soft tissue mass is not a feature of an AVM. When the characteristic MRI features of an infantile hemangioma are present in the appropriate clinical context, the diagnosis is usually secure. However, if the margins are unsharp or the lesion is multicompartmental, hemangioendothelioma, rhabdomyosarcoma, fibrosarcoma, and infantile myofibromatosis may also be considered.

The involuting phase of infantile hemangiomas lasts 1 to 7 years and complete regression is seen in most patients by 10 years of age (1). During this period, hemangiomas become progressively less cellular and undergo fibrofatty replacement. During the involuting phase, the tumor enhancement pattern decreases and the lesion appears more heterogeneous. There is also increased fibrous tissue, and calcification and blood breakdown products may be apparent. Once involution is completed, fibrofatty tissue may be the only remnant of a hemangioma identifiable by MRI.

Congenital hemangiomas are clinically and biologically different from infantile hemangiomas (Table 40A). They (Figures 40F, 40G) are present from birth and rapidly decrease rather than increase in size at presentation compared with infantile hemangiomas. They are often associated with dilated draining channels, arterial aneurysms, and/or arteriovenous fistulas (5). Unlike infantile hemangiomas, congenital hemangiomas may be infiltrative and occupy multiple compartments. Most of these are classified as a rapidly involuting congenital hemangioma (RICH) and will completely involute by approximately one year of age. Rarely, congenital hemangiomas do not

Table 40A. Comparison of infantile hemangioma, RICH, and KHE.

	Margins	Clinical Course	Other
Infantile Hemangioma	Well-defined	Initially enlarges during first year of life, then slowly regresses.	Glucose transporter protein isoform 1 (GLUT1) positive (8).
Congenital Hemangioma (RICH, NICH)	Infiltrative	Presents at birth, then rapidly involutes. If it grows with the child or does not involute by approximately 1 year of age, then NICH (non-involuting congenital hemangioma) (6).	May develop heart failure due to AV shunting (typically seen with liver lesions); may develop mild thrombocytopenia.
Kaposiform Hemangioendothelioma (KHE)	Infiltrative	50% present at birth, rapidly enlarges, may spontaneously shrink, but never disappears.	No heart failure, but may develop significant thrombocytopenia.

involute and grow proportionately with the child. These lesions are classified as a non-involuting congenital hemangioma (NICH) (6).

Kaposiform hemangioendothelioma (KHE) (Figures 40H, 40I) is a unique entity that is distinct clinically and biologically from infantile and congenital hemangioma. Kaposiform hemangioendothelioma is usually larger than infantile hemangiomas at presentation, and does not completely involute. It may be associated with Kasabach-Merrit Syndrome, a triad of cutaneous vascular lesion, anemia, and thrombocytopenia at birth (7). Approximately 50% of KHE are present at birth and the remainder present during the first year of life. These tumors may have infiltrative margins, cross fascial planes, and show a diffuse and inhomogeneous enhancement pattern. Although KHE may present at birth, it disproportionately increases in size, rather than rapidly involutes like a congenital hemangioma.

Since the clinical and imaging findings in this patient were characteristic of an infantile hemangioma, the lesion was not in a critical region, and it showed no evidence of overlying skin necrosis, no therapy was given.

What the Clinician Needs to Know

1. Distinguish an infantile hemangioma from infantile fibrosarcoma or hemangioendothelioma.
2. Is the lesion isolated to the subcutaneous fat, does it violate the superficial fascia, what muscle compartments are involved, and are neurovascular structures encased?

Answers

1. True.
2. Proliferative, involuting, and involuted.

Additional Examples

Infantile Hemangioma of the Trunk and Hepatic Hemangiomas

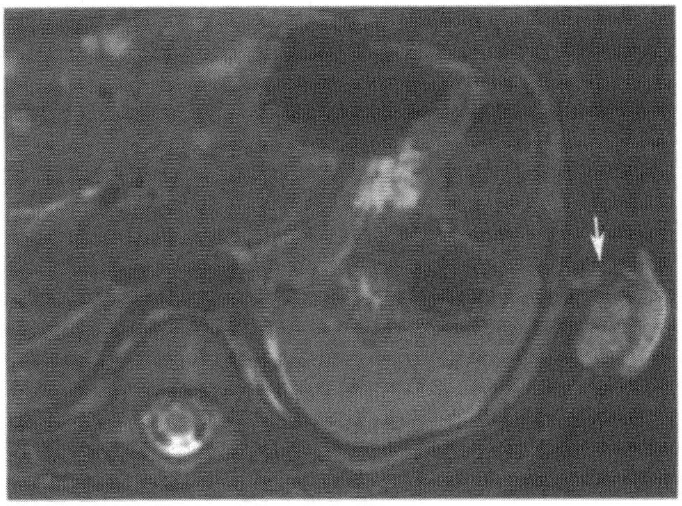

Figure 40D. Axial T2 FS.

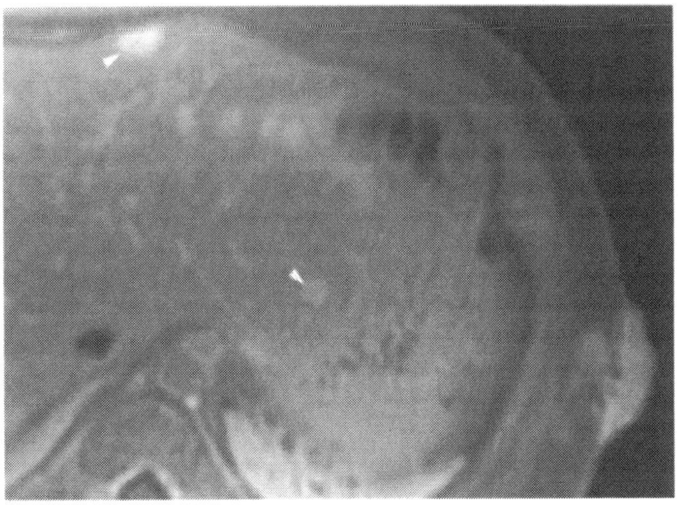

Figure 40E. Axial T1 post-Gd FS (different level).

Findings

This is a 3-month-old boy with a history of multiple cutaneous infantile hemangiomas.

Figures 40D, 40E. A left flank hemangioma is identified. It is multilobulated, markedly hyperintense on T2, and shows diffuse enhancement. A draining vein is identified **(arrow)**. There are also multiple round lesions within the liver that are moderately hyperintense on T2 and show homogeneous enhancement **(arrowheads)** compatible with liver hemangiomas.

Congenital Hemangioma

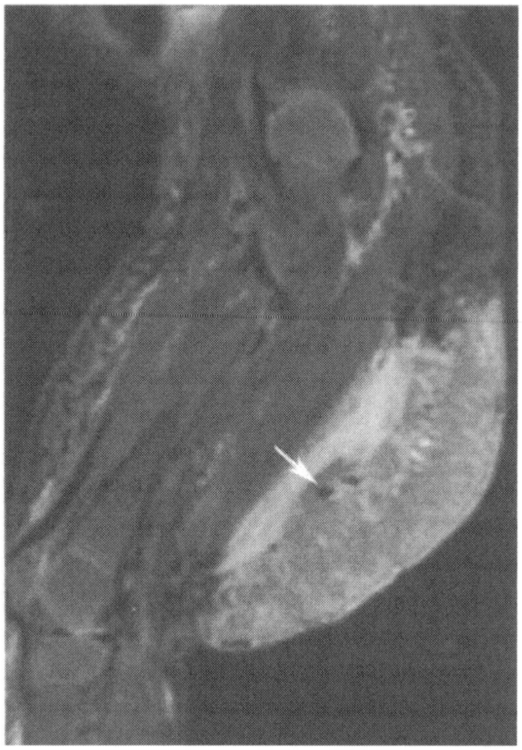

Figure 40F. Sagittal STIR of the right thigh.

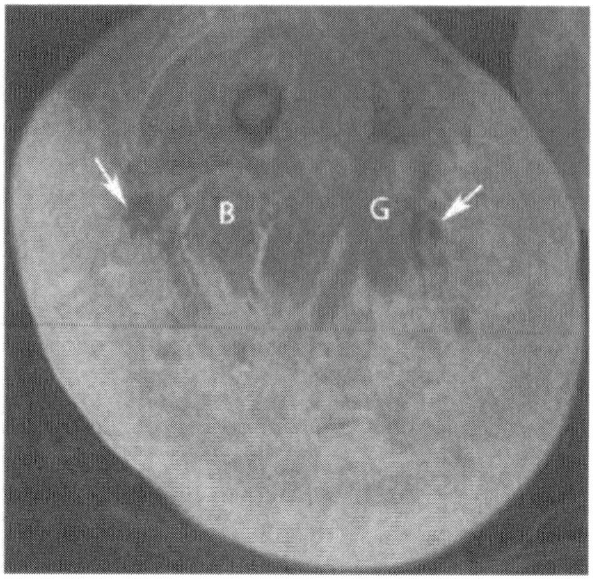

Figure 40G. Axial T1 post-Gd FS of the right proximal thigh.

Findings

This is a 6-day-old boy born with a large right thigh mass and congestive heart failure.

Figures 40F, 40G. This mass exhibits transfascial extension and occupies multiple soft tissue compartments. It is mainly subcutaneous and invades the biceps femoris (B), semimembranosus, semitendinosus, and gracilis (G) muscles. Numerous flow voids are evident **(arrows)**. At 11 weeks of age, the mass had nearly completely involuted.

Kaposiform Hemangioendothelioma

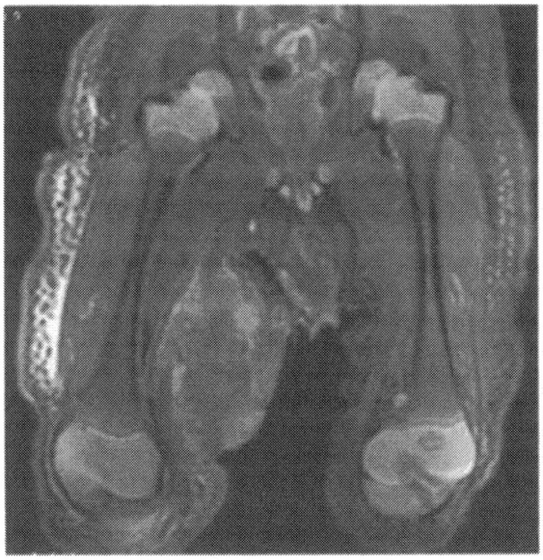

Figure 40H. Coronal STIR.

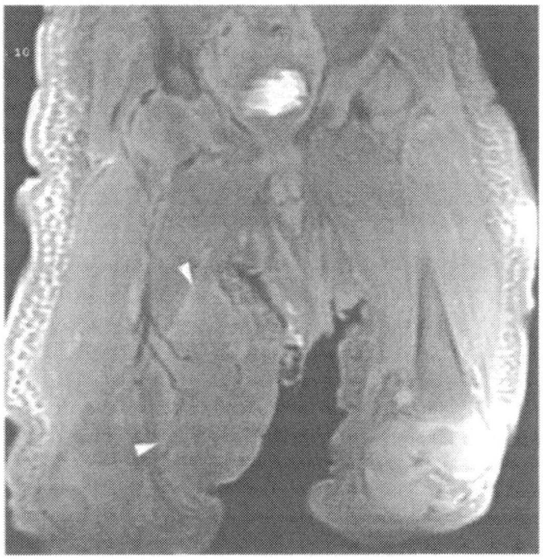

Figure 40I. Coronal T1 post-Gd FS.

Findings

This is a 7-day-old boy born with a large right thigh mass, thrombocytopenia, and no signs of heart failure.

Figures 40H, 40I. The right medial thigh mass has heterogeneously mild increased STIR SI and shows moderate enhancement **(arrowheads)**. Flow voids are identified arising from the femoral profunda artery. The margins of this mass are indistinct and there is significant subcutaneous edema. The clinical and laboratory findings and subsequent pathologic specimen confirmed the diagnosis of kaposiform hemangioendothelioma.

Pitfalls and Pearls

Infantile hemangiomas are sharply defined with respect to adjacent soft tissues in all imaging planes. Consider kaposiform hemangioendothelioma or infantile fibrosarcoma if the margins are unsharp.

References

1. Mueller BU, Mulliken JB. The infant with a vascular tumor. *Semin Perinatol* 1999; 23:332–340.
2. Hughes JA, Hill V, Patel K, Syed S, Harper J, De Bruyn R. Cutaneous haemangioma: Prevalence and sonographic characteristics of associated hepatic haemangioma. *Clin Radiol* 2004; 59:273–280.
3. Mulliken JB, Glowacki J. Hemangiomas and vascular malformations in infants and children: A classification based on endothelial characteristics. *Plast Reconstr Surg* 1982; 69:412–422.
4. Burrows PE, Laor T, Paltiel H, Robertson RL. Diagnostic imaging in the evaluation of vascular birthmarks. *Dermatol Clin* 1998; 16:455–488.
5. Berenguer B, Mulliken JB, Enjolras O, et al. Rapidly involuting congenital hemangioma: clinical and histopathologic features. *Pediatr Dev Pathol* 2003; 6:495–510.

6. Mulliken JB, Enjolras O. Congenital hemangiomas and infantile hemangioma: Missing links. *J Am Acad Dermatol* 2004; 50:875–882.
7. Mulliken JB, Anupindi S, Ezekowitz RA, Mihm MC, Jr. Case records of the Massachusetts General Hospital. Weekly clinicopathological exercises. Case 13-2004: A newborn girl with a large cutaneous lesion, thrombocytopenia, and anemia. *N Engl J Med* 2004; 350:1764–1775.
8. North PE, Waner M, Mizeracki A, Mihm MC, Jr. GLUT1: A newly discovered immunohistochemical marker for juvenile hemangiomas. *Hum Pathol* 2000; 31:11–22.

Case 41

History

This is a 5-year-old girl who has had a bump over her left, anterior mid-tibia for 5 months. Plain radiographs were normal (not shown).

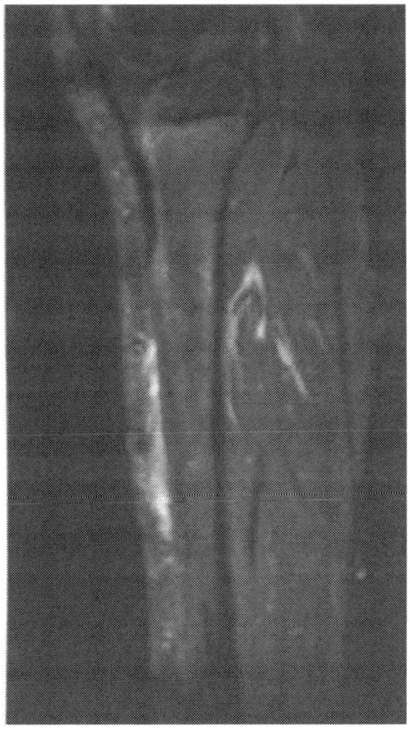

Figure 41A. Sagittal STIR of the left tibia.

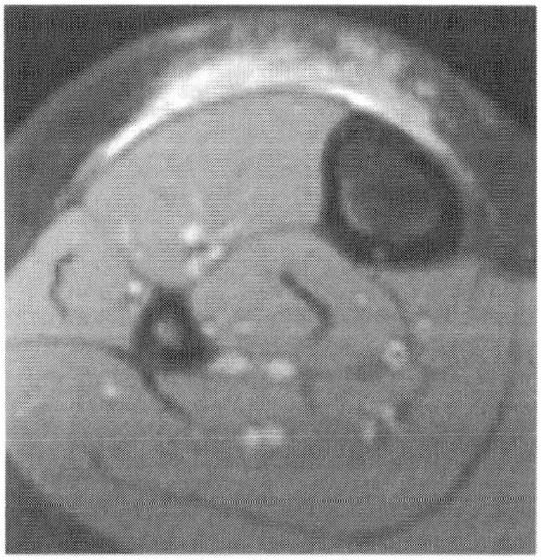

Figure 41B. Axial PD FS.

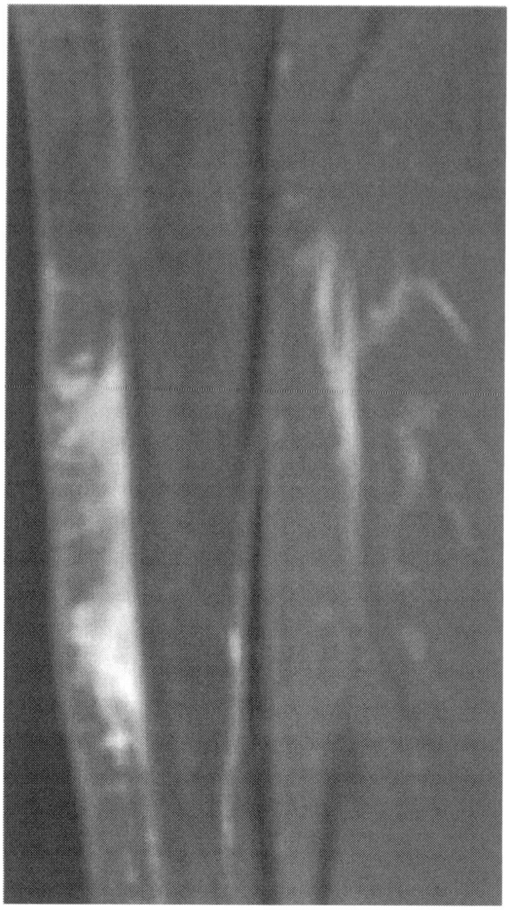

Figure 41C. Sagittal T1 post-Gd FS.

Figures 41A, 41B, 41C. There is an ill-defined subcutaneous mass noted along the anterior surface of the mid-tibia and tibialis anterior muscle. The mass is hyperintense on fluid sensitive sequences and enhances after gadolinium administration.

Diagnosis

Subcutaneous granuloma annulare

Questions

1. What is the most common location for subcutaneous granuloma annulare?
2. T/F: Subcutaneous granuloma annulare often extends into the adjacent muscles.

Discussion

Subcutaneous granuloma annulare (SGA) is a benign inflammatory process that is confined to the subcutaneous tissue layers. The suspected etiologies include post-traumatic, immune-complex reaction, and infectious (1). The four subtypes are localized, generalized, perforating, and subcutaneous (2). The subcutaneous form of granuloma annulare is noted almost exclusively in childhood. The cutaneous forms of granuloma annulare are apparent clinically and usually are not referred for radiologic imaging.

Subcutaneous granuloma annulare is typically seen in patients 2 to 5 years of age with a range of 6 months to 15 years (1, 3). In children under 5, it is considered the most common benign soft tissue mass that leads to biopsy in the lower extremity (23–30% of all biopsies) (4). The most common location for SGA is pretibial but it may occur elsewhere including the upper extremities, buttocks, face, and scalp (Answer to Question 1). The lesions are usually unilateral and solitary, but are occasionally multifocal (Figures 41D, 41E).

The MRI features of SGA are variable and often nonspecific. Patients usually present for imaging because of a rapidly growing mass. In one series, the usual size ranged from 1 to 4 cm (2). Subcutaneous granuloma annulare is usually hypointense to hyperintense on T1, shows variable SI on fluid-sensitive sequences, and enhances intensely with gadolinium (Figures 41C, 41E) (1, 2, 5). The borders are variable and may be either ill-defined or nodular. Lesions are usually confined to the subcutaneous fat with a broad base along the superficial fascia, and do not invade muscle (Answer to Question 2).

The differential diagnosis includes, but is not limited to, post-traumatic fat necrosis, hematoma, foreign body reaction (Figures 41F, 41G), nodular fasciitis, microcystic lymphatic malformation, cellulitis, rheumatoid nodules, fibrous hamartoma, and even malignant conditions such as subcutaneous sarcomas (e.g., myxoid fibrous histiocytoma) and granulocytic sarcoma. When the mass is confined to the subcutaneous pretibial space, is painless, and the patient is from 3 to 5 years old, the diagnosis of SGA should be suggested.

MRI evaluation was considered in this patient only after other etiologies, particularly trauma, had been excluded. Biopsy confirmed the diagnosis of SGA.

Orthopedic Perspective

The primary concern of the orthopedist in this context is to exclude sarcoma. Most pretibial lesions are self-limiting benign, traumatic, or reactive processes and are of little clinical significance. Subcutaneous sarcomas rarely occur in children and require identification because of the obvious treatment differences. Diagnosis is usually possible based on characteristic clinical findings, occasionally supplemented by MRI. When concern for malignancy persists, a definitive diagnosis may require a biopsy (either open or needle).

What the Clinician Needs to Know

1. Is the lesion restricted to the subcutaneous tissues, or does it cross tissue planes?
2. Is there osseous involvement?

Answers

1. Pretibial.
2. False.

Additional Examples

Subcutaneous Granuloma Annulare, Multifocal

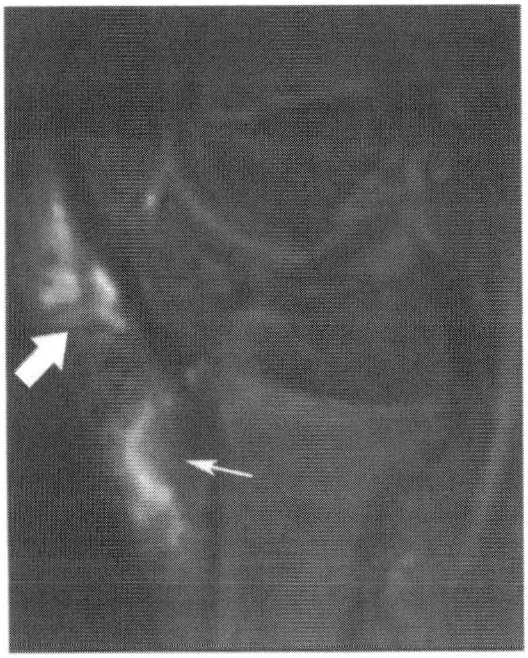

Figure 41D. Sagittal STIR.

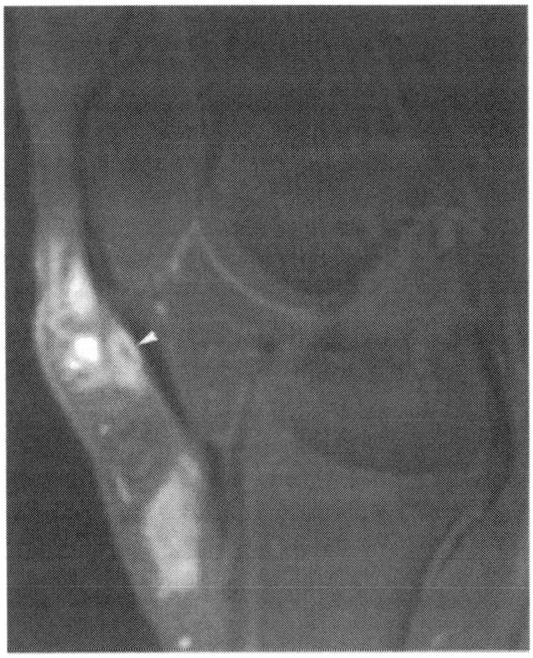

Figure 41E. Sagittal T1 post-Gd FS.

Findings

This is a 5-year-old girl with painless swelling over the anterior knee for over one year.

Figures 41D, 41E. There are two subcutaneous foci identified, one pretibial **(thin arrow)** and the other prepatellar **(thick arrow)**. The prepatellar lesion shows multiple nodules with variable STIR SI with peripheral enhancement **(arrowhead)**. These lesions were surgically confirmed to represent SGA.

Foreign Body Granuloma

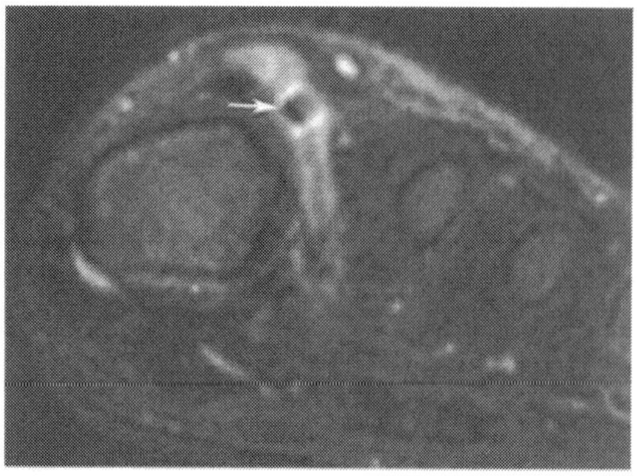

Figure 41F. Coronal T2 FS of the left foot.

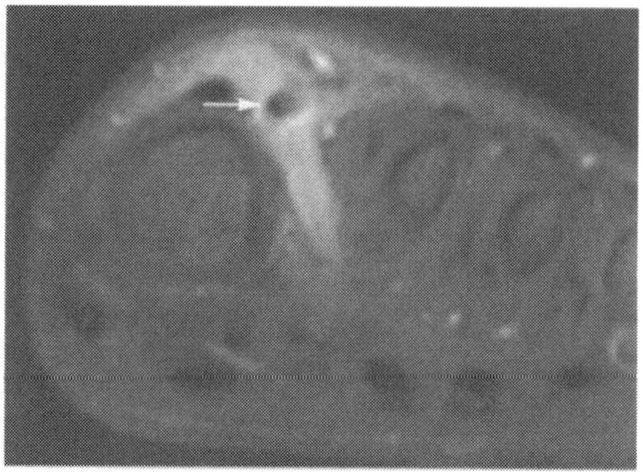

Figure 41G. Coronal T1 post-Gd FS.

Findings

This is a 12-year-old boy who sustained a puncture wound with a pencil to the dorsum of his foot three years earlier. He has had persistent swelling and redness, but no drainage. A radiograph showed swelling over the first and second metatarsal, but no foreign body (not shown).

Figures 41F, 41G. T2 and post-Gd sequences show a hypointense object within the deep soft tissues **(arrow)**. T2 hyperintensity and enhancement extend from the foreign body to the subcutaneous tissues and to the interosseous muscles of the first and second metatarsals. A pencil tip with foreign body granuloma were confirmed surgically.

Pitfalls and Pearls

1. The recurrence rate of subcutaneous granuloma annulare is high, ranging from 40% to 75% (6).
2. Although a benign process, granuloma annulare may rapidly enlarge and show marked contrast enhancement, raising the concern of a soft tissue malignancy.

References

1. De Maeseneer M, Vande Walle H, Lenchik L, Machiels F, Desprechins B. Subcutaneous granuloma annulare: MR imaging findings. *Skeletal Radiol* 1998; 27:215–217.
2. Kransdorf MJ, Murphey MD, Temple HT. Subcutaneous granuloma annulare: radiologic appearance. *Skeletal Radiol* 1998; 27:266–270.
3. Laor T. MR imaging of soft tissue tumors and tumor-like lesions. *Pediatr Radiol* 2004; 34:24–37.
4. Kransdorf MJ. Benign soft-tissue tumors in a large referral population: Distribution of specific diagnoses by age, sex, and location. *AJR Am J Roentgenol* 1995; 164:395–402.
5. Chung S, Frush DP, Prose NS, Shea CR, Laor T, Bisset GS. Subcutaneous granuloma annulare: MR imaging features in six children and literature review. *Radiology* 1999; 210:845–849.
6. Vandevenne JE, Colpaert CG, De Schepper AM. Subcutaneous granuloma annulare: MR imaging and literature review. *Eur Radiol* 1998; 8:1363–1365.

Case 42

History

This 11-year-old girl fell on her left upper arm during gymnastics 2 weeks ago. One day prior to the MRI, she developed severe arm pain and redness. She had a temperature of 39 C, and was confused. A radiograph of the arm was normal (not shown).

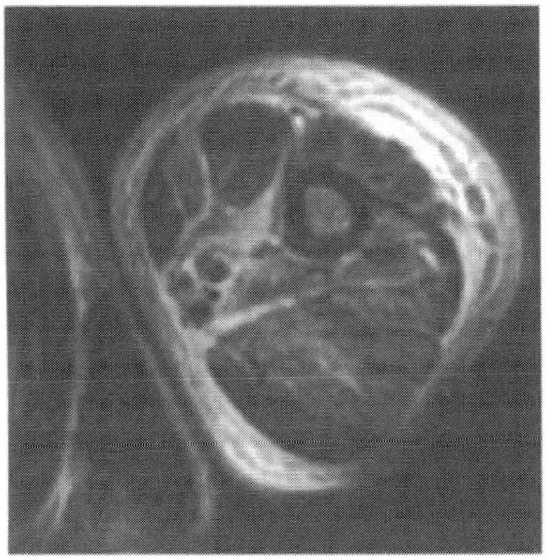

Figure 42A. Axial T2 FS of the left upper arm.

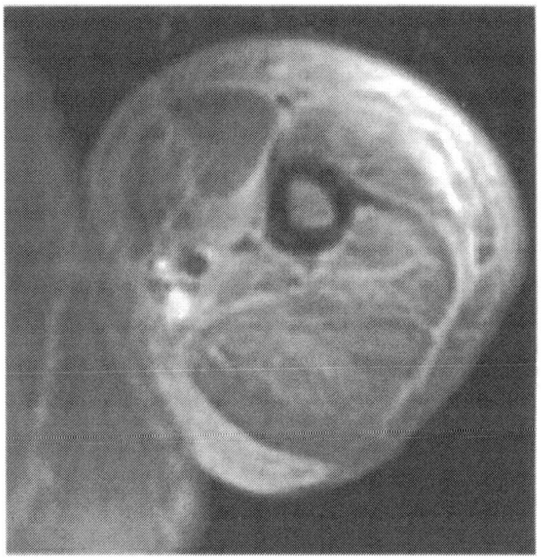

Figure 42B. Axial T1 post-Gd FS.

Figures 42A, 42B, (42A with annotations). There is striking T2 hyperintensity and enhancement within the subcutaneous tissues as well as the superficial and deep fascia **(arrows)** of the upper arm. There is muscle inflammation manifested by mild increased T2 SI (*) and conspicuous enhancement.

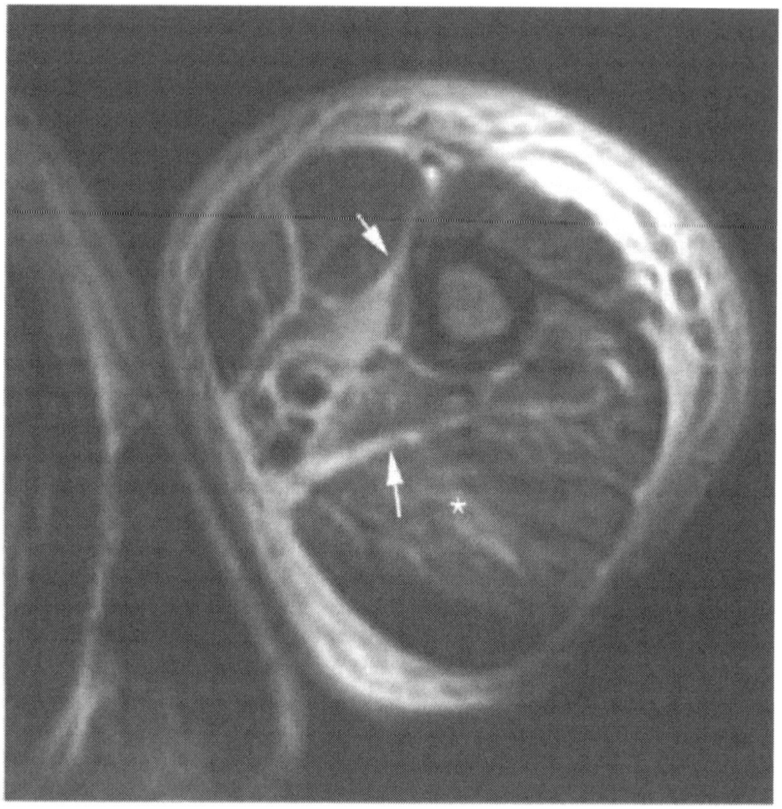

Figure 42A* Annotated.

Diagnosis

Necrotizing fasciitis

Questions

1. What is the proposed etiology for muscle enhancement in the setting of necrotizing fasciitis?
2. How does MRI differentiate necrotizing fasciitis from cellulitis?

Discussion

Necrotizing fasciitis is a potentially fatal soft tissue infection involving the subcutaneous fat as well as superficial and deep fascia (1). The etiology is often polymicrobial with both aerobic and anaerobic organisms. In one study, beta-hemolytic streptococci was isolated from 31/73 surgically proven cases (2). The risk factors for necrotizing fasciitis include infection, trauma, surgery, chronic illnesses, and immunosuppression.

A diffuse edema pattern involving the subcutaneous fat, superficial and deep fascia, and adjacent muscles may be seen on MRI. Rarely, gas bubbles are evident, and are hypointense on all imaging sequences (3). Because of increased pressures generated by the fascia edema, a compartment syndrome may develop, leading to muscle necrosis. The initial MRI features of muscle necrosis are increased T2 SI and diffuse heterogeneous enhancement. The enhancement pattern in muscle may reflect capillary extravasation of contrast from necrotic tissue (Answer to Question 1) (4). In late stages, liquefaction may occur, resulting in areas of nonenhanced musculature (Figures 42C, 42D), and eventually intramuscular abscess formation. These MRI features are very sensitive for necrotizing fasciitis, but not specific. Similar changes may be seen in patients with dermatomyositis, other vasculitidies, trauma, and pyomyositis (Figures 42E–42G) (5). However, given the potential complications and occasional lethal outcome of this disorder, the diagnosis of necrotizing fasciitis should be suggested in the appropriate clinical context.

In contrast to necrotizing fasciitis, isolated cellulitis demonstrates inflammatory changes that are limited to the subcutaneous tissues (Figures 42H, 42I) (1). There is usually skin thickening as well as reticulated and ill-defined extension of inflammatory changes along the subcutaneous fat and fascia. Deep fascial edema is usually absent (Answer to Question 2). The most common causative organism isolated is *Staphylococcus aureus* (6). Cellulitis is generally apparent clinically, but MRI may be useful to rule out complications including pyomyositis and abscess. Bacterial pyomyositis is most often seen in patients who live in tropical regions and *Staphylococcus aureus* is the most often isolated (7). Transfascial muscle edema with increased T2 SI and enhancement may be seen and be indistinguishable from necrotizing fasciitis.

Blood cultures were positive for group A *Streptococcus* in this patient and necrotizing fasciitis was confirmed at fasciectomy. She was given high dose penicillin and clindamycin and was discharged home on day 9 of her hospital stay.

Orthopedic Perspective

The main issue for the orthopedist is to distinguish superficial cellulitis from a deep infection involving muscle with abscess formation. The former can frequently be treated with rest, elevation, and broad-spectrum antibiotics, but the latter frequently requires aggressive surgical debridement. Necrotizing fasciitis is especially alarming because of its rapid progression and the need to distinguish it from gas gangrene. Delay in treatment can lead to significant morbidity, and occasionally mortality, if extensive myonecrosis and compartment syndrome occur. MR imaging also gives direction to the surgeon regarding the extent of involvement and the compartments that will require debridement.

What the Clinician Needs to Know

1. Distinguish superficial cellulitis from necrotizing fasciitis because the latter requires urgent surgical debridement.
2. Define the extent of involvement to guide surgical debridement.

Answers

1. Capillary extravasation of gadolinium from necrotic muscle tissue.
2. Necrotizing fasciitis demonstrates edema in the subcutaneous soft tissues, superficial and deep fascia, and muscle edema. Cellulitis shows similar changes in the superficial tissues, but deep fascial involvement is usually absent.

Additional Examples

Advanced Necrotizing Fasciitis

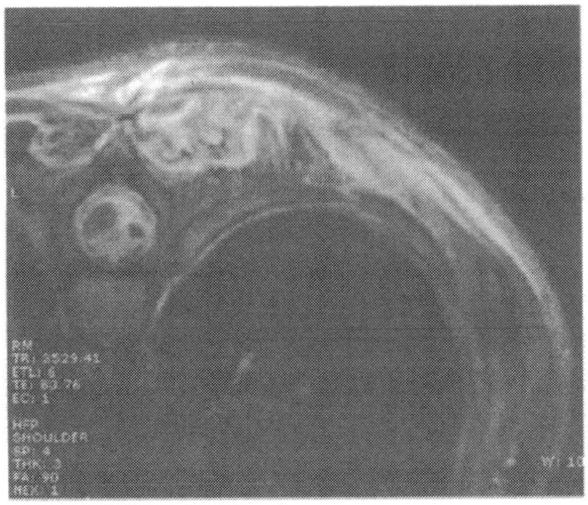

Figure 42C. Axial T2 FS of the mid-back.

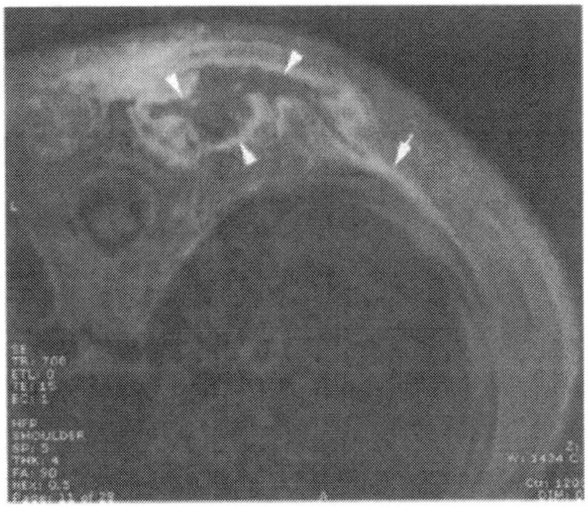

Figure 42D. Axial T1 post-Gd FS.

Findings

This is a 2-year-old girl who presented with a 9-day history of high fevers, neck pain, and an abnormal gait.

Figure 42C. There is diffuse heterogeneous increased T2 SI within the markedly enlarged right erector spinae muscle and deep fascia bilaterally.

Figure 42D. There is rim enhancement around the poorly enhancing erector spinae muscle **(arrowheads)** as well as diffuse enhancement within the deep fascia **(arrow)**. Necrotizing fasciitis was found surgically and myonecrosis was confirmed pathologically.

Pyomyositis with Abscess

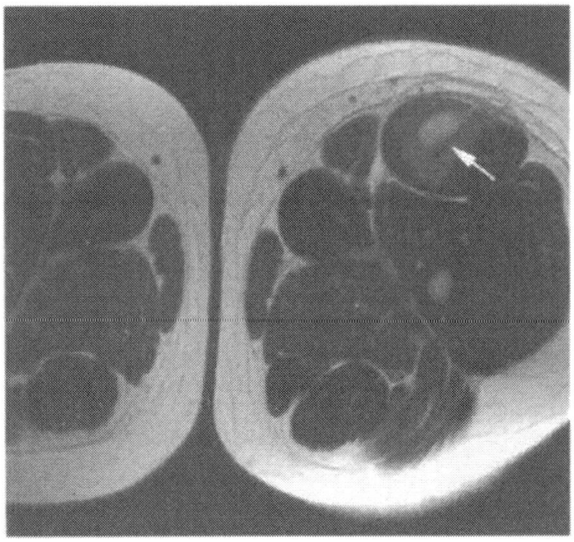

Figure 42E. Axial T2 (no FS) of the thighs.

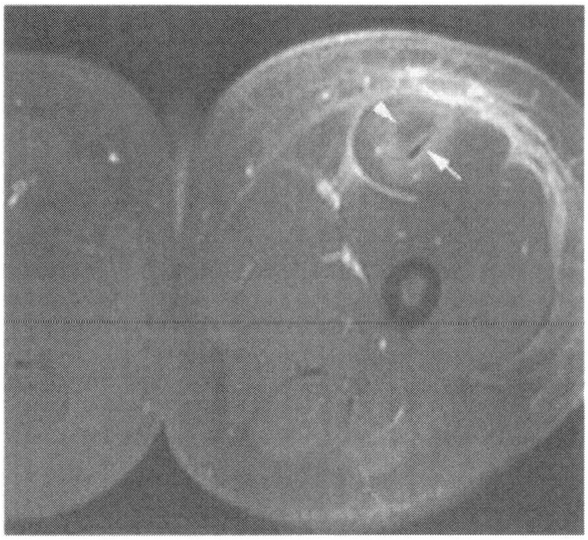

Figure 42F. Axial T1 post-Gd FS.

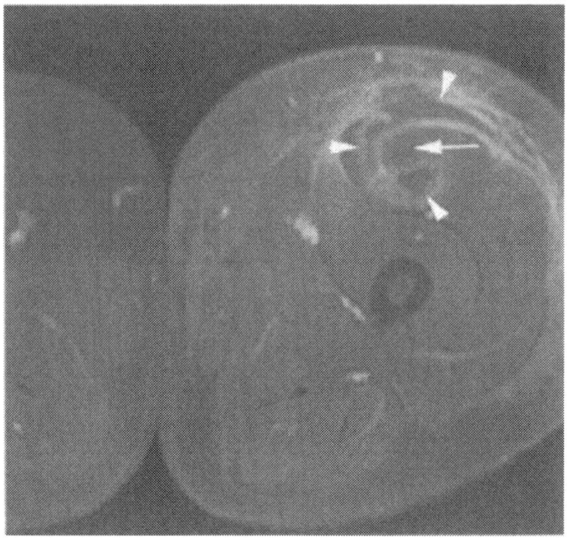

Figure 42G. Axial T1 post-Gd FS.

Findings

This is a 16 year old with a 1-week history of left thigh pain.

Figures 42E, 42F, 42G. There is a T2 hyperintense collection at the myotendinous junction of the rectus femoris **(arrows)**. There are intramuscular enhancing fluid collections **(arrowheads)** that track anteriorly to the superficial fascia. Pus was drained operatively and grew gram positive cocci. Presumably, this patient had a rectus femoris myotendinous tear with a secondarily infected hematoma.

Diffuse Forearm Cellulitis

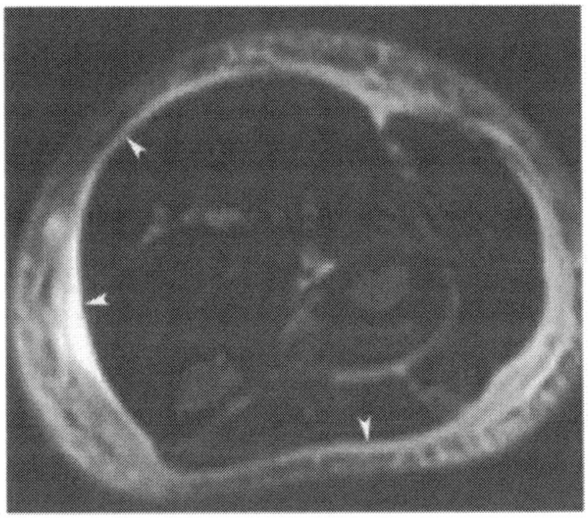

Figure 42H. Axial T2 FS of the left forearm.

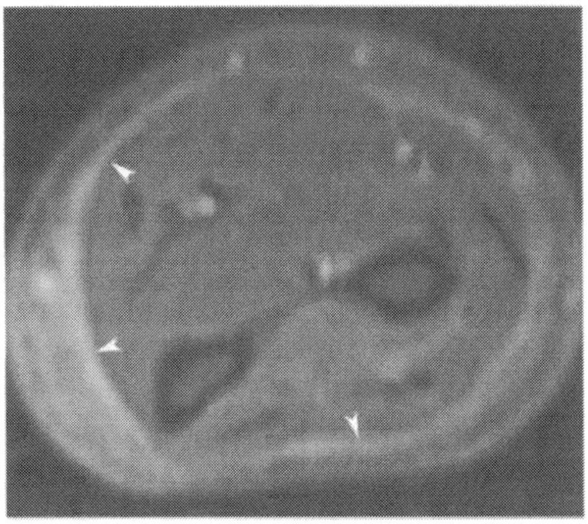

Figure 42I. Axial T1 post-Gd FS.

Findings

This 11-year-old boy had progressive forearm swelling following a 5 days earlier.

Figures 42H, 42I. There is diffuse reticular and confluent increased T2 SI and enhancement present within the subcutaneous soft tissues. There is diffuse thickening of the skin and enhancement of the superficial fascia **(arrowheads)**. No deep fascia or muscle edema is present to suggest necrotizing fasciitis.

Pitfalls and Pearls

The MRI features of necrotizing fasciitis are often nonspecific, and may overlap with trauma and systemic conditions such as dermatomyositis and vasculitis. However, when the typical MRI findings are present in the appropriate clinical context, the diagnosis of this potentially life-threatening disorder should be suggested.

References

1. Struk DW, Munk PL, Lee MJ, Ho SG, Worsley DF. Imaging of soft tissue infections. *Radiol Clin North Am* 2001; 39:277–303.
2. Bakleh M, Wold LE, Mandrekar JN, Harmsen WS, Dimashkieh HH, Baddour LM. Correlation of histopathologic findings with clinical outcome in necrotizing fasciitis. *Clin Infect Dis* 2005; 40:410–414.
3. Fugitt JB, Puckett ML, Quigley MM, Kerr SM. Necrotizing fasciitis. *Radiographics* 2004; 24:1472–1476.
4. Schmid MR, Kossmann T, Duewell S. Differentiation of necrotizing fasciitis and cellulitis using MR imaging. *AJR Am J Roentgenol* 1998; 170:615–620.
5. Arslan A, Pierre-Jerome C, Borthne A. Necrotizing fasciitis: Unreliable MRI findings in the preoperative diagnosis. *Eur J Radiol* 2000; 36:139–143.
6. Ladhani S, Garbash M. Staphylococcal skin infections in children: Rational drug therapy recommendations. *Paediatr Drugs* 2005; 7:77–102.
7. Yu CW, Hsiao JK, Hsu CY, Shih TT. Bacterial pyomyositis: MRI and clinical correlation. *Magn Reson Imaging* 2004; 22:1233–1241.

Case 43

History

This is a 6-day-old girl with an enlarging right calf mass.

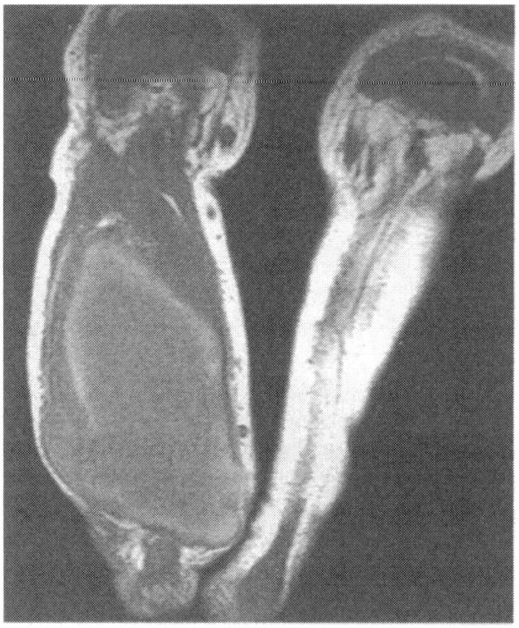

Figure 43A. Coronal T1.

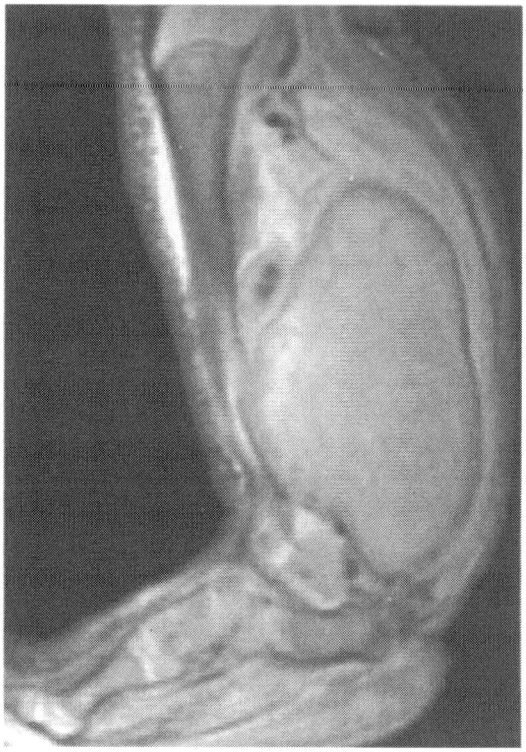

Figure 43B. Sagittal STIR.

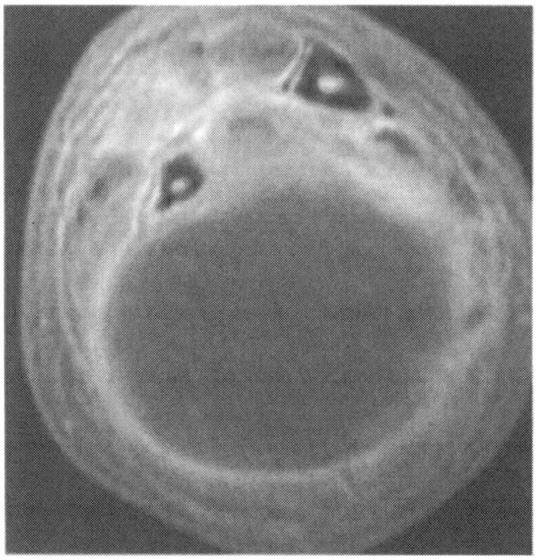

Figure 43C. Axial T1 post-Gd FS.

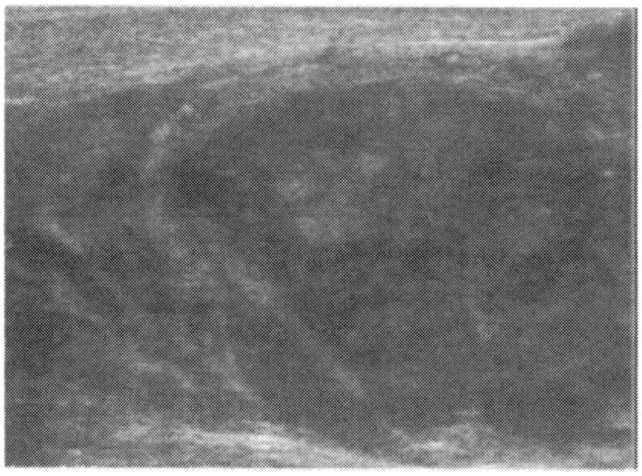

Figure 43D. Sagittal ultrasound with color Doppler.

Figures 43A, 43B. A large sharply marginated mass is identified in the posterior right calf. The center is hypointense on T1 and markedly hyperintense on STIR, particularly near the fibula.

Figure 43C. The center of the mass remains hypointense, but the periphery enhances, especially in the interosseous region.

Figure 43D. Ultrasound shows a complex mass without flow by color Doppler.

Figure 43E. Central matrix calcifications are noted **(arrowheads)**, as well as tibial metaphyseal and diaphyseal radiolucent lesions **(arrows)**. Posterior tibial and fibular remodeling deformity is evident.

Figure 43F. Sharply marginated radiolucent lesions are identified in the proximal femurs **(arrows)**.

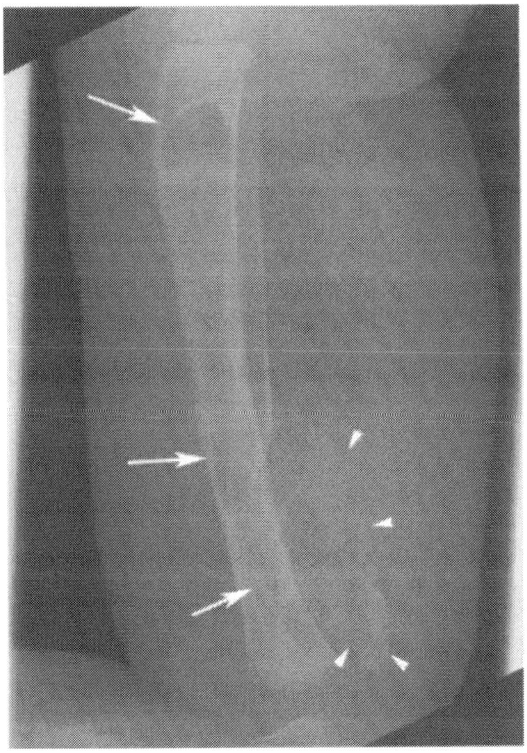

Figure 43E. Lateral radiograph of the right lower leg, 7 months later.

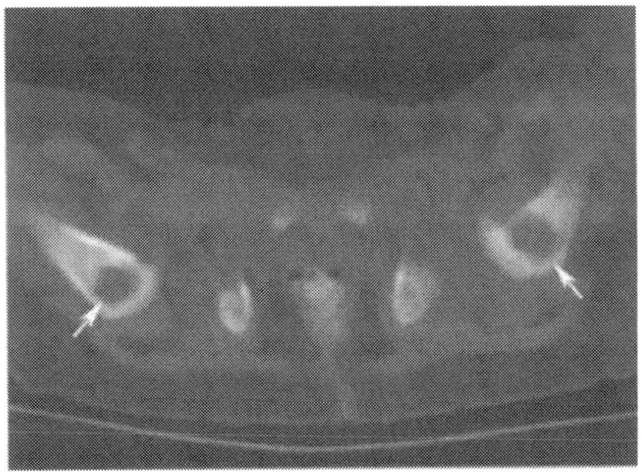

Figure 43F. Axial CT through the hips.

Diagnosis

Infantile myofibromatosis, multicentric

Questions

1. What feature of infantile myofibromatosis carries a poor prognosis?
2. What term describes the peripheral enhancement pattern noted following gadolinium in some cases of infantile myofibromatosis?

Discussion

Fibrous tumors of infancy are rare. The most common subtype is infantile myofibromatosis (AKA congenital generalized fibromatosis) (1). It may involve the subcutaneous soft tissues, muscles, bone, viscera, and rarely the nervous system. About half of these tumors present at birth and 88% are found by 2 years of age (2). More than half of these lesions are solitary and the remainder are multicentric, with or without visceral involvement. It is a benign, nonmetastasizing tumor with the majority of lesions undergoing spontaneous involution within one to two years. The exception is multicentric myofibromatosis with visceral involvement, which has a poor prognosis related to involvement of the cardiopulmonary or gastrointestinal system (Answer to Question 1) (3).

The MRI features of infantile myofibromatosis mirror the histology and biologic behavior of the lesion. Myofibroblasts are usually located peripherally, whereas more primitive hemangiopericytoma-like cells are located centrally (2). These latter cells tend to undergo necrosis, hemorrhage, and calcification. In general, these tumors are hypo- to isointense on T1 and hyperintense on fluid sensitive sequences (2, 4, 5). The center of the tumor may show T2 hyperintensity, depending on the degree of cellularity and necrosis present. The enhancement pattern of these lesions is also variable. The target appearance has been applied to the peripheral enhancement pattern without central enhancement noted following intravenous gadolinium (Figures 43G–43I) (Answer to Question 2) (2, 5).

Calcification may be seen centrally within these tumors (Figure 43E) (3). Osseous deformity related to extrinsic compression by the tumor is evident in some cases. Intraosseous infantile myofibromatosis usually involves the metaphysis, and is not necessarily associated with a soft tissue mass (Figures 43E, 43F). These lesions may be well defined, purely lytic, and show a peripheral sclerotic rim (2).

Infantile myofibromatosis should not be confused with an intramuscular desmoid tumor, a lesion that is usually found in older children and young adults and tends to recur (5). The MRI features of infantile myofibromatosis may mimic abscess (with rim enhancement), other fibrous tumors, neurofibromas, soft tissue sarcomas with central tumoral necrosis, such as infantile fibrosarcoma and infantile rhabdomyosarcoma, congenital and infantile hemangioma, hemangiopericytoma, and vascular anomalies such as lymphatic malformations. When there is a soft tissue mass with multifocal osseous involvement as seen in this newborn, the differential is limited to myofibromatosis, congenital leukemia/lymphoma, or an aggressive sarcoma with metastasis; therefore biopsy is required.

The initial diagnosis was a macrocystic lymphatic malformation with internal hemorrhage. When ultrasound-guided sclerotherapy was attempted, the lesion was found to be solid, but avascular (Figure 43D). Percutaneous needle biopsy was performed and the diagnosis of infantile myofibromatosis was established. Surgery was not performed and the patient did not receive chemotherapy. The skeletal and soft tissue lesions were followed clinically and have spontaneously regressed. The child is asymptomatic and has no developmental or growth abnormalities.

Orthopedic Perspective

The orthopedist needs to distinguish this lesion from the sarcomas and vascular lesions noted in the above discussion. The diagnosis is made after biopsy. Skeletal survey and, in some cases, abdominal CT are obtained to search for other lesions, since the presence of visceral involvement portends a worse prognosis. Since the lesions usually spontaneously involute, myofibromatosis is usually observed, without chemotherapy or surgical intervention.

What the Clinician Needs to Know

1. The clinician needs to distinguish this entity from malignant, vascular, and infectious processes.
2. Is the process multicentric?
3. Is there visceral involvement?

Answers

1. Multicentric disease with visceral involvement.
2. Target appearance.

Additional Example

Solitary Infantile Myofibromatosis with Target Sign

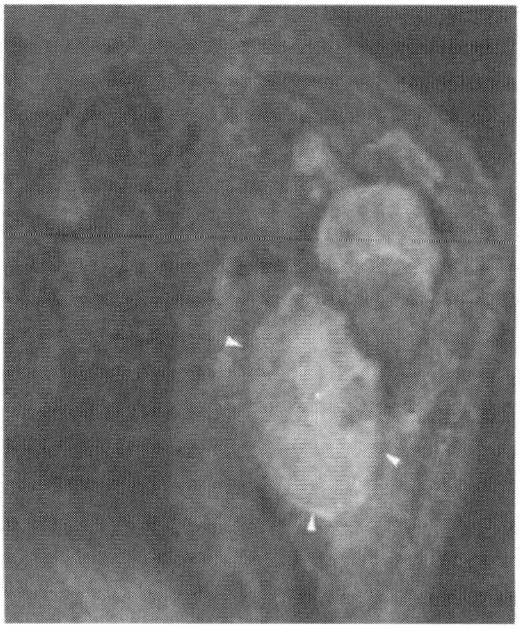

Figure 43G. Oblique coronal STIR of the left shoulder.

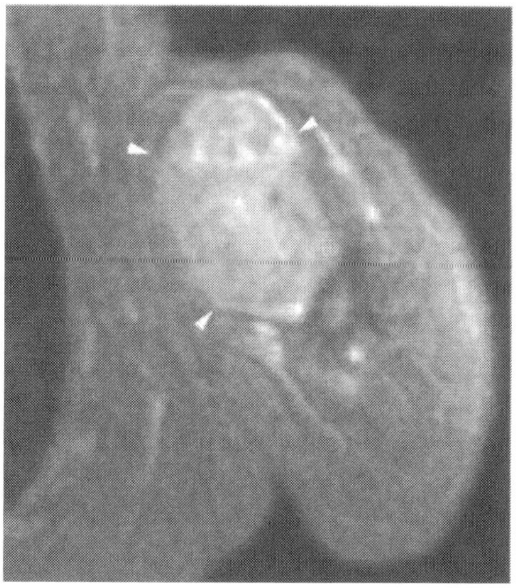

Figure 43H. Axial T2 FS.

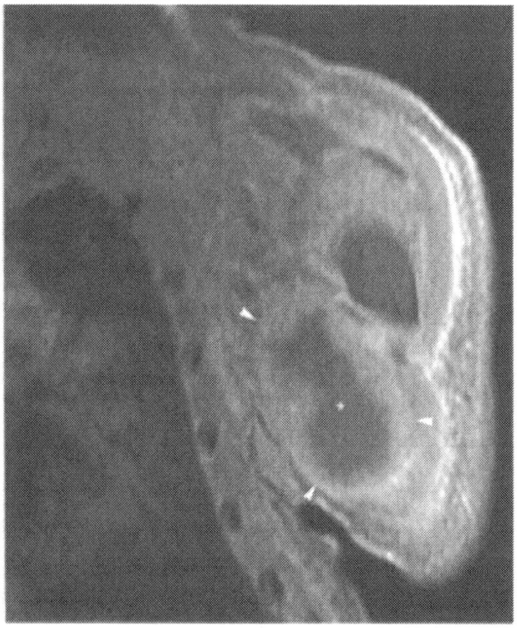

Figure 43I. Axial T1 post-Gd FS.

Findings

This is a 5-day-old otherwise healthy boy, who was found to have a mass along the proximal left arm at birth.

Figures 43G, 43H. There is a large well-defined, heterogeneously hyperintense soft tissue mass **(arrowheads)** abutting the medial aspect of the humerus. The center of the mass (*) is slightly more hyperintense than the periphery.

Figure 43I. There is peripheral enhancement **(arrowhead)** of the mass, around a nonenhancing center (*) (target sign).

Pitfalls and Pearls

1. The imaging features of infantile myofibromatosis are nonspecific, but when a target sign is present and there are multiple sharply defined metaphyseal lytic bone lesions, the diagnosis of multicentric myofibromatosis is likely.
2. Since the pathologic features of myofibromatosis may suggest a hemangiopericytoma, a lesion with a very different biologic behavior, the pathologist should be alerted to the possibility of infantile myofibromatosis.

References

1. Spadola L, Anooshiravani M, Sayegh Y, Jequier S, Hanquinet S. Generalised infantile myofibromatosis with intracranial involvement: Imaging findings in a newborn. *Pediatr Radiol* 2002; 32:872–874.
2. Koujok K, Ruiz RE, Hernandez RJ. Myofibromatosis: imaging characteristics. *Pediatr Radiol* 2004.
3. Davies RS, Carty H, Pierro A. Infantile myofibromatosis: A review. *Br J Radiol* 1994; 67:619–623.
4. Johnson GL, Baisden BL, Fishman EK. Infantile myofibromatosis. *Skeletal Radiol* 1997; 26:611–614.
5. Eich GF, Hoeffel JC, Tschappeler H, Gassner I, Willi UV. Fibrous tumours in children: Imaging features of a heterogeneous group of disorders. *Pediatr Radiol* 1998; 28:500–509.

Case 44

History

This is a 9-year-old boy with volar wrist swelling for the last 2 months.

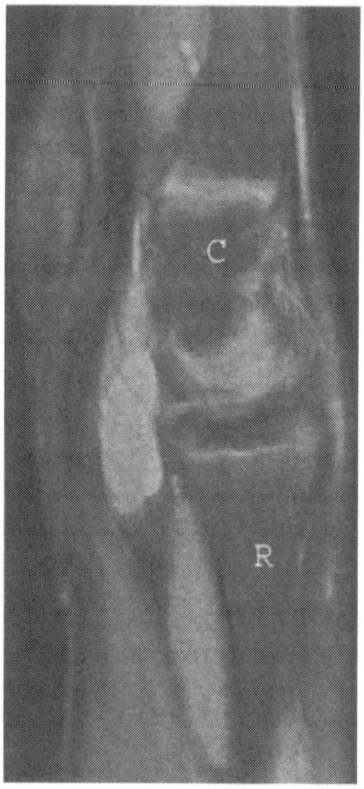

Figure 44A. Sagittal PD FS.

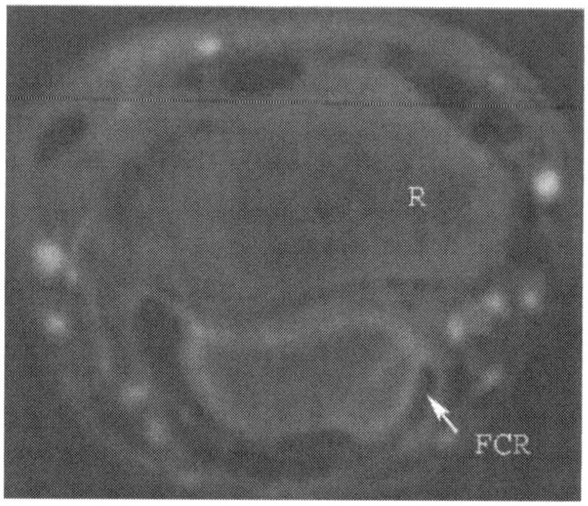

Figure 44B. Axial T1 post-Gd FS.

Figures 44A, 44B, (44B with annotations). There is a sharply defined mass situated deep to the flexor tendons of the wrist. The mass displaces the tendons and is intimately associated with the flexor carpi radialis tendon sheath (FCR). The mass is hyperintense on the PD FS sequence and demonstrates rim **(arrowheads)** and septal **(arrow)** enhancement. Capitate (C), radius (R).

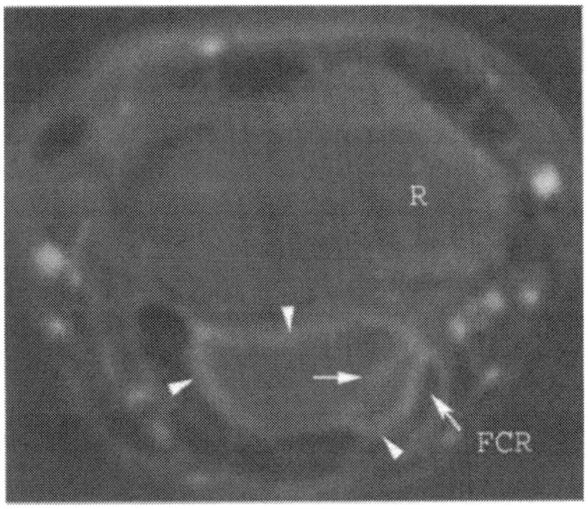

Figure 44B* Annotated.

Diagnosis

Ganglion cyst

Questions

1. T/F: Volar wrist ganglion cysts are more common than dorsal ganglion cysts in children under 10 years of age.
2. Ganglion or synovial cysts of the knee generally arise in relation to what anatomic structure?

Discussion

Ganglion cysts are benign, fluid or mucinous filled structures, usually resulting from repetitive trauma. Ganglion cysts may be found related to the joint capsule, ligaments, tendon sheaths, bursae, or subchondral bone (1). Ganglion cysts have a thin fibrous lining, whereas synovial cysts have a thin epithelial lining. On MRI, it is impossible to differentiate these two entities. By convention, both are simply referred to as ganglion cysts, unless pathologic proof is available.

Ultrasound may be the initial modality in the evaluation of ganglion cysts. Uncomplicated ganglion cysts are anechoic or hypoechoic, compressible, and demonstrate increased through transmission. US may help distinguish a simple ganglion cyst from a solid soft tissue lesion as well as vascular malformations.

Ganglion cysts are usually hypointense on T1, markedly hyperintense on fluid-sensitive sequences, and may demonstrate thin rim enhancement (1). When ganglion cysts are predominantly mucinous, they may demonstrate mild hyperintensity on T1W images. Ganglion cysts may be round or multilobular with well-defined margins. Internal septations may be identified related to prior trauma or infection, and these can enhance with gadolinium.

The relationship to adjacent tendons, joint, and neurovascular structures is important in the evaluation of ganglion cysts. It is critical to identify any communication of a ganglion cyst with adjacent tendon sheath because successful resection, after failed nonoperative treatment, requires that the entire tract and cyst be removed to prevent recurrence (2).

In the adult population and in children over 10 years of age, ganglion cysts of the wrist are usually located dorsally, and typically arise from the scapholunate interosseous membrane (2, 3). In children under 10 years of age, a higher incidence of volar compartment ganglion cysts has been noted (Answer to Question 1) (3). Wang et al. evaluated 14 children under the age of 10 years (average: 38 months) with ganglion cysts of the wrist and found 7 retinacular cysts (ganglia arising from the flexor tendon sheath), 5 volar wrist ganglia, and 2 dorsal wrist ganglia (4). Volar wrist ganglion cysts may arise from the radiocarpal joint, scaphotrapezial joint, and ulnocarpal joint (2). Ulnocarpal joint ganglion cysts may be associated with tears of the triangular fibrocartilage. In the foot, dorsal ganglion cysts (Figures 44C–44E) are more common than plantar ganglion cysts (5).

In the knee, ganglion cysts are often related to the cruciate ligaments (Figure 44F) (Answer to Question 2), and less commonly arise from a torn meniscus. In one study of 85 knees, 49 cysts arose from the anterior cruciate ligament, 16 from the posterior cruciate ligament, and 15 from menisci (6). Ganglion cysts within Hoffa's fat pad should also be distinguished from infrapatellar bursitis, which has a characteristic tear-drop appearance (Figures 44G, 44H) (7).

Not all extraosseous soft tissue cysts are ganglion cysts. Other considerations include bursa, soft tissue myxoma, macrocystic lymphatic malformation, epidermoid cysts (Figures 44I–44K), and soft tissue sarcoma (Figures 44L–44N). Soft tissue myxomas may demonstrate marked T2 hyperintensity, show variable T1 SI, and be indistinguishable from a ganglion cyst before contrast. Contrast administration usually distinguishes these two entities by demonstrating diffuse or nodular enhancement with myxomas, as apposed to thin rim enhancement with ganglion cysts (8). Epidermoid cysts are benign lesions that are found in the subcutaneous tissue that may occur from abnormal migration of epidermal tissue during embryogenesis or trauma (9). These lesions are usually iso- to mildly hyperintense on T1, hyperintense on fluid-sensitive sequences, and lack central enhancement. Of the soft tissue sarcomas, synovial cell sarcoma is most variable in appearance and may contain cystic change (10). Whenever a solid soft tissue component is identified in a cystic lesion, a sarcoma should always be considered.

In this patient, the cystic mass did not resolve after 6 months of conservative therapy. Complete excision was performed and the lesion was pathologically proven to represent a ganglion cyst.

Orthopedic Perspective

The primary issue from the orthopedist's perspective when a mass presents in the soft tissues of the extremities or torso is whether it is benign or malignant. In general, cystic lesions in children adjacent to tenosynovial structures can be observed and often resolve spontaneously, whereas solid lesions need to be evaluated more carefully. Clinically, one can attempt to make this distinction by transillumination of the cyst, but ultrasound is much more reliable and is a relatively easy and inexpensive modality to follow ganglia. If they become symptomatic, the MRI will guide the surgical approach to ensure complete excision, and, as noted above, if there is communication with a joint or tendon sheath, this should be excised as well.

What the Clinician Needs to Know

1. Is the lesion solid or cystic? Cystic lesions near tenosynovial structures can usually be observed and may resolve spontaneously. Solid lesions are of more concern.
2. What is the precise anatomic location of the ganglion cyst and is there potential communication with the adjacent joint or tendon sheaths? This is helpful to plan surgical excision when needed.

Answers

1. True.
2. Cruciate ligaments.

Additional Examples

Ganglion Cyst of the Dorsum of the Foot

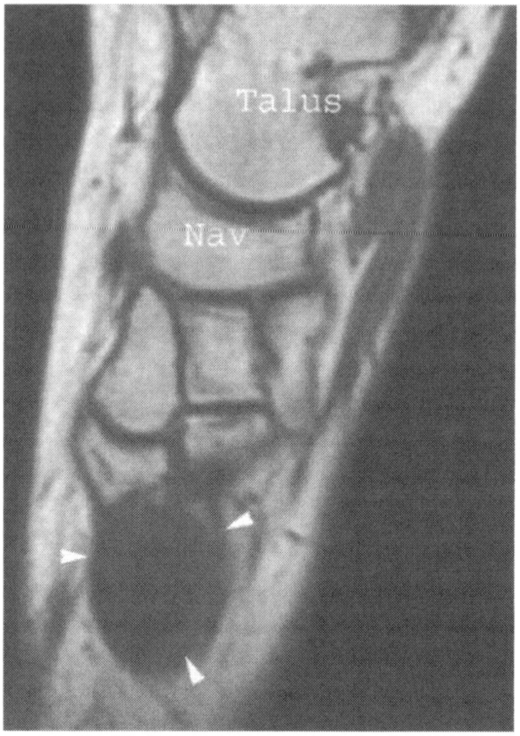

Figure 44C. Axial (footprint) T1.

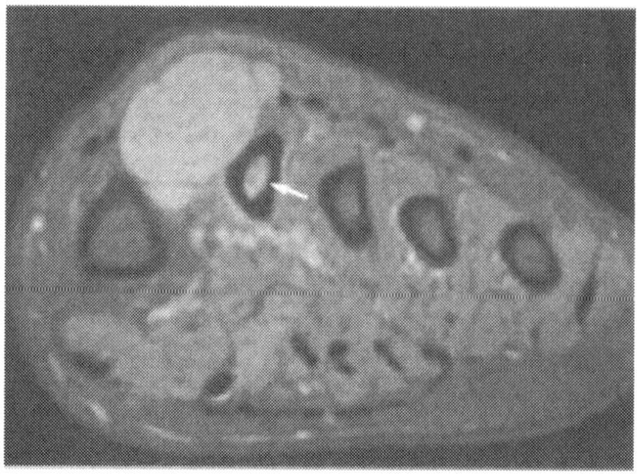

Figure 44D. Coronal T2 FS.

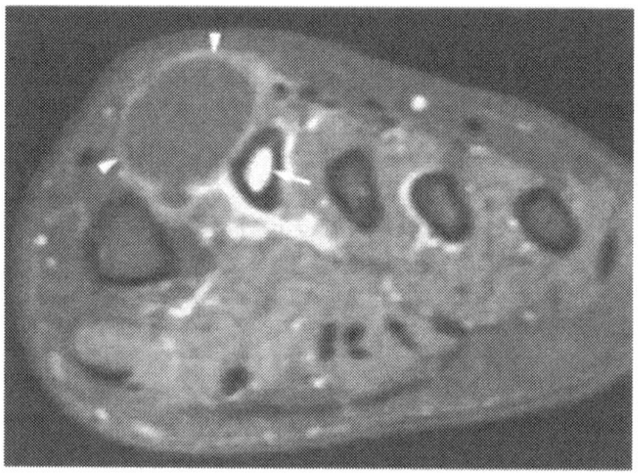

Figure 44E. Coronal T1 post-Gd FS.

Findings

This is a 13-year-old girl with an enlarging mass over the dorsum of the first and second metatarsal heads.

Figures 44C, 44D, 44E. The mass is hypointense on T1, hyperintense with lobulations on T2, and demonstrates thin rim enhancement **(arrowheads)**. This was a pathologically proven ganglion cyst. There is also marrow edema of the second metatarsal **(arrows)** compatible with stress reaction.

Ganglion Cyst in Hoffa's Fat Pad

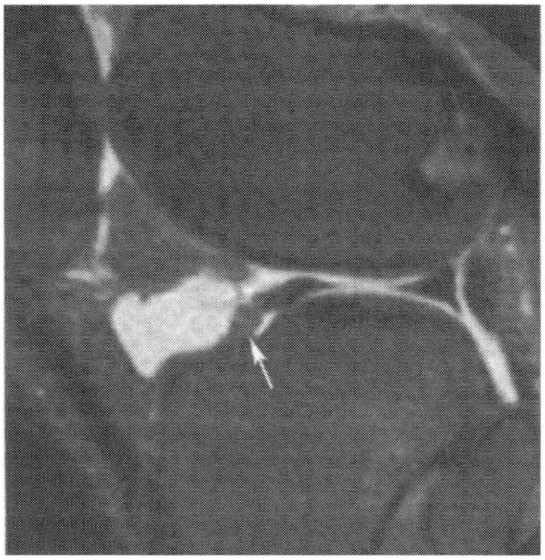

Figure 44F. Sagittal STIR.

Findings

This is a 20-year-old male with chronic knee pain.

Figure 44F. There is a globular appearing cystic structure extending into Hoffa's fat pad that is associated with the anterior cruciate ligament. There is also mildly increased SI within the anterior meniscal root **(arrow)**.

Infrapatellar Bursitis

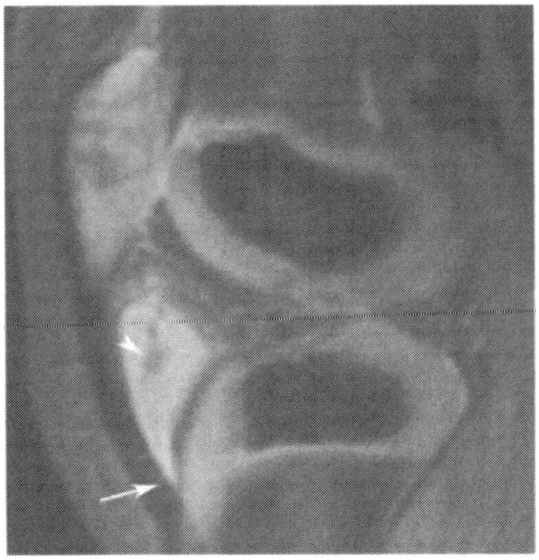

Figure 44G. Sagittal PD FS.

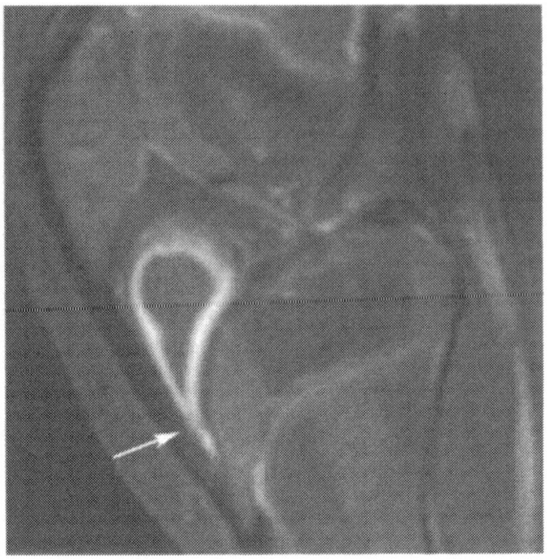

Figure 44H. Sagittal T1 post-Gd FS.

Findings

This is a 3-year-old boy with an infrapatellar mass that had been present for several months.

Figures 44G, 44H. A cystic infrapatellar mass is present that is predominantly hyperintense on the PD FS sequence and demonstrates a thick enhancing rim following contrast. Foci of low SI are present centrally on the PD FS sequence **(arrowhead)** that were found to represent rice bodies (fibrin and/or collagen encapsulated necrotic cellular debris) at surgery. Note the characteristic tear-drop appearance **(arrow)**, which distinguishes infrapatellar bursitis from a Hoffa's fat pad ganglion cyst.

Epidermoid Cyst

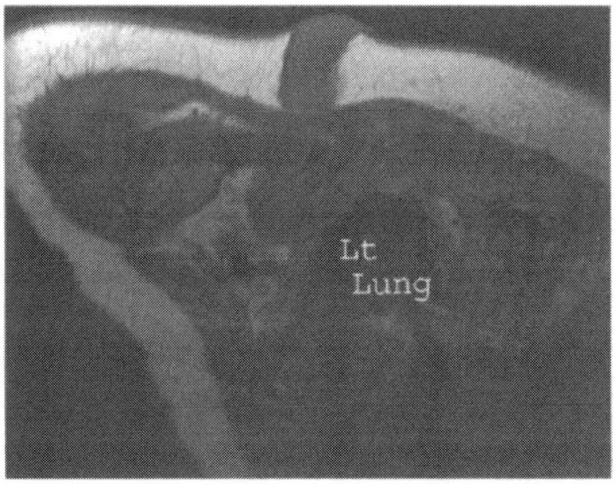

Figure 44I. Axial T1 of the left upper chest (prone positioning).

Figure 44J. Axial T2 FS.

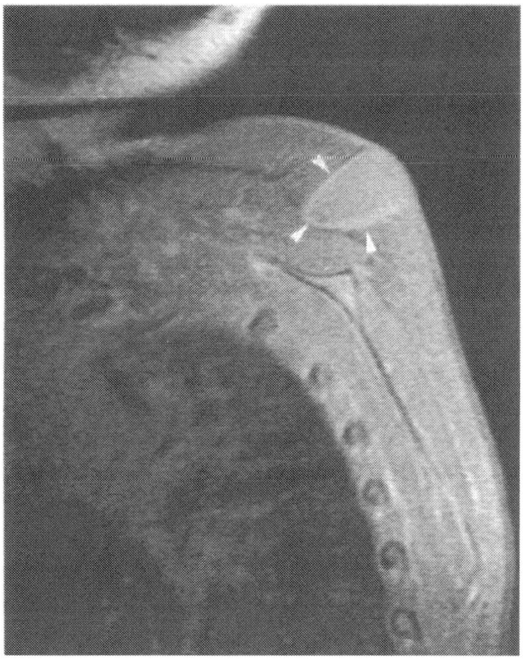

Figure 44K. Sagittal T1 post-Gd FS.

Findings

This is a 6-month-old with a palpable mass over his left back.

Figures 44I, 44J, 44K. There is a large well-defined subcutaneous lesion over the left upper back. The lesion is mildly hyperintense on T1, markedly hyperintense on T2, and demonstrates thin rim enhancement **(arrowheads)**. This was a pathologically proven epidermoid cyst.

Extraskeletal Ewing's Sarcoma

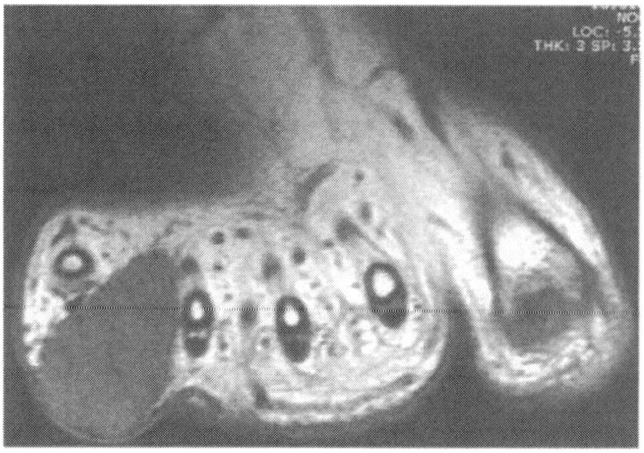

Figure 44L. Coronal T1 of the right foot.

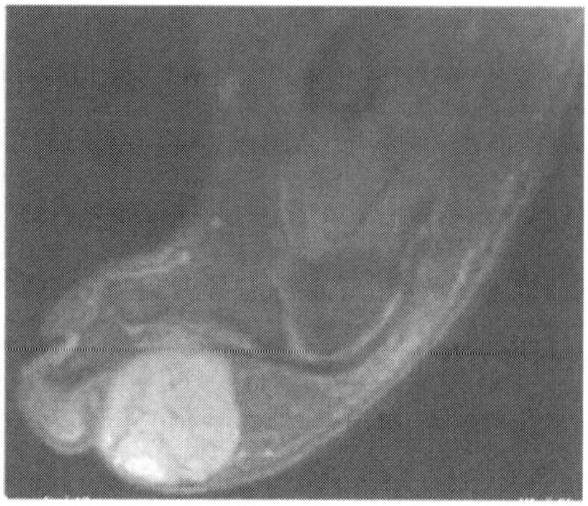

Figure 44M. Sagittal PD FS.

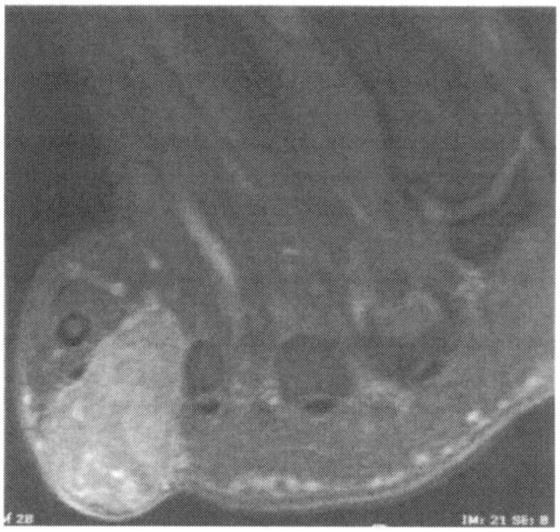

Figure 44N. Coronal T1 post-Gd FS.

Findings

This is a 15-year-old boy who discovered a nonpainful reddish mass over the plantar aspect of his right fifth toe.

Figures 44L, 44M, 44N. There is a large subcutaneous mass extending between the 4th and 5th metatarsals. It is mildly hyperintense on T1, heterogeneously hyperintense on PD FS, and shows heterogeneous enhancement. The bones are normal. This was pathologically proven to be an extraskeletal Ewing's sarcoma.

Pitfalls and Pearls

Although typical ganglion cysts are usually evaluated without gadolinium, atypical features that suggest a cystic neoplasm or inflammatory arthropathy should prompt contrast administration.

References

1. Ma LD, McCarthy EF, Bluemke DA, Frassica FJ. Differentiation of benign from malignant musculoskeletal lesions using MR imaging: Pitfalls in MR evaluation of lesions with a cystic appearance. *AJR Am J Roentgenol* 1998; 170:1251–1258.
2. Nahra ME, Bucchieri JS. Ganglion cysts and other tumor related conditions of the hand and wrist. *Hand Clin* 2004; 20:249–260, v.
3. Satku K, Ganesh B. Ganglia in children. *J Pediatr Orthop* 1985; 5:13–15.
4. Wang AA, Hutchinson DT. Longitudinal observation of pediatric hand and wrist ganglia. *J Hand Surg [Am]* 2001; 26:599–602.
5. Llauger J, Palmer J, Monill JM, Franquet T, Bague S, Roson N. MR imaging of benign soft-tissue masses of the foot and ankle. *Radiographics* 1998; 18:1481–1498.
6. Krudwig WK. Schulte KK, Heinemann C. Intra-articular ganglion cysts of the knee joint: A report of 85 cases and review of the literature. *Knee Surg Sports Traumatol Arthrosc* 2004; 12:123–129.
7. Helpert C, Davies AM, Evans N, Grimer RJ. Differential diagnosis of tumours and tumour-like lesions of the infrapatellar (Hoffa's) fat pad: Pictorial review with an emphasis on MR imaging. *Eur Radiol* 2004; 14:2337–2346.
8. Murphey MD, McRae GA, Fanburg-Smith JC, Temple HT, Levine AM, Aboulafia AJ. Imaging of soft-tissue myxoma with emphasis on CT and MR and comparison of radiologic and pathologic findings. *Radiology* 2002; 225:215–224.
9. Shibata T, Hatori M, Satoh T, Ehara S, Kokubun S. Magnetic resonance imaging features of epidermoid cyst in the extremities. *Arch Orthop Trauma Surg* 2003; 123:239–241.
10. Nakanishi H, Araki N, Sawai Y, et al. Cystic synovial sarcomas: Imaging features with clinical and histopathologic correlation. *Skeletal Radiol* 2003; 32:701–707.

Case 45

History

This is a 12-year-old boy who had trauma to his left calf while playing baseball 12 days ago. He presents after increasing mid-calf pain and 3 days of fever. A radiograph of the lower leg and ankle was normal (not shown).

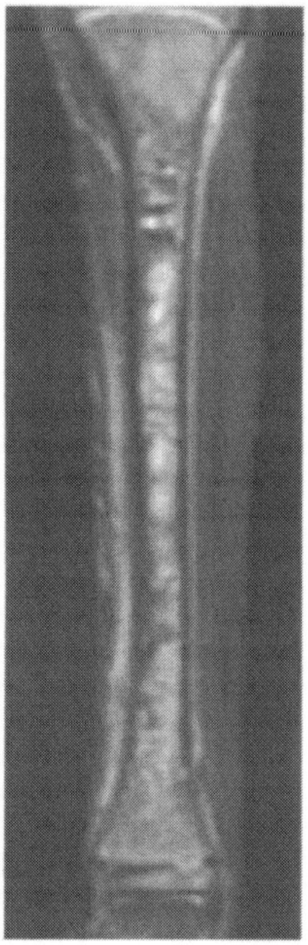

Figure 45A. Coronal STIR of the left tibia.

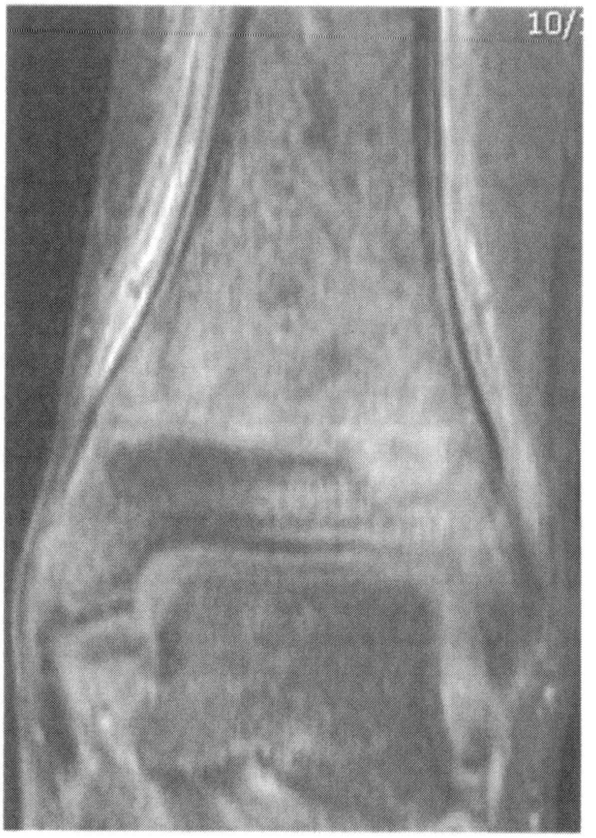

Figure 45B. Coronal T1 post-Gd FS.

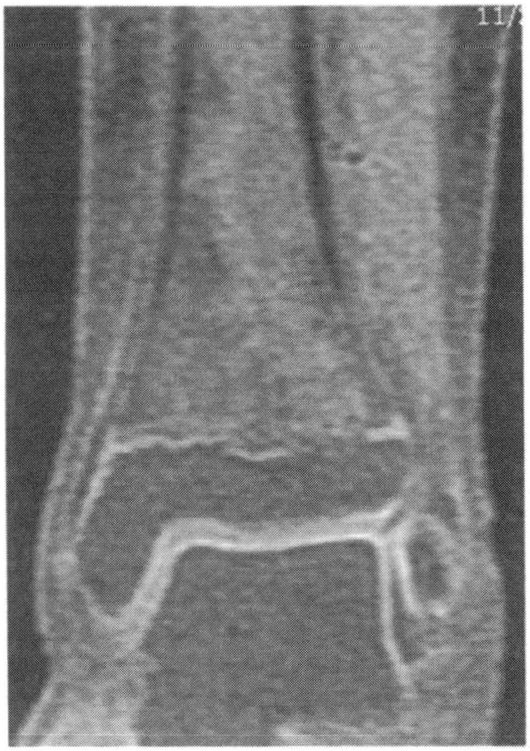

Figure 45C. Coronal 3D SPGR FS, one month later.

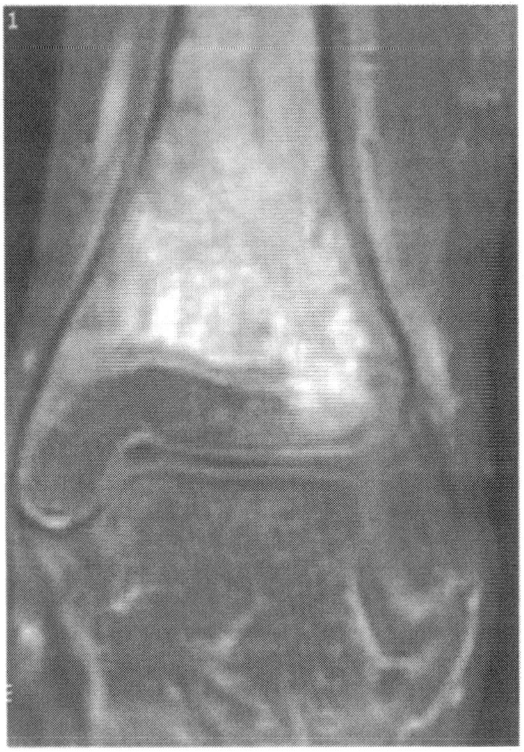

Figure 45D. Coronal T1 post-Gd FS.

Figure 45A. There is diffuse increased SI within the marrow cavity of the distal tibial metaphysis and patchy hyperintensity throughout the diaphysis.

Figures 45B, (45B with annotations). Lamellated periosteal hyperintensity and enhancement **(thick arrows)** is also seen. The abnormal SI and enhancement cross the physis **(arrowhead)** and extend into the epiphysis. Multiple small pockets of nonenhancing marrow are present within the metaphysis (*).

Figures 45C, 45D, (45C with annotations). The physeal cartilage SI is disrupted laterally **(arrowhead)**. Transphyseal enhancement is seen in a similar distribution.

Figure 45E. No physeal bony bridge is seen.

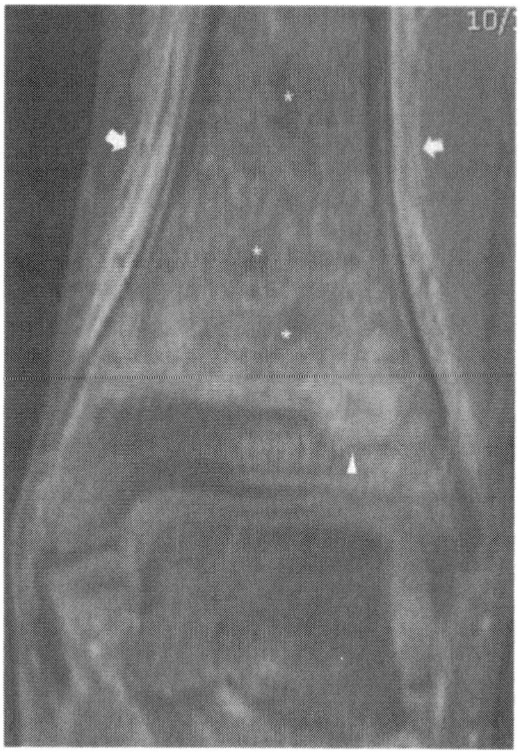

Figure 45B* Annotated.

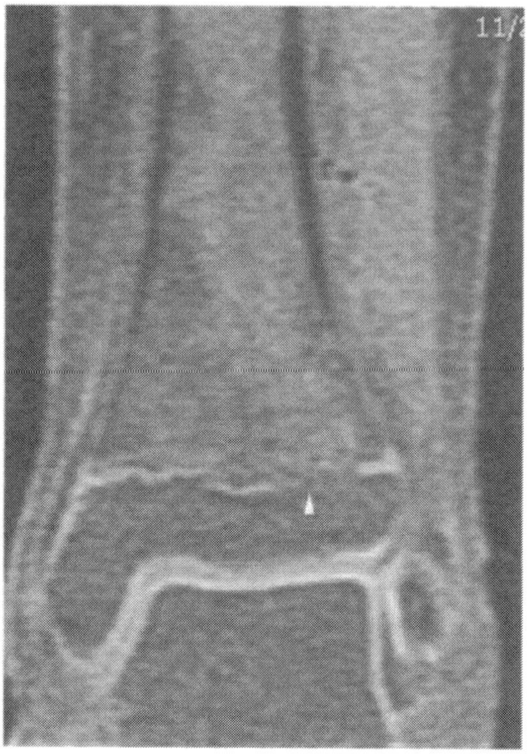

Figure 45C* Annotated.

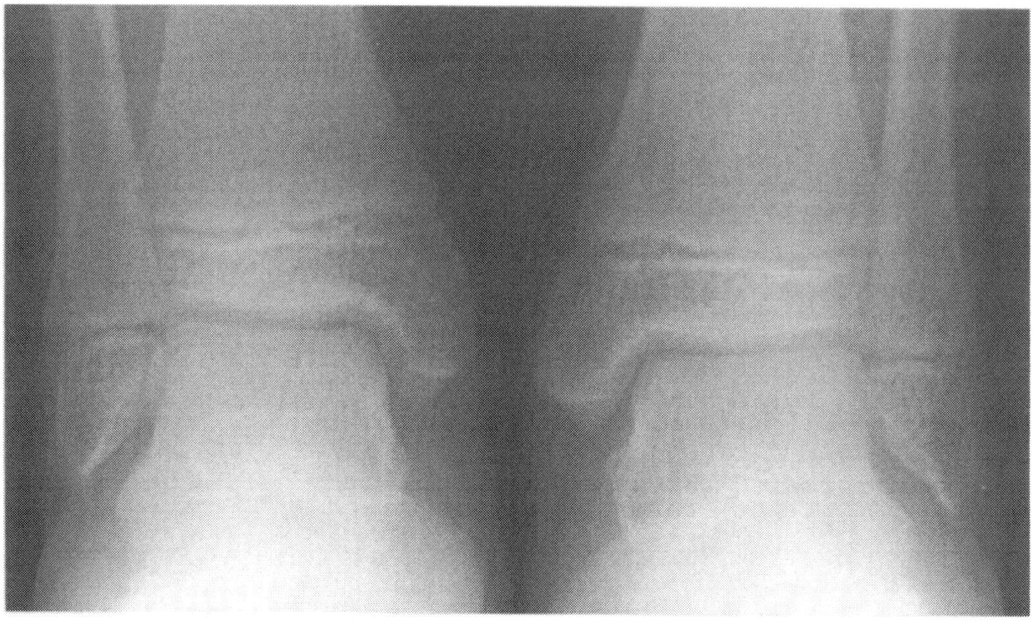

Figure 45E. AP radiograph, 9 months later.

Diagnosis

Acute tibial osteomyelitis with transphyseal extension

Questions

1. T/F: Osteomyelitis is more common in the lower extremity than in the upper extremity.
2. What is the most common cause of pyogenic osteomyelitis?

Discussion

Osteomyelitis occurs more commonly in the lower extremity than the upper extremity (Answer to Question 1) (1). The metaphyseal and metaphyseal equivalent regions are most often involved with hematogenous osteomyelitis and *Staphylococcus aureus* is the most frequently isolated organism (Answer to Question 2) (2, 3). The incidence of *Streptococcus pneumoniae* and *Hemophilus influenza* osteomyelitis is in decline due to routine vaccination against these agents.

The radiographic findings of acute osteomyelitis may be normal during the first two weeks of infection. The indication for MRI is to confirm the diagnosis and to assess for complications that may require interventions, such as intramedullary, subperiosteal, or soft tissue abscess formation.

Extension of metaphyseal osteomyelitis to the epiphysis or diaphysis may occur at any age. Traditionally, metaphyseal osteomyelitis with extension to the epiphysis has been described prior to 18 months of age and after skeletal maturity (4). However, even after the age of 18 months, the open physis in children provides an incomplete barrier against transphyseal osteomyelitis. Crossing vessels persist, although the majority of the metaphyseal nutrient arterioles involute and become terminal sinusoids before reaching the physis.

Historically, hematogenous diaphyseal osteomyelitis was considered uncommon in childhood. MRI and scintigraphy have shown that diaphyseal extension is often seen with metaphyseal osteomyelitis and focal diaphyseal signal abnormalities may be observed separate from the main site of infection (Figure 45F). In one study involving 136 children with acute osteomyelitis, multifocality was detected in 19% of patients by bone scintigraphy (5).

Alternative diagnoses for osteomyelitis, particularly if there is transphyseal extension and/or multifocality, include chronic recurrent multifocal osteomyelitis (CRMO), infarction and neoplastic processes such as Langerhans cell histiocytosis (LCH), lymphoma/leukemia, osteosarcoma (Figures 45G, 45H), Ewing's sarcoma, and metastases. Osteosarcoma and Ewing's sarcoma usually have discrete soft tissue masses with less marrow and juxtacortical soft tissue edema compared with osteomyelitis. In the absence of abscess, differentiation of osteomyelitis from LCH, leukemia/lymphoma, and metastatic neuroblastoma on MRI grounds is often problematic, and requires correlation with clinical and laboratory studies. Radiographs are mandatory, and CT may be useful in selected cases, particularly if osteosarcoma is a strong consideration (Figure 45H).

Follow-up MR imaging is helpful for evaluating treatment failure and growth plate complications from transphyseal osteomyelitis. 3D SPGR FS or similar imaging has been shown to be useful in the evaluation of suspected premature physeal fusion (6). However, imaging too early can be misleading as evident in this patient. An SPGR sequence performed one month after the diagnosis of osteomyelitis showed loss of the normal physeal cartilage SI (Figure 45C), suggesting a physeal bar. However, loss of the normal physeal cartilage SI on SPGR images within a few months of the diagnosis of osteomyelitis is nonspecific and may simply reflect persistent inflammatory change and granulation tissue.

In this patient, blood cultures and bone biopsy grew *Staphylococcus aureus*. He was treated with systemic antibiotics and did well. No leg length discrepancy was noted on clinical follow-up.

Orthopedic Perspective

In cases of osteomyelitis, MRI is often deemed necessary to assess for intraosseous or subperiosteal abscesses that may require surgical drainage. The location and the extent are important in planning guided CT or surgical drainage. If the abscess is less than 1.5 cm, antibiotics alone may be tried. The possibility of a subsequent premature growth arrest should be considered when osteomyelitis crosses the physis, and follow-up MRI assists in planning surgical interventions to minimize growth disturbance.

What the Clinician Needs to Know

1. What is the full extent of osteomyelitis? Does it cross the physis, how far does it extend in the diaphysis, is there an intra-articular component, and is there a large extraosseous component?
2. Is there a drainable intra- or extraosseous abscess?
3. Is there premature physeal fusion or growth disturbance on follow-up imaging?

Answers

1. True.
2. *Staphylococcus aureus*.

Additional Examples

Femur Osteomyelitis, Multifocal

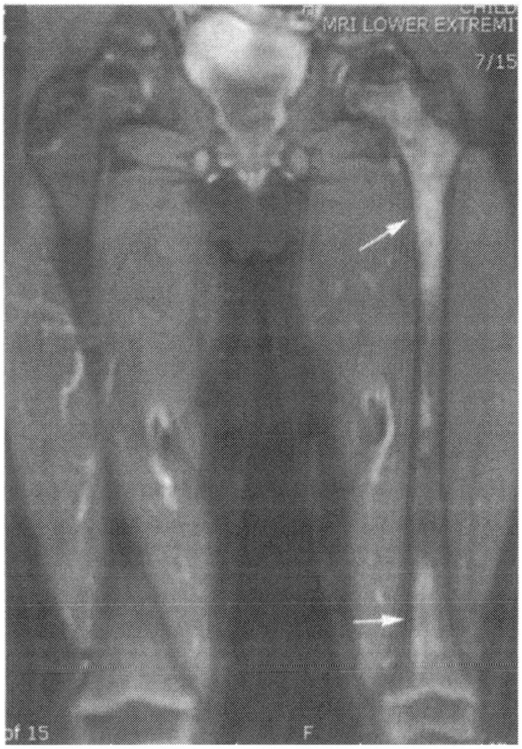

Figure 45F. Coronal STIR.

Findings

This 5-year-old girl had positive blood cultures and clinical signs of osteomyelitis.

Figure 45F. Separate areas of increased SI on fluid sensitive sequences are present in the proximal metaphysis/diaphysis and the mid-diaphysis of the femur **(arrows)**, compatible with either multifocal osteomyelitis or metaphyseal osteomyelitis with a secondary area of reactive hyperemia. Although the STIR SI in the distal left femur could conceivably represent residual red marrow, the striking hyperintensity and asymmetry as compared to the right side is more consistent with an additional focus of osteomyelitis.

Tibia Osteosarcoma with Transphyseal Extension

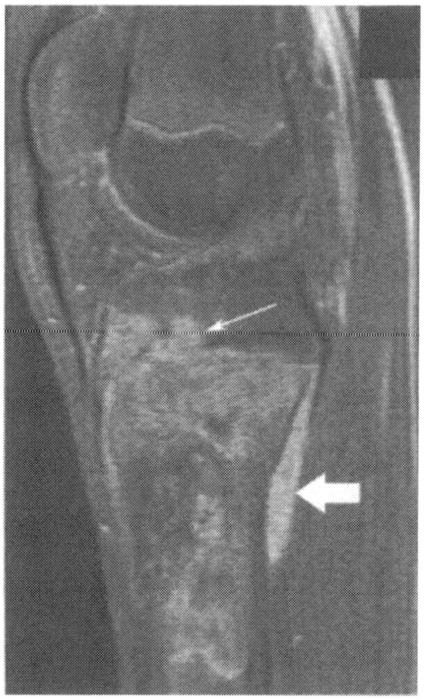

Figure 45G. Sagittal T1 post-Gd FS of the proximal left tibia.

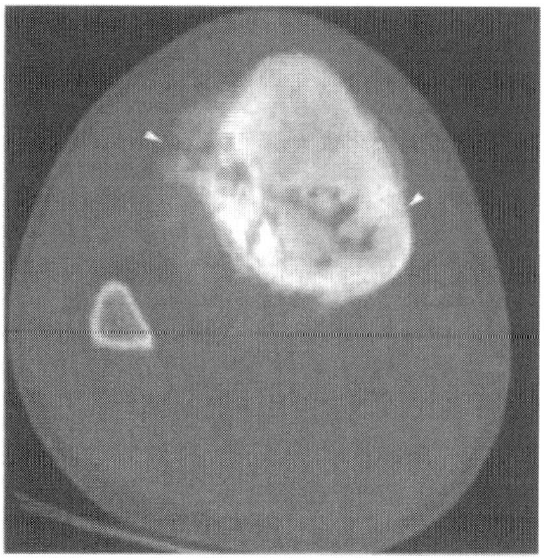

Figure 45H. CT of the proximal tibia.

Findings

This is a 14-year-old boy with chronic knee pain.

Figure 45G. There is abnormal, heterogeneous marrow enhancement that extends across the physis **(arrow)**. There is also an extraosseous soft tissue mass present **(thick arrow)**.

Figure 45H. Extensive osteoid producing intramedullary tumor, pathologic fracture **(arrowheads)**, extraosseous soft tissue component, and aggressive appearing periosteal reaction are seen. This lesion was pathologically confirmed conventional osteosarcoma.

Pitfalls and Pearls

1. Transphyseal osteomyelitis can still occur after 18 months of age because of incomplete involution of crossing physeal vessels.
2. The alternative causes for diffuse marrow infiltration with transphyseal spread includes neoplastic processes such as leukemia, osteosarcoma, metastases, and Langerhans cell histiocytosis. Extensive osteomyelitis can mimic neoplasm and a biopsy may be required for differentiation.

References

1. Waagner DC. Musculoskeletal infections in adolescents. *Adolesc Med* 2000; 11:375–400.
2. Luhmann JD, Luhmann SJ. Etiology of septic arthritis in children: An update for the 1990s. *Pediatr Emerg Care* 1999; 15:40–42.

3. Goergens ED, McEvoy A, Watson M, Barrett IR. Acute osteomyelitis and septic arthritis in children. *J Paediatr Child Health* 2005; 41:59–62.
4. Oudjhane K, Azouz EM. Imaging of osteomyelitis in children. *Radiol Clin North Am* 2001; 39:251–266.
5. Howman-Giles R, Uren R. Multifocal osteomyelitis in childhood: Review by radionuclide bone scan. *Clin Nucl Med* 1992; 17:274–278.
6. Lohman M, Kivisaari A, Vehmas T, Kallio P, Puntila J, Kivisaari L. MRI in the assessment of growth arrest. *Pediatric Radiology* 2002; 32:41–45.

Case 46

History

This is an 11-year-old boy who injured his right knee while skiing.

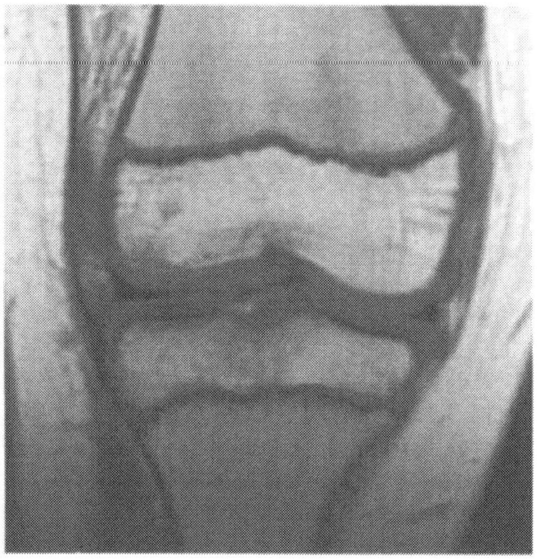

Figure 46A. Coronal T1 of the right knee.

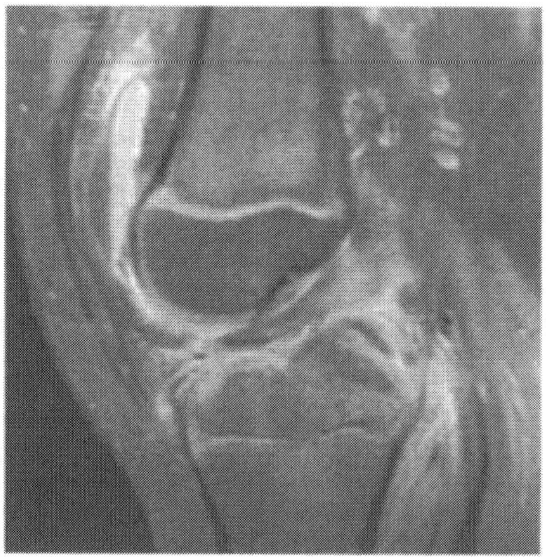

Figure 46B. Sagittal PD FS.

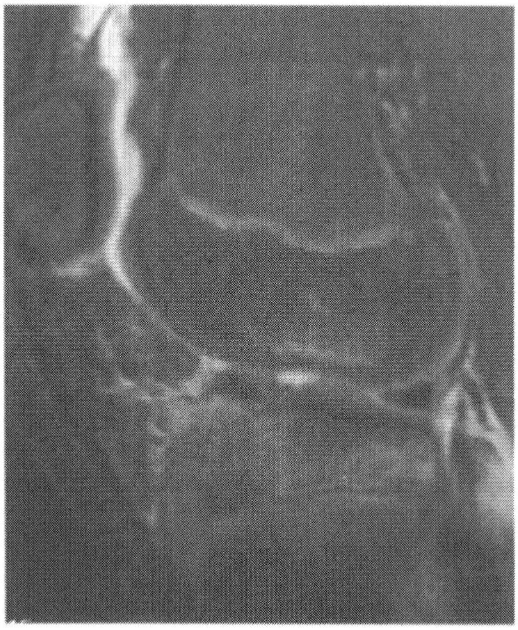

Figure 46C. Sagittal STIR through the lateral femoral condyle.

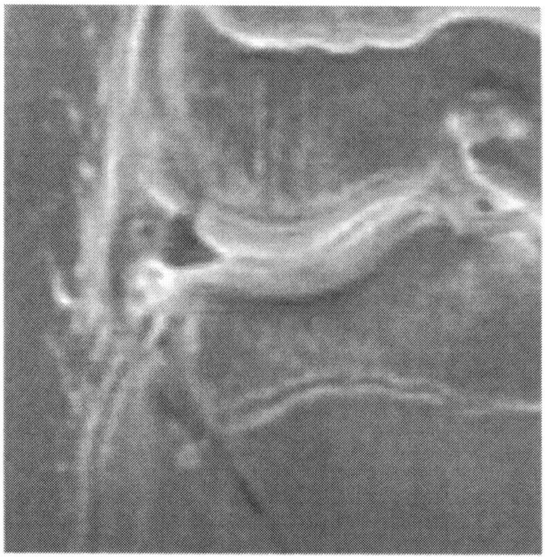

Figure 46D. Coronal PD FS.

Figures 46A, 46B, (46A, 46B with annotations). A tibial eminence fracture fragment is identified **(arrows)**. The anterior cruciate ligament (ACL) fibers insert onto the fracture fragment. There is mild intrasubstance ACL hyperintensity with preservation of ligamentous morphology. A lipohemarthrosis is present (F, fat; B, blood).

Figures 46C, (46C with annotations). There is mild increased SI seen in the subchondral epiphyseal marrow of the lateral femoral condyle and the posterolateral tibia (*). There is mild intrasubstance globular SI at the root of the anterior horn of the lateral meniscus, near the ACL attachment **(arrow)**.

Figures 46D, (46D with annotations). There is diffuse increased SI within the soft tissues along the margin of lateral compartment. A focal avulsion of the perichondrium is present (S) at the attachment of the lateral capsular ligament and fibers of the iliotibial band **(arrowheads)**.

Figure 46E. A tibial eminence fracture is evident **(arrow)**. The lateral tibial epiphyseal avulsion injury was not visible on the AP radiograph (not shown).

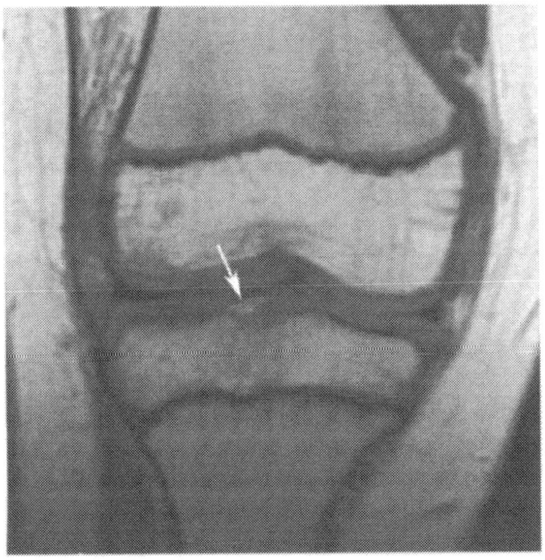

Figure 46A* Annotated.

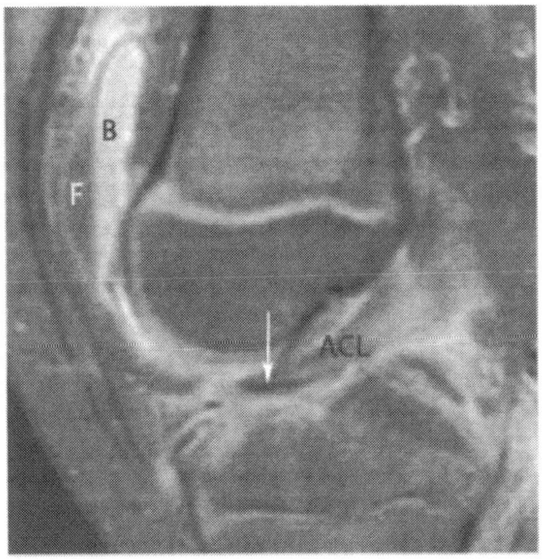

Figure 46B* Annotated.

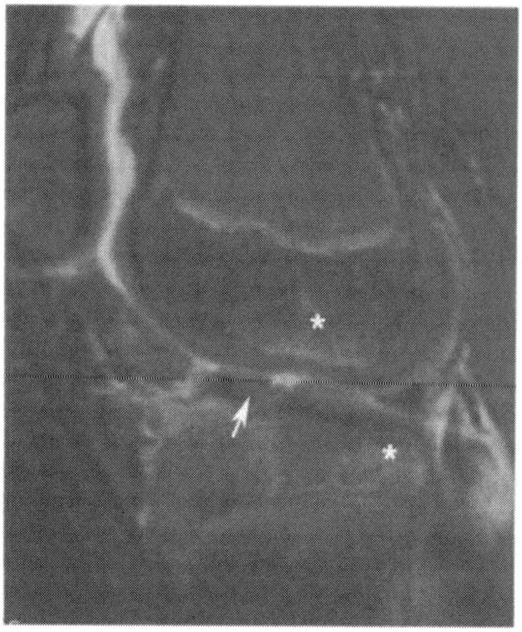

Figure 46C* Annotated.

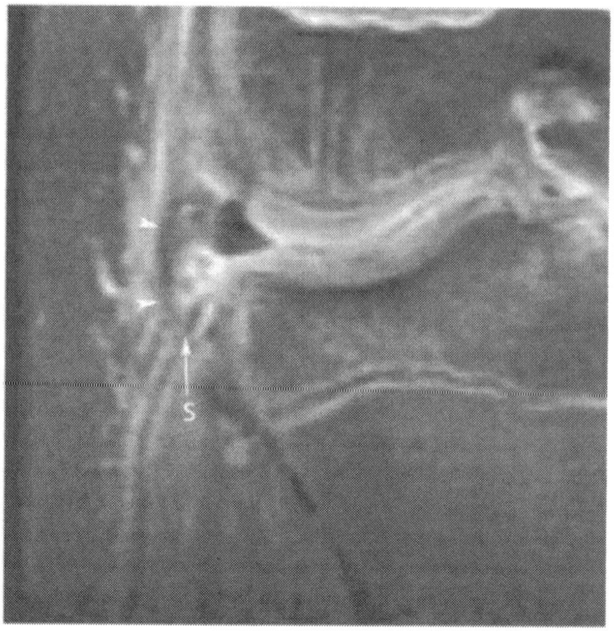

Figure 46D* Annotated.

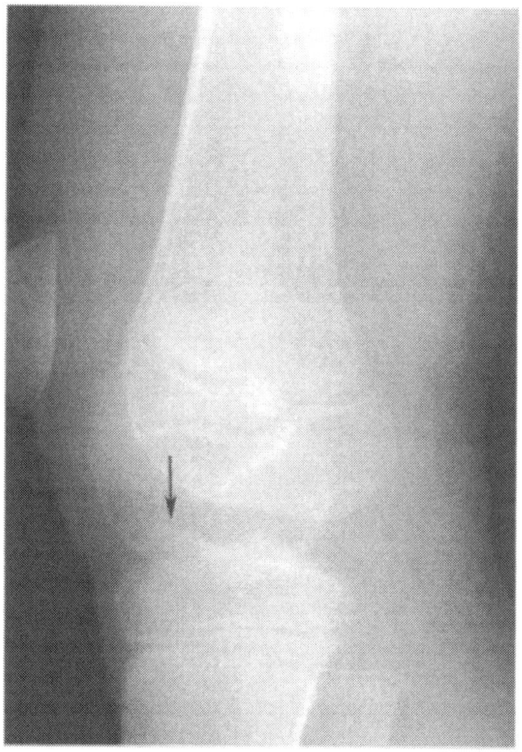

Figure 46E. Lateral radiograph.

Diagnosis

ACL complex injury with tibial eminence fracture (with lateral capsular ligament avulsion)

Questions

1. What is considered the weakest component of the ACL complex in children with an open physis?
2. When an ACL tear is present, are patients with open, or closed physes more likely to have an associated meniscal tear?

Discussion

ACL complex injuries in children differ based on age (preadolescent versus adolescent), gender, and anatomic variations. ACL midsubstance tears are observed more commonly in adolescent children and adult patients, whereas preadolescent children have been observed to have a higher incidence of tibial eminence fractures (1). In the child with an open physis, the weakest component of the ACL complex is the chondro-osseous tibial eminence; consequently, this region is predisposed to avulsion injury (Answer to Question 1). Partial ACL tears often coexist with tibial eminence avulsion injuries (2). Children with narrow intercondylar notches tend to suffer midsubstance ACL tears, whereas children with wider intercondylar notches tend to suffer tibial eminence fractures (3).

Identification of an associated meniscal tear is important since it may alter operative management of ACL injuries. In one study, ACL injuries were associated with a meniscal tear in 23% and 41% of patients with open and closed physis, respectively (Answer to Question 2) (4). The authors noted that lateral meniscal tears were more common than medial meniscal tears in both patient populations. However, other studies have observed that ACL and medial meniscal tears are more common in patients with open physis (5, 6). In the setting of traumatic hemarthrosis, one study observed that the majority of children had underlying ACL or meniscal pathology (preadolescents 7–12 yo: 47% ACL tear, 47% meniscal tear, 6% both; adolescents 13–18 yo: 55% ACL tear, 27% meniscal tear, 18% both) (7). In this study, tibial avulsion injuries were three times more common in the preadolescent population. Significantly, the authors also found coexisting osteochondral injuries in 13% of preadolescent patients and 5% of adolescent patients.

Historically, a Segond fracture describes a lateral capsular ligament avulsion fracture of the middle 1/3 of the lateral margin of the tibial epiphysis (Figures 46F, 46G). More recent cadavaric studies have shown that the anterior oblique band of the fibular collateral ligament and the iliotibial band may also contribute to these fractures (8). A Segond fracture may occur with excessive varus stress and/or internal rotation of the knee. Segond injuries are usually apparent on radiography (Figure 46F), but may be quite subtle on MRI (Figure 46G). Familiarity with this MRI appearance is critical because of its high association with ACL complex injuries. In this case, the avulsion injury occurred at the ligamentous attachment, and since there was no significant bony avulsion fragment, the injury was only apparent on MRI (Figure 46D).

A lateral notch sign is an osteochondral fracture of the anterior to middle aspect of the lateral femoral condyle (9). Like the Segond fracture, it has a high association with ACL complex injuries when the cortical depression is >1.5 mm (Figure 46H). Anterior translation of the tibia is another finding that may be seen with ACL complex tears. With complete ACL complex disruption, the tibia is free to translate anteriorly. With anterior tibial translation, a redundant and wavy PCL may be seen. The inferior articular margin uncovering may also be seen with anterior tibial translation.

This patient had an arthroscopic reduction of the tibial eminence avulsion fracture. There was mild intrasubstance hemorrhage within the intact ACL at its insertion onto the tibial eminence fracture fragment. The heterogeneous SI in the anterior horn of the lateral meniscus was not a tear at arthroscopy and was likely a normal variant (Figure 46C). The speckled appearance of the central aspect of the anterior horn of the lateral meniscus is related to the normal ACL attachment (10).

Orthopedic Perspective

Tibial eminence fracture was once considered the pediatric injury equivalent of the ACL injury in adults. However, ACL midsubstance tears are being seen with increased frequency in children and adolescents. Tibial eminence fractures represent an ACL avulsion fracture through the relatively weaker bone of the tibial eminence. Treatment is based on the degree of displacement of the fragment, with cast treatment of nondisplaced fractures (type 1). Hinged (type 2) and displaced (type 3) fractures may reduce with knee extension allowing closed treatment. However, most fractures do not reduce, thus requiring reduction and fixation that can usually be performed arthroscopically. Although MRI is usually not required for diagnosis, it is useful to identify associated injuries, such as meniscal tears or an entrapped meniscus. The entrapped meniscus may block closed reduction of the displaced tibial eminence and has been identified in 26% of type 2 fractures and 65% of type 3 fractures (11). The ACL remains morphologically intact in tibial eminence fractures. However, plastic deformation with slightly increased anterior laxity is the norm.

What the Clinician Needs to Know

1. What is the degree of displacement of the tibial eminence fracture?
2. Are there associated findings, such as chondral injury, meniscal tear, or entrapped meniscus?

Answers

1. Chondro-osseous junction of the tibial eminence.
2. Patients with closed physis are more likely to have a meniscal tear in the presence of a torn ACL.

Additional Examples

Segond Fracture

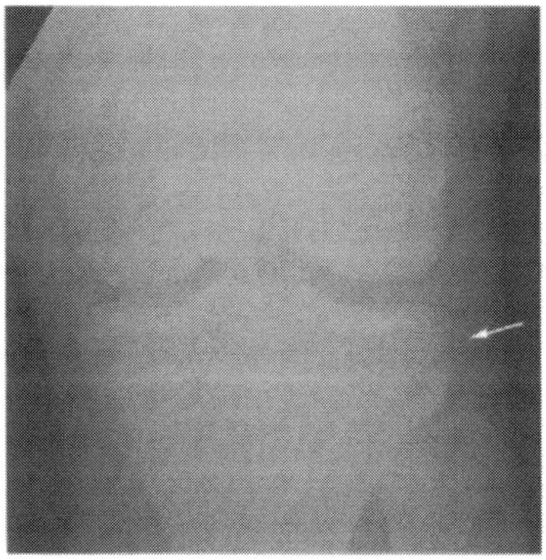

Figure 46F. AP radiograph of the left knee.

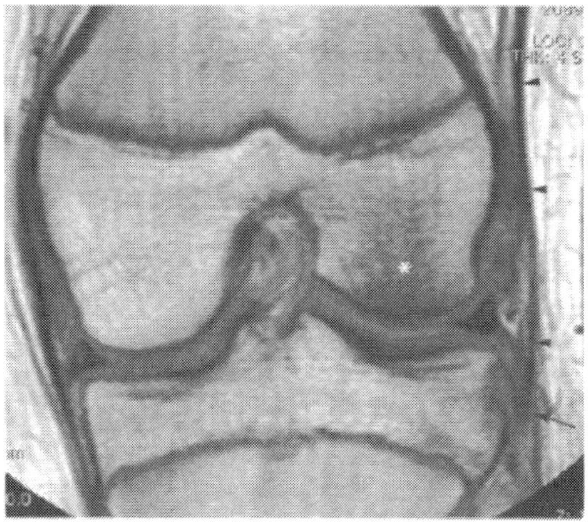

Figure 46G. Coronal T1.

Findings

This 14-year-old boy injured his knee playing basketball.

Figures 46F, 46G. The Segond fracture is demonstrated on both the radiograph and MRI **(arrows)**. Note the relationship of the iliotibial band **(arrowheads)** to the Segond fracture. Subchondral edema is present in the lateral femoral condyle (*). Tears of the ACL and the posterior horn of the medial meniscal were also present (not shown).

Lateral Notch Sign

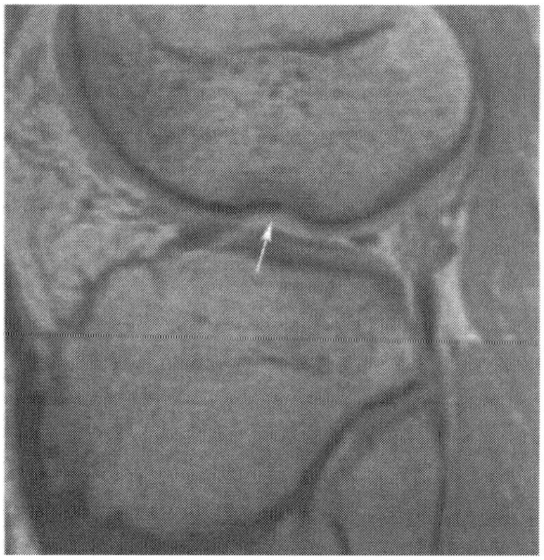

Figure 46H. Sagittal PD.

Findings

This 17-year-old boy injured his knee playing basketball.

Figure 46H. A depressed fracture of the lateral femoral condyle is present, measuring greater than 1.5 mm **(arrow)**, and is compatible with a lateral notch sign. There is also subchondral epiphyseal marrow edema and articular cartilage thinning associated with the depressed fracture. An ACL and a lateral meniscal tear were also present (not shown).

Pitfalls and Pearls

1. Carefully inspect the overlying articular cartilage whenever subchondral edema is identified. Osteochondral injuries are frequently found in patients with subchondral edema (4). Use PD FS, 3D SPGR FS or similar sequences to evaluate the overlying articular cartilage.
2. Segond fractures and the lateral condylar notch signs have high associations with ACL complex injuries.
3. Tibial eminence and Segond fractures have important clinical implications, but may be inconspicuous on MRI. Don't forget the radiographs!

References

1. Johnston DR, Ganley TJ, Flynn JM, Gregg JR. Anterior cruciate ligament injuries in skeletally immature patients. *Orthopedics* 2002; 25:864–871; quiz 872–863.
2. Accousti WK, Willis RB. Tibial eminence fractures. *Orthop Clin North Am* 2003; 34:365–375.
3. Kocher MS, Mandiga R, Klingele K, Bley L, Micheli LJ. Anterior cruciate ligament injury versus tibial spine fracture in the skeletally immature knee: A comparison of skeletal maturation and notch width index. *J Pediatr Orthop* 2004; 24:185–188.
4. Oeppen RS, Connolly SA, Bencardino JT, Jaramillo D. Acute injury of the articular cartilage and subchondral bone: A common but unrecognized lesion in the immature knee. *AJR Am J Roentgenol* 2004; 182:111–117.

5. Lee K, Siegel MJ, Lau DM, Hildebolt CF, Matava MJ. Anterior cruciate ligament tears: MR imaging-based diagnosis in a pediatric population. *Radiology* 1999; 213:697–704.
6. Zobel MS, Borrello JA, Siegel MJ, Stewart NR. Pediatric knee MR imaging: Pattern of injuries in the immature skeleton. *Radiology* 1994; 190:397–401.
7. Stanitski CL, Harvell JC, Fu F. Observations on acute knee hemarthrosis in children and adolescents. *J Pediatr Orthop* 1993; 13:506–510.
8. Campos JC, Chung CB, Lektrakul N, et al. Pathogenesis of the Segond fracture: Anatomic and MR imaging evidence of an iliotibial tract or anterior oblique band avulsion. *Radiology* 2001; 219:381–386.
9. Pao DG. The lateral femoral notch sign. *Radiology* 2001; 219:800–801.
10. Shankman S, Beltran J, Melamed E, Rosenberg ZS. Anterior horn of the lateral meniscus: Another potential pitfall in MR imaging of the knee. *Radiology* 1997; 204:181–184.
11. Kocher MS, Micheli LJ, Gerbino P, Hresko MT. Tibial eminence fractures in children: Prevalence of meniscal entrapment. *Am J Sports Med* 2003; 31:404–407.

Case 47

History

This is a 14-month-old boy with acute onset of left lower extremity non-weightbearing. He is afebrile, has an ESR of 31, and a normal WBC. There is no history of trauma. His parents had taken him camping in New Hampshire two months earlier. Radiographs revealed a knee joint effusion, but were otherwise normal (not shown).

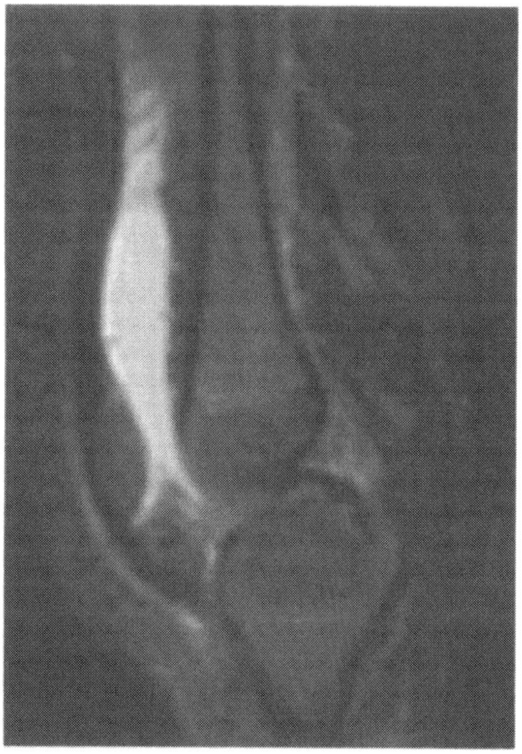

Figure 47A. Sagittal STIR of the left knee.

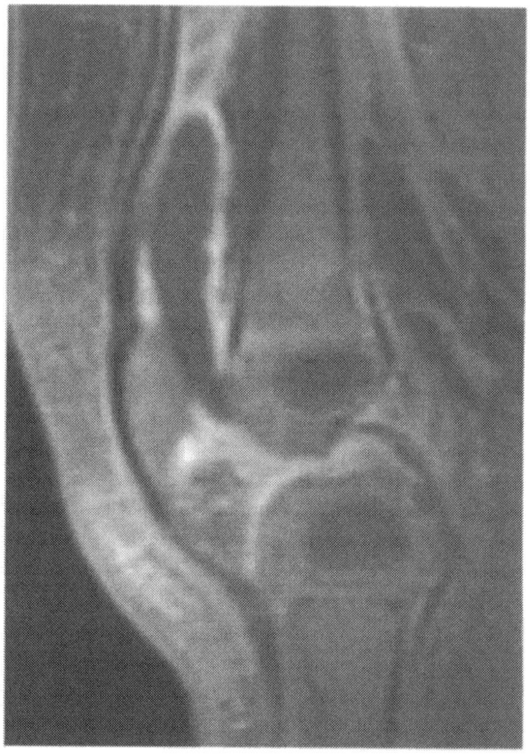

Figure 47B. Sagittal T1 post-Gd FS.

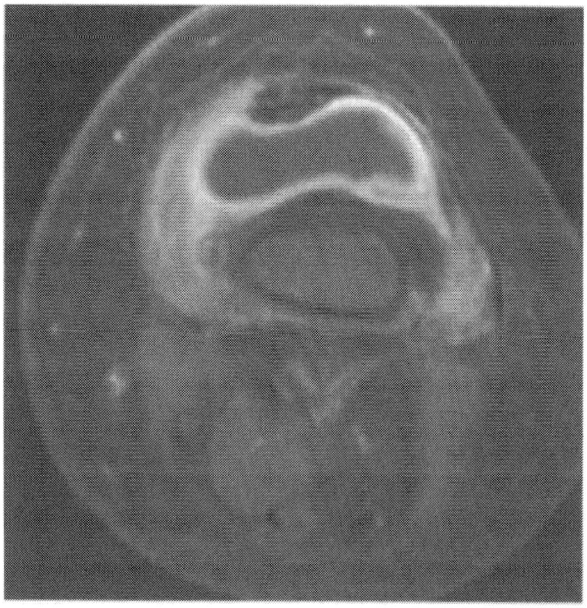

Figure 47C. Axial T1 post-Gd FS.

Figures 47A, 47B, 47C. There is a moderate to large joint effusion with synovial thickening and enhancement. There is also increased STIR SI and enhancement of the infrapatellar subcutaneous tissues, Hoffa's fat pad, and quadriceps muscle. There is normal marrow SI.

Diagnosis

Lyme arthritis

Questions

1. What joint is most commonly affected by Lyme arthritis?
2. T/F: MRI readily distinguishes between oligoarticular JRA, septic arthritis, and Lyme arthritis.

Discussion

Lyme disease is the result of a systemic infection by the spirochete *Borrelia burgdorferi*, which is transmitted by the deer tick *Ixodes daminii* (1). Arthritis is generally a late manifestation of Lyme disease and may present weeks to years after initial infection. In the United States, the majority of cases of Lyme disease occur in the New England and Mid-Atlantic states (2). Approximately 25% of cases of Lyme arthritis occur in children under 15 years of age. The arthritis is usually oligoarticular and the knee is affected 80% of the time (Answer to Question 1) (3). Other sites of involvement include the shoulders, elbows, ankles, hips (Figures 47D, 47E), wrists, temporomandibular joint, and small bones of the extremities. Lyme arthritis may present acutely with knee pain that can be indistinguishable clinically from septic arthritis or new onset oligoarticular juvenile rheumatoid arthritis (JRA).

Like the clinical exam, the MRI features of Lyme arthritis may be indistinguishable from other infectious and noninfectious arthritidies (Answer to Question 2). Some authors have suggested a role for MRI in the differentiation between Lyme and septic arthritis. Ecklund et al. observed that patients with Lyme arthritis more frequently had myositis, adenopathy, and lacked subcutaneous edema compared with septic arthritis (4). Myositis most often involved the popliteus muscle in their series of patients. However, septic arthritis, transient synovitis, trauma, and JRA cannot be excluded on MRI grounds alone (5). Although subcutaneous edema appears to be more common in septic arthritis, it has been described with Lyme arthritis (6) and the nonspecific nature of this finding is also illustrated by its presence in this case. Intra-articular blood products have been observed with Lyme arthritis (7) but also occur with trauma, pigmented villonodular synovitis (PVNS), synovial venous malformation, and hemophilic arthropathy. Additional MRI features common to Lyme arthritis as well as septic and nonseptic arthritis include joint effusion, synovial thickening/enhancement, and juxta-articular marrow edema.

The MRI was ordered in this patient because of the concern for osteomyelitis and septic arthritis. A subsequent joint aspiration was negative for septic arthritis and a Lyme titer was positive. The patient was treated with amoxicillin for one month.

Orthopedic Perspective

Lyme arthritis of the knee is often difficult to differentiate from septic arthritis both clinically and by MRI. Both may present with a painful, swollen knee with limp and limited motion. Definitive Lyme titers are not back in time to aid with the initial differentiation and the use of rapid Lyme tests is controversial. Aspiration is usually performed to differentiate the two; however, some cases of Lyme arthritis may have high joint fluid white blood cell counts. MRI is not typically ordered to aid in this differentiation; its role in this context is to exclude osteomyelitis, abscess, or neoplasm.

What the Clinician Needs to Know

Have alternative considerations for a nontraumatic joint effusion been excluded, such as PVNS or osteomyelitis with secondary septic arthritis?

Answers

1. Knee.
2. False.

Additional Example

Lyme Arthritis Mimicking Septic Arthritis/Osteomyelitis

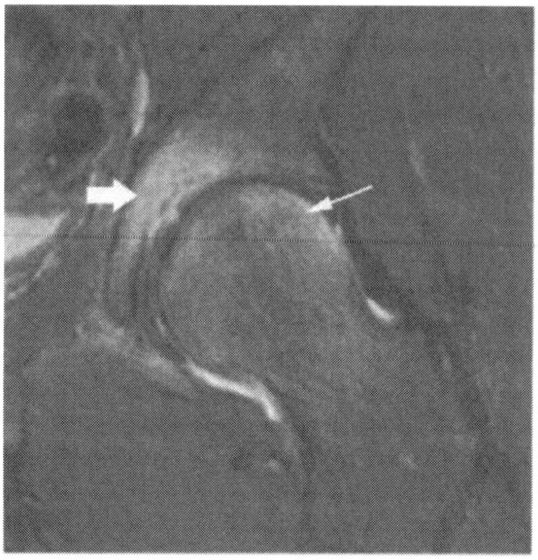

Figure 47D. Coronal STIR of the left hip.

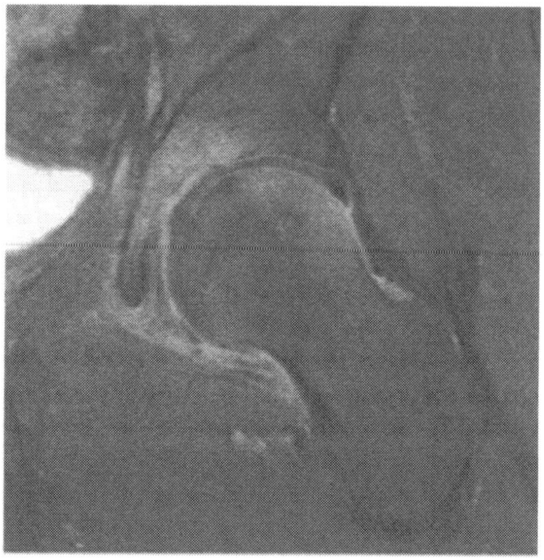

Figure 47E. Coronal T1 post-Gd FS.

Findings

This is a 15-year-old boy with left hip pain for two weeks. He was afebrile, and had an ESR of 69 and a normal WBC.

Figures 47D, 47E. There is increased STIR SI and enhancement within the acetabulum **(thick arrow)**, femoral head **(thin arrow)**, and synovium. A joint aspiration and blood cultures were negative for infection, and subsequent Lyme titers were positive.

Pitfalls and Pearls

1. Although most common in the knee, Lyme disease does affect most other joints, and should be placed in the differential diagnosis of mono/oligoarticular arthritis.
2. The radiologist may be the first to suggest the correct diagnosis of Lyme arthritis on MRI performed in the setting of joint pain, with clinical features atypical for septic arthritis, osteomyelitis, or oligoarticular JRA.

References

1. Massarotti EM. Lyme arthritis. *Med Clin North Am* 2002; 86:297–309.
2. Willis AA, Widmann RF, Flynn JM, Green DW, Onel KB. Lyme arthritis presenting as acute septic arthritis in children. *J Pediatr Orthop* 2003; 23:114–118.
3. Cavallaro A, Harrer T, Richter H, Bautz W, Fellner FA. MR findings in acute Lyme disease affecting the knee: A case report. *Rontgenpraxis* 2002; 54:210–213.
4. Ecklund K, Vargas S, Zurakowski D, Sundel RP. MRI features of lyme arthritis in children. *AJR Am J Roentgenol* 2005; 184:1904–1909.

5. Graif M, Schweitzer ME, Deely D, Matteucci T. The septic versus nonseptic inflamed joint: MRI characteristics. *Skeletal Radiol* 1999; 28:616–620.
6. Schmitz G, Vanhoenacker FM, Gielen J, De Schepper AM, Parizel PM. Unusual musculoskeletal manifestations of Lyme disease. *Jbr-Btr* 2004; 87:224–228.
7. Seldes R, Glasgow SG, Torg JS. Atraumatic spontaneous hemarthrosis associated with Lyme arthritis: A case report. *Clin Orthop* 1993:269–271.

Case 48

History

This is a 15-year-old boy, an avid basketball player with left knee pain for the past few weeks.

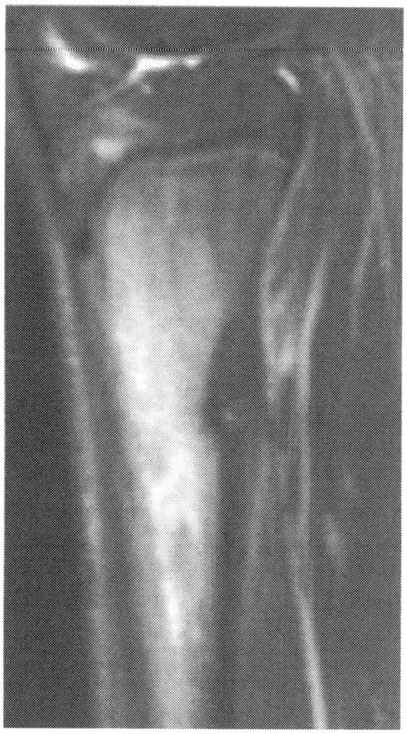

Figure 48A. Sagittal STIR of the left knee.

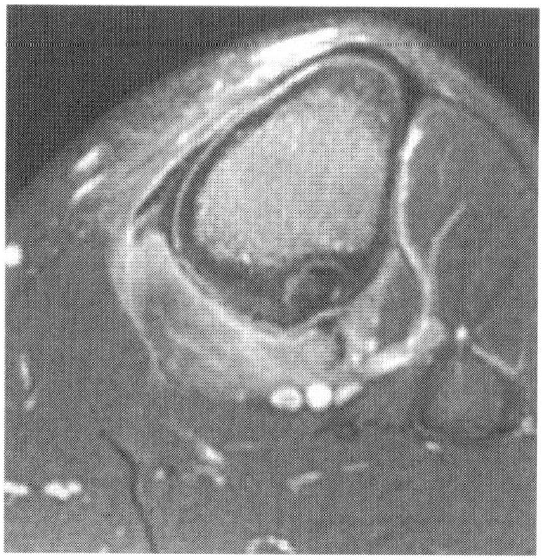

Figure 48B. Axial T1 post-Gd FS.

Figures 48A, 48B. There is a hypointense lesion along the posterior cortex of the tibia associated with STIR hyperintensity and enhancement in the adjacent marrow and juxtacortical soft tissues. The cortical lesion demonstrates mild rim enhancement.

Figure 48C. This cortical lesion has sclerotic, well-defined margins. The cortex overlying the lesion is thinned, particularly proximally **(thick arrow)**, but a pathologic fracture is not identified. A septated appearance is noted in the caudal aspect of this lesion **(arrowhead)**.

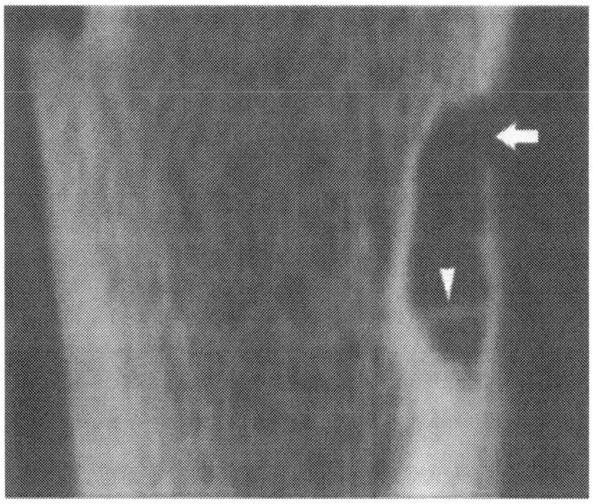

Figure 48C. Sagittal CT reformation of the left proximal tibia.

Diagnosis

Nonossifying fibroma (NOF)

Questions

1. What syndromes are associated with multiple NOFs?
2. What is the most common signal intensity of an NOF on T2W sequences?

Discussion

Nonossifying fibroma (NOF) (AKA fibroxanthoma, fibrous cortical defect) is the most common benign bone lesion in children and adolescence. In a selected patient population of 154 children, Caffey observed that NOF occurred in approximately 36% of patients, was absent in children under 24 months of age, and was most frequently observed in children between 4 and 12 years of age (1). Histologically, NOF is composed of fibroblasts, giant cells, and xanthoma cells, and contains little collagen unless there has been prior fracture (2). NOF is most commonly encountered in the metaphyses of the proximal and distal tibia and the distal femur. It is generally silent, and is discovered incidentally, unless a pathologic fracture has occurred. Multiple NOFs may be seen in Jaffe-Campanacci syndrome (multiple NOFs, multiple café au lait spots, and mental retardation) and occasionally in neurofibromatosis type 1 (NF 1) (Answer to Question 1) (3–5). There are rare reports of rickets associated with NOFs, with resolution of the endocrinopathy following surgical treatment of the osseous lesion (6).

On radiography, these lesions are often multilobulated with a soap-bubble appearance, have well-defined sclerotic margins, and are cortically based within the metaphysis or metadiaphysis (Figures 48D–48F). Larger lesions may cause bony expansion and may appear medullary based, mimicking fibrous dysplasia or even a malignancy (Figures 48G–48I). Thinning or even absence of overlying cortex may also raise concerns of an aggressive neoplasm.

The plain radiographic features of NOF are characteristic and further imaging is rarely necessary. Unfortunately, the diagnostic features of the lesion may not be appreciated and patients may undergo MRI because of pain and concerns of a more aggressive process. In one study involving 19 patients, NOFs were hypointense on T1 of variable T2 SI (79% hypointense, 21% hyperintense), and all demonstrated enhancement after intravenous gadolinium administration (Answer to Question 2) (7). In this study, areas of T2 hypointensity corresponded with areas of hypercellular fibrous tissue and hemosiderin, whereas areas of T2 hyperintensity corresponded to massive aggregation of foamy histiocytes with little hypercellular fibrous tissue at pathology. All lesions were noted to have a hypointense rim (Figures 48D–48I), corresponding with marginal sclerosis on plain radiography. Internal septations (Figure 48C) and less commonly blood products may also be seen. Extraosseous soft tissue or marrow signal abnormality may be seen beyond the confines of the NOF (21%), particularly after pathologic fracture (7).

The edema and contrast enhancement associated with an occult or healing fracture of NOF may suggest an aggressive neoplasm, and a careful review of the plain radiographs or CT may resolve this confusion. On occasion, a fracture line is not identified on plain radiography or CT, and it can only be presumed based on the associated imaging features (Figures 48A–48C). Sclerosis associated with NOF healing tends to progress from the diaphyseal end of the lesion, with corresponding hypointensity on all MR imaging sequences.

When NOF has a bubbly character and encroaches on the medullary cavity, alternative considerations include fibrous dysplasia, metaphyseal giant cell tumor, cortical desmoid, chondromyxoid fibroma, and aneurysmal bone cyst. Cortical desmoids (AKA avulsive cortical irregularity) are characteristically located in the distal femoral metaphyseal cortex, at the insertion site of either the medial head of the gastrocnemius or adductor magnus muscles (see Case 77). Giant cell tumors in the skeletally immature arise almost exclusively in the metaphysis (8). Chondromyxoid fibromas are benign cartilaginous tumors located in the metaphysis and are characterized by multiple delicate bubbly lucencies with a thinner rim of peripheral sclerosis compared with NOFs, and occur in older patients (average of 28 years in a series of 34 patients) (9).

This patient was referred to MRI because of the suspicion of osteomyelitis. The features of a NOF were seen on the initial radiograph. The significant marrow and soft tissue edema on MRI, as well as the lack of fracture definition on the subsequent CT, raised the alternative considerations of osteomyelitis and osteoid osteoma. However, the constellation of imaging findings in this case is still best explained by an ordinary NOF, with stress reaction. This lesion was pathologically proven NOF.

Orthopedic Perspective

The main issue is to distinguish NOF from other lesions. Often they are found incidentally when the child sustains an injury to a knee or ankle. Pain is often attributed to the presence of NOF, although in most instances NOFs are asymptomatic and the pain is due to another cause. Occasionally, NOFs present following pathologic fracture. The major dilemma for the orthopedist is to decide whether surgical treatment is needed to prevent pathologic fracture, especially in an athletic young patient. There are no clinical or radiographic guidelines that help with this decision, but recently biomechanical analysis of CT scans of both the normal and involved bone using density phantoms has provided us some ability to judge the axial, torsional, and bending rigidities of the bone, which aids in advising parents how to proceed (10).

What the Clinician Needs to Know

1. Is this a NOF or is it another neoplasm or infection?
2. Is the involved bone likely to fracture with activities of daily living or sports activity?

Answers

1. Jaffe-Campanacci syndrome and NF 1.
2. Hypointense on T2.

Additional Examples

NOF

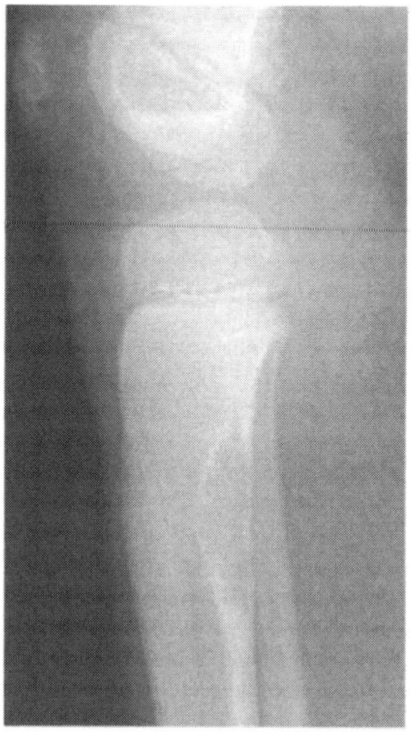

Figure 48D. Lateral radiograph of the left knee.

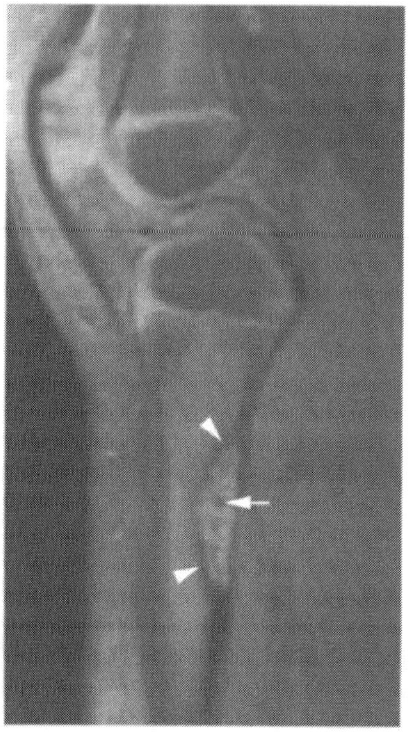

Figure 48E. Sagittal PD FS.

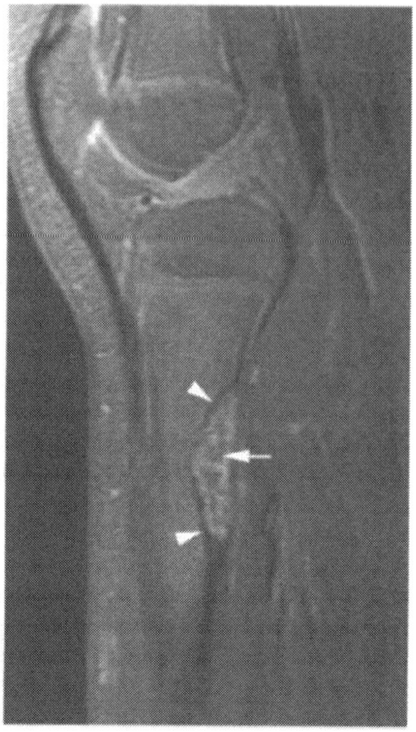

Figure 48F. Sagittal T1 post-Gd FS.

Findings

This is a 3-year-old girl with left knee pain.

Figure 48D. There is an osteolytic, multilobulated lesion present in the posterior aspect of the proximal tibia metadiaphysis with sharply defined, sclerotic margins.

Figures 48E, 48F. This lesion demonstrates mixed SI on the PD FS and patchy enhancement with multiple hypointense foci **(arrows)**. A thin hypointense rim surrounds the lesion **(arrowheads)**. These features were characteristic of an NOF; therefore, the patient was observed.

Atypical NOF

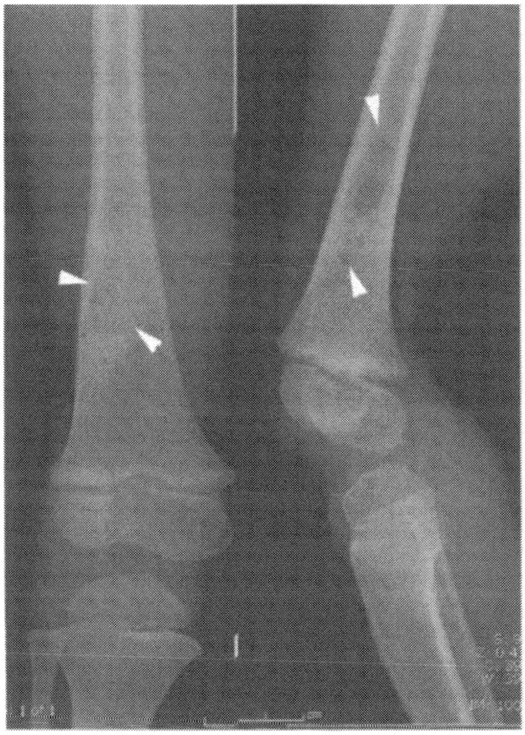

Figure 48G. Plain radiographs of the right femur.

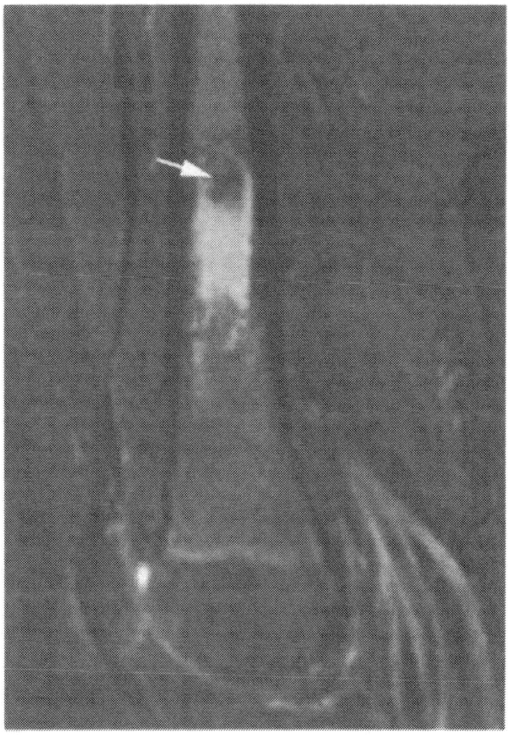

Figure 48H. Sagittal STIR.

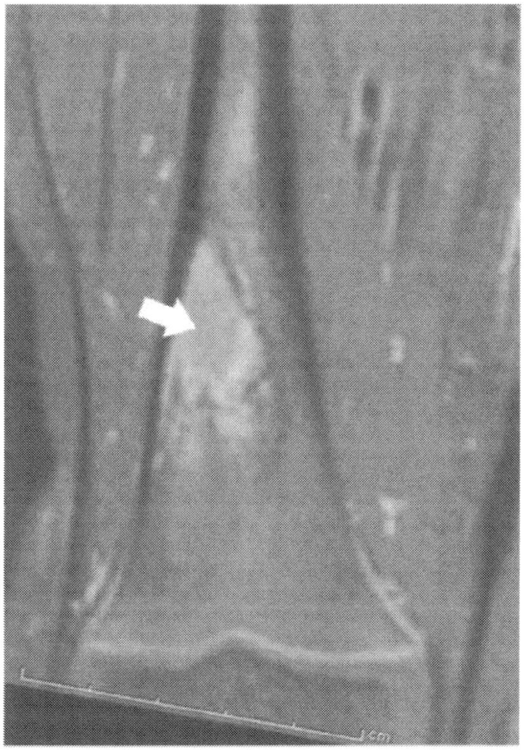

Figure 48I. Coronal T1 post-Gd FS.

Findings

This is a 3-year-old boy with a new right-sided limp.

Figure 48G. A slightly expansile osteolytic lesion with poorly defined margins is identified in the distal femoral diaphysis **(arrowheads)**.

Figure 48H. This lesion is predominantly hyperintense with a more focal area of hypointensity located superiorly **(thin arrow)**.

Figure 48I. Mild central enhancement **(thick arrow)** surrounded by moderate rim enhancement is evident. Because of the patient's age and the concern for Langerhans cell histiocytosis, or less likely an aggressive neoplasm, biopsy was performed. The lesion was pathologically proven NOF.

Pitfalls and Pearls

1. Both NOF and fibrous dysplasia may contain fibrous tissue that may be hypointense on all imaging sequences and enhance.
2. MRI evaluation of NOFs should be reserved for those lesions with atypical plain radiographic features. If an occult fracture is suspected, carefully review plain radiographs and consider CT.
3. Rarely NOF is associated with rickets, and surgical treatment of the osseous lesion corrects the endocrinopathy.

References

1. Caffey J. On fibrous defects in cortical walls of growing tubular bones: Their radiologic appearance, structure, prevalence, natural course, and diagnostic significance. *Adv Pediatr* 1955; 7:13–51.
2. Kransdorf MJ, Utz JA, Gilkey FW, Berrey BH. MR appearance of fibroxanthoma. *J Comput Assist Tomogr* 1988; 12:612–615.
3. Colby RS, Saul RA. Is Jaffe-Campanacci syndrome just a manifestation of neurofibromatosis type 1? *Am J Med Genet A* 2003; 123:60–63.
4. Hau MA, Fox EJ, Cates JM, Brigman BE, Mankin HJ. Jaffe-Campanacci syndrome: A case report and review of the literature. *J Bone Joint Surg Am* 2002; 84-A:634–638.
5. Faure C, Laurent JM, Schmit P, Sirinelli D. Multiple and large non-ossifying fibromas in children with neurofibromatosis. *Ann Radiol (Paris)* 1986; 29:369–373.
6. Pollack JA, Schiller AL, Crawford JD. Rickets and myopathy cured by removal of a nonossifying fibroma of bone. *Pediatrics* 1973; 52:364–371.
7. Jee WH, Choe BY, Kang HS, et al. Nonossifying fibroma: Characteristics at MR imaging with pathologic correlation. *Radiology* 1998; 209:197–202.
8. Kransdorf MJ, Sweet DE, Buetow PC, Giudici MA, Moser RP, Jr. Giant cell tumor in skeletally immature patients. *Radiology* 1992; 184:233–237.
9. Brien EW, Mirra JM, Kerr R. Benign and malignant cartilage tumors of bone and joint: Their anatomic and theoretical basis with an emphasis on radiology, pathology and clinical biology. I. The intramedullary cartilage tumors. *Skeletal Radiol* 1997; 26:325–353.
10. Hong J, Cabe GD, Tedrow JR, Hipp JA, Snyder BD. Failure of trabecular bone with simulated lytic defects can be predicted non-invasively by structural analysis. *J Orthop Res* 2004; 22:479–486.

Case 49

History

This is a 5-year-old boy with a 5-month history of left arm pain.

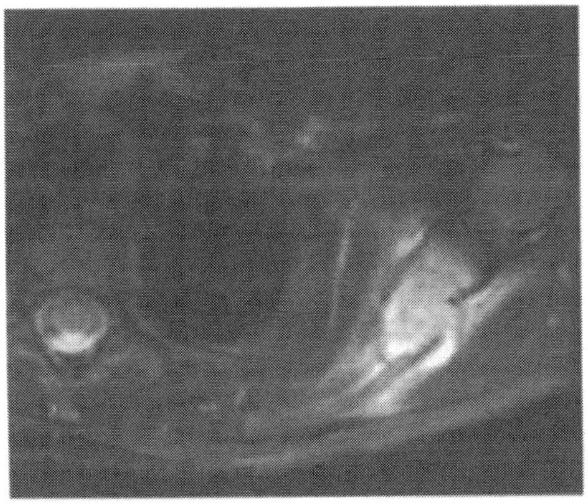

Figure 49A. Axial T2 FS.

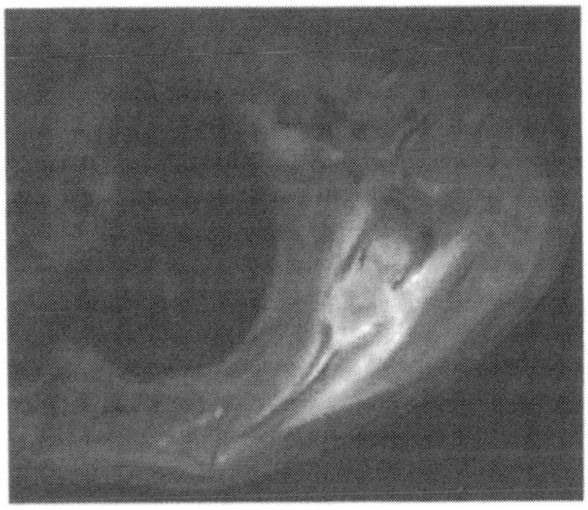

Figure 49B. Axial T1 post-Gd FS.

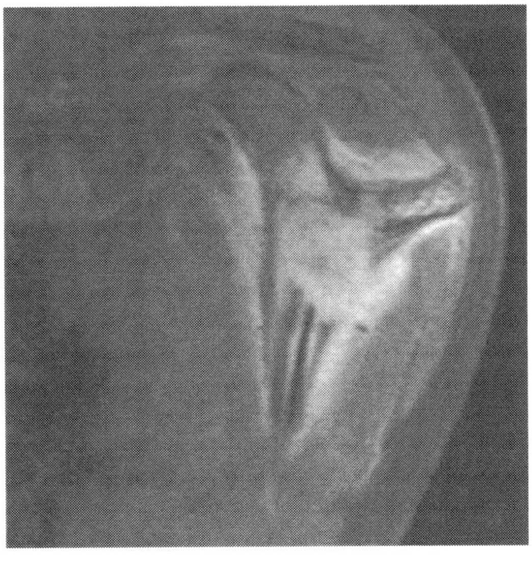

Figure 49C. Oblique sagittal T1 post-Gd FS.

Figures 49A, 49B, 49C, (49B, 49C with annotations). There is a mass arising from the scapula. The margins are sharply defined **(black arrowheads)**. It is T2 hyperintense and demonstrates an enhancement pattern that is greater peripherally than centrally. The lesion transgresses the cortex (C) and lifts the periosteum (P). There is extension of the mass into the infraspinatus muscle **(white arrowhead)**. There is edema and enhancement of the adjacent marrow, periosteum, and juxtacortical soft tissues. The glenohumeral joint is spared.

Figure 49D. No calcified matrix is seen within the lesion. Note sharply marginated cortical destruction **(black arrowhead)** correlating with a narrow zone of transition seen on MRI.

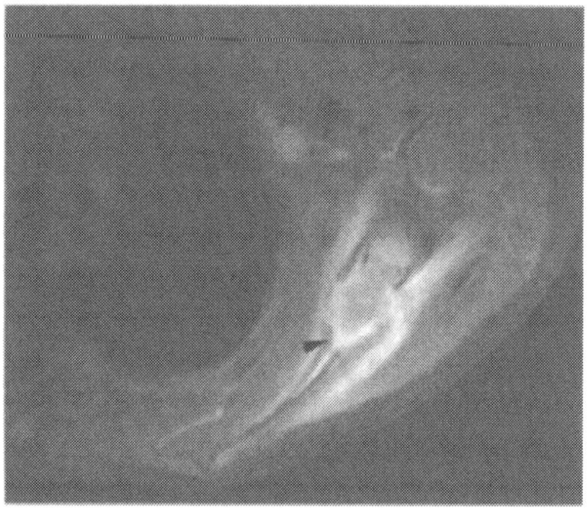

Figure 49B* Annotated.

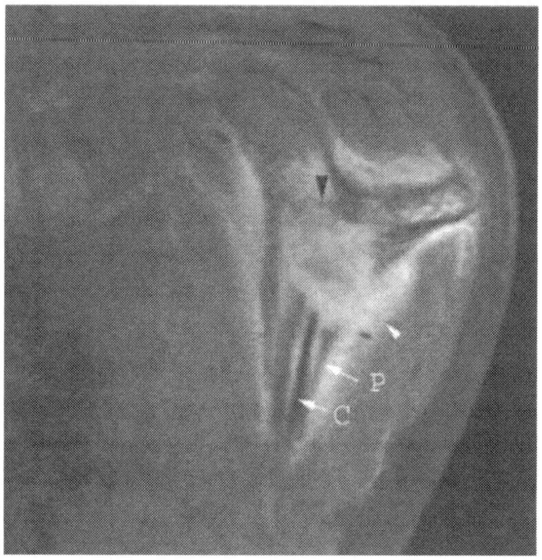

Figure 49C* Annotated.

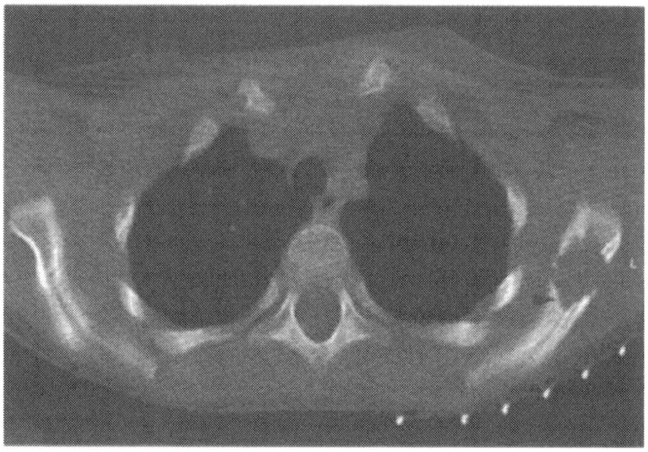

Figure 49D. CT.

Diagnosis

Langerhans cell histiocytosis (LCH)

Questions

1. What are the three classically described clinical forms of LCH?
2. T/F: The incidence of solitary LCH of bone that eventually becomes multifocal is approximately 75%.

Discussion

The three classically described clinical forms of Langerhans cell histiocytosis (LCH) (AKA eosinophilic granuloma or histiocytosis X) are: eosinophilic granuloma (5–15 yo, disease limited to lung or bone), Hand-Schuller-Christian disease (1–5 yo, triad of osseous lesions, diabetes insipidus, exophthalmos), and Letterer-Siwe disease (<2 yo, fulminant LCH with osseous and solid organ involvement) (Answer to Question 1) (1). Of the three, eosinophilic granuloma is the most common and benign form. Letterer-Siwe disease is least common and is often associated with multiorgan involvement followed by death within 1 to 2 years after presentation. An alternative clinical categorization of LCH is tumor confined to bone (local or multifocal), which has a good prognosis, and LCH with solid organ involvement, which has a poor prognosis (2).

A solitary bone lesion accounts for approximately 70% of all cases of LCH (3).The majority of LCH bone lesions occur in the skull (Figure 49E), vertebral column (Figures 49F, 49G), pelvis (Figures 49H–49J), scapula, and ribs. Approximately 1/3 of LCH bone lesions occur within the long tubular bones. The most common location in long bones in order of frequency is the femur, humerus (Figures 49K–49N), and tibia. Within long bones, LCH may occur anywhere but diaphyseal involvement is most common. The incidence of solitary LCH of bone that eventually becomes multifocal (with or without solid organ involvement) is approximately 10% (Answer to Question 2) (1).

The plain radiographic features of LCH of bone depend on disease activity (4). Active LCH bone lesions may appear aggressive, with a nonsclerotic geographic, moth-eaten or occasionally permeative osteolytic pattern. Aggressive periosteal reaction, endosteal scalloping, and cortical disruption with extraosseous soft tissue mass may be seen. A calcified matrix is absent. Inactive or healing LCH lesions may have a more benign appearance (Figures 49H–49J), with variable radiolucency, a sclerotic margin, and thick or wavy mature periosteal reaction.

Various descriptive terms have been applied to the distinctive radiographic features of LCH of bone. Active LCH lesions often demonstrate a "hole-within-a-hole" appearance that represents a confluence of osteolytic lesions typically seen in flat bones. A button sequestrum may be identified, particularly within the skull (5). In the spine, a vertebral plana deformity may result. In the calvarium, beveled-edge and geographic osteolytic lesions are classic findings (1). The "floating tooth" sign may be seen with mandibular or maxillary bone lesions, once the process engulfs the root of the tooth.

The MRI features of LCH are nonspecific during the active phase and resemble other aggressive appearing processes such as osteomyelitis and occasionally Ewing's sarcoma or lymphoma. LCH involving tubular bones arises from the medullary cavity and may demonstrate cortical destruction with extraosseous soft tissue extension. A clue that favors the diagnosis of LCH on MRI is sharply marginated cortical destruction, correlating with a narrow zone of transition on plain radiography and CT (Figure 49D). LCH is usually hypointense to isointense on T1, hyperintense on fluid sensitive

Table 49A. Selected aggressive bone tumors without calcific matrix (7, 8).

	Age at presentation (years)	Location	Location in long tubular bones
LCH	5–15	Axial > appendicular	diaphysis
Ewing's sarcoma	10–20	Axial = appendicular	metadiaphysis
Lymphoma	50–60; rare before 10	Appendicular > axial	metadiaphysis

sequences, and diffusely enhances after gadolinium administration (6). There is often adjacent marrow and juxtacortical soft tissue edema, even in the absence of pathologic fracture. Low SI on both T1 and fluid-sensitive sequences may be seen during the healing phase of LCH (6). Rarely, fluid-fluid levels and secondary aneurysmal bone cyst may develop (5).

Alternative diagnoses in this case include infection and less likely leukemia/lymphoma and Ewing's sarcoma (Table 49A). Ewing's sarcoma was considered less likely because of the patient's age, geographic margins, and modest extraosseous soft tissue mass. This lesion was pathologically proven to be LCH by CT-guided biopsy. After the provisional diagnosis of LCH was made, methylprednisolone was injected into the lesion at the time of biopsy. A skeletal survey showed no other lesions. Because this was localized LCH, the patient was not given chemotherapy and was symptom-free 9 months from presentation.

Orthopedic Perspective

LCH is a great mimic and should be considered in the differential diagnosis of all unusual radiolucent lesions of childhood. Because LCH may present with a destructive appearance, periosteal reaction, and perilesional edema or soft tissue mass, the referring diagnosis is frequently Ewing's sarcoma (especially in diaphyseal lesions of the long bones). Differentiating LCH from leukemia, lymphoma, and Ewing's sarcoma is also problematic in the spine since all can present as vertebra plana. A simple method to distinguish LCH from other lesions is to obtain a skull radiograph. If there is a "punched-out" lesion, the diagnosis is virtually established. If there is not, then further diagnostic studies are indicated. CT-directed needle biopsies are useful to clarify this distinction if diagnostic tissue can be obtained. Infection should also be considered in the differential diagnosis, and cultures should be obtained at time of biopsy. The other difficulty with LCH is assessing for the presence of multiple lesions. Some may not be readily apparent on radionuclide bone scans (although most are) and there is no general agreement on whether a bone scan, skeletal survey, or both should be performed. After excluding other systemic involvement, treatment of the solitary LCH lesion has become relatively straightforward recently with the use of intralesional injections of steroids. This almost always leads to resolution of the lesion and plain radiographs are sufficient to follow the patient. In lesions that require an open biopsy, simple curettage is usually curative, although in long bones of the lower extremity, casting or internal fixation may be required. Multicentric bone lesions, recurrences, and organ involvement often require systemic treatment with chemotherapeutic drugs. Patients should be assessed for diabetes insipidus with first void urine specific gravity measurement.

What the Clinician Needs to Know

1. Is the lesion solitary or multifocal?
2. Is this LCH or other lesion such as infection, leukemia, lymphoma, or Ewing's sarcoma?

3. Can the lesion be diagnosed by needle biopsy and treated by intralesional steroid administration?

Answers

1. Eosinophilic granuloma, Hand-Schuller-Christian disease, and Letterer-Siwe disease.
2. False; 10%.

Additional Examples

Solitary LCH of the Calvarium

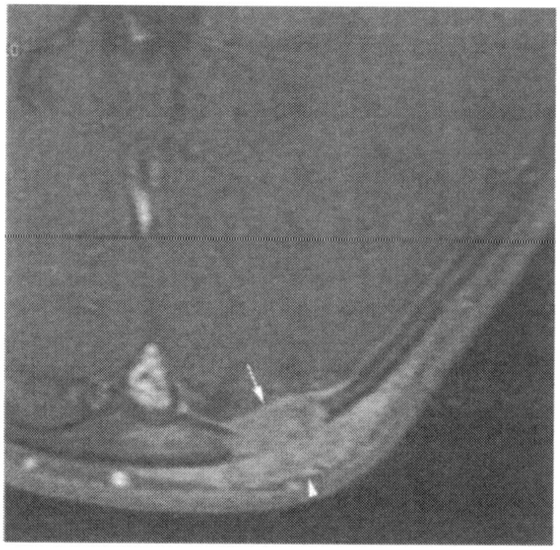

Figure 49E. Axial T1 post-Gd FS.

Findings

This 13-year-old boy had a slowly growing, painful lump over his left occiput.

Figure 49E. An enhancing tumor is seen to destroy inner and outer tables of the occipital calvarium, with extension to the epidural **(arrow)** and subgaleal spaces **(arrowhead)**. The presence of central enhancement rules out an epidermoid cyst. This was pathologically proven LCH.

Vertebrae Plana Due to LCH

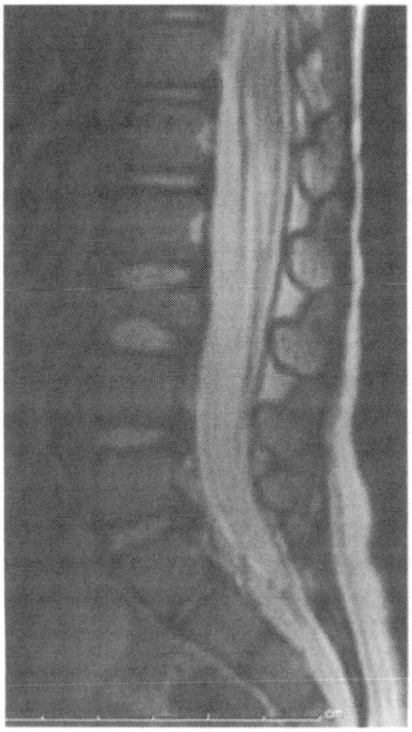

Figure 49F. Sagittal T2 of the lumbar spine.

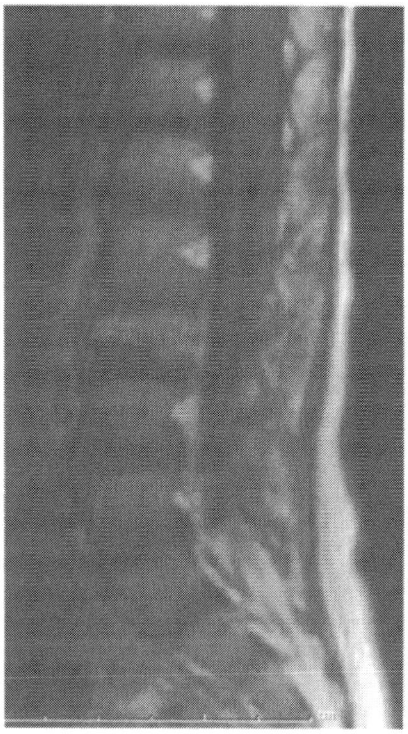

Figure 49G. Sagittal T1 post-Gd FS.

Findings

This is a $1^1/_2$-year-old boy with back pain who fell 2 weeks prior to presentation.

Figures 49F, 49G. There is L3 vertebral plana and diffuse enhancement. The adjacent intervertebral disks show normal SI and are increased in height, thereby ruling out osteomyelitis-disciitis. No epidural component is seen. The patient underwent CT-guided biopsy of the lesion, and LCH was confirmed. Other considerations for vertebral plana based on the patient's age would include neuroblastoma metastasis and leukemia/lymphoma.

LCH, Healing Phase

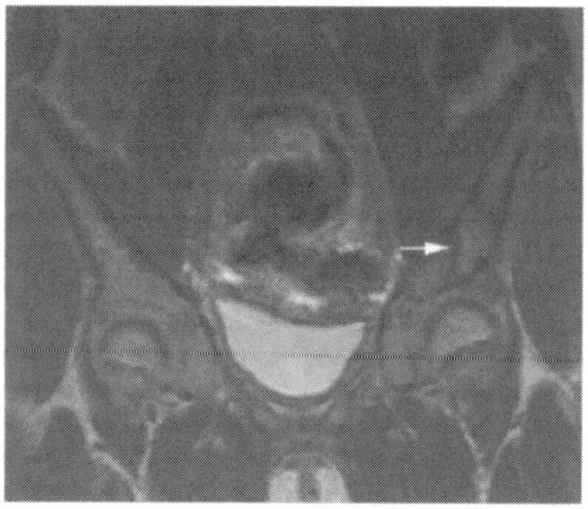

Figure 49H. Coronal T1.

Figure 49I. Coronal T2 FS.

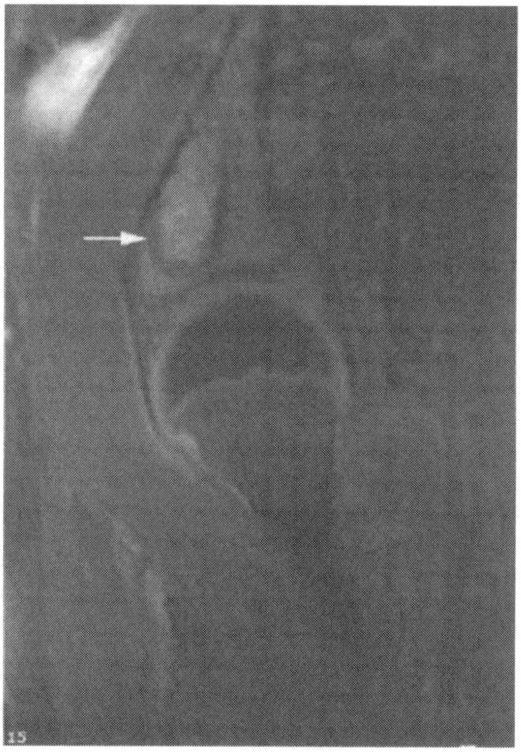

Figure 49J. Sagittal T1 post-Gd FS.

Findings

This is an 8-year-old girl with chronic left hip pain for several months.

Figures 49H, 49I, 49J. There is a well-defined lesion in the ilium that has a sharply demarcated hypointense margin **(arrows)**. The center of the lesion is hyperintense on T2 and enhances. There is also a moderate degree of adjacent marrow edema/enhancement and minimal juxtacortical soft tissue edema. This was biopsy proven LCH.

Pathologic Fracture from LCH

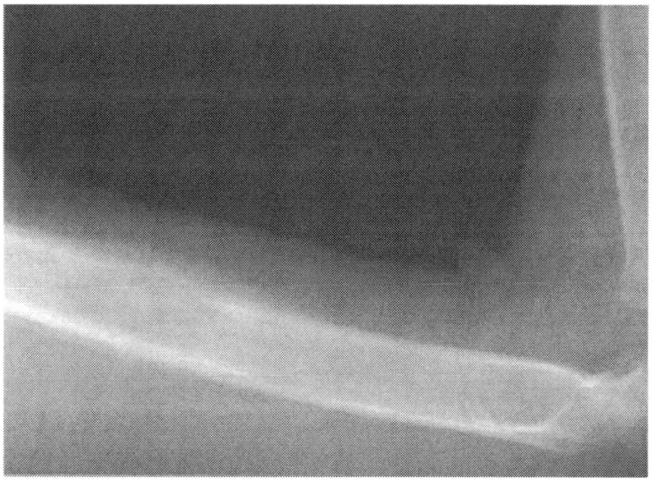

Figure 49K. Lateral radiograph of the left humerus.

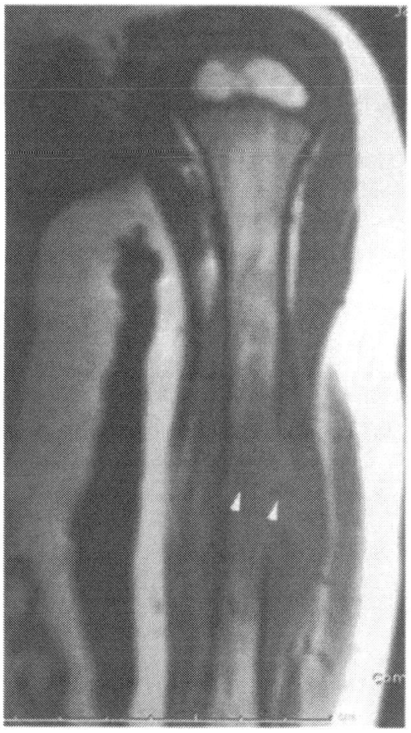

Figure 49L. Coronal T1.

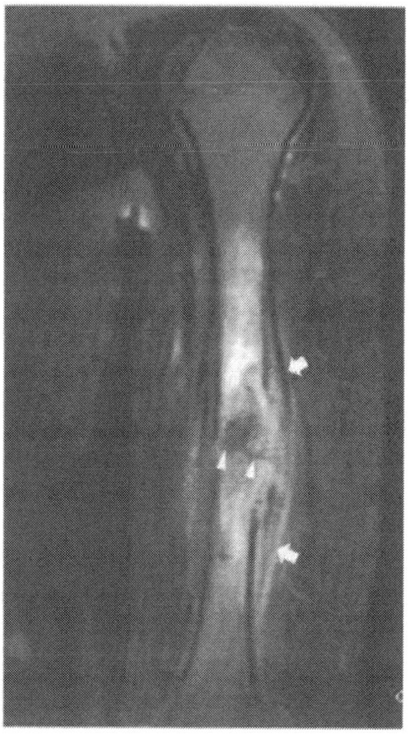

Figure 49M. Coronal T2 FS.

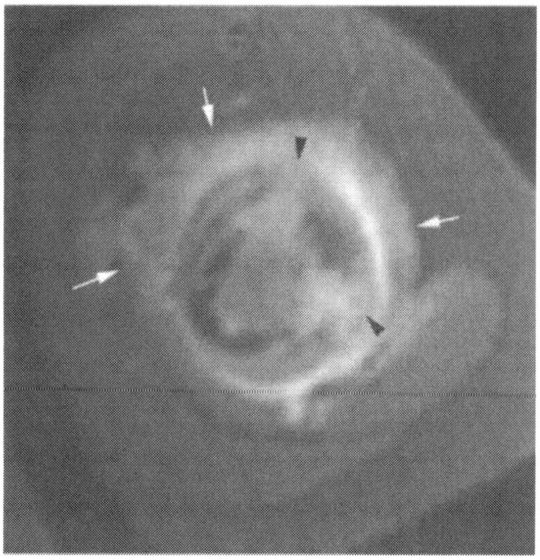

Figure 49N. Axial T1 post-Gd FS.

Findings

This is a 6-year-old boy with left arm pain following a fall 1 month ago.

Figure 49K. A moth-eaten osteolytic lesion is present in the mid-diaphysis of the humerus.

Figures 49L, 49M. A thin hypointense line on T1 **(white arrowheads)** and T2W sequences is identified, representing a nondisplaced pathologic fracture. There is aggressive, laminated periosteal reaction present **(thick arrows).** There is also extensive edema in the adjacent marrow and extraosseous soft tissues.

Figure 49N. Soft tissue extension **(white thin arrows)** and cortical breech **(black arrowheads)** is better appreciated on an axial section. This was biopsy proven LCH.

Pitfalls and Pearls

1. Multifocal LCH tends to occur in patients under 5 years, whereas solitary osseous LCH usually occurs in patients from 5 to 15 years.
2. One advantage of MRI is its ability to detect LCH lesions in symptomatic regions before they are apparent radiographically (6).

References

1. Stull MA, Kransdorf MJ, Devaney KO. Langerhans cell histiocytosis of bone. *Radiographics* 1992; 12:801–823.
2. Meyer JS, Harty MP, Mahboubi S, et al. Langerhans cell histiocytosis: Presentation and evolution of radiologic findings with clinical correlation. *Radiographics* 1995; 15:1135–1146.
3. Kransdorf MJ, Smith SE. Lesions of unknown histogenesis: Langerhans cell histiocytosis and Ewing sarcoma. *Semin Musculoskelet Radiol* 2000; 4:113–125.
4. Azouz EM, Saigal G, Rodriguez MM, Podda A. Langerhans' cell histiocytosis: Pathology, imaging and treatment of skeletal involvement. *Pediatr Radiol* 2005; 35:103–115.
5. Hindman BW, Thomas RD, Young LW, Yu L. Langerhans cell histiocytosis: Unusual skeletal manifestations observed in thirty-four cases. *Skeletal Radiol* 1998; 27:177–181.
6. George JC, Buckwalter KA, Cohen MD, Edwards MK, Smith RR. Langerhans cell histiocytosis of bone: MR imaging. *Pediatr Radiol* 1994; 24:29–32.
7. Krishnan A, Shirkhoda A, Tehranzadeh J, Armin AR, Irwin R, Les K. Primary bone lymphoma: Radiographic-MR imaging correlation. *Radiographics* 2003; 23:1371–1383; discussion 1384–1377.
8. Arndt CA, Crist WM. Common musculoskeletal tumors of childhood and adolescence. *N Engl J Med* 1999; 341:342–352.

Case 50

History

These three patients have an underlying malignancy. What type of supportive medication did they all receive while undergoing chemotherapy?

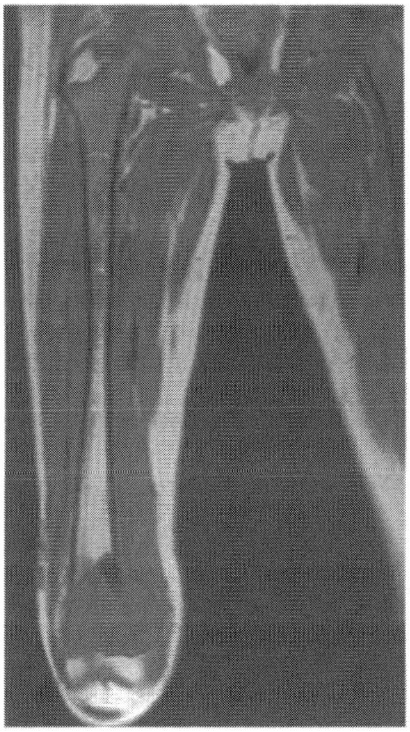

Figure 50A. Patient 1, 10-year-old boy. Coronal T1 of the right femur.

Figure 50B. Patient 1, 10-year-old boy. Coronal STIR.

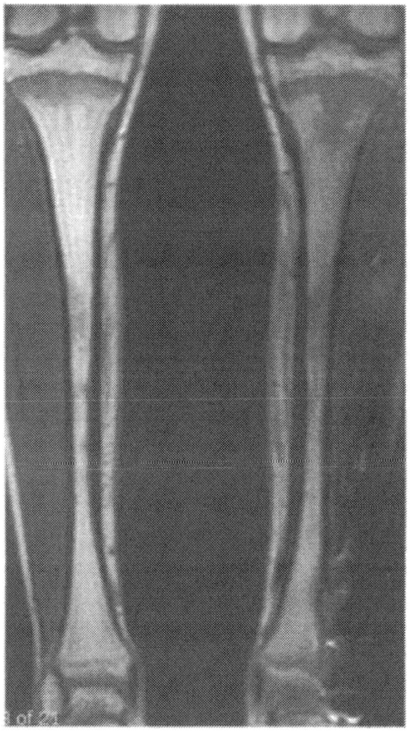

Figure 50C. Patient 2, 16-year-old boy. Coronal T1.

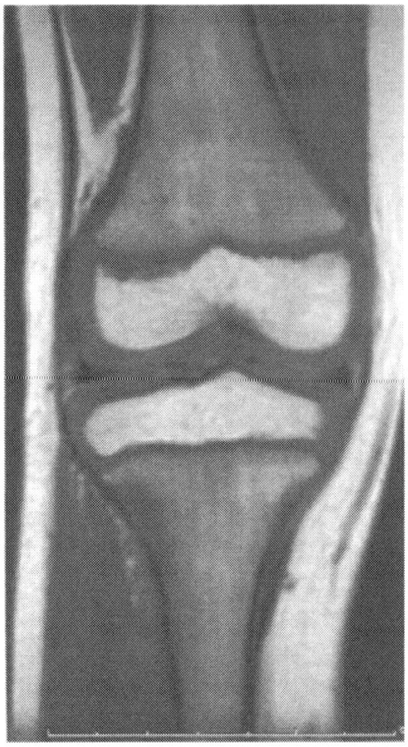

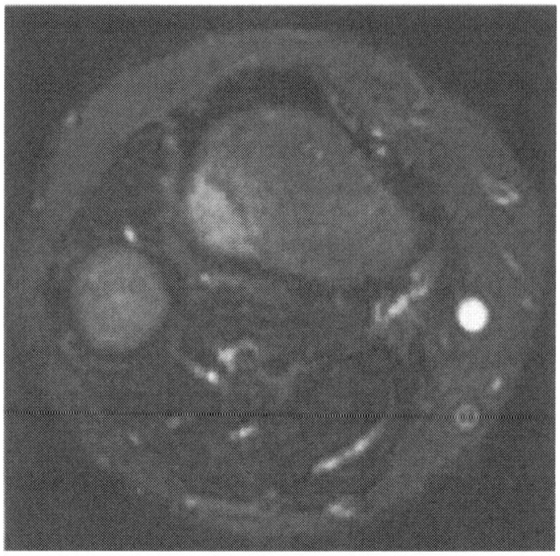

Figure 50E. Patient 3, 7-year-old boy. Axial STIR through the proximal tibia.

Figure 50D. Patient 3, 7-year-old boy. Coronal T1 of the right knee.

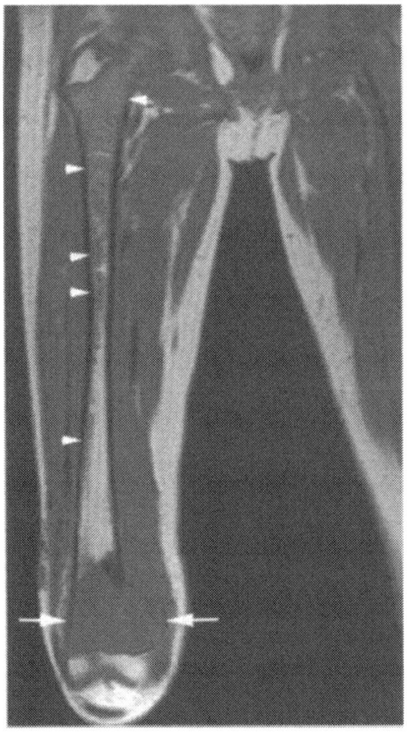

Figure 50A* Annotated.

Figures 50A, 50B, (50A with annotations). There is a complex mass in the distal femur that is centered in the metaphysis **(arrows)**. There are confluent and nodular areas in the marrow of the proximal and mid-femur that are slightly hypointense on T1 **(arrowheads)** and hyperintense on STIR.

Figure 50C. There are subphyseal and mid-diaphyseal areas of hypointensity identified in both tibias that do not cause mass effect. Susceptibility artifact related to a fibular resection is seen in the left calf.

Figures 50D, 50E. Along the lateral aspect of the proximal tibia metaphysis, there is an ovoid focus that is hypointense on T1 and moderately hyperintense on STIR. There is also diffuse marrow hypointensity on T1 that spares the epiphyses.

Diagnosis

Granulocyte colony-stimulating factor (G-CSF) effect on marrow

Questions

1. T/F: Yellow marrow conversion occurs within 6–8 weeks after radiation therapy.
2. By what age is yellow marrow conversion in the diaphysis usually complete?

Discussion

Granulocyte colony-stimulating factor (G-CSF) is an immunomodulatory agent that stimulates bone marrow production of neutrophils. Among several therapeutic applications, it is administered to patients undergoing cancer chemotherapy to accelerate recovery from neutropenia (1). In children treated with G-CSF, red marrow reconversion may be diffuse or confined to the metaphysis (2). Red marrow reconversion rarely may appear globular and mass-like and mimic neoplastic disease. On MRI, red marrow reconversion related to G-CSF therapy follows similar SI to red marrow and is isointense on T1 and mildly hyperintense on fluid-sensitive sequences.

Radiation therapy administered for childhood malignancies may also produce marrow signal alterations on MRI. Bone marrow may become hyperintense on fluid-sensitive sequences for the first 1–2 weeks after radiation. This may be followed by fatty marrow conversion after 6–8 weeks (Answer to Question 1) (3). In the vertebral bodies of children, return of red marrow by 11–30 months may be seen (4), whereas persistent fat SI in the radiation field may be seen in the extremities. Bone complications related to radiation include osteonecrosis, insufficiency fractures, physeal growth arrest, and the development of tumors including osteochondromas and bone/soft tissue malignancies.

It is important to be familiar with the appearance and age distribution of normal red marrow to permit differentiation from pathologic states (Diagram 50A). In the extremity, yellow marrow conversion occurs from distal to proximal. Yellow marrow conversion occurs in a predictable pattern in a given tubular bone. When the epiphyseal ossification center forms, yellow marrow conversion usually occurs within a few months (5). Yellow marrow conversion subsequently occurs in the diaphysis followed by the distal metaphysis (6). Yellow marrow conversion within the proximal metaphyses is variable, and persistent red marrow may be seen throughout adulthood. Yellow marrow conversion should be complete in the diaphysis by approximately 5 years of age (Answer to Question 2). With anemia, red marrow reconversion occurs

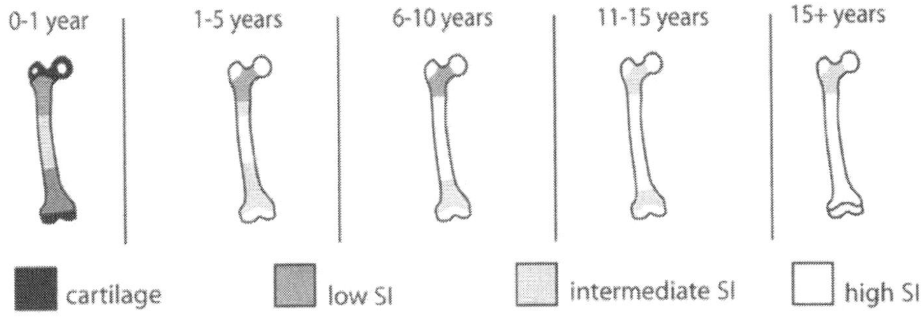

Diagram 50A. Normal marrow conversion with age. Normal fatty marrow distribution changes depending on age and are best depicted on T1W sequences as above. (Adapted by permission from Laor T, Jaramillo J, Oestreich AE. Muskuloskeletal System. In: Practical Pediatric Imaging, eds. Kirks DR, Griscom NT et al. Philadelphia: Lippincott-Raven; 1998: 331. Copyright, 1998, Lippincott Williams & Wilkins.)

in the reverse order: proximal metaphysis followed by distal metaphysis, diaphysis, and finally the epiphysis. In older children and adolescents, red marrow in the epiphysis should always be considered pathologic, and underlying metabolic or hematologic conditions as well as malignancy should be excluded.

Yellow marrow is usually moderately hyperintense with respect to muscle on T1 and hypointense on fluid sensitive sequences (8). Red marrow is usually isointense on T1 and isointense to mildly hyperintense on fluid sensitive sequences with respect to muscle. The exception to this rule is in the newborn (Figures 50F, 50G). Red marrow may appear isointense with respect to muscle on fluid sensitive sequences and may be relatively more hypointense compared with older infants and toddlers. A similar appearance has been observed in late gestational normal fetal mice and pigs (9, 10). Although the reason for the MRI features of red marrow in the newborn is unclear, it may be related to generalized physiologic osteosclerosis of the newborn (11). Even when yellow marrow conversion is evident on a T1W sequence, marrow SI may still be increased on fluid sensitive sequence in older infants and toddlers (Figures 50H, 50I). This observation presumably reflects the normal high water content within yellow marrow and/or the contribution from residual red marrow.

The differential diagnosis for a focal marrow replacement process includes metastases, osteomyelitis, and stress reaction. Metastatic disease, from etiologies such as neuroblastoma and leukemia, tend to deposit in the same locations as residual red marrow (i.e., metaphyseal regions) (Figures 50D, 50E) (12). Benign and malignant entities may cause trabecular disruption, mass effect, periosteal reaction, and/or juxtacortical soft tissue edema/mass. Residual red marrow, on the other hand, is typically flame shaped or has a paintbrush appearance, with no disruption of the normal trabecular pattern. T1W or PDW high resolution images should be carefully inspected and correlated with the area of signal abnormality on fluid sensitive sequences to assess for trabecular disruption and mass. Alternative strategies to differentiate normal marrow from tumoral infiltration includes chemical shift and echoplanar diffusion imaging (Table 50A) (13). Gadolinium enhanced sequences may be confusing since residual red marrow and pathologic processes may both demonstrate variable enhancement.

Patient 1 had biopsy proven conventional osteosarcoma of the distal femur. The MRI was ordered after chemotherapy and G-CSF therapy. The proximal and mid-diaphyseal changes were consistent with red marrow conversion.

Patient 2 had a nonmetastatic Ewing's sarcoma of the left fibula, which was resected (hence the susceptibility artifact). He was treated with G-CSF and chemotherapy after tumor resection. The signal changes in the proximal tibias were presumably related to red marrow reconversion.

Patient 3 had stage 4 neuroblastoma and was on chemotherapy and G-CSF. One year prior to the MRI, he received whole body radiation. The marrow changes were

Table 50A. Various MRI appearance of yellow and red marrow and tumor.

Sequence	Yellow Marrow	Red Marrow	Tumor
T1	Moderately hyperintense to muscle	Mildly hyperintense to isointense to muscle	Hypo- to isointense to muscle
T2 FS or STIR	Hypointense to muscle	Iso- to slightly hyperintense to muscle	Hyperintense to muscle
Diffusion (echoplanar)	– restricted diffusion	– restricted diffusion	+ restricted diffusion
Chemical shift (out of phase)	Hypointense	Hypointense	Hyperintense

Source: Data from Vanel D, et al. (13).

attributable to both G-CSF and whole body radiation. In addition to red marrow reconversion, the proximal tibial metaphyseal lesion was consistent with metastasis based on a MIBG (iodine-131-metaiodobenzylguanidine) study (not shown).

What the Clinician Needs to Know

1. Are the MRI features characteristic for radiation change or G-CSF marrow effect or does the patient require short term MRI follow-up, bone scintigraphy, PET, or even biopsy?
2. Is the red marrow distribution and SI appropriate for age? Is there complete marrow replacement (rule out leukemia/lymphoma) or is there too much yellow marrow for age?

Answers

1. True.
2. Approximately 5 years.

Additional Examples

Normal Red Marrow in a Newborn

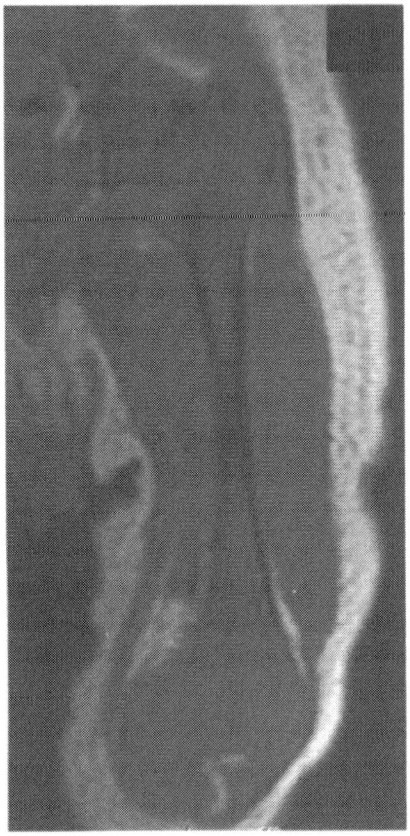

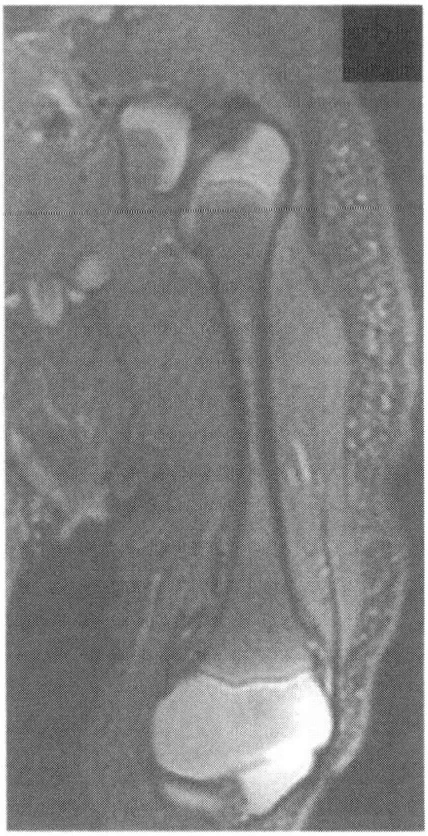

Figure 50F. Coronal T1 of the left femur.

Figure 50G. Coronal STIR.

Findings

This is a 7-day-old boy who was being evaluated for a soft tissue mass (not shown).

Figures 50F, 50G. The normal femoral bone marrow is diffusely hypointense on T1 and isointense on STIR with respect to muscle.

Normal Red Marrow in an 11 Month Old

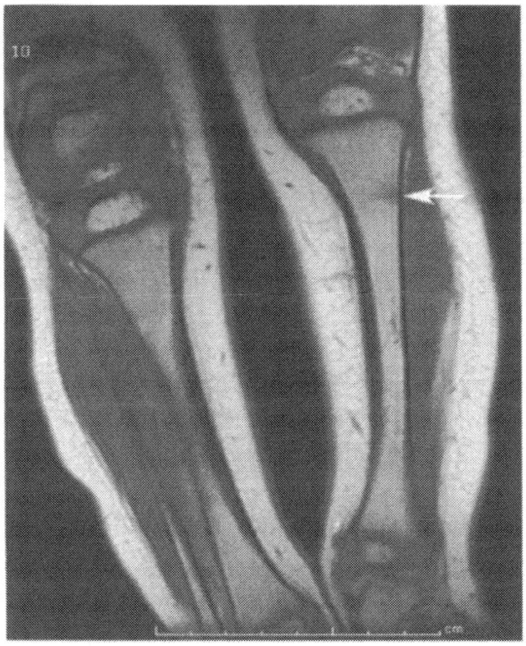

Figure 50H. Coronal T1 of the tibias.

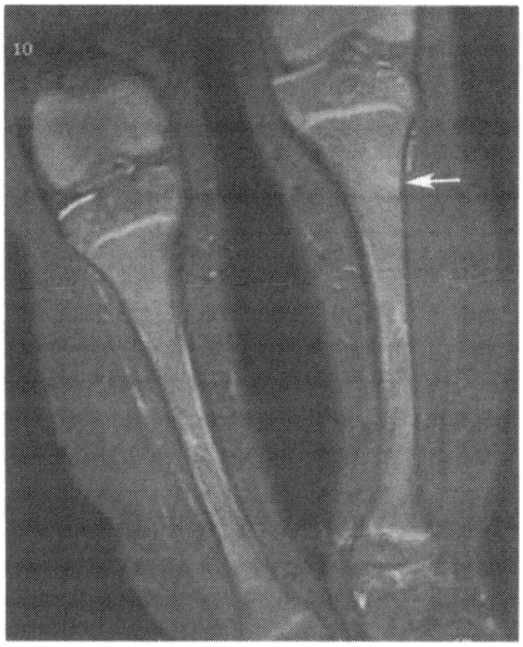

Figure 50I. Coronal STIR.

Findings

Figures 50H, 50I. This is an 11-month-old boy with stress reaction in the left proximal tibia **(arrows)**. There is partial yellow marrow conversion present, with the marrow moderately hyperintense on T1 and mild-moderately hyperintense on STIR. Compare this patient with the newborn (Figures 50F, 50G). Despite the presence of more yellow marrow in this 11 month old, the STIR sequence is relatively more hyperintense compared with the newborn.

Pitfalls and Pearls

1. Epiphyseal or mid-diaphyseal red marrow SI should always be considered abnormal in the older child.
2. Causes for diffuse marrow replacement include: tumor packing from metastatic neuroblastoma, sarcomas, leukemia/lymphoma, severe anemia, and metabolic disorders such as hemochromatosis, Gaucher's disease, and mucopolysaccharidosis (14).

References

1. Hubel K, Engert A. Clinical applications of granulocyte colony-stimulating factor: An update and summary. *Ann Hematol* 2003; 82:207–213.
2. Fletcher BD, Wall JE, Hanna SL. Effect of hematopoietic growth factors on MR images of bone marrow in children undergoing chemotherapy. *Radiology* 1993; 189:745–751.
3. Fletcher BD. Effects of pediatric cancer therapy on the musculoskeletal system. *Pediatr Radiol* 1997; 27:623–636.

4. Cavenagh EC, Weinberger E, Shaw DW, White KS, Geyer JR. Hematopoietic marrow regeneration in pediatric patients undergoing spinal irradiation: MR depiction. *AJNR Am J Neuroradiol* 1995; 16:461–467.
5. Laor T, Jaramillo J, Oestreich AE. Muskuloskeletal System. In *Practical Pediatric Imaging*, eds Kirks DR, Griscom NT, et al. Philadelphia: Lippincott-Raven; 1998; 331.
6. Jaramillo D, Laor T, Hoffer FA, et al. Epiphyseal marrow in infancy: MR imaging. *Radiology* 1991; 180:809–812.
7. Moore SG, Dawson KL. Red and yellow marrow in the femur: Age-related changes in appearance at MR imaging. *Radiology* 1990; 175:219–223.
8. Moulopoulos LA, Dimopoulos MA. Magnetic resonance imaging of the bone marrow in hematologic malignancies. *Blood* 1997; 90:2127–2147.
9. Connolly SA, Jaramillo D, Hong JK, Shapiro F. Skeletal development in fetal pig specimens: MR imaging of femur with histologic comparison. *Radiology* 2004; 233:505–514.
10. Ichikawa Y, Sumi M, Ohwatari N, et al. Evaluation of 9.4-T MR microimaging in assessing normal and defective fetal bone development: Comparison of MR imaging and histological findings. *Bone* 2004; 34:619–628.
11. Kottamasu SR. Bone changes associated with systemic disease. In: *Caffey's Pediatric Diagnostic Imaging,* 10th ed. Kuhn JP, Slovis TL, Haller J, ed. Philadelphia: Mosby, 2004; 2416–2454.
12. Herman TE, Siegel MJ. Case 30: Neoplastic marrow infiltration due to neuroblastoma. *Radiology* 2001; 218:91–94.
13. Vanel D, Dromain C, Tardivon A. MRI of bone marrow disorders. *Eur Radiol* 2000; 10:224–229.
14. States LJ. Imaging of metabolic bone disease and marrow disorders in children. *Radiol Clin North Am* 2001; 39:749–772.

Case 51

History

This is a 7-month-old boy with multiple liver lesions noted on a prior CT scan performed elsewhere following a reported fall.

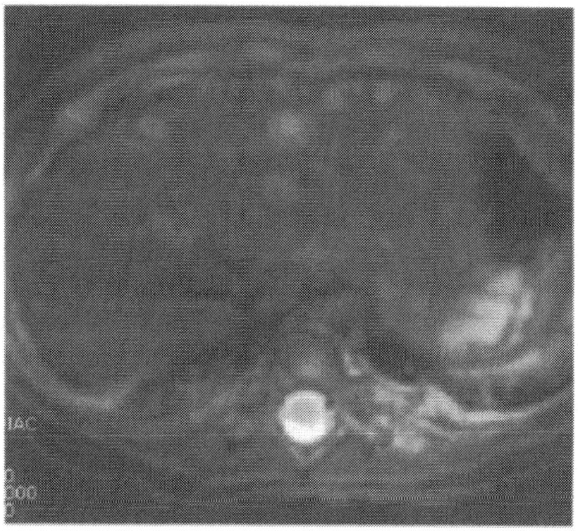

Figure 51A. Axial T2 FS.

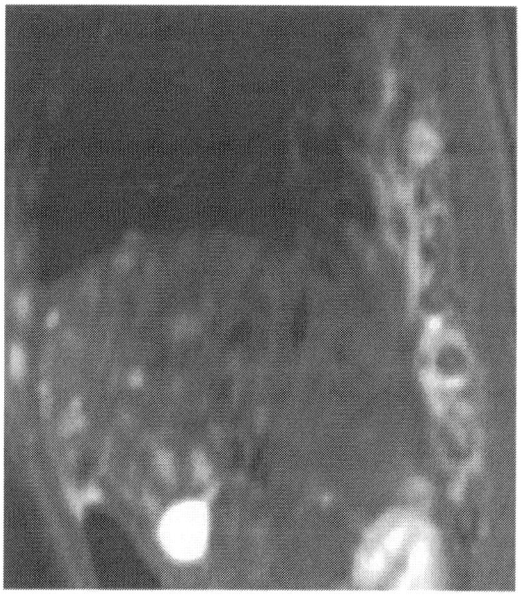

Figure 51B. Sagittal STIR, to the right of midline.

Figures 51A, (51A with annotations). There are innumerable, well-defined hyperintense lesions within the liver, consistent with hemangiomas. In addition, there is disruption of the architecture of the left rib, near its costovertebral articulations. There is hyperintensity within the rib head and neck **(thick black arrow)** as well as the adjacent posterior soft tissues **(white arrow)**. There is slight hyperintensity within the contralateral rib neck **(thin black arrow)**.

Figures 51B, (51B with annotations). The right parasagittal image shows hyperintensity surrounding two rib necks **(arrows)**.

Figure 51C. The MRI findings correspond to fractures of the rib necks on the CT **(arrows)**.

Diagram 51A. With anteroposterior compression of the chest, there is excessive leverage of the posterior ribs over the fulcrum of the transverse processes. This places tension along the inner aspects of the rib head and neck regions, resulting in fractures at these sites **(arrows)**. This mechanism is also consistent with the morphologic patterns of injury occurring at other sites along the rib arcs and at the costochondral junctions **(arrows)**.

Figure 51D. A classic metaphyseal lesion with a bucket-handle pattern is also evident in the proximal tibia **(arrow)**.

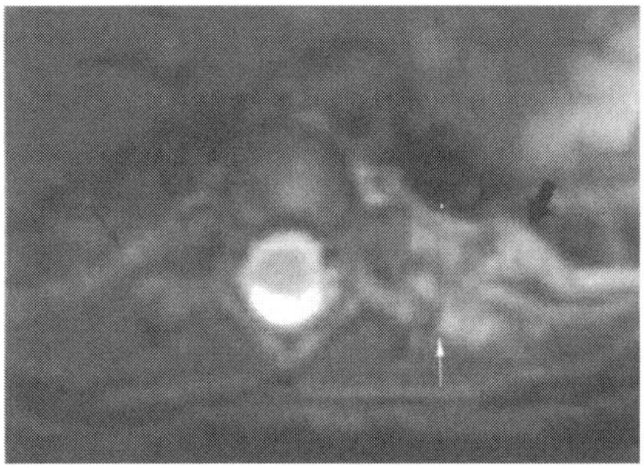

Figure 51A* Annotated.

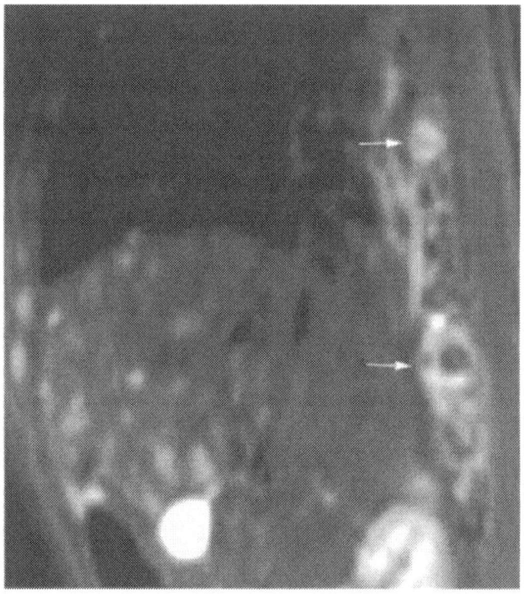

Figure 51B* Annotated.

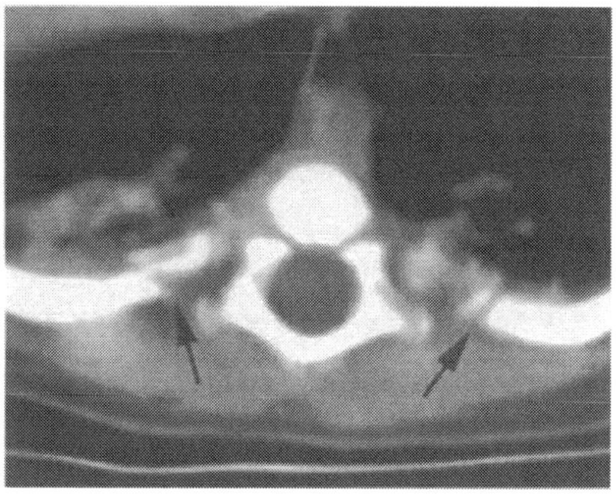

Figure 51C. CT of the abdomen performed 10 days earlier.

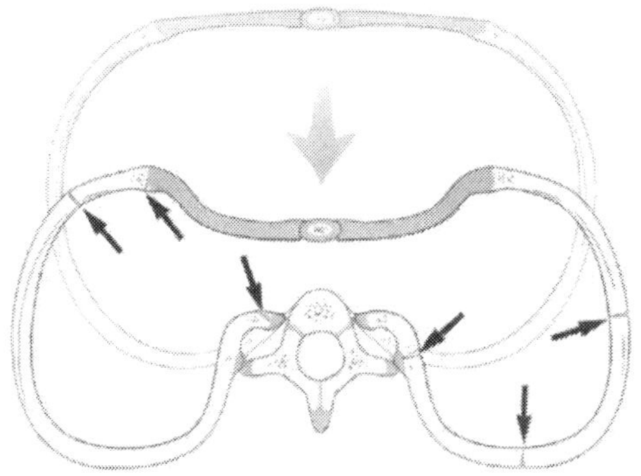

Diagram 51A. Mechanism of injury. (Reprinted from Kleinman P (1) with permission from Elsevier.)

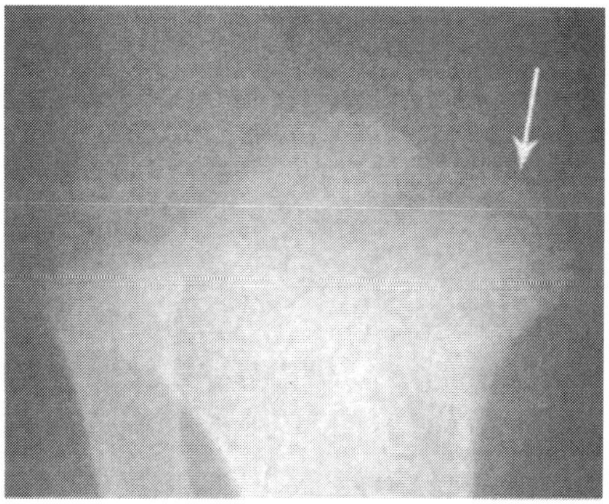

Figure 51D. AP of the right proximal tibia.

Diagnosis

Posterior rib fractures due to child abuse

Questions

1. T/F: Fractures of the rib head and neck in child abuse are caused by direct impact.
2. T/F: Posterior rib fractures are common with cardiopulmonary resuscitation (CPR).

Discussion

Approximately one million children are found to be maltreated annually, and 1500 are victims of fatal abuse. Children at greatest risk are below one year of age, and it is in this age group that skeletal injuries are most common and distinctive (1, 2). Rib fractures are the most common fracture noted in fatally abused infants, and most occur near the costovertebral articulations at the rib head or neck. Clinical, postmortem, and experimental studies indicate that fractures in this region occur with excessive leverage of the posterior rib over the fulcrum of the transverse process with anteroposterior chest compression (Diagram 51A) (Answer to Question 1). The fracture morphology is best appreciated in the axial plane, and thus CT tends to be more sensitive than radiography, particularly in the acute phase (1, 3). Although MRI of the thorax is not routinely done in the context of suspected child abuse, posterior rib fractures may be displayed on MRI done for other reasons, as in this case.

Rib fractures are commonly encountered along with characteristic injuries involving the metaphyses of the long bones, particularly at the knees and ankles. The so-called classic metaphyseal lesion (CML) has distinctive radiologic features, and in infants, the corner fracture and bucket-handle fracture patterns carry a high specificity for inflicted trauma (Figure 51D) (1). Although the more extensive metaphyseal injuries may be visible on MRI, plain radiography remains the gold standard for the detection of these often subtle injuries.

MRI is quite useful in assessing physeal injuries and epiphyseal separations in the context of suspected abuse. These are most commonly reported in the humerus and femurs, and although the findings may be apparent when the involved epiphyses are ossified, the findings may be subtle and the features may mistakenly suggest a dislocation, rather than fracture (1). In contrast to the CML, epiphyseal separations are primarily cartilaginous injuries, and are usually associated with a subtle metaphyseal fragment (Salter-Harris type 2), which may be the first clue to a fracture (Figures 51E–51H). MRI should include cartilage-sensitive sequences, usually a combination of gradient echo and PD with fat saturation. A 3D SPGR FS or similar sequence provides the option to reformat images in the optimal plane to assess displacement and angular deformity. Because abuse may not be suspected when a physeal separation is encountered in an infant with a reported fall, initial radiography should include views of the entire ipsilateral extremity, since additional fractures may be encountered. As with other unexplained injuries in the infant and toddler, a skeletal survey is usually indicated.

There are few differential considerations for posteromedial rib fractures in infants. They have not been documented with customary cardiopulmonary resuscitation (Answer to Question 2), and rarely occur with severe accidents in children (1). Posterior rib fractures occasionally occur during difficult vertex deliveries of large infants, but these fractures involve the mid-posterior arcs, rather that the costovertebral articulation (4). As with any unexplained fracture in young children, bone density and morphology should be visually assessed, and correlated with the physical examination and history, to exclude an underlying metabolic disorder (1).

In addition to the above injuries, skeletal survey demonstrated fractures of the right femoral shaft as well as a right frontoparietal skull fracture (not shown). Old vitreous hemorrhage was found in the right eye. Abuse was supported by the Department of Social Services, and the patient was placed in protective custody after discharge. The presumed liver hemangiomas were followed without intervention.

Orthopedic Perspective

Fractures are the second most common presentation of physical abuse after skin lesions, such as bruises and burns. Thus, abused patients commonly present for musculoskeletal injury evaluation and diagnostic imaging. There is no fracture that definitively makes the diagnosis of physical abuse; however, there are fracture patterns that carry a high specificity for abuse, including classic metaphyseal lesions and posteromedial rib fractures. Long bone fractures in children <1 year old and multiple fractures in various stages of healing are commonly seen with abuse, but must be viewed in conjunction with the history, physical exam, and other imaging studies. The orthopedist needs to know if these fractures are present and whether a skeletal survey is indicated. Imaging is also vital in excluding a variety of diseases and normal variants that may mimic abuse. Differentiation of abuse from mild forms of osteogenesis imperfecta is a common problem, and is usually accomplished on clinical and imaging grounds. However, there is an increasing tendency to rely on collagen analysis and genetic testing for osteogenesis imperfecta in problematic cases.

What the Clinician Needs to Know

1. What are the patterns, number, and distribution of fractures?
2. What are the ages of the fractures?
3. Are there skeletal injuries present that carry a high specificity for abuse?
4. Are there any features suggestive of osteogenesis imperfecta?

Answers

1. True.
2. False.

Additional Example

Distal Humeral Epiphyseal Separation in Child Abuse

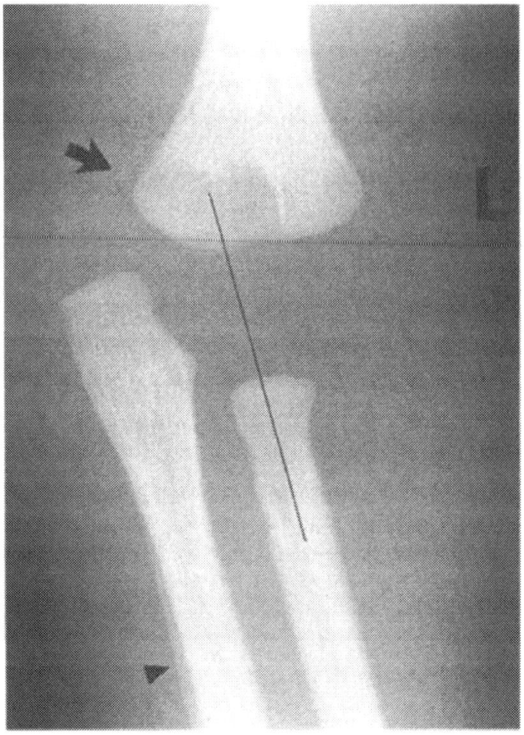

Figure 51E. Oblique radiograph of the right elbow.

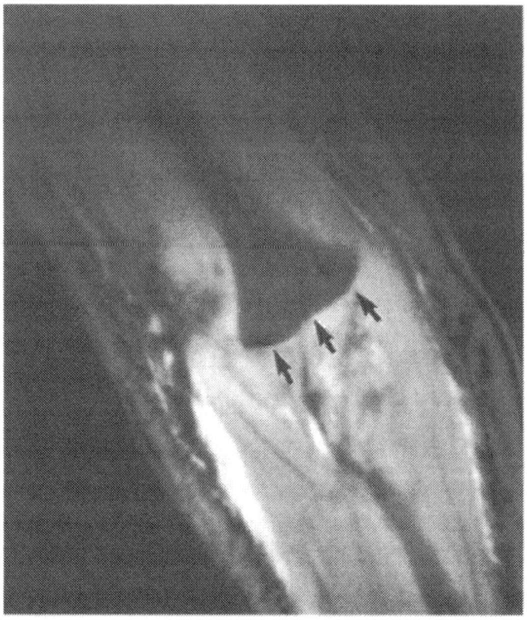

Figure 51F. Coronal MPGR through the distal humerus.

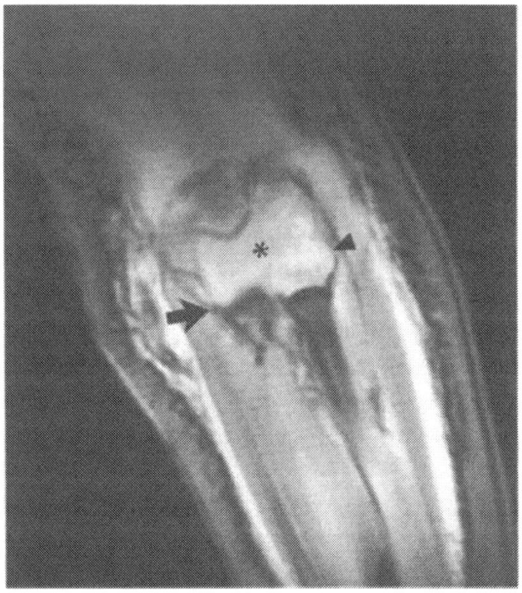

Figure 51G. Coronal MPGR just posterior to Figure 51F.

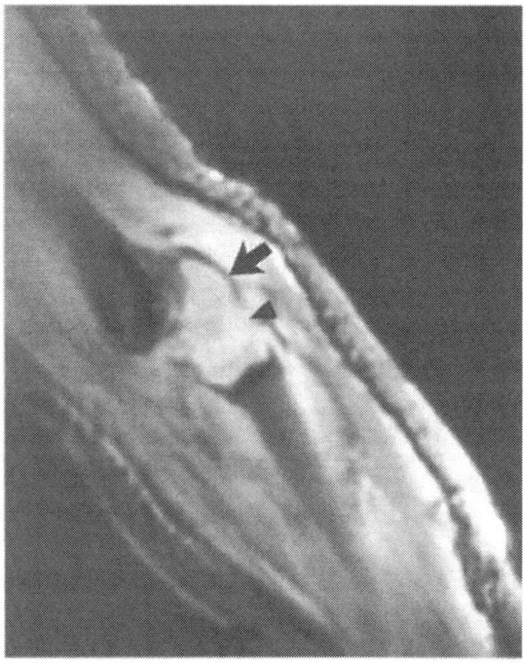

Figure 51H. Sagittal MPGR through the lateral elbow.

Findings

This 3-month-old infant presented with a swollen right elbow and unwillingness to use the right arm.

Figure 51E. The alignment of the radius is with the medial metaphysis **(straight black line)** rather than the lateral metaphysis of the distal humerus. A small metaphyseal fragment is present medially **(arrow)**. Posttraumatic subperiosteal new bone formation is noted along the ulnar diaphysis **(arrowhead)** without underlying fracture.

Figure 51F. A "naked" metaphysis **(arrows)** is evident. The epiphysis is not imaged in this plane.

Figure 51G. A more posterior image shows a normal relationship of the distal humeral epiphysis (*) with the proximal radius **(arrowhead)** and ulna **(arrow)**, but no contact with the distal humerus.

Figure 51H. A sagittal image through the radiocapitellar articulation confirms posterior displacement of the distal humeral epiphysis **(arrow)** and a normal relationship between the capitellum and the radius **(arrowhead)**. (Figures 51E–51H reprinted with permission from Springer (5).)

Pitfalls and Pearls

1. Abused children may present with clinical findings suggesting natural illness and rib fractures may be identified incidentally on MRI or CT.
2. When abuse is suspected, a skeletal survey should be obtained with a suitable high-detail imaging system.
3. Whenever abuse is suspected, close consultation with the referring clinician is essential, and referral to a child protection team may be indicated.

References

1. Kleinman PK. *Diagnostic Imaging of Child Abuse*, 2nd ed. St. Louis, MO: Mosby Year Book Inc., 1998.
2. Summary Child Maltreatment 2003. U.S. Department of Health and Human Services, Administration for Children and Families (accessed 2005): http://www.acf.hhs.gov/programs/cb/publications/cm03/summary.htm.
3. Barsness KA, Cha ES, et al. The positive predictive value of rib fractures as an indicator of nonaccidental trauma in children. *J Trauma* 2003; 54(6):1107–1110.
4. Hartmann RW. Radiological case of the month: Rib fractures produced by birth trauma. *Arch Pediatr Adolesc Med* 1997; 151(9):947–948.
5. Nimkin K, Kleinman PK, Teeger S, Spevak MR. Distal humeral physeal injuries in child abuse: MR imaging and ultrasonography findings. *Pediatr Radiol* 1995; 25:562–565.

Case 52

History

This is a 14-year-old boy with shoulder pain.

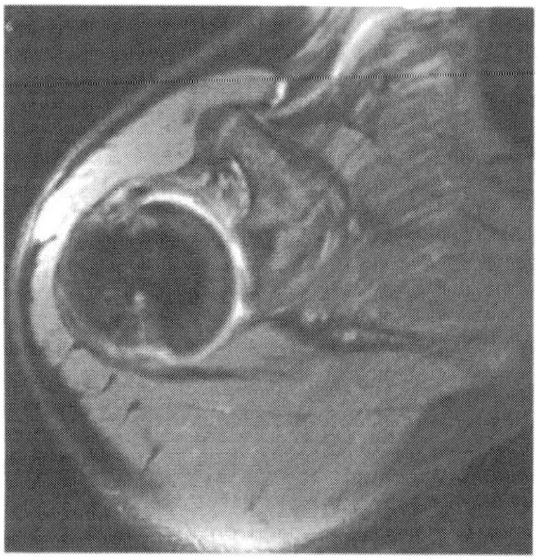

Figure 52A. Axial PD FS of the right shoulder.

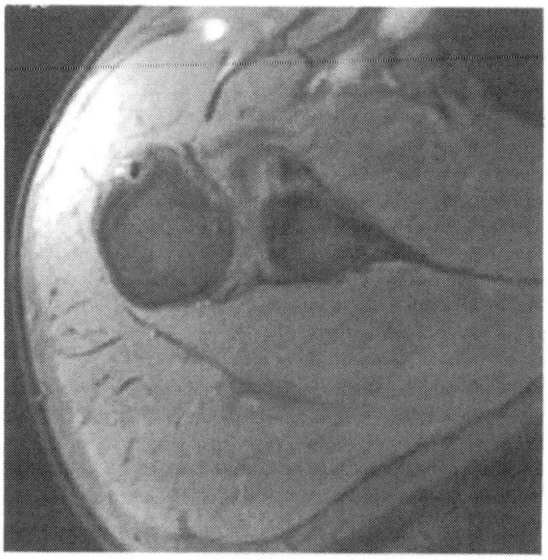

Figure 52B. Axial MPGR (caudal to Figure 52A).

Figures 52A, (52A with annotations). There is a focal concavity and adjacent marrow hyperintensity of the posterolateral margin of the humeral head **(thick arrow)** consistent with a compression fracture.

Figures 52B, (52B with annotations). The anteroinferior labrum is avulsed and anteromedially displaced **(thin arrow)**. The anteroinferior labrum remains attached by an intact scapular periosteum **(arrowhead)**.

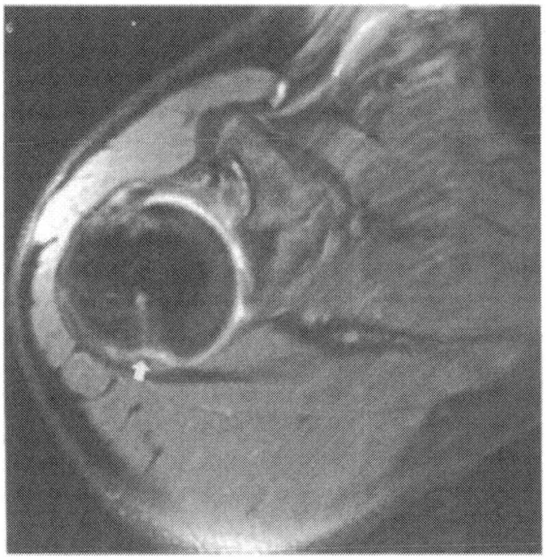

Figure 52A* Annotated.

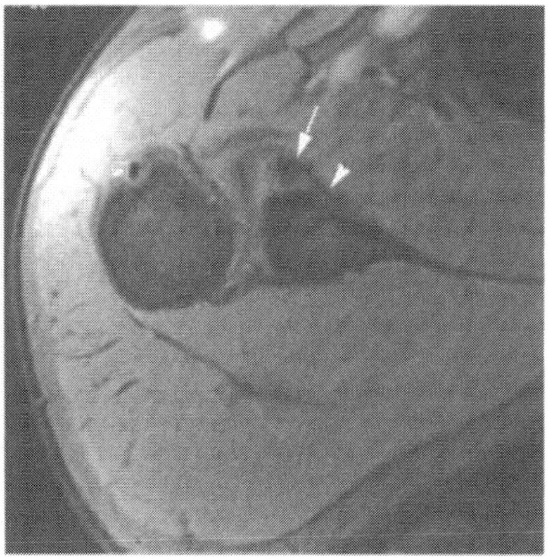

Figure 52B* Annotated.

Diagnosis

Anterior labroligamentous periosteal sleeve avulsion (ALPSA) and Hills-Sachs deformity

Questions

1. What is the most common cause of posterior glenohumeral dislocation in infancy?
2. What is a classic Bankart lesion?

Discussion

Shoulder instability is divided into atraumatic and traumatic (most common) etiologies. The most common direction of instability is anterior for both types. In one study comparing 70 shoulders in 66 children with shoulder instability, atraumatic instability presented in younger patients (9.3 years compared with 14.2 years for patients with traumatic instability), tended to be multidirectional, and frequently occurred voluntarily (1). Glenoid hypoplasia and ligamentous laxity may contribute to atraumatic shoulder instability (2). Recurrent dislocation has been observed more frequently in younger patients. McLaughlin et al. observed that recurrent dislocation occurred in 90% of patients presenting with their first dislocation under the age of 20, 60% in patients presenting between 20 and 40, and 10% in patients over 40 years of age (3). Traumatic anterior shoulder dislocation tends to occur with a fall on an outstretched hand. Proximal physeal humeral fractures may occur in children with forces similar to those that result in shoulder dislocation in skeletally mature patients (4). Chronic shoulder instability may also result from repetitive microtrauma and is most often seen in throwing athletes, but may also occur in gymnasts and swimmers (5).

Posterior shoulder dislocations are rare in the immature skeleton. The principal exception is found in the infant and child with a brachial plexus birth injury resulting in glenohumeral dysplasia (see Case 28). With brachial plexus birth injury, posterior subluxation or dislocation occurs due to muscle imbalance and unopposed internal rotation forces of the humerus (Answer to Question 1). In one study that followed 134 children with neonatal brachial plexus birth palsy, 8% were found to have posterior shoulder dislocations at one year of age (6).

The imaging findings of anterior shoulder instability include: Hill-Sachs deformity (superolateral compression fracture of the humeral head) and a classic Bankart lesion (anteroinferior labral tear/separation and a tear of the scapular periosteum) (Answer to Question 2) (2). Bankart lesions may be cartilaginous or osseous. There are several variations of the Bankart lesion (Table 52A). The Bankart lesion and Bankart variations disrupt the inferior glenohumeral ligament complex, the most important ligament for passive stabilization of the glenohumeral joint (7). Normal variations, including a sublabral foramen and a Buford complex, affect the appearance of the anterosuperior glenoid labrum and should not be confused with a Bankart or superior labral anterior-posterior (SLAP) lesion. The sublabral foramen is located between the bony glenoid and anterosuperior glenoid labrum, and a Buford complex is a hypertrophied middle glenohumeral ligament with an absent anterosuperior glenoid labrum.

There are several important normal MRI features of the immature glenoid. The normal anterosuperior glenoid cartilage has a slightly concave shape. However, the osseous glenoid is slightly convex and rounded and slopes medially to join the coracoid (Figure 52C). A potential diagnostic pitfall is to mistake this normally rounded anterosuperior glenoid contour for glenoid dysplasia, a condition that may lead to instability. Inferiorly, the glenoid may appear diminutive and flattened, with a conspicuous elongated labrum (Figure 52D).

Table 52A. Bankart lesions and variations.

Type	Description
Bankart lesion	Anteroinferior labral tear with a torn scapular periosteum
Anterior labroligamentous periosteal sleeve avulsion (ALPSA)	Anteroinferior labral tear with medial displacement and an intact scapular periosteum
Perthes	Anteroinferior labral tear without displacement with an intact scapular periosteum
Humeral avulsion glenohumeral ligament (HAGL)	Avulsion of the inferior glenohumeral ligament at its humeral insertion
Bony avulsion glenohumeral ligament (BAGHL)	HAGL lesion, but with bony avulsion at the humeral insertion
Floating anterior inferior glenohumeral ligament (AIGHL)	Combined Bankart and HAGL lesion
Glenolabral articular disruption (GLAD)	Anteroinferior labral tear associated with a glenoid chondral defect

Source: Data from Beltran J et al. (8).

This patient subsequently underwent arthroscopy and an anterior labral tear without bony Bankart lesion was identified. A labral repair and capsulorrhaphy were performed. At one-year follow-up, he was symptom free without evidence of recurrent shoulder dislocation or subluxation.

Orthopedic Perspective

Instability is a common shoulder disorder. The unconstrained geometry of the glenohumeral joint allows for great mobility; however, this is at the expense of intrinsic stability. In general, instability falls into two patterns: atraumatic multidirectional (posterior, inferior) instability, often in a ligamentously lax adolescent female, and traumatic anterior instability, often in an adolescent male contact athlete.

Multidirectional instability is treated with prolonged physical therapy and rehabilitation. Surgery is performed as a last resort and involves capsulorraphy, usually done open. Traumatic anterior instability has a high risk of recurrent instability. Some have advocated surgical stabilization of all first-time adolescent dislocators. Others recommend surgical stabilization only once recurrent instability occurs. Traumatic anterior instability is often associated with anteroinferior capsulolabral separation, the Bankart lesion. This leads to incompetence of the inferior glenohumeral ligament and recurrent instability. This author stabilizes first time dislocators if there is a Bankart lesion. MRI can help with determining the presence of the Bankart lesion. In the acute setting, conventional MRI often has an arthrogram effect from the traumatic hemarthrosis. Arthroscopic fixation of labral detachment has become the preferred method of repair.

What the Clinician Needs to Know

Distinguish secondary anatomic causes of shoulder instability. For instance, is there a Bankart-type lesion associated with instability or is instability due to ligamentous laxity alone?

Answers

1. Brachial plexus birth injury with acquired glenohumeral dysplasia.
2. Anteroinferior labral tear and a tear of the scapular periosteum.

Additional Example

Normal Glenoid in a Young Child

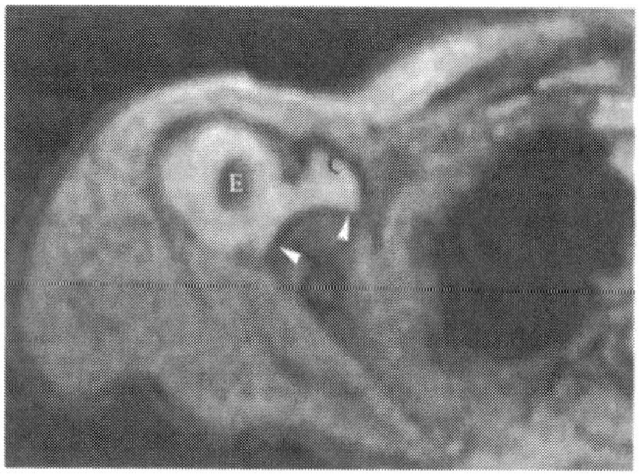

Figure 52C. Axial MPGR of the right shoulder.

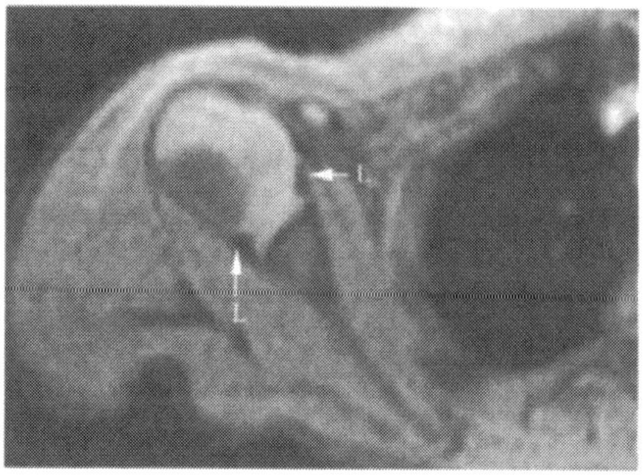

Figure 52D. Axial MPGR.

Findings

This is a 6-month-old girl with a normal right shoulder.

Figure 52C. The anterosuperior hyaline cartilage of the glenoid is continuous with the cartilaginous coracoid (C). Note the normal convexity of the anterosuperior osseous glenoid **(arrowheads)**. Secondary ossification center of the humerus (E).

Figure 52D. More inferiorly, the glenoid articular surface is flat and the labrum is prominent and elongated. Fibrocartilaginous labrum (L).

Pitfalls and Pearls

Abduction external rotation (ABER) imaging may be performed as an additional sequence to specifically evaluate the anteroinferior capsulolabral complex. This shoulder position places additional tension on the anterior inferior glenohumeral ligament and will accentuate Bankart and Bankart-variant injuries.

References

1. Lawton RL, Choudhury S, Mansat P, Cofield RH, Stans AA. Pediatric shoulder instability: Presentation, findings, treatment, and outcomes. *J Pediatr Orthop* 2002; 22:52–61.
2. Rafii M. Non-contrast MR imaging of the glenohumeral joint: Part II. Glenohumeral instability and labrum tears. *Skeletal Radiol* 2004; 33:617–626.
3. McLaughlin HL, Cavallaro WU. Primary anterior dislocation of the shoulder. *Am J Surg* 1950; 80:615–621; passim.
4. Dameron TB, Jr., Reibel DB. Fractures involving the proximal humeral epiphyseal plate. *J Bone Joint Surg Am* 1969; 51:289–297.
5. Kocher MS, Waters PM, Micheli LJ. Upper extremity injuries in the paediatric athlete. *Sports Med* 2000; 30:117–135.
6. Moukoko D, Ezaki M, Wilkes D, Carter P. Posterior shoulder dislocation in infants with neonatal brachial plexus palsy. *J Bone Joint Surg Am* 2004; 86-A:787–793.
7. Shankman S, Bencardino J, Beltran J. Glenohumeral instability: Evaluation using MR arthrography of the shoulder. *Skeletal Radiol* 1999; 28:365–382.
8. Beltran J, Kim DH. MR imaging of shoulder instability injuries in the athlete. *Magn Reson Imaging Clin N Am* 2003; 11:221–238.

Case 53

History

This is an 11-year-old girl with right arm pain for 3 months that has become worse over the last few weeks.

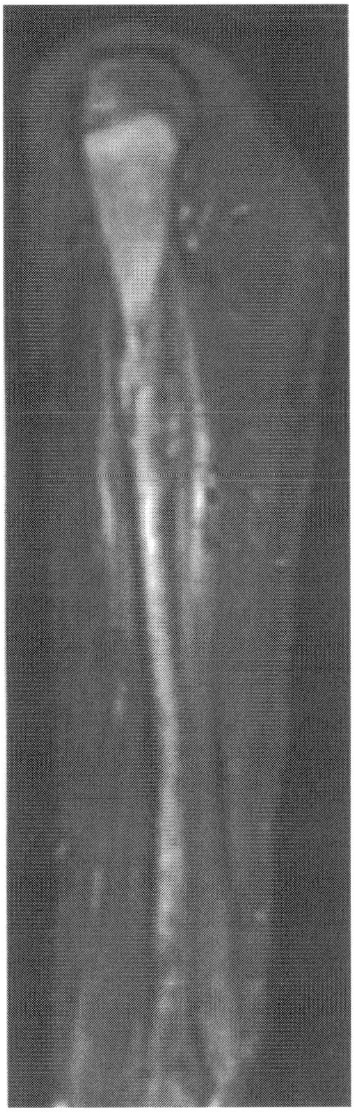

Figure 53A. Coronal STIR of the right humerus.

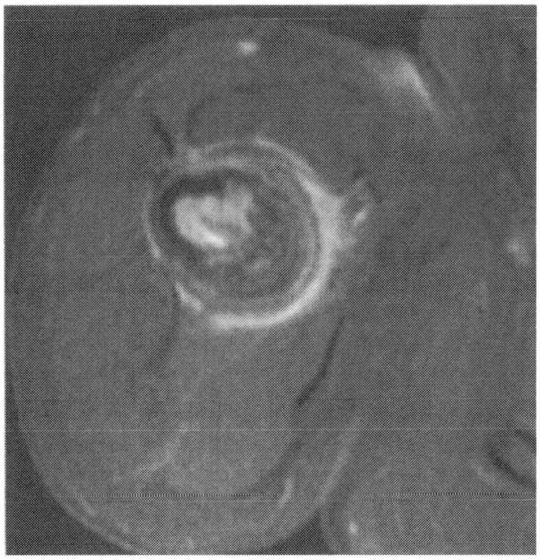

Figure 53B. Axial T1 post-Gd FS.

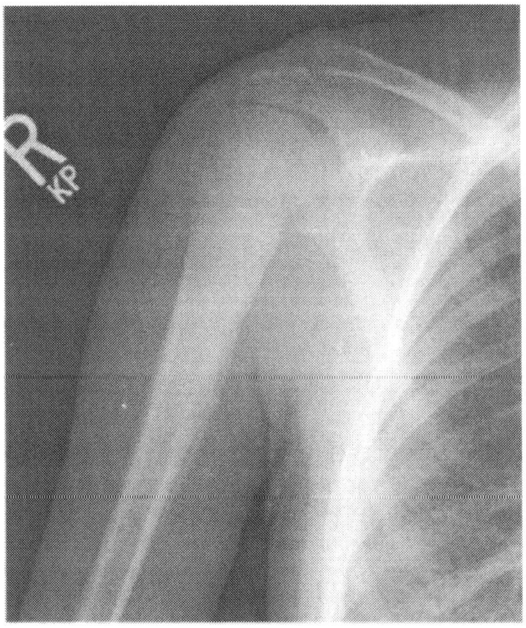

Figure 53C. AP radiograph.

Figures 53A, 53B. There is diffuse marrow edema present centered in the metadiaphysis of the humerus. There is focal cortical destruction medially, laminated periosteal reaction, and adjacent soft tissue edema and enhancement. No discrete soft tissue mass is identified.

Figure 53C. A permeative and moth-eaten mixed osteolytic and sclerotic lesion is present within the proximal humeral metadiaphysis. Laminated periosteal reaction, most evident over an area of medial cortical destruction, and a wide zone of transition are present.

Diagnosis

Osteomyelitis mimicking Ewing's sarcoma

Questions

1. T/F: MRI easily distinguishes infectious and post-traumatic granulation tissue from an underlying tumor.
2. T/F: Bone remodeling is a unique feature of slow growing, benign bone tumors and is not seen with bone sarcomas.

Discussion

The clinical and radiographic features of infection or trauma may overlap with malignancies such as Ewing's sarcoma, osteosarcoma, as well as Langerhans cell histiocytosis (LCH). In a series of 21 patients with biopsy and/or blood culture positive subacute osteomyelitis who were initially thought to have bone tumors, the primary radiographic abnormalities were osteolysis with a sclerotic border in 14 cases, osteolysis without definite borders in six, and onion skin periosteal reaction in one (1). With osteomyelitis or trauma, aggressive appearing periosteal reaction and intra- and extraosseous granulation tissue may mimic tumor (Answer to Question 1). Distinguishing trauma from tumor may be problematic when a discrete fracture line is not identified (Figures 53D–53F) (2).

Bone sarcomas may also mimic osteomyelitis. High grade osteosarcomas and Ewing's sarcoma may outgrow their blood supply leading to tumoral necrosis, mimicking abscess. The diagnosis of Ewing's sarcoma may be particularly problematic when minimal extraosseous soft tissue disease is associated with the intraosseous tumor (Figures 53G, 53H). Chemotherapy or radiation therapy may lead to either tumoral necrosis or cystic change that may mimic an abscess as well.

MRI evaluation after pathologic fracture is often requested when plain radiographic features suggest an underlying bone neoplasm. The presence of a discrete intra- or extraosseous soft tissue mass should suggest an underlying neoplasm (Figures 53I–53L). However, the presence of a juxtacortical soft tissue mass does not invariably point to a malignant neoplasm. Benign lesions such as fibrous dysplasia and unicameral bone cysts that have sustained a pathologic fracture may also demonstrate intra- and extraosseous soft tissue masses on MRI (3, 4).

Although malignant tumors tend to destroy, rather than expand, cortical bone, sarcomas may occasionally manifest indolent features (Figures 53M–53P). Because of a slow growth pattern, bone remodeling and expansion rather than cortical destruction may be seen, mimicking fibrous dysplasia (Answer to Question 2). Furthermore, the lack of a significant extraosseous component may contribute to the benign appearance of a slow growing sarcoma.

In this patient, biopsy was performed due to plain radiographic and MRI features suggesting Ewing's sarcoma. The diagnosis of chronic osteomyelitis was made, although no organism was isolated. The patient was treated empirically for osteomyelitis and responded. A follow-up radiograph at 3 months showed healing osteomyelitis (with no evidence of malignancy).

Orthopedic Perspective

One must always consider mimics such as osteomyelitis, LCH, fibrous dysplasia, and fatigue fracture when evaluating a patient with a suspected aggressive bone tumor. Often the plain radiograph is the most helpful study to establish the diagnosis. Benign tumors such as aneurysmal bone cyst can appear quite aggressive and mimic a malignant bone tumor. The periosteal reaction of a cortical osteoid osteoma may be extensive and mimic Ewing's sarcoma, and as shown here, osteomyelitis can be indistinguishable from sarcoma or leukemia until the biopsy is done. Of more concern is a sarcoma that mimics a benign lesion. Osteosarcomas may have areas with fluid-fluid levels that lead the clinician to suspect an aneurysmal bone cyst. When a pathological fracture has occurred, the correct diagnosis can be extremely difficult to discern because the usual clues are lost in the fracture hematoma. A biopsy does not always clarify the issue because of the callus formation and the difficulty of obtaining adequate tissue after fracture. The obfuscation that can occur with MRI after stress reaction and fracture is illustrated (Figures 53D–53F, 53I–53L), and this predicament can make efforts to obtain an accurate diagnosis treacherous. A high index of suspicion is necessary and a careful history is important. Was the trauma sufficient to cause the fracture, or was there little or no trauma, suggesting that the fracture was pathologic? Were there symptoms such as pain or limp prior to the fracture, again suggesting a pre-existing lesion? A correct diagnosis is greatly facilitated by a careful history, physical examination, and review of all recent and old imaging studies.

What the Clinician Needs to Know

1. Is this a benign, infectious, or traumatic condition mimicking a malignant bone tumor?
2. Is this a malignant tumor masquerading as a benign condition? Infection is particularly difficult to distinguish from a malignant bone tumor.
3. A pathological fracture makes the diagnosis of an underlying malignant or benign bone tumor extremely difficult to establish. A carefully taken history is essential.

Answers

1. False.
2. False.

Additional Examples

Stress Injury Mimicking Ewing's Sarcoma

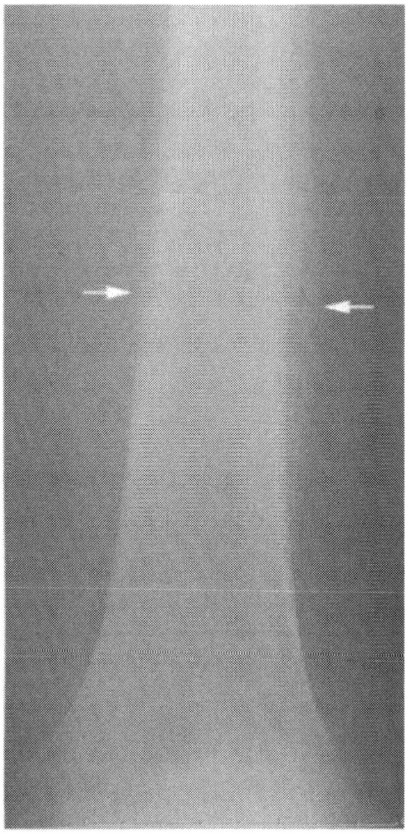

Figure 53D. AP radiograph of the distal left femur.

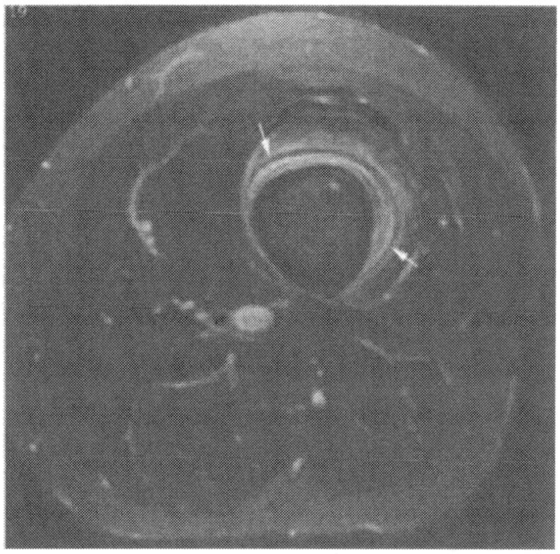

Figure 53E. Axial T2 FS distal diaphysis.

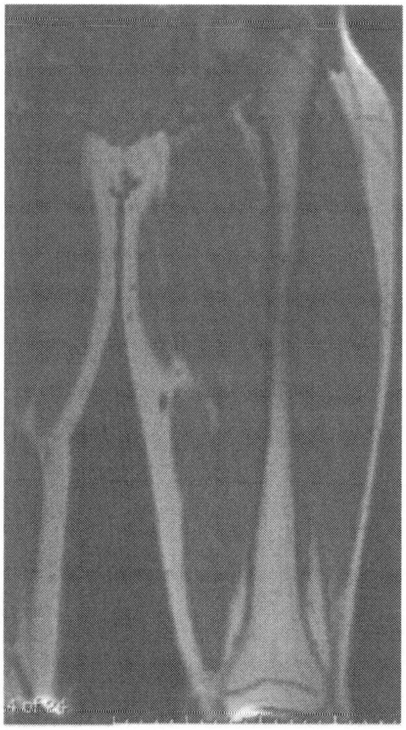

Figure 53F. Coronal T1.

Findings

This is a 15-year-old boy with bilateral hip pain.

Figure 53D. Aggressive appearing, laminated periosteal reaction is evident in the distal diaphysis of the femur **(arrows)**.

Figure 53E. T2 hyperintense laminated periosteal reaction **(arrows)** is evident without underlying soft tissue mass or fracture line.

Figure 53F. There is normal marrow SI seen within the distal diaphysis of the femur. A follow-up plain radiograph and MRI 9 weeks later (not shown) showed decreased periosteal reaction and no underlying soft tissue mass. The diagnosis of stress reaction without occult fracture was made. Short-interval follow-up radiographs and MRI averted the need for biopsy.

Ewing's Sarcoma Without Juxtacortical Soft Tissue Mass

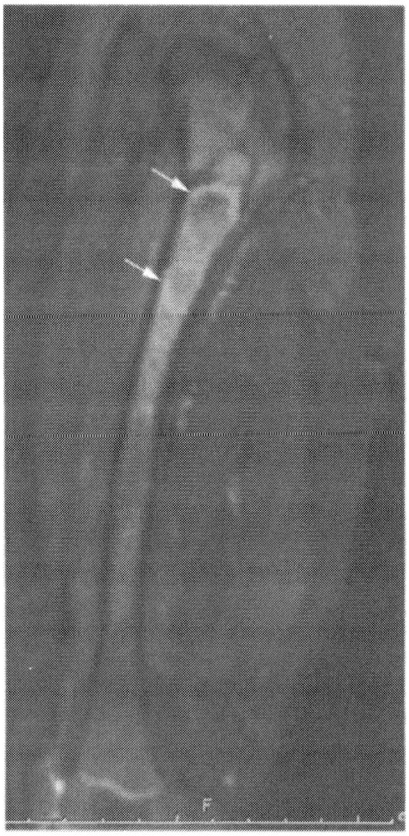

Figure 53G. Sagittal STIR of the right femur.

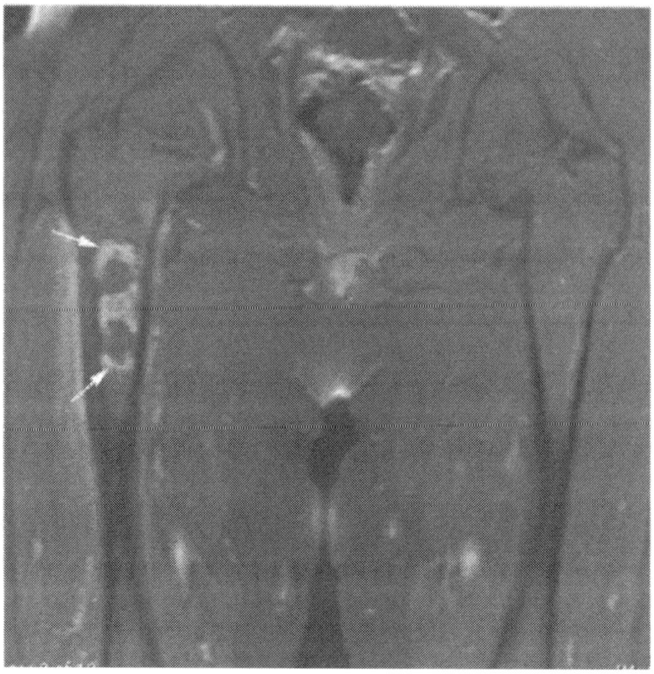

Figure 53H. Coronal T1 post-Gd FS.

Findings

This is a 3-year-old boy with right thigh pain. The radiograph showed a permeative lesion involving the proximal right femur (not shown).

Figures 53G, 53H. There is extensive heterogeneous signal abnormality in the marrow of the proximal femur with ring-like hyperintensities and enhancement **(arrows)**. There is minimal juxtacortical soft tissue enhancement, but no discrete mass is seen. Because of the patient's age and lack of soft tissue mass, other diagnoses including LCH, fibrous dysplasia, and atypical osteomyelitis were considered. This was pathologically proven to be a Ewing's sarcoma.

Osteosarcoma with Pathologic Fracture

Findings

This is a 14-year-old boy who sustained a pathologic fracture through the mid-diaphysis of his femur after skiing.

Figure 53I. A mixed osteolytic-sclerotic lesion with a wide transition zone is identified.

Figures 53J, 53K, 53L. Diffuse marrow and extraosseous soft tissue edema is present. On axial sequences, a juxtacortical intermediate T2 and enhancing soft tissue mass is identified **(arrowheads)**. This soft tissue mass is easily distinguishable from the adjacent feathery, soft tissue edema. This was pathologically proven to be a conventional osteosarcoma located in the diaphysis.

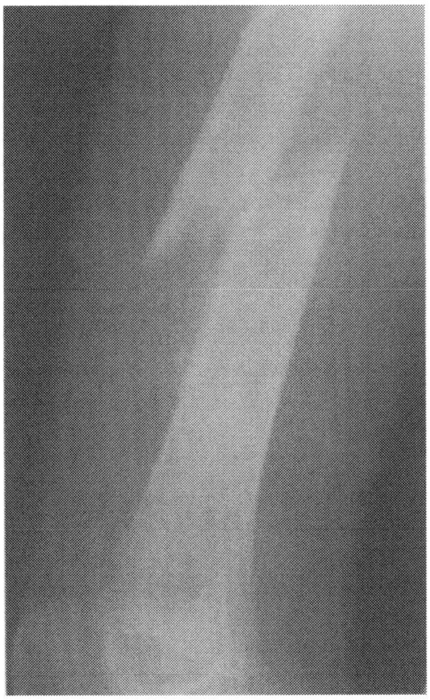

Figure 53I. Lateral radiograph of the left femur.

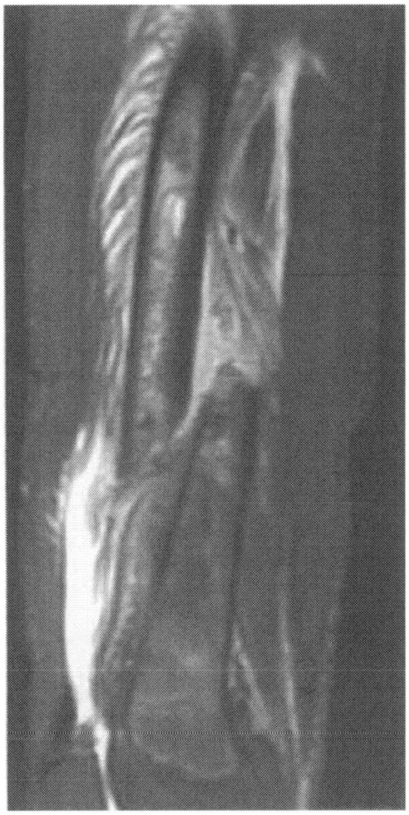

Figure 53J. Sagittal STIR.

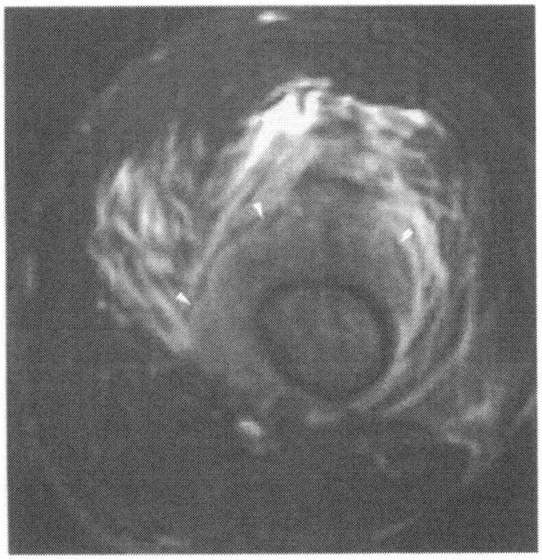

Figure 53K. Axial T2 FS mid-diaphysis.

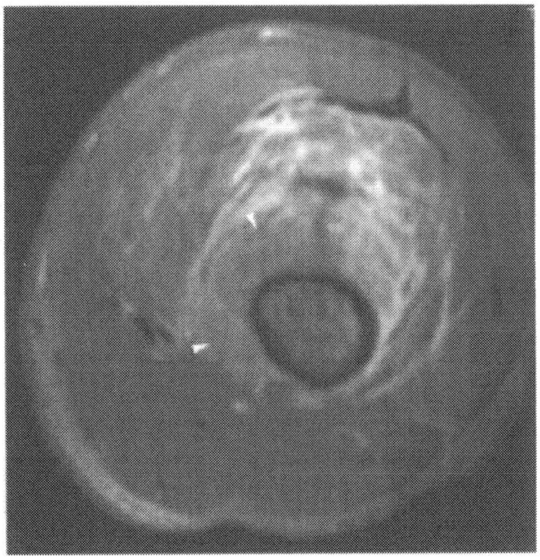

Figure 53L. Axial T1 post-Gd FS mid-diaphysis.

Ewing's Sarcoma Mimicking a Benign Lesion

Findings

This is a 17-year-old boy with right leg pain and an enlarging calf mass. Three years earlier he was casted for a fracture through what was considered to be a unicameral bone cyst.

Figure 53M. An expansile mass is identified **(arrowheads)** with ground-glass attenuation and an appearance suggesting an aneurysmal bone cyst, giant cell tumor, or less likely fibrous dysplasia.

Figures 53N, 53O, 53P. There is a heterogeneous mass with variable T2 SI and enhancement associated with fluid-fluid levels **(arrows)**. There is minimal extraosseous soft tissue mass. This was pathologically proven Ewing's sarcoma.

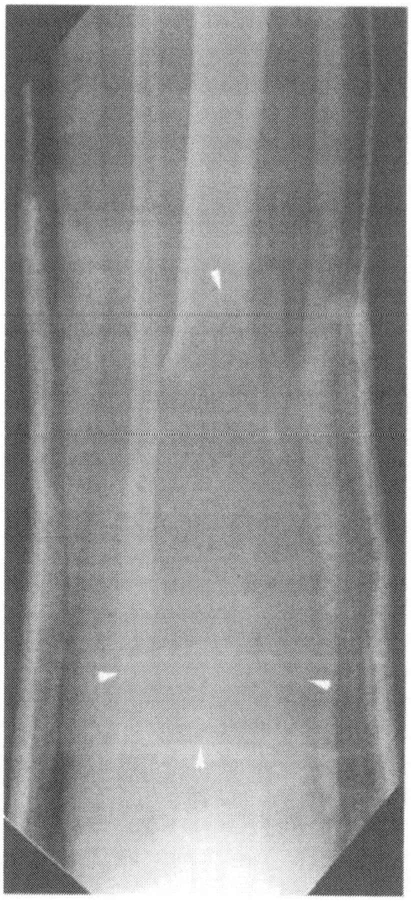

Figure 53M. AP radiograph of the right tibia/fibula.

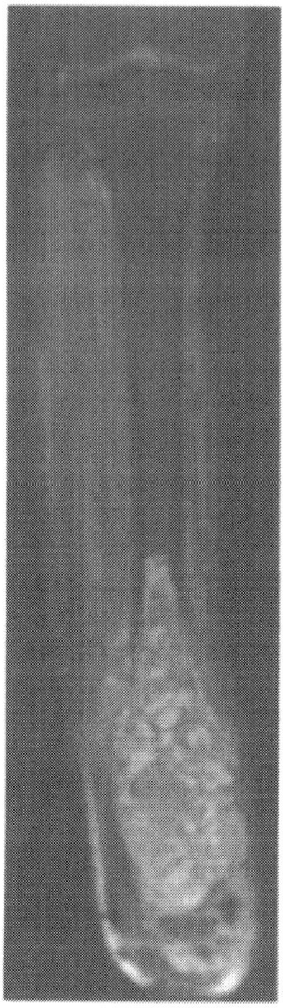

Figure 53N. Coronal STIR.

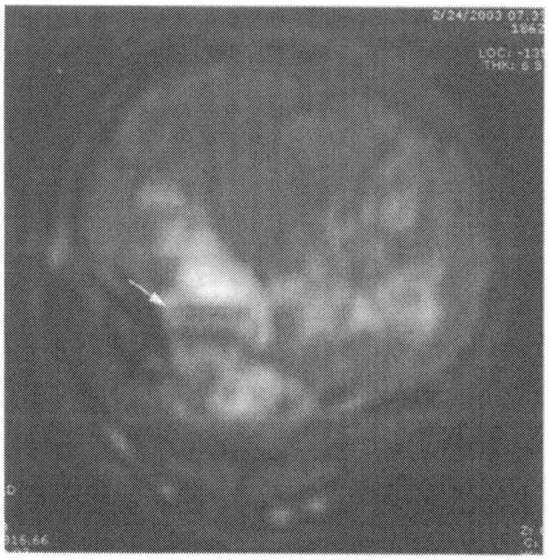

Figure 53O. Axial T2 FS.

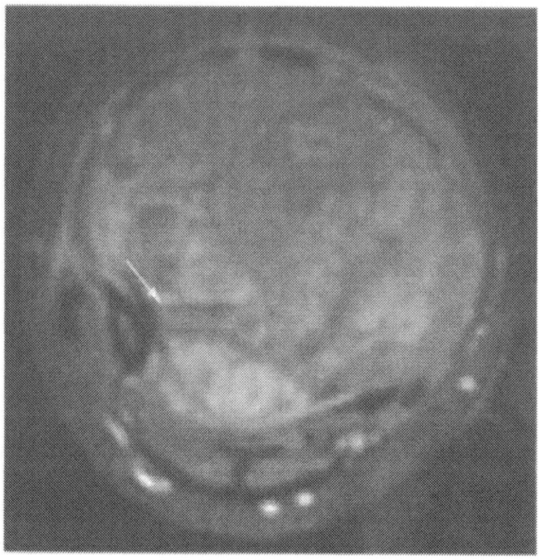

Figure 53P. Axial T1 post-Gd FS.

Pitfalls and Pearls

1. A juxtacortical soft tissue mass is usually, *but not invariably*, associated with primary malignant bone tumors in children.
2. Healing after a pathologic fracture through a unicameral bone cyst, nonossifying fibroma, or fibrous dysplasia may mimic a malignant bone lesion because of the presence of juxtacortical soft tissue component (granulation tissue and callus).

References

1. Cottias P, Tomeno B, Anract P, Vinh TS, Forest M. Subacute osteomyelitis presenting as a bone tumour: A review of 21 cases. *Int Orthop* 1997; 21:243–248.
2. Fayad LM, Kamel IR, Kawamoto S, Bluemke DA, Frassica FJ, Fishman EK. Distinguishing stress fractures from pathologic fractures: A multimodality approach. *Skeletal Radiol* 2005; 34:245–259.
3. Jee WH, Choi KH, Choe BY, Park JM, Shinn KS. Fibrous dysplasia: MR imaging characteristics with radiopathologic correlation. *AJR Am J Roentgenol* 1996; 167:1523–1527.
4. Margau R, Babyn P, Cole W, Smith C, Lee F. MR imaging of simple bone cysts in children: Not so simple. *Pediatr Radiol* 2000; 30:551–557.

Case 54

History

This is a 16-year-old girl with right knee popping and instability.

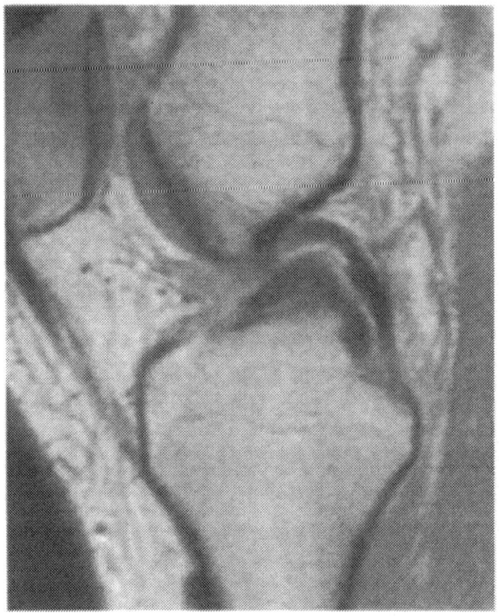

Figure 54A. Sagittal PD through the right medial knee.

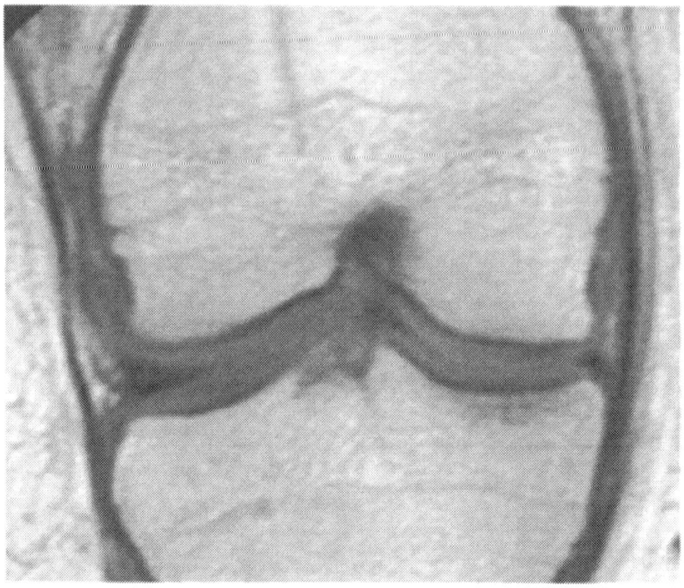

Figure 54B. Coronal T1.

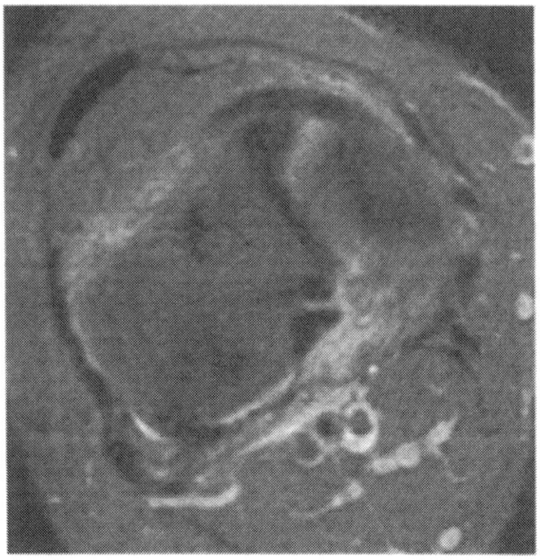

Figure 54C. Axial STIR through the meniscus.

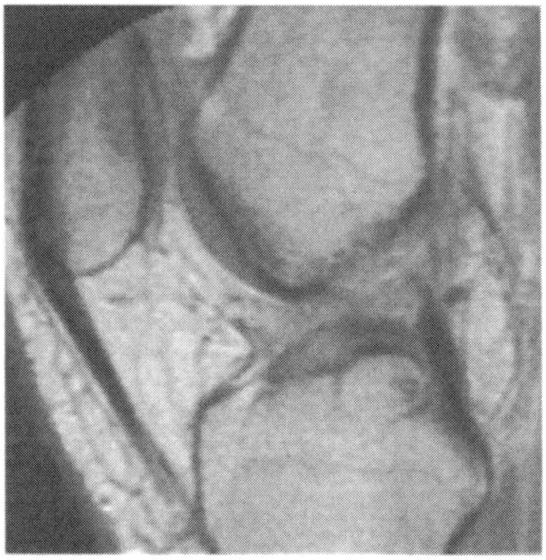

Figure 54D. Sagittal PD.

Figures 54A, (54A with annotations). A "double PCL" sign is present. The false PCL (*) is located anterior to the true PCL (P).

Figures 54B, (54B with annotations). The anterior horn of the medial meniscus is shortened, but maintains a triangular configuration **(arrowhead)**. A displaced meniscal fragment is present in the intercondylar notch **(black arrow)**.

Figures 54C, (54C with annotations). The torn inner edge of the medial meniscus is displaced medially (*).

Figures 54D, (54D with annotations). The ACL is also torn and the hypointense fibers are horizontally oriented within the intercondylar notch **(white arrow)**.

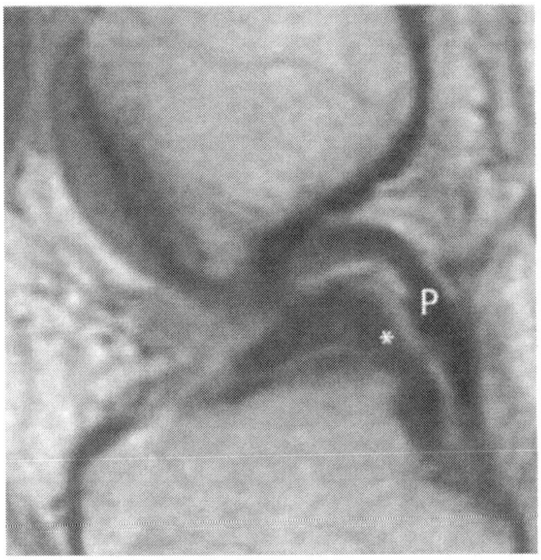

Figure 54A* Annotated.

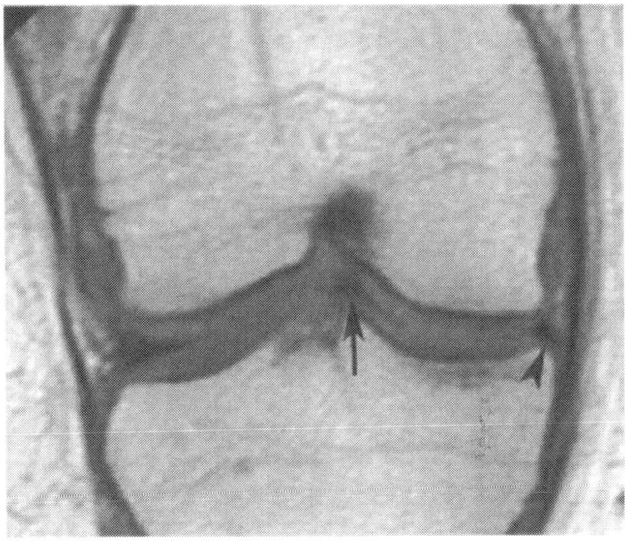

Figure 54B* Annotated.

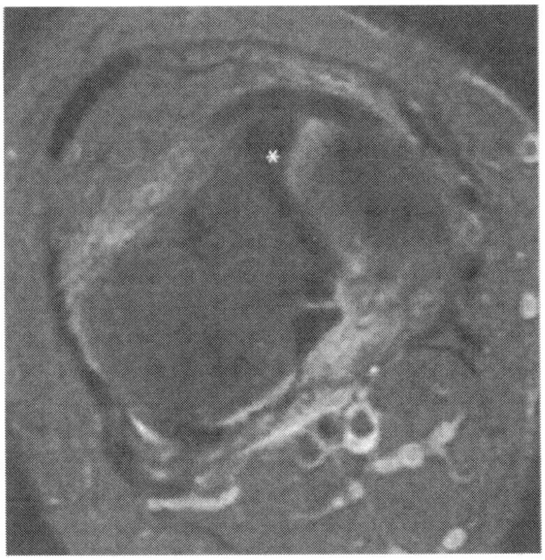

Figure 54C* Annotated.

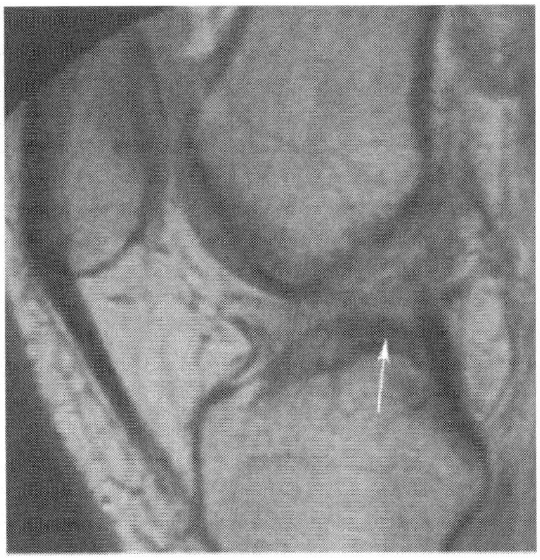

Figure 54D* Annotated.

Diagnosis

Bucket-handle medial meniscal tear and ACL tear

Questions

1. What are 5 MRI signs seen with bucket-handle tears?

Discussion

There are three types of displaced meniscal tears: bucket-handle tear, flap tear with displacement, and a free fragment with displacement (1). Of the three, the bucket-handle tear is most common, usually involving the medial meniscus. A bucket-handle tear represents a displaced longitudinal vertical tear of the inner edge of the meniscus, which remains attached anteriorly and posteriorly (Diagram 54A). The displaced fragment assumes the configuration of the handle of a bucket. A simultaneous tear of the ACL may be seen in 11% to 48% of cases (2).

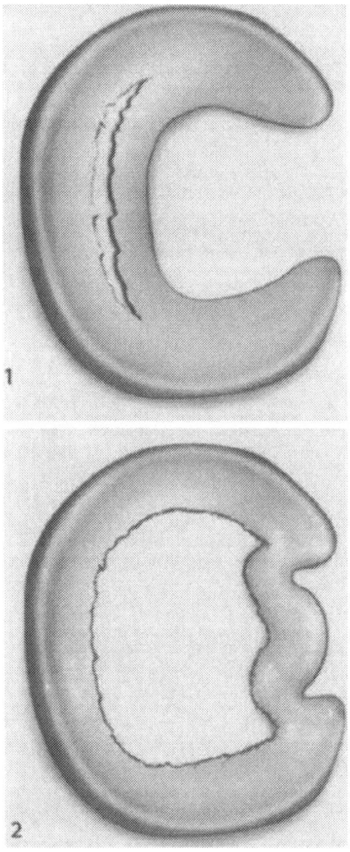

Diagram 54A. Bucket-handle tear. (1) Nondisplaced longitudinal vertical meniscal tear. (2) Longitudinal vertical meniscal tear with a displaced inner fragment (bucket-handle tear). (Reprinted from Dorsay TA et al. with permission from Springer (3).)

Five MRI imaging findings of a bucket-handle tear are: double PCL sign, intercondylar notch fragment sign, double anterior horn sign, a flipped meniscus sign, and absent bow-tie sign (Answer to Question 1) (1).

1. The double PCL sign is seen almost exclusively with medial meniscal tears. The displaced fragment locates anterior to the PCL (Figure 54A). This should not be confused with the normal posterior meniscal root insertion or with the meniscofemoral ligament of Humphrey.
2. The intercondylar notch fragment sign is present when the displaced meniscal fragment is found within the intercondylar notch (Figure 54B). This should not be confused with the normal meniscal root.
3. The double anterior horn sign is present when the displaced meniscal fragment is found anteriorly in the ipsilateral knee compartment. The anterior meniscal horn is located anteriorly, and the second triangle is the displaced posterior horn fragment (Figures 54E, 54F). The transverse meniscal ligament may be prominent and closely apposed to the normal anterior horn of the menisci. This normal finding should not be mistaken for the double anterior horn sign.
4. The flipped meniscus sign is present when the displaced meniscal fragment is found cephalad to the ipsilateral meniscal horn, usually the anterior horn. This increases the height of the meniscus. This should be differentiated from a tall, asymmetric discoid meniscus with degeneration.
5. The absent bow-tie sign is positive when the normal bow tie configuration of the meniscus is absent, or is present on only one sagittal image. This may be falsely positive after prior meniscectomy, meniscal hypoplasia, and in very young children. This may be falsely negative when a bucket-handle tear occurs in a discoid meniscus.

This patient had a knee injury while playing basketball. The bucket-handle and ACL tears were confirmed at surgery. The bucket-handle tear fragment was reduced and secured with sutures and the ACL reconstruction was performed using the pes anserine tendon.

Orthopedic Perspective

The MRI diagnosis of a bucket-handle meniscus tear is important in cases of isolated meniscal injury and in cases associated with ACL injury. A bucket-handle meniscus tear can present as a locked knee, which is a relative surgical emergency. Arthroscopy is performed to confirm the diagnosis (Figure 54G), reduce the tear, and repair the meniscus. In cases of meniscal tear associated with ACL injury, some surgeons prefer to acutely repair the meniscus and then stage the ACL reconstruction approximately three months later. Bucket-handle meniscus tears usually encompass a large percentage of the meniscus tissue and occur in a zone with good vascularity. Both of these factors advocate for acute meniscal repair.

What the Clinician Needs to Know

1. Location and degree of displacement of bucket-handle tears.
2. The presence of coexisting ACL and osteochondral injuries.

Answer

1. Double PCL sign, intercondylar notch fragment sign, double anterior horn sign, a flipped meniscus sign, and absent bow-tie sign.

Additional Examples

Bucket-Handle Tear of the Lateral Meniscus—Double Anterior Horn Sign

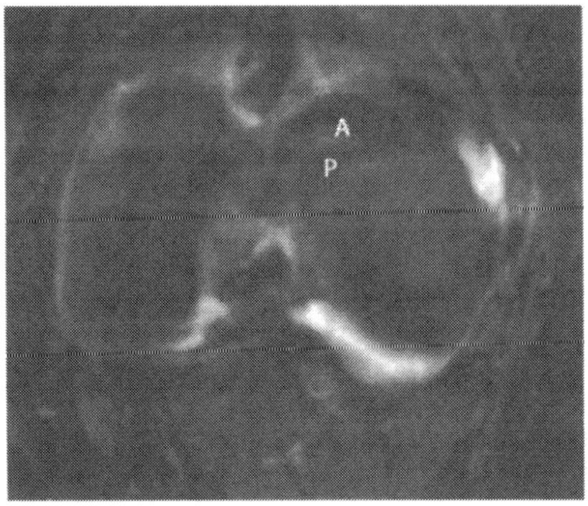

Figure 54E. Axial T2 FS.

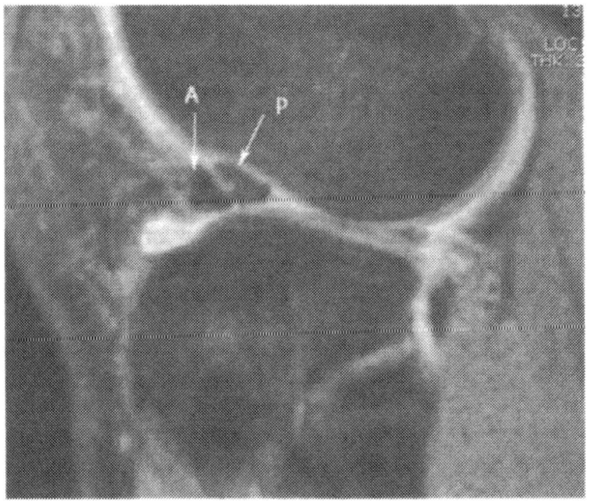

Figure 54F. Sagittal PD FS.

Findings

This is a 17-year-old girl who presented with persistent left knee pain after falling on her knee one month ago.

Figures 54E, 54F. Axial and sagittal images show a displaced posterior horn of the lateral meniscus. The displaced, inner fragment of the posterior horn (P) is located immediately posterior to the anterior horn (A). This is known as the double anterior horn sign. At surgery, this patient had bucket-handle tears of both menisci. The ACL fibers were intact but there was a proximal avulsion injury (not shown).

Bucket-Handle Tear, Arthroscopy

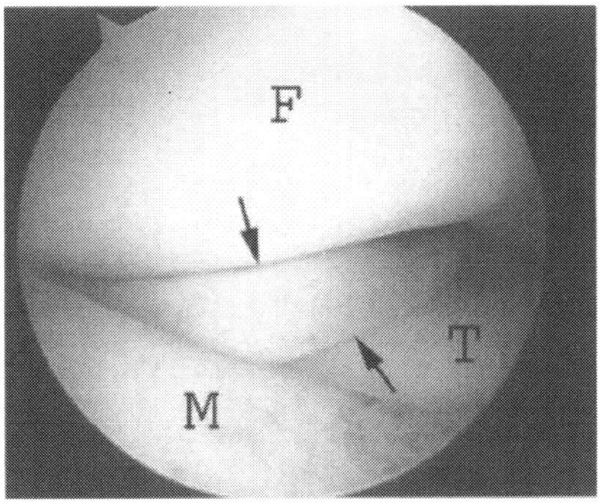

Figure 54G. Medial compartment.

Findings

Figure 54G. Arthroscopy demonstrates a medially displaced posterior horn medial meniscal fragment **(arrows)**. The periphery of the meniscus (M) is intact. Medial femoral condyle (F) and tibia (T).

Pitfalls and Pearls

1. Isolated ACL tears may be treated conservatively in children, whereas ACL tears associated with a meniscal tear are often treated surgically.
2. The absent bowtie sign does not apply to a discoid meniscus. A discoid meniscus may have two normal appearing bowties on contiguous sagittal images and still have a bucket-handle tear.
3. Do not confuse the posterior horn meniscal root insertion for an intercondylar notch sign.

References

1. Lieberman KA. The absent bow tie sign. *Radiology* 2000; 215:263–265.
2. Aydingoz U, Firat AK, Atay OA, Doral MN. MR imaging of meniscal bucket-handle tears: A review of signs and their relation to arthroscopic classification. *Eur Radiol* 2003; 13:618–625.
3. Dorsay TA, Helms CA. Bucket-handle meniscal tears of the knee: Sensitivity and specificity of MRI signs. *Skeletal Radiol* 2003; 32:266–272.

Case 55

History

This is an 11-year-old boy with back, left buttock, and knee pain.

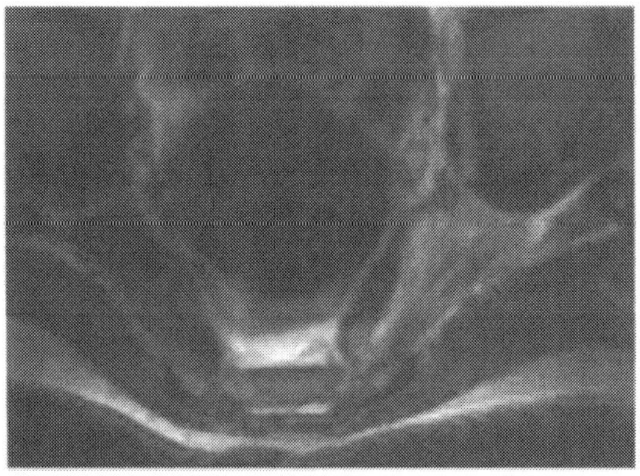

Figure 55A. Axial T2.

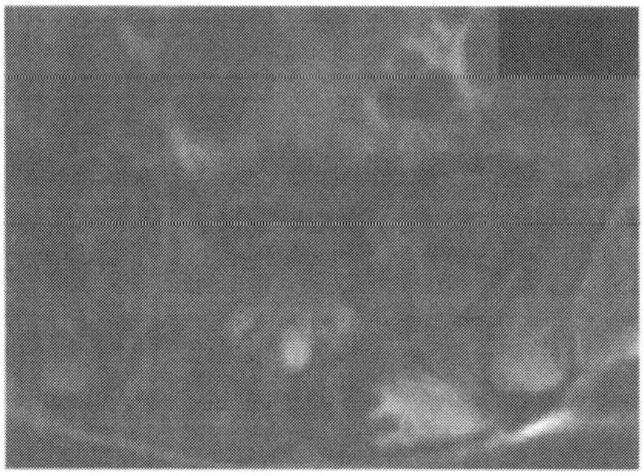

Figure 55B. Axial T2.

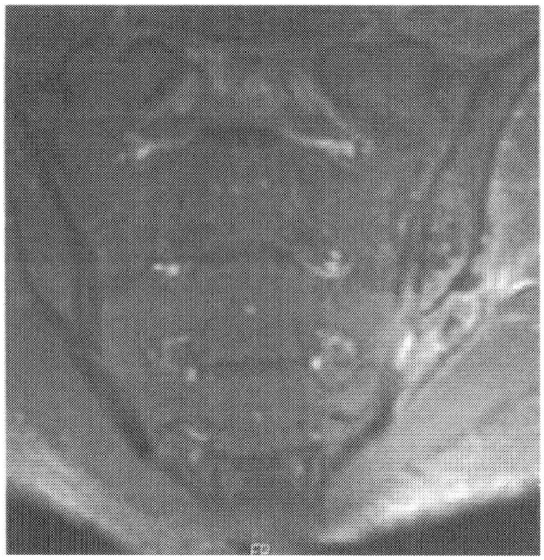

Figure 55C. Oblique coronal T1 post-Gd FS.

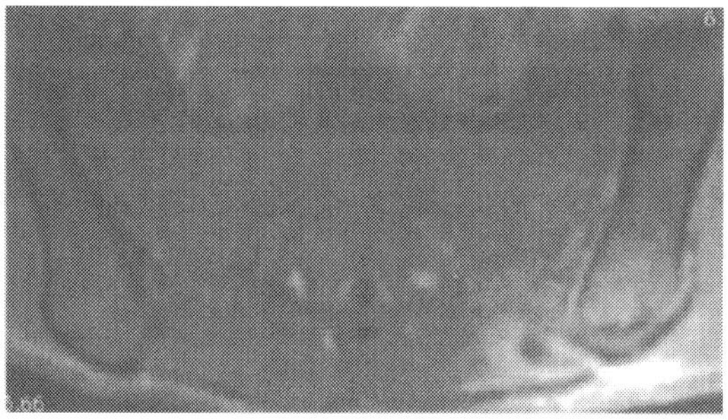

Figure 55D. Axial T1 post-Gd FS.

Figures 55A, 55B. There is increased SI within the left piriformis, erector spinae, and gluteus maximus muscles. In addition, there is increased SI within the metaphyseal equivalent region of the posterior inferior iliac spine (PIIS).

Figures 55C, 55D, (55C, 55D with annotations). Following gadolinium administration, there is abnormal enhancement of the metaphyseal equivalent (M), physis (P), and apophysis (A) of the left PIIS. The abnormal enhancement extends into the left SI joint, gluteus maximus muscle, and erector spinae muscle. Nonenhancing fluid collections are present in the physis of the PIIS **(thin arrow)** and left erector spinae muscle **(arrowhead)**. A posterior iliac subperiosteal fluid collection is also evident **(thick arrow)**.

Figure 55E. Note normal secondary ossification center of the PIIS apophysis **(arrow)**.

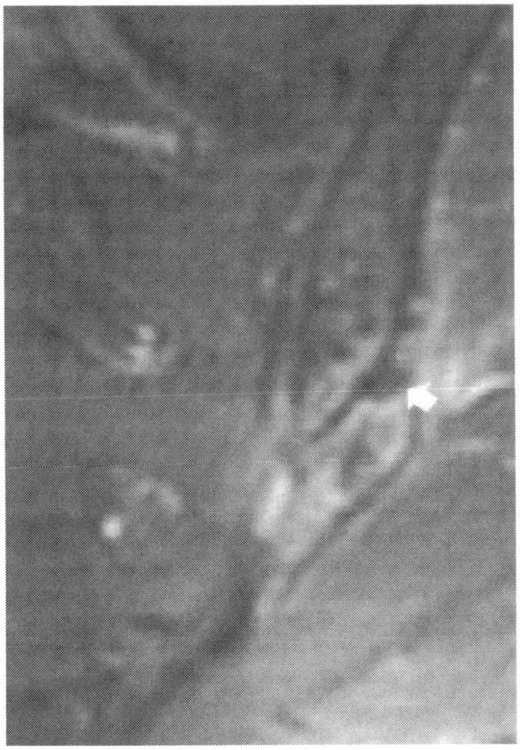

Figure 55C* Annotated.

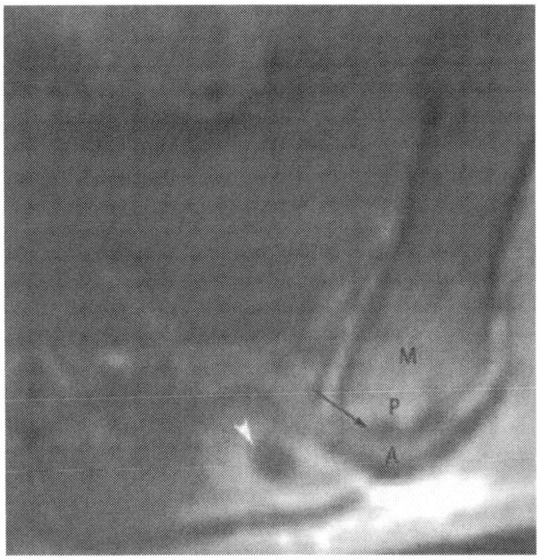

Figure 55D* Annotated detail.

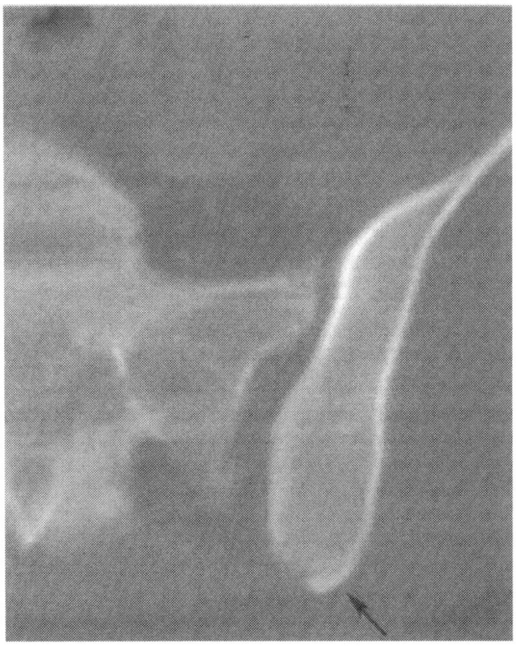

Figure 55E. Normal CT anatomy of the posterior inferior iliac spine (PIIS) in a 15-year-old male.

Diagnosis

Iliac osteomyelitis and sacroiliitis

Questions

1. What pelvic bone is most often affected by osteomyelitis?
2. What is the differential diagnosis for asymmetric sacroiliitis?

Discussion

The iliac bone is the most common location for pelvic osteomyelitis (Answer to Question 1) (1). It is the largest of the three pelvic bones and contains the greatest number of metaphyseal equivalent zones. In early childhood, the iliac crest and the anterior and posterior iliac spines form a broad based apophysis that undergoes secondary ossification during adolescence. In preadolescent patients, the separate cartilaginous apophyses are well shown by MRI, and with ossification, the various apophyses become visible by CT (Figure 55E).

The pathogenesis of iliac osteomyelitis mirrors the development of osteomyelitis in the long bones in the immature skeleton. Infection begins in the metaphyseal equivalent region and can secondarily involve the physis, apophysis, and adjacent muscles. As seen in this case, juxta-articular osteomyelitis can also lead to secondary sacroiliitis. A similar process can be seen in the acetabulum leading to a septic hip (Figure 55F) and the pubic symphysis (Figure 55G). The most common organism isolated in cases of pelvic osteomyelitis is *Staphylococcus aureus* (1).

Alternative diagnoses for nonpyogenic sacroiliac joint inflammation include juvenile rheumatoid arthritis (JRA), seronegative spondyloarthropathies, Lyme arthritis, post-immunization arthritis, and trauma (Answer to Question 2). The inflammatory changes related to nonpyogenic sacroiliitis usually begin in the joint. Therefore, the MRI findings include increased sacroiliac joint fluid, enhancing synovium, and juxta-articular marrow edema that may symmetrically affect both sides of the joint. Some authors have noted that it is difficult to distinguish between pyogenic and nonpyogenic sacroiliitis since both share similar sacroiliac joint inflammatory changes (2). However, when sacroiliac inflammatory changes result from adjacent osteomyelitis, the diagnosis of pyogenic sacroiliitis is usually straightforward. Typical features of osteomyelitis such as intraosseous abscess and cloaca formation may be present in the adjacent metaphyseal equivalent–apophyseal complex. In addition, soft tissue and anterior and posterior iliac subperiosteal fluid collections may be present (3). These osseous and soft tissue changes are usually not seen with noninfectious sacroiliitis.

This patient initially had fevers and left knee pain that spontaneously resolved. Two weeks later, he again had fevers and new pain in the left lower back, buttock, and along a sciatic nerve distribution. He initially underwent hip ultrasound and plain radiographs that were normal. Blood cultures grew *Staphylococcus aureus* and he was subsequently referred for MRI. The piriformis muscle involvement noted on this study (Figure 55A) correlated with his sciatic symptoms. He was treated conservatively with IV antibiotics and did well.

Orthopedic Perspective

Pelvic osteomyelitis can present with vague symptoms that are less well localized, such as back, hip, pelvic, and flank pain, generalized malaise, or even findings that are suspicious for appendicitis. Due to delay in diagnosis, cases may progress to more severe infections with pelvic abscesses, septic arthritis, bacteremia, septic emboli, and even systemic sepsis.

In addition to localizing the sites and extent of involvement, MRI can also guide treatment by identifying abscesses. Pelvic osteomyelitis is typically treated medically with antibiotics. However, cases with large (>1.5 cm) intraosseous, subperiosteal, or intramuscular abscesses (such as a psoas abscess) require drainage. If the abscess is located in an area of the pelvis that is difficult or morbid to reach surgically, CT-guided drainage is preferred. Small abscesses and sacroiliac joint involvement can be initially managed medically with antibiotics; however, if there is no improvement, drainage may be required.

Pelvic and spinal osteomyelitis generally have worse prognoses than appendicular osteomyelitis, likely related to delay in diagnosis and progression to advanced infection. Uncomplicated appendicular osteomyelitis is typically treated with intravenous antibiotics for 3 to 5 days until clinical and laboratory improvement is noted, followed by oral antibiotics for 3 to 4 weeks. On the other hand, pelvic and spinal osteomyelitis are treated with intravenous antibiotics for 3 to 4 weeks (see Case 30, osteomyelitis CPG). Follow-up MRI is indicated in cases without expected clinical and laboratory improvement, despite an appropriate intravenous antibiotic regimen.

What the Clinician Needs to Know

1. Distinguishing between pyogenic and nonpyogenic sacroiliitis.
2. The presence and extent of an abscess or phlegmon.
3. The presence of articular extension.

Answers

1. Iliac bone.
2. JRA, seronegative spondyloarthropathies, Lyme arthritis, post-immunization arthritis, and trauma.

Additional Examples

Triradiate Osteomyelitis and Septic Hip

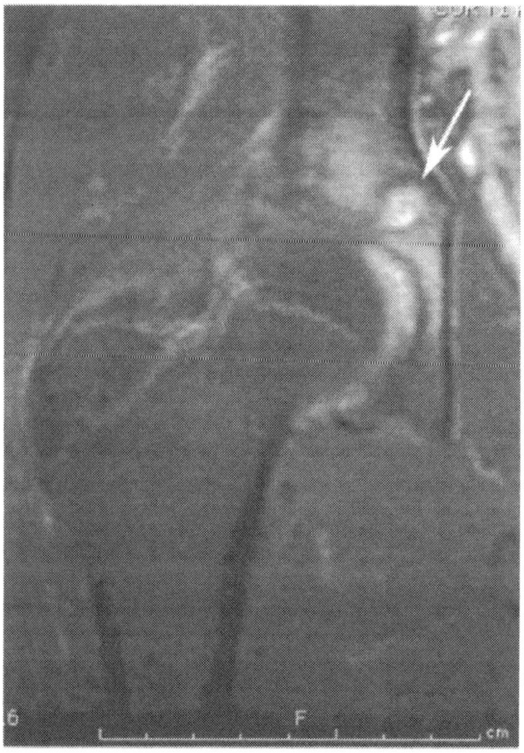

Figure 55F. Coronal T1 post-Gd FS of the right hip.

Findings

This is 14-year-old male with right hip pain and fever.

Figure 55F. There is abnormal enhancement of the triradiate cartilage **(arrow)**, iliac and ischial bone marrow, as well as synovial enhancement in the right hip. Blood cultures were positive for *Staphylococcus aureus* and he was treated with a 6-week course of IV cefazolin after a peripherally inserted central catheter (PICC) was placed.

Pubic Symphysis Osteomyelitis and Synovitis

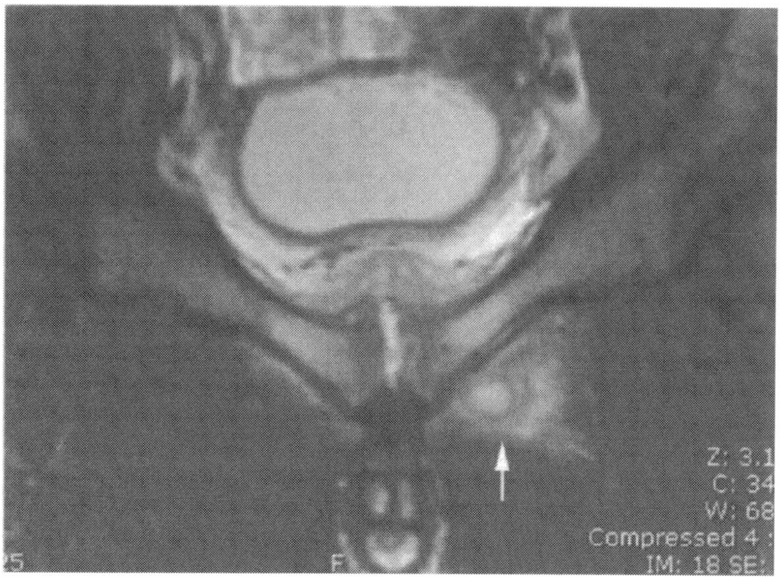

Figure 55G. Coronal STIR.

Findings

This is a 14-year-old male with fever and point tenderness over the left pubic bone.

Figure 55G. There is abnormal SI within the medial portions of the superior pubic rami and pubic symphysis. There is diffuse abnormal SI within the left adductor brevis muscle with a small focal fluid collection **(arrow)**. The pubic symphysis was subsequently biopsied under fluoroscopic guidance, and bone cultures were positive for *Staphylococcus aureus*. A PICC was placed and a six week course of IV oxacillin was given.

Pitfalls and Pearls

1. Keep in mind that a spondyloarthropathy, such as psoriatic arthritis, may occasionally manifest impressive bone marrow edema and enhancement along an inflamed sacroiliac joint, suggesting a pyogenic etiology.
2. Early MR imaging of pelvic osteomyelitis may show minimal osseous signal abnormality. Soft tissue edema may be the most conspicuous finding initially. Post-contrast images and early follow-up MRI may be required for definitive diagnosis.

References

1. Davidson D, Letts M, Khoshhal K. Pelvic osteomyelitis in children: A comparison of decades from 1980–1989 with 1990–2001. *J Pediatr Orthop* 2003; 23:514–521.
2. Haliloglu M, Kleiman MB, Siddiqui AR, Cohen MD. Osteomyelitis and pyogenic infection of the sacroiliac joint: MRI findings and review. *Pediatr Radiol* 1994; 24:333–335.
3. Sturzenbecher A, Braun J, Paris S, Biedermann T, Hamm B, Bollow M. MR imaging of septic sacroiliitis. *Skeletal Radiol* 2000; 29:439–446.

Case 56

History

This is a 5-year-old girl who fell onto her right elbow.

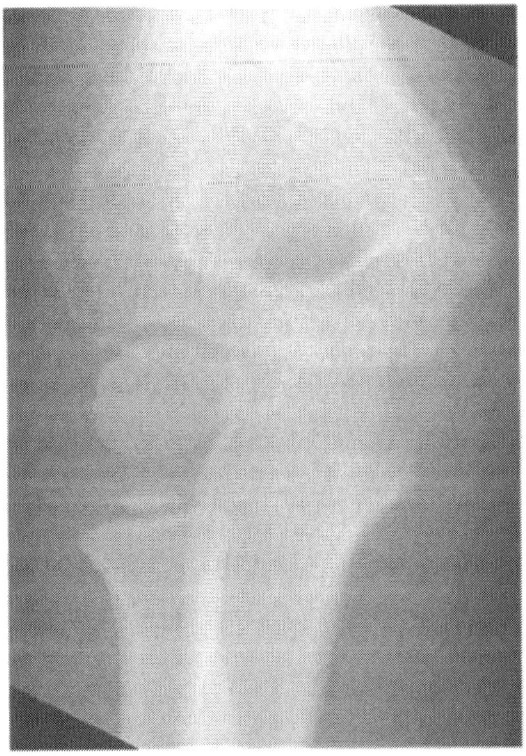

Figure 56A. AP radiograph of the right elbow.

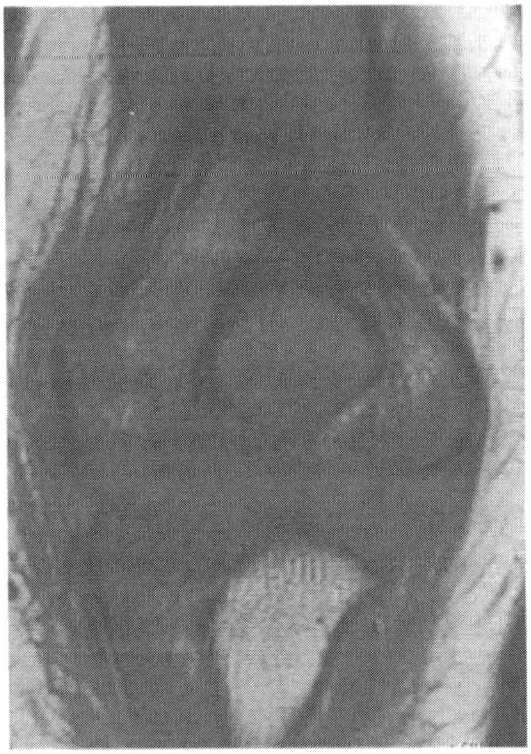

Figure 56B. Coronal T1.

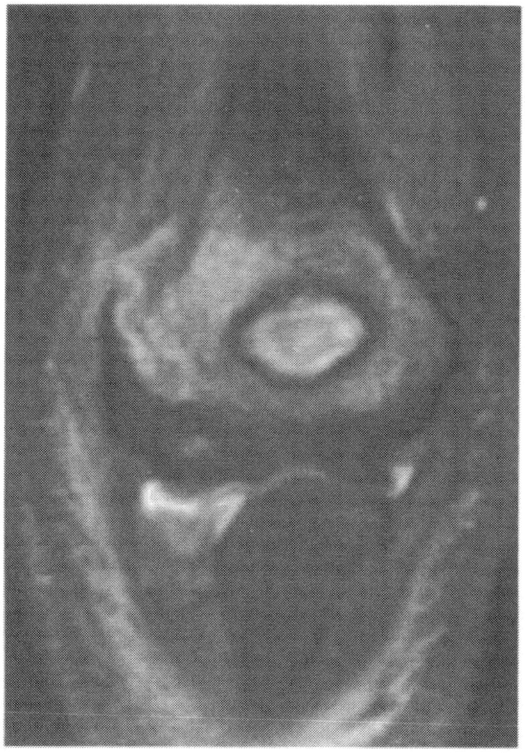

Figure 56C. Coronal STIR.

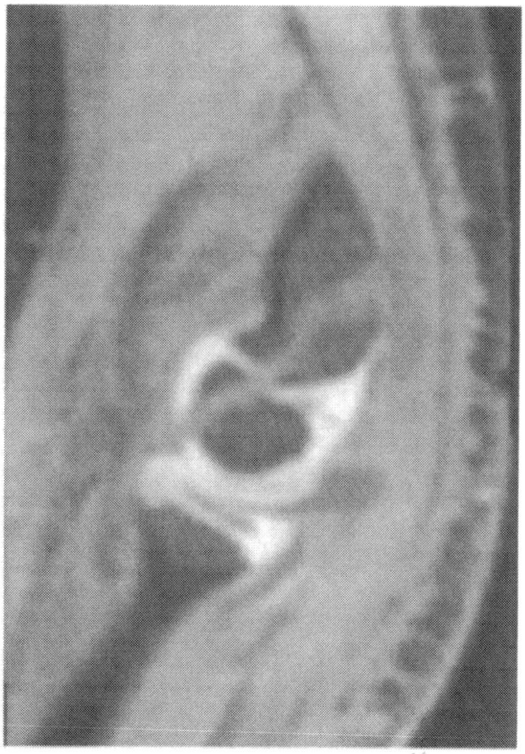

Figure 56D. Sagittal 3D SPGR FS.

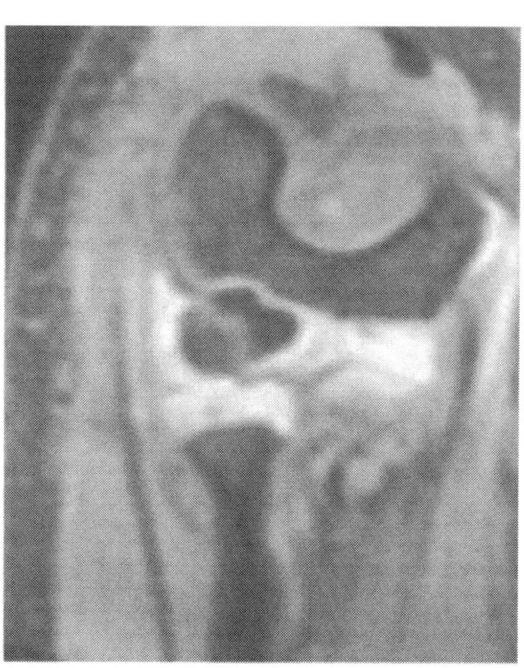

Figure 56E. Coronal 3D SPGR FS.

Figures 56A, (56A with annotations). There is a fracture through the lateral supracondylar region **(thick arrow)** extending to the distal humeral physis. There is the suggestion of fracture extension through the capitellum **(arrowheads).**

Figures 56B, 56C. There is diffuse T1 hypointensity and STIR hyperintensity in the distal humeral metaphysis. The lateral condylar fracture line extends to the physis, but epiphyseal extension of the fracture is not clearly evident.

Figures 56D, 56E (56D with annotations). With 3D reformatted images, the lateral condylar fracture is elegantly shown to extend from the lateral metaphysis **(arrowhead)** across the physis, splitting the capitellum and disrupting the articular surface **(thin arrow).**

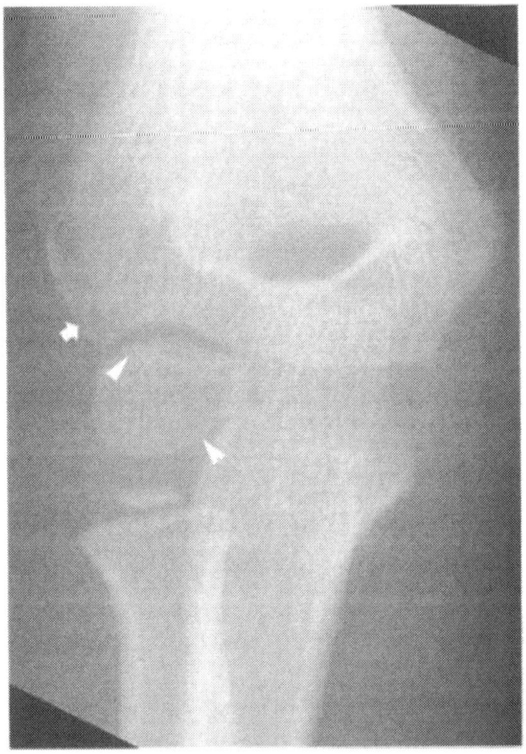

Figure 56A* Annotated.

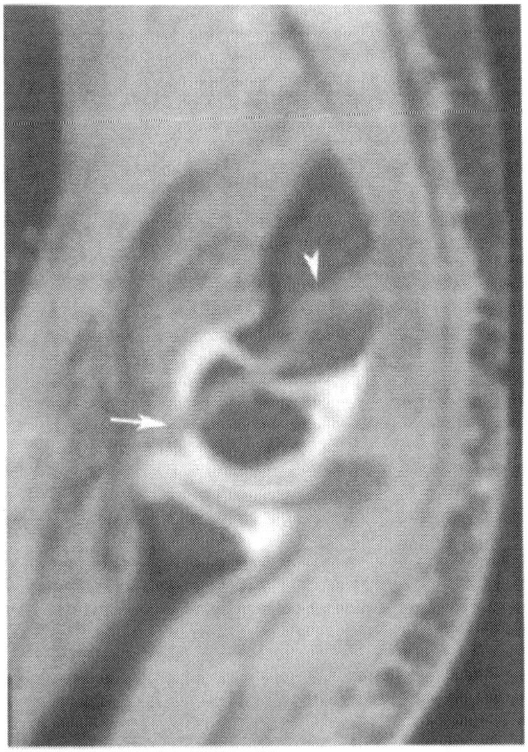

Figure 56D* Annotated.

Diagnosis

Lateral condylar fracture

Questions

1. Into which Salter-Harris (SH) fracture category do most lateral condylar fractures fall?
2. What amount of displacement defines an unstable lateral condylar fracture fragment?

Discussion

Lateral condylar fractures are the second most common type of distal humerus fracture in children. The most common pattern is supracondylar and the third most common is the medial epicondylar type (Diagram 56A) (1). Lateral condylar fractures usually occur with a fall on an outstretched hand leading to forced varus angulation when the forearm is supinated. These forces result in a fracture of the lateral aspect of the distal humerus metaphysis that usually propagates across the physis to involve a variable portion of the lateral condyle. In most instances, lateral condylar fractures are SH 4 injuries (Answer to Question 1) (2).

Lateral condylar fractures are considered unstable when the fragment is displaced by greater than 2 mm (Answer to Question 2) (3). Unstable fractures usually require surgical management whereas fractures with 2 mm or less displacement are considered stable and may be treated by casting. A lateral condylar fracture may also be considered unstable if the fracture extends to the articular surface (2).

The utility of MRI assessment of pediatric elbow fractures is unclear because some studies have suggested that the additional information changes management, while others indicate that it does not (4, 5). In a study of 12 patients, Kamegaya et al. showed that MRI may help determine which patients required pin fixation or closed reduction therapy by identifying unstable intra-articular fractures that were initially considered stable nonarticular fractures on plain radiography (6).

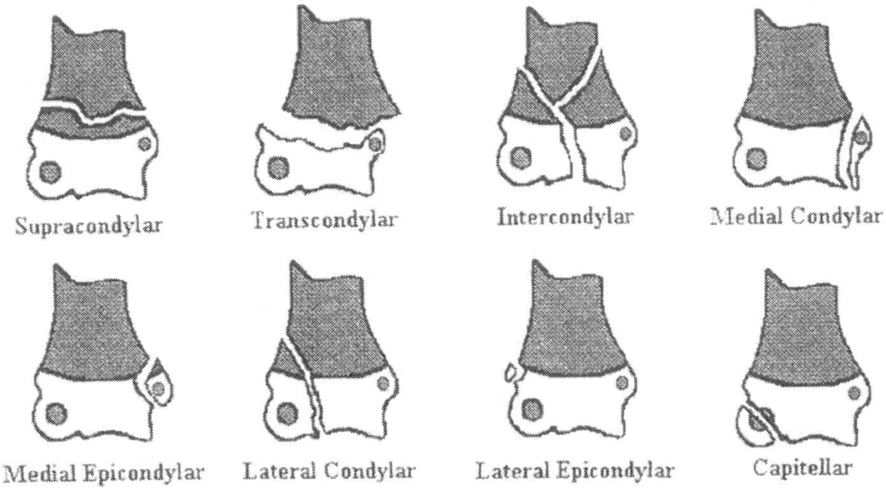

Diagram 56A. Distal humerus fractures. (Reprinted from Bettran J et al. with permission from Elsevier (7).)

MRI evaluation of lateral condylar fractures should include assessment for loose bodies, extent of fracture displacement, fracture extension to the capitellum and articular surface, and any additional fractures. Cartilage sensitive sequences such as GRE or PD with FS are essential. A 3D SPGR FS or equivatent sequence is particularly useful to assess the physis, cartilaginous epiphysis, and the articular cartilage. With healing fractures, intravenous gadolinium administration may be used to evaluate ununited lateral condylar fractures that have been treated nonoperatively for the presence of granulation tissue along the fracture line and osteonecrosis of the fracture fragment (Figures 56F–56H).

This patient was treated conservatively with casting. The fracture was considered stable since the fracture fragment showed less than 2 mm of distraction. A follow-up plain radiograph (not shown) showed healing and bony incorporation of the fracture fragment.

Orthopedic Perspective

Lateral condylar fractures in children are managed based on the extent of displacement. Nondisplaced fractures are treated with cast immobilization. Slightly displaced fractures are usually treated with percutaneous pinning because of the risk of further displacement. Significantly displaced fractures require open reduction and internal fixation. Displacement can usually be adequately assessed on plain radiographs, particularly the oblique view. However, in young patients with small lateral condylar flake fractures, MRI may be indicated to assess the amount of displacement, as well as articular surface involvement.

What the Clinician Needs to Know

1. The amount of displacement of the fracture in millimeters.
2. Does the fracture extend to the articular surface? Is the fracture a Salter-Harris type 2 or type 4 injury?
3. Are there any other fractures or ligamentous/tendinous injuries?

Answers

1. Salter-Harris type 4 fractures.
2. >2 mm.

Additional Example

Healing Lateral Condylar Fracture

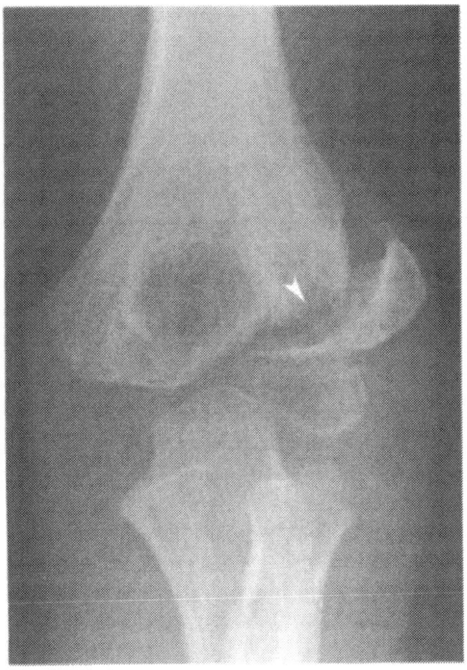

Figure 56F. AP radiograph of the left elbow.

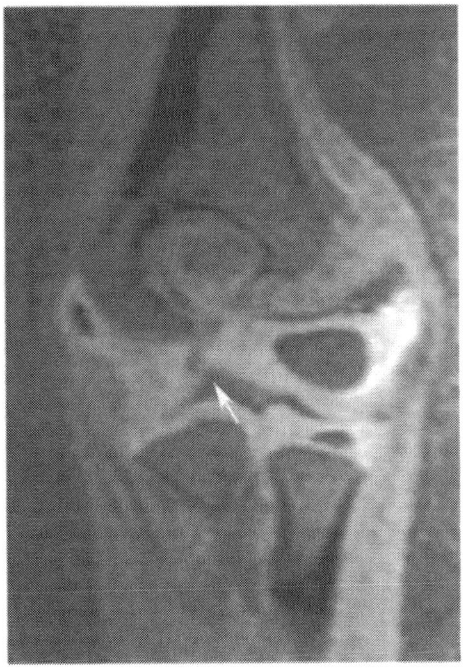

Figure 56G. Coronal 3D SPGR FS.

Findings

This 6-year-old boy fell and injured his left elbow.

Figure 56F. This lateral condylar fracture was initially treated conservatively with casting, but this radiograph at 1 month showed no bony union. Note periarticular osteopenia, diastatic fracture line **(arrowhead),** and relative sclerosis of the lateral condylar fracture margin.

Figure 56G. There is displacement and fracture extension through trochlea to the articular surface **(arrow).**

Figure 56H. Homogeneous enhancement of granulation tissue along the fracture line is evident **(arrowhead).** There is no evidence of osteonecrosis. Because the fracture fragment was approximately 4 mm distracted and there was articular incongruity the patient underwent open reduction.

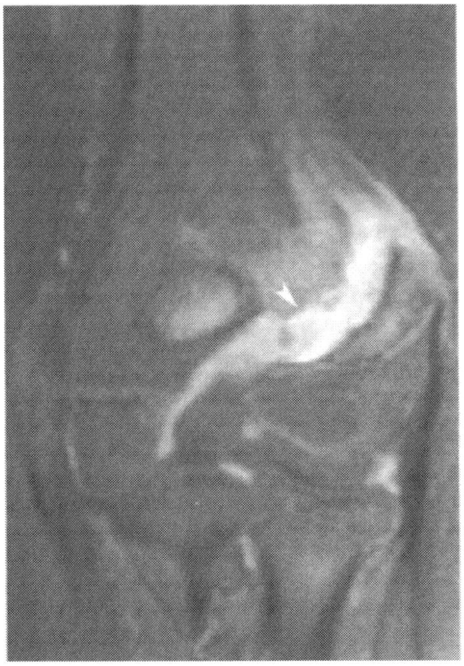

Figure 56H. Coronal T1 post-Gd FS.

Pitfalls and Pearls

Nondisplaced lateral condylar fractures are easily overlooked on radiographs, and MRI may be ordered unnecessarily. Plain radiographs should always be reviewed prior to the MRI.

References

1. Laor T, Jaramillo D, Oestreich AE. Musculoskeletal system. In: *Practical Pediatric Imaging: Diagnostic Radiology of Infants and Children*, 3rd ed., Chapter 5, Kirks DR, ed. Philadelphia: Lippincott-Raven, 1998: 327–510.
2. Rutherford A. Fractures of the lateral humeral condyle in children. *J Bone Joint Surg Am* 1985; 67:851–856.
3. Thonell S, Mortensson W, Thomasson B. Prediction of the stability of minimally displaced fractures of the lateral humeral condyle. *Acta Radiol* 1988; 29:367–370.
4. Beltran J, Rosenberg ZS, Kawelblum M, Montes L, Bergman AG, Strongwater A. Pediatric elbow fractures: MRI evaluation. *Skeletal Radiol* 1994; 23:277–281.
5. Griffith JF, Roebuck DJ, Cheng JC, et al. Acute elbow trauma in children: Spectrum of injury revealed by MR imaging not apparent on radiographs. *AJR Am J Roentgenol* 2001; 176:53–60.
6. Kamegaya M, Shinohara Y, Kurokawa M, Ogata S. Assessment of stability in children's minimally displaced lateral humeral condyle fracture by magnetic resonance imaging. *J Pediatr Orthop* 1999; 19:570–572.
7. Beltran J, Rosenberg ZS. MR imaging of pediatric elbow fractures. *Magn Reson Imaging Clin N Am* 1997; 5:567–578.

Case 57

History

This is a 7-year-old girl with a right popliteal fossa mass. Her right leg is 5 cm longer than the left.

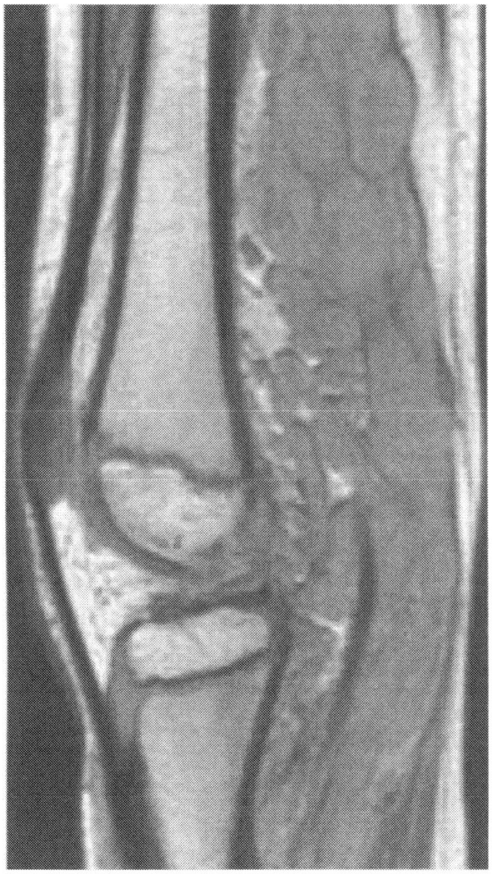

Figure 57A. Sagittal PD of the right knee.

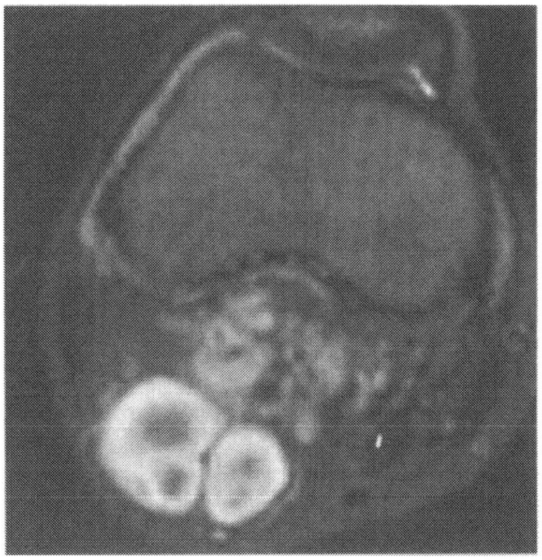

Figure 57B. Axial T2 FS.

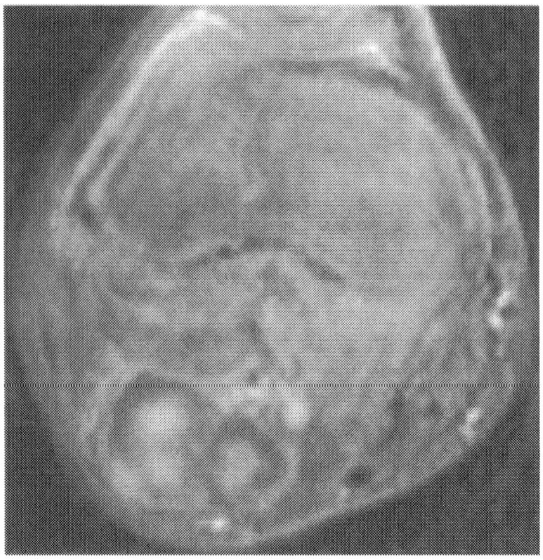

Figure 57C. Axial T1 post-Gd FS.

Figure 57A. There are multiple well-defined branching masses posterior to the knee with a "bag of worms" pattern. The masses are isointense with respect to adjacent muscle.

Figure 57B. On T2, these sharply marginated, round masses are relatively hypointense centrally and hyperintense peripherally (target sign).

Figure 57C. The corresponding post-Gd image shows prominent central enhancement.

Diagnosis

Plexiform neurofibroma

Questions

1. What are the three morphologic types of neurofibroma?
2. What is the explanation for the target sign?

Discussion

Neurofibromatosis (NF) I and II are distinct entities that are both autosomal dominant inherited. The incidence of NF I is approximately 1:3000–1:5000 and NF II is approximately 1:50,000 (1). The clinical manifestations of NF I include: two or more neurofibromas, optic glioma, Lisch nodules, osseous lesions, café au lait spots, and axillary or inguinal freckling. The clinical manifestations of NF II include: vestibular schwannomas (usually bilateral), meningiomas, cutaneous neurofibromas, and cataracts.

There are three morphologic types of neurofibroma: localized, diffuse, and plexiform (Answer to Question 1) (2). Ninety percent of patients with localized or diffuse neurofibromas do not have neurofibromatosis type I. In contrast, plexiform neurofibromas are virtually pathognomonic of neurofibromatosis type I (2). Localized and plexiform neurofibromas tend to be tubular in shape, and follow the course of the involved nerve. Localized neurofibromas are confined to a single nerve that is usually superficially located. Plexiform neurofibromas usually involve a major deep nerve and extend into its branches. Diffuse neurofibromas, on the other hand, are infiltrative lesions that are usually found within the subcutaneous soft tissues and a distinct relationship with a nerve or nerve branch is usually not identifiable by MRI.

In general, neurofibromas are hypo- to isointense on T1, hyperintense on fluid sensitive sequences, and demonstrate variable enhancement. The fat around the neurofibroma may be displaced creating the so-called split fat pad sign, a finding best appreciated on T1W sequences (Figure 57D) (3). A thickened nerve may be followed into and away from a neurofibroma (Figure 57E). The split fat pad sign is nonspecific and may also be seen in patients with schwannomas and malignant peripheral nerve sheath tumors (MPNST). The target sign may be seen with both localized and plexiform neurofibromas. On T2W sequences, the center of the neurofibroma is hypointense and the peripheral rim is hyperintense (Figure 57E). The center of the target sign enhances more than the peripheral rim (Figures 57B, 57C). Histologically, the center is composed of fibrous tissue and the hyperintense periphery is composed of densely cellular myxoid material (Answer to Question 2) (2). A bag of worms appearance is seen with plexiform neurofibromas and reflects involvement of a major nerve with nodular branch extension (Figures 57F, 57G). The fascicular sign describes multiple ring-like structures on T2W sequences that are sectioned axially through a peripheral nerve sheath tumor (neurofibroma, schwannoma, and MPNST).

Muscle atrophy may occur when an efferent nerve is involved by neurofibromas. On the other hand, plexiform neurofibromas of NF I may be a component of a diffuse mesodermal dysplasia that produces gigantism of an affected extremity, leading to elephantiasis neuromatosa (Figures 57H–57J). In patients with this pattern, diffuse mesodermal overgrowth occurs with extensive vascular malformations and underlying bone dysplasia in the affected limb (4). There are several other causes of limb overgrowth, which are outlined in Table 57A. Other skeletal manifestations of neurofibromatosis type I include kyphoscoliosis (most common), ribbon-shaped ribs, costovertebral

Table 57A. Causes of limb overgrowth (7–9).

Beckwith-Wiedeman syndrome: Diffuse limb overgrowth without underlying soft tissue abnormality with an association with solid organ neoplasms such as Wilms tumor.
Macrodystrophia lipomatosa: Diffuse soft tissue overgrowth, particularly involving the subcutaneous fat.
NF I: Diffuse mesodermal overgrowth and vascular malformations.
Klippel Trenaunay syndrome: Soft tissue overgrowth associated with varicose veins, deep venous anomalies, venous and lymphatic malformations, and dermal capillary-lymphatic vesicles.
Proteus syndrome: Overgrowth of vascular, connective, tissue, adipose, and neural tissue associated with low and high flow vascular malformations.
Bannayan-Riley-Ruvalcaba syndrome: Similar features with Proteus syndrome with the additional features of macrocephaly and gastrointestinal hamartomatous polyposis.
Diffuse lymphatic or venolymphatic malformations
Parkes-Weber syndrome: Arteriovenous malformations and arteriovenous fistulas associated with diffuse soft tissue overgrowth.

dislocation, rib notching, anterior tibia bowing, pseudoarthrosis, lambdoid suture defects, sphenoid wing dysplasia, multiple nonossifying fibromas, and posterior vertebral scalloping secondary to dural ectasia (Figure 57J) (5). A swan-neck deformity may occur associated with a severe S-shaped scoliosis.

In one study comparing 10 NF I patients with MPNST with 40 patients with NF I without malignancy, tumoral heterogeneity on MRI was consistently found with MPNST, whereas tumoral homogeneity was consistently seen with neurofibromas without evidence of malignancy (6). New focal and persistent pain or neuropathy, enlarging mass, and MRI features including tumor size greater than 5 cm and ill-defined tumoral margins are additional features suggesting malignancy (Figures 57K, 57L) (3). On the other hand, a well-defined tumor margin does not imply benignity, since MPNST may form a pseudocapsule.

The knee mass in this patient with known neurofibromatosis type I was compatible with a plexiform neurofibroma. Since there were no malignant characteristics, the lesion was not resected. Epiphysiodesis of the right lower extremity is planned after her right leg reaches an adult length in order to correct her leg length discrepancy.

Orthopedic Perspective

Neurofibromatosis, for the most part, is a nonsurgical disease. The orthopedist may at times need to treat limb-length discrepancies with epiphysiodeses or lengthening procedures and angular deformities of the extremities with osteotomies. NF patients may also develop significant scoliosis requiring spinal fusion and some patients require surgical management of pseudoarthrosis of the tibia. However, when the primary presenting symptom is a mass, the obvious concern is distinguishing benign schwannomas and neurofibromas from MPNST. The distinction of benign from malignant masses on MRI grounds alone is often difficult, and correlation with the clinical findings is important. If a nerve sheath tumor is painful, and the pain is constant and awakens the patient from sleep, it is concerning. If the mass has grown in size, especially if the change is recent and dramatic, it also raises the suspicion of malignancy. In most instances, the MRI will be the most crucial imaging tool to help distinguish benign from malignant disease, but it is not definitive in this regard. MRI is often able to reliably discern plexiform neurofibromas, as noted above, but when there is a solitary solid mass, it must be assessed for possible malignancy. PET scanning shows promise as a potential tool to distinguish a malignancy from the frequently multiple benign neurofibromas in patients with NF.

Biopsies are not always definitive. Needle biopsies and open biopsies suffer from problems with sampling error. If a diagnosis of malignancy is rendered, then the diagnosis is definitive, but a large, solid appearing mass that is histologically benign may have areas of malignant change elsewhere in the lesion, which are not detected until the lesion is excised. The challenge is to know when to excise and when to observe, because excision may lead to neurologic deficit. After establishing a diagnosis, the MRI information is useful in planning the surgical approach. For benign schwannomas, the orthopedist wants to know the relationship of the lesion to major nerves and some indication about whether the lesion can be separated from the nerve to preserve function. Usually schwannomas can be dissected away from the nerve so that the function can be preserved. Neurofibromas are not as well circumscribed and may be intimately attached or intertwined with the nerve. Decisions then need to be made relative to the merits of resecting the lesion versus preserving function.

Malignant lesions require a wide resection. In some centers, preoperative radiotherapy is used especially for large lesions in order to limit the likelihood of recurrence if the tumor is peeled away from adjacent neurovascular structures. MRI is helpful in elucidating the structures in the vicinity of the tumor. The surgeon needs to know if bone will need to be resected or if there is a plane between the tumor and the bone. Similarly, the extent of the soft tissue resection and the muscles that might need to be resected in whole or in part can be determined. MRI offers the best anatomic information and is extremely useful in planning surgical resections. It is also helpful in detecting recurrence over time. For large malignant schwannomas it is sometimes wise to obtain a baseline postoperative MRI (after waiting for the postoperative edema to dissipate) so that if a recurrence is suspected, it can be distinguished from postoperative change.

What the Clinician Needs to Know

1. What is the probability that the nerve sheath tumor is benign versus malignant?
2. What is the extent of the lesion in the soft tissues and its relationship to the bone?
3. Can the adjacent nerves and vessels be preserved while still getting a negative margin?
4. Is it a lesion that can be safely dissected away from a major peripheral nerve?
5. Is there a local recurrence?

Answers

1. Localized, diffuse, and plexiform.
2. This is seen in localized and plexiform neurofibromas on T2W sequences. The hypointense center is composed of fibrous tissue and the hyperintense periphery is composed of myxoid material.

Additional Examples

Localized Neurofibroma with Split Fat Pad Sign and Target Sign

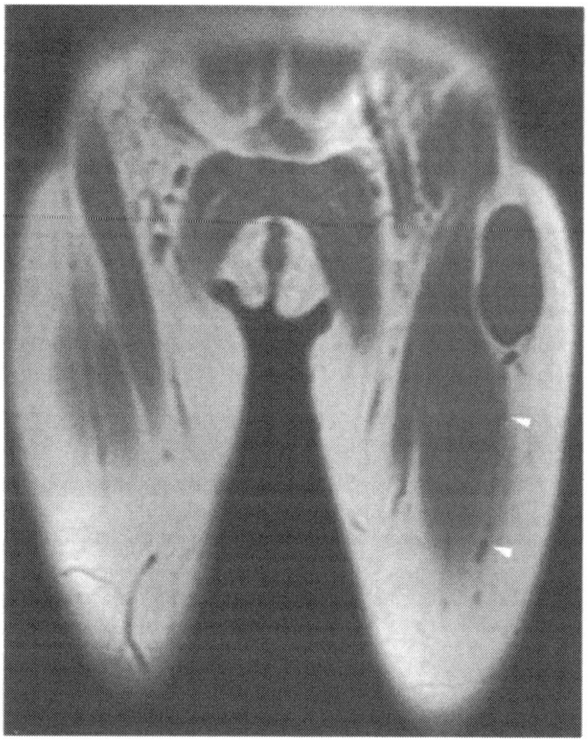

Figure 57D. Coronal T1.

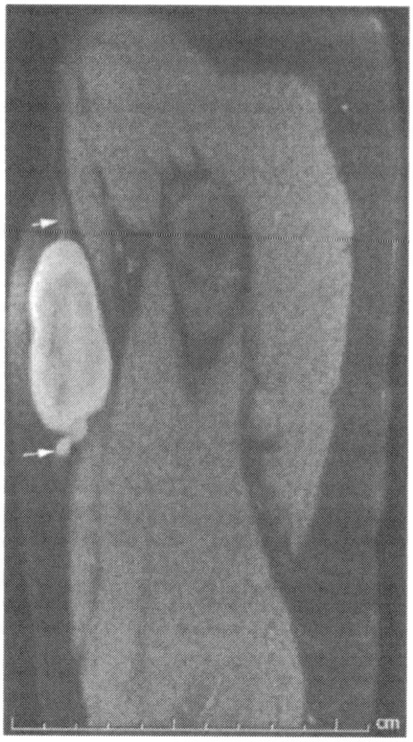

Figure 57E. Sagittal PD FS of the left thigh.

Findings

This is a 12-year-old girl with known NF I with an enlarging left anterior thigh mass.

Figure 57D. There is a large mass within the anterior thigh that displaces the adjacent fat (split fat pad sign). Multiple smaller neurofibromas are evident **(arrowheads).**

Figure 57E. The proximal and distal superficial nerve branches are identified with respect to the large neurofibroma **(arrows).** This mass has a hypointense center and hyperintense rim, compatible with the target sign.

Plexiform Neurofibroma Mimicking a Venous Malformation

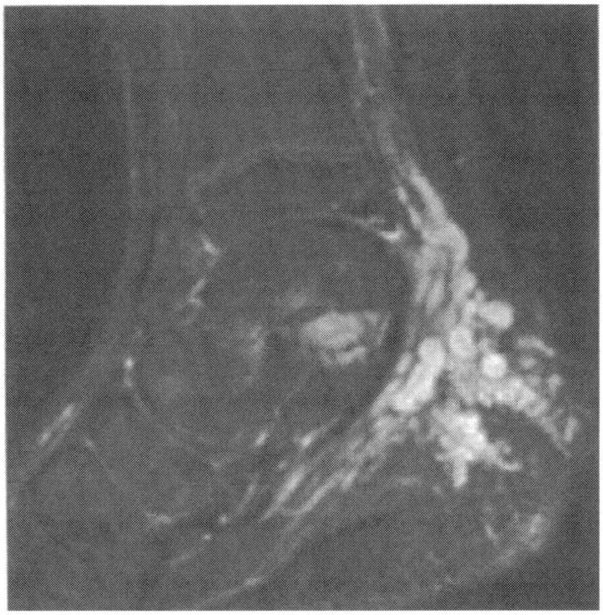

Figure 57F. Sagittal STIR.

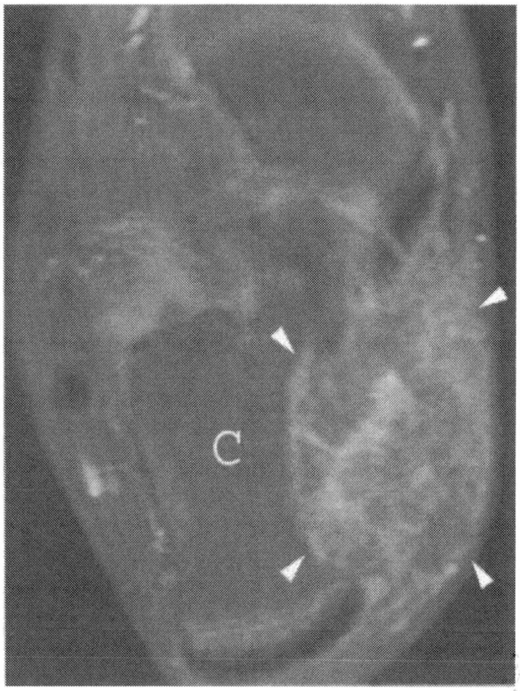

Figure 57G. Axial T1 post-Gd FS.

Findings

This is a 6-year-old boy with NF I and a posterior ankle mass.

Figure 57F. Branching hyperintense masses with a "bag of worms" appearance are identified along the posteromedial ankle. The lesion was initially thought to represent a venous malformation.

Figure 57G. Post-Gd image confirms that this is a plexiform soft tissue neoplasm rather than a vascular malformation **(arrowheads).** Axial image also demonstrates remodeling of the medial calcaneal body (C) due to the tumor.

Elephantiasis Neuromatosa and Dural Ectasia

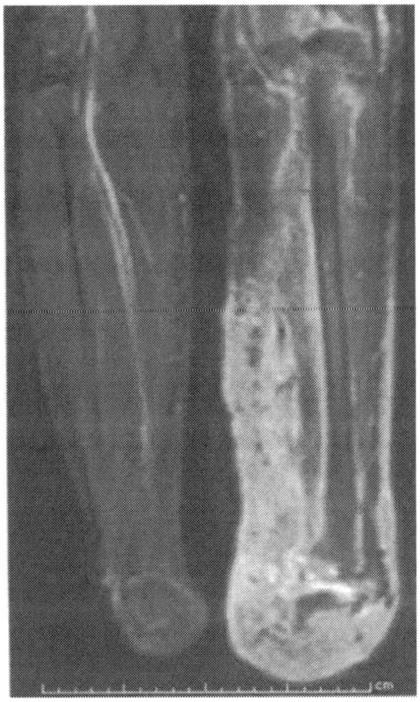

Figure 57H. Coronal STIR of the lower legs.

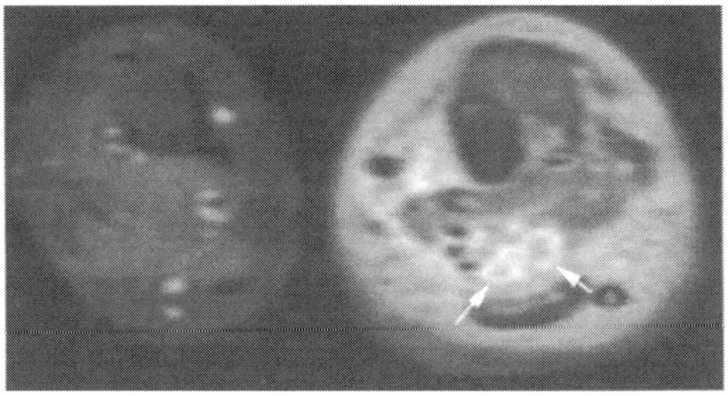

Figure 57I. Axial STIR.

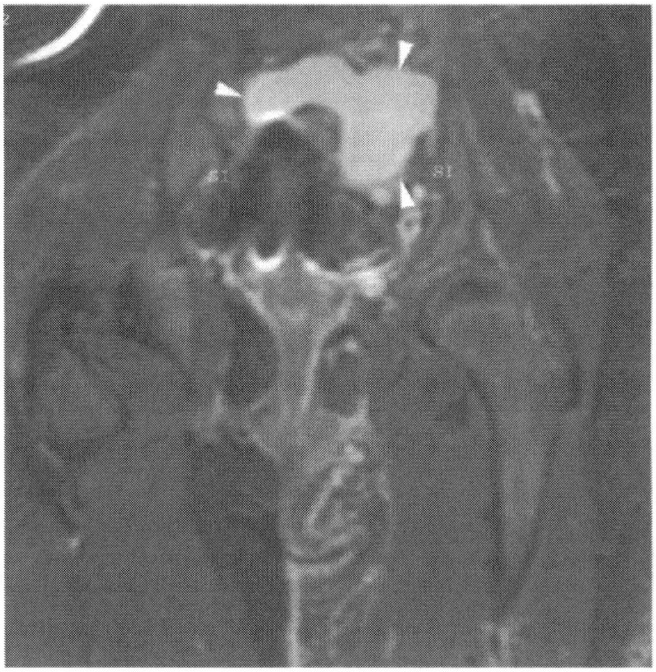

Figure 57J. Coronal STIR through the sacral region.

Findings

This is a 13-year-old boy with NF I.

Figures 57H, 57I. The left calf is markedly enlarged compared with the right. On the left, there is increased STIR SI and replacement of the normal subcutaneous fat and soft tissues. The overlying skin is diffusely thickened. The muscles on the left show mild diffuse increased STIR SI compared with the normal right side. The margins of the process are indistinct and multiple fascial planes are crossed. Within the mid-calf, there are round masses with a target sign **(arrows).**

Figure 57J. Dural ectasia and bony erosion is evident at the level of S1 **(arrowheads).** Sacroiliac joint (SI).

Neurofibromatosis I with Low Grade Malignant Peripheral Nerve Sheath Tumor (MPNST)

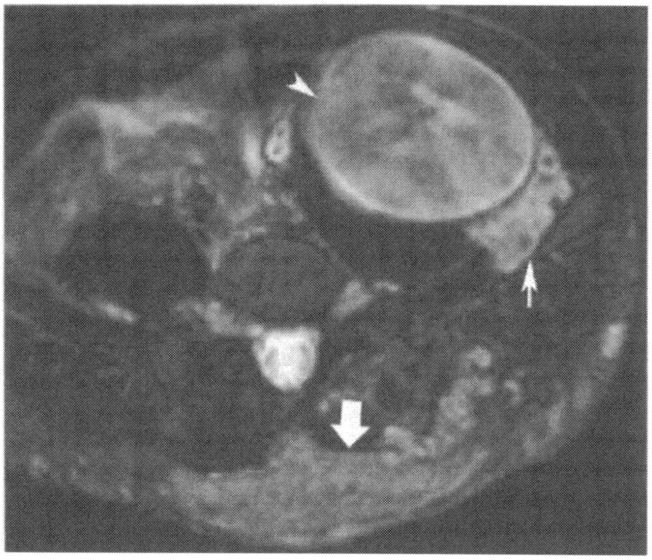

Figure 57K. Axial T2 FS.

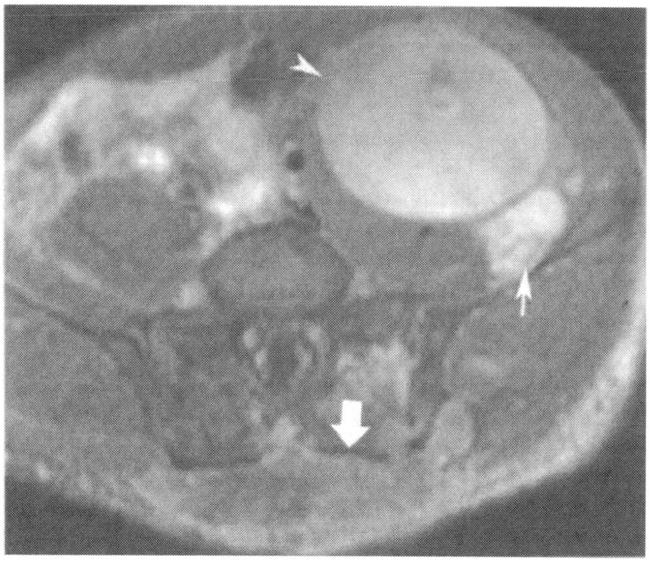

Figure 57L. Axial T1 post-Gd FS.

Findings

This is a 16-year-old girl boy with a known history of NF I.

Figures 57K, 57L. There is a >6 cm well-defined mass located within the extraperitoneal space **(arrowhead)** that is heterogeneously hyperintense on T2 and shows inhomogeneous enhancement. A conglomerate of plexiform neurofibromas with target signs is identified in the region of the femoral nerve **(thin arrows).** Note diffuse neurofibromas involving the subcutaneous tissues of the back **(thick arrows).** Because of its size, the mass was resected and found to represent a low grade MPNST.

Pitfalls and Pearls

1. The most important clinical sign of malignant degeneration of a neurofibroma is new focal pain or neuropathy.
2. The two risk factors for pediatric neurofibrosarcoma are prior radiation and a history of NF I (10).

References

1. Barkovich AJ. The phakomatoses. In: *Pediatric Neuroimaging*, 3rd ed. Philadelphia: Lippincott Williams & Wilkins, 2000; 383–442.
2. Murphey MD, Smith WS, Smith SE, Kransdorf MJ, Temple HT. From the archives of the AFIP. Imaging of musculoskeletal neurogenic tumors: Radiologic-pathologic correlation. *Radiographics* 1999; 19:1253–1280.
3. Lin J, Martel W. Cross-sectional imaging of peripheral nerve sheath tumors: Characteristic signs on CT, MR imaging, and sonography. *AJR Am J Roentgenol* 2001; 176:75–82.
4. Stevens KJ, Ludman CN, Sully L, Preston BJ. Magnetic resonance imaging of elephantiasis neuromatosa. *Skeletal Radiol* 1998; 27:696–701.
5. Bernauer TA, Mirowski GW, Caldemeyer KS. Neurofibromatosis type 1: Part II. Non-head and neck findings. *J Am Acad Dermatol* 2001; 44:1027–1029.
6. Mautner VF, Friedrich RE, von Deimling A, et al. Malignant peripheral nerve sheath tumours in neurofibromatosis type 1: MRI supports the diagnosis of malignant plexiform neurofibroma. *Neuroradiology* 2003; 45:618–625.
7. Gorlin RJ, Cohen MM, Jr., Condon LM, Burke BA. Bannayan-Riley-Ruvalcaba syndrome. *Am J Med Genet* 1992; 44:307–314.
8. Levine C. The imaging of body asymmetry and hemihypertrophy. *Crit Rev Diagn Imaging* 1990; 31:1–80.
9. Burrows PE, Laor T, Paltiel H, Robertson RL. Diagnostic imaging in the evaluation of vascular birthmarks. *Dermatol Clin* 1998; 16:455–488.
10. Neville H, Corpron C, Blakely ML, Andrassy R. Pediatric neurofibrosarcoma. *J Pediatr Surg* 2003; 38:343–346; discussion 343–346.

Case 58

History

This is a 10-year-old girl with new right hip pain.

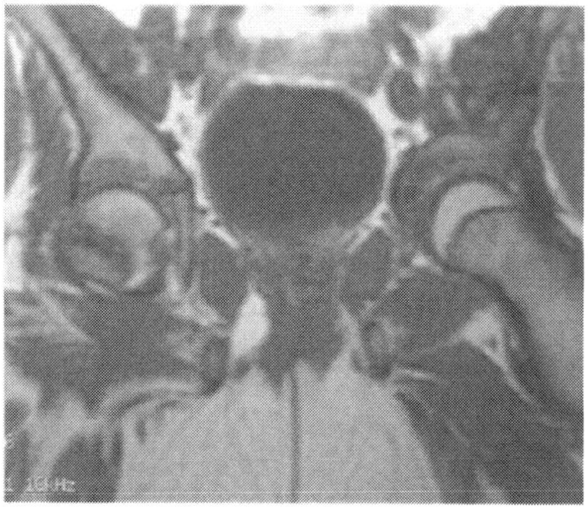

Figure 58A. Coronal T1.

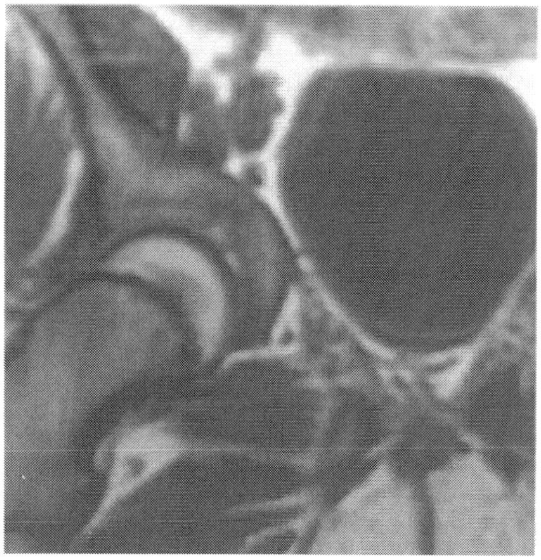

Figure 58B. Coronal T1.

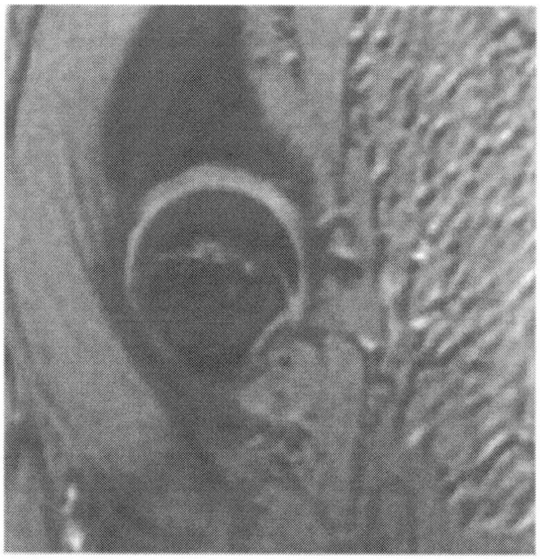

Figure 58C. Sagittal 3D SPGR FS, Right.

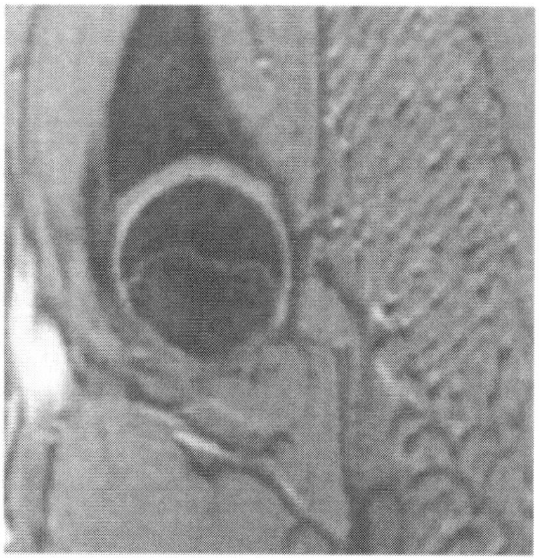

Figure 58D. Sagittal 3D SPGR FS, Left.

Figures 58A, 58B (58B with annotations). On the right, there is mild physeal widening and diminished SI in the epiphyseal marrow adjacent to the physis **(arrow).** The right capital epiphysis is slightly medially displaced.

Figures 58C, 58D. There is mild physeal irregularity and widening on the right. Compare the subtle right posterior epiphyseal displacement (Figure 58C) with the normal left hip (Figure 58D).

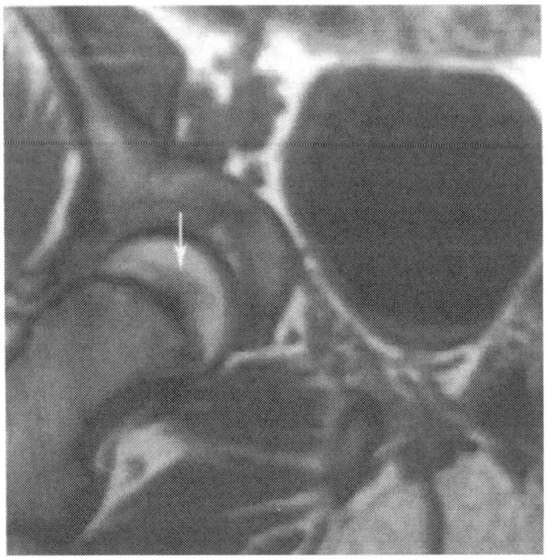

Figure 58B* Annotated.

Diagnosis

Slipped capital femoral epiphysis (SCFE)

Questions

1. T/F: The majority of surgically confirmed cases of SCFE have adjacent periphyseal marrow edema demonstrated on MRI.
2. What are some complications of SCFE?

Discussion

The etiology for slipped capital femoral epiphysis (SCFE) is controversial. Some authors have labeled SCFE as an epiphysiolysis, physiolysis, or a Salter-Harris type 1 injury (1, 2). The etiology for SCFE is probably a combination of delayed metaphyseal endochondral ossification and repetitive microtrauma predisposing to epiphyseal displacement. The risk factors for SCFE include: obesity, African descent, endocrinopathies such as hypothyroidism and growth hormone deficiency, metabolic disorders such as rickets and malnutrition, renal failure, radiation therapy, and developmental dysplasia of the hip (DDH) (3–5). In a large multicenter study, the average age for girls and boys was 12 years and 13.5 years, respectively, and was bilateral in 22% of patients (3). Underlying endocrinopathies or metabolic disorders should be strongly considered when SCFE occurs in children less than 10 years of age.

The early radiographic findings of SCFE include: widened femoral physis, small capital femoral epiphysis, localized osteopenia of the affected hip, and loss of Klein's line. Klein's line is an extension of the superior margin of the femoral neck. It should bisect roughly 20% of the capital femoral epiphysis on the AP projection. The direction of the epiphyseal displacement is most commonly posteromedially. Since the posterior component predominates, the slippage will be most apparent on the lateral projection, but will also be evident on the frontal view if the involved hip is held in external rotation. If the frog lateral view is limited, the diagnosis may be missed unless a true lateral projection is obtained. In such cases, the patient may be unnecessarily referred for MRI.

The value of MRI is to delineate SCFE in the preslip or early slip stage in those patients who are at highest risk (Figures 58E–58G) (6). Patients in the pre-slip stage present radiographs that are normal or show apparent physeal widening without epiphyseal displacement. In patients with obvious SCFE at presentation, MRI contributes little additional diagnostic information beyond that which is apparent on radiographs. Additional MRI features of SCFE include: focal or diffuse physeal widening, periphyseal edema, joint effusion, and synovitis. However, not all cases of SCFE manifest periphyseal marrow edema. In Uman et al.'s series, only 2 of 15 patients showed periphyseal marrow edema (Answer to Question 1) (7). The absence of periphyseal edema may indicate that SCFE is related to chronic physeal stress rather than an acute, posttraumatic epiphyseal separation.

T1 and PD and water sensitive sequences should be obtained to assess anatomy and bone marrow edema. Images should be obtained in a combination of coronal, sagittal, and axial planes to best assess the direction and extent of slippage. The axial oblique plane through the long axis of the femoral neck is particularly useful to identify subtle posterior slippage (Figure 58F).

MRI may also assess for the complications of SCFE and its treatment. These include: premature degenerative changes, labral tears related to CAM-type femoroacetabular impingement (FAI) (see Case 71), ischemic necrosis (Figures 58H, 58I), chondrolysis

(Figure 58J), and hardware failure. Patients may eventually develop coxa vara deformity, related to medial growth arrest (Answer to Question 2).

Based on the MRI findings of a preslip stage of SCFE, an in situ transphyseal epiphyseal pinning of the femoral head was performed.

Orthopedic Perspective

SCFE usually presents as hip, thigh, or knee pain in the obese child. SCFE is classified based on chronicity: acute (<3 weeks) versus chronic (>3 weeks); stability: stable (able to bear any weight) versus unstable (unable to bear any weight); and severity based on slip angle (mild, moderate, severe). The unstable SCFE is a better predictor of ischemic necrosis rather than the severity of the slip. A metabolic work-up is usually performed in atypical cases (e.g., very young patients, nonobese patients, or bilateral SCFE). Treatment is typically hip pinning in situ (both stable and unstable SCFE) to prevent further slippage and vascular disruption to the capital femoral epiphysis. After pinning, patients are usually kept 6 weeks in crutches. They are then allowed to weight bear the following 6 weeks but are not allowed to play sports. Chondrolysis and ischemic necrosis after SCFE are observed; these patients may eventually need hip fusion or total hip replacement.

What the Clinician Needs to Know

1. The diagnosis of SCFE in preslip cases or very mild slips.
2. The presence and extent of physeal edema.
3. Evidence of ischemic necrosis or chondrolysis.
4. Involvement of the opposite hip.

Answers

1. False.
2. Premature degenerative changes, labral tears related to CAM-type femoroacetabular impingement, ischemic necrosis, chondrolysis, and coxa vara.

Additional Examples

SCFE

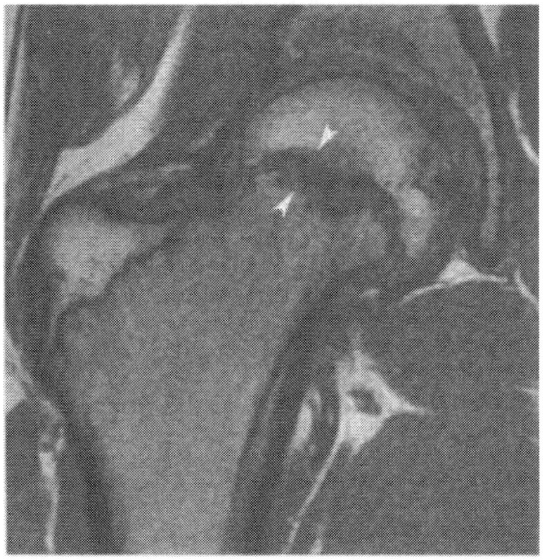

Figure 58E. Coronal T1 of the right hip.

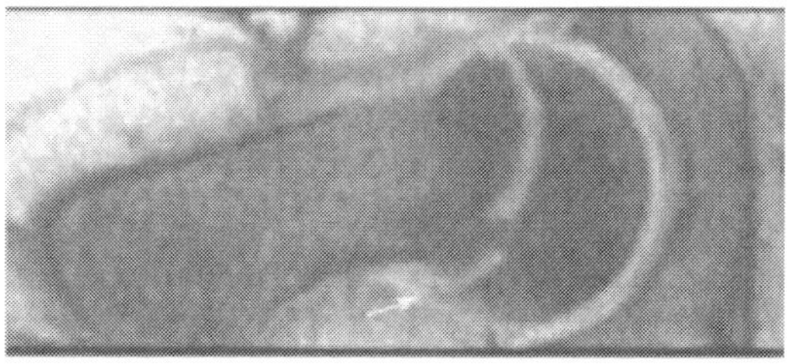

Figure 58F. Oblique-axial 3D SPGR FS.

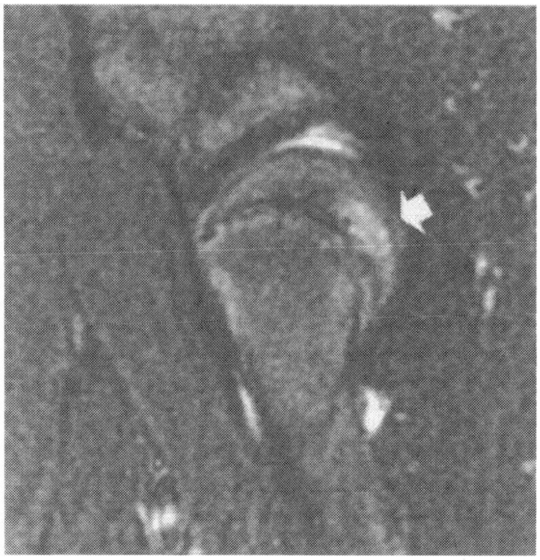

Figure 58G. Oblique-sagittal STIR.

Findings

This 13-year-old girl complained of right hip and knee pain. She had a history of a left SCFE that was pinned. Because she was at risk for right SCFE, and radiographs were considered normal (not shown), she was referred for MRI.

Figure 58E. An irregular widened physis is evident **(arrowheads)**.

Figure 58F. Subtle posterior epiphyseal displacement is seen with increased posterior femoral head neck offset **(thin arrow)** and diminished anterior offset.

Figure 58G. There is mild periphyseal marrow edema within the epiphysis **(thick arrow)**.

Ischemic Necrosis After SCFE

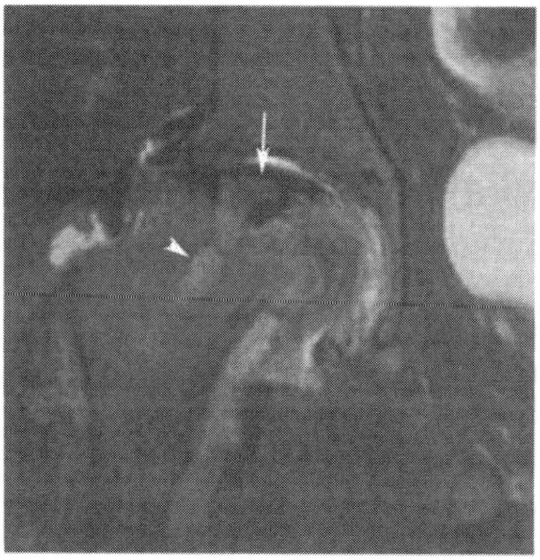

Figure 58H. Coronal T2 FS of the right hip.

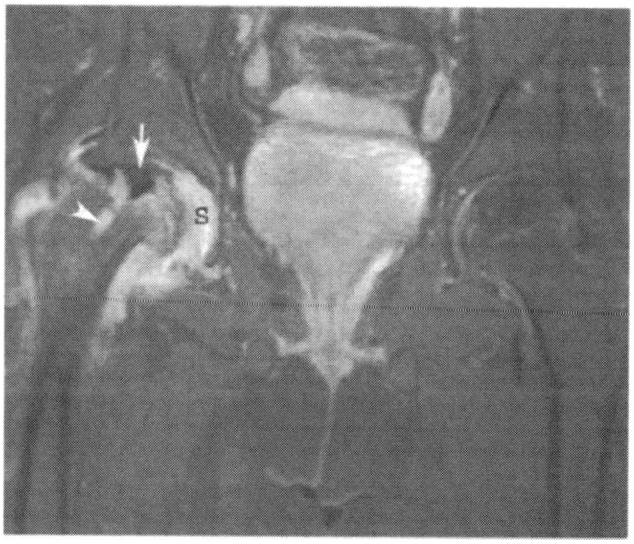

Figure 58I. Coronal T1 post-Gd FS.

Findings

This 13-year-old girl had SCFE and underwent transphyseal pinning in the past. The pin was removed 3 months previously.

Figures 58H, 58I. The femoral head is flattened and offset medially. There is markedly decreased SI in the lateral half of the femoral epiphysis with corresponding non-perfusion following gadolinium **(arrows)**. The remaining viable epiphysis shows intense enhancement. There is marked synovial enhancement (S) consistent with synovitis, but there were no clinical or laboratory signs of infectious arthritis. Note ghost track enhancement **(arrowheads)**.

Chondrolysis and Premature Degenerative Changes After SCFE

Figure 58J. Coronal T1 FS direct Gd arthrography.

Findings

This 16-year-old girl had persistent left hip pain after SCFE that was treated with pinning.

Figure 58J. The superolateral joint space distance measures less than 3 mm, suggesting chondrolysis (loss of articular cartilage) (5). There is mild increased T2 SI within the acetabular roof (A), with a subchondral cyst **(thin arrow)** reflecting early degenerative changes. Susceptibility artifact from prior transphyseal screw is present **(arrowhead)**. A rim of juxtaphyseal osteophytes is also present **(thick arrow)**, consistent with CAM-type FAI impingement (see Case 71).

Pitfalls and Pearls

1. If SCFE is identified in a preadolescent patient, consider metabolic causes such as rickets or pseudohypoparathyroidism.
2. There is a significant incidence of bilaterality in patients with SCFE. Image the opposite hip when the diagnosis of SCFE is initially made.
3. Do not perform MRI to diagnose SCFE until good quality radiographs are reviewed.

References

1. Kumar K. Should slipped capital femoral epiphysis (SCFE), a misnomer, be renamed as idiopathic capital femoral physiolysis (ICFP). *J Bone Joint Surg Br* 2002; 84:932.
2. Loder RT. Unstable slipped capital femoral epiphysis. *J Pediatr Orthop* 2001; 21:694–699.
3. Loder RT. The demographics of slipped capital femoral epiphysis: An international multicenter study. *Clin Orthop Relat Res* 1996; 8–27.
4. Loder RT, Wittenberg B, DeSilva G. Slipped capital femoral epiphysis associated with endocrine disorders. *J Pediatr Orthop* 1995; 15:349–356.

5. Boles CA, el-Khoury GY. Slipped capital femoral epiphysis. *Radiographics* 1997; 17:809–823.
6. Lalaji A, Umans H, Schneider R, Mintz D, Liebling MS, Haramati N. MRI features of confirmed "pre-slip" capital femoral epiphysis: A report of two cases. *Skeletal Radiol* 2002; 31:362–365.
7. Umans H, Liebling MS, Moy L, Haramati N, Macy NJ, Pritzker HA. Slipped capital femoral epiphysis: A physeal lesion diagnosed by MRI, with radiographic and CT correlation. *Skeletal Radiol* 1998; 27:139–144.

Case 59

History

This is a 1-day-old, 30-week gestational boy with a right arm and chest wall mass.

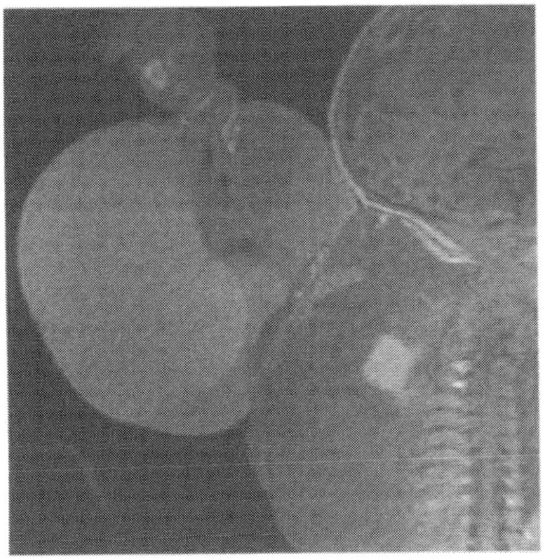

Figure 59A. Coronal T1 of the right arm.

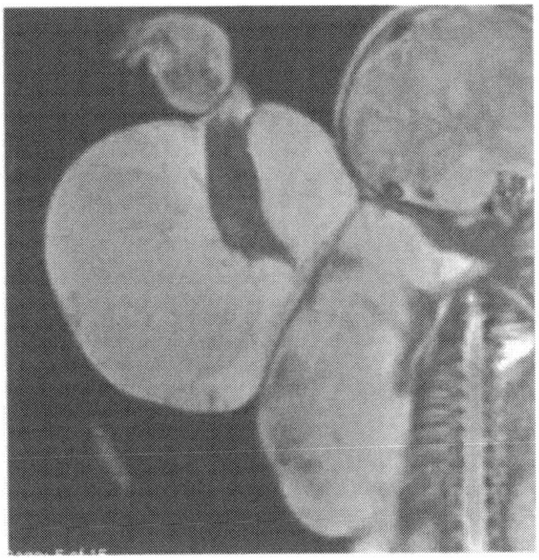

Figure 59B. Coronal STIR.

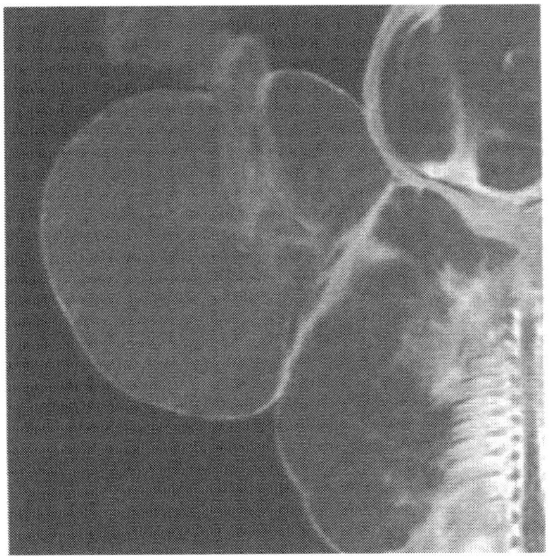

Figure 59C. Coronal T1 post-Gd FS.

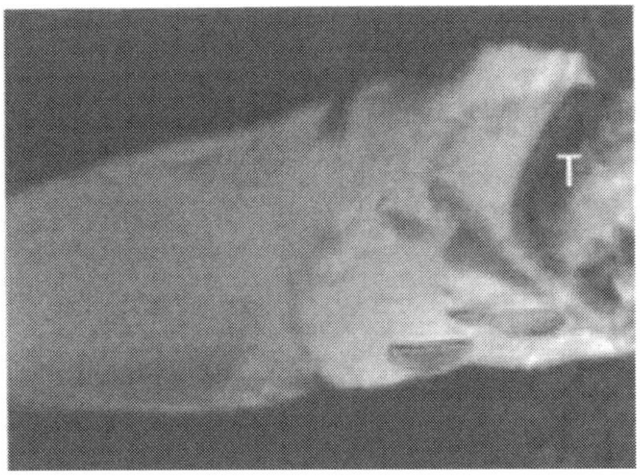

Figure 59D. Axial MPGR.

Figures 59A, 59B, (59B with annotations). A large septated mass arises from the right arm and right chest wall. The mass is of mixed SI, but is predominantly hypointense on T1 and hyperintense on STIR. A well-defined component located inferiorly shows intermediate SI on T1 and STIR **(arrowheads)**.

Figure 59C. There is mild rim and septal enhancement. No solid soft tissue component or high flow vessels are identified.

Figures 59D, (59D with annotations). There is blooming artifact with a layering effect within the dependent (posterior) aspect of this mass, consistent with blood products **(arrows)**. Thorax (T).

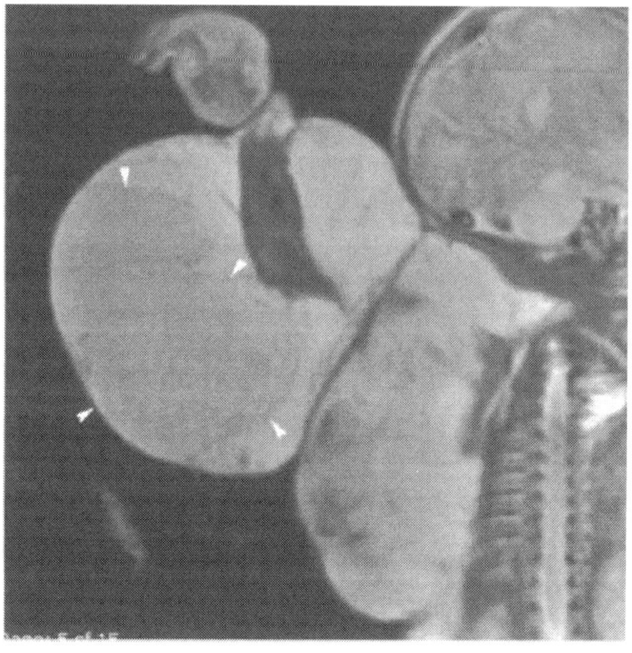

Figure 59B* Annotated.

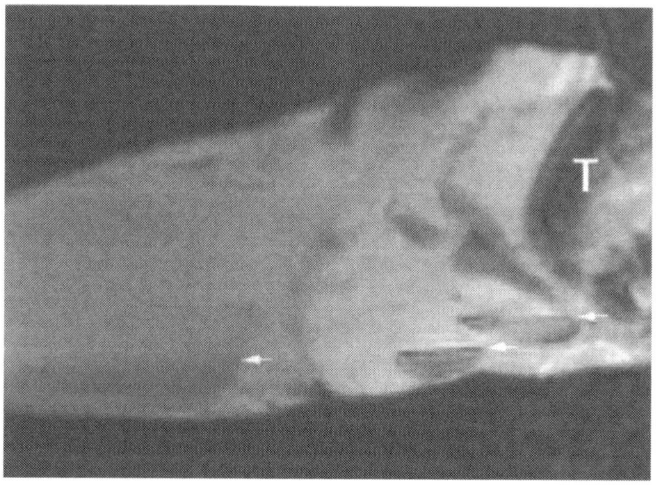

Figure 59D* Annotated.

Diagnosis

Macrocystic lymphatic malformation

Questions

1. What is the most common location for lymphatic malformations?
2. T/F: Microcystic lymphatic malformations may mimic soft tissue neoplasms.

Discussion

The most common location for lymphatic malformations is in the head and neck region followed by the axilla (Answer to Question 1) (1). The extremities and torso are less commonly affected. Lymphatic malformations may be classified as macrocystic or microcystic. The most common genetic condition associated with macrocystic lymphatic malformations is Turner's syndrome (45,X) (2). Other genetic disorders include, but are not limited to Down's and other trisomy syndromes, Noonan's, and Klinefelter's syndromes. They may also occur in generalized vascular malformation syndromes such as Klippel-Trenaunay, Proteus, and Bannayan-Riley-Ruvalcaba syndrome (3).

Macrocystic lymphatic malformations are usually hypointense on T1 and markedly hyperintense on fluid sensitive sequences. After intravenous gadolinium administration, septal and rim enhancement may be seen outlining large cysts of lymphatic fluid. Hemorrhagic components and mineralization may be seen within these lesions, particularly after trauma or infection. They are best depicted with gradient echo sequences and will demonstrate blooming artifact.

Microcystic lymphatic malformations are also hypointense on T1 and hyperintense on fluid sensitive sequences. With intravenous gadolinium, tiny confluent septa may enhance. Distinguishing capillary malformations from microcystic lymphatic malformations may be impossible both on MRI and histology (4). In addition, microcystic lymphatic and capillary malformations often coexist in the same region. Since the individual septa may not be resolved on MRI, microcystic lymphatic malformations may mimic a soft tissue neoplasm (Figures 59E–59G) (Answer to Question 2) (5).

Gorham disease is characterized by lymphatic malformation involving bone. The imaging manifestations of Gorham disease represent a spectrum, from permeative osteolysis (Figures 59E–59G) to vanishing bone (Figures 59H–59J). On MRI, diffuse intraosseous enhancement may be seen, representing a combination of vascular channels including microcystic lymphatic malformation and hypervascular fibrous tissue (6). Extraosseous extension into adjacent soft tissues is often seen (Figures 59E–59G). Diffuse lymphatic malformation may be considered a defect of mesenchymal cells affecting multiple compartments of a body region (4). This is in contrast to sarcomas, which represent an aggressive process that begins in one anatomic space and grows into adjacent compartments.

Alternative diagnoses for a large extremity mass in a newborn include infantile fibrosarcoma, infantile myofibromatosis, congenital hemangioma, hemangioendothelioma, and infantile rhabdomyosarcoma. These lesions are easily distinguished from macrocystic lymphatic malformations on clinical exam and on MRI. Unlike lymphatic malformations, these alterative diagnoses are generally characterized by a dense soft tissue mass that enhances after gadolinium, rather than large lymphatic cysts with septal and wall enhancement.

This patient's extensive upper extremity lymphatic malformation was initially thought to represent gastroschisis on prenatal ultrasound. The patient was delivered by c-section at 30 weeks gestation because of fetal distress. He had no other anomalies and had a normal karyotype. At four months of age, a radical debulking surgery of his lymphatic malformation was performed with good results.

What the Clinician Needs to Know

1. What is the size and extent (unicompartmental or multicompartmental) of the lymphatic malformation?
2. Is the lymphatic malformation predominantly macrocystic or microcystic? Which components of the lymphatic malformation appear amenable for sclerotherapy?
3. Is the lesion an isolated lymphatic malformation, or is it a mixed malformation with additional venous or capillary components?

Answers

1. Head and neck.
2. True.

Additional Examples

Gorham Disease

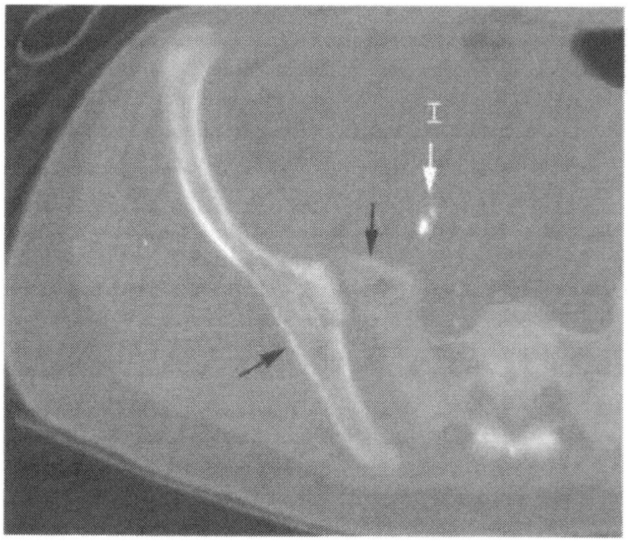

Figure 59E. CT of the right pelvis.

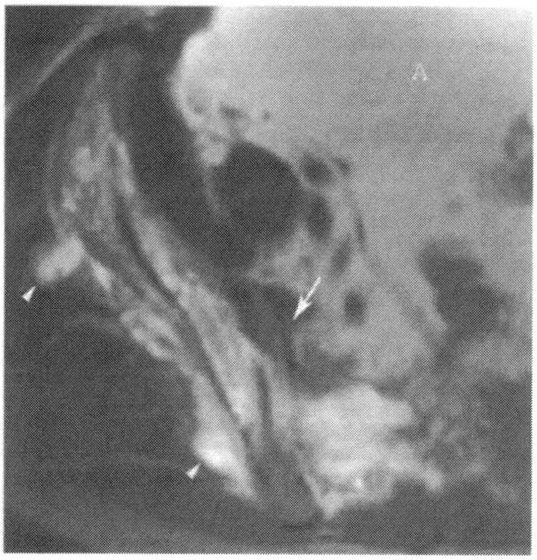

Figure 59F. Axial T2 FS.

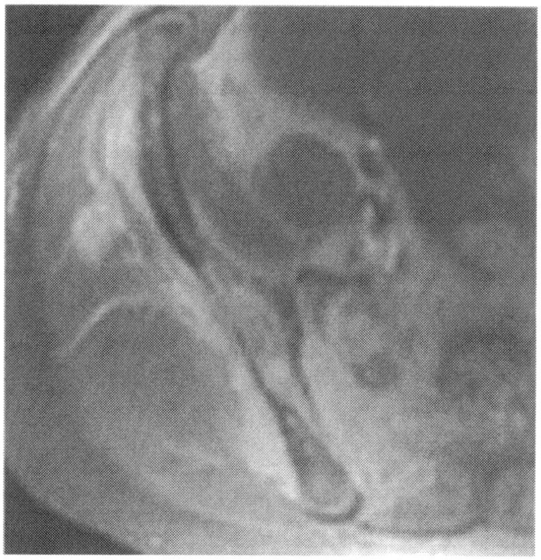

Figure 59G. Axial T1 post-Gd FS.

Findings

This is a 12-year-old girl with an extensive microcystic lymphatic malformation involving the right hemipelvis.

Figure 59E. Permeative destruction of the right ilium and sacrum is seen **(black arrows)**. Contrast in the deep iliac lymph node chains **(arrow, I)** is from prior lymphangiography.

Figures 59F, 59G. An extensive multicompartmental lesion is present that is hyperintense on T2 and shows diffuse enhancement. It involves the sacrum, right sacroiliac joint **(white arrow)**, ilium, and gluteal muscles **(arrowheads)**. In addition, diffuse extraperitoneal involvement anteromedial to the iliopsoas muscle is seen. Note massive ascites (A).

Gorham Disease

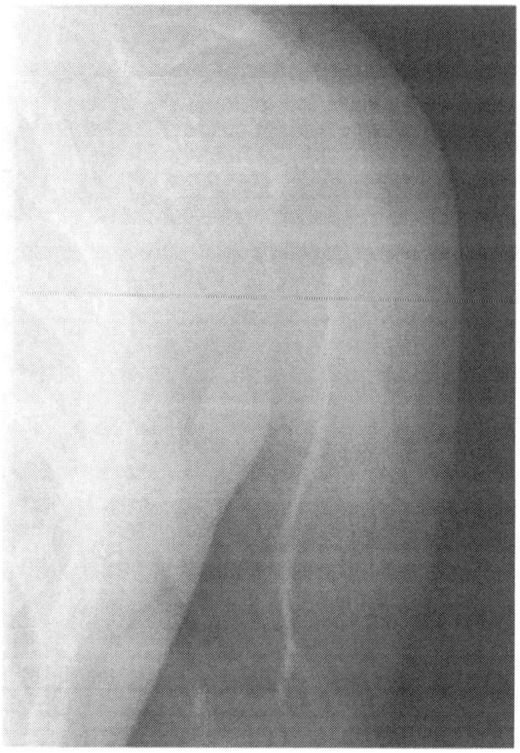

Figure 59H. AP radiograph of the left humerus.

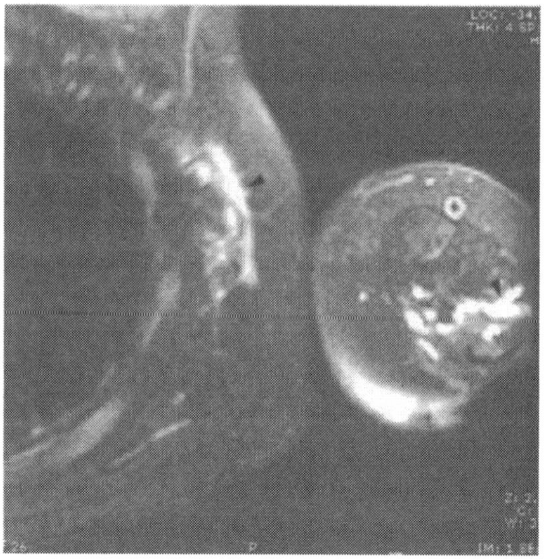

Figure 59I. Axial T2 FS through the mid-left humerus.

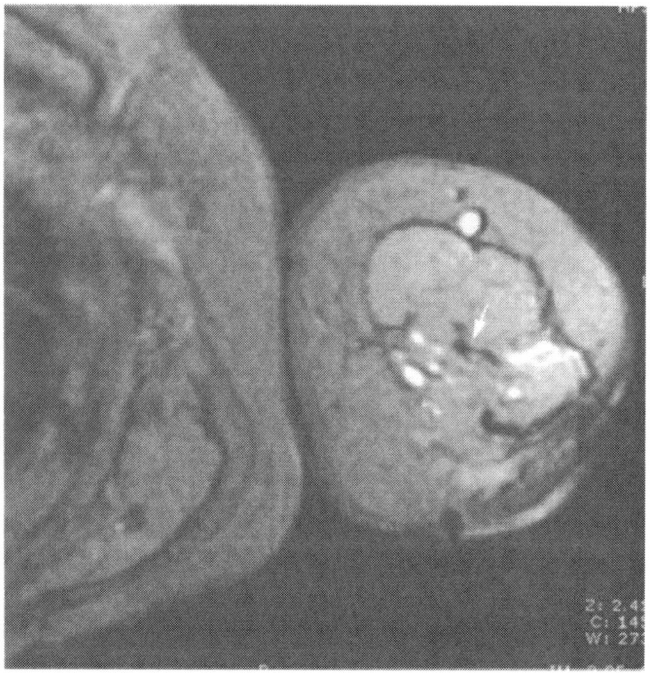

Figure 59J. Axial MPGR.

Findings

This is a 12-year-old girl with a known history of Gorham disease. She had prior incomplete surgical debulking of a complex lymphatic malformation.

Figure 59H. There is extensive osteolysis of the humerus and scapula.

Figures 59I, 59J. Residual soft tissue microcystic lymphatic malformation abuts the shaft of the humerus and left chest wall **(arrowheads)**. MPGR sequence better delineates the residual humeral shaft cortex **(arrow)**.

Pitfalls and Pearls

1. Capillary malformations and microcystic lymphatic malformations may be indistinguishable on MRI. Both can mimic a soft tissue neoplasm.
2. Lymphatic malformations are congenital lesions that are present at birth. However, they may not be diagnosed until later in life when they change in size due to trauma or infection.

References

1. Dubois J, Garel L. Imaging and therapeutic approach of hemangiomas and vascular malformations in the pediatric age group. *Pediatr Radiol* 1999; 29:879–893.
2. Gallagher PG, Mahoney MJ, Gosche JR. Cystic hygroma in the fetus and newborn. *Semin Perinatol* 1999; 23:341–356.
3. Burrows PE, Laor T, Paltiel H, Robertson RL. Diagnostic imaging in the evaluation of vascular birthmarks. *Dermatol Clin* 1998; 16:455–488.
4. Burrows PE. Personal communication. 2005.
5. Konez O, Burrows PE. Magnetic resonance of vascular anomalies. *Magn Reson Imaging Clin N Am* 2002; 10:363–388, vii.
6. Manisali M, Ozaksoy D. Gorham disease: Correlation of MR findings with histopathologic changes. *Eur Radiol* 1998; 8:1647–1650.

Case 60

History

This is a 15-year-old girl with chronic lower back pain.

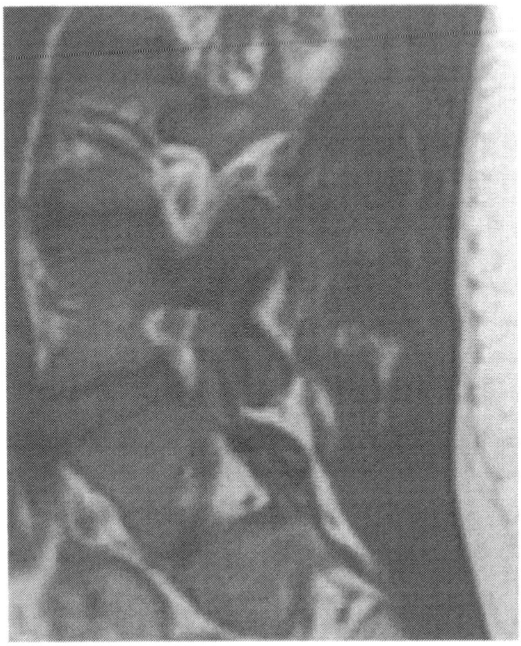

Figure 60A. Sagittal T1 left parasagittal.

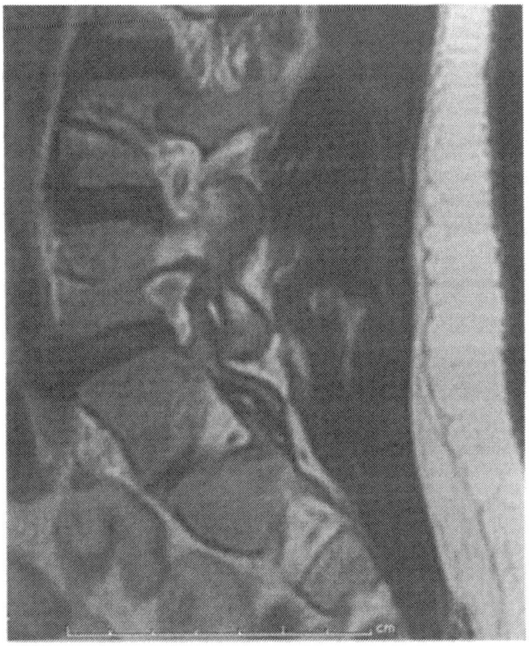

Figure 60B. Sagittal T2 left parasagittal.

Figures 60A, 60B, (60A, 60B with annotations). Intermediate T1 and T2 SI is identified along a cleavage plane in the left L5 pars interarticularis **(arrows)**. No evidence of spondylolesthesis is seen. Mild T2 hyperintensity is seen within the pedicle and superior and inferior articular facets of L5, consistent with marrow edema **(arrowheads)**. A similar appearance was also seen on the right (not shown).

Figures 60C, 60D, 60E. There is bilateral, marked radiotracer uptake at the L5 pars interarticularis. On CT, bilateral, spondylolysis is identified **(arrows)**. The margins of the pars are well defined and sclerotic, indicating chronicity.

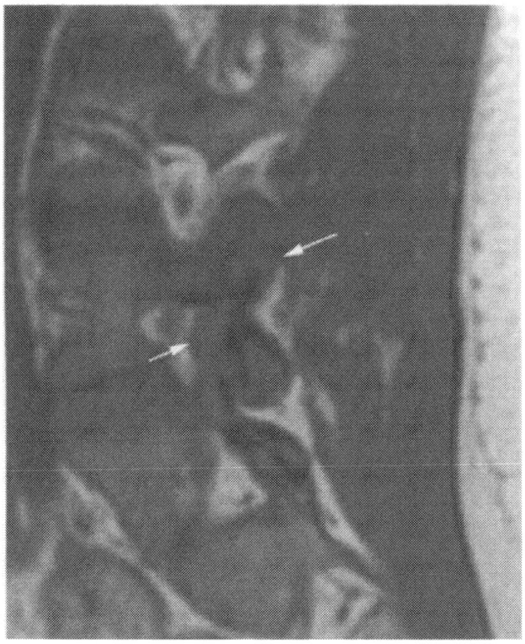

Figure 60A* Annotated.

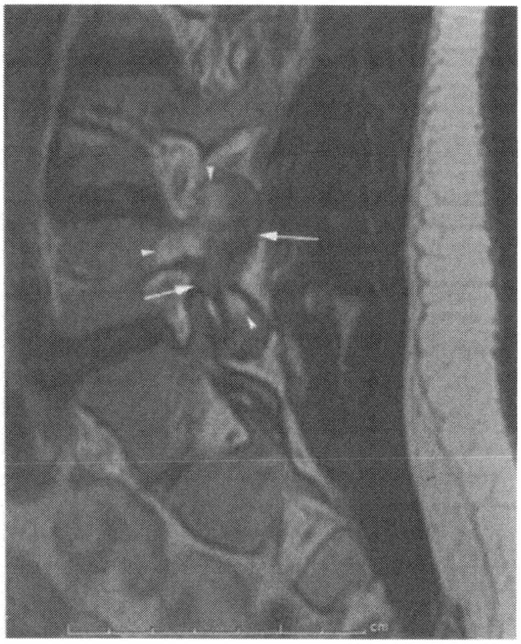

Figure 60B* Annotated.

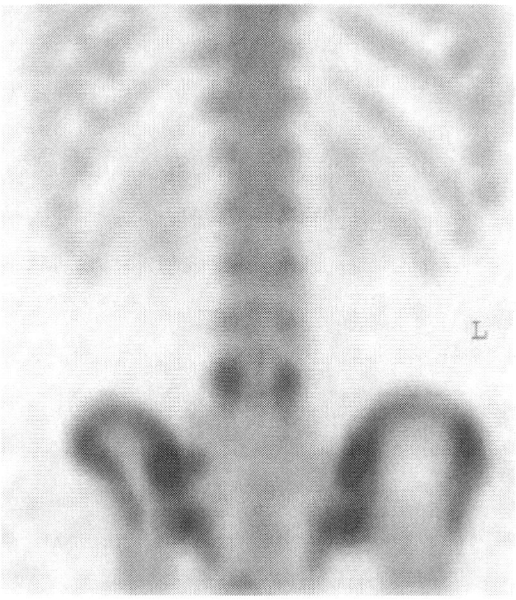

Figure 60C. Tc-99m HDP bone scan.

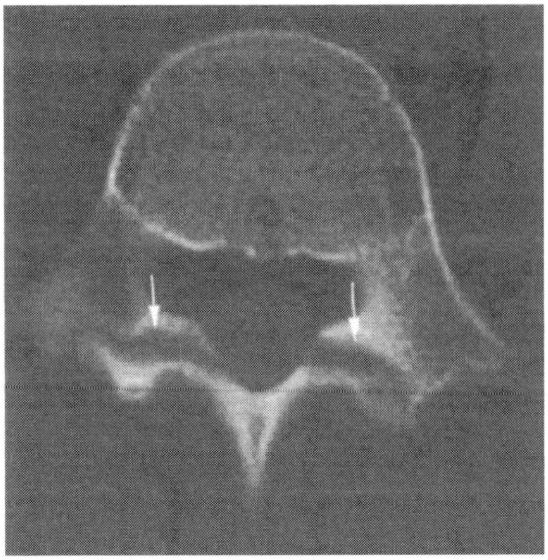

Figure 60D. CT at L5.

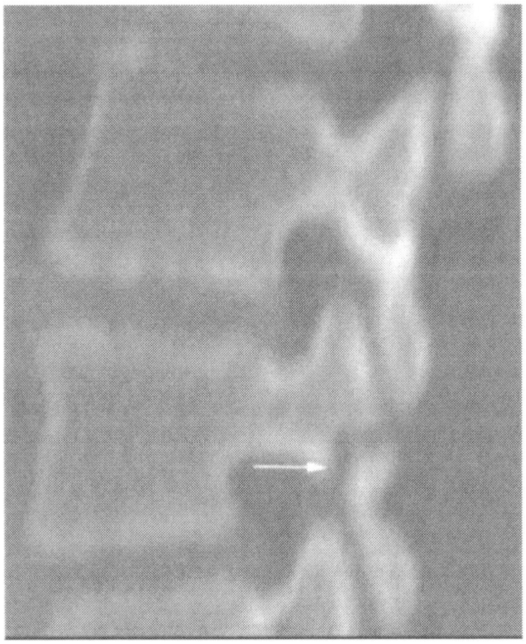

Figure 60E. Sagittal reformat CT, left parasagittal.

Diagnosis

Bilateral spondylolysis of L5

Questions

1. What is the most common location for spondylolysis?
2. T/F: Spondylolesthesis may occur in the absence of spondylolysis.
3. What are some congenital vertebral column deformities that may predispose patients to spondylolesthesis?

Discussion

Spondylolysis usually results from a stress fracture of the pars interarticularis between the superior and inferior articular facets. Bilateral pars defects are seen in 80% of cases (1). Spondylolysis occurs most commonly at the L5 level since this is where the greatest amount of mechanical stress related to flexion and extension occurs (Answer to Question 1) (2). The second most common level is at L4.

Spondylolesthesis may develop when the spondylolysis is bilateral and it is more likely to occur during periods of rapid growth. In a study by Fredrickson et al., the incidence of spondylolysis was 4.4% in a group of 500 unselected 6-year-old children (3). A genetic predisposition exists, with Alaskan Native Americans at highest risk (26%) and black females at lowest risk (1.1%) (2). In a study of 100 athletic children who presented with back pain, Micheli et al. noted that 47% had spondylolysis with or without spondylolesthesis, compared with an adult control population where the incidence was 5% (4).

Patients with congenital vertebral column deformities may develop spondylolesthesis. A spondylolysis does not have to be present for spondylolesthesis to occur (Answer to Question 2). Spondylolesthesis from congenital deformities may eventually lead to spondylolysis, however, due to altered axial loading. The spectrum of congenital deformities that may cause spondylolesthesis include: hypoplasia of the upper surface of the first sacral vertebral body, hypoplasia/aplasia of the facets, elongation of the pars interarticularis, spina bifida, and lumbosacral kyphosis (Answer to Question 3) (5).

The MRI diagnosis of spondylolysis is made if a line or band through the pars is identified on both T1 and T2W sequences, which disrupts normal cortical and marrow SI. The cleavage plane may be hypointense on T1 and variable SI on T2. If there is also significant spondylolesthesis, the pars defect may be occupied by deep paraspinal fat. Routine lumbar spine MRI with straight axial and sagittal sequences or imaging planes angled to the intervertebral disk may not be adequate for evaluation of spondylolysis. In one study, routine lumbar spine MRI had a positive predictive value of 14% to 18% for a pars defect, but had a negative predictive value of 97–99% (6). In another study, MRI missed 30% (20/66) of spondylolyses, based on direct visualization of a pars defect, that were later confirmed by plain radiography or CT (7). However, in this study, sensitivity increased to 97% (64/66) when indirect signs of spondylolysis were evaluated. These features included: increased anteroposterior diameter of the spinal canal, wedging of the posterior aspect of the vertebral body (Figure 60F), spondylolesthesis, and reactive marrow or osseous degenerative changes of the affected pars and neighboring facets. With unilateral spondylolysis, contralateral pars stress (Figures 60G, 60H) and subsequent degenerative changes may develop (8).

MRI may confidently exclude a pars defect if normal marrow SI is identified throughout the pars. MRI may be falsely positive because volume averaging may mimic a pars defect. In addition, sclerosis at the pars may mimic a defect because the pars may be hypointense on all imaging sequences. To better conform to the orientation of the pars, 3D acquisition or reverse angle oblique axial images may provide better delineation of the pars compared with standard orthogonal planes (9).

This patient was initially treated with bracing and strengthening exercises, which improved her back pain. Since her back pain improved, surgery was not being considered.

Orthopedic Perspective

Spondylolysis is the most common cause of back pain in the active child and adolescent. Repetitive hyperextension activities, such as gymnastics, dancing, ice-skating, and football (e.g., lineman), may subsequently lead to spondylolysis of the pars interarticularis. The diagnosis is suggested clinically by low back pain that worsens with hyperextension. The diagnosis is usually established by radiography, CT, and/or bone scintigraphy. Primary evaluation by MRI requires additional tailored sequences to assess the pars interarticularis and this must be communicated to the radiologist and technologist. Primary evaluation by MRI of spondylolysis is indicated if symptoms are atypical and there is concern for pathology of the cord and intervertebral disks. Treatment of spondylolysis with up to grade 2 spondylolesthesis (<50% displacement) requires immobilization in an antilordotic overlap brace, restriction of activity, and hamstrings stretching. Surgical fixation is considered when there is established nonunion, failed brace treatment, and the patient is symptomatic with functional limitations. Grade 3–4 spondylolesthesis (>50% displacement) is often referred for surgical fixation initially. Reduction may be considered, particularly in patients with spondyloptosis (>100% displacement). The complications of reduction of spondylolesthesis include nerve root injury in a minority of patients.

What the Clinician Needs to Know

1. What is the percentage of anterolesthesis, if present?
2. Is there an underlying congenital deficiency of the facet or posterior elements?
3. Are there secondary degenerative changes present at other vertebral levels, facets, and the intervertebral disk?

Answers

1. L5.
2. True.
3. Hypoplasia of the upper surface of the first sacral vertebral body, hypoplasia/aplasia of the facets, elongation of the pars interarticularis, spina bifida, and lumbrosacral kyphosis.

Additional Examples

Indirect Signs and Secondary Degenerative Changes from Spondylolysis

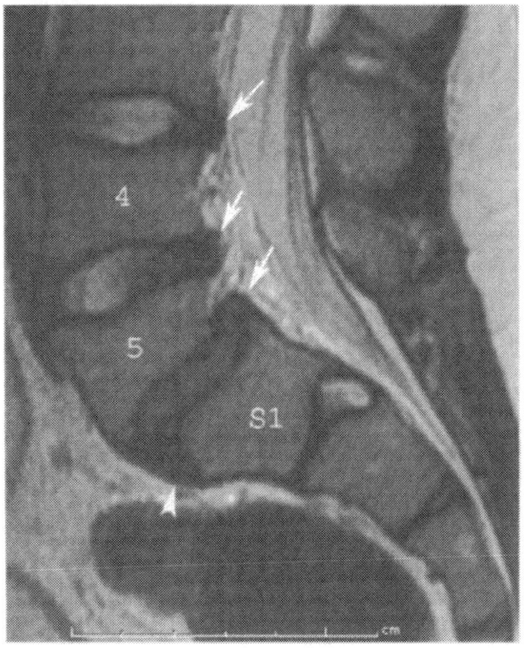

Figure 60F. Sagittal T2.

Findings

This is a 12-year-old boy with chronic back pain. He had bilateral L5 pars defects confirmed by MRI and CT (not shown).

Figure 60F. The indirect signs of spondylolysis seen in this case include: grade 2 anterolesthesis of L5/S1, increased anterior-posterior diameter at L5, and posterior wedge deformity of L5. Multilevel disk bulges are seen **(arrows)**. A limbus-like deformity related to a growth disturbance of the superior S1 endplate **(arrowhead)** is also seen.

Unilateral Left L5 Spondylolysis

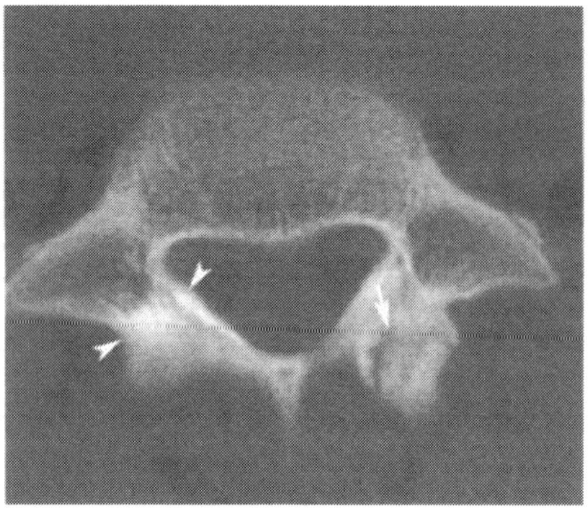

Figures 60G. Axial CT through the L5 pars interarticularis.

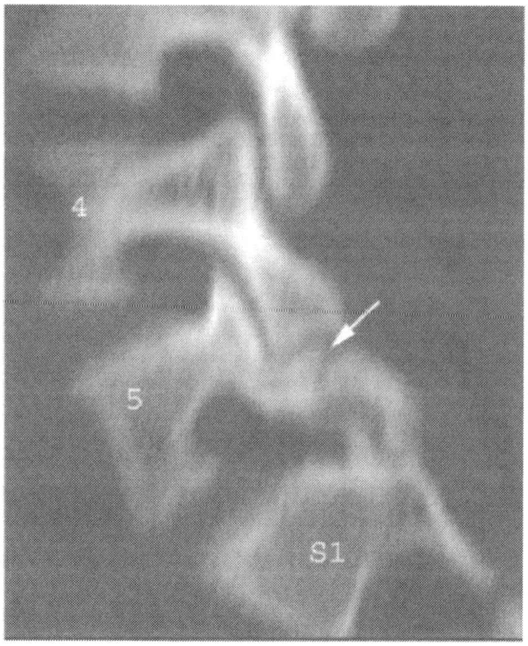

Figure 60H. Sagittal reformat CT through the left L5 pars interarticularis.

Findings

This is a 13-year-old girl with chronic back pain.

Figures 60G, 60H. There is a left L5 pars defect identified without displacement **(arrow)**. There is stress related sclerosis and hypertrophy of the intact right L5 pedicle and pars (**between arrowheads)**.

Pitfalls and Pearls

1. Intact pars interarticularis marrow SI confidently excludes spondylolysis.
2. It is not uncommon to encounter an incidental spondylolysis in the pediatric and adolescent population. The pars interarticularis should always be carefully inspected for this ubiquitous defect on spinal MRI performed for any reason.

References

1. Herring JA. Back pain. In: *Tachdjian's Pediatric Orthopaedics*, 3rd ed., Chapter 7. Philadelphia: W.B. Saunders, 2002; 95–108.
2. Lim MR, Yoon SC, Green DW. Symptomatic spondylolysis: Diagnosis and treatment. *Curr Opin Pediatr* 2004; 16:37–46.
3. Fredrickson BE, Baker D, McHolick WJ, Yuan HA, Lubicky JP. The natural history of spondylolysis and spondylolisthesis. *J Bone Joint Surg Am* 1984; 66:699–707.
4. Micheli LJ, Wood R. Back pain in young athletes: Significant differences from adults in causes and patterns. *Arch Pediatr Adolesc Med* 1995; 149:15–18.

5. Logroscino G, Mazza O, Aulisa G, Pitta L, Pola E, Aulisa L. Spondylolysis and spondylolisthesis in the pediatric and adolescent population. *Childs Nerv Syst* 2001; 17:644–655.
6. Saifuddin A, Burnett SJ. The value of lumbar spine MRI in the assessment of the pars interarticularis. *Clin Radiol* 1997; 52:666–671.
7. Ulmer JL, Mathews VP, Elster AD, Mark LP, Daniels DL, Mueller W. MR imaging of lumbar spondylolysis: The importance of ancillary observations. *AJR Am J Roentgenol* 1997; 169:233–239.
8. Sirvanci M, Ulusoy L, Duran C. Pedicular stress fracture in lumbar spine. *Clin Imaging* 2002; 26:187–193.
9. Campbell RS, Grainger AJ. Optimization of MRI pulse sequences to visualize the normal pars interarticularis. *Clin Radiol* 1999; 54:63–68.

Case 61

History

This is a 7-year-old boy with right hip pain and limp for several days.

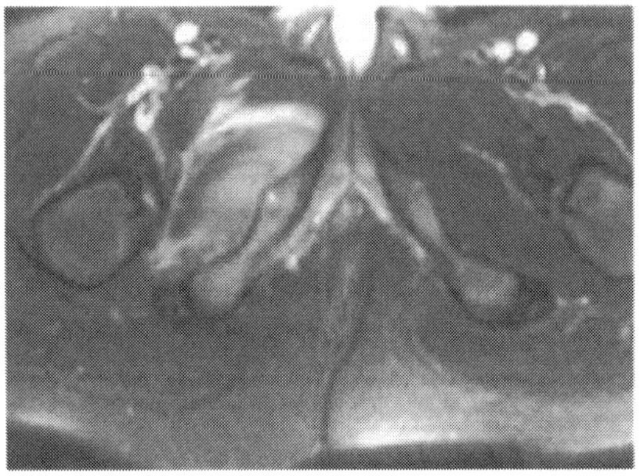

Figure 61A. Axial T2 FS of the right hip.

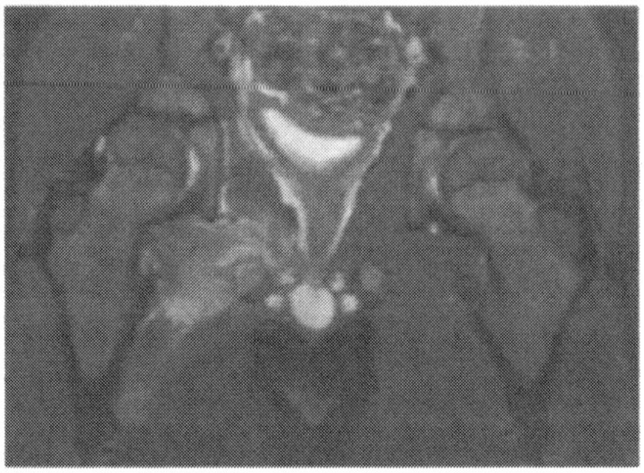

Figure 61B. Coronal STIR.

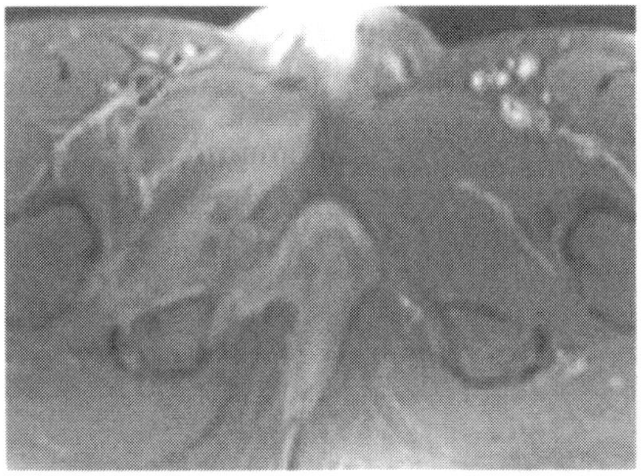

Figure 61C. Axial T1 post-Gd FS.

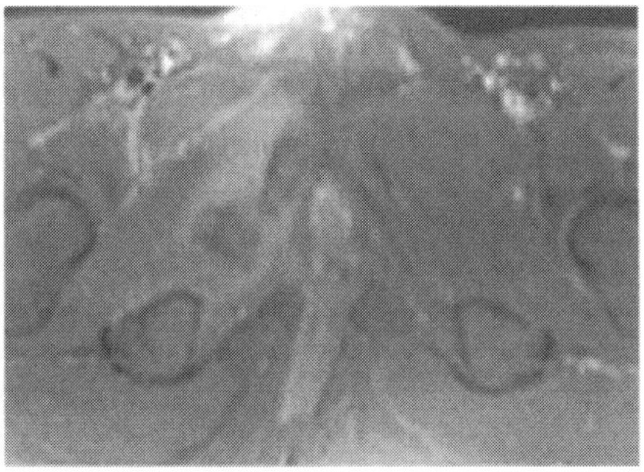

Figure 61D. Axial T1 post-Gd FS.

Figures 61A, 61B, (61A with annotations). There is diffuse increased SI within the right adductor magnus, obturator externus, obturator internus, and within the ischiopubic synchondrosis and adjacent bone. A more discrete, rounded focus of fluid SI is located within the synchondrosis **(arrowhead)**. Note the appearance of the normal left synchondrosis **(thin arrow)**.

Figures 61C, 61D, (61D with annotations). A focal rim-enhancing fluid collection is present surrounding the right ischiopubic synchondrosis **(thick arrow)**. There is also abnormal enhancement within the adductor and obturator externus muscles.

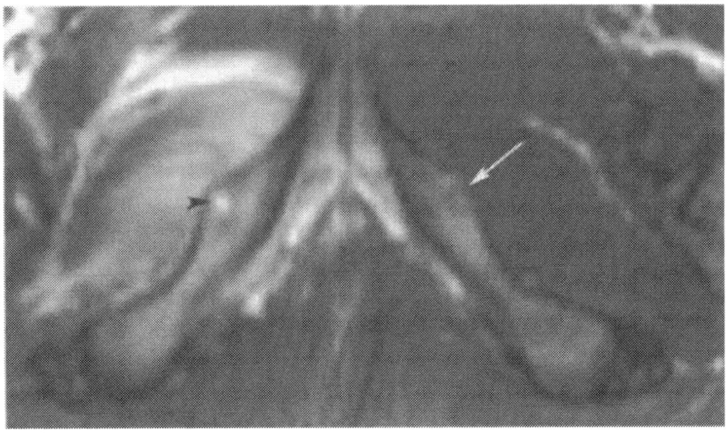

Figure 61A* Annotated detail.

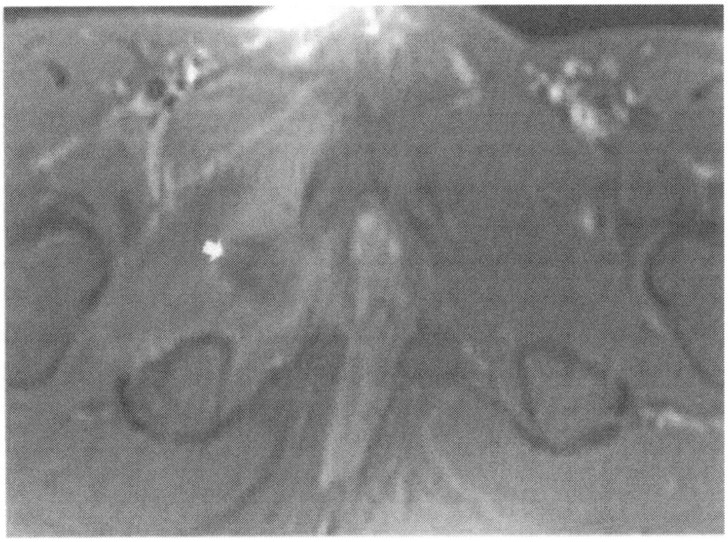

Figure 61D* Annotated.

Diagnosis

Ischiopubic synchondrosis osteomyelitis

Questions

1. Over what age range does the ischiopubic synchondrosis fuse?
2. T/F: A normal unfused ischiopubic synchondrosis may show adjacent soft tissue T2 hyperintensity and enhancement on MRI.

Discussion

The ischiopubic synchondrosis (IPS) shares similar anatomic features with the metaphyses of long bones (1). The IPS has metaphyseal equivalent regions on either side of the cartilaginous synchondrosis. Osteomyelitis, as seen in this case, as well as other pathologic metaphyseal lesions may begin at the metaphyseal equivalent region of either the pubic or ischial side of the IPS.

The plain radiographic evaluation of the ischiopubic synchondrosis (IPS) is challenging because its appearance changes with skeletal maturity. Prior to osseous fusion, the cartilaginous IPS is radiolucent on plain radiography. Osseous fusion occurs between 4 and 12 years of age and is commonly asymmetric (Answer to Question 1) (2). Immediately prior to fusion, the normal IPS may have a bulbous, mass-like appearance. When unilateral, this finding may be alarming on plain radiography and MRI (Figures 61E–61G). The IPS T2 SI and enhancement will vary depending on the stage of development and this MRI pattern may be mistaken for pathology such as neoplasm, infection, or stress injury.

A clue that may help differentiate the normal IPS from a pathologic process is the appearance of the adjacent soft tissues. Avulsion injuries from the adductor muscles, neoplasm, and osteomyelitis often have prominent adjacent soft tissue masses, inflammation, or fluid collections. The variable MRI appearance of the normal IPS is confined to cartilage and bone, and there should be only minor adjacent soft tissue signal alteration (Figure 61G). In one study evaluating the IPS in 28 asymptomatic children, 57% demonstrated minor adjacent soft tissue edema near the IPS (Answer to Question 2) and the majority demonstrated T1 hypointensity, T2 hyperintensity, and enhancement within the IPS (3).

A second clue to distinguish between the normal variations of the IPS and pathology is to determine if the signal abnormality is centered at, or lies at a distance from, the IPS. If the IPS is secondarily involved from a primary pubic or ischial lesion, then differentiation from the normal IPS is straightforward (Figures 61H, 61I).

The patient's right hip pain and limp were initially evaluated by plain radiograph and US, which were inconclusive. His symptoms were explained by the extensive IPS osteomyelitis with inflammatory changes involving several muscles that insert onto the proximal femur (obturator externus, obturator internus, and adductor muscles). Blood cultures grew *Staphylococcus aureus* and he did well following a course of antibiotics.

Orthopedic Perspective

The clinical indication to pursue advanced imaging of asymmetries of the IPS includes focal tenderness or when the asymmetry appears mass-like. Asymmetries of the normal IPS are common and true pathology uncommon. Pathologic considerations are similar to those affecting metaphyseal or metaphyseal-equivalent regions elsewhere, including stress reaction, fracture, and rarely neoplasm.

What the Clinician Needs to Know

Is there an intraosseous or soft tissue collection, and if so, what is the best approach for drainage? Is the collection amenable to CT-guided aspiration?

Answers

1. 4–12 years.
2. True.

Additional Examples

Normal Asymmetry of the Ischiopubic Synchondrosis

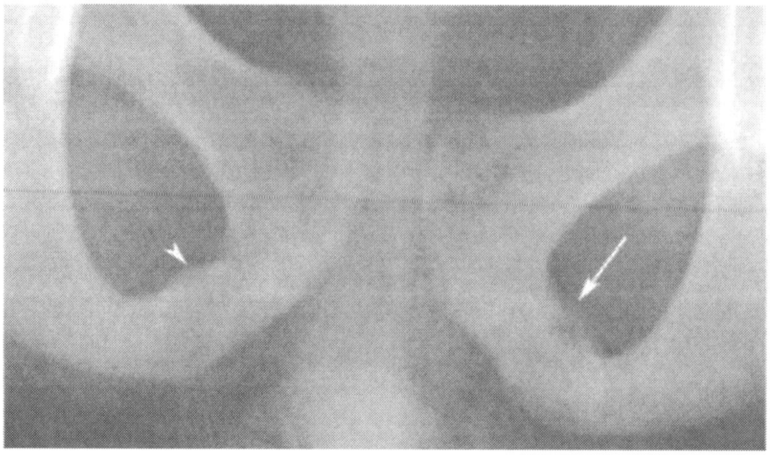

Figure 61E. AP radiograph.

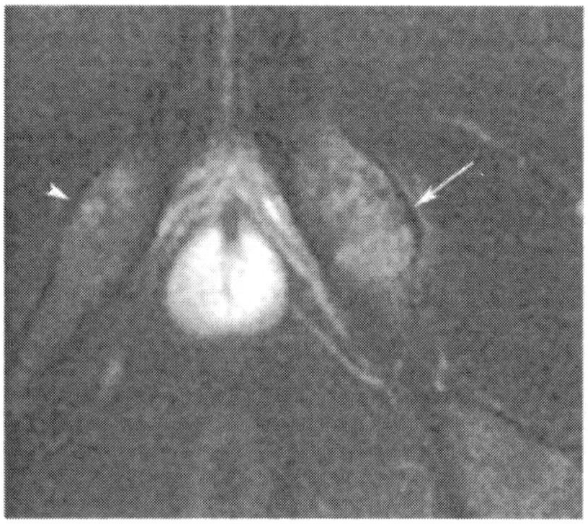

Figure 61F. Axial T2 FS.

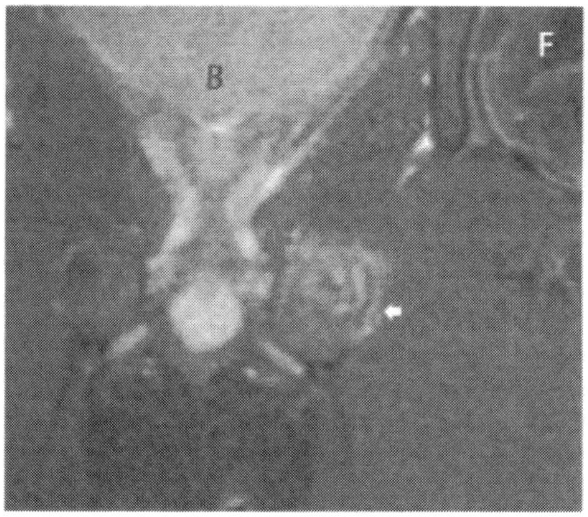

Figure 61G. Coronal T1 post-Gd FS (through both IPS).

Findings

This is an 11-year-old boy who suffered a minor injury to his pelvis and had no symptoms referrable to the IPS.

Figure 61E. The normal left IPS **(thin arrow)** was mistaken for a neoplasm on plain radiographs and the patient was referred for MRI.

Figure 61F. The right IPS **(arrowhead)** is less bulbous than the left **(thin arrow)**, as it approaches fusion. The left sided finding represents normal asymmetric fusion of the IPS.

Figure 61G. There is mild normal extraosseous soft tissue enhancement adjacent to the IPS **(thick arrow)**. B, bladder; F, femoral head.

Langerhans Cell Histiocytosis (LCH) of the Ischium

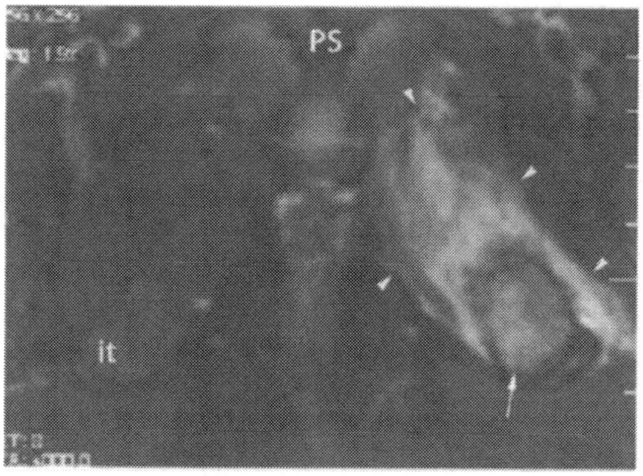

Figure 61H. Axial T2 FS.

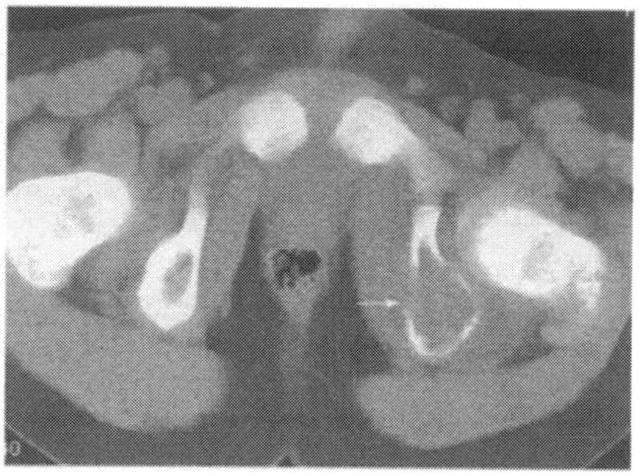

Figure 61I. CT.

Findings

This is a 4-year-old boy who presented with left hip and buttock pain.

Figure 61H. There is increased T2 SI throughout the left ischium **(arrow)** and adjacent soft tissues **(arrowheads)**. Pubic symphysis (PS); Normal right ischial tuberosity (it).

Figure 61I. CT confirms an osteolytic lesion **(arrow)** that is centered in the ischium and not the IPS. This was pathologically confirmed LCH.

Pitfalls and Pearls

1. Beware of isolated, increased SI on T2W sequences within the ischiopubic synchondrosis, without associated mass or gross soft tissue signal abnormality. This may be the normal appearance of the ischiopubic synchondrosis (3).
2. Interestingly, delayed IPS fusion tends to occur on the opposite side of foot preference (4).
3. Obturator phlegmon syndrome refers to pyomyositis of the obturator internus muscle. This may result from an isolated pyomyositis or may be secondary to pelvic osteomyelitis (5).
4. Make sure to image the entire pelvis in patients who are referred to evaluate the IPS, since the culprit for the patient's symptoms may very well lie elsewhere in the pelvis.

References

1. Iqbal A, McKenna D, Hayes R, O'Keeffe D. Osteomyelitis of the ischiopubic synchondrosis: Imaging findings. *Skeletal Radiol* 2004; 33:176–180.
2. Ogden JA. Pelvis. In: *Skeletal Injury in the Child*, 3rd ed., Chapter 19. New York: Springer, 2000; 790–830.
3. Herneth AM, Trattnig S, Bader TR. et al. MR imaging of the ischiopubic synchondrosis. *Magn Reson Imaging* 2000; 18:519–524.
4. Herneth AM, Philipp MO, Pretterklieber ML, Balassy C, Winkelbauer FW, Beaulieu CF. Asymmetric closure of ischiopubic synchondrosis in pediatric patients: Correlation with foot dominance. *AJR Am J Roentgenol* 2004; 182:361–365.
5. Orlicek SL, Abramson JS, Woods CR, Givner LB. Obturator internus muscle abscess in children. *J Pediatr Orthop* 2001; 21:744–748.

Case 62

History

This is a 14-year-old boy with a 2-month history of left leg pain.

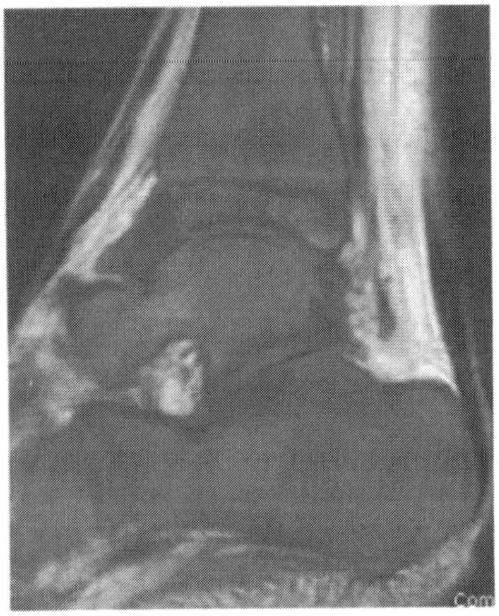

Figure 62A. Sagittal T1 of the left ankle.

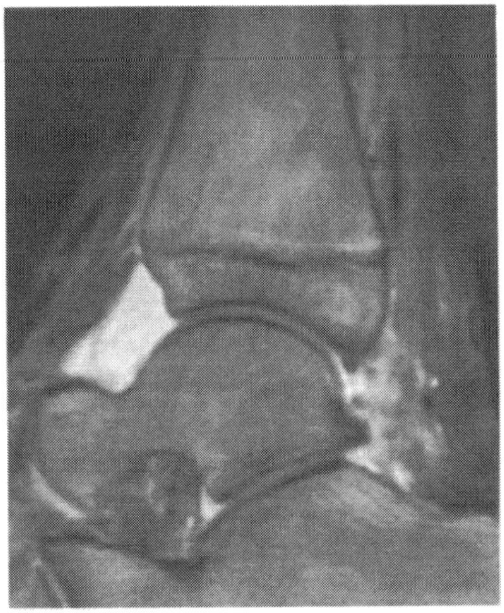

Figure 62B. Sagittal STIR.

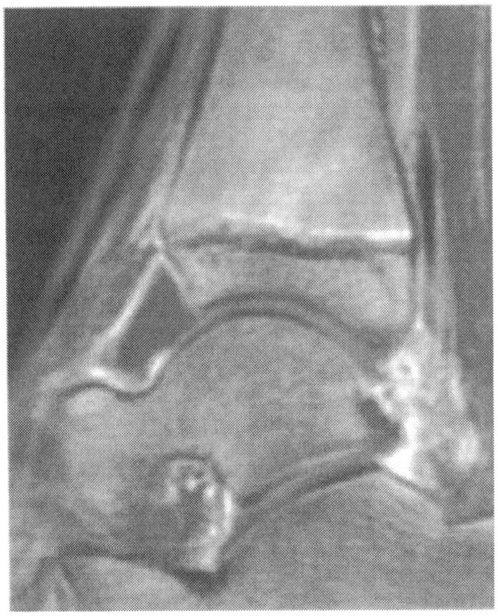

Figure 62C. Sagittal T1 post-Gd FS.

Figures 62A, 62B. There is a moderate size ankle joint effusion associated with synovial thickening and enhancement extending into Kager's fat pad. The marrow is diffusely hypointense on T1 and hyperintense on STIR images.

Figure 62C. There is diffuse marrow enhancement as well as a transverse linear enhancing zone in the subphyseal metaphysis of the tibia.

Figure 62D. There is subphyseal radiolucency in the distal tibia and, to a lesser extent, the fibula **(arrows)**, corresponding to hyperintensity in Figure 62B and enhancement in Figure 62C.

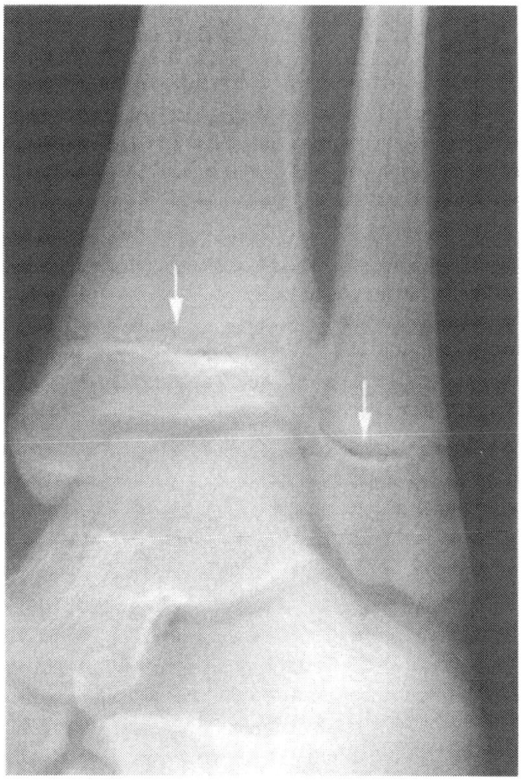

Figure 62D. Oblique radiograph.

Diagnosis

Acute lymphoblastic leukemia, leukemic lines, and candida arthritis

Questions

1. T/F: MRI readily differentiates leukemic arthritis from infectious arthritis.
2. T/F: Residual red marrow may be slightly hyperintense to muscle on fluid-sensitive sequences.

Discussion

Leukemia is the most common malignancy in the pediatric population (1). Acute lymphoblastic leukemia (ALL) comprises 80% and acute myeloid leukemia (AML) 15–20% (1, 2). In the pediatric population, the peak incidence of ALL is 2–5 years. AML occurs most frequently in the first two years of life, with a smaller peak in later childhood.

The musculoskeletal manifestations of acute leukemia and complications of treatment include: marrow packing/infiltration, permeative lytic bone destruction, periosteal new bone formation, granulocytic sarcoma, insufficiency and pathologic fracture, osteonecrosis, spontaneous hemorrhage, osteomyelitis, infectious arthritis, leukemic arthritis, and secondary neoplasms.

Clinically, leukemic arthritis may mimic infectious arthritis, Lyme disease, and other inflammatory arthridities. Leukemic arthritis occurs in 12% to 65% of cases of childhood leukemia, whereas the incidence in adult leukemia patients is 4% to 13% (3). Proposed etiologies for leukemic arthritis includes leukemic synovial infiltration, intra-articular hemorrhage related to thrombocytopenia, and/or immune complex induced synovitis (3). Infectious arthritis may be suggested when the synovitis is associated with juxta-articular marrow edema, but the findings are by no means diagnostic (4). When there is diffuse marrow replacement, as in this case, it is impossible to differentiate between leukemic and septic arthritis by MRI (Answer to Question 1).

On radiography, diffuse osteopenia of ALL may be due to bone marrow packing by leukemic cells or altered mineral metabolism (5, 6). Leukemic lines represent metaphyseal radiolucent bands with variably sclerotic margins that parallel the physis (Figures 62E–62H). Radiodense metaphyseal bands and growth recovery lines may be seen adjacent to leukemic lines (5). Stunted endochondral ossification from stress is the assumed mechanism for leukemic lines (7). Fractures may accompany leukemic lines due to osseous weakening (Figures 62E–62H). Periostitis may be seen after fracture or when leukemic infiltration is present.

On MRI, marrow packing may be distinguished from residual red marrow by comparing marrow contents with respect to muscle. Marrow packing by leukemic cells is hypointense on T1 and markedly hyperintense on fluid-sensitive sequences with respect to muscle (8). Normal red marrow may be slightly hyperintense on fluid-sensitive sequences, has a typical flame shaped or paint brush appearance, and follows a characteristic distribution that changes with age (see Case 50) (Answer to Question 2). Cortical bone infiltration by leukemic cells may elevate the periosteum and a juxtacortical soft tissue mass may be seen mimicking other small round blue cell tumors such as Ewing's sarcoma (6).

Granulocytic sarcomas (AKA chloromas) are focal, solid tumors composed of immature myeloid cells. They almost always occur in the setting of AML (9). These tumors are more common in children (10.9% incidence) than adults with AML (10, 11). Granulocytic sarcomas most commonly occur in the subcutaneous soft tissues followed by the orbital or periorbital regions. Other reported locations include, but are not limited to: lymph nodes, bone/periosteum (Figures 62I–62K), gingiva, and testis (12). On MRI, these tumors have a nonspecific appearance and are isointense on T1, hyperintense on fluid-sensitive sequences, and demonstrate variable enhancement after gadolinium (12). Granulocytic sarcomas may demonstrate rim enhancement with a hypointense center, mimicking abscess (11).

The diagnosis of candida septic arthritis was made after joint aspiration. One month prior to the MRI, the patient was diagnosed with acute lymphoblastic lymphoma, T-cell phenotype.

Orthopedic Perspective

Leukemia can present with a variety of differing musculoskeletal manifestations as noted above. Usually the presentation mimics osteomyelitis or septic arthritis and a high index of suspicion is needed to make the correct diagnosis. A complete blood count and smear will usually settle the issue, but it is important to appreciate that these may be normal and the diagnosis will hinge on astute suspicion and a bone marrow aspiration. It is important to inquire about whether the child has been ill, appears anemic, or has had bruising, all of which may point to the correct diagnosis.

At times the presentation may mimic a sarcoma as noted in the last case. In these circumstances, an open or needle biopsy may be necessary to establish the correct diagnosis.

What the Clinician Needs to Know

1. Is there is a diffuse marrow replacement?
2. Is there a joint effusion?
3. What is the etiology for the periosteal reaction? Is the periosteal reaction due to leukemic infiltration or is there an underlying pathologic or insufficiency fracture present?
4. Are the findings classic for infection or trauma, or are these findings due to leukemia, which requires a different evaluation to include a CBC with smear and a bone marrow biopsy?
5. A high index of suspicion is needed. Are there other clinical signs such as paleness, anemia, and bruising to lead to the consideration of the correct diagnosis?

Answers

1. False.
2. True.

Additional Examples

ALL with Leukemic Lines

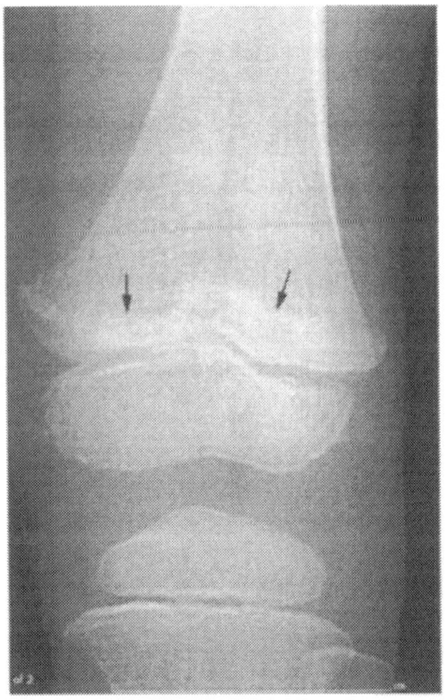

Figure 62E. AP radiograph, left knee.

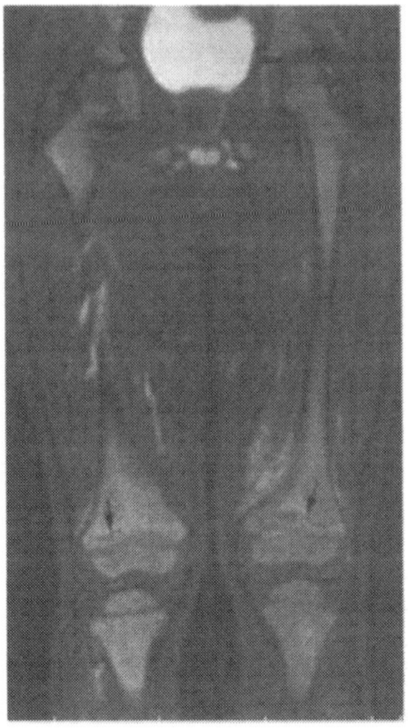

Figure 62F. Coronal STIR.

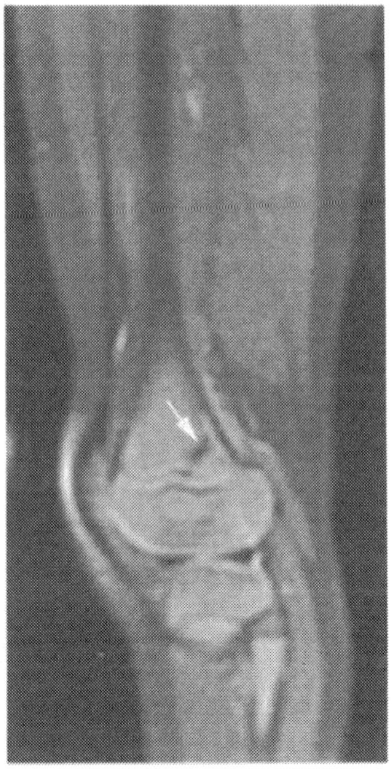

Figure 62G. Sagittal PD FS.

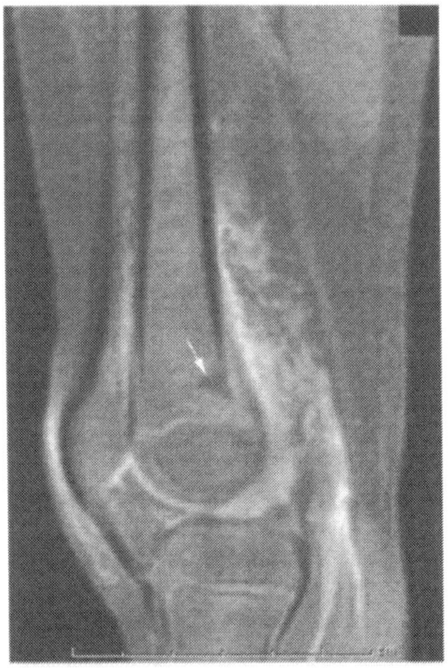

Figure 62H. Sagittal T1 post-Gd FS.

Findings

This is a 4-year-old boy with acute left knee pain and swelling.

Figure 62E. Parallel radiolucent and radiodense metaphyseal bands are present compatible with leukemic lines **(black arrows)**.

Figure 62F. There is diffuse medullary hyperintensity consistent marrow replacement. Bilateral hypointense horizontal metaphyseal bands **(black arrows)** with marginal hyperintensity correlate with the leukemic lines seen on plain radiography. On the left, there is also juxtacortical soft tissue edema.

Figures 62G, 62H. An oblique fracture line **(white arrows)** is present with adjacent juxtacortical diffuse soft tissue enhancement. The findings on MRI prompted a hematologic work-up that confirmed ALL.

Granulocytic Sarcoma of Bone Mimicking Ewing's Sarcoma

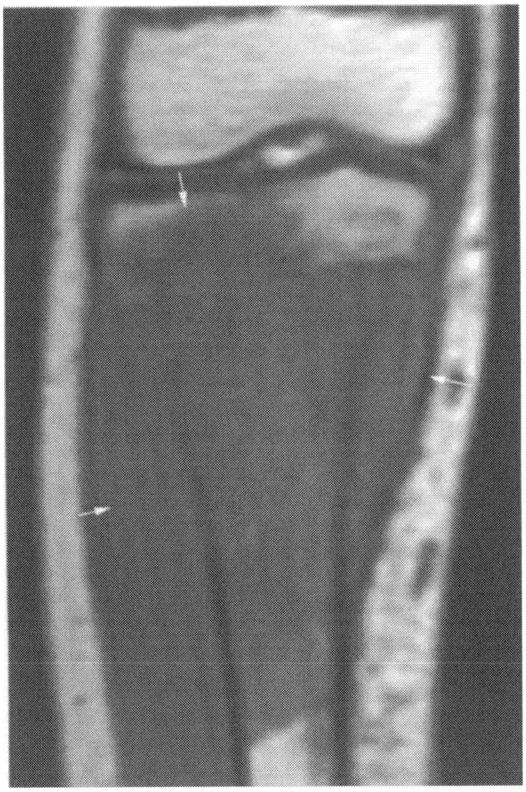

Figure 62I. Coronal T1 of the right tibia.

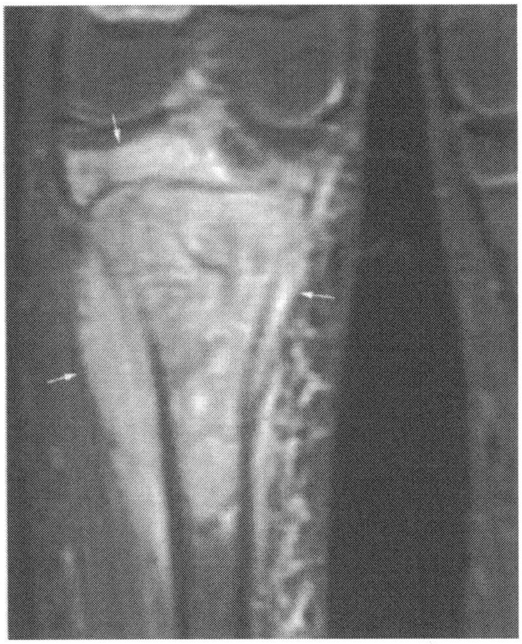

Figure 62J. Coronal STIR.

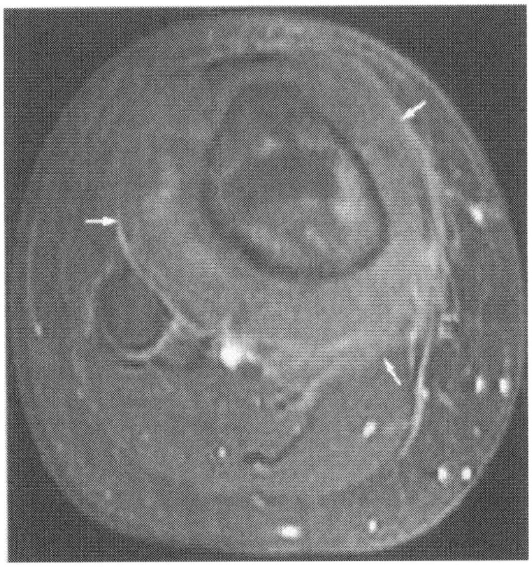

Figure 62K. Axial T1 post-Gd FS.

Findings

This is a 14-year-old boy with 4 weeks of right leg pain. He has a history of AML that was diagnosed 4 years ago.

Figure 62I. There is a well circumscribed zone of T1 hypointensity located in the proximal tibial metaphysis extending into the epiphysis. Note subperiosteal extension of metaphyseal disease **(arrows)**.

Figures 62J, 62K. The process has variable SI on STIR and demonstrates peripheral enhancement. It infiltrates through the cortex producing a large circumferential extraosseous mass **(arrows)**. This was pathologically proven to be a granulocytic sarcoma.

Pitfalls and Pearls

1. Bone/joint pain is a common presentation for childhood leukemia. In the child with a joint effusion, the radiologist may be the first to suggest the correct diagnosis based on leukemic lines, demineralization, permeative bone destruction, periosteal reaction, or pathologic fracture.
2. Oncologic causes of monoarticular synovitis include leukemia, intra-articular osteoid osteoma, and chondroblastoma.

References

1. Chan KW. Acute lymphoblastic leukemia. *Curr Probl Pediatr Adolesc Health Care* 2002; 32:40–49.
2. Aquino VM. Acute myelogenous leukemia. *Curr Probl Pediatr Adolesc Health Care* 2002; 32:50–58.
3. Evans TI, Nercessian BM, Sanders KM. Leukemic arthritis. *Semin Arthritis Rheum* 1994; 24:48–56.
4. Lee SK, Suh KJ, Kim YW, et al. Septic arthritis versus transient synovitis at MR imaging: Preliminary assessment with signal intensity alterations in bone marrow. *Radiology* 1999; 211:459–465.
5. Resnick D, Haghighi P. Lymphoproliferative and myeloproliferative disorders. In: *Diagnosis of Bone & Joint Disorders*, 4th ed. Philadelphia: W.B. Saunders, 2002; 2291–2345.
6. Parker BR. Leukemia and lymphoma in childhood. *Radiol Clin North Am* 1997; 35: 1495–1516.
7. Laor T, Jaramillo D, Oestreich AE. Musculoskeletal system. In: *Practical Pediatric Imaging: Diagnostic Radiology of Infants and Children*, 3rd ed., Chapter 5, Kirks DR, ed. Philadelphia: Lippincott-Raven, 1998; 327–510.
8. Moulopoulos LA, Dimopoulos MA. Magnetic resonance imaging of the bone marrow in hematologic malignancies. *Blood* 1997; 90:2127–2147.
9. Binder C, Tiemann M, Haase D, Humpe A, Kneba M. Isolated meningeal chloroma (granulocytic sarcoma): A case report and review of the literature. *Ann Hematol* 2000; 79:459–462.
10. Dusenbery KE, Howells WB, Arthur DC, et al. Extramedullary leukemia in children with newly diagnosed acute myeloid leukemia: A report from the Children's Cancer Group. *J Pediatr Hematol Oncol* 2003; 25:760–768.
11. Ooi GC, Chim CS, Khong PL, et al. Radiologic manifestations of granulocytic sarcoma in adult leukemia. *AJR Am J Roentgenol* 2001; 176:1427–1431.
12. Pui MH, Fletcher BD, Langston JW. Granulocytic sarcoma in childhood leukemia: Imaging features. *Radiology* 1994; 190:698–702.

Case 63

History

This is a 17-month-old girl with an enlarging right buttock mass. She is otherwise healthy. There is no history of recent buttock injections or trauma.

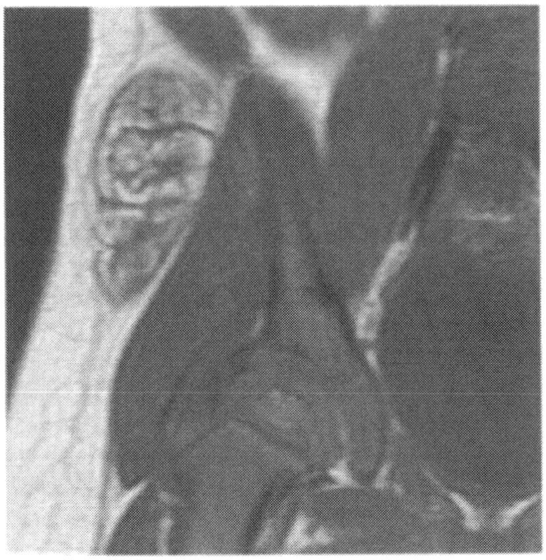

Figure 63A. Coronal T1.

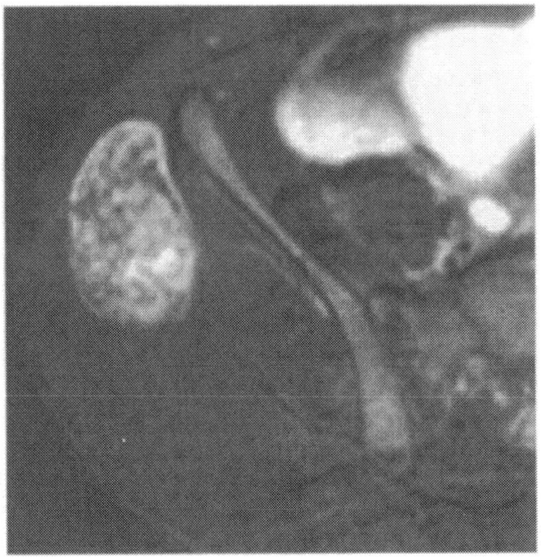

Figure 63B. Axial T2 FS.

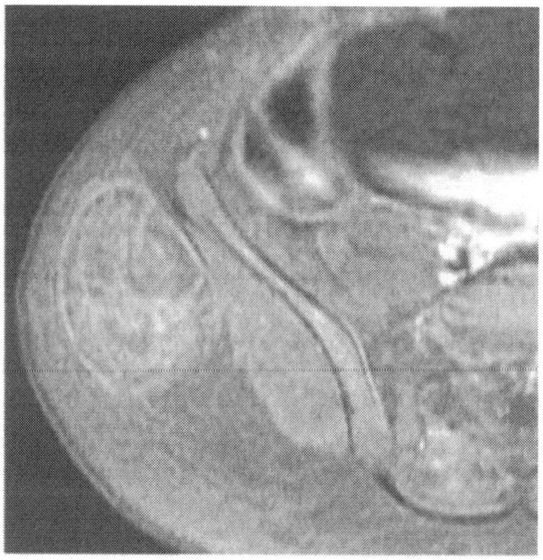

Figure 63C. Axial T1 post-Gd FS.

Figure 63A. There is a heterogeneous soft tissue mass confined to the subcutaneous soft tissues in the superior aspect of the right buttock. T1 hyperintense areas consistent with mature fat are interspersed with hypointense tissue and fibrous septae.

Figures 63B, 63C. The mass abuts, but does not invade the gluteus maximus muscle. It shows heterogeneous increased SI on T2 and it enhances. No flow voids are present. No calcifications were seen on plain radiographs (not shown).

Diagnosis

Lipoblastoma

Questions

1. T/F: Lipoblastomas have a high incidence of metastasis at presentation.
2. What is the differential diagnosis of an extremity soft tissue mass in an infant that contains both soft tissue and fat elements?

Discussion

Lipoblastoma is an uncommon benign tumor, usually seen in children under 3 years of age, with just over half presenting before age one (1). They are painless, growing masses and are encountered most often in the lower extremities. The tumors are composed of immature and mature adipocytes, mesenchymal cells, and fibrous septae (2). The natural history of these tumors is differentiation into simple lipomas with no reported cases of malignant transformation or metastasis (Answer to Question 1) (2, 3).

Lipoblastomas are usually confined to the subcutaneous soft tissues and have well-defined margins. The term lipoblastomatosis has been used when tumors are non-encapsulated, infiltrative, and involve both the subcutaneous soft tissues and deep muscles (4). Lipoblastomas may have variable soft tissue and fat SI present. They are rich in mature adipocytes and may be indistinguishable from simple lipomas. Mature adipocytes within these lesions follow the SI of subcutaneous fat on all imaging sequences. The immature cellular and fibrous elements demonstrate hypointensity on T1, hyperintensity on fluid sensitive sequences, and often enhance (4, 5). Predictably, lipoblastomas may mimic a soft tissue sarcoma when there is a paucity of mature adipocytes (Figures 63D–63F).

In the infant, alternative considerations for a mass with both soft tissue and fat elements include a complicated lipoma (secondary to fat necrosis/infection/trauma), involuting congenital or infantile hemangioma, venous malformation with adjacent fatty atrophy of muscle, teratoma, and fibrous hamartoma (Answer to Question 2). Fibrous hamartomas (Figures 63G–63I) most commonly occur in male children under one year of age and tend to involve the axillary fold (6). Fibrous hamartomas contain three elements: mature adipocytes, fibrous tissue, and primitive mesenchymal cells.

In older children, additional considerations for a tumor with both fat and soft tissue elements include liposarcoma, hibernoma, myolipoma, angiolipoma, chondroid lipoma, and lipofibromatosis (4). The MRI features of lipoblastoma may be indistinguishable from a liposarcoma (Figures 63J–63L), but the key differentiating point for these two soft tissue tumors is the age of the patient. Liposarcomas are extremely rare in children under 10 years of age, whereas lipoblastomas most frequently occur in children under 3 years of age.

At surgery, this tumor showed no evidence of underlying gluteal muscle invasion and the patient had complete excision of the buttock mass. The diagnosis of lipoblastoma was confirmed pathologically.

Orthopedic Perspective

The main issue for the orthopedist is to differentiate this lesion from the other tumors listed above. A biopsy may be necessary. These lesions are usually excised, and surgical considerations are similar to lipomas (see Case 72). As with lipomas, lipoblastomas may be observed if they are asymptomatic.

What the Clinician Needs to Know

1. The clinician needs to differentiate lipoblastoma from other benign and rarely malignant tumors that contain fat.
2. The extent of the lesion needs to be defined by MRI to aid in surgical planning since these are usually excised.

Answers

1. False.
2. Lipoblastoma, complicated lipoma, involuting congenital or infantile hemangioma, venous malformation with adjacent fatty atrophy of muscle, teratoma, and fibrous hamartoma.

Additional Examples

Lipoblastoma Mimicking a Sarcoma

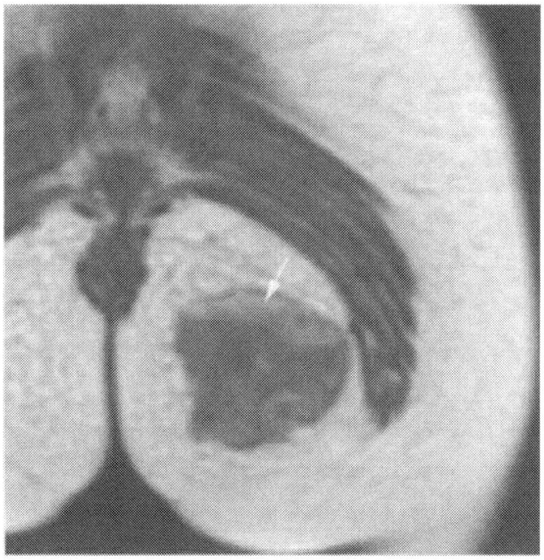

Figure 63D. Coronal T1.

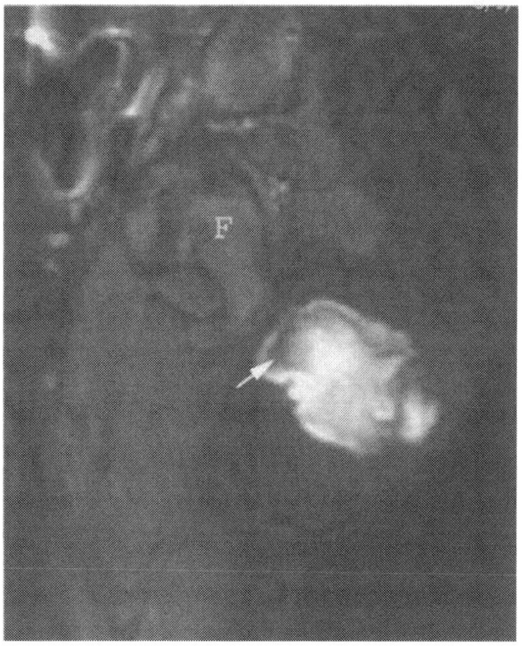

Figure 63E. Sagittal STIR.

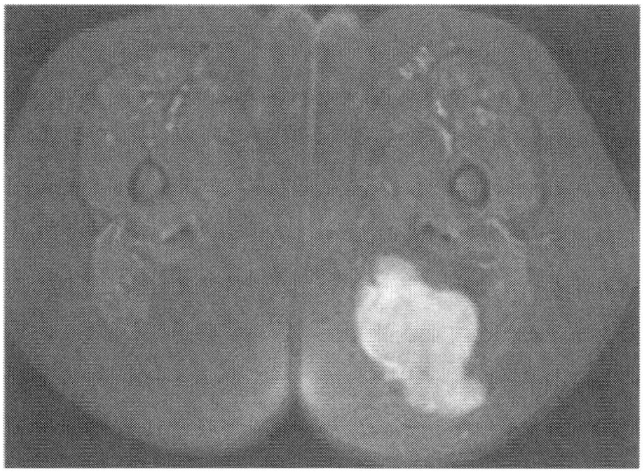

Figure 63F. Axial T1 post-Gd FS.

Findings

This is a 14-month-old girl with an enlarging left buttock mass.

Figures 63D, 63E, 63F. There is a well-defined mass in the subcutaneous soft tissues of the left buttock without deep invasion. There is a minimal amount of fat SI noted within the lesion **(arrows)**. The mass is markedly hyperintense on STIR and shows diffuse enhancement. This was pathologically proven to be a lipoblastoma. Femur (F).

Fibrous Hamartoma

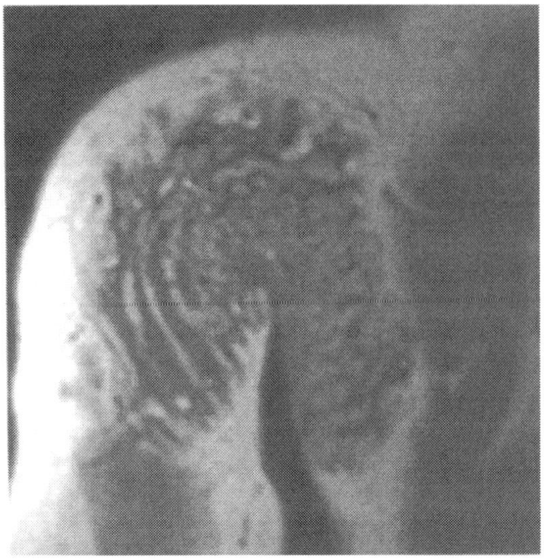

Figure 63G. Coronal T1 of the right shoulder.

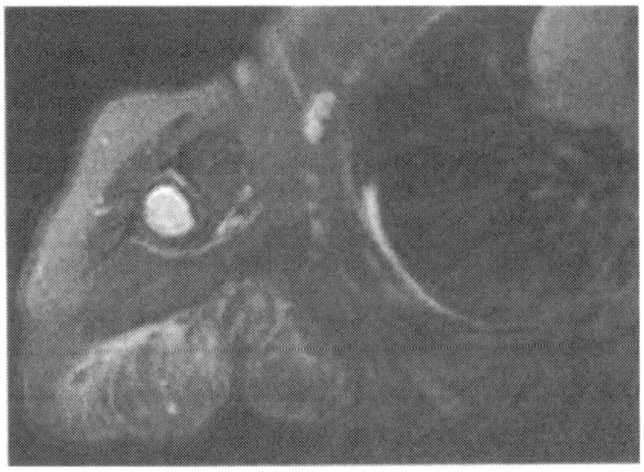

Figure 63H. Axial T2 FS.

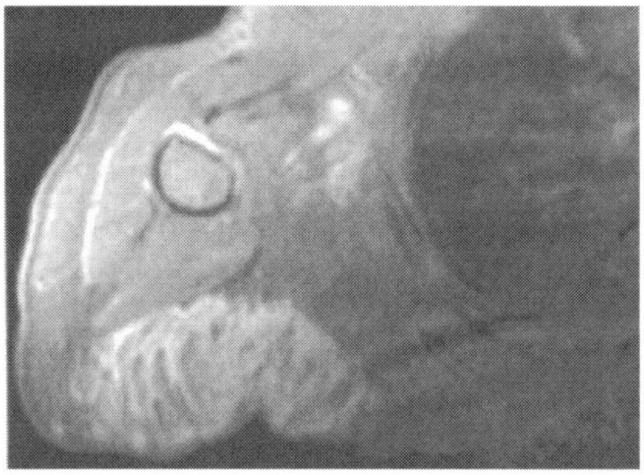

Figure 63I. Axial T1 post-Gd FS.

Findings

This is a 9-month-old boy with a painless, firm mass within the right axilla.

Figures 63G, 63H, 63I. The mass has characteristic organized stripes of mixed intermediate T1 and increased T2 SI separated by fat SI (7) and demonstrates heterogeneous enhancement. This was a pathologically proven fibrous hamartoma.

Myxoid Liposarcoma

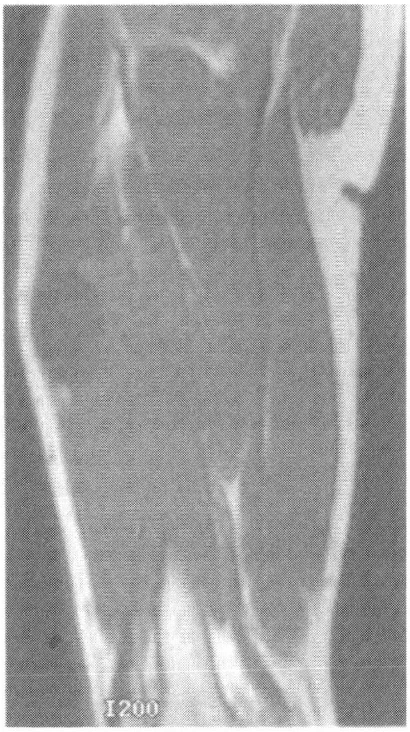

Figure 63J. Sagittal T1 of the right thigh.

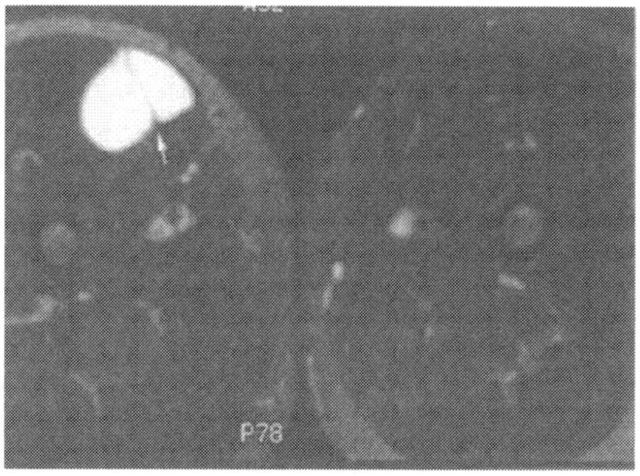

Figure 63K. Axial T2 FS.

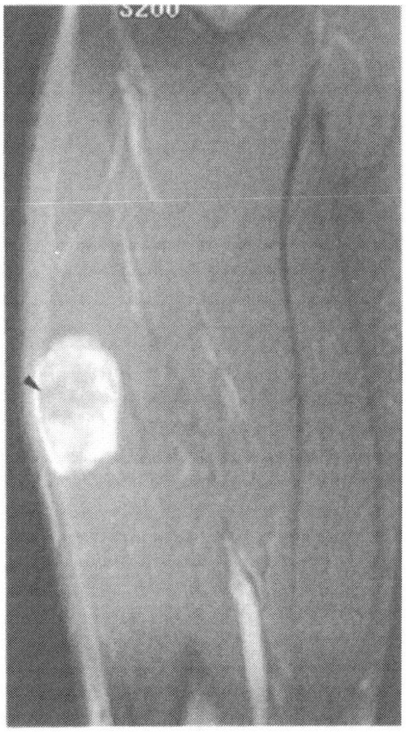

Figure 63L. Sagittal T1 post-Gd FS.

Findings

This is a 14-year-old boy with an enlarging well-defined mass in the vastus medialis muscle.

Figure 63J. This lesion is hypointense on T1 without evidence of fat signal.

Figure 63K. This lesion is strikingly hyperintense on T2 with a single septation **(arrow)**.

Figure 63L. There is intense peripheral and heterogeneous central enhancement **(arrowhead)**. This nonspecific appearance may be seen with a wide variety of soft tissue sarcomas. This was pathologically proven to be a myxoid liposarcoma.

Pitfalls and Pearls

Do not confuse the term lipoblastoma with a malignant lesion. Lipoblastomas are benign lesions usually found in children under 3 years of age, whereas liposarcomas are rare in children (4).

References

1. Harrer J, Hammon G, Wagner T, Bolkenius M. Lipoblastoma and lipoblastomatosis: A report of two cases and review of the literature. *Eur J Pediatr Surg* 2001; 11:342–349.
2. O'Donnell KA, Caty MG, Allen JE, Fisher JE. Lipoblastoma: Better termed infantile lipoma? *Pediatr Surg Int* 2000; 16:458–461.
3. Ha TV, Kleinman PK, Fraire A, et al. MR imaging of benign fatty tumors in children: Report of four cases and review of the literature. *Skeletal Radiol* 1994; 23:361–367.
4. Murphey MD, Carroll JF, Flemming DJ, Pope TL, Gannon FH, Kransdorf MJ. From the archives of the AFIP: Benign musculoskeletal lipomatous lesions. *Radiographics* 2004; 24:1433–1466.
5. Reiseter T, Nordshus T, Borthne A, Roald B, Naess P, Schistad O. Lipoblastoma: MRI appearances of a rare paediatric soft tissue tumour. *Pediatr Radiol* 1999; 29:542–545.
6. Dickey GE, Sotelo-Avila C. Fibrous hamartoma of infancy: Current review. *Pediatr Dev Pathol* 1999; 2:236–243.
7. Loyer EM, Shabb NS, Mahon TG, Eftekhari F. Fibrous hamartoma of infancy: MR-pathologic correlation. *J Comput Assist Tomogr* 1992; 16:311–313.

Case 64

History

This is a healthy 6-year-old girl with a leg-length discrepancy.

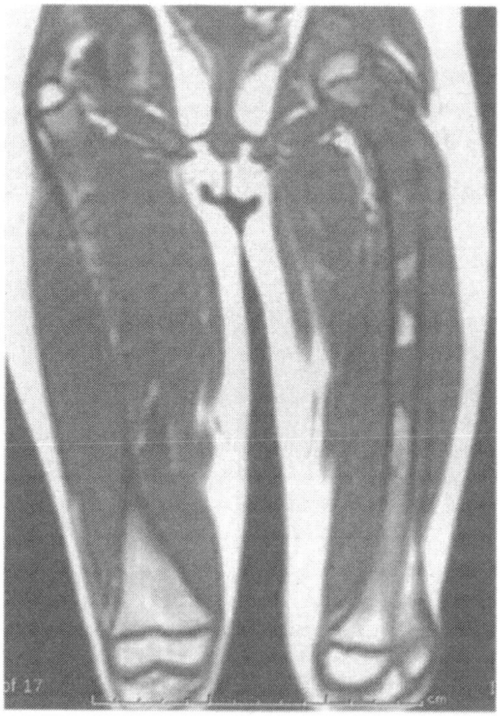

Figure 64A. Coronal T1.

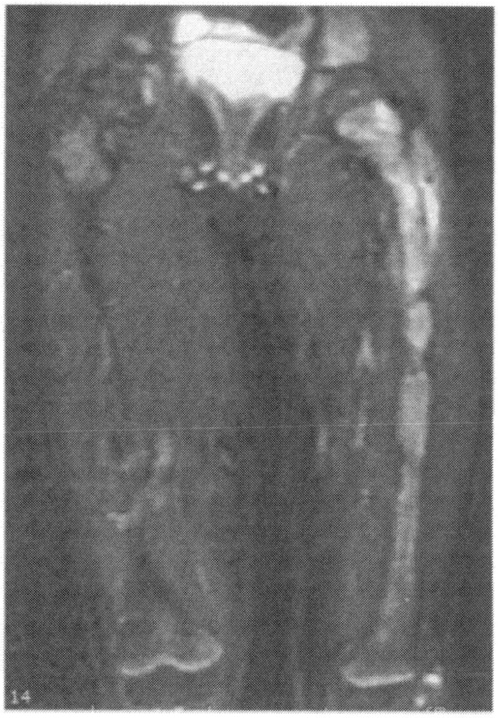

Figure 64B. Coronal STIR.

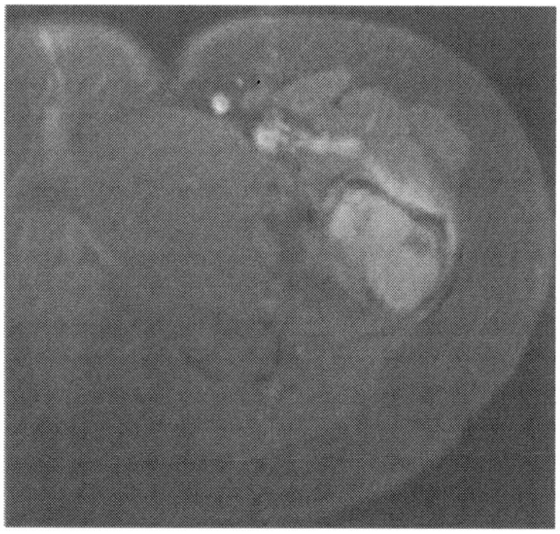

Figure 64C. Axial T1 post-Gd FS of the proximal left femur.

Figure 64A. There is slight medullary expansion and cortical thinning of the left femoral neck and mid-diaphysis.

Figures 64B, 64C, 64D. There is diffuse increased STIR SI and enhancement throughout the marrow of the femur as well as the lower ilium. There is increased SI and enhancement within the juxtacortical soft tissues lateral to the proximal femur. There is also a focus of increased enhancement in the capital femoral epiphysis, similar in character to the other lesions.

Figure 64E. There is a Shepard's crook deformity. Variable ground glass opacity and marrow expansion is seen involving the proximal $^{3}/_{4}$ of the femur. Cortical thinning is seen, particularly along the medial aspect of the femoral neck **(arrowhead)**.

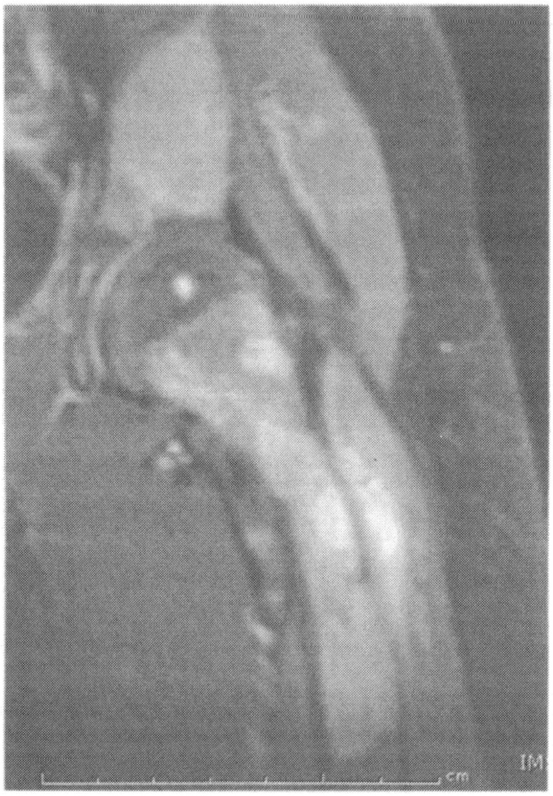

Figure 64D. Coronal T1 post-Gd FS.

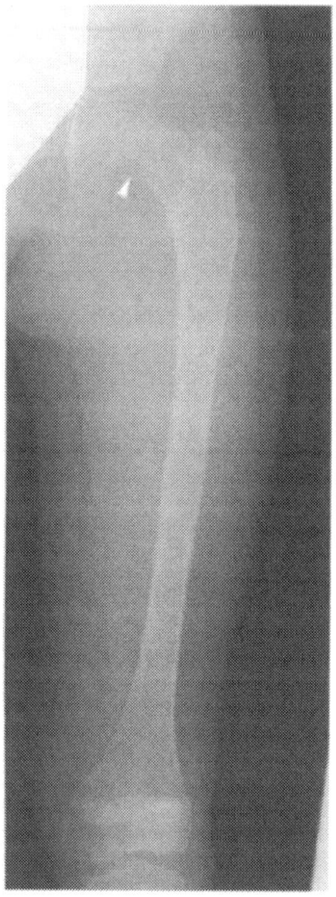

Figure 64E. AP radiograph.

Diagnosis

Polyostotic fibrous dysplasia

Questions

1. What is the most common location for monostotic fibrous dysplasia?
2. What is the incidence of malignant transformation of fibrous dysplasia?

Discussion

Fibrous dysplasia is a common, benign fibro-osseous lesion of bone. Eighty percent of cases are monostotic. Monostotic fibrous dysplasia most commonly affects the ribs, followed by the proximal femur (Answer to Question 1) (1). Polyostotic fibrous dysplasia most often affects the femur, tibia, pelvis, and foot. Monostotic fibrous dysplasia usually presents in the first and second decade of life, whereas polyostotic fibrous dysplasia usually presents during the first decade of life. Polyostotic fibrous dysplasia is associated with two rare syndromes. McCune-Albright syndrome is usually seen in females, and is manifest by the triad of café au lait skin pigmentation, precocious puberty, and polyostotic fibrous dysplasia. Mazabraud's syndrome is characterized by intramuscular myxomas associated with polyostotic fibrous dysplasia (2).

Radiographically, fibrous dysplasia usually arises centrally or slightly eccentrically within the medullary cavity of the diaphysis or metadiaphysis of long bones. The lesions may demonstrate well-defined sclerotic or nonsclerotic margins, endosteal scalloping, cortical thinning, and mild bony expansion. Lesions have variable opacity and may be osteolytic, ground glass, sclerotic, or contain a grossly mineralized matrix (3). Fibrous dysplasia may appear multilobulated with variable internal septation. Angular and modeling deformity may result from weightbearing stresses, such as the Shepard's crook deformity of the femur (Figure 64E).

Fibrous dysplasia has a variable appearance on MRI. In one study examining 13 biopsy proven cases, fibrous dysplasia was hypointense on T1, variable SI on T2 (38% hypointense, 62% hyperintense) (Figures 64F–64H), and all showed enhancement (central > peripheral). Extraosseous soft tissue extension was noted in 31% of cases (4). Hyperintensity on T2 correlated pathologically with areas containing fewer bony trabeculae, less cellularity, and fewer collagen fibers compared with areas that were hypointense on T2. A hypointense rim on all imaging sequences was also identified in most cases, presumably representing endosteal bone reaction. Cystic change, fluid-fluid levels, blood products, and secondary aneurysmal bone cyst formation may also be seen (4, 5).

The radiographic features of fibrous dysplasia usually point to the diagnosis. MRI evaluation is performed if the plain radiographic features are atypical (Figures 64I–64K) or if there is unexplained pain. Etiologies for focal pain include occult pathologic fracture and superimposed malignancy. Sarcomatous transformation occurs in approximately 0.5% of cases; osteosarcoma is most common, followed by fibrosarcoma (Answer to Question 2) (6). The incidence of sarcomatous transformation is higher in patients with polyostotic fibrous dysplasia.

In this patient, fibrous dysplasia was also seen in the left tibia (not shown). The patient carried the diagnosis of polyostotic fibrous dysplasia, but did not have McCune-Albright or Mazabraud's syndrome. The pathologic diagnosis of fibrous dysplasia was confirmed when the patient underwent curretage and packing of the proximal left femur lesion.

Orthopedic Perspective

The primary issues for the orthopedist are to distinguish fibrous dysplasia from other lesions and to decide if prophylactic fixation is necessary to prevent pathological fracture. Fibrous dysplasia can mimic several other neoplasms (unicameral bone cyst, nonossifying fibroma, osteofibrous dysplasia, adamantinoma and others). The decision about whether to undertake prophylactic treatment is difficult. Short of CT imaging for biomechanical analysis, the prediction of fracture risk is not precise. Lesions in the lower extremity, especially in the intertrochanteric region of the proximal femur, are at highest risk. It is easier to treat an impending varus deformity before it occurs. Treatment of fibrous dysplasia in the major weightbearing bones of the lower extremity includes curettage, bone graft packing of the lesion, and internal fixation. In the upper extremity, internal fixation is usually not required. Unless there is pain or an impending fracture, surgical treatment is not necessary. Patients with polyostic fibrous dysplasia should be evaluated for endocrinopathies or myxomas.

What the Clinician Needs to Know

1. Distinguish fibrous dysplasia from other benign and malignant neoplasms.
2. The clinician needs to assess fracture risk, especially in the long bones of the lower extremity, and treat accordingly.
3. Assess for polyostotic disease and evaluate for endocrinopathies or other lesions such as myxoma.

Answers

1. Ribs.
2. 0.5%. The incidence is higher in patients with the polyostotic fibrous dysplasia.

Additional Examples

Fibrous Dysplasia

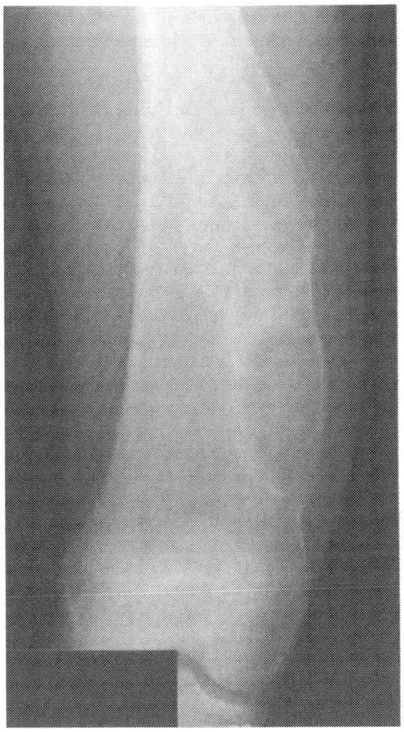

Figure 64F. AP radiograph of the distal right femur.

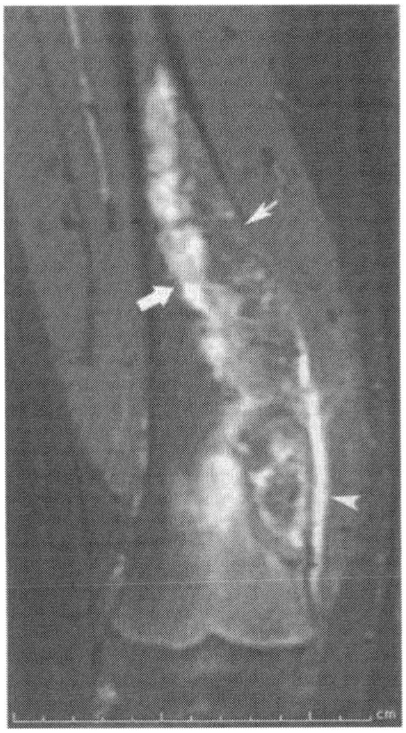

Figure 64G. Coronal STIR.

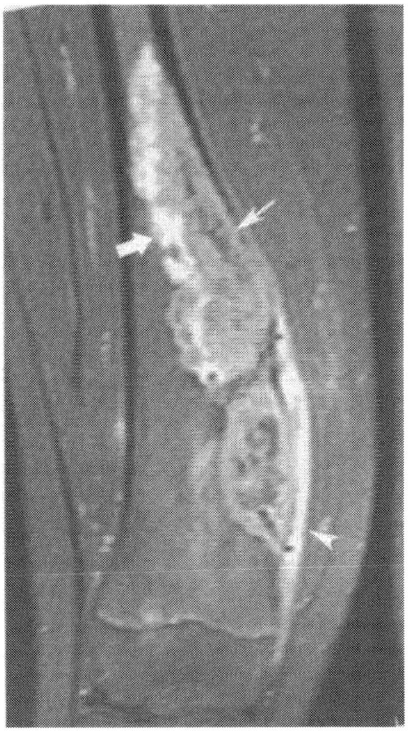

Figure 64H. Coronal T1 post-Gd FS.

Findings

This is a 13-year-old girl with known fibrous dysplasia.

Figure 64F. An eccentric expansile lesion is present with a sclerotic sharp transition zone and variable ground glass opacity.

Figures 64G, 64H. The lesion is heterogeneous with areas of hypointensity and moderate enhancement medially **(thin arrows)**, and hyperintensity and marked peripheral enhancement laterally **(thick arrows)**. There is also periosteal hyperintensity and enhancement medially **(arrowheads)**. The lesion was curettaged and packed with both fibular bone graft and bone chips. This lesion was pathologically proven to be fibrous dysplasia.

Fibrous Dysplasia with Pathologic Fracture

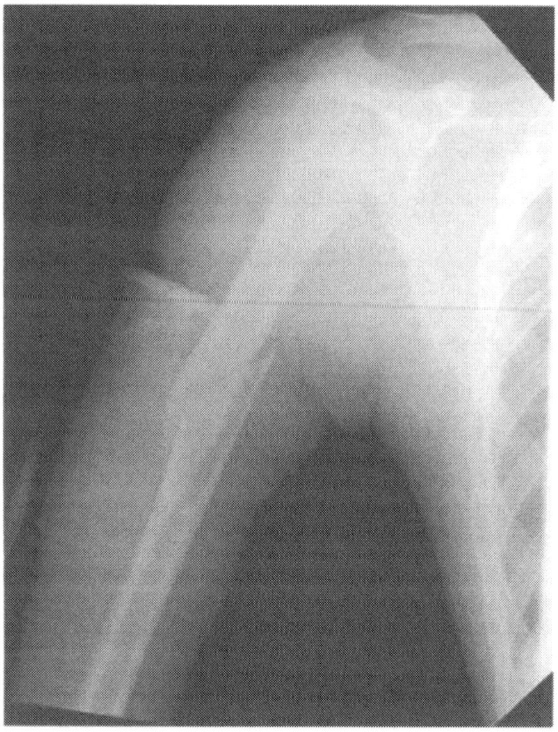

Figure 64I. AP radiograph of the right humerus.

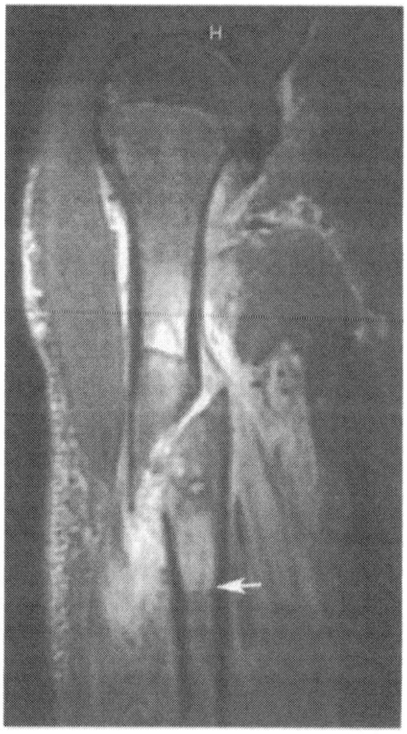

Figure 64J. Sagittal STIR.

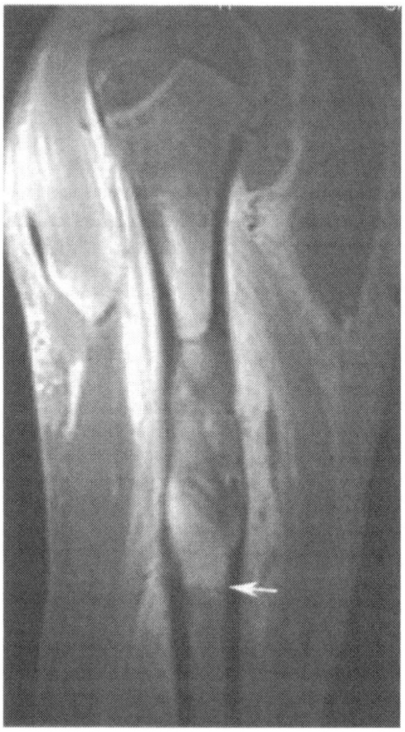

Figure 64K. Sagittal T1 post-Gd FS.

Findings

This 14-year-old boy sustained a pathologic fracture of his right humerus while throwing a baseball.

Figure 64I. Based on location, age, and appearance, the most likely diagnosis was a unicameral bone cyst.

Figures 64J, 64K. This lesion shows diffuse hyperintensity in the marrow and adjacent soft tissues. A sharply defined zone of increased marrow enhancement is present (**arrows**, distal margin), suggesting a tumor matrix, rather than blood or cyst fluid. This was pathologically proven to be fibrous dysplasia. The lesion was curettaged and packed with cancellous autograft and demineralized bone paste. Open reduction and internal fixation of the fracture with four titanium nails was performed.

Pitfalls and Pearls

1. The presence of ill-defined extraosseous soft tissue extension in the setting of fibrous dysplasia is not uncommon and should not be misinterpreted as malignant transformation.
2. A variant of fibrous dysplasia is osteofibrous dysplasia, which typically arises from the diaphysis of the anterior tibia cortex (see Case 73).

References

1. Kransdorf MJ, Moser RP, Jr., Gilkey FW. Fibrous dysplasia. *Radiographics* 1990; 10:519–537.
2. Iwasko N, Steinbach LS, Disler D, et al. Imaging findings in Mazabraud's syndrome: Seven new cases. *Skeletal Radiol* 2002; 31:81–87.
3. Smith SE, Kransdorf MJ. Primary musculoskeletal tumors of fibrous origin. *Semin Musculoskelet Radiol* 2000; 4:73–88.
4. Jee WH, Choi KH, Choe BY, Park JM, Shinn KS. Fibrous dysplasia: MR imaging characteristics with radiopathologic correlation. *AJR Am J Roentgenol* 1996; 167:1523–1527.
5. Okada K, Yoshida S, Okane K, Sageshima M. Cystic fibrous dysplasia mimicking giant cell tumor: MRI appearance. *Skeletal Radiol* 2000; 29:45–48.
6. Schwartz DT, Alpert M. The malignant transformation of fibrous dysplasia. *Am J Med Sci* 1964; 247:1–20.

Case 65

History

This is an 11-year-old girl with a known leg-length discrepancy (the right leg is 1 cm longer than the left) and new left-sided knee pain.

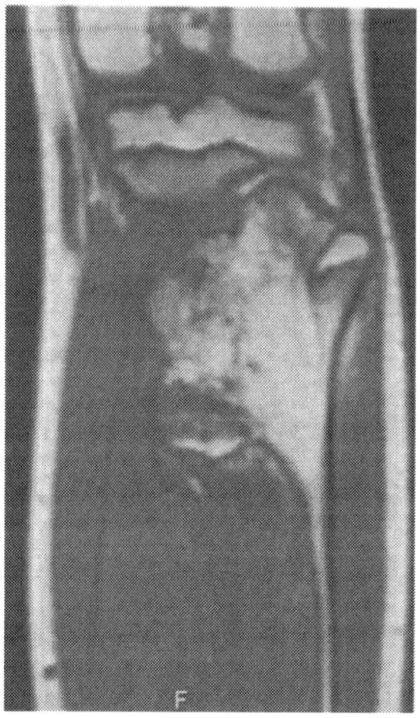

Figure 65A. Coronal T1 of the left proximal lower leg.

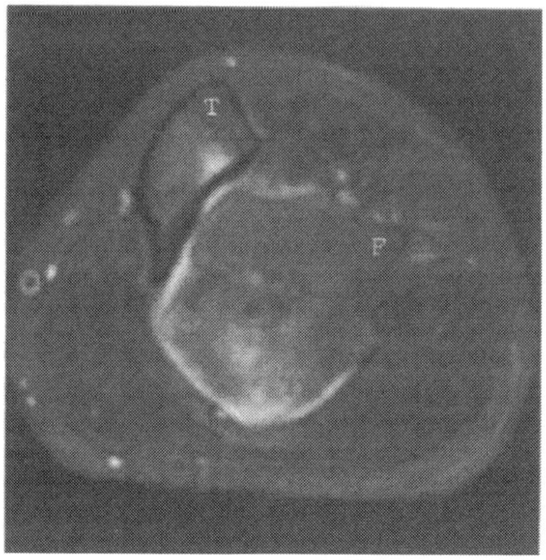

Figure 65B. Axial T2 FS.

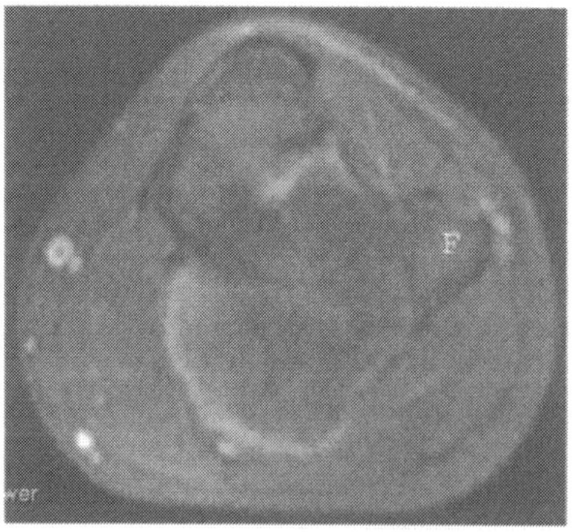

Figure 65D. Axial T1 post-Gd FS.

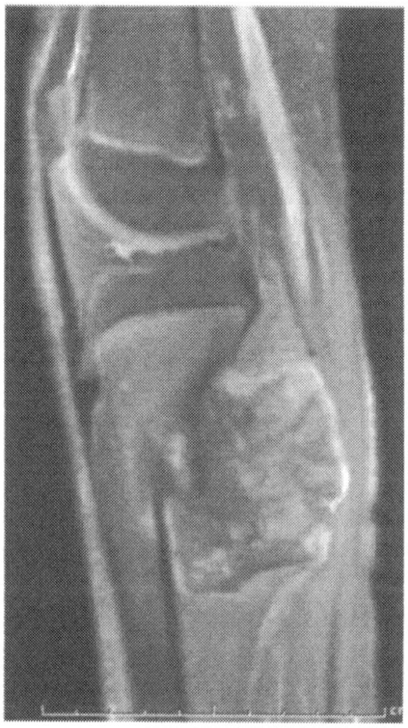

Figure 65C. Sagittal PD FS.

Figure 65A. There is a large, broad based mass arising from the medial aspect of the proximal fibula. The fibular marrow cavity and cortex are continuous with this mass.

Figures 65B, 65C, (65B with annotations). A thin cartilaginous cap is evident that is markedly hyperintense on fluid-sensitive sequences (**thin arrows**). A hypointense band overlies the cartilaginous cap, compatible with perichondrium (**arrowheads**). Ill-defined T1 hypointense areas located beneath the cap become hyperintense with a ring-and-arc pattern on fluid-sensitive sequences (*). These regions correspond to cartilaginous elements with variable water content and ossification. There is bony remodeling and edema (**thick arrow**) within the adjacent tibia.

Figure 65D. There is enhancement of the ring-and-arc pattern within the lesion matrix, as well as enhancement of the thin cartilage cap, corresponding to the T2 hyperintensity in Figure 65B. Fibula (F).

Figure 65E. Plain radiograph shows that this mass arises from the fibula. The margins are sclerotic, the matrix is mineralized, and the marrow cavity and cortex are continuous with the mass.

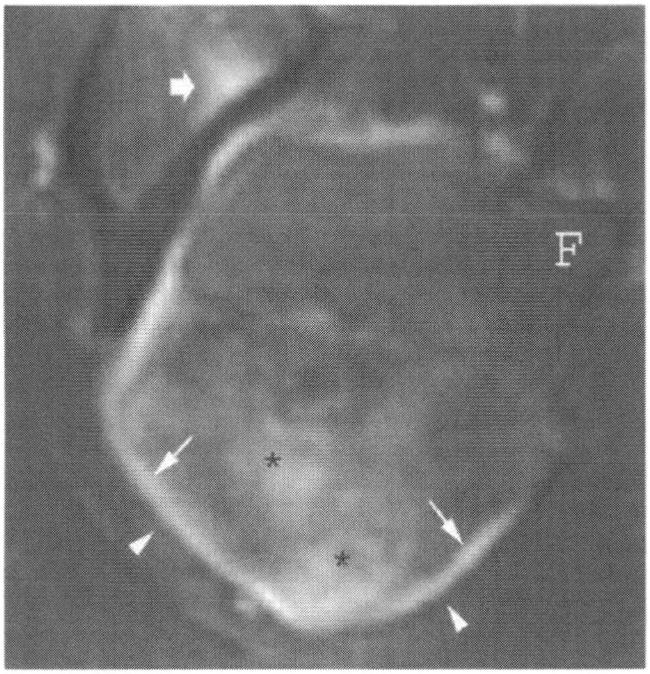

Figure 65B* Annotated detail.

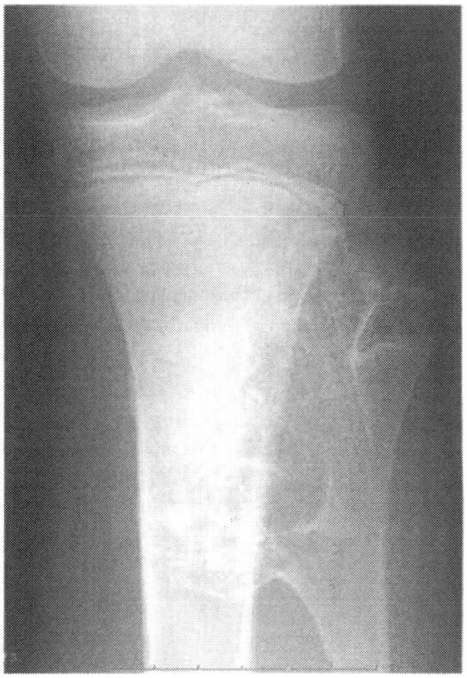

Figure 65E. AP radiograph.

Diagnosis

Sessile osteochondroma of the proximal fibula

Questions

1. What percentage of solitary osteochondromas undergoes malignant transformation?
2. T/F: Focal pain over an osteochondroma in a child is usually indicative of malignant transformation.

Discussion

Osteochondromas are commonly encountered benign tumors of bone. The lesions usually arise sporadically, but may also occur in an autosomal dominant pattern, multiple hereditary exostoses (AKA osteochondromatosis, diaphyseal aclasis) (Figures 65F–65H). They can develop following trauma or radiation. The lesions are typically metaphyseal, usually affecting the lower extremities, and are most common around the knees (1). Once skeletal maturity is reached, the vast majority of these benign tumors cease growing.

On radiography, the cortex and medullary cavity of the parent bone are continuous with the osteochondroma. The cartilaginous components of the osteochondroma may demonstrate chondroid matrix calcification. The apex of osteochondromas typically points away from the nearby joint. Osteochondromas may be pedunculated (narrow stalk) (Figure 65I) or sessile (broad based) (Figures 65A–65E). The radiographic features are usually diagnostic and no further imaging is required for asymptomatic lesions.

On MRI, osteochondromas may appear completely cartilaginous prior to the appearance of an ossification center (Figures 65J–65L). The cartilaginous cap of osteochondromas is usually hypo- to isointense on T1 and hyperintense on fluid-sensitive sequences related to its high water content (Figure 65B) (2, 3). A thin, low SI layer overlying the cartilage cap is usually seen correlating with the perichondrium pathologically (1). Areas of chondroid matrix within the lesion may manifest variable signal intensity on fluid sensitive sequences related to differences in water composition and stage of ossification (Figure 65B). On post-Gd sequences, septal and thin peripheral enhancement may also be seen overlying the cartilaginous cap. Enhancement of the ring-and-arc pattern of the cartilaginous component may be seen as well and does not necessarily imply chondrosarcomatous transformation (2).

MRI best assesses the soft tissue and osseous complications related to osteochondromas. In the soft tissues, tenosynovitis, neurovascular compression, and reactive bursa formation over the osteochondroma may occur (4). Pseudoaneurysm formation and distal muscle atrophy may occur related to neurovascular compression. The spectrum of skeletal complications related to osteochondromas include physeal growth alterations, angular deformities (e.g., tibiotalar slant, pseudo-Madelung deformity), and erosion or modeling disturbance of adjacent osseous structures. Direct impact, stress injury, or pathologic fracture of an osteochondroma may also occur. Fractures are more common with pedunculated than sessile osteochondromas (5). The complex morphology of these lesions is not always revealed adequately on standard orthogonal imaging, and oblique reformatted images generated from a 3D fat saturated spoiled gradient echo or similar sequence may better show the relationship of the cartilage cap of the osteochondroma to adjacent neurovascular structures.

Chondrosarcomatous transformation occurs in approximately 1% of patients with solitary osteochondromas and 5% to 25% of patients with multiple hereditary exostoses (Answer to Question 1) (2). However, chondrosarcomatous transformation usually occurs in patients around 40 years of age, and is rare in children (6). The suspicion of malignancy should be raised when there is focal pain at the site of an existing osteochondroma and the cartilaginous cap is over 2cm thick in adult patients and over 3cm in children (2). However, focal pain is far more likely to result from mechanical contact of an osteochondroma with adjacent structures, than from malignant transformation (Answer to Question 2).

Alternative diagnostic considerations for an osteochondroma include, but are not limited to: bizarre parosteal osteochondromatous proliferation (BPOP) (Figures 65M–65O), periosteal chondroma, traction exostosis, myositis ossificans, metachondromatosis, and parosteal osteosarcomas. BPOP is an osteochondroma variant that most commonly occurs in the hands and feet, but approximately 25% occur in the tubular long bones (7). Osteochondromas may mimic periosteal chondromas, particularly when they are imaged before ossification has begun (Figures 65J–65L). Traction exostosis may occur at ligamentous or tendinous attachments to bone (1). The diagnosis of metachondromatosis should be considered in the setting of osteochondromas with enchondromas. With metachondromatosis, the osteochondromas point toward the joint rather than away. Osteochondromas should be differentiated from parosteal osteosarcomas, which most commonly arise from the posterior metaphysis of the distal femur in patients 30–40 years of age (8).

In this patient with a typical sessile osteochondroma, the knee pain was likely related to altered mechanics due to the large size of the lesion and abutment of the osteochondroma against the tibia, causing marrow edema.

Orthopedic Perspective

The diagnosis is usually evident from the radiograph, and for solitary osteochondromas, observation is the rule unless they cause symptoms due to fracture, irritation of a nerve, or bursitis. For large osteochondromas that require excision, such as the fibular osteochondroma shown here, the identification of the relationship of the lesion relative to the major vessels and nerves is crucial. These can be challenging cases due to the size and location of the osteochondroma. MR angiography may be useful. Particularly difficult sites for osteochondromas are the medial aspect of the proximal humerus, and the posterior aspect of the proximal tibia and fibula. Preoperative planning is essential to avoid complications since these are benign lesions and excision is in a sense elective. The treatment principle is to attempt to remove the entire cartilaginous cap along with the stalk to avoid recurrence and MRI is very helpful in preoperative planning. Patients with multiple osteochondromas may develop angular deformities and limb-length inequality, which may require surgical correction.

What the Clinician Needs to Know

1. Is there continuity of the cortex of the host bone and the osteochondroma?
2. What is the extent of the osteochondroma relative to the adjacent major nerves and blood vessels?

Answers

1. 1%.
2. False.

Additional Examples

Multiple Hereditary Exostoses

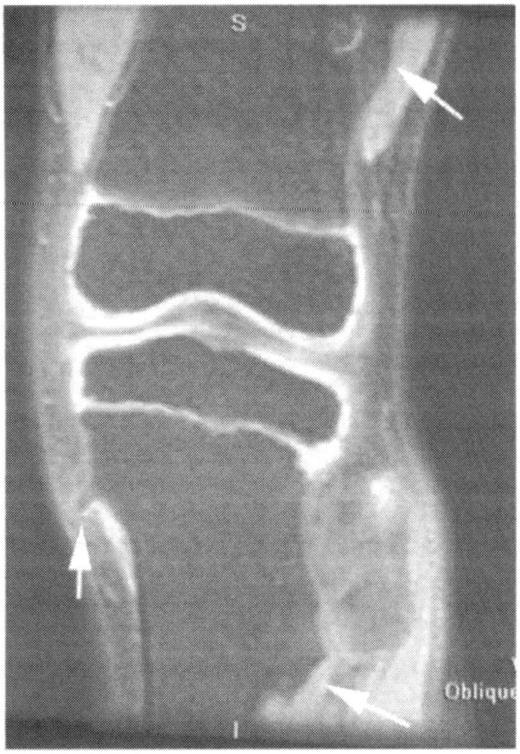

Figure 65F. Coronal 3D SPGR FS of the left knee.

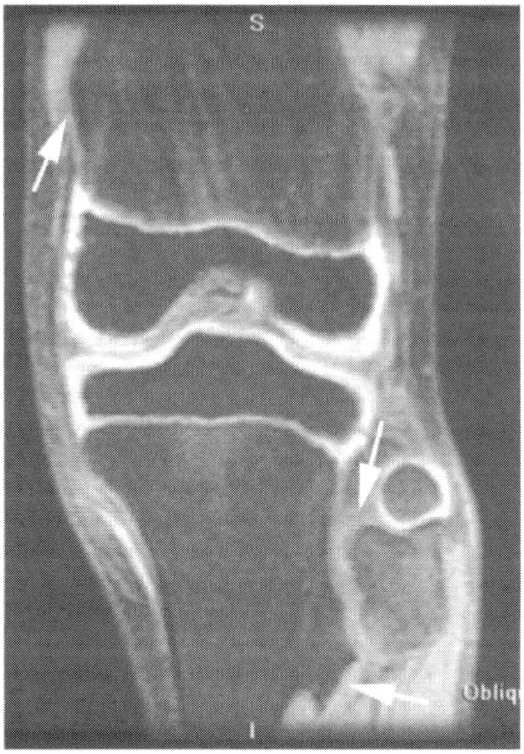

Figure 65G. Coronal 3D SPGR FS.

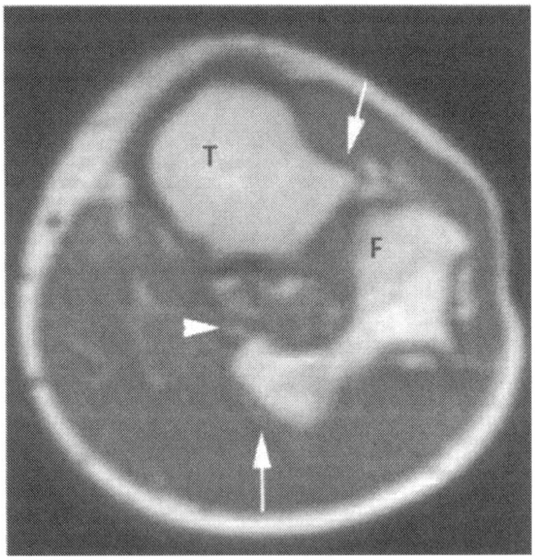

Figure 65H. Axial PD (proximal tibia-fibula).

Findings

This is an 8-year-old boy with known multiple hereditary exostoses and worsening posterior knee pain.

Figures 65F, 65G, 65H. There are multiple sessile and pedunculated osteochondromas arising from the distal femur and proximal tibia and fibula **(arrows)**. On the axial sequence, a pedunculated osteochondroma arising from the fibula impinges upon the tibial nerve and posterior tibial artery/vein complex **(arrowhead)**. Tibia (T), fibula (F).

Pedunculated Osteochondroma

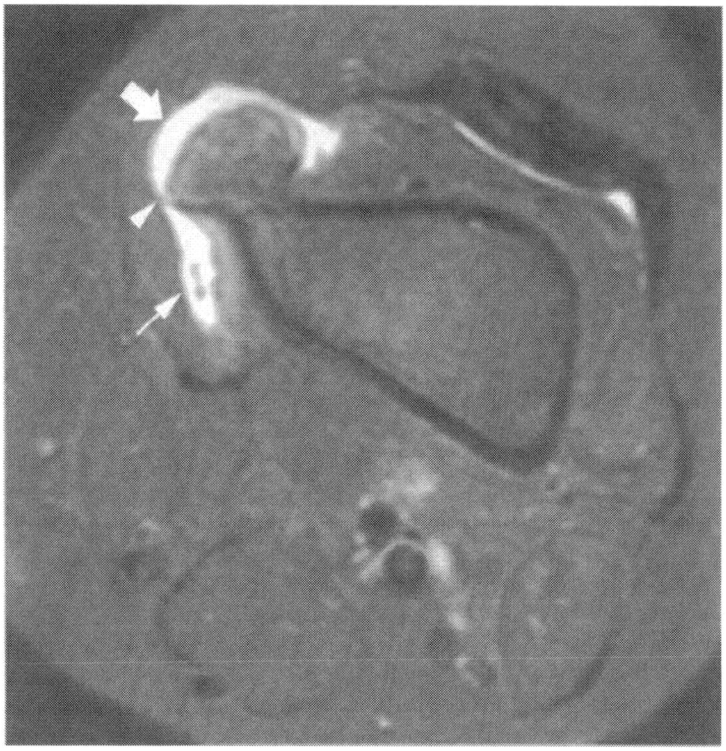

Figure 65I. Axial STIR of the left distal femur.

Findings

This is a 10-year-old girl who had acute medial knee pain after running.

Figure 65I. There is a pedunculated osteochondroma identified with a bulbous cartilaginous cap **(thick arrow)**. The perichondrium **(arrowhead)** separates the cartilaginous cap from adjacent bursa-like fluid collection **(thin arrow)** and edema extending into the vastus medialis. An osteochondroma with a large cartilage cap was confirmed pathologically.

Sessile Osteochondroma Mimicking a Periosteal Chondroma

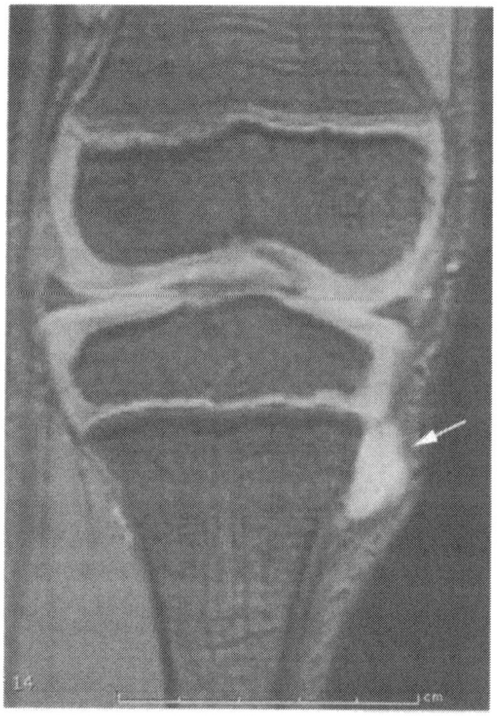

Figure 65J. Coronal MPGR of the right knee.

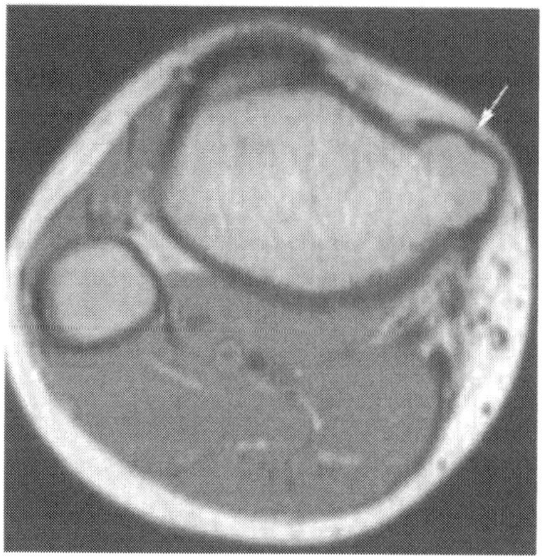

Figure 65K. Axial PD.

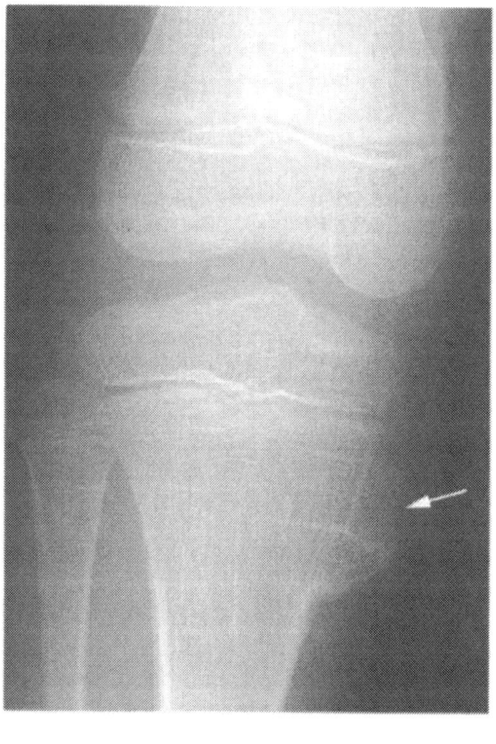

Figure 65L. AP radiograph 2 years later.

Findings

This is a 6-year-old boy whose mother noticed a painless bump over the medial aspect of his knee.

Figures 65J, 65K. There is a broad-based mass along the medial tibia metaphysis that abuts the physis and follows cartilage SI **(arrows)**.

Figure 65L. The radiograph obtained 2 years later demonstrates ossification of this mass **(arrow)**, and the diagnosis of a sessile osteochondroma can now be made with confidence.

Bizarre Parosteal Osteochondromatous Proliferation (BPOP)

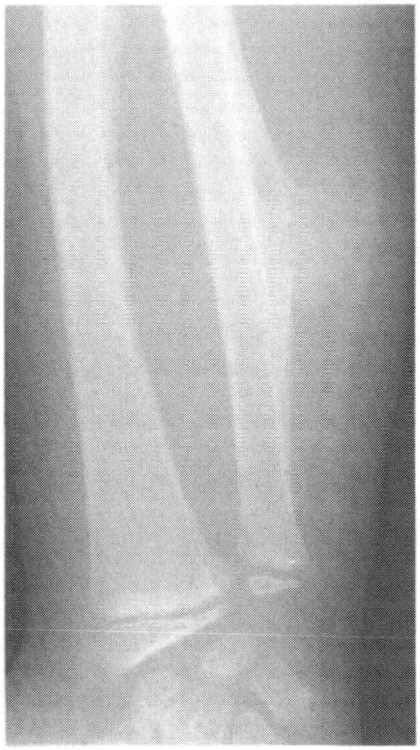

Figure 65M. AP radiograph of the left forearm. (Reprinted from Zambrano E et al. with permission (9).)

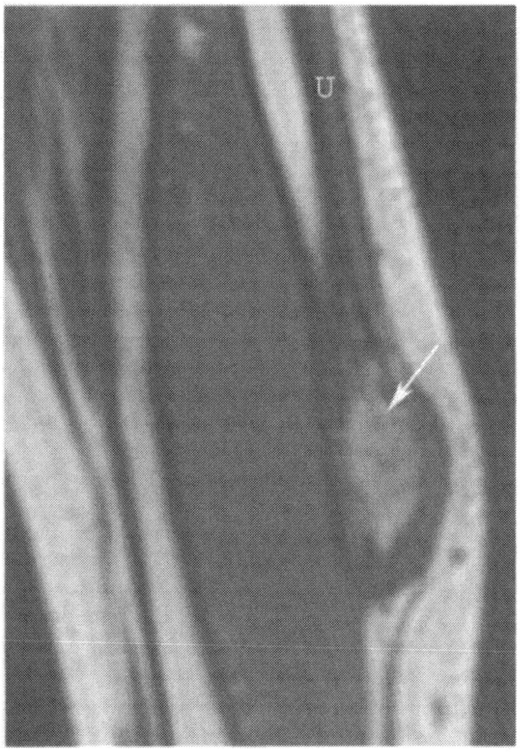

Figure 65N. Coronal T1 of the left forearm.

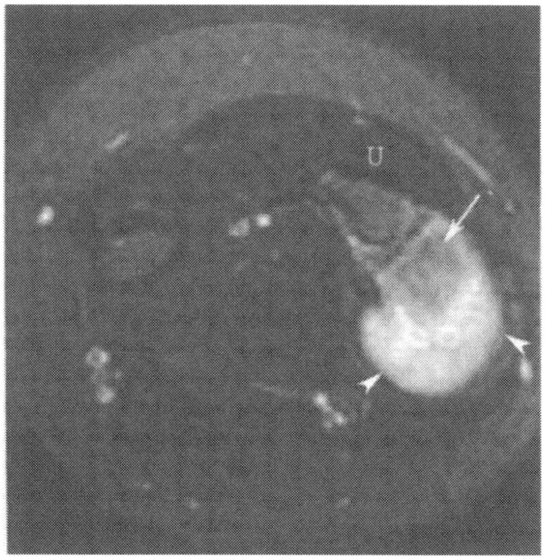

Figure 65O. Axial PD FS through the ulnar lesion.

Findings

This is a 7-year-old girl with an enlarging mass in her left forearm.

Figure 65M. Plain radiograph demonstrates an exophytic mass arising from the mid-diaphysis of the ulna. There is expansion and radiolucency of the adjacent medullary cavity.

Figures 65N, 65O. Fatty marrow is evident within the lesion **(arrows)** and a thick cartilaginous component is identified at the vertex **(arrowheads)**. This was a pathologically proven BPOP. Ulna (U).

Pitfalls and Pearls

1. Chondrosarcomatous transformation of osteochondromas is very rare in children. In a child, the most likely cause for pain over an osteochondroma is pressure effect on adjacent soft tissues or direct injury to the osteochondroma.
2. Osteochondromas are readily diagnosed on plain radiographs. MRI evaluation is performed to assess for bone and soft tissue complications.
3. Osteochondromas may enlarge rapidly during growth spurts, prompting a referral for MRI because of concerns of an aggressive neoplasm.

References

1. Murphey MD, Choi JJ, Kransdorf MJ, Flemming DJ, Gannon FH. Imaging of osteochondroma: Variants and complications with radiologic-pathologic correlation. *Radiographics* 2000; 20:1407–1434.
2. Woertler K, Lindner N, Gosheger G, Brinkschmidt C, Heindel W. Osteochondroma: MR imaging of tumor-related complications. *Eur Radiol* 2000; 10:832–840.
3. Robbin MR, Murphey MD. Benign chondroid neoplasms of bone. *Semin Musculoskelet Radiol* 2000; 4:45–58.
4. Brien EW, Mirra JM, Luck JV, Jr. Benign and malignant cartilage tumors of bone and joint: Their anatomic and theoretical basis with an emphasis on radiology, pathology and clinical biology. II. Juxtacortical cartilage tumors. *Skeletal Radiol* 1999; 28:1–20.
5. Karasick D, Schweitzer ME, Eschelman DJ. Symptomatic osteochondromas: imaging features. *AJR Am J Roentgenol* 1997; 168:1507–1512.
6. Lee KC, Davies AM, Cassar-Pullicino VN. Imaging the complications of osteochondromas. *Clin Radiol* 2002; 57:18–28.
7. Torreggiani WC, Munk PL, Al-Ismail K, et al. MR imaging features of bizarre parosteal osteochondromatous proliferation of bone (Nora's lesion). *Eur J Radiol* 2001; 40:224–231.
8. Murphey MD, Robbin MR, McRae GA, Flemming DJ, Temple HT, Kransdorf MJ. The many faces of osteosarcoma. *Radiographics* 1997; 17:1205–1231.
9. Zambrano E, Nose V, Perez-Atayde AR, et al. Distinct chromosomal rearrangements in subungual (Dupuytren) exostosis and bizarre parosteal osteochondromatous proliferation (Nora lesion). *Am J Surg Pathol* 2004; 28:1033–1039.

Case 66

History

This is a 13-year-old girl with right hip pain and fever.

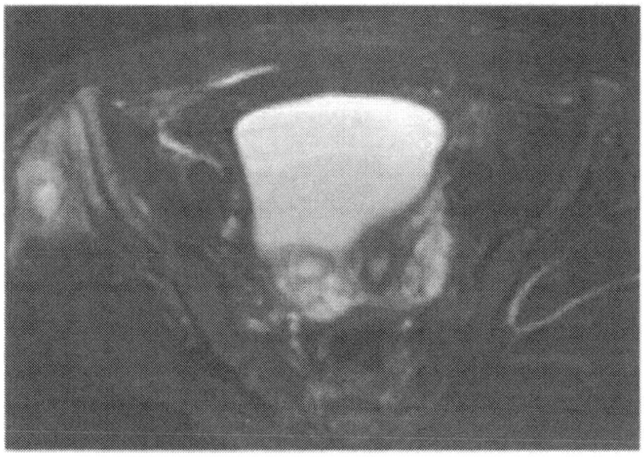

Figure 66A. Axial T2 FS.

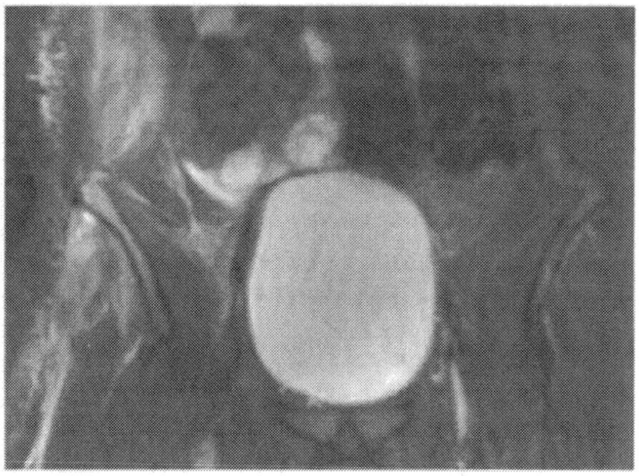

Figure 66B. Coronal STIR.

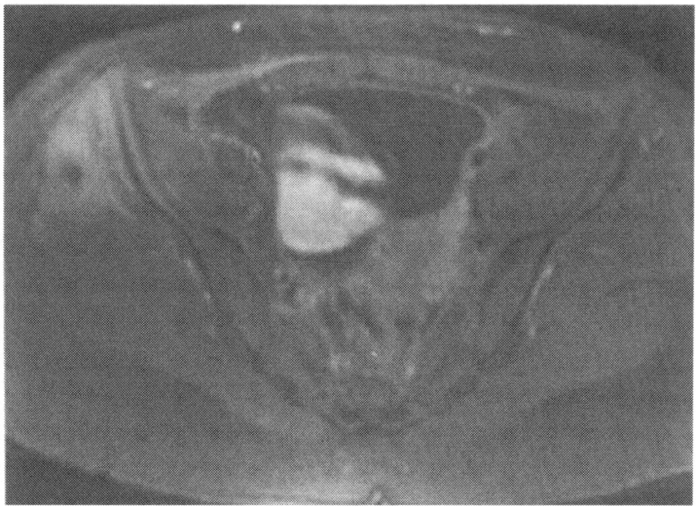

Figure 66C. Axial T1 post-Gd FS.

Figures 66A, 66B, 66C, (66B, 66C with annotations). There is increased SI on fluid-sensitive sequences and enhancement in the right iliac bone, gluteus minimus, gluteus medius, and subcutaneous tissues of the anterior abdominal wall and upper thigh. A multiloculated fluid collection is noted at the gluteal attachment **(thin arrow)**. A small amount of intraperitoneal free fluid is also seen **(arrowhead)**. On the post-Gd sequence, a small loculated fluid collection is present in the gluteus medius muscle **(thick arrow)**.

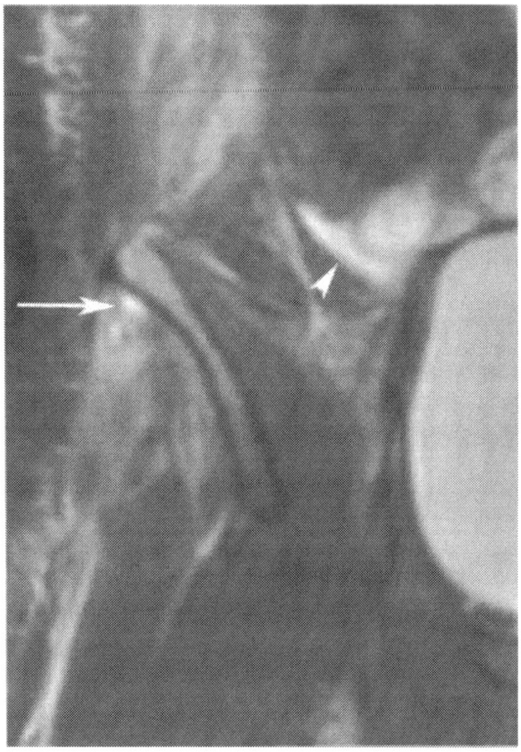

Figure 66B* Annotated detail.

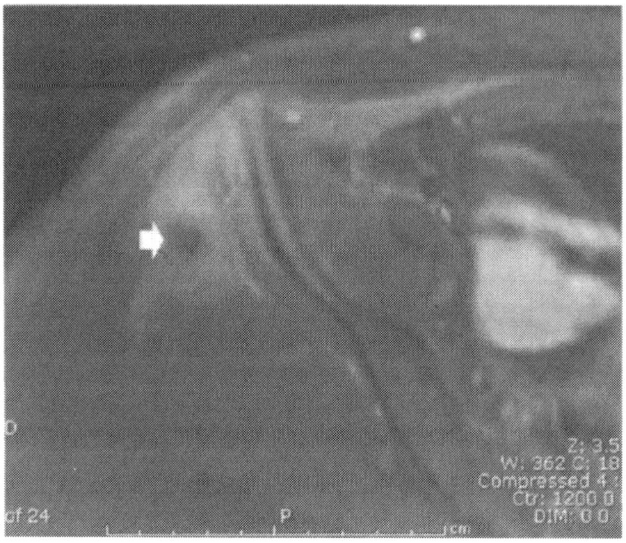

Figure 66C* Annotated detail.

Diagnosis

Iliac crest osteomyelitis

Questions

1. Which muscles attach to the iliac crest?
2. T/F: Hip pain is the most common presentation for pelvic osteomyelitis.

Discussion

The iliac crest is a broad-based apophysis that serves as the attachment site of the abdominal musculature (transverse abdominis, internal oblique, and external oblique) and the quadratus lumborum muscle (Answer to Question 1) (1). The iliac crest is predisposed to similar avulsive injuries and metaphyseal equivalent infections that are seen elsewhere in the pelvis, such as the rectus femoris muscle attachment to the anterior inferior iliac spine (Figure 66D), sartorius attachment to the anterior superior iliac spine, and hamstring attachment to the ischial tuberosity. When pelvic bone apophyses are near a joint, secondary septic arthritis may develop (see Case 55) (2).

Imaging of pediatric pelvic osteomyelitis is often delayed because symptoms are nonspecific and children may be poor historians. The most common clinical presentation is hip pain (Answer to Question 2) (3). These patients may also present with peritoneal signs once inflammation extends to the peritoneal cavity (Figure 66B). Therefore, abdominal and hip imaging with ultrasonography is often explored before the diagnosis of pelvic osteomyelitis is made.

The osseous and soft tissue inflammatory changes seen with pelvic osteomyelitis may be mistaken for an avulsive injury or aggressive neoplasm. Post-gadolinium T1W sequences are helpful to distinguish loculated fluid collections and nonspecific muscle edema related to avulsive or muscle injury and a discrete solid enhancing mass due to a primary bone tumor. The primary imaging feature of pelvic sarcomas is mass effect with absent or minor infiltrative edema compared with osteomyelitis (Figure 66E). Distinguishing early osteomyelitis versus an avulsive injury in the absence of juxtacortical inflammatory changes or fluid collections may be difficult (Figures 66F, 66G). The presence of nonenhancing fluid collections may also be nonspecific when pelvic myotendinous or avulsive injury have been severe, leading to a rim enhancing juxtacortical muscle hematoma or seroma. These post-traumatic collections may mimic a soft tissue abscess associated with pelvic osteomyelitis. Therefore, correlation with blood cultures and clinical exam is essential.

This patient initially presented to an ER with abdominal pain and a low-grade fever. A KUB and abdomen CT were normal. She was discharged and given nonsteroidal anti-inflammatory medications. She returned later with a 4-day history of right hip pain and a fever. A pelvic MRI was ordered after hip radiographs and an ultrasound were normal. Blood cultures grew *Staphylococcus aureus*. Due to the small size of the gluteus medius muscle abscess, the patient was treated nonoperatively with antibiotics and did well.

Orthopedic Perspective

Pelvic osteomyelitis can be diagnostically challenging because patients may present with nonspecific signs and symptoms. Abdominal and lower extremity pain may lead to imaging of the wrong body part and a delayed diagnosis. When symptoms appear to be localized to the hip, MRI of the entire pelvis is usually advisable to insure diagnosis of extra-articular pelvic osteomyelitis. The presence, location, size of abscess and phlegmon should be described as well as their relationship with neurovascular structures, joint and physes.

What the Clinician Needs to Know

1. Define both the intra- and extrapelvic soft tissue extent of pelvic osteomyelitis.
2. Exclude any discrete abscess collections within the intra- and extrapelvic spaces.

Answers

1. Abdominal wall muscles: transverse abdominis, internal oblique, external oblique, and the quadratus lumborum.
2. True.

Additional Examples

Anterior Inferior Iliac Spine (AIIS) Osteomyelitis

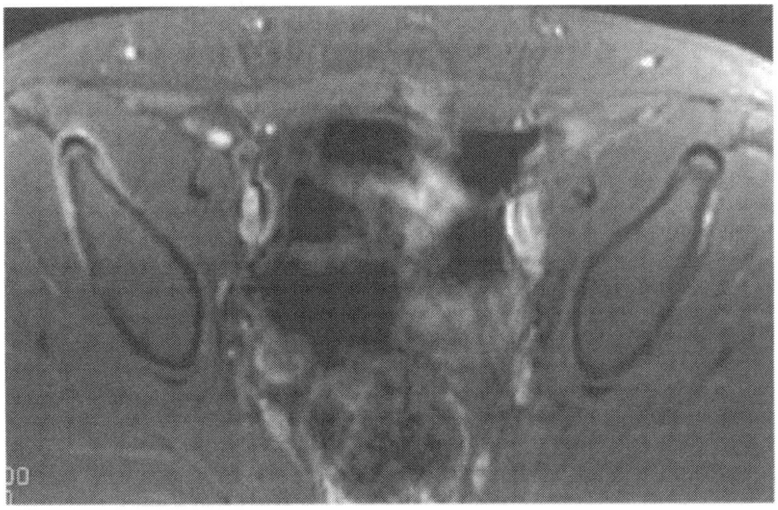

Figure 66D. Axial T1 post-Gd FS.

Findings

This is a 9-year-old girl with right hip pain for one day after playing soccer.

Figure 66D. The AIIS periosteal, apophyseal, and medullary bone enhancement is nonspecific, and could represent either a rectus femoris AIIS avulsion injury or osteomyelitis. Blood cultures grew *Staphylococcus aureus*. She was treated for presumed osteomyelitis.

Ewing's Sarcoma of the Iliac Wing

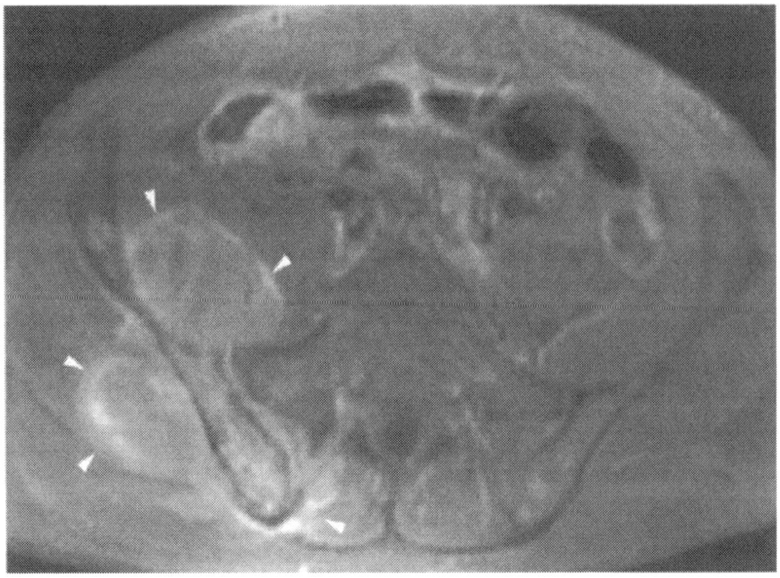

Figure 66E. Axial T1 post-Gd FS.

Findings

This is a 16-year-old boy with a 3-month history of worsening right hip pain.

Figure 66E. There is a large enhancing soft tissue mass centered in the right iliac wing, extending posteriorly to the sacroiliac joint and paraspinal musculature. The mass displaces the iliacus muscle anteriorly and infiltrates and displaces the gluteal muscles posteriorly **(arrowheads)**. The dominant feature is the large mass without the typical inflammatory changes of an infectious process.

Iliac Crest Apophysitis

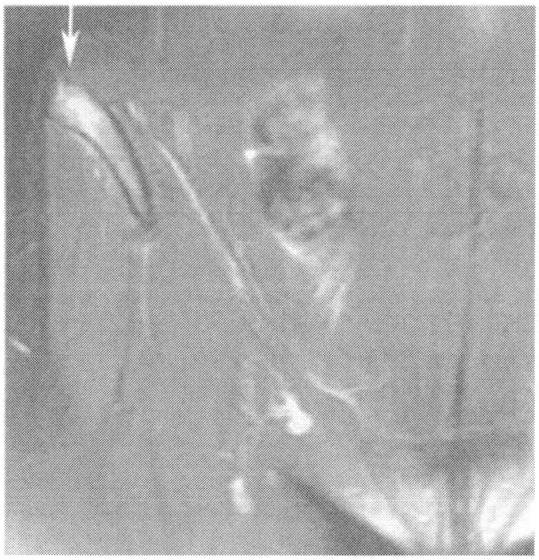

Figure 66F. Coronal T2 FS.

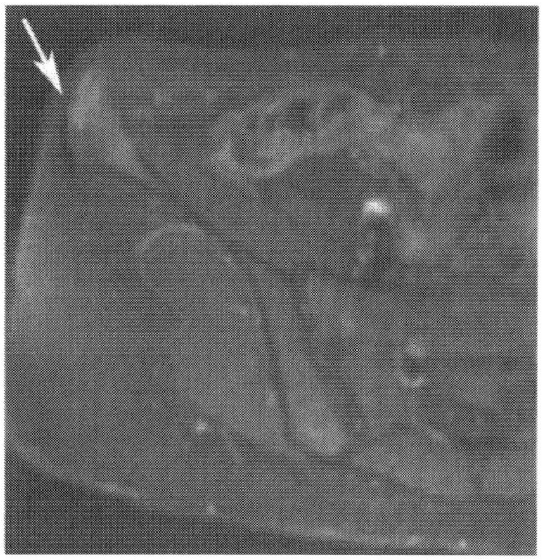

Figure 66G. Axial T1 post-Gd FS.

Findings

This is a 16-year-old girl with right iliac crest pain related to running.

Figures 66F, 66G. There is increased T2 SI and enhancement in the iliac crest apophysis and metaphyseal equivalent region **(arrows)**. There are no adjacent soft tissue inflammatory changes. This patient had no clinical or laboratory evidence of osteomyelitis. Therefore, she was treated with resting and not given antibiotics.

Pitfalls and Pearls

Iliac crest avulsion injuries and early osteomyelitis may be indistinguishable by MRI if there is no associated abscess. Short term radiologic follow-up and clinical correlation may be required to distinguish these two entities.

References

1. Stevens MA, El-Khoury GY, Kathol MH, Brandser EA, Chow S. Imaging features of avulsion injuries. *Radiographics* 1999; 19:655–672.
2. Sturzenbecher A, Braun J, Paris S, Biedermann T, Hamm B, Bollow M. MR imaging of septic sacroiliitis. *Skeletal Radiol* 2000; 29:439–446.
3. Davidson D, Letts M, Khoshhal K. Pelvic osteomyelitis in children: A comparison of decades from 1980–1989 with 1990–2001. *J Pediatr Orthop* 2003; 23:514–521.

Case 67

History

This is a 3-year-old boy with an enlarging right great toe mass. Radiographs of the toe demonstrated enlargement of the soft tissues without evidence of calcifications or osseous abnormality (not shown).

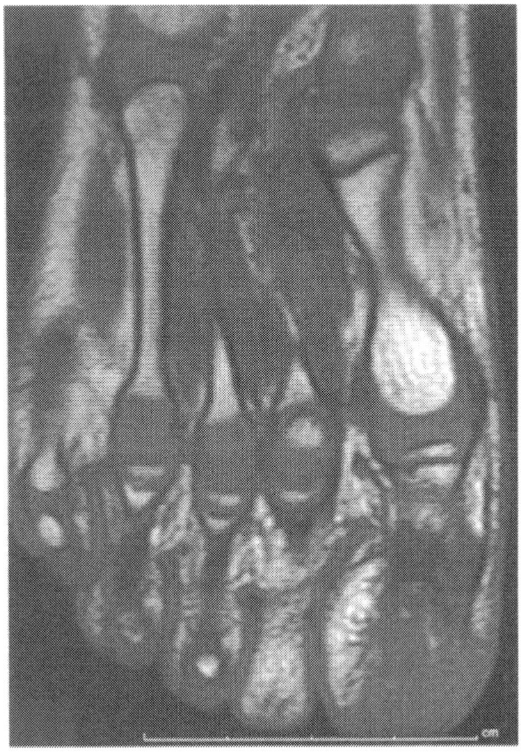

Figure 67A. Axial (footprint) T1 of the right forefoot.

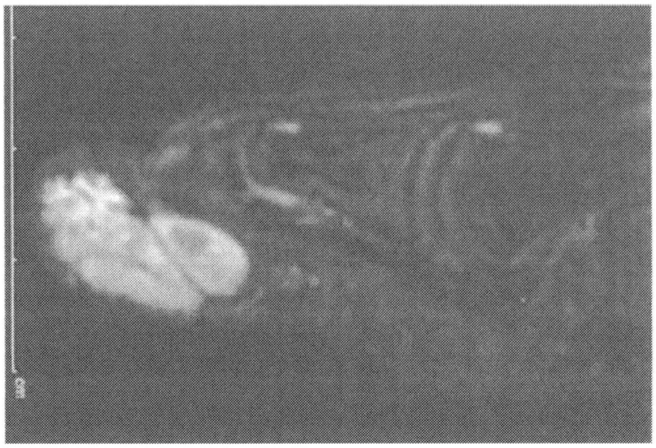

Figure 67B. Sagittal T2 FS of the great toe.

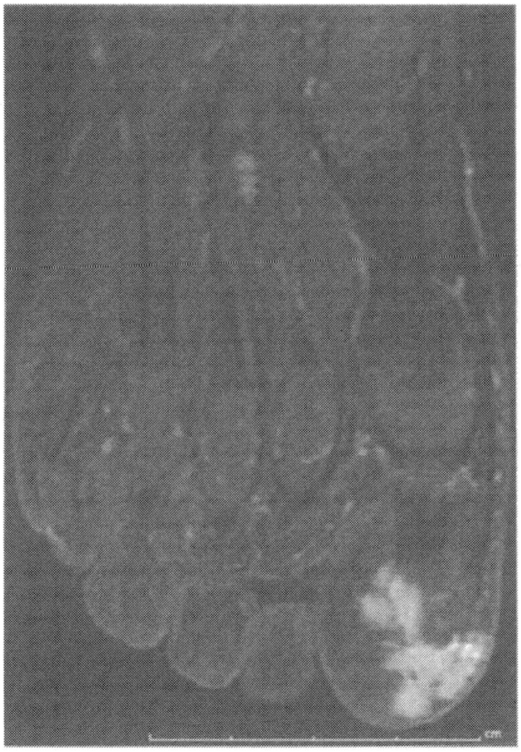

Figure 67C. Axial (footprint) T1 post-Gd FS.

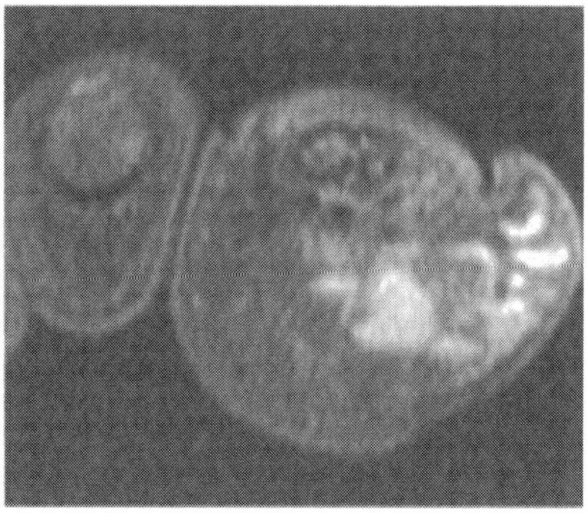

Figure 67D. Coronal T1 post-Gd FS.

Figures 67A, 67B, 67C, 67D, (67B with annotations). Within the subcutaneous soft tissues of the plantar aspect of the great toe, there is a multilobulated mass that is hypointense on T1, markedly hyperintense on T2, and shows patchy, tubular enhancement. The osseous structures are not involved. A hypointense intraluminal structure is identified **(arrow)**, compatible with thrombus. A fluid-fluid level is also present **(arrowhead)**.

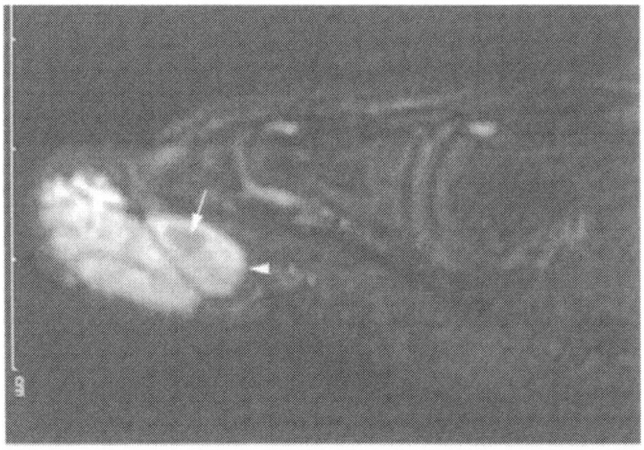

Figure 67B* Annotated.

Diagnosis

Venous malformation of the great toe

Questions

1. T/F: Venous malformations often spontaneously involute.
2. T/F: Pure venous malformations are always associated with hemihypertrophy.

Discussion

Vascular anomalies are common soft tissue masses in the pediatric population. Studies based on pathologic confirmation underrepresent the prevalence of vascular anomalies (1, 2); these lesions usually do not come to biopsy since the diagnosis is frequently made based on clinical and imaging features.

Mulliken and Glowacki have proposed a classification of vascular anomalies that separates vascular malformations from vascular tumors based on their biologic and clinical behavior (3). This system has been widely accepted and adopted by the International Society for the Study of Vascular Anomalies (ISSVA) (Table 67A) (4). Based on this classification system, the vast majority of symptomatic vascular anomalies found in children after the age of 12 months, adolescents, and adults are vascular malformations, not hemangiomas as often expressed in the literature. An alternative classification for benign vascular anomalies divides hemangioma into five different groups based on their pathologic characteristics. These include capillary, cavernous, arteriovenous, venous, and mixed lesions (5). This pathologic classification system is potentially confusing because it does not distinguish between infantile hemangiomas that spontaneously involute and venous malformations that may be progressively disfiguring with age and do not involute (Answer to Question 1).

Venous malformations (VM) are frequently imaged with MRI and are characteristically isointense to muscle on T1 and markedly hyperintense on fluid-sensitive sequences. The margins are often lobulated and fluid-fluid levels may be seen. Focal T1 and T2 hypointense intraluminal filling defects may be identified within venous mal-

Table 67A. Vascular anomalies.

Vascular tumors	Hemangioma	Infantile (usually present at approximately 3 months) Congenital hemangioma (present at birth) • RICH (rapidly involuting congenital hemangioma) • NICH (noninvoluting congenital hemangioma: persists after 1 year of age)
	Other vascular tumors	Kaposiform hemangioendothelioma Tufted angioma Angiosarcoma
Vascular malformations	High flow	Arteriovenous fistula (AVF) Arteriovenous malformation (AVM)
	Low flow	Capillary malformation Venous malformation Lymphatic malformation • Macrocystic • Microcystic Combined

Source: Data from Burrow PE et al. (11).

formations, representing thrombi (Figures 67E). When a thrombus within a venous structure calcifies, it is called a phlebolith (Figure 67F). The utility of MRI in this context is to define soft tissue extent: is the lesion restricted to one anatomic space or is it multicompartmental? Is the lesion purely subcutaneous or is there also involvement of muscle, bone (Figures 67G, 67H), or viscera?

On post-Gd sequences, VMs usually show patchy, tubular enhancement, reflecting slow flow physiology (6). Imaging venous malformations too early after gadolinium administration may simulate a lymphatic or venolymphatic malformation (Figures 67I, 67J). With appropriate delay, venous malformations will enhance, except where there are thrombi. On delayed post-Gd dynamic MR angiography, contrast delineation of tubular and nodular venous sinusoids as well as anomalous tortuous draining veins may be seen (7).

With pure venous malformations, limb hypoplasia may be seen (Answer to Question 2) (8). Limb enlargement is often seen with lymphatic, capillary, mixed venolymphatic, and arteriovenous malformations (9).

Venous malformations should not be confused with arteriovenous malformations (AVM) (Figures 67K–67N). An AVM is composed of a complex network of high flow vascular channels that directly connect arterial and venous systems (6). Soft tissue mass, arterioles, and a capillary bed are absent. AVM should be distinguished from an arteriovenous fistula (AVF), which represents a single arterial-venous channel without intervening arteriole and capillary bed. MRI of AVMs demonstrates a confluence of multiple tubular flow voids on T1, T2, and T1 post-Gd sequences because of high flow characteristics. Dynamic post-gadolinium MR angiography is useful to delineate both feeding and early draining vascular channels (Figure 67N). AVMs should be distinguished from lesions with AVM-like channels such as infantile and congenital hemangioma (see Case 40) and soft tissue sarcomas such as infantile fibrosarcoma (see Case 95).

If MRI is indeterminate, ultrasound may be used as a supplementary imaging tool to asses flow characteristics. Ultrasound may also be used for screening soft tissue lesions so that the MRI may be tailored to the principal diagnostic possibilities. Tubular slow flow vessels are seen with venous spectral Doppler in venous malformations, whereas high flow vessels with arterial spectral Doppler are usually apparent with AVMs (10).

In this patient, the right great toe mass was resected and surgically confirmed to represent a venous malformation. Surgical resection was preferred over sclerotherapy because the venous malformation was both focal and superficial.

What the Clinician Needs to Know

1. Lesions that are amenable for resection (discrete lesions, unicompartmental, and superficial) versus lesions that are better suited for sclerotherapy (diffuse lesions, multicompartmental, and deep).
2. Is the lesion a pure venous malformation or is the lesion mixed?
3. What are the main draining veins of the venous malformation?

Answers

1. False.
2. False.

Additional Examples

Extensive VM with Phleboliths

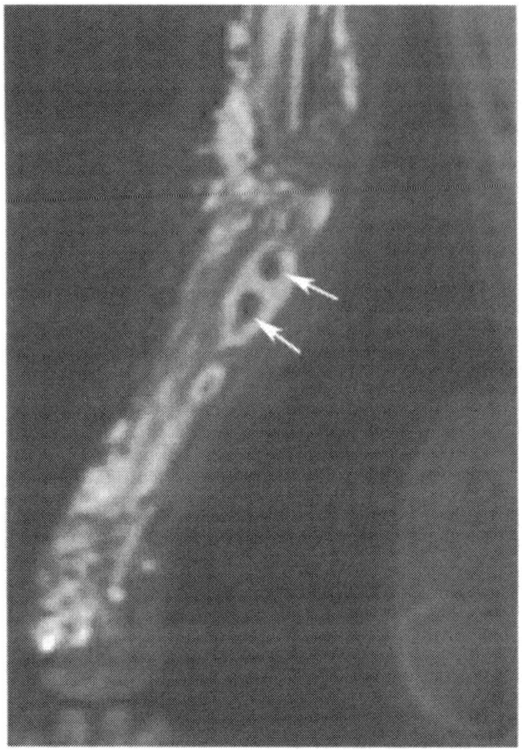

Figure 67E. Coronal STIR of the right arm.

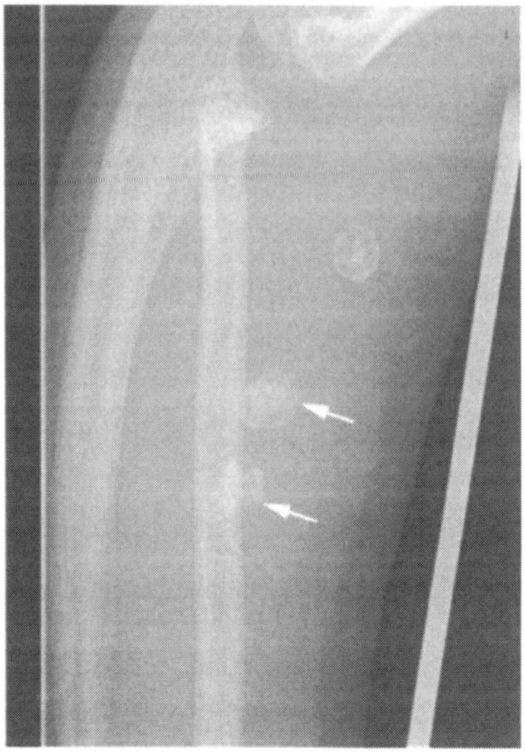

Figure 67F. Lateral radiograph of the right arm.

Findings

This is a 5-year-old girl with an extensive subcutaneous and intramuscular venous malformation of the right arm.

Figures 67E, 67F. Two rounded areas of decreased SI **(arrows)** correspond with phleboliths seen on the plain radiograph.

Extensive VM with Osseous Involvement

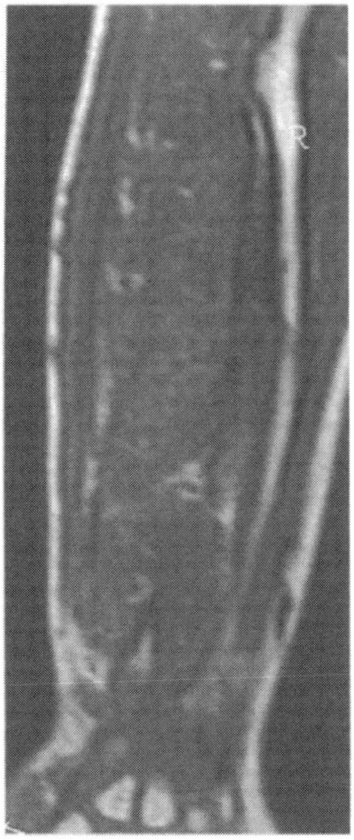

Figure 67G. Coronal T1 of the left forearm.

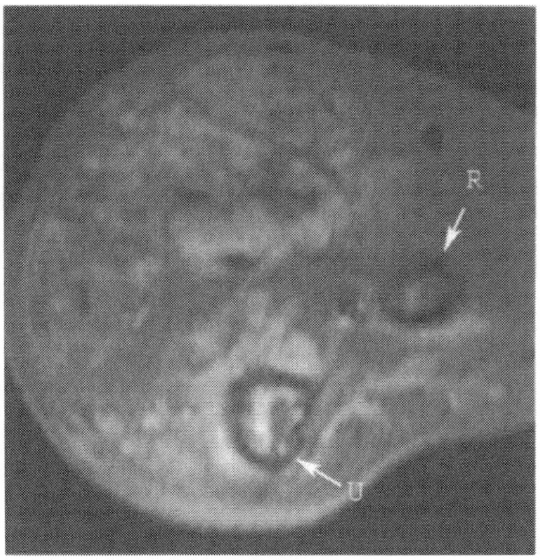

Figure 67H. Axial T1 post-Gd FS.

Findings

This is an 8-year-old boy with a known left forearm venous malformation.

Figure 67G. Lace-like fatty change is present in the forearm muscles.

Figure 67H. Extensive intramuscular and osseous patchy tubular enhancement is present. The margins are lobulated. These features are characteristic of a diffuse VM. Radius (R), Ulna (U).

Venous Malformation Initially Diagnosed as a Lymphatic Malformation

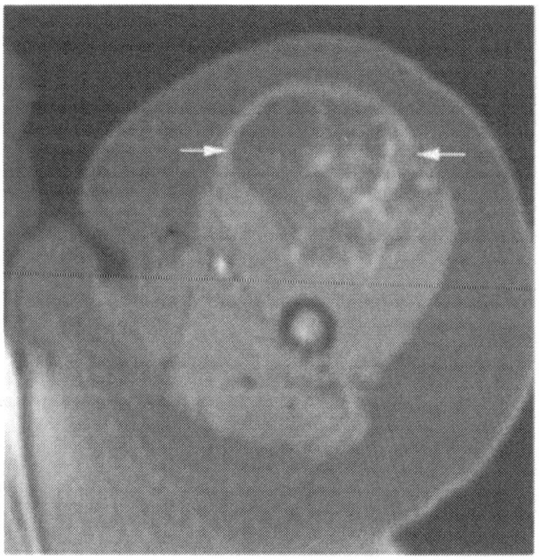

Figure 67I. Axial T1 post-Gd FS of the left thigh (at 5 months of age).

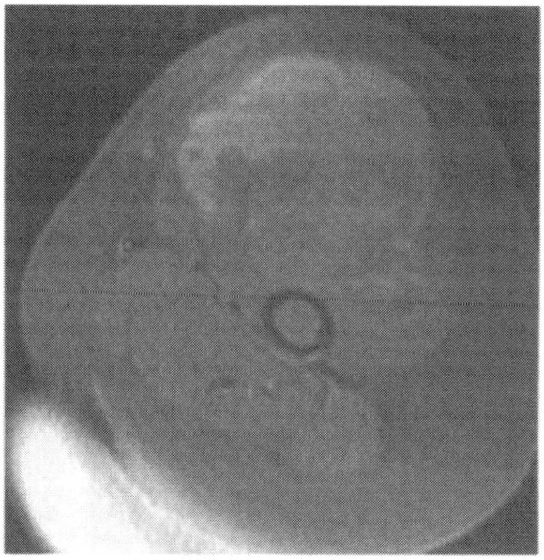

Figure 67J. Axial T1 post-Gd FS of the left thigh (at 20 months of age).

Findings

This 5-month-old girl had an enlarging left thigh mass.

Figure 67I. There is a large lobulated mass identified in the proximal quadriceps muscle. Post contrast imaging that was obtained too early following intravenous gadolinium administration gives the false impression of septal enhancement of a multilocular cystic mass, mimicking a macrocystic lymphatic malformation. Fluid-fluid levels are also present **(arrows)**.

Figure 67J. Post contrast imaging at 20 months of age with a longer post contrast delay shows enhancement of multiple confluent tubular structures, characteristic features of a venous malformation.

Recurrent Arteriovenous Malformation

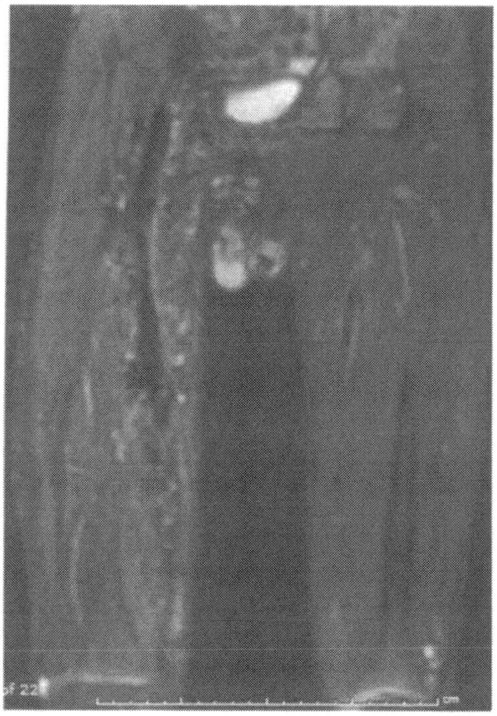

Figure 67K. Coronal STIR.

Figure 67L. Coronal STIR.

Figure 67M. Coronal T1 post-Gd FS.

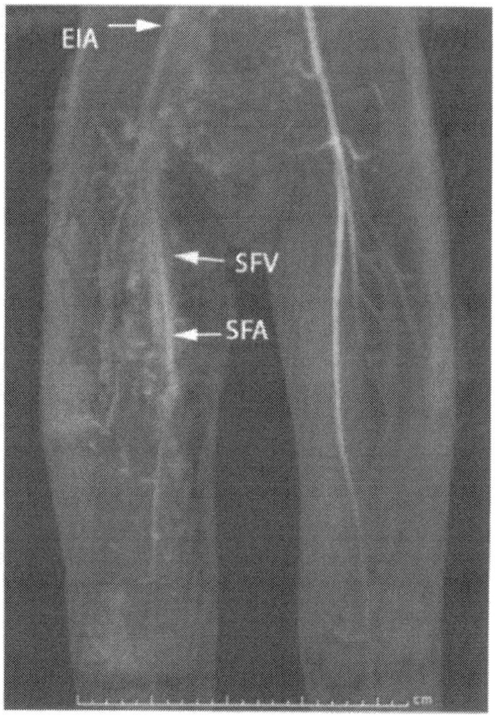

Figure 67N. Coronal gadolinium enhanced dynamic MRA, first pass.

Findings

This is a 13-year-old boy with progressive swelling of his right thigh and heart failure.

Figures 67K, 67L, 67M. On the right, there are confluent flow voids present in the medial thigh. There is mild increased STIR SI and enhancement in the adjacent muscles but a discrete mass is not identified.

Figure 67N. Dynamic Gd MRA demonstrates confluent vascular channels along the right medial thigh. The external iliac artery (EIA), superficial femoral artery (SFA), and a large superficial femoral vein (SFV) simultaneously enhance. This study was ordered for pre-embolization planning for a diffuse proximal thigh AVM.

Pitfalls and Pearls

1. Extensive venous malformations are associated with limb hypoplasia, whereas mixed, lymphatic, and arteriovenous malformations are associated with limb enlargement (8).
2. Contrast-enhanced imaging may falsely suggest a lymphatic malformation if scanning is done too soon following intravenous gadolinium administration.

References

1. Kransdorf MJ. Benign soft-tissue tumors in a large referral population: Distribution of specific diagnoses by age, sex, and location. *AJR Am J Roentgenol* 1995; 164:395–402.
2. Kransdorf MJ. Malignant soft-tissue tumors in a large referral population: Distribution of diagnoses by age, sex, and location. *AJR Am J Roentgenol* 1995; 164:129–134.
3. Mulliken JB, Glowacki J. Hemangiomas and vascular malformations in infants and children: A classification based on endothelial characteristics. *Plast Reconstr Surg* 1982; 69:412–422.
4. Enjolras O, Mulliken JB. Vascular tumors and vascular malformations (new issues). *Adv Dermatol* 1997; 13:375–423.
5. Murphey MD, Fairbairn KJ, Parman LM, Baxter KG, Parsa MB, Smith WS. From the archives of the AFIP. Musculoskeletal angiomatous lesions: Radiologic-pathologic correlation. *Radiographics* 1995; 15:893–917.
6. Konez O, Burrows PE. Magnetic resonance of vascular anomalies. *Magn Reson Imaging Clin N Am* 2002; 10:363–388, vii.
7. Herborn CU, Goyen M, Lauenstein TC, Debatin JF, Ruehm SG, Kroger K. Comprehensive time-resolved MRI of peripheral vascular malformations. *AJR Am J Roentgenol* 2003; 181:729–735.
8. Enjolras O, Ciabrini D, Mazoyer E, Laurian C, Herbreteau D. Extensive pure venous malformations in the upper or lower limb: A review of 27 cases. *J Am Acad Dermatol* 1997; 36:219–225.
9. Enjolras O, Chapot R, Merland JJ. Vascular anomalies and the growth of limbs: A review. *J Pediatr Orthop B* 2004; 13:349–357.
10. Paltiel HJ, Burrows PE, Kozakewich HP, Zurakowski D, Mulliken JB. Soft-tissue vascular anomalies: Utility of US for diagnosis. *Radiology* 2000; 214:747–754.
11. Burrows PE, Laor T, Paltiel H, Robertson RL. Diagnostic imaging in the evaluation of vascular birthmarks. *Dermatol Clin* 1998; 16:455–488.

Case 68

History

This is a 9-year-old boy with a history of recurrent left knee swelling and pain.

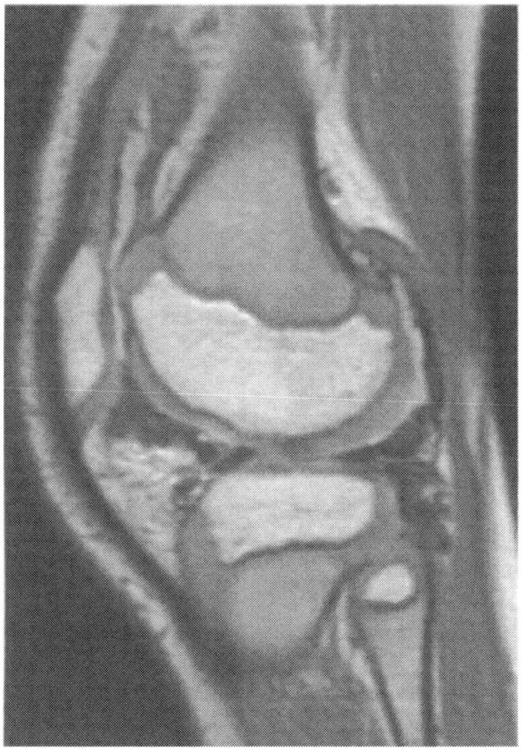

Figure 68A. Sagittal PD of the left knee.

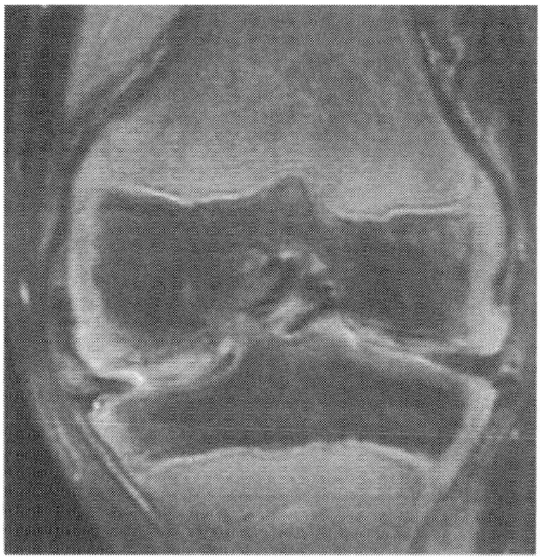

Figure 68B. Coronal PD FS.

Figures 68A, (68A with annotations). The synovium is thickened and contains a hypointense rim **(arrowhead)** representing hemosiderin. Confluent hemosiderin staining is also present in Hoffa's fat pad and the posterior joint region **(arrows)** near the popliteus tendon.

Figure 68B. Cartilage thinning, subchondral cortical irregularity, and joint space narrowing are evident.

Figures 68C, 68D, 68E. The synovium is thickened and demonstrates diffuse enhancement **(arrow)**. There is a moderate-size joint effusion present (E). A hypointense inner rim is identified consistent with hemosiderin staining **(arrowhead)**. Layering blood products are also present along the medial aspect of the suprapatellar bursa **(thick arrow)**.

Figure 68F. Dense effusion is present.

Figure 68G. There is widening of the intercondylar notch, femoral condylar flattening, and cortical irregularity.

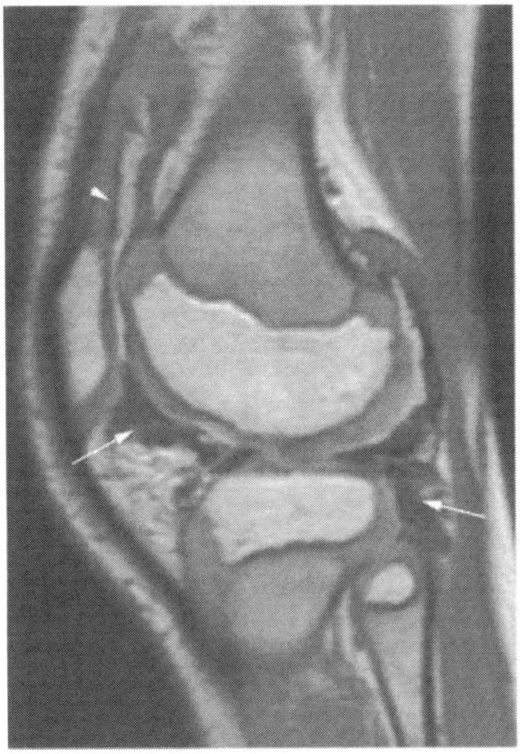

Figure 68A* Annotated.

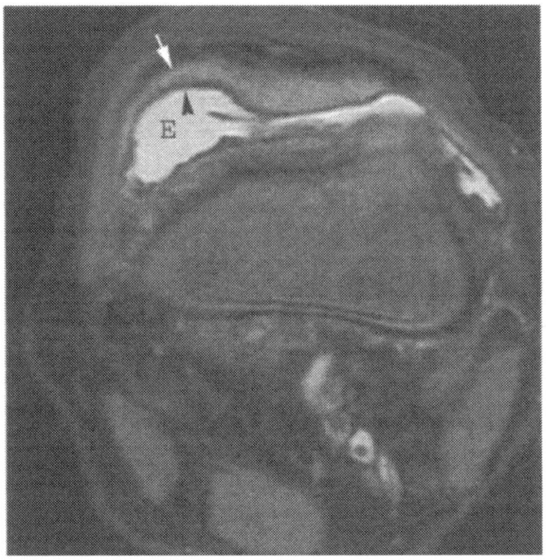

Figure 68C. Axial PD FS.

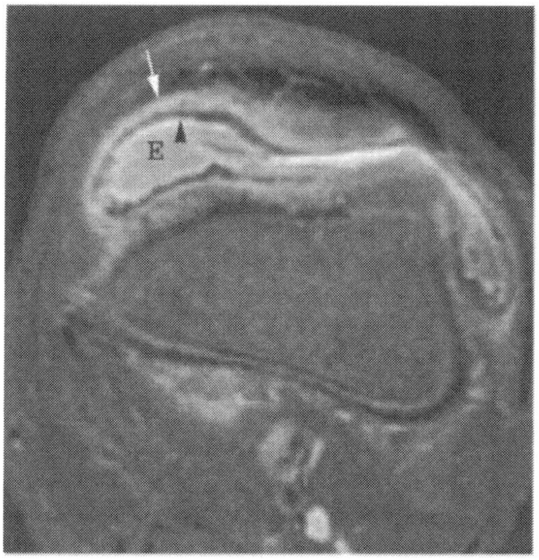

Figure 68D. Axial T1 post-Gd FS.

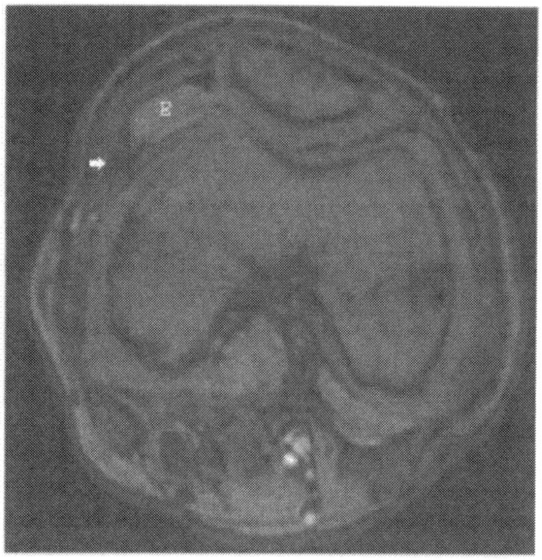

Figure 68E. Axial MPGR.

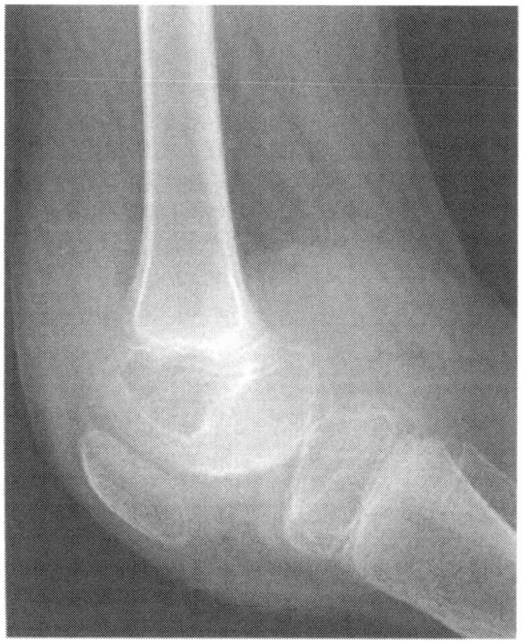

Figure 68F. Lateral radiograph of the left knee.

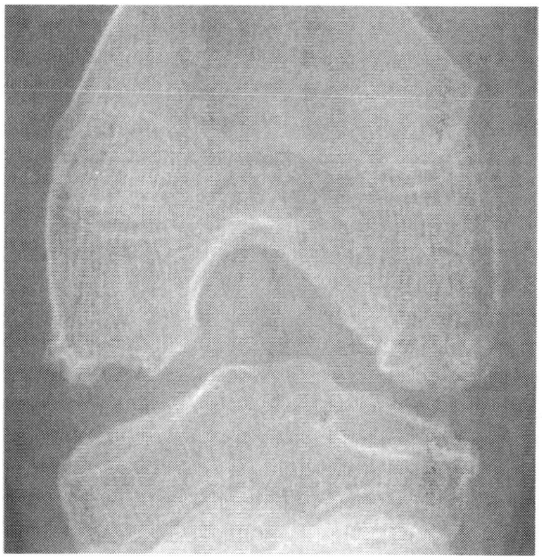

Figure 68G. Tunnel radiograph.

Diagnosis

Hemophilic arthropathy

Questions

1. T/F: Pannus formation is a nonspecific finding and may be seen in any inflammatory arthropathy.
2. What is the differential diagnosis of conditions associated with intra-articular blood products?

Discussion

Hemophilia is a bleeding disorder related to deficient plasma thromboplastin factor 8 (hemophilia A) or 9 (hemophilia B or Christmas disease) (1). It is an X-linked recessive condition with variable disease severity. The musculoskeletal manifestations of hemophilia include hemophilic arthropathy and intraosseous and soft tissue pseudotumors.

Hemophilic arthropathy describes an inflammatory arthritis that results from repeated hemarthrosis. The most commonly affected joint is the knee followed by the ankle, elbow, and shoulder (2). With recurrent bouts of hemarthrosis, synovial hyperplasia, hyperemia, and granulation tissue formation, or pannus, eventually develop, mimicking both infectious and noninfectious inflammatory arthropathies (Answer to Question 1). Chronicity eventually leads to juxta-articular osteopenia, articular cartilage destruction, subchondral cysts, marginal and central bone erosions, and secondary degenerative changes (3). Chronic hyperemia leads to epiphyseal overgrowth and early physeal closure.

Distinctive radiographic changes in the knee in patients with hemophilic arthropathy include inferior patella pole squaring, widening of the intercondylar notch, and increased periarticular density representing intra-articular blood (3). Other characteristic plain radiographic features include broadening of the trochlear notch in the elbow and tibiotalar slant (medial downward slanting tibiotalar joint with ankle valgus deformity) (Figure 68H).

The value of MRI in hemophilic arthropathy lies in its ability to detect degenerative changes before they are radiographically visible. An MRI grading system for hemophilic arthropathy has been proposed (Table 68A) (4). An MRI feature of hemophilic arthropathy that is not typically seen in other inflammatory arthropathies is the abundance of blooming artifact (signal loss) related to repeated hemarthrosis leading to hemosiderin deposition. This susceptibility artifact is best appreciated on GRE sequences (5). Other causes of hemarthrosis include: trauma, infection, inflammatory

Table 68A. MRI grading of hemophilic arthropathy.

MRI Findings
0. Normal.
I. Slight synovial hypertrophy with hemosiderin present within the joint.
II. Synovial hypertrophy, abundant hemosiderin within the joint, cartilaginous erosions.
III. Diffuse cartilaginous destruction, joint space narrowing, osseous erosions, and/or subchondral cysts.
IV. Severe joint derangement, secondary osteoarthritis, and/or ankylosis.

Source: From Soler R et al. with permission from Springer (4).

arthritis, pigmented villonodular synovitis (PVNS), and synovial venous malformation (AKA synovial hemangioma) (Answer to Question 2) (6).

Hemophilic pseudotumors (hematomas) may occur within the soft tissues or bone (3). Within bone, they may be intramedullary or subperiosteal. Soft tissue pseudotumors usually present as a mass after minor trauma or exertion and are usually located at the myotendinous junction (Figures 68I, 68J).

This patient with a history of hemophilia B (Christmas disease) had an open synovectomy 2 months after the MRI to manage chronic knee pain and swelling.

Orthopedic Perspective

The clinician and patient generally know the diagnosis of hemophilia long before a referral to MRI. The study is performed to determine if the patient will benefit from synovectomy. The degree of synovial inflammation as well as articular cartilage loss are important facts for the clinician. When the affected joint has a chronically painful effusion with limited motion, synovectomy may be performed, to slow the progression of joint destruction.

What the Clinician Needs to Know

What is the status of the articular cartilage and synovium? Are bony erosions present?

Answers

1. True.
2. Trauma, infection, inflammatory arthritis, hemophilia, PVNS, and synovial venous malformation (AKA synovial hemangioma).

Additional Examples

Hemophilic Arthropathy with Tibiotalar Slant and Erosions

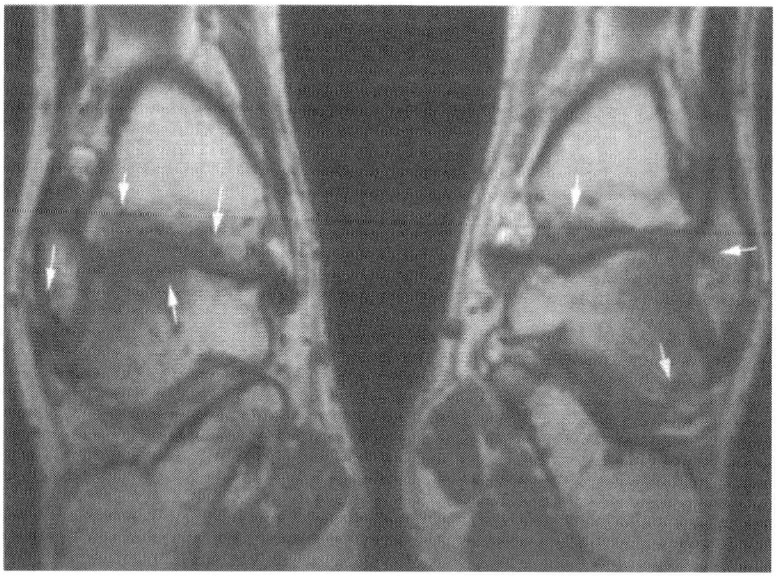

Figure 68H. Coronal T1.

Findings

This is a 17-year-old boy with known hemophilia with worsening bilateral ankle pain.

Figure 68H. There is bilateral tibiotalar slant (ankle valgus) associated with innumerable erosions **(arrows)** and joint narrowing. There is marrow edema in the lateral talus and subchondral tibial plafond bilaterally.

Hemophilic Pseudotumor at the Biceps Myotendinous Junction

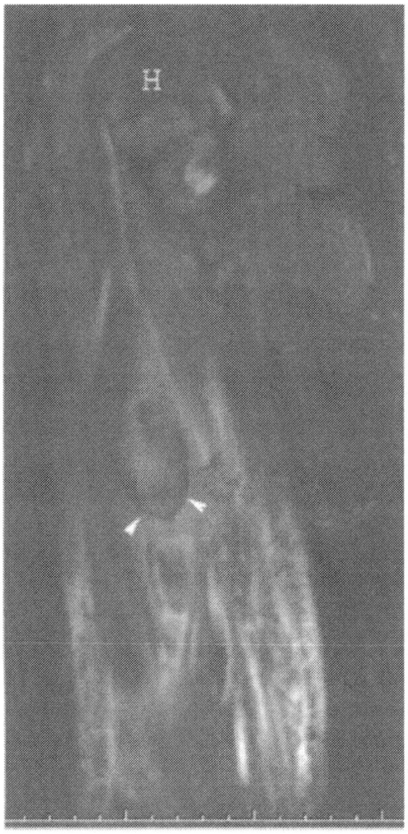

Figure 68I. Sagittal STIR of the right arm.

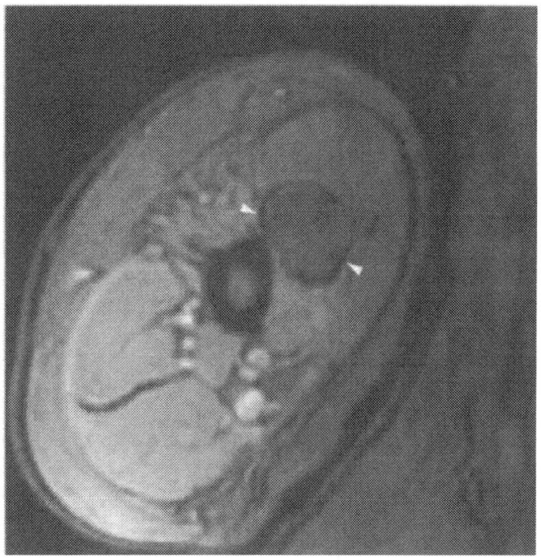

Figure 68J. Axial MPGR mid-right arm.

Findings

This is a 17-year-old boy with right arm pain.

Figures 68I, 68J. On STIR, inflammatory changes are present within the fascia, muscles, and subcutaneous soft tissues. At the myotendinous junction of the biceps, a pseudotumor with a hemosiderin rim **(arrowheads)** is identified. Humeral head (H).

Pitfalls and Pearls

With the exception of intra-articular blood products, the MRI and radiographic features of hemophilic arthropathy may be indistinguishable from other infectious and noninfectious inflammatory arthropathies.

References

1. Lan HH, Eustace SJ, Dorfman D. Hemophilic arthropathy. *Radiol Clin North Am* 1996; 34:446–450.

2. Llauger J, Palmer J, Roson N, Bague S, Camins A, Cremades R. Nonseptic monoarthritis: Imaging features with clinical and histopathologic correlation. *Radiographics* 2000; 20 Spec No: S263–278.
3. Kerr R. Imaging of musculoskeletal complications of hemophilia. *Semin Musculoskelet Radiol* 2003; 7:127–136.
4. Soler R, Lopez-Fernandez F, Rodriguez E, Marini M. Hemophilic arthropathy: A scoring system for magnetic resonance imaging. *Eur Radiol* 2002; 12:836–843.
5. Rand T, Trattnig S, Male C, et al. Magnetic resonance imaging in hemophilic children: Value of gradient echo and contrast-enhanced imaging. *Magn Reson Imaging* 1999; 17:199–205.
6. Azouz EM. Arthritis in children: Conventional and advanced imaging. *Semin Musculoskelet Radiol* 2003; 7:95–102.

Case 69

History

This is a 12-month-old boy with left third digit swelling. There was no history of trauma or fever.

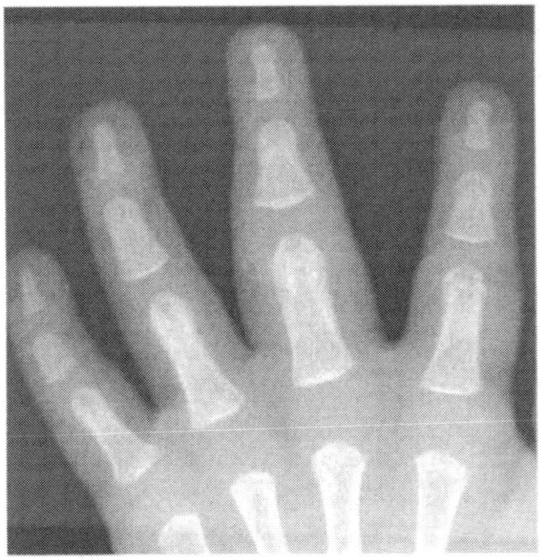

Figure 69A. PA radiograph of the digits of the left hand.

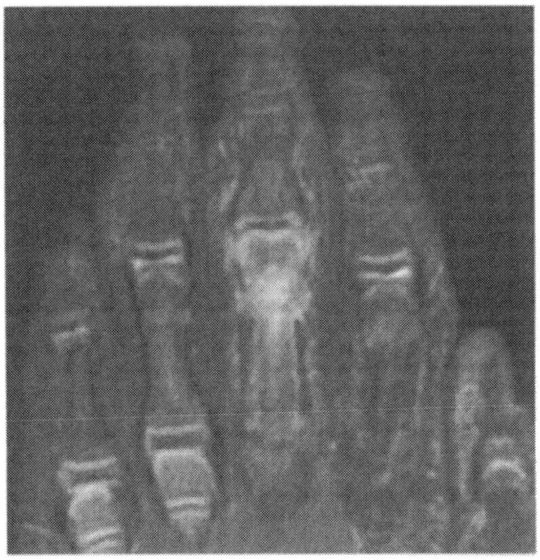

Figure 69B. Coronal STIR.

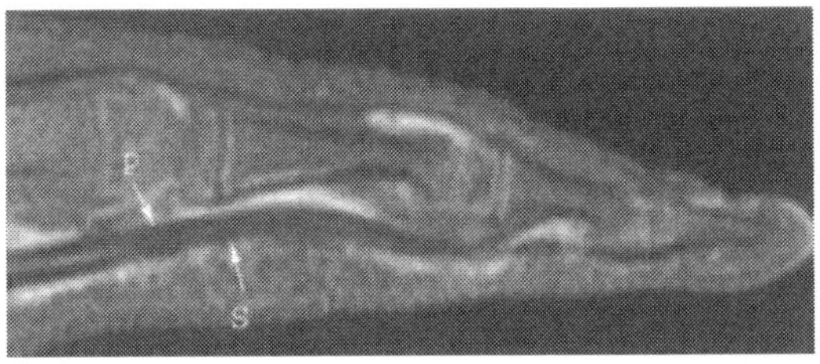

Figure 69C. Sagittal T1 post-Gd FS.

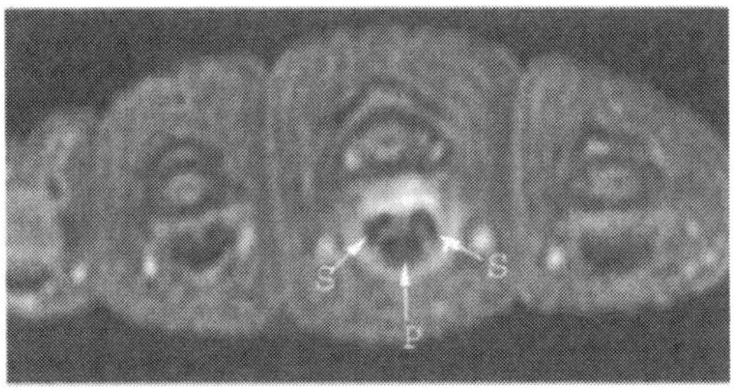

Figure 69D. Axial T1 post-Gd FS through the proximal phalanx of the third digit.

Figure 69A. The third digit is enlarged with fusiform swelling. There is cortical thickening of the proximal phalanx and spurring at the base of the middle phalanx. Bony mineralization is normal.

Figures 69B, 69C, 69D, (69C with annotations). There is increased STIR SI in the proximal phalanx of the third digit. There is synovial hypertrophy and enhancement of the distal interphalangeal (DIP) joint **(arrow)** and tendon sheaths **(arrowheads)** of the flexor digitorum profundus (P) and superficialis (S).

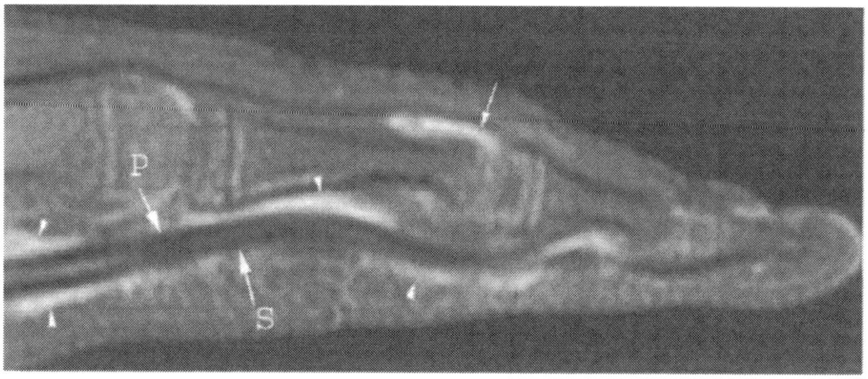

Figure 69C* Annotated.

Diagnosis

Psoriatic dactylitis

Questions

1. T/F: Tenosynovitis is a dominant feature of dactylitis seen in patients with juvenile psoriatic arthritis.
2. T/F: The diagnosis of psoriatic arthritis may be made in the absence of dermatologic lesions.

Discussion

Psoriatic dactylitis (sausage digit) results from a combination of tenosynovitis and intra-articular synovitis, leading to a swollen and painful digit (1). The dominant feature of dactylitis of psoriatic arthritis and other seronegative spondyloarthropathies is tenosynovitis (especially the flexor tendons) (Answer to Question 1). Psoriatic arthritis tends to have significant extra-articular inflammation compared with rheumatoid arthritis but these two diseases may be indistinguishable when there is only synovitis (Figure 69E) (2).

There are two classification systems for juvenile psoriatic arthritis: International League of Associations for Rheumatology (ILAR) and the Vancouver Criteria (Table 69A) (3). The criteria for the diagnosis of juvenile psoriatic arthritis do not require skin changes in either of these classification systems (Answer to Question 2).

In psoriatic arthritis, the DIP joint is more frequently involved, the distribution may be asymmetric, and bony mineralization is usually normal to increased. Marginal erosions may be combined with bony proliferation, leading to juxta-articular spurring (4). Juvenile rheumatoid arthritis (JRA), on the other hand, usually affects the carpal, metacarpophalangeal (MCP), and proximal interphalangeal (PIP) joints, and may show juxta-articular or global osteopenia and marginal erosions. Juxta-articular and central erosions, as well as periosteal reaction, may occur with both JRA and juvenile psoriatic arthritis. Central erosions and eventual fibrous or bony ankylosis may occur with JRA and psoriatic arthritis, once the overlying protective hyaline cartilage has been destroyed.

On MRI, the inflammatory changes of psoriatic arthritis may be detected earlier when viewed in conjunction with radiographs. Tenosynovitis, synovitis, pre-erosive osteitis, erosions, and articular destruction are assessed with a combination of T1, water sensitive, and post-Gd sequences. Pre-erosive osteitis (subchondral edema) may predict future sites of erosions and synovial hyperplasia may predict the future number of erosions in a given joint (5). Pre-erosive osteitis may be considered when radiogra-

Table 69A. Vancouver criteria for juvenile psoriatic arthritis.

1. Definite
   A. Arthritis and psoriatic rash
   Or
   B. Arthritis with 3 of 4 minor criteria:
   Dactylitis, nail pitting/oncholysis, psoriasis-like rash, and family history of psoriasis in a first- or second-degree relative
2. Probable
   Arthritis with 2 of the 4 minor criteria (in B)

Source: Data from Southwood TR, Petty RE, Malleson PN, et al. Psoriatic arthritis in children. Arthritis Rheum 1989; 32:1007.

phy and T1W sequences demonstrate no cortical destruction in the setting of decreased SI on T1 or increased T2 SI in subchondral bone.

The differential diagnosis for an acutely swollen digit includes trauma, infectious tenosynovitis, and osteonecrosis related to sickle cell disease. Tumoral causes of an enlarged digit include macrodystrophia lipomatosa, neurofibromatosis type 1, Proteus syndrome, and underlying lymphatic or mixed vascular malformations.

The diagnosis of juvenile psoriatic arthritis was made in this patient because he had both skin lesions of psoriasis and arthritis. Although the diagnosis was established, the MRI was requested to determine disease activity prior to aggressive methotrexate therapy.

Orthopedic Perspective

Psoriatic dactylitis is generally a clinical diagnosis and advanced imaging is usually not performed. However, MRI may be useful to guide therapy, based on the extent of synovial hypertrophy, marrow edema, and extraosseous inflammatory change. When the presentation of dactylitis is subacute, osteomyelitis/septic arthritis may be a consideration, and certain MRI features may provide strong support for an infectious etiology (see Case 18).

What the Clinician Needs to Know

1. What is the extent of disease activity?
2. Is there an alternative etiology for the digital swelling, such as osteomyelitis/septic arthritis or an occult fracture?

Answers

1. True.
2. True.

Additional Example

Oligoarticular Juvenile Rheumatoid Arthritis

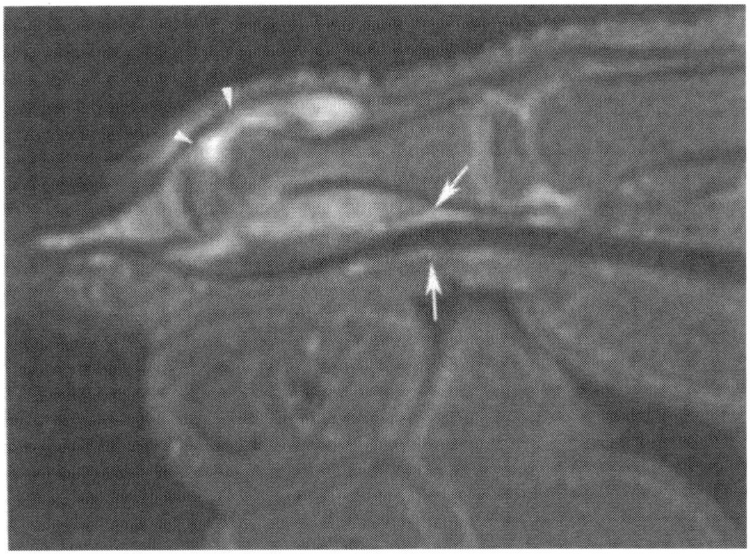

Figure 69E. Sagittal T1 post-Gd FS of the right thumb.

Findings

This is a 9-year-old girl who had chronic swelling of her right thumb.

Figure 69E. There is diffuse synovial thickening and enhancement of the interphalangeal joint of the thumb **(arrowheads)**. There is also mild tendon sheath enhancement along the flexor pollicis longus **(arrows)**, compatible with tenosynovitis. She was subsequently diagnosed with oligoarticular juvenile rheumatoid arthritis.

Pitfalls and Pearls

1. Dactylitis may occur with both juvenile seronegative spondyloarthropathies and JRA. However, inflammatory changes with juvenile seronegative spondyloarthropathies tend to affect the extra-articular soft tissues whereas JRA tends to be confined to the joint.
2. When symptoms are localized to a specific digit, MR imaging of the individual digit and metacarpal should be performed rather than the entire hand and wrist. The affected digit should be imaged with a high resolution matrix, small field of view, and gadolinium should be given.

References

1. Olivieri I, Barozzi L, Favaro L, et al. Dactylitis in patients with seronegative spondylarthropathy: Assessment by ultrasonography and magnetic resonance imaging. *Arthritis Rheum* 1996; 39:1524–1528.
2. Jevtic V, Watt I, Rozman B, Kos-Golja M, Demsar F, Jarh O. Distinctive radiological features of small hand joints in rheumatoid arthritis and seronegative spondyloarthritis demonstrated

by contrast-enhanced (Gd-DTPA) magnetic resonance imaging. *Skeletal Radiol* 1995; 24: 351–355.

3. Southwood TR, Petty RE, Malleson PN, Delgado EA, Hunt DW, Wood B, Schroeder ML. Psoriatic arthritis in children. *Arthritis Reheum.* 1989 Aug; 32(8):1007–1013.
4. Azouz EM, Duffy CM. Juvenile spondyloarthropathies: Clinical manifestations and medical imaging. *Skeletal Radiol* 1995; 24:399–408.
5. Savnik A, Malmskov H, Thomsen HS, et al. MRI of the wrist and finger joints in inflammatory joint diseases at 1-year interval: MRI features to predict bone erosions. *Eur Radiol* 2002; 12:1203–1210.

Case 70

History

This is a 4-year-old boy with left elbow pain.

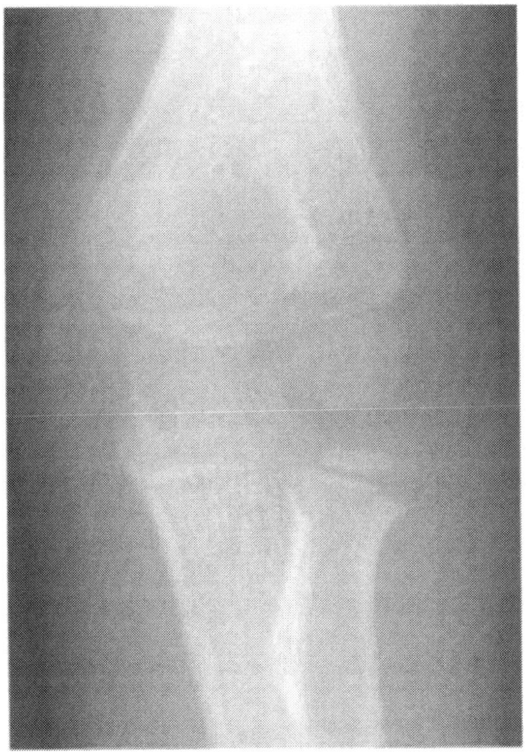

Figure 70A. AP radiograph of the left elbow.

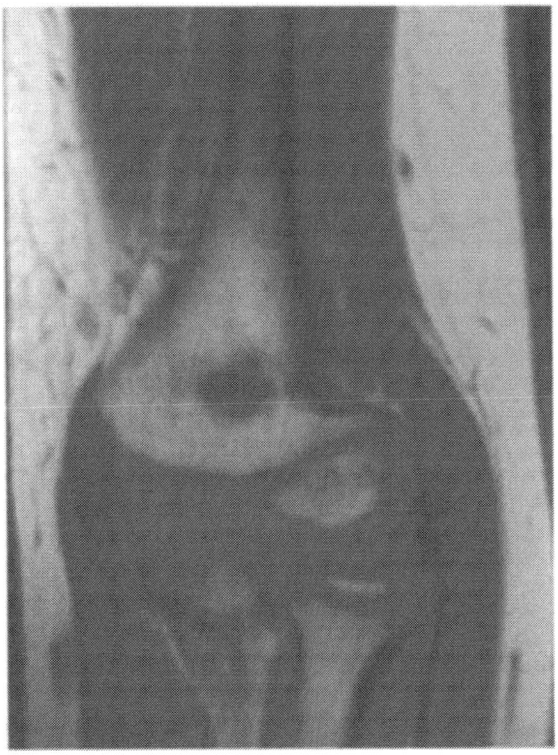

Figure 70B. Coronal T1.

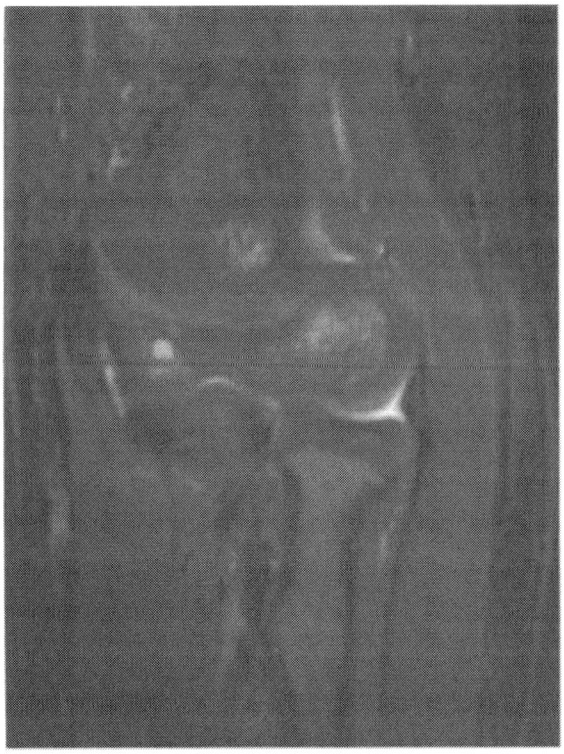

Figure 70C. Coronal STIR.

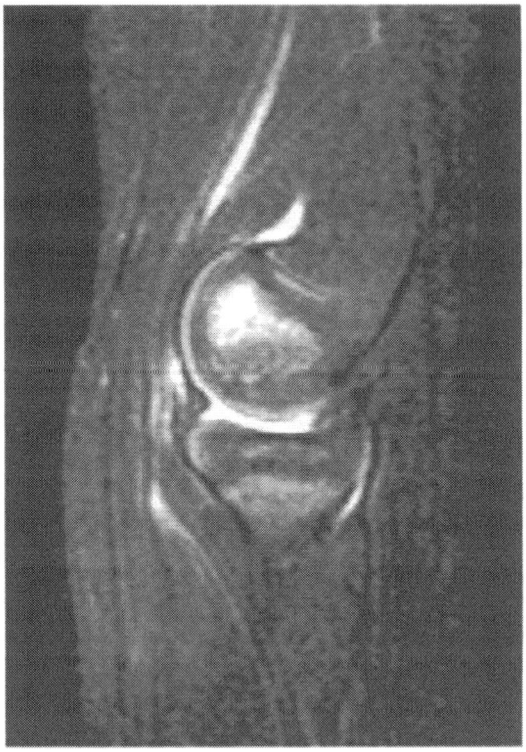

Figure 70D. Sagittal PD FS through the capitellum.

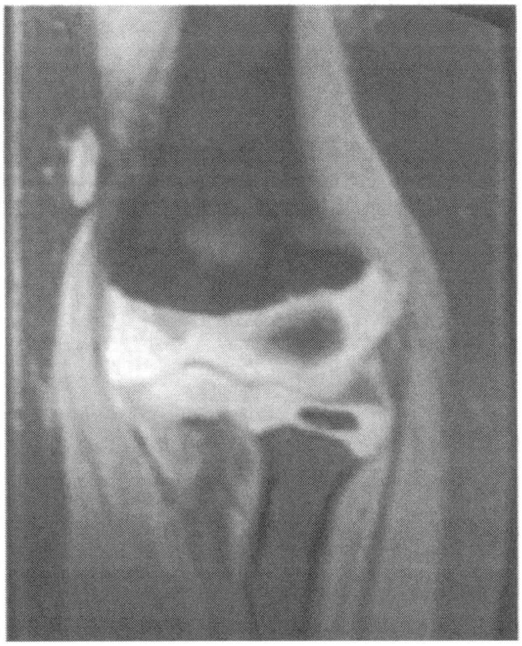

Figure 70E. Coronal 3D SPGR FS reformat.

Figure 70A. The radiograph shows diminished density of the capitellum.

Figures 70B, 70C, 70D, (70C with annotations). There is T1 hypointensity and corresponding hyperintensity on fluid-sensitive sequences affecting the capitellar ossification center **(arrow)**. There is also focal T1 hypointensity and corresponding hyperintensity on STIR within the pre-ossification center of the trochlea **(arrowhead)**.

Figure 70E. There is normal cartilage SI within the capitellum and trochlea on this cartilage-sensitive sequence.

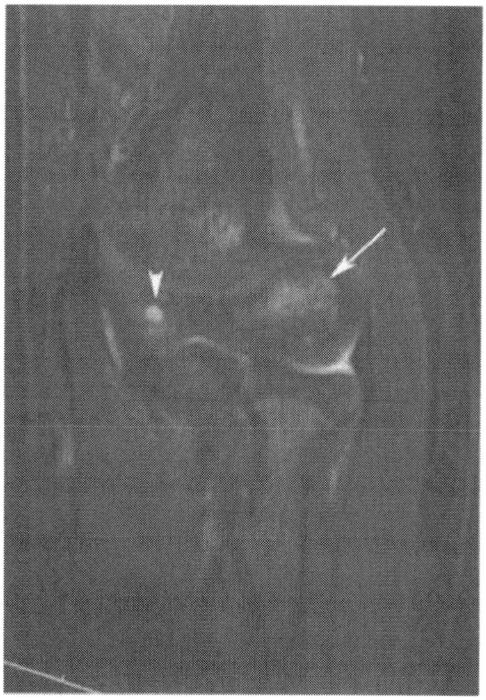

Figure 70C* Annotated.

Diagnosis

Panner's disease and normal trochlear pre-ossification center

Questions

1. What is the difference in the marrow edema patterns between osteochondritis dissecans (OCD) and Panner's disease?
2. What is the cause of increased T2 hyperintensity in the pre-ossification center of the trochlea?

Discussion

Panner's disease is an osteochondrosis of the capitellum occurring in children age 4–10 years with a male predominance (1). For an explanation of osteochondrosis, please see Case 99. Unlike osteochondritis dissecans of the capitellum (OCD), Panner's disease usually lacks a clear history of trauma. The etiology of Panner's disease is unknown, although some have theorized that it may be related to chronic microtrauma and even ischemic necrosis (2). It is a self-limited condition and does not lead to secondary degenerative changes and loose bodies. Panner's disease tends to involve most of the capitellar ossification center, whereas OCD is more focal and is typically located anterolaterally (Answer to Question 1) (3).

The plain radiographic findings of Panner's disease include sclerosis, fragmentation, trabecular rarefaction, and poorly defined cortical margins of the capitellum. With MRI, signal abnormalities are restricted to the capitellar ossification center, and cartilage sensitive sequences show intact epiphyseal and articular cartilage. The ossification center shows nonspecific, decreased SI on T1 and corresponding increased T2 SI, reflecting marrow edema. In contrast to this typical pattern, Stoane et al. reported a case in which the capitellum was hypointense on both T1 and T2 sequences, and they concluded that the findings were consistent with trauma induced ischemic necrosis (2).

The distinctive trochlear signal characteristics seen in this case of Panner's disease (Figures 70B, 70C) should not be mistaken for a pathologic process. The pre-ossification center of the epiphysis may demonstrate transient hypointensity on T1 and hyperintensity on fluid-sensitive sequences. This phenomenon likely reflects increased free water associated with chondrocyte hypertrophy that occurs just prior to the appearance of secondary ossification centers (Answer to Question 2) (4).

In contrast to Panner's disease, OCD of the capitellum occurs in older children (11–15 years) and focally affects the anterolateral subchondral bone plate. It is felt to be related to chronic microtrauma due to compression from valgus stress, as is typically seen in throwing athletes (3). Like Panner's disease, OCD of the capitellum more commonly occurs in boys. The radiographic features of OCD of the capitellum include anterolateral flattening, lucency, sclerosis, subchondral cortical defects, and eventually loose body development in some instances (5). While Panner's disease is self-limited, capitellar OCD may eventually lead to secondary degenerative joint disease.

The utility of MRI is to assess for OCD stability since this guides therapy (Figures 70F–70H). MRI features of instability include: loose bodies, fluid signal between the parent bone and OCD fragment, cyst formation between the parent bone and OCD fragment, articular cartilage thinning over the osteochondral lesion, and fluid signal traversing the articular cartilage and subchondral bone plate extending into the bed of the OCD (6). With direct gadolinium arthrography, unstable lesions demonstrate contrast insinuating between the lesion and parent bone (7). With IV gadolinium, enhancement between the parent bone and the OCD fragment is compatible with granulation tissue and suggests instability (8). A lack of enhancement at the parent bone–OCD interface suggests stability. Intravenous gadolinium may also help assess the vascularity of the OCD fragment (9).

This patient was treated with casting and rest and was doing well at short term follow-up.

Orthopedic Perspective

Panner's disease and OCD of the capitellum can usually be distinguished based on clinical grounds. Panner's disease occurs in younger children and there is usually no history of repetitive valgus overuse. MRI evaluation of OCD may provide information regarding integrity of the articular surface, stability of the lesion, presence of loose bodies, and healing of the lesion. This information is useful in guiding treatment. Early stage OCD without articular surface fissuring is treated nonoperatively. Nonhealing lesions and lesions with articular cartilage disruption are treated with arthroscopic drilling. Loose bodies are treated with arthroscopic excision.

What the Clinician Needs to Know

1. Distinguish among Panner's disease, OCD, and synovial processes secondarily affecting the capitellum.
2. For OCD of the capitellum, determine stability as well as the presence of loose bodies.

Answers

1. Panner's disease tends to affect most of the capitellum, whereas osteochondritis dissecans tend to be more focal and involve the anterolateral capitellum.
2. Increased free water associated with chondrocyte hypertrophy just prior to the appearance of the ossification center.

Additional Examples

Osteochondritis Dissecans of the Capitellum with Loose Body

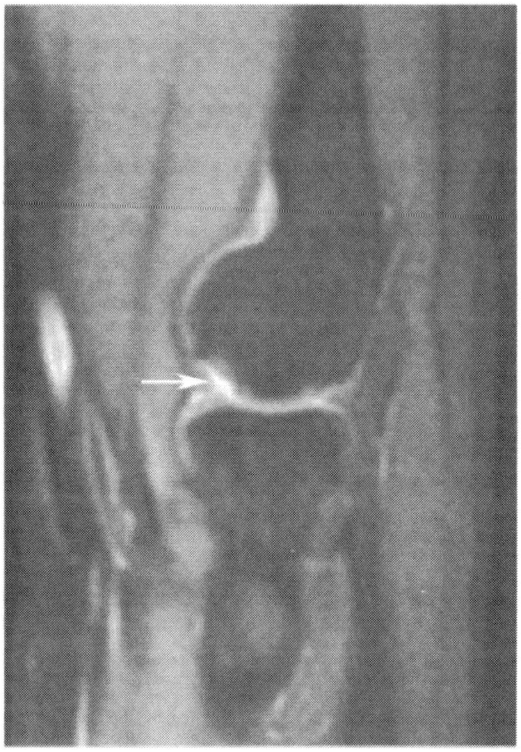

Figure 70F. Sagittal PD FS of the right elbow.

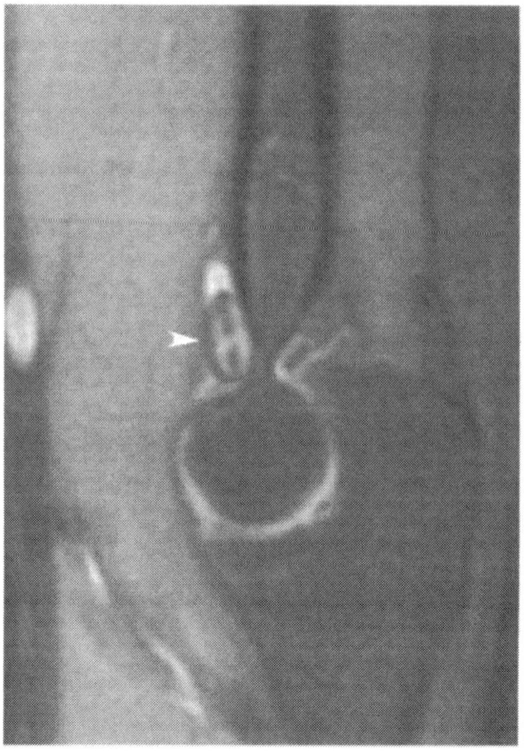

Figure 70G. Sagittal PD FS.

Findings

This 14-year-old, right-handed gymnast had persistent popping and pain in her elbow.

Figure 70F. There is a small osteochondral lesion located along the anterolateral capitellum **(arrow)**.

Figure 70G. An osteochondral loose body is identified **(arrowhead)** in the anterior elbow joint surrounded by a small joint effusion. This was subsequently confirmed surgically.

Osteochondritis Dissecans of the Capitellum with Loose Body

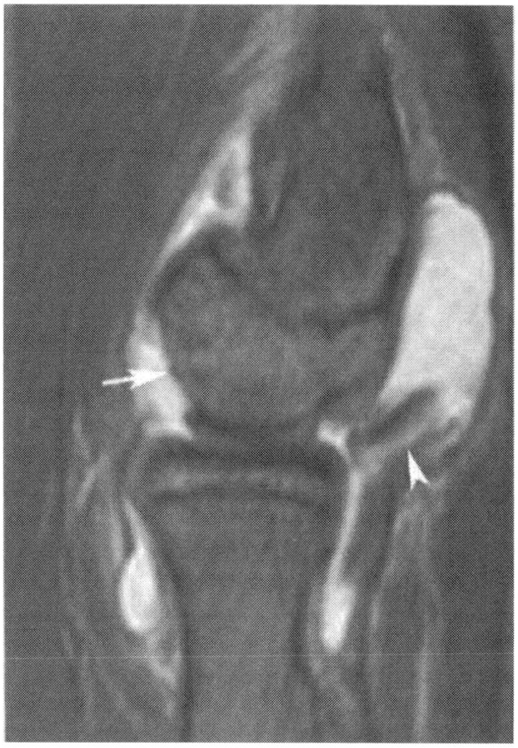

Figure 70H. Sagittal STIR of the left elbow.

Findings

This 12-year-old boy had chronic diminished elbow flexion with no reported history of trauma.

Figure 70H. There is a large joint effusion. There is an osteochondral lesion located along the anterolateral capitellum **(arrow)**, without underlying marrow edema. An osteochondral loose body is present in the posterior elbow joint **(arrowhead)**.

Pitfalls and Pearls

1. Like most osteochondroses, Panner's disease cannot be differentiated from a normal developmental variant on radiographic grounds. MRI has potential utility when clinical findings point to the diagnosis.
2. OCD of the capitellum may occur secondary to disorders associated with an abnormal radiocapitellar relationship. Look for signs of a prior radial neck/head fracture, bone dysplasia, or radiocapitellar subluxation/dislocation.

References

1. Klein EW. Osteochondrosis of the capitellum (Panner's disease): Report of a case. *Am J Roentgenol Radium Ther Nucl Med* 1962; 88:466–469.

2. Stoane JM, Poplausky MR, Haller JO, Berdon WE. Panner's disease: X-ray, MR imaging findings and review of the literature. *Comput Med Imaging Graph* 1995; 19:473–476.
3. Bradley JP, Petrie RS. Osteochondritis dissecans of the humeral capitellum: Diagnosis and treatment. *Clin Sports Med* 2001; 20:565–590.
4. Chapman VM, Nimkin K, Jaramillo D. The pre-ossification center: Normal CT and MRI findings in the trochlea. *Skeletal Radiol* 2004; 33:725–727.
5. Takahara M, Ogino T, Sasaki I, Kato H, Minami A, Kaneda K. Long term outcome of osteochondritis dissecans of the humeral capitellum. *Clin Orthop* 1999:108–115.
6. De Smet AA, Ilahi OA, Graf BK. Reassessment of the MR criteria for stability of osteochondritis dissecans in the knee and ankle. *Skeletal Radiol* 1996; 25:159–163.
7. Fritz RC. MR imaging of osteochondral and articular lesions. *Magn Reson Imaging Clin N Am* 1997; 5:579–602.
8. Bohndorf K. Osteochondritis (osteochondrosis) dissecans: A review and new MRI classification. *Eur Radiol* 1998; 8:103–112.
9. Peiss J, Adam G, Casser R, Urhahn R, Gunther RW. Gadopentetate-dimeglumine-enhanced MR imaging of osteonecrosis and osteochondritis dissecans of the elbow: Initial experience. *Skeletal Radiol* 1995; 24:17–20.

Case 71

History

This is a 14-year-old boy with a history of left slipped capital femoral epiphysis (SCFE) with new left hip pain, decreased range of motion, and hip clicking. Similar but lesser symptoms were also present on the right.

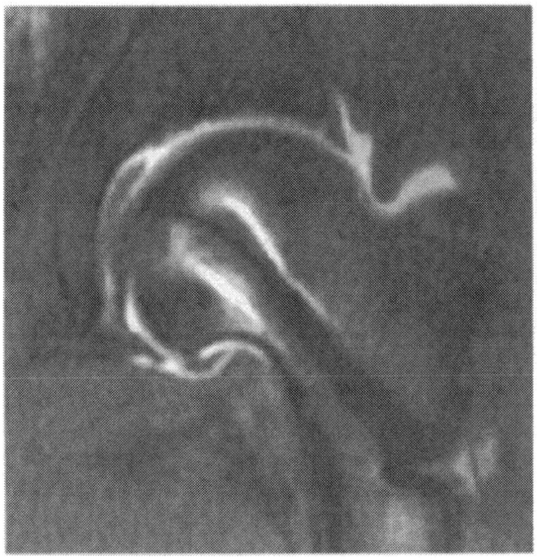

Figure 71A. Coronal T1 FS direct Gd arthrography.

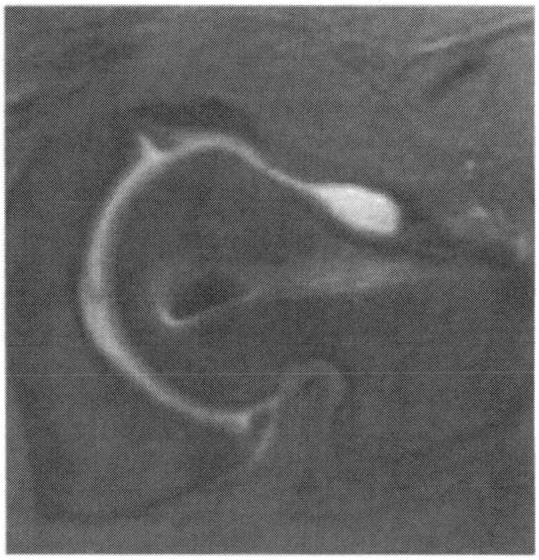

Figure 71B. Axial T1 FS direct Gd arthrography.

Figures 71A, 71B, (71B with annotations). The anterosuperior femoral head/neck offset is reduced **(thin arrow)** due to posteromedial epiphyseal displacement. An osteolabral tear/separation is present **(arrowhead)**. A normal posterior sublabral sulcus is present **(thick arrow)** (see Case 84). Transphyseal screw (*).

Figure 71C. Left-sided SCFE and transphyseal screw are seen. The femoral neck is short, indicating growth disturbance. The right hip is normal.

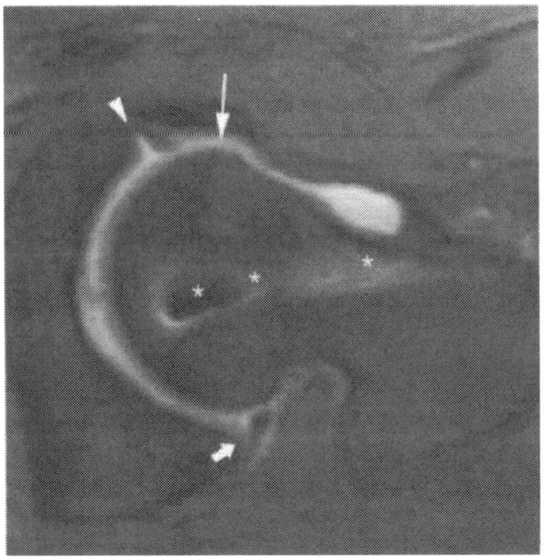

Figure 71B* Annotated.

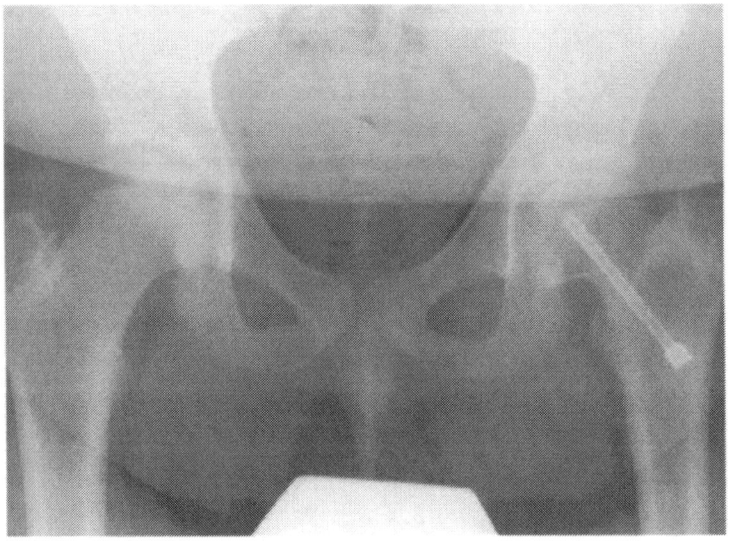

Figure 71C. AP pelvis.

Diagnosis

CAM-type femoroacetabular impingement (FAI) with anterosuperior osteolabral tear/separation related to SCFE

Question

1. What anatomic abnormalities may lead to femoroacetabular impingement?

Discussion

Femoroacetabular impingement (FAI) describes abnormal abutment of the femoral head and acetabulum with hip motion that may lead to early labral degeneration. FAI may result from proximal femoral (CAM-type) and acetabular causes (pincer-type) (Answer to Question 1) (1). CAM-type impingement includes: reduced femoral head/neck offset, juxtaphyseal osteophytes (or "physeal bump"), and an aspherical femoral head (Figure 71D) (2, 3). Pincer-type impingement is a result of femoral head overcoverage (coxa profunda), acetabular protrusio, and acetabular retroversion. Acetabular retroversion is present when the acetabulum is more posteriorly oriented than normal. This may lead to a prominent anterosuperior acetabular ridge that becomes an obstacle for flexion and internal rotation of the hip. With pincer-type impingement, the femoral head is usually normal, although both pincer-type and CAM-type impingement may coexist.

Although FAI is increasingly recognized in the adult population, experience with this disorder in the pediatric population is limited. Developmental dysplasia of the hip (DDH) and slipped capital femoral epiphysis (SCFE) are potential causes of FAI. Other etiologies include: Legg-Calve-Perthes (LCP) disease, normal anatomical variations of the femoral head/neck junction, spondyloepiphyseal dysplasia (SED), infection, nonpyogenic synovitis (e.g., JRA), and trauma.

Acetabular chondral and labral injury related to femoroacetabular impingement have been observed along the anterosuperior rim in patients with isolated CAM-type impingement, whereas circumferential labral and cartilage damage tend to occur in patients with isolated pincer-type impingement (4, 5). Direct MR arthrography is useful to assess for labral tears, osteolabral separation, loose bodies, and articular cartilage degeneration (6). Adequate distension of the joint capsule with contrast delineates the undersurface of the labrum, osteolabral junction, and articular cartilage, where the earliest degenerative changes of FAI may be seen. Indirect MR arthrography may also be performed, but delineation of labral pathology is suboptimal (see Case 37). In addition to T1W FS images, T2W FS sequence should be included in the arthrogram to assess for subchondral cysts (ganglion cysts), marrow edema, and other soft tissue findings related to altered weightbearing and impingement.

As a consequence of prior SCFE in this patient, the femoral head/neck offset was reduced, which predisposed to FAI. A right hip arthrogram also showed labral pathology similar to that on the left (not shown). The labral tear/separation was confirmed and debrided during arthroscopy. Postoperatively, his hip pain and impingement symptoms resolved.

Orthopedic Perspective

FAI is a newly described condition that is being actively studied. FAI may be responsible for a large proportion of cases of early degenerative joint disease of the hip.

CAM-type FAI may be idiopathic (normal spherical head with a head/neck junction bump) or secondary to another hip disease process (e.g., DDH, SCFE, LCP). In SCFE, the proximal femoral epiphysis displaces posteromedially, which leads sometimes to a subtle anteromedial bump that may cause anterior impingement with hip flexion. Over time, this may lead to labral-chondral injury of the acetabulum and the development of degenerative joint disease. An aspherical femoral head resulting from various hip entities including LCP or trauma may cause earlier, more severe degenerative changes and multidirectional impingement. The orthopedist needs to know the degree of CAM-type impingement and the degree of degenerative changes before surgical correction. Patients are less likely to benefit from correction of FAI if there are advanced degenerative changes present and may ultimately require a total hip replacement. The orthopedist also needs to know if loose bodies are present and the presence of chondral and osteochondral fractures. Newer open and arthroscopic techniques are being developed to remove the bony prominence of the anterior femoral head/neck junction to relieve the impingement.

What the Clinician Needs to Know

1. What type of femoroacetabular impingement is present: CAM-type, pincer-type, or both?
2. For CAM-type impingement, is the deformity a subtle femoral-head/neck bump or is it due to an aspherical femoral head?
3. With CAM-type impingement, what is the degree of acetabular chondral/labral injury?
4. Based on the distribution of degenerative changes and shape of the femoral head, what is the underlying etiology that is causing femoroacetabular impingement (e.g., DDH, SCFE, LCP, post-traumatic deformity, prior infection)?

Answer

1. CAM-type impingement: reduced femoral head/neck offset, juxtaphyseal osteophytes (or "physeal bump"), and an aspherical femoral head. Pincer-type impingement: acetabular protrusio, coxa profunda, and acetabular retroversion.

Additional Example

Cam-Type FAI with Loose Body and Labral Tear/Degeneration In Spondyloepiphyseal Dysplasia

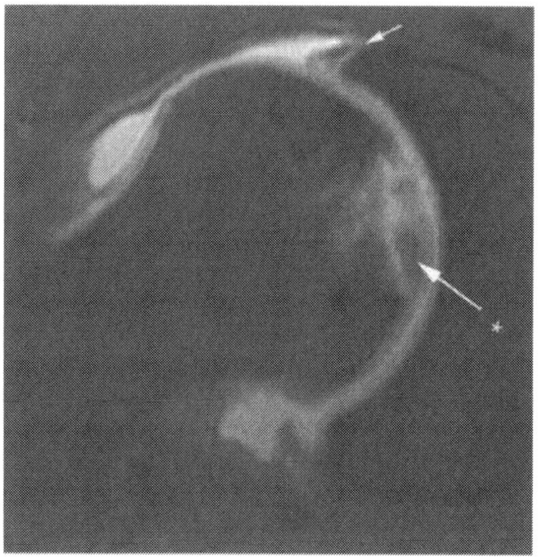

Figure 71D. Axial T1 FS direct Gd arthrography, left hip.

Findings

This 16-year-old boy with a history of spondyloepiphyseal dysplasia (SED) had increasing left hip pain.

Figure 71D. There is loss of the normal triangular shape of the anterior labrum with undersurface irregularity as well as contrast insinuating between the osteolabral junction **(arrow)**, consistent with labral tear/degeneration. There is also contrast undercutting of an osteochondral fragment that was surgically proven to be a loose body (*). CAM-type FAI in this patient was due to an aspherical femoral head related to SED.

Pitfalls and Pearls

1. The most common location for labral degeneration from femoroacetabular impingement is anterosuperior.
2. The imaging findings associated with FAI must be viewed in the clinical context. There is much to be learned about this newly described condition, so a cautious approach to radiologic diagnosis is warranted.

References

1. Ganz R, Parvizi J, Beck M, Leunig M, Notzli H, Siebenrock KA. Femoroacetabular impingement: A cause for osteoarthritis of the hip. *Clin Orthop* 2003:112–120.
2. Jager M, Wild A, Westhoff B, Krauspe R. Femoroacetabular impingement caused by a femoral osseous head-neck bump deformity: Clinical, radiological, and experimental results. *J Orthop Sci* 2004; 9:256–263.

3. Siebenrock KA, Wahab KH, Werlen S, Kalhor M, Leunig M, Ganz R. Abnormal extension of the femoral head epiphysis as a cause of cam impingement. *Clin Orthop* 2004:54–60.
4. Beck M, Kalhor M, Leunig M, Ganz R. Hip morphology influences the pattern of damage to the acetabular cartilage: Femoroacetabular impingement as a cause of early osteoarthritis of the hip. *J Bone Joint Surg Br* 2005; 87:1012–1018.
5. Leunig M, Podeszwa D, Beck M, Werlen S, Ganz R. Magnetic resonance arthrography of labral disorders in hips with dysplasia and impingement. *Clin Orthop* 2004:74–80.
6. Schmid MR, Notzli HP, Zanetti M, Wyss TF, Hodler J. Cartilage lesions in the hip: Diagnostic effectiveness of MR arthrography. *Radiology* 2003; 226:382–386.

Case 72

History

This 13-year-old girl noticed a non-tender mass in her left thigh 1 month ago.

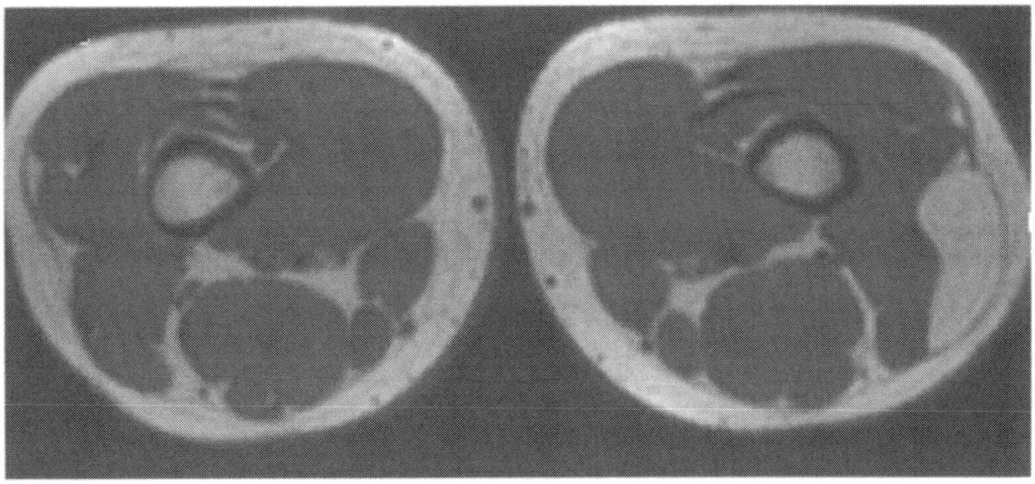

Figure 72A. Axial T1 of both distal thighs.

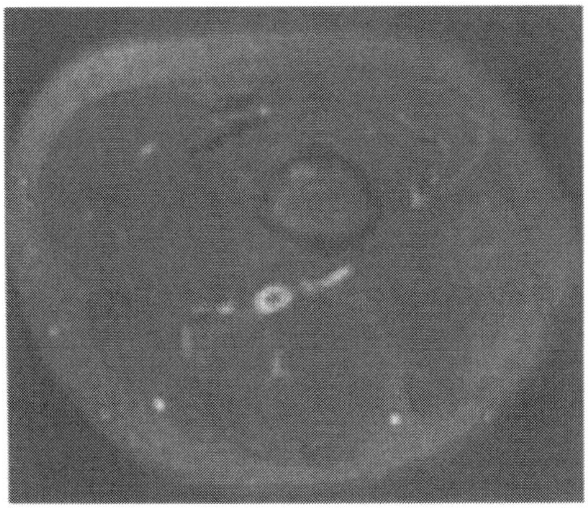

Figure 72B. Axial STIR of the left distal thigh.

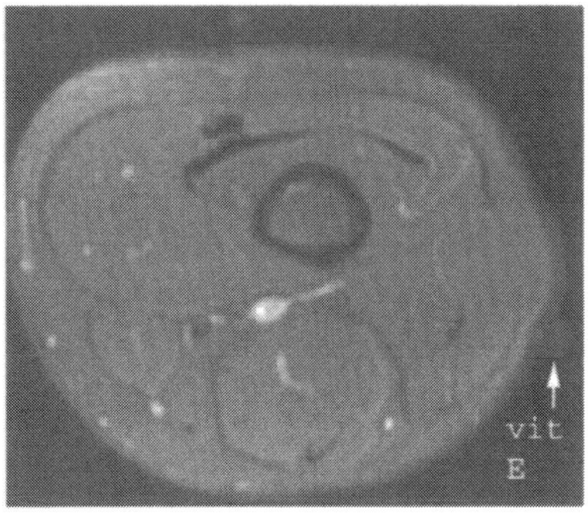

Figure 72C. Axial T1 post-Gd FS of the left distal thigh.

Figures 72A, 72B, 72C. On the left, there is a well-defined, encapsulated mass located between the superficial fascia and the vastus lateralis muscle. It follows fat SI on T1 and STIR and does not enhance. Vitamin E tablet (vit E) is located over the palpable mass.

Diagnosis

Lipoma

Questions

1. T/F: MRI easily distinguishes between mature lipoblastoma and lipoma.
2. T/F: Septations within a lipomatous lesion exclude the diagnosis of a lipoma.

Discussion

Lipomas or other fat containing lesions constitute approximately 6% of all soft tissue tumors in children (1). Most lipomas are superficial and tend to be well-defined and encapsulated (Figure 72D). Deep lipomas tend to be larger with infiltrative margins that may involve muscle (2).

On radiography, up to 11% of lipomas show calcification (2). On CT and MRI, simple lipomas are homogeneous tumors that follow fat on all imaging sequences. On post-Gd sequences, there may be mild capsular enhancement. There should be no non-fatty soft tissue elements present in an uncomplicated lipoma. A simple lipoma is usually indistinguishable from a mature lipoblastoma (see Case 63) (Answer to Question 1) (3). Since lipoblastomas may transform into benign lipomas, distinguishing a mature lipoblastoma from a lipoma is probably not clinically relevant.

Trauma, infection, or fat necrosis may complicate the appearance of a lipoma. Blood products, internal septation, and nodular soft tissue SI may alter the characteristic benign appearance of lipomas on MRI (Figures 72E, 72F). If septations are >2 mm thick, septal nodularity is present, the tumor is greater than 10 cm in size, or there is >25% nonfat SI soft tissue component within the tumor, the diagnosis of a lipoma should not be made on MR imaging grounds alone (4). Conversely, if septations are under <2 mm and septal nodularity and nonadipose soft tissue mass are absent, an MRI diagnosis of a lipoma is appropriate (Answer to Question 2).

Differential considerations for a lipoma include well-differentiated liposarcoma (if patient is >10 years old), lipoblastoma (if the patient is <3 years old), involuted infantile hemangioma, and venous malformation with adjacent fatty atrophy. Lipoma is one of the few soft tissue lesions that can usually be diagnosed with confidence on MRI grounds alone. Other soft tissue masses that MRI may provide a presumptive diagnosis include: some vascular tumors, vascular malformations, certain fibrous tumors, neurogenic tumors, mature myositis ossificans, ganglion cysts, and giant cell tumors (5).

Despite a confident diagnosis of a lipoma on MRI grounds, surgical removal was elected for cosmetic reasons. The patient was considered a good candidate for operative resection because the lipoma was well encapsulated and the lesion did not involve muscle or adjacent neurovascular structures.

Orthopedic Perspective

Once the diagnosis of a lipoma is secure, the decision for the orthopedist and patient will be whether to resect the lesion. This depends on the symptoms, the cosmetic appearance, and the potential morbidity of excision. The MRI is very useful in defining the tissue planes involved, the relationship to neurovascular structures, and the extent of the lesion. These characteristics make it possible for the orthopedist to advise the patient on the pros and cons of excision. In most cases lipomas "shell out" quite easily, although intramuscular lipomas may require some muscle resection.

What the Clinician Needs to Know

1. Is the lesion composed of mature fat and diagnostic of a lipoma, or is a biopsy needed to exclude other lesions?
2. Is this a well-defined lipoma that is easily excised, or is it an intramuscular lipoma that will require some sacrifice of muscle to excise?
3. To determine resectability, the clinician needs to know the relationship of the lipoma to the neurovascular structures and muscle.

Answers

1. False.
2. False.

Additional Examples

Superficial Lipoma

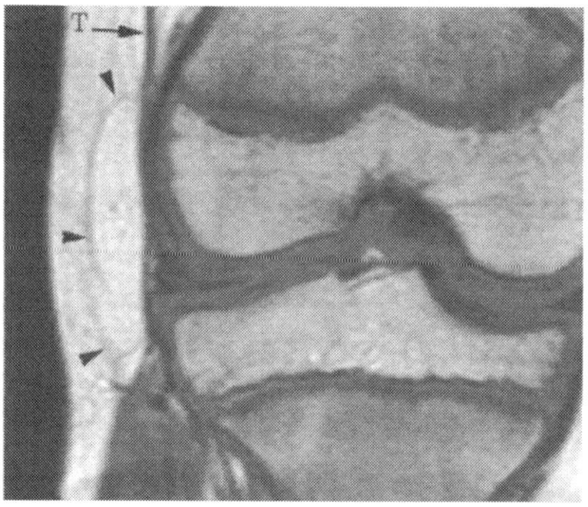

Figure 72D. Coronal T1 of the right knee.

Findings

This 9-year-old girl was referred for posterior right knee pain and a lateral soft tissue mass.

Figure 72D. There is a well-defined homogeneously T1 hyperintense mass **(arrowheads)** superficial to the tensor fascia lata (T) that follows the SI of subcutaneous fat. This lesion was consistent with a superficial lipoma.

Intramuscular Lipoma of the Vastus Lateralis

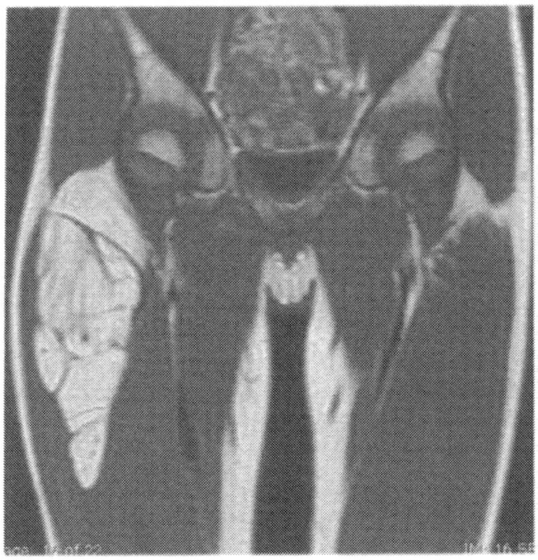

Figure 72E. Coronal T1.

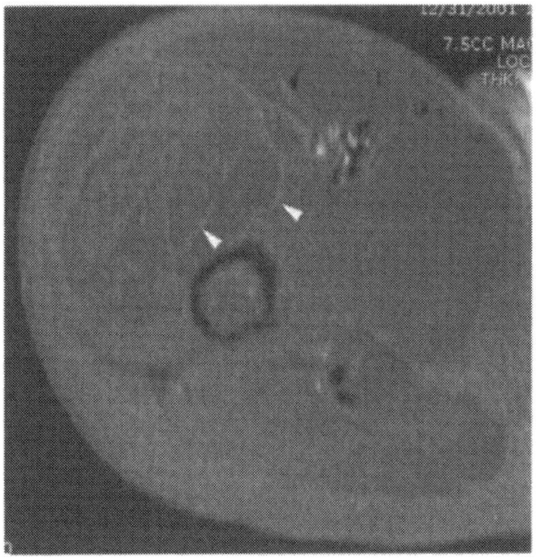

Figure 72F. Axial T1 post-Gd FS right thigh.

Findings

This 11-year-old boy recently noted a dramatic increase in the diameter of his right thigh.

Figure 72E. There is a large septated T1 hyperintense mass (similar to subcutaneous fat) within the right vastus lateralis muscle. STIR images (not shown) showed no abnormal increased SI within the lesion.

Figure 72F. There is mild capsular and septal enhancement **(arrowheads)**. Because these septa are <2 mm thick, the MRI features are all compatible with a lipoma. The diagnosis of lipoma was confirmed surgically.

Pitfalls and Pearls

1. Multiple lipomatous tumors may be seen in Proteus syndrome, Bannayan-Riley-Ruvalcaba syndrome, and Cowden's disease (2).
2. When midline lipomas are identified over the back, rule out occult spinal dysraphism.

References

1. Aflatoon K, Aboulafia AJ, McCarthy EF, Jr., Frassica FJ, Levine AM. Pediatric soft-tissue tumors. *J Am Acad Orthop Surg* 2003; 11:332–343.
2. Murphey MD, Carroll JF, Flemming DJ, Pope TL, Gannon FH, Kransdorf MJ. From the archives of the AFIP: Benign musculoskeletal lipomatous lesions. *Radiographics* 2004; 24:1433–1466.
3. Reiseter T, Nordshus T, Borthne A, Roald B, Naess P, Schistad O. Lipoblastoma: MRI appearances of a rare paediatric soft tissue tumour. *Pediatr Radiol* 1999; 29:542–545.
4. Kransdorf MJ, Bancroft LW, Peterson JJ, Murphey MD, Foster WC, Temple HT. Imaging of fatty tumors: Distinction of lipoma and well-differentiated liposarcoma. *Radiology* 2002; 224:99–104.
5. Kransdorf MJ, Murphey MD. Radiologic evaluation of soft-tissue masses: A current perspective. *AJR Am J Roentgenol* 2000; 175:575–587.

Case 73

History

This is a 3-year-old boy who refuses to bear weight on his left leg after a fall.

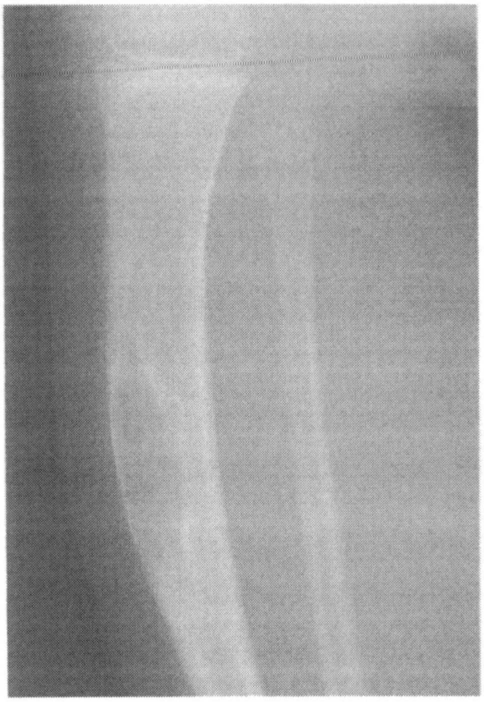

Figure 73A. Lateral radiograph of the left tibia and fibula.

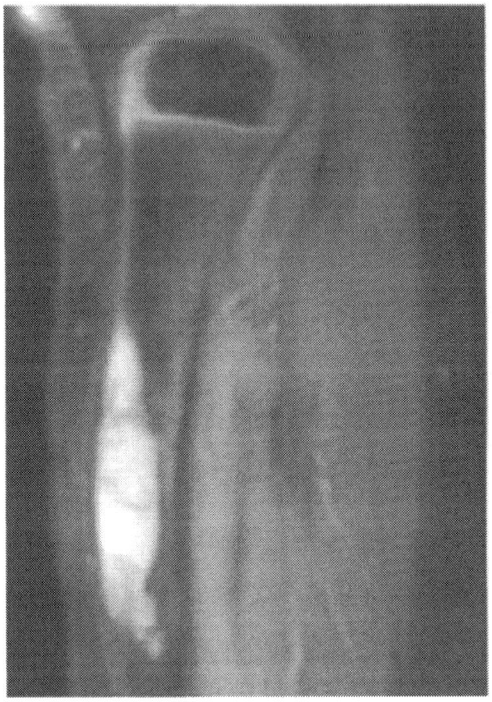

Figure 73B. Sagittal STIR.

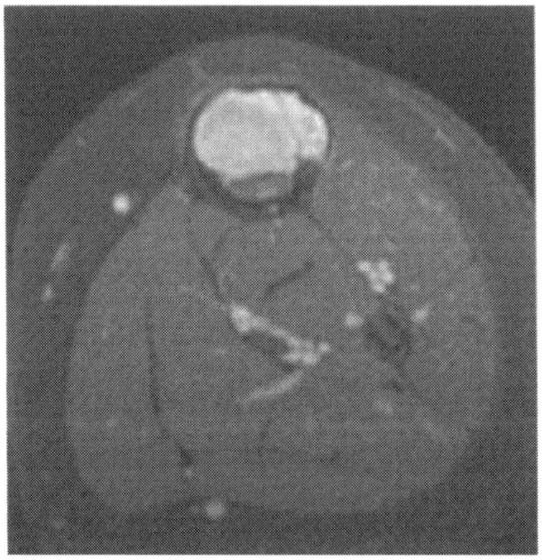

Figure 73C. Axial T1 post-Gd FS.

Figures 73A, (73A with annotations). There is a pathologic fracture **(arrow)** through a slightly expansile osteolytic lesion that thins the anterior cortex of the proximal tibia diaphysis. The margins are well defined and sclerotic. There is the suggestion of ground glass matrix opacity. There is anterior apex bowing deformity.

Figures 73B, 73C. The cortically based lesion is sharply marginated, containing scattered hypointense septations. It demonstrates heterogeneous increased STIR SI and mild heterogeneous enhancement. There is no extraosseous soft tissue mass.

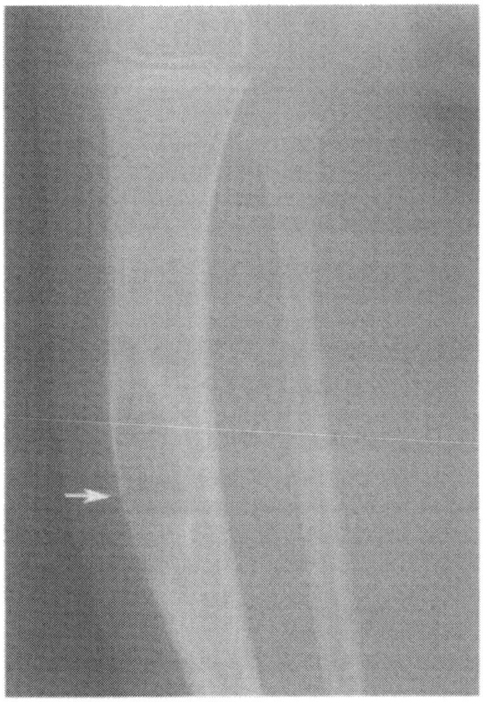

Figure 73A* Annotated.

Diagnosis

Osteofibrous dysplasia with pathologic fracture

Questions

1. What are the usual ages of presentation for osteofibrous dysplasia and adamantinoma?
2. T/F: Adamantinoma more often demonstrates internal hemorrhagic products and extraosseous extension than osteofibrous dysplasia.

Discussion

Osteofibrous dysplasia and adamantinoma (Figures 73D–73F) are two pathologic entities that classically occur in the anterior cortex of the mid-tibial diaphysis. Infrequently, a synchronous lesion may be identified within the fibula. These lesions rarely occur elsewhere. An intermediate lesion has been described, differentiated adamantinoma, which contains elements of both osteofibrous dysplasia and adamantinoma (1).

Osteofibrous dysplasia is a benign fibro-osseous lesion that superficially resembles fibrous dysplasia by histology (1). This lesion has no metastatic potential. Adamantinoma, on the other hand, is a low-grade epithelial neoplasm that may be locally aggressive and occasionally metastasizes, particularly to the lung. These two lesions appear to be related because both have been observed to have cytokeratin-positive cells at histology (2). Adamantinoma may be distinguished from osteofibrous dysplasia by the presence of characteristic hyperchromatic epithelial islands found at histology (2). Osteofibrous dysplasia tends to occur in children under 10 years of age, whereas adamantinoma usually occurs in adults, with a mean age of 30 years. However, adamantinoma displays a wide age spectrum (depending on diagnostic criteria) that extends into childhood (Answer to Question 1) (1). Therefore, strict age criteria should not be used for diagnoses of anterior tibial cortex lesions, because these entities overlap demographically.

The radiographic features of osteofibrous dysplasia and adamantinoma are similar. On radiography, a cortically based osteolytic lesion is identified that may have a ground-glass matrix. The orientation of the lesion is usually longitudinal with respect to the tibia shaft. The margins are usually well-defined and often sclerotic, indicating a slow growing, indolent process. Bloem et al. compared the radiographic features of 46 patients with fibrous dysplasia and osteofibrous dysplasia to those of 22 patients with adamantinoma. The authors observed that younger age at presentation, anterior bowing deformity, ground-glass matrix, absence of multilayered periosteal reaction, and moth-eaten destruction favored the diagnosis of osteofibrous dysplasia rather than adamantinoma (3).

On MRI, these lesions are usually isointense on T1, with variable hyperintensity on fluid sensitive sequences, and they enhance. In a study comparing osteofibrous dysplasia and adamantinoma, Arcara et al. observed that osteofibrous dysplasia was usually more homogeneous on T1, T2, and post-Gd sequences compared with adamantinoma (4). Adamantinoma, on the other hand, was more often heterogeneous on both T1 and T2W sequences and demonstrated marked heterogeneous enhancement with vascular-like structures. Adamantinoma was also more likely to have hemorrhagic products and demonstrate extraosseous extension compared with osteofibrous dysplasia (Answer to Question 2). Both osteofibrous dysplasia and adamantinoma were observed to have variable intramedullary extension. Van der Woude et al. observed that classic adamantinoma was more often composed of multiple separate foci, whereas differentiated adamantinoma tended to present as a single focus (5).

Since radiography and MRI cannot definitively distinguish between these three lesions, the diagnosis is usually made after biopsy. These tumors should be distinguished from fibrous dysplasia, which is usually located centrally or eccentrically within the medullary cavity, as well as Langerhans cell histiocytosis and small round blue cell malignancies, such as Ewing's sarcoma (Figures 73G–73I). When there is a pseudoarthrosis or anterior apex angulation associated with a dysplastic cystic mid-diaphyseal lesion, neurofibromatosis type 1 should also be considered.

In this patient, the tibial cortical lesion was pathologically proven to be osteofibrous dysplasia. Curettage and packing with bone graft was performed. Follow-up MRI and radiography demonstrated no evidence of recurrence.

Orthopedic Perspective

Tibial shaft lesions involving the anterior cortex often pose extremely challenging diagnostic problems. The primary differential diagnosis includes fibrous dysplasia, osteofibrous dysplasia and adamantinoma, although Ewing's sarcoma may also present in the tibial shaft as illustrated here (Figures 73G–73I). Osteofibrous dysplasia is a benign lesion that weakens the bone and frequently presents with prominent tibial bowing. Pathologic fracture is common. It would seem that treatment would be simple—curettage and bone graft packing of the lesion—but the recurrence rate from intralesional treatment is virtually 100%. If left alone, however, there are data to suggest that these lesions regress over time. The recently described entity termed differentiated adamantinoma has made treatment decisions difficult because it assumes that osteofibrous dysplasia is a precursor lesion to classic adamantinoma, a low-grade neoplasm. Whether osteofibrous dysplasia is a precursor lesion (like carcinoma in situ) is unclear.

The only known definitive treatment of osteofibrous dysplasia is resection, but this usually involves an intercalary resection of the tibia. Although reconstructive options of allografts, bone transport, and vascularized fibular grafts are available, they are fairly aggressive treatments for a benign lesion and are associated with potentially serious complications. These reconstructions at times develop complications that result in amputation, which is a radical result for a benign lesion. Conservative treatment involves the use of a brace and the attempt to prevent pathologic fracture with the hope that the lesion will involute. The most difficult dilemma for the orthopedist is to distinguish osteofibrous dysplasia from adamantinoma. The location in the anterior cortex of the tibial shaft associated with anterior tibial bowing is reassuring, but an open biopsy is almost always necessary to be certain of the diagnosis. Osteofibrous dysplasia can be observed with attention directed to prevention of fracture, but adamantinoma requires complete surgical excision.

The entity differentiated adamantinoma is an unfortunate misnomer. This entity has not been clearly shown to develop metastases (unlike classical adamantinoma) but the term adamantinoma may encourage the surgeon to treat this like classic adamantinoma with a radical resection or amputation. The proper treatment of differentiated adamantinoma is unclear, but resection, if it does not result in significant morbidity, or close observation is reasonable.

What the Clinician Needs to Know

1. Distinguish osteofibrous dysplasia from fibrous dysplasia, adamantinoma, and Ewing's sarcoma. This almost always requires a biopsy.
2. Is there extraosseous extension?
3. Is there a pathologic fracture?
4. Are there satellite lesions in the tibia, or a second lesion in the fibula?

Answers

1. Osteofibrous dysplasia usually occurs in children under 10 years of age, whereas adamantinoma tends to occur in adult patients, with a mean age of presentation of 30 years.
2. True.

Additional Examples

Adamantinoma

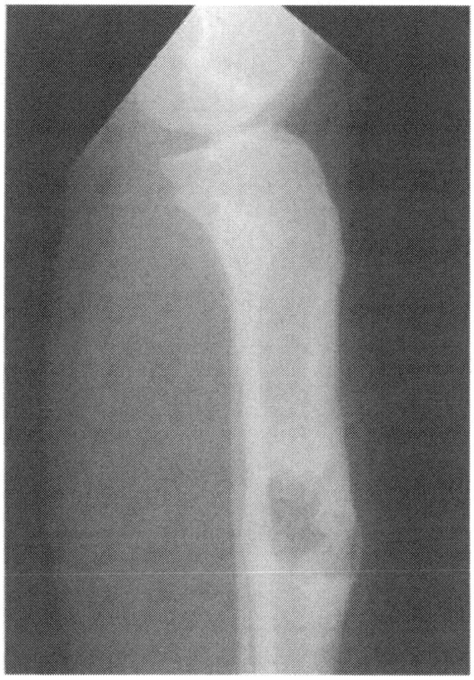

Figure 73D. Lateral radiograph of the left tibia.

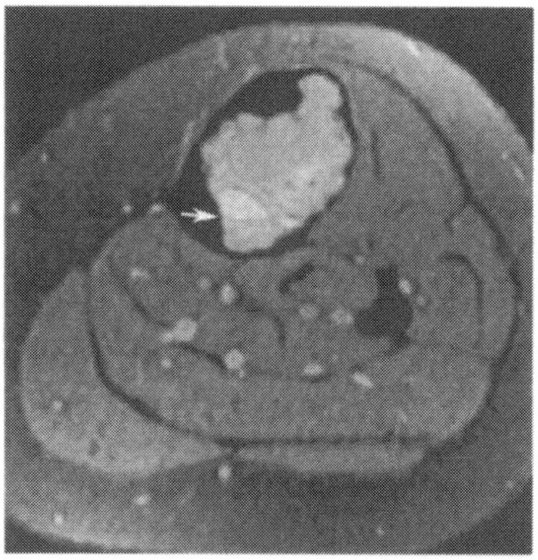

Figure 73E. Axial PD FS.

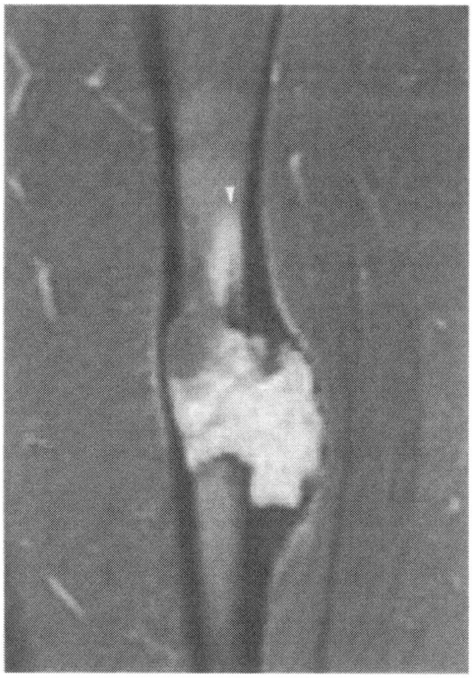

Figure 73F. Coronal T1 post-Gd FS.

Findings

This is an 18-year-old female runner who carried the diagnosis of shin splints for several months.

Figure 73D. There is a well-defined multilobular expansile osteolytic lesion in the mid-tibia diaphysis. It thins the anterior cortex, and does not have ground glass opacity.

Figure 73E. A fluid-fluid level is present **(arrow)**. There is no extraosseous extension.

Figure 73F. There is near complete tumoral enhancement seen. There is also edema seen within the adjacent marrow cavity **(arrowhead)**. There is no extraosseous soft tissue component. Note that this lesion is more heterogeneous in appearance compared with osteofibrous dysplasia (Figures 73A–73C). This was a pathologically proven adamantinoma.

Ewing's Sarcoma with Anterior Tibial Cortex Destruction

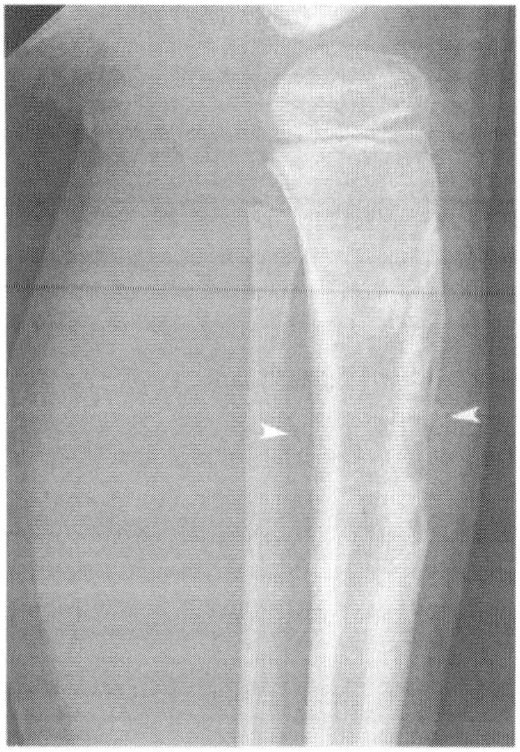

Figure 73G. Lateral radiograph of the left tibia.

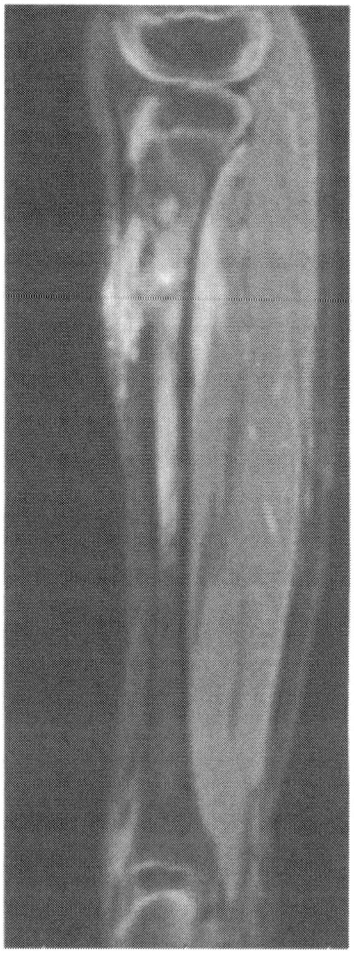

Figure 73H. Sagittal PD FS.

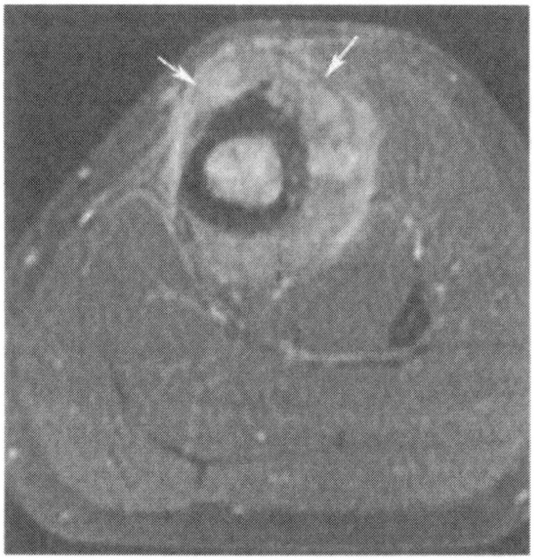

Figure 73I. Axial T1 post-Gd FS.

Findings

This 8-year-old boy had a 5-week history of swelling of his left lower leg.

Figure 73G. A broad based nonexpansile osteolytic process is evident involving the anterior tibia cortex. There is irregular saucerization of the cortex, aggressive periosteal reaction **(arrowheads)**, and permeative medullary bone destruction.

Figures 73H, 73I. A circumferential juxtacortical soft tissue mass is identified that elevates the periosteum **(arrows)**. This was a pathologically proven Ewing's sarcoma.

Pitfalls and Pearls

1. Anterior bowing in a child less than 10 years of age favors osteofibrous dysplasia whereas an older patient with no bowing favors adamantinoma or Ewing's sarcoma.
2. If a lesion is encountered resembling fibrous dysplasia, but is based in the anterior cortex of the tibia in a young child, consider osteofibrous dysplasia. Fibrous dysplasia is usually an intramedullary lesion.

References

1. Kahn LB. Adamantinoma, osteofibrous dysplasia and differentiated adamantinoma. *Skeletal Radiol* 2003; 32:245–258.
2. Sweet DE, Vinh TN, Devaney K. Cortical osteofibrous dysplasia of long bone and its relationship to adamantinoma: A clinicopathologic study of 30 cases. *Am J Surg Pathol* 1992; 16:282–290.
3. Bloem JL, van der Heul RO, Schuttevaer HM, Kuipers D. Fibrous dysplasia vs. adamantinoma of the tibia: Differentiation based on discriminant analysis of clinical and plain film findings. *AJR Am J Roentgenol* 1991; 156:1017–1023.
4. Arcara LK, Murphey MD, Gannon FH, Jelinek JS, Flemming DJ, Dinauer PA. Adamantinoma and osteofibrous dysplasia: Radiologic differentiation [abstract]. *Skeletal Radiol* 2005; 34:559–566.
5. Van der Woude HJ, Hazelbag HM, Bloem JL, Taminiau AH, Hogendoorn PC. MRI of adamantinoma of long bones in correlation with histopathology. *AJR Am J Roentgenol* 2004; 183:1737–1744.

Case 74

History

This is a 4-year-old previously well girl with a 1-month history of severe constipation associated with abdominal pain and fever. She is now unable to walk.

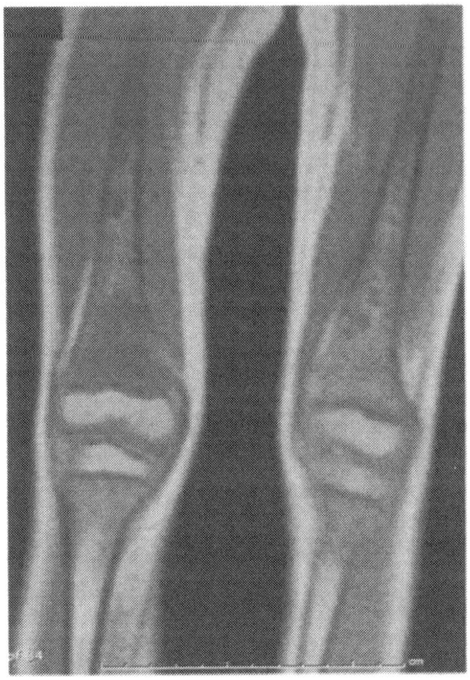

Figure 74A. Coronal T1.

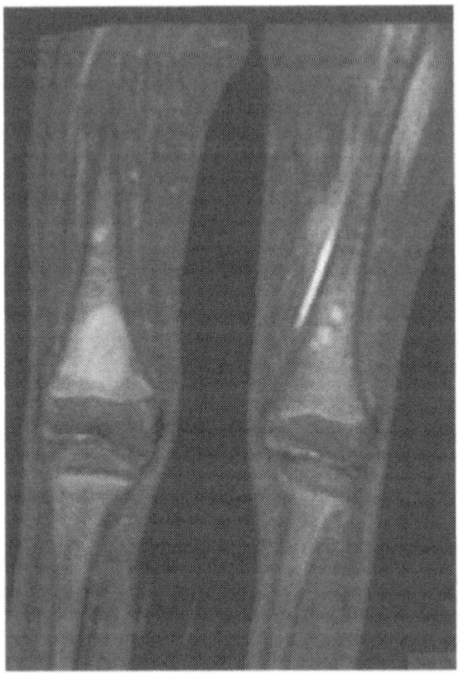

Figure 74B. Coronal STIR.

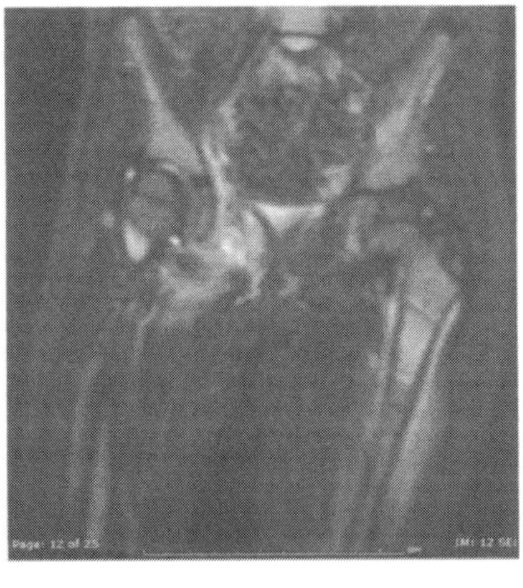

Figure 74C. Coronal STIR.

Figures 74A, 74B, 74C, (74C with annotations). There is abnormal nodular T1 hypointensity and STIR hyperintensity within the marrow of both femurs. Increased STIR SI is also present in the periosteum and juxtacortical soft tissues **(thick arrow)**, associated with conspicuous marrow disease in the left proximal femur **(thin arrow)**. A large heterogeneous SI mass is present in the right suprarenal region **(arrowheads)**.

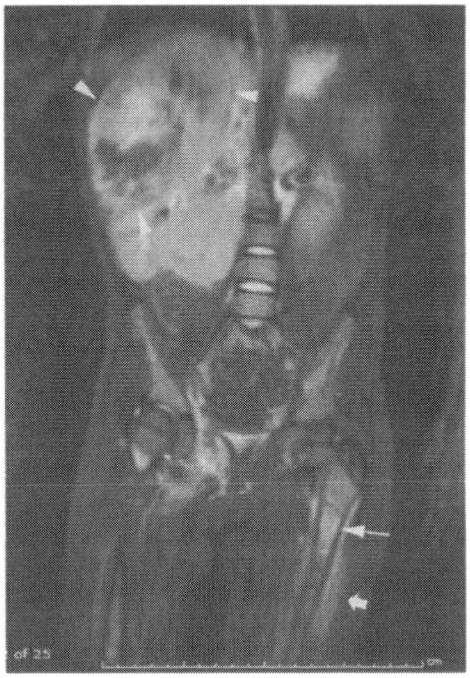

Figure 74C* Annotated.

Diagnosis

Neuroblastoma with bone metastases

Questions

1. T/F: Neuroblastoma osseous metastases occur most often in the metaphyses.
2. What other neoplastic processes fall into the differential diagnosis of osseous metastatic neuroblastoma?

Discussion

Neuroblastoma is the most common extracranial solid neoplasm of childhood (1). Over 90% of cases present before the age of 5 years with the peak incidence at 2–3 years. These tumors arise most commonly from the adrenal medulla, but may occur anywhere there is sympathetic neural tissue including paraspinal sympathetic ganglia, organ of Zuckerhandl, celiac plexus, posterior mediastinum, and neck. The most common site of metastasis is the skeleton, occurring in up to 50–60% of patients at presentation (2). Other sites of metastases include lymph nodes, liver, and skin. Bony metastases may be seen in the axial (Figures 74D, 74E) and the appendicular skeleton (Figures 74F–74I).

On MRI, neuroblastoma metástatic to bone tends to be diffuse and less commonly focal (2). The lesions are hypointense on T1, hyperintense on fluid sensitive sequences, and enhance with gadolinium. They may be associated with cortical destruction, periosteal reaction, and extraosseous extension. Within the long bones, metastatic disease tends to occur within the proximal, and less commonly the distal metaphyses (Answer to Question 1) and follows the same distribution as red marrow reconversion (3). There are several MRI features that permit differentiation of metastatic neuroblastoma from red marrow. Red marrow tends to have a paintbrush or flame shaped pattern without discrete mass effect or trabecular destruction. Neuroblastoma metastases, on the other hand, tend to have a discrete mass effect with infiltrative or nodular margins and destroy trabecular architecture. Furthermore, red marrow is usually less hyperintense than metastatic disease on fluid-sensitive sequences. Technetium-99m methylene-diphosphonate bone scintigraphy (Tc-99m MDP), I-131 metaiodobenzylguanidine (MIBG), and fluorodeoxyglucose positron emission tomography (FDG-PET) uptake in these lesions may be used to further characterize the biologic activity of neuroblastoma metastases and to exclude the diagnosis of residual red marrow (1).

The determination of stage 4S skeletal metastases is currently determined by blind iliac bone marrow biopsies (<10% of tumor cells on bone marrow biopsy) in the absence of distant osseous metastases (4). If osseous metastases are present (with involvement of the cortex), the patient is upgraded to stage 4 disease, which has a worse prognosis. In general, children under 1 year of age have stage 4S disease. Distant osseous metastasis may be determined by nuclear scintigraphy, plain radiography, CT, and/or MRI (1, 5, 6). Whole body MRI using a combination of STIR, T1W, and in- and out-of-phase GRE sequences may be used as a screening tool, but large studies comparing its accuracy with respect to scintigraphy are needed (5). The combination of MRI and scintigraphy may increase accuracy of detecting metastatic disease (6).

Alternative considerations for metastatic neuroblastoma to bone include leukemia/lymphoma, Langerhans cell histiocytosis (LCH) (Answer to Question 2), bacterial osteomyelitis, chronic recurrent multifocal osteomyelitis (see Case 2), chronic anemia, and normal residual red marrow. Differentiation between metastatic neuroblastoma and leukemia/lymphoma or LCH is problematic since all may demonstrate diffuse or multifocal bone lesions and all may occur in children under 5 years of age (7, 8). Pyogenic osteomyelitis is a relatively common entity in children and may be multifocal in as many as 19% of cases (9). Therefore, in the absence of a primary neoplasm or diffuse osseous replacement, pyogenic osteomyelitis should be considered in the setting of metaphyseal or metaphyseal equivalent osseous disease.

In this patient, biopsy of the suprarenal mass and blind iliac bone biopsy confirmed the diagnosis of stage 4 neuroblastoma. She received vincristine, adriamycine, and cyclophosphamide. She subsequently underwent autologous peripheral blood stem cell transplant. She was well after 1.5 months of follow-up. Her clinical care was subsequently transferred back to a pediatric oncologist located close to her home.

Orthopedic Perspective

The issues related to making the diagnosis of metastatic neuroblastoma are similar to lymphoma of bone (see Case 24). Urinary catecholamines, if present, are helpful to establish the diagnosis. Once the diagnosis is evident, the role of the orthopedist is to decide whether a patient is at risk for pathological fracture and will require prophylactic fixation. If fracture occurs, operative stabilization may be necessary to provide comfort and relief, although these patients usually have end stage disease by this time.

What the Clinician Needs to Know

1. Is this metastatic neuroblastoma, other small round blue cell malignancy, LCH, or osteomyelitis?
2. Is the patient at risk for pathologic fracture? CT and biomechanical modeling of the data may be useful.
3. Treat fractures when they occur to provide palliation.

Answers

1. True.
2. Leukemia/lymphoma and Langerhans cell histiocytosis.

Additional Examples

Neuroblastoma Presenting with Calvarial Metastasis

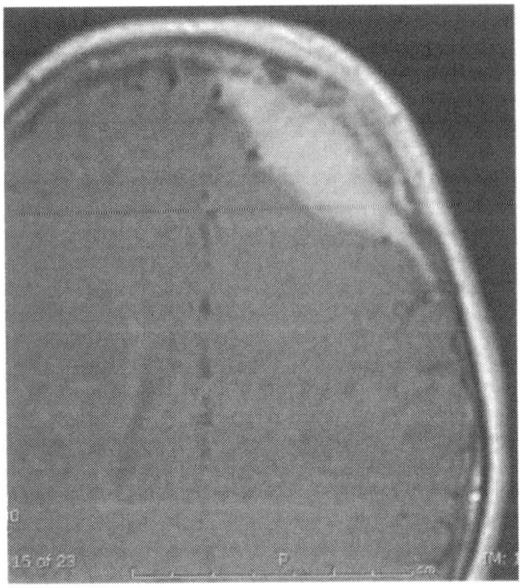

Figure 74D. Axial T1 post-Gd FS.

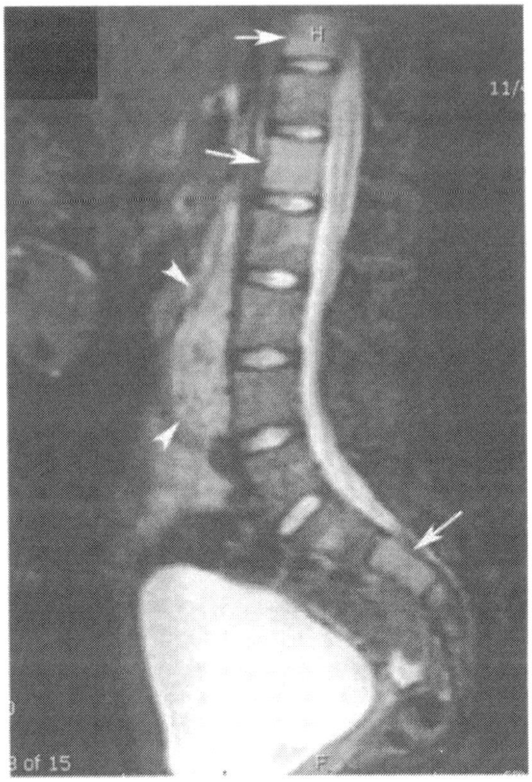

Figure 74E. Sagittal STIR.

Findings

This is a 15-year-old girl who fell off her motorbike with loss of consciousness. An outside CT (not shown) reportedly demonstrated an epidural hematoma.

Figure 74D. There is an extra-axial enhancing mass originating from the left frontal calvarium.

Figure 74E. Multiple metastases are seen in the lower thoracic and upper lumbar spine and the sacrum **(arrows)**. A retroperitoneal, extra-adrenal mass is identified **(arrowheads)**. Complete work-up led to the diagnosis of stage 4 neuroblastoma. This is an uncommon age for presentation of neuroblastoma.

Neuroblastoma with Bony Relapse

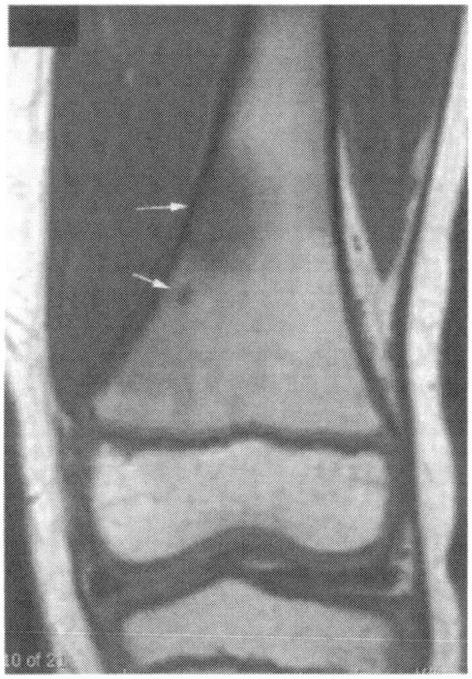

Figure 74F. Coronal T1 of the left knee.

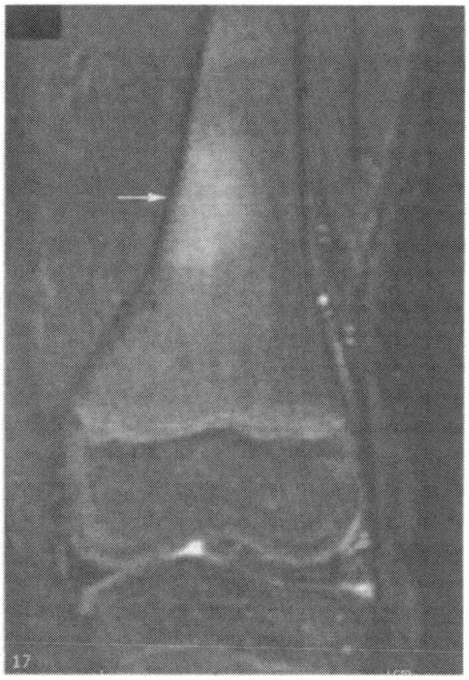

Figure 74G. Coronal STIR.

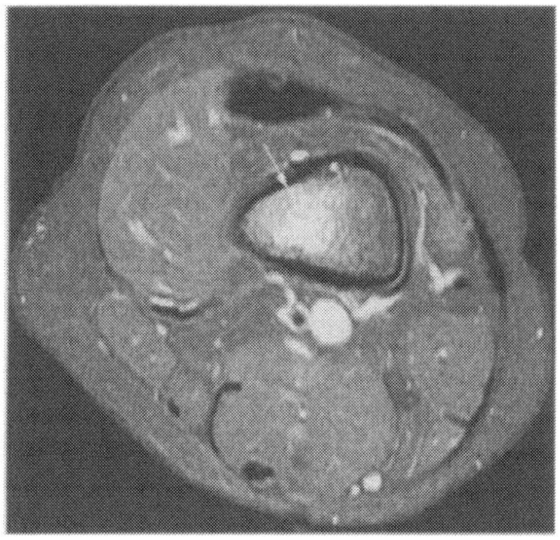

Figure 74H. Axial T1 post-Gd FS.

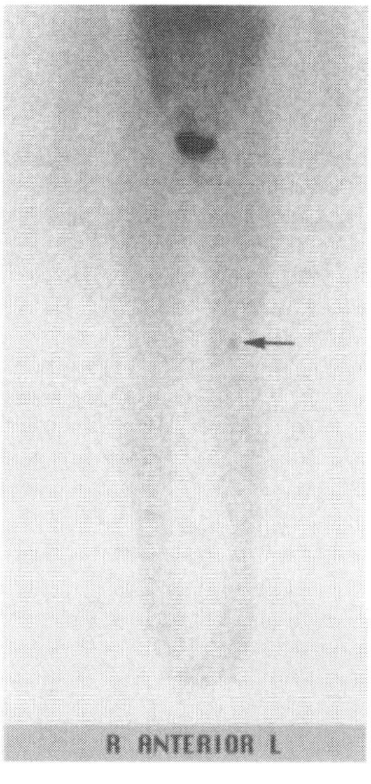

Figure 74I. MIBG (I-131-meta-iodobenzylguanidine).

Findings

This is an 8-year-old girl with stage 4 neuroblastoma with left knee pain.

Figures 74F,74G, 74H, 74I. There is a poorly defined, infiltrative mass in the distal femoral metadiaphysis that is hypointense on T1, hyperintense on STIR, and demonstrates homogeneous enhancement **(arrows)**. This correlates with a focal area of uptake on the MIBG exam **(arrow)**, compatible with metastatic neuroblastoma.

Pitfalls and Pearls

1. Neuroblastoma skeletal metastases in the young child (<5 years old) often present with atypical clinical features resembling osteomyelitis or arthritis.
2. Although neuroblastoma typically presents in young children, the diagnosis should be considered in the differential of multiple infiltrative skeletal lesions throughout childhood, particularly if extraosseous extension is present.

References

1. Kushner BH. Neuroblastoma: A disease requiring a multitude of imaging studies. *J Nucl Med* 2004; 45:1172–1188.
2. Hiorns MP, Owens CM. Radiology of neuroblastoma in children. *Eur Radiol* 2001; 11:2071–2081.
3. Herman TE, Siegel MJ. Case 30: Neoplastic marrow infiltration due to neuroblastoma. *Radiology* 2001; 218:91–94.
4. McHugh K, Pritchard J. Problems in the imaging of three common paediatric solid tumours. *Eur J Radiol* 2001; 37:72–78.
5. Laffan EE, O'Connor R, Ryan SP, Donoghue VB. Whole-body magnetic resonance imaging: A useful additional sequence in paediatric imaging. *Pediatr Radiol* 2004; 34:472–480.
6. Siegel MJ, Ishwaran H, Fletcher BD, et al. Staging of neuroblastoma at imaging: Report of the radiology diagnostic oncology group. *Radiology* 2002; 223:168–175.
7. Parker BR. Leukemia and lymphoma in childhood. *Radiol Clin North Am* 1997; 35:1495–1516.
8. Meyer JS, Harty MP, Mahboubi S, et al. Langerhans cell histiocytosis: Presentation and evolution of radiologic findings with clinical correlation. *Radiographics* 1995; 15:1135–1146.
9. Howman-Giles R, Uren R. Multifocal osteomyelitis in childhood: Review by radionuclide bone scan. *Clin Nucl Med* 1992; 17:274–278.

Case 75

History

This is a 4-year-old boy with left ankle pain and swelling.

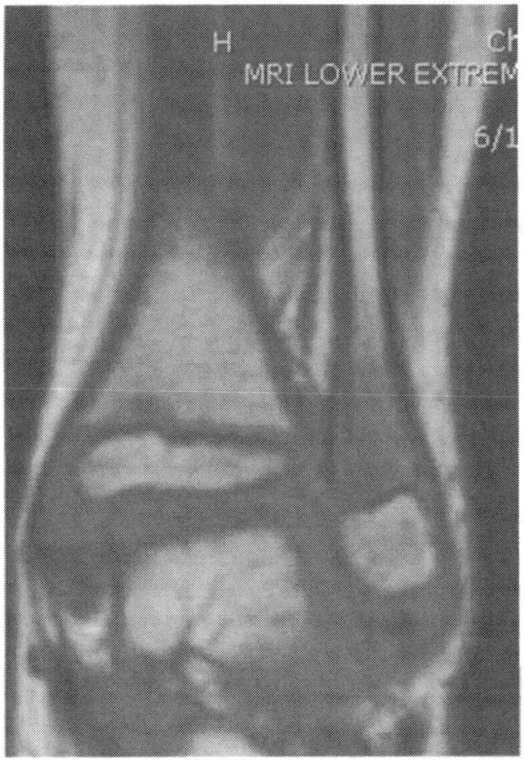

Figure 75A. Coronal T1 of the left ankle.

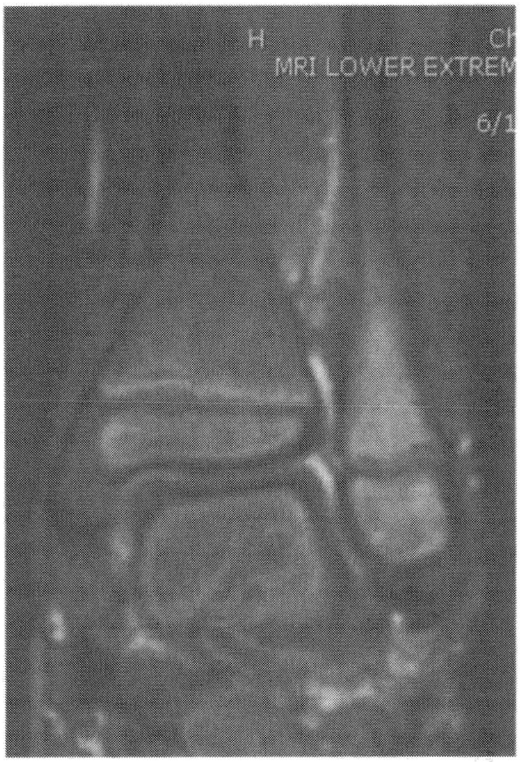

Figure 75B. Coronal STIR.

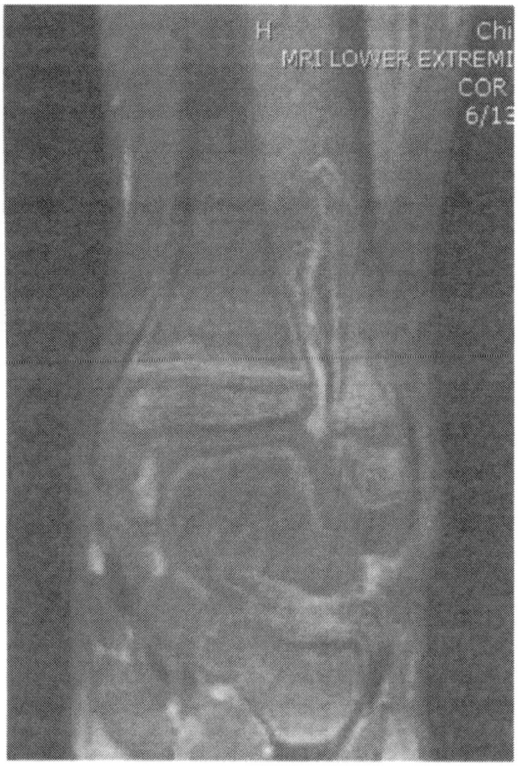

Figure 75C. Coronal T1 post-Gd FS.

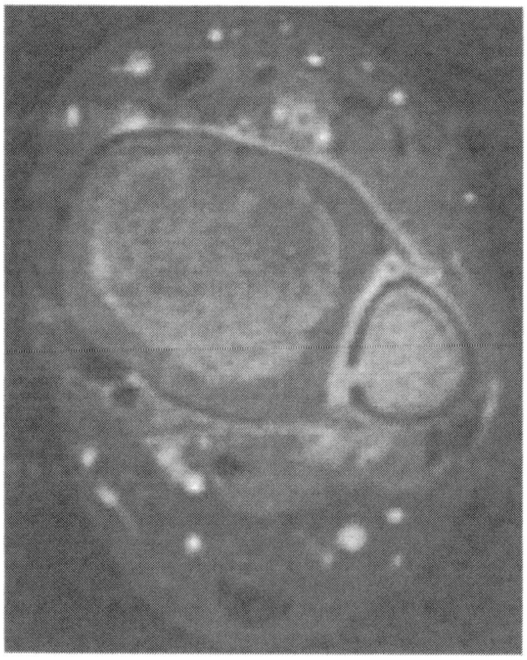

Figure 75D. Axial T1 post-Gd FS (through the fibula metaphysis).

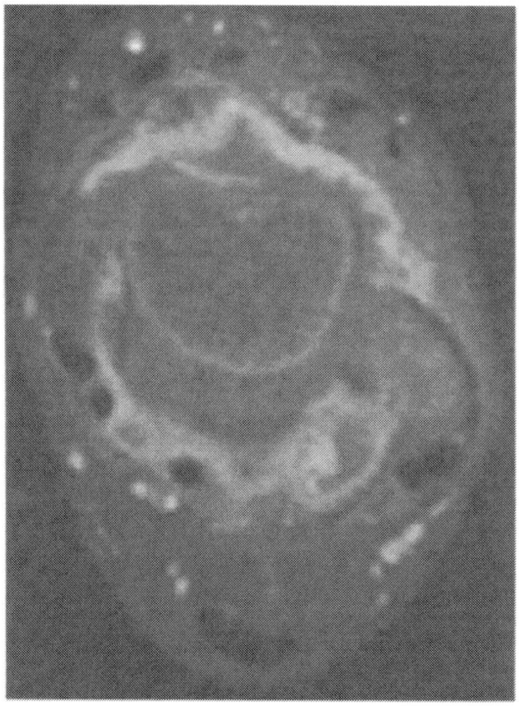

Figure 75E. Axial T1 post-Gd FS (through the fibula epiphysis).

Figures 75A, 75B, 75C, 75D, 75E, (75D, 75E with annotations). Metaphyseal marrow edema is evident within the distal fibula on T1 and STIR sequences with corresponding enhancement following gadolinium. Epiphyseal edema is visible on STIR. Normal physeal SI is seen. The intra-articular metaphyseal cortex is breached medially **(arrowhead)**. There is a moderate joint effusion surrounded by marked synovial enhancement **(arrow)**.

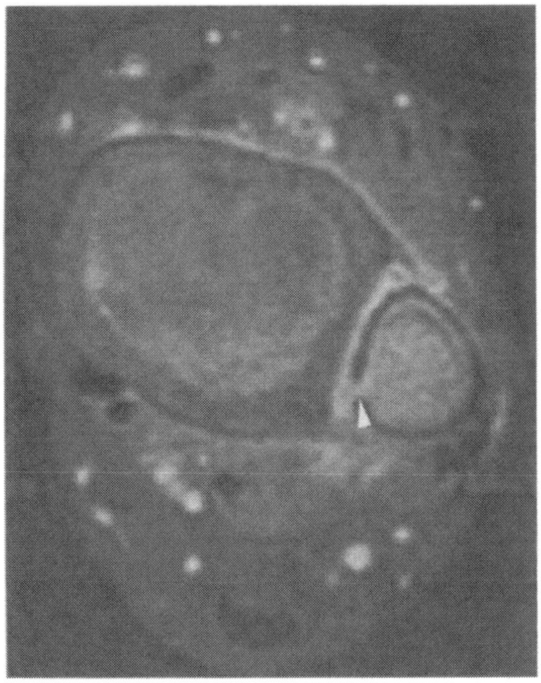

Figure 75D* Annotated.

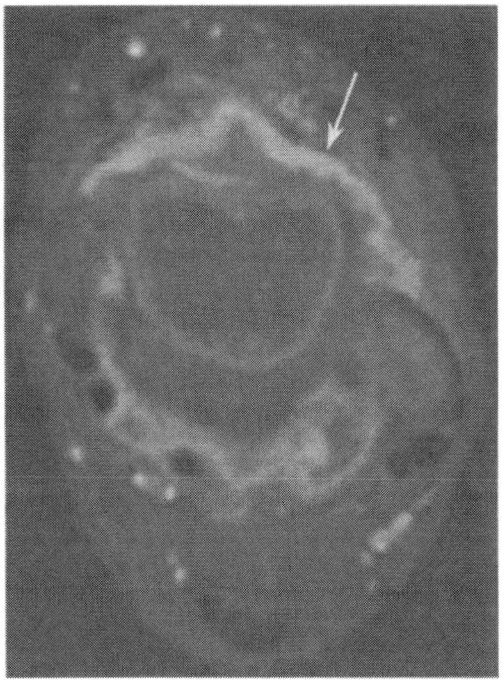

Figure 75E* Annotated.

Diagnosis

Fibular osteomyelitis and secondary septic arthritis

Questions

1. What are two explanations for a joint effusion in the setting of intra-articular osteomyelitis?
2. T/F: Secondary septic arthritis in the setting of osteomyelitis is a rare occurrence.

Discussion

A joint effusion in the setting of intra-articular osteomyelitis may be due to a sympathetic effusion or secondary septic arthritis (Answer to Question 1). Post-gadolinium images in this setting are useful. Modest thin synovial enhancement is expected in the setting of a sympathetic effusion. Features suggesting secondary septic arthritis include: thick and irregular synovial enhancement and the presence of a cloaca, a direct intra-articular communication with osteomyelitis via a cortical breach.

Secondary septic arthritis in the setting of osteomyelitis is not uncommon (Answer to Question 2). The incidence is variable depending on the organism: 22% for Group A beta hemolytic *Streptococcus*, 28% for *Staphylococcus aureus*, and 61% for *Streptococcus pneumonia* (1). Osteomyelitis leading to septic arthritis occurs by two routes: hematogeneous inoculation from metaphyseal or epiphyseal osteomyelitis or by direct extension through a cloaca. Subperiosteal abscesses may also develop after the cortex is breached (Figures 75F, 75G). Subperiosteal abscesses may be quite dramatic in children since the periosteum is less adherent to the cortex compared with adults.

The alternative diagnoses of a joint effusion with juxta-articular marrow edema are primary septic arthritis, Lyme arthritis, juvenile rheumatoid arthritis (JRA), spondyloarthropathy such as psoriatic arthritis, or trauma. Transient synovitis is usually unassociated with juxta-articular marrow edema (2). Primary and secondary septic arthritis may be indistinguishable and both may demonstrate cortical breach. Although juxta-articular marrow edema and enhancement can be seen with JRA, isolated fibular marrow edema would be unusual. The presence of a fracture line or appropriate clinical history would be required to diagnose traumatic causes of effusion with juxta-articular marrow edema.

This patient presented with a sore throat and ankle pain and swelling. He had blood and throat cultures that were positive for Group A Streptococcus. The patient was begun on azithromycin at an outside institution and transferred to our institution for an MRI of the ankle. The patient's ankle pain and swelling were already improving by the time the MRI was obtained. Therefore, the patient was carefully followed and a joint debridement was not considered because his symptoms were improving with antibiotic treatment alone. The patient was subsequently treated with clindamycin for 4 weeks for osteomyelitis and secondary septic arthritis.

Orthopedic Perspective

When confronted with a joint effusion in proximity to long bone osteomyelitis, timely and accurate differentiation between septic arthritis and a sympathetic effusion is essential. Septic arthritis is treated as an emergency, with prompt surgical drainage for large joints such as the shoulder, hip, and knee. Surgical drainage is also frequently performed for elbow and ankle septic arthritis. On the other hand, a sympathetic effusion is observed and typically resolves with effective management of the underlying osteomyelitis. Septic arthritis resulting from cortical breaching of metaphyseal osteomyelitis should be considered for joints where the metaphysis is intracapsular. Although MRI may be useful in differentiating a septic arthritis from sympathetic effusion, when in doubt a joint aspiration should be performed. Primary septic arthritis should be distinguished from secondary septic arthritis due to osteomyelitis since the extent of surgical debridement and the course of antibiotics may vary.

What the Clinician Needs to Know

1. Is the effusion in proximity to long bone osteomyelitis a sympathetic effusion or septic arthritis?
2. Are there other soft tissue, subperiosteal, or intraosseous collections that require drainage?

Answers

1. Septic arthritis or sympathetic effusion.
2. False.

Additional Example

Osteomyelitis and Septic Arthritis

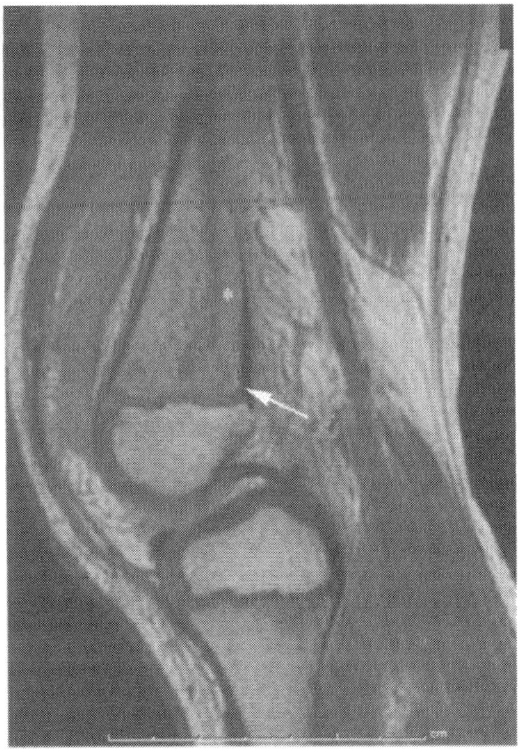

Figure 75F. Sagittal PD of the right knee.

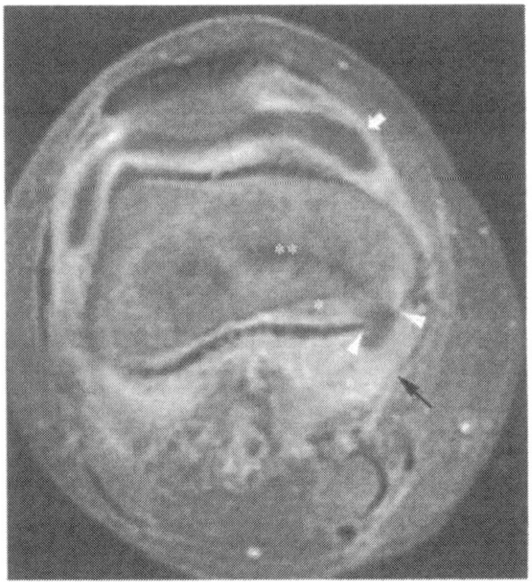

Figure 75G. Axial T1 post-Gd FS.

Findings

This is an 8-year-old boy with a one-week history of right knee pain and swelling.

Figure 75F. A subperiosteal fluid collection is present in the distal femoral metaphysis (*). The subperiosteal fluid collection terminates at the perichondrial attachment **(white arrow)**.

Figure 75G. An ill-defined intramedullary collection is evident in the distal femur (**). The distal femoral intra-articular periosteum **(arrowheads)** is breached by pus. Juxtacortical diffuse soft tissue enhancement extends posteriorly **(black arrow)**. A moderate sized joint effusion as well as synovial thickening and enhancement are present **(thick arrow)**. This patient subsequently underwent right knee arthrocentesis, partial synovectomy, and distal femoral metaphysis irrigation and debridement with creation of a cortical window.

Pitfalls and Pearls

Primary septic arthritis usually presents with symmetric marrow edema on both sides of the joint. Secondary septic arthritis due to intra-articular osteomyelitis presents with marrow edema that may involve both sides of the joint, but the marrow changes tend to be asymmetric.

References

1. Ibia EO, Imoisili M, Pikis A. Group A beta-hemolytic streptococcal osteomyelitis in children. *Pediatrics* 2003; 112:e22–26.
2. Lee SK, Suh KJ, Kim YW, et al. Septic arthritis versus transient synovitis at MR imaging: Preliminary assessment with signal intensity alterations in bone marrow. *Radiology* 1999; 211:459–465.

Case 76

History

This is a 3-week-old boy born at 36 weeks gestation, who recently stopped moving his left arm. A radiograph is normal and the proximal humoral secondary ossification center has not appeared (not shown).

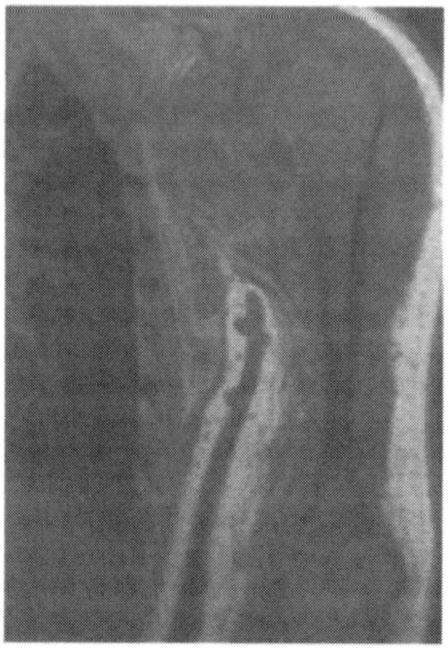

Figure 76A. Coronal T1 of the left humerus.

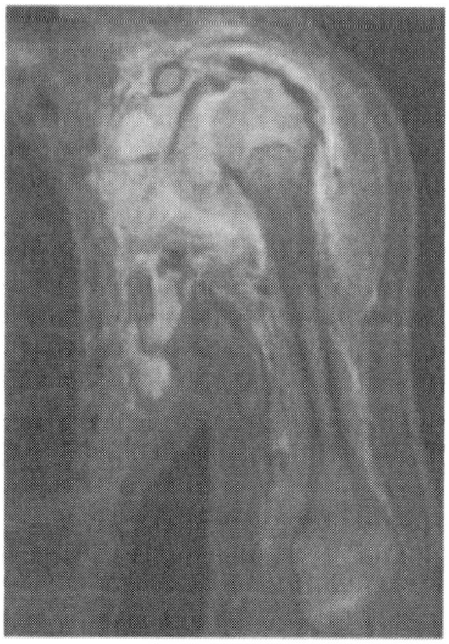

Figure 76B. Coronal STIR.

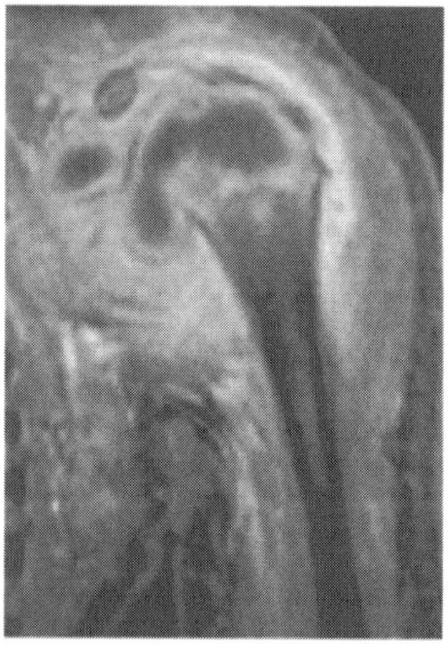

Figure 76C. Coronal T1 post-Gd FS.

Figures 76A, 76B. There is diffuse hyperintensity on the STIR sequence within the soft tissues of the shoulder, extending to the arm and axilla. There is focal hyperintensity in the subcoracoid region and along the medial margins of the cartilaginous proximal humeral epiphysis. There is disruption of the lateral aspect of the physis, and increased SI is evident in the juxtaphyseal margins of the epiphysis and metaphysis.

Figures 76C, (76C with annotations). There is a proximal humerus epiphyseal rim enhancing loculated fluid collection that extends into the glenohumeral joint. A separate smaller rim enhancing fluid collection is also seen medially in the region of the subcoracoid bursae (**arrow**). There is juxtaphyseal, epiphyseal, and metaphyseal enhancement (**arrowheads**), corresponding to the hyperintensity on the STIR sequence.

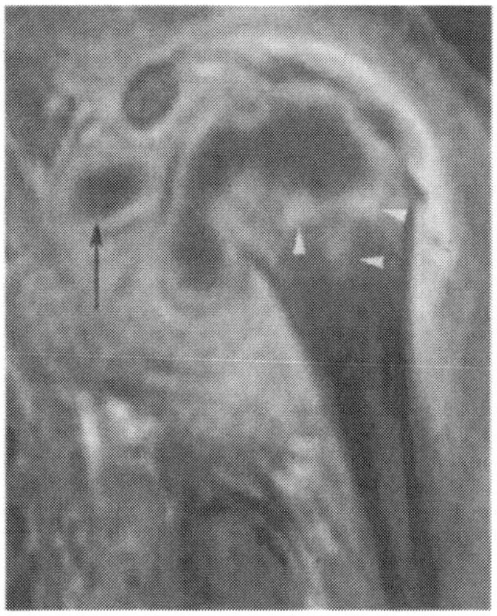

Figure 76C* Annotated.

Diagnosis

Neonatal osteomyelitis and septic arthritis from group B *Streptococcus*

Questions

1. What are the two most common organisms isolated in neonatal osteomyelitis?
2. What is the most common location for neonatal osteomyelitis from Group B *Streptococcus*?

Discussion

The two clinical presentations of neonatal osteomyelitis are: (1) septicemia associated with disseminated infection and (2) focal osteomyelitis with symptoms related to a single site, with minimal or no systemic manifestations. Most cases of neonatal osteomyelitis present with focal disease without septicemia (1). In focal cases, the presentation may be confused with a birth fracture or nerve palsy. Cases with septicemia and disseminated infection often present in the neonatal intensive care unit in premature infants with multiple medical problems. Risk factors for neonatal osteomyelitis include prematurity, complications from pregnancy, or a history of invasive procedures such as vascular access. Various studies have demonstrated either *Staphylococcus aureus* or group B *Streptococcus* as the most common organism isolated in patients with neonatal osteomyelitis (Answer to Question 1). Less commonly, neonatal osteomyelitis may result from group A *Streptococcus*, *Gonococcus*, syphilis, and gram negative organisms such as *Escherichia coli*, *Klebsiella*, and *Enterobacter* (1). For patients with group B *Streptococcus* osteomyelitis, the most common location is the proximal humerus and the second most common is the hip (Answer to Question 2).

In children under 18 months of age, the most common location for hematogeneous osteomyelitis of the long bones is the epiphysis (2). The pathophysiology of epiphyseal osteomyelitis in the newborn is determined by the vascular supply of the secondary ossification center. In children under 18 months, transphyseal vessels directly connect the nutrient arteries of the diaphysis to the epiphyseal circulation (3). Therefore, hematogenous bacterial seeding may bypass the metaphysis and physis, and implant and colonize the epiphysis. The modest metaphyseal disease in this case is consistent with transphyseal spread from a primary epiphyseal focus of infection.

In children over 18 months with open physes, the most common location for hematogeneous osteomyelitis of the long bones is the metaphysis (2). The pathophysiology of osteomyelitis in older children differs from that in infants. At approximately 18 months of age, transphyseal vessels gradually regress and become terminal sinusoids in the metaphysis. Bacterial colonization now occurs at the sinusoidal level near the physis where blood flow is sluggish (see Case 45). Although less common, primary hematogeneous epiphyseal osteomyelitis still occurs in older children and is usually related to the epiphyseal spherical growth plate (see Case 30) (4).

When osteomyelitis develops in an unossified epiphysis, the plain radiographic findings may be confusing. A septic physeal separation with lateral displacement of the metaphysis from the articular surface may suggest a dislocation (5). MRI or ultrasound is required in these cases to define the anatomy. Ultrasonography may also be useful to evaluate for pseudosubluxation and for septic arthritis. When the epiphysis is ossified, the infection and hyperemia may result in a "vanishing epiphysis" on follow-up images (6).

The role of MRI in the evaluation of suspected neonatal osteomyelitis is to determine the presence of isolated septic arthritis, septic arthritis with coexisting osteomyelitis, and the presence of drainable abscess collections. The administration of intravenous gadolinium is essential in this context. Although the soft tissue changes are quite dramatic in this case, the cartilaginous and osseous alterations are relatively modest on T1 and STIR sequences. Delineation of the epiphyseal centered abcess that extends beyond the humeral head is possible only on the post-Gd sequence.

The differential diagnosis for this case is limited. The patient's age, clinical presentation, location of the lesion, extensive inflammatory changes of the adjacent soft tissues, lack of a solid mass, and the presence of a rim enhancing epiphyseal fluid collection are all features that point strongly to neonatal osteomyelitis. Alternative but unlikely considerations include birth trauma or child abuse, nonpyogenic synovitis, and congenital leukemia.

This infant had an unremarkable delivery and neonatal course. He was well at home until his parents noticed that he was no longer moving his left arm. He was initially thought to have a brachial plexus palsy until an abnormal erythrocyte sedimentation rate (ESR) value was obtained. After the MRI, the left shoulder was surgically debrided. Group B *Streptococcus* was cultured from a surgically confirmed humeral head abscess. He was treated with IV antibiotics. At follow-up 2 years later, his left arm function was normal and no growth arrest was evident by radiographs (not shown) or clinical examination.

Orthopedic Perspective

MRI guides the management of neonatal osteomyelitis by identifying intraosseous/soft tissue abscesses and associated septic arthritis. Large abscesses (>1.5 cm) require drainage, either surgically or by interventional radiology. Concomitant septic arthritis may be seen in cases with epiphyseal involvement, or in cases with metaphyseal involvement in which the metaphysis is intracapsular, such as the hip or shoulder. Septic arthritis is typically treated with surgical drainage. Serial aspirations may be effective in small joints.

The sequelae of neonatal osteomyelitis may be profound. Because of epiphyseal involvement, transphyseal extension, and delayed diagnosis with more advanced disease stage, metaphyseal and epiphyseal growth disturbance may occur. These growth disturbances are often consequential given the young age of the patient, and may range from limb-length inequality to destruction of the involved area. When secondary septic arthritis is also present, destruction of the articular cartilage may lead to degenerative arthritis. Early diagnosis and effective treatment are the keys to avoid later sequelae from neonatal osteomyelitis.

What the Clinician Needs to Know

1. The presence of soft tissue/intraosseous abscess and septic arthritis.
2. Is the physis intact, or is there an epiphyseal separation?

Answers

1. *Staphylococcus aureus* and group B *Streptococcus*.
2. Humerus.

Pitfalls and Pearls

The lack of epiphyseal ossification may make the diagnosis of neonatal osteomyelitis difficult. Pseudosubluxation and rarefaction of the metaphyseal trabeculae are subtle early radiographic findings of neonatal osteomyelitis. Unlike osteomyelitis in older children, radiographic findings in neonates are usually abnormal at presentation (1).

References

1. Overturf GD, Marcy SM. Bacterial infections of the bones and joints. In: *Infectious Diseases of the Fetus and Newborn Infant*, 5 ed., Remington JS, Klein JO, eds. Philadelphia: W.B. Saunders, 2001; 1019–1034.
2. Marin C, Sanchez-Alegre ML, Gallego C, et al. Magnetic resonance imaging of osteoarticular infections in children. *Curr Probl Diagn Radiol* 2004; 33:43–59.
3. Oudjhane K, Azouz EM. Imaging of osteomyelitis in children. *Radiol Clin North Am* 2001; 39:251–266.
4. Jaramillo D, Hoffer FA. Cartilaginous epiphysis and growth plate: Normal and abnormal MR imaging findings. *AJR Am J Roentgenol* 1992; 158:1105–1110.
5. Kaye JJ, Winchester PH, Freiberger RH. Neonatal septic "dislocation" of the hip: True dislocation or pathological epiphyseal separation? *Radiology* 1975; 114:671–674.
6. Wood BP. The vanishing epiphyseal ossification center: A sequel to septic arthritis of childhood. *Radiology* 1980; 134:387–389.

Case 77

History

This is a 5-year-old girl with left knee pain. At an outside hospital, an osteolytic lesion was incidentally found on radiographs.

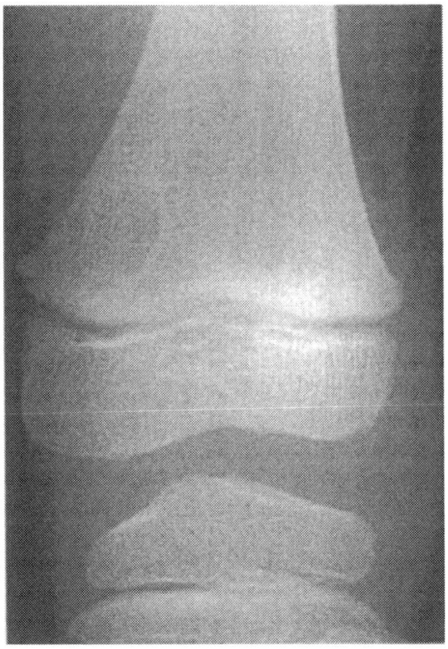

Figure 77A. AP radiograph of the left knee.

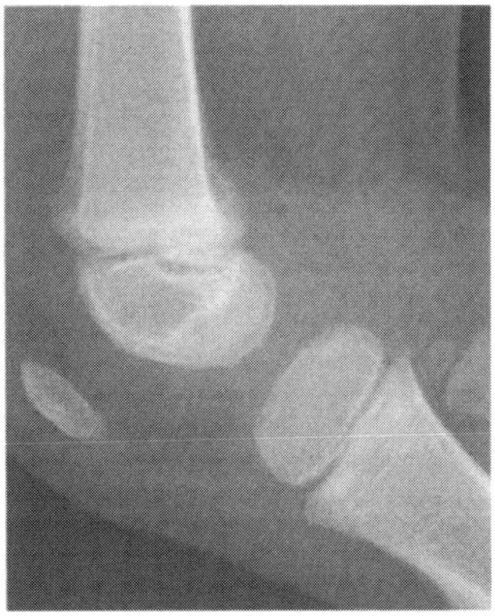

Figure 77B. Lateral radiograph.

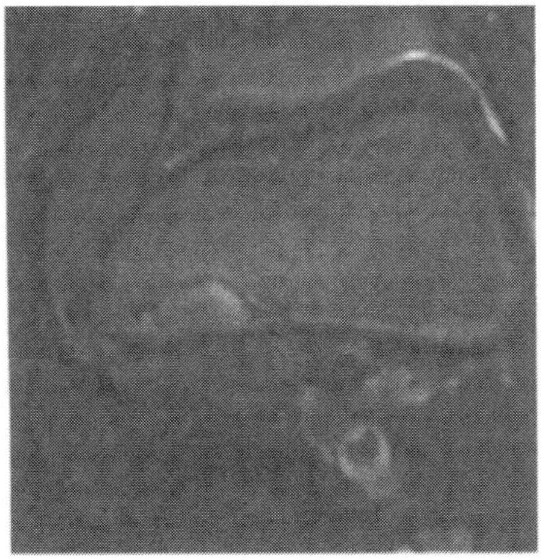

Figure 77C. Axial T2 FS of the distal femur.

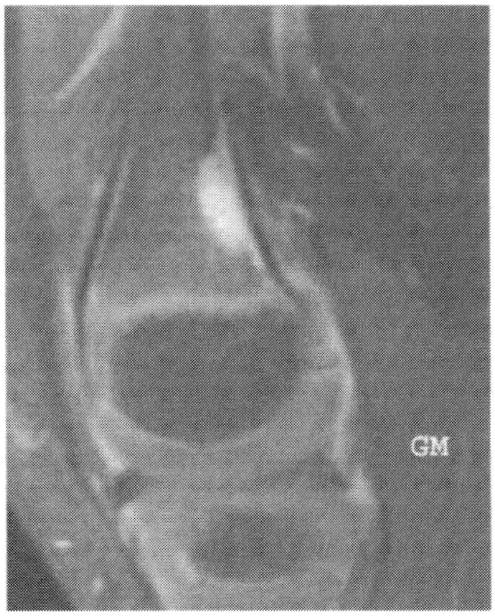

Figure 77D. Sagittal T1 post-Gd FS through the medial femoral condyle.

Figures 77A, 77B, (77A with annotations). There is a round lucent lesion with a well-defined sclerotic rim **(arrowheads)** located in the medial aspect of the distal femoral metaphysis. There is focal irregular periosteal new bone formation posteriorly on the lateral projection.

Figures 77C, 77D. A cortically based lesion is identified at the origin of the medial head of the gastrocnemius muscle (GM). It is mildly hyperintense on T2 and shows homogeneous enhancement. A thin hypointense rim is identified, best seen on the axial sequence, that corresponds to the sclerotic rim on the plain radiograph. The adjacent soft tissues and marrow signal are normal.

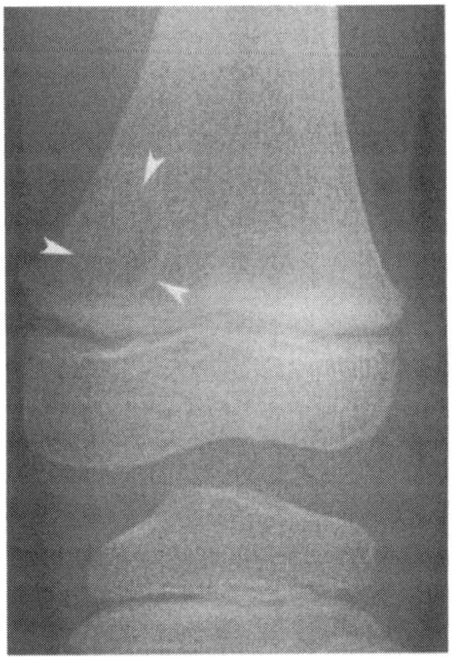

Figure 77A* Annotated.

Diagnosis

Avulsive cortical irregularity

Questions

1. Where are avulsive cortical irregularities of the femur most commonly found?
2. T/F: Avulsive cortical irregularities are commonly asymptomatic.

Discussion

Avulsive cortical irregularities (AKA cortical or periosteal desmoid) are benign lesions that are characteristically located along the posteromedial aspect of the distal femoral metaphysis (Answer to Question 1). Chronic periosteal avulsive stress related to the medial head of the gastrocnemius or the adductor magnus muscle is the most commonly accepted explanation for these lesions (1, 2). They are found almost exclusively in children and adolescents and more commonly affect boys. They are usually asymptomatic in children (Answer to Question 2).

On radiography, avulsive cortical irregularities have a variable appearance, but are usually found when they are osteolytic. The margins are usually sharp and sclerotic, but may occasionally be indistinct. The lesion is eccentric and located along the cortical margin of the distal femur metaphysis. It may assume a concave, convex, or divergent (both concave and convex) appearance (3). There may be variable periosteal new bone formation seen on plain radiography that may mimic an aggressive lesion (Figures 77E–77G).

On MRI, avulsive cortical irregularities are usually hypointense on T1, hyperintense on fluid-sensitive sequences, and occasionally show enhancement, as in this case. In one series, 3 of 12 patients with avulsive cortical irregularities showed enhancement (3). In this same study, the authors noticed that all their cases had a hypointense rim surrounding the avulsive cortical irregularity on all imaging sequences. There may also be mild increased SI on fluid sensitive sequences within the adjacent marrow that may be related to chronic stress reaction (4). Overuse injuries may also lead to tendinopathy with myxoid degeneration of the medial head of the gastrocnemius muscle and tendon, in the absence of osseous abnormalities (Figure 77H).

Alternative considerations for an avulsive cortical irregularity include nonossifying fibroma (NOF), osteoid osteoma, and aggressive lesions such as osteomyelitis, or even an osteosarcoma. NOFs, like avulsive cortical irregularities, are often eccentric lesions with a cortical defect. However, NOFs are typically located in the metaphysis and migrate to the diaphysis with age. An avulsive cortical irregularity may be favored over a NOF if the medial head of the gastrocnemius or adductor magnus muscle can be traced to the lesion. The sclerotic margin, cortical thickening, and central osteolysis of an avulsive cortical irregularity may mimic an osteoid osteoma (Figures 77E–77G); however, the dramatic marrow edema characteristic of osteoid osteoma is generally lacking. Aggressive processes such as osteomyelitis or osteosarcoma may be suggested on radiographs due to the presence of thick irregular periosteal reaction. MRI evaluation may be helpful in these circumstances by demonstrating typical features of infection (e.g., intra-/extraosseous soft tissue abscess/phlegmon, cloaca, sequestrum) or extraosseous soft tissue mass.

The patient's work-up terminated after MRI evaluation. The lesion was not biopsied and the patient was referred back to the care of her primary physician.

Orthopedic Perspective

Avulsive cortical irregularities of the knee are observed frequently in the skeletally immature knee. These are considered incidental findings; the patient's knee pain symptoms are probably completely unrelated. The most important clinical consideration and pitfall of these lesions is not to mistake these lesions for a neoplastic process that may lead to unnecessary biopsy.

What the Clinician Needs to Know

An unequivocal diagnosis of an avulsive cortical irregularity to avoid unnecessary biopsy.

Answers

1. Posteromedial aspect of the distal femoral metaphysis.
2. True.

Additional Examples

Avulsive Cortical Irregularity Simulating an Osteoid Osteoma

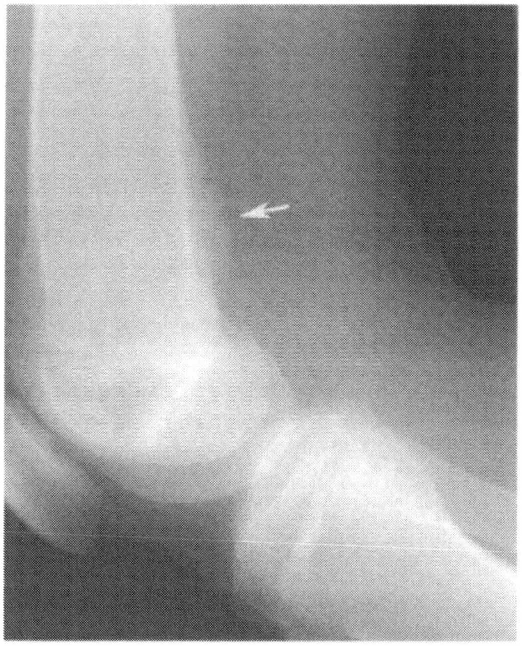

Figure 77E. Lateral radiograph of the right knee.

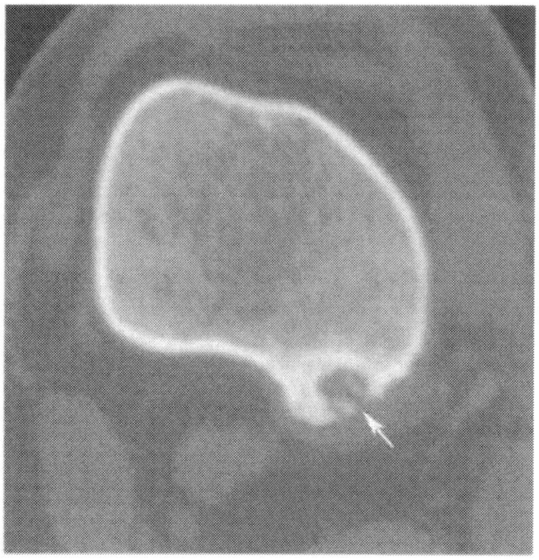

Figure 77F. CT.

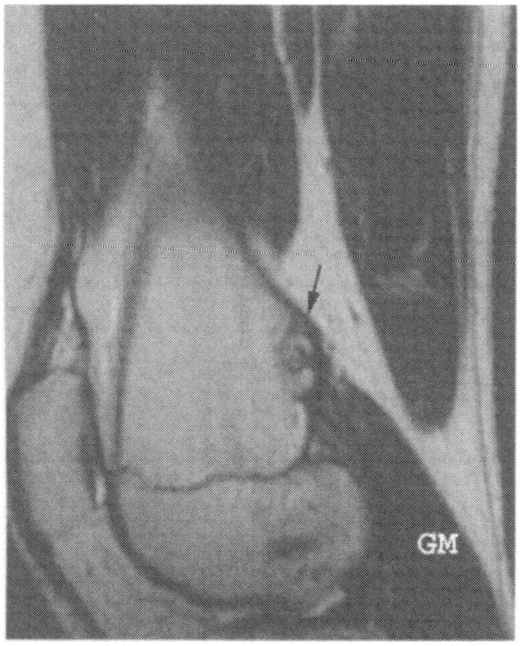

Figure 77G. Sagittal T2.

Findings

This 14-year-old boy had right-sided knee pain.

Figure 77E. Thick mature periosteal new bone is noted along the distal femoral diaphysis **(arrow)**.

Figure 77F. CT shows a rounded cortically based lucent lesion with a mineralized matrix and a densely sclerotic margin **(arrow)**.

Figure 77G. The MRI shows the lesion **(arrow)** to be closely related to the origin of the medial head of the gastrocnemius muscle (GM). The adjacent marrow SI is normal. Although the CT appearance is suggestive of an osteoid osteoma, the radiograph is more in keeping with an avulsive cortical irregularity.

Tendinopathy with Myxoid Degeneration

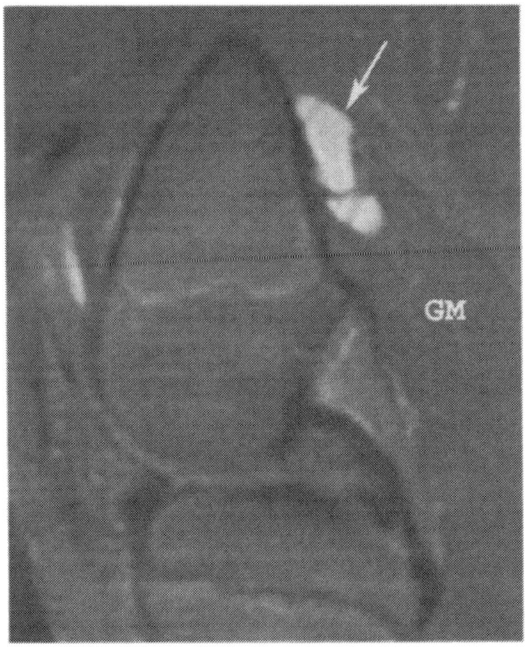

Figure 77H. Sagittal STIR of the medial left knee.

Findings

This 11-year-old girl had chronic medial knee pain.

Figure 77H. There is a multilobulated hyperintense lesion **(arrow)** at the origin of the medial head of the gastrocnemius muscle (GM). The adjacent marrow and cortex of the distal femur are normal. This lesion was felt to represent myxoid degeneration (ganglion cyst) of the tendon origin of the medial head of the gastrocnemius muscle.

Pitfalls and Pearls

An avulsive cortical irregularity is a radiographic diagnosis. MRI should be utilized for problem solving when the lesion has atypical features.

References

1. Yamazaki T, Maruoka S, Takahashi S, et al. MR findings of avulsive cortical irregularity of the distal femur. *Skeletal Radiol* 1995; 24:43–46.
2. Resnick D, Greenway G. Distal femoral cortical defects, irregularities, and excavations. *Radiology* 1982; 143:345–354.
3. Suh JS, Cho JH, Shin KH, et al. MR appearance of distal femoral cortical irregularity (cortical desmoid). *J Comput Assist Tomogr* 1996; 20:328–332.
4. Posch TJ, Puckett ML. Marrow MR signal abnormality associated with bilateral avulsive cortical irregularities in a gymnast. *Skeletal Radiol* 1998; 27:511–514.

Case 78

History

This is a 2-year-old boy who sustained a fall on his left knee. He had swelling that resolved after 3 days. However, the swelling returned a month later and has persisted. He does not have a fever, is able to weight bear on his left knee, and his WBC is normal.

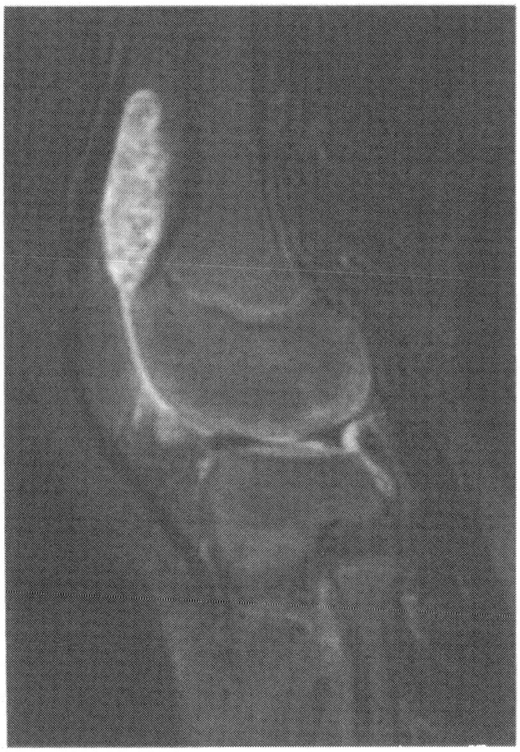

Figure 78A. Sagittal T2 FS of the left knee.

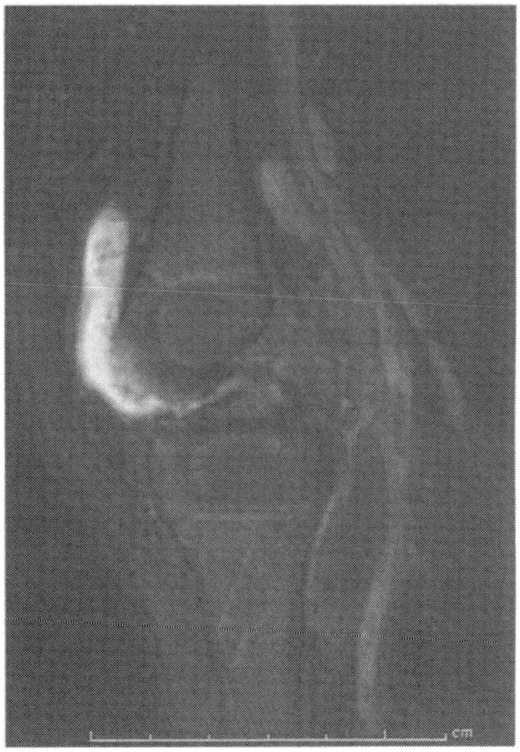

Figure 78B. Sagittal T2 FS, medial to Figure 78A.

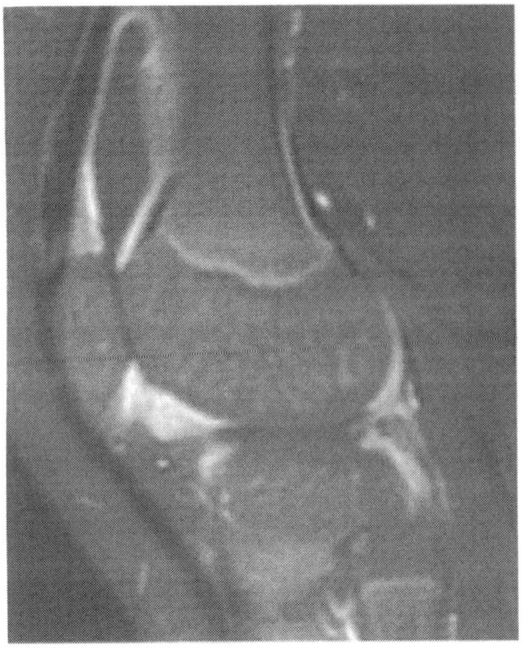

Figure 78C. Sagittal T1 post-Gd FS.

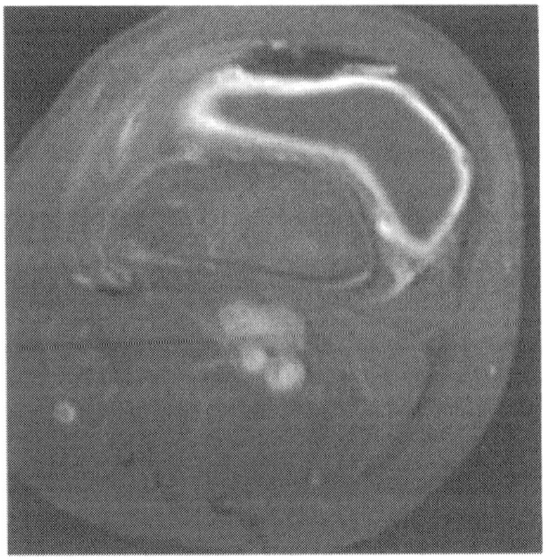

Figure 78D. Axial T1 post-Gd FS.

Figure 78A. There is a moderate size joint effusion containing innumerable low SI particles.
Figure 78B. Reactive lymph nodes are seen in the popliteal fossa.
Figures 78C, 78D. Following contrast, there is enhancement of thickened synovium and Hoffa's fat pad.

Diagnosis

Juvenile rheumatoid arthritis (JRA) with rice bodies

Questions

1. What joint is most commonly affected in JRA?
2. What is the differential diagnosis for rice bodies?

Discussion

The knee is the most common joint affected in patients with juvenile rheumatoid arthritis (JRA) (Answer to Question 1) (1). MRI features specific to the knee in JRA include meniscal hypoplasia, degeneration and tear, and inflammation of Hoffa's fat pad (Figure 78E) (1, 2). Other changes of JRA that may be seen in any joint include synovial hyperplasia and enhancement, tenosynovitis, myositis, rice bodies, central and peripheral subchondral erosions, pannus, intra-articular hemorrhage, epiphyseal overgrowth, periostitis, adjacent marrow edema, joint subluxation, and advanced bone age (3, 4). Periostitis is a feature of JRA that is not seen in the adult type (seropositive) rheumatoid arthritis (RA) (4).

Rice bodies are fibrin and/or collagen encapsulated necrotic cellular debris located within the joint. They are usually hypointense on both T1 and fluid-sensitive sequences (5, 6). On radiography, rice bodies are noncalcified, whereas synovial osteochondromatosis and post-traumatic loose bodies are often radio-opaque. Rice bodies may also be seen with RA, seronegative spondyloarthropathies, and unusual entities such as tuberculous arthritis (Answer to Question 2) (3, 7).

Pannus is composed of reactive granulation tissue within the joint. Fibrous pannus is hypointense on T1, hypointense to intermediate on T2, and demonstrates minimal enhancement. Hypervascular pannus is hypointense on T1, intermediate to hyperintense on fluid-sensitive sequences, and demonstrates significant enhancement (5). Pannus is not specific for JRA and may be seen with other rheumatologic disorders and infectious arthritis.

Longstanding JRA may lead to lipoma arborescens (Figures 78F–78H) and severe degenerative changes with cartilaginous loss, subchondral erosions, and joint space narrowing (Figures 78I–78K). Lipoma arborescens represents synovial metaplasia that may be primary, or secondary to trauma or inflammatory arthritis (8). Fat saturated spin echo or GRE sequences help differentiate lipoma arborescens from intra-articular hemorrhage related to JRA. Lipomatous arborescens follows fat on all imaging sequences, whereas intra-articular hemorrhage shows variable SI depending on the age of the blood products. Intra-articular hemorrhage demonstrates blooming artifact on GRE sequences, whereas lipoma arborescens does not.

This patient was diagnosed with oligoarticular JRA, negative rheumatoid factor but positive ANA. His only symptomatic joint was his knee. He was started on naproxen and later switched to methotrexate, which controlled his symptoms. Because he was ANA positive, routine ophthalmologic visits were scheduled because of the increased risk of developing uveitis.

Orthopedic Perspective

When a history of trauma is lacking, morning stiffness, boggy appearance of the affected joint, normal mechanical examination, polyarticular involvement, and abnormal laboratory values all point to the diagnosis of JRA. However, as seen in this case, JRA may have an atypical presentation and the clinician may refer these patients for MRI, particularly when only a single joint is affected. After osteomyelitis/septic arthritis have been excluded, the clinical and laboratory work-up continues, since Lyme arthritis, other unusual infectious arthridities, and occult trauma are clinical and radiologic mimickers of JRA.

What the Clinician Needs to Know

Determine disease activity, based on the degree of synovial hypertrophy, subchondral edema, and active erosions.

Answers

1. Knee.
2. JRA, RA, seronegative spondyloarthropathies, and tuberculous arthritis.

Additional Examples

Inflammation of Hoffa's Fat Pad

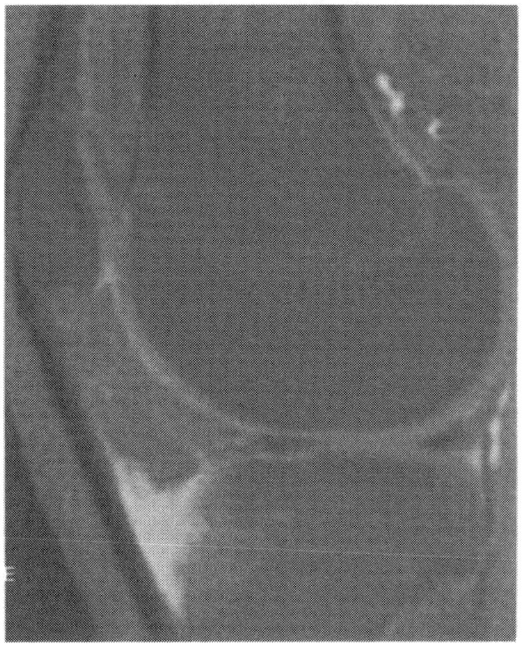

Figure 78E. Sagittal T1 post-Gd FS of the right knee.

Findings

This is a 15-year-old girl with a known history of JRA. She recently had new right knee swelling.

Figure 78E. There is focal enhancement of Hoffa's fat pad in the absence of a joint effusion or enhancing synovium. These findings may be seen in the setting of JRA, but also occur with the seronegative spondyloarthropathies and trauma.

Chronic JRA with Lipoma Arborescens

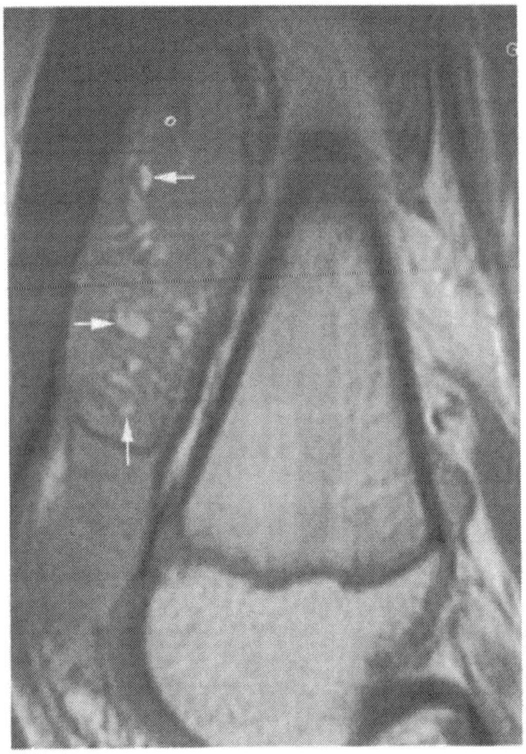

Figure 78F. Sagittal PD of the left knee.

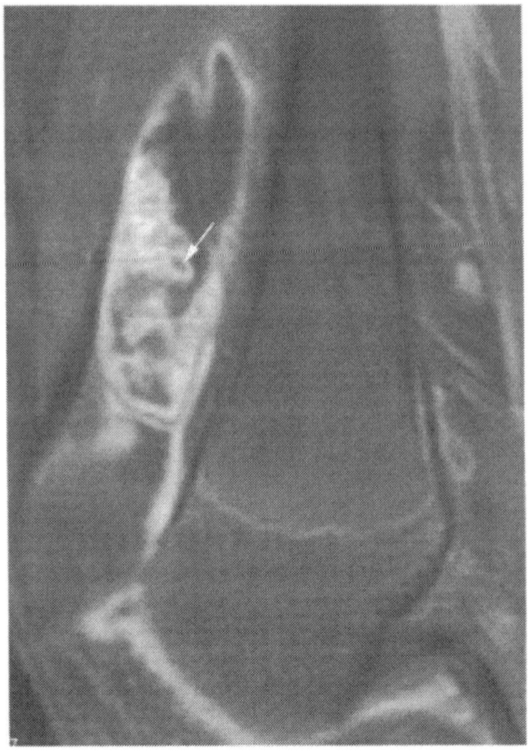

Figure 78G. Sagittal T1 post-Gd FS.

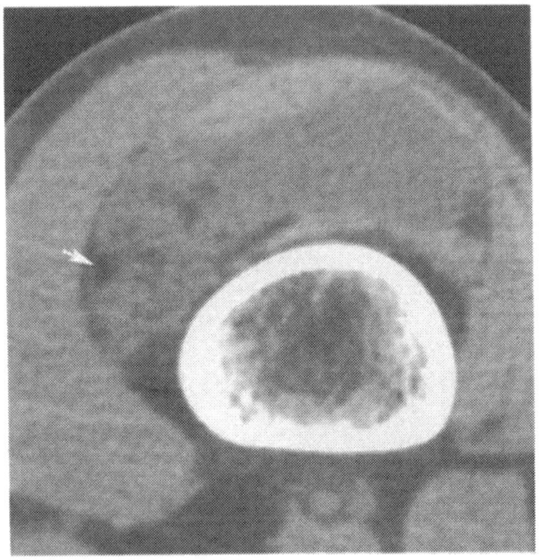

Figure 78H. CT through the suprapatellar joint space.

Findings

This 13-year-old boy had new onset of left knee pain without a history of trauma.

Figures 78F, 78G. There is a large joint effusion with synovial hypertrophy and enhancement. In addition, there are scattered areas of fat SI present within the leaves of the synovium **(arrows)**, compatible with lipoma arborescens.

Figure 78H. CT confirms the presence of synovial fat deposits **(arrow)**. Lipoma arborescens was confirmed at synovectomy.

Longstanding Advanced JRA

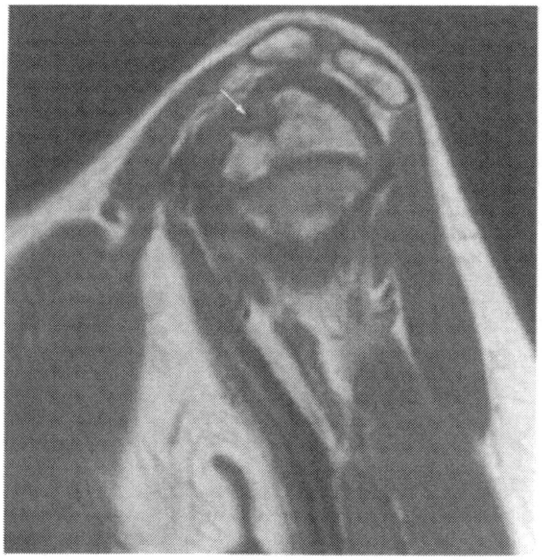

Figure 78I. T1 sagittal of the left shoulder.

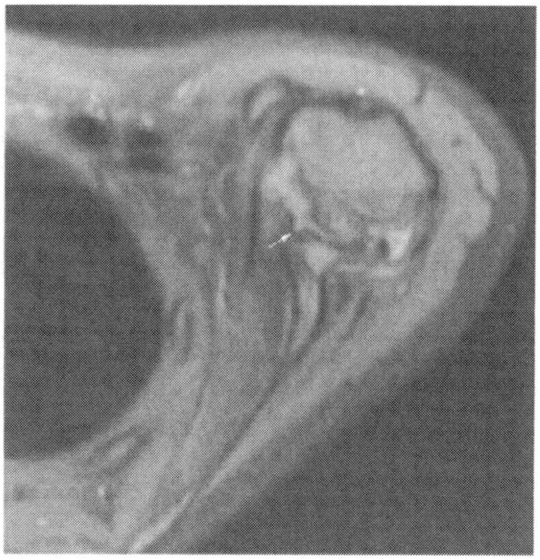

Figure 78J. Axial PD FS.

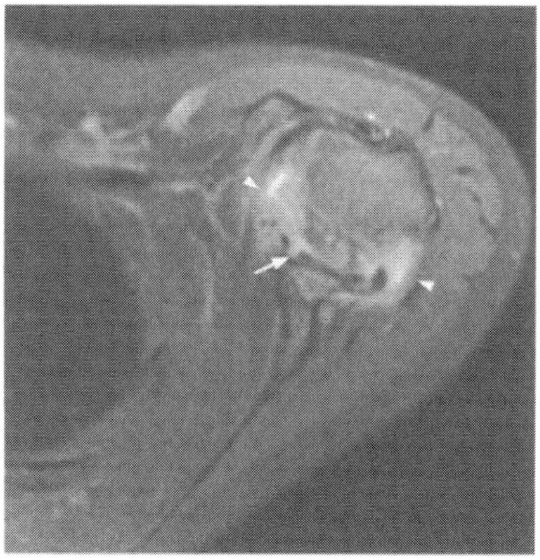

Figure 78K. axial T1 post-Gd FS.

Findings

This is a 9-year-old girl with known systemic JRA since the age of 2 years.

Figures 78I, 78J, 78K. There is severe left glenohumeral joint narrowing, erosions **(arrows)**, and diffuse synovial thickening and enhancement **(arrowheads)**, all compatible with advanced changes of JRA. Note enhancement within central erosions of the glenoid **(arrows)**.

Pitfalls and Pearls

Inflammation of Hoffa's fat pad is a nonspecific finding and may also result from trauma, infection, Osgood-Schlatter's disease, and seronegative spondyloarthropathies (see Case 85).

References

1. Gylys-Morin VM, Graham TB, Blebea JS, et al. Knee in early juvenile rheumatoid arthritis: MR imaging findings. *Radiology* 2001; 220:696–706.
2. Winalski CS, Palmer WE, Rosenthal DI, Weissman BN. Magnetic resonance imaging of rheumatoid arthritis. *Radiol Clin North Am* 1996; 34:243–258, x.
3. Azouz EM. Arthritis in children: Conventional and advanced imaging. *Semin Musculoskelet Radiol* 2003; 7:95–102.
4. Cohen PA, Job-Deslandre CH, Lalande G, Adamsbaum C. Overview of the radiology of juvenile idiopathic arthritis (JIA). *Eur J Radiol* 2000; 33:94–101.
5. Narvaez JA, Narvaez J, Roca Y, Aguilera C. MR imaging assessment of clinical problems in rheumatoid arthritis. *Eur Radiol* 2002; 12:1819–1828.
6. Chung C, Coley BD, Martin LC. Rice bodies in juvenile rheumatoid arthritis. *AJR Am J Roentgenol* 1998; 170:698–700.
7. Chau CL, Griffith JF. Musculoskeletal infections: ultrasound appearances. *Clin Radiol* 2005; 60:149–159.
8. Murphey MD, Carroll JF, Flemming DJ, Pope TL, Gannon FH, Kransdorf MJ. From the archives of the AFIP: Benign musculoskeletal lipomatous lesions. *Radiographics* 2004; 24:1433–1466.

Case 79

History

This is a 13-year-old boy with a left buttock mass present since the age of 3 years that has recently increased in size. Plain radiographs of the pelvis were normal (not shown).

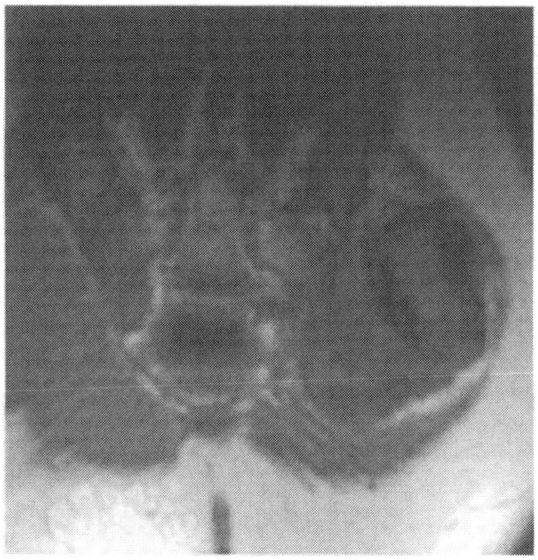

Figure 79A. Coronal T1.

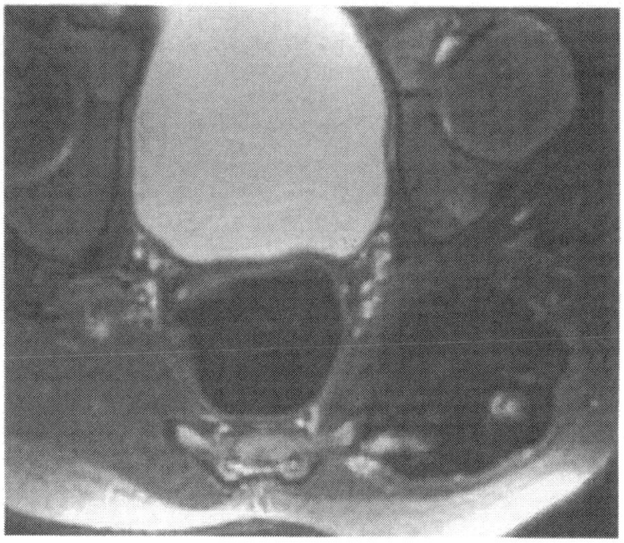

Figure 79B. Axial T2 FS.

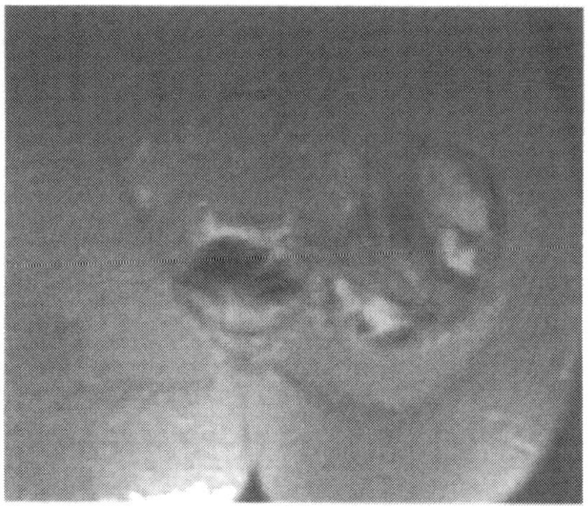

Figure 79C. Coronal T1 post-Gd FS.

Figures 79A, 79B, (79A with annotations). There is a well-defined mass within the left gluteus maximus muscle **(arrowheads)**. It is isointense to slightly hyperintense on T1 and is markedly hypointense with foci of hyperintensity on the T2W image.

Figure 79C. The mass demonstrates scattered areas of mild enhancement.

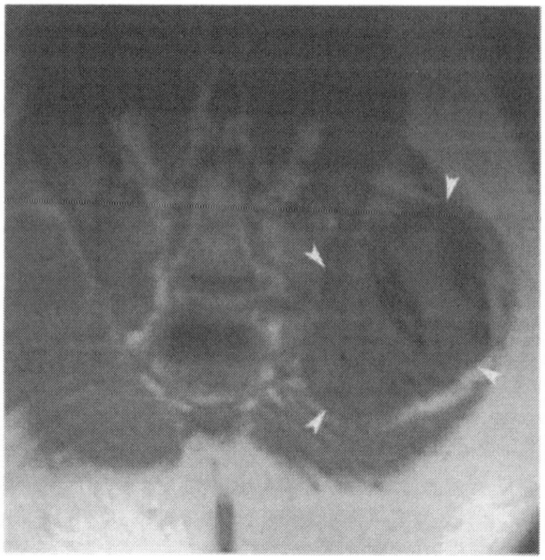

Figure 79A* Annotated.

Diagnosis

Desmoid fibromatosis

Questions

1. T/F: Desmoid fibromatosis that is hyperintense on T2W sequences has a higher chance of recurrence after therapy.
2. T/F: Pediatric desmoid fibromatosis tends to have more infiltrative margins, a greater tendency to involve the extremities, and a higher local recurrence rate than its adult counterpart.

Discussion

Desmoid fibromatosis (AKA aggressive/infantile/musculoaponeurotic fibromatosis) is one of several histologically benign, but locally recurring fibrous tumors of childhood. Other terms used to describe desmoid fibromatosis include: desmoid tumor, aggressive fibromatosis, infantile fibromatosis, musculoaponeurotic fibromatosis, and low grade well differentiated fibrosarcoma (1, 2). Although the majority of desmoid fibromatoses occur in patients in their twenties and thirties, up to 30% of cases are in children and adolescents (1). They may be subclassified on the basis of location: extra-abdominal, abdominal wall, and intra-abdominal. Intra-abdominal desmoid fibromatosis is usually seen in patients with familial adenomatous polyposis syndrome (FAPS) (Figures 79D, 79E) and is rarely seen in isolation (3). When desmoid fibromatosis is found in bone, it is called a desmoplastic fibroma.

The MRI features of desmoid fibromatosis are variable. These tumors may be hypointense on all imaging sequences due to an abundance of collagen and fibrous tissue and decreased cellularity. When there is a highly cellular component, tumors may be isointense to muscle on T1 and hyperintense on fluid sensitive sequences. Tumoral enhancement is commonly seen following gadolinium. Desmoids that are predominantly hyperintense on T2W sequences are more likely to grow and recur compared with desmoids that are predominantly hypointense (Answer to Question 1) (4). However, the presence or degree of enhancement has not been shown to predict local recurrence or growth of desmoids (2).

In one study evaluating 8 children and 32 adult patients with desmoid fibromatosis, the lesions in children more frequently had infiltrative margins (63%), as compared with a nodular pattern in adults (81%) (2). They were more likely to involve the extremities (Figures 79F, 79G), and had a higher local recurrence rate (87%) than in adults (50%) (Answer to Question 2). The extensor side of the upper extremity and the flexor side of the lower extremity tend to be affected in children (5).

Other benign fibrous tumors seen in children include: fibrous hamartoma, myofibromatosis, fibromatosis colli, digital fibromatosis, infantile desmoid fibromatosis, calcifying aponeurotic fibroma, and hyaline fibromatosis (6).

This patient subsequently underwent surgical resection of the left buttock mass, which was pathologically confirmed to represent desmoid fibromatosis. At 4-month follow-up, the patient was well without clinical evidence of recurrence, but was subsequently lost to follow-up.

Orthopedic Perspective

Desmoid fibromatoses are primarily treated by surgical excision and, although benign, they have a high propensity for recurrence. The borders tend to be infiltrative, making it difficult to achieve negative margins. Imaging is important to the surgeon for planning the surgical excision. The attempt is to achieve a wide margin of normal tissue around the lesion to prevent recurrence. Knowing precisely where the tumor is with respect to surrounding major nerves, blood vessels, and bones is crucial in planning local treatment options. At times, adjuvant therapies, such as radiotherapy, are used to reduce the likelihood of local recurrence. In skeletally immature patients, it is important to assess the relationship of the tumor to the growth plates to determine if they will be in a treatment field. Chemotherapy is used for treatment of recurrent lesions and those desmoids where a resection might lead to functional morbidity (e.g., desmoids near major nerves or multiple desmoids). The masses usually do not shrink following chemotherapy administration because of the collagen content of the tumor, but there are often signs of response seen on MRI (i.e., the tumor becomes more hypointense on fluid sensitive sequences), which are helpful to the clinician.

In the case of the buttock mass shown above, it should be possible to achieve a wide surgical margin without injuring the sciatic nerve, although some sacrum might need to be included. The foot lesion (Figures 79G, 79H) does not have a local resection option; so a reasonable treatment approach would include partial foot amputation, radiation, chemotherapy, or observation, depending on symptoms. Following resection, the MRI is useful for monitoring for recurrence and it is often reasonable to get a baseline postoperative MRI after the edema from the operation has resolved to compare with future studies.

What the Clinician Needs to Know

1. The clinician needs to know the anatomic extent of the lesion to plan surgical excision.
2. Are the growth plates in a potential radiation field in skeletally immature patients?
3. Has the lesion responded to chemotherapy?
4. Is there evidence of recurrence?

Answers

1. True.
2. True.

Additional Examples

Intra-abdominal Desmoid Fibromatosis

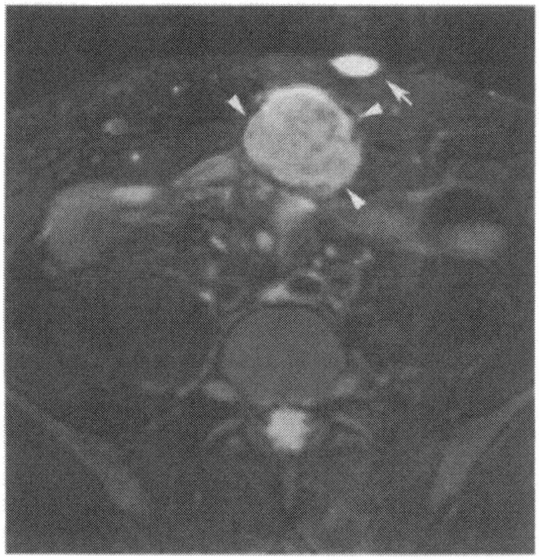

Figure 79D. Axial T2 FS.

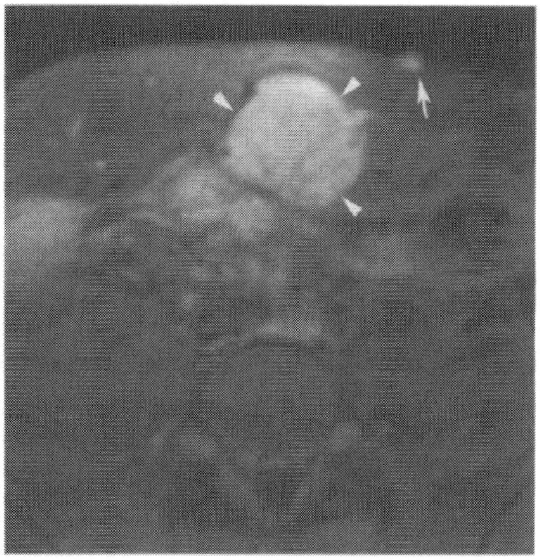

Figure 79E. Axial T1 post-Gd FS.

Findings

This is an 18-year-old male with a history of FAPS who had a total colectomy 2 years earlier.

Figures 79D, 79E. A round, well-defined anterior abdomen mass is identified **(arrowheads)** that is hyperintense on T2 and shows homogeneous enhancement. At surgery, the lesion was noted to be related to the earlier incision and intra-abdominal desmoid fibromatosis was confirmed pathologically. Vitamin E marker **(arrow)**.

Desmoid Fibromatosis of the Foot

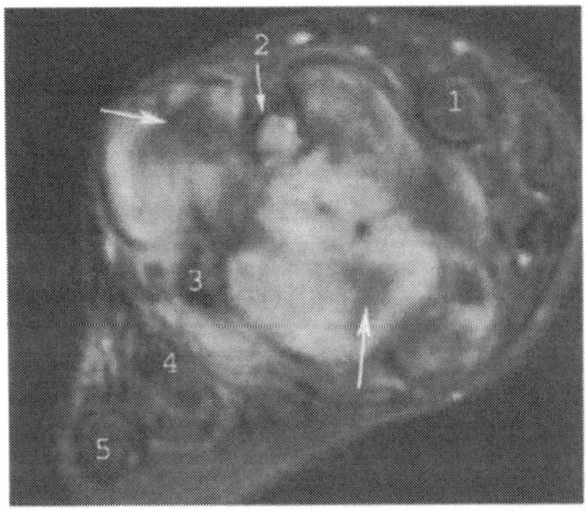

Figure 79F. Coronal T2 FS distal metatarsals of the right foot.

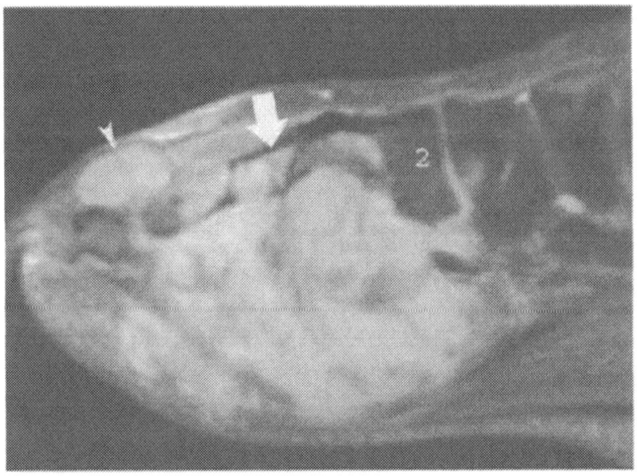

Figure 79G. Sagittal T1 post-Gd FS.

Findings

This is a 12-year-old boy who had a growing mass in his right foot first noticed 5 months ago.

Figures 79F, 79G. A multicompartmental, infiltrative mass is identified that is predominantly hyperintense on T2 and shows intense enhancement. There are focal areas of T2 hypointensity consistent with fibrous tissue **(thin arrows)**. The mass invades the second metatarsal **(thick arrow)** and extends through the second metatarso-phalangeal joint to the dorsum of the forefoot **(arrowhead)**. Biopsy confirmed desmoid fibromatosis. The appearance of desmoid fibromatosis with a significant cellular component is indistinguishable from soft tissue sarcomas. Metatarsals labeled 1–5.

Pitfalls and Pearls

1. Desmoid fibromatosis with a high cellular content has a nonspecific MRI appearance that is hypointense on T1, hyperintense on T2, with diffuse enhancement on post-Gd sequences. It may be indistinguishable from soft tissue sarcomas and other fibrous tumors.
2. Desmoid fibromatosis is seen in 9–18% of patients with FAPS and has a predilection for abdominal surgical incision sites (4).

References

1. Petchprapa CN, Haller JO, Schraft S. Imaging characteristics of aggressive fibromatosis in children. *Comput Med Imaging Graph* 1996; 20:153–158.
2. Romero JA, Kim EE, Kim CG, Chung WK, Isiklar I. Different biologic features of desmoid tumors in adult and juvenile patients: MR demonstration. *J Comput Assist Tomogr* 1995; 19:782–787.
3. Kingston CA, Owens CM, Jeanes A, Malone M. Imaging of desmoid fibromatosis in pediatric patients. *AJR Am J Roentgenol* 2002; 178:191–199.

4. Healy JC, Reznek RH, Clark SK, Phillips RK, Armstrong P. MR appearances of desmoid tumors in familial adenomatous polyposis. *AJR Am J Roentgenol* 1997; 169:465–472.
5. Spiegel DA, Dormans JP, Meyer JS, et al. Aggressive fibromatosis from infancy to adolescence. *J Pediatr Orthop* 1999; 19:776–784.
6. Eich GF, Hoeffel JC, Tschappeler H, Gassner I, Willi UV. Fibrous tumours in children: Imaging features of a heterogeneous group of disorders. *Pediatr Radiol* 1998; 28:500–509.

Case 80

History

This is an 11-month-old girl who presented with focal enlargement of the second toe.

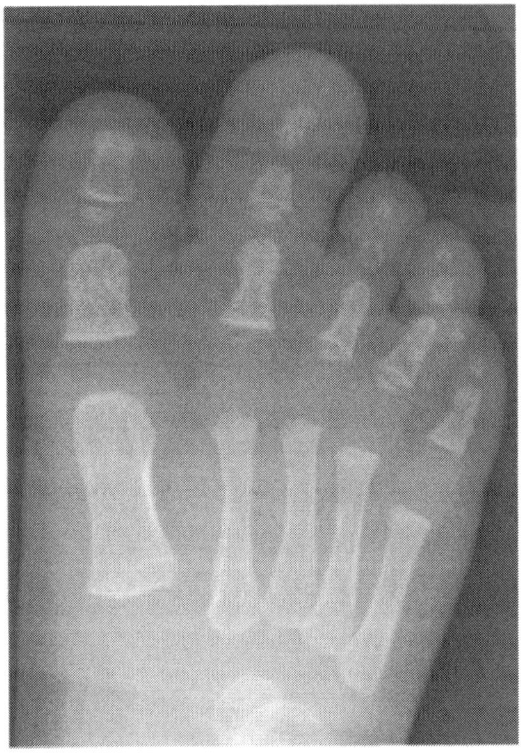

Figure 80A. AP radiograph.

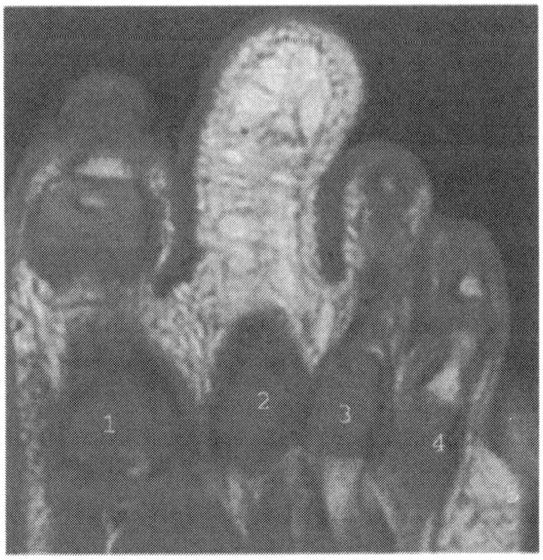

Figure 80B. T1 axial (footprint).

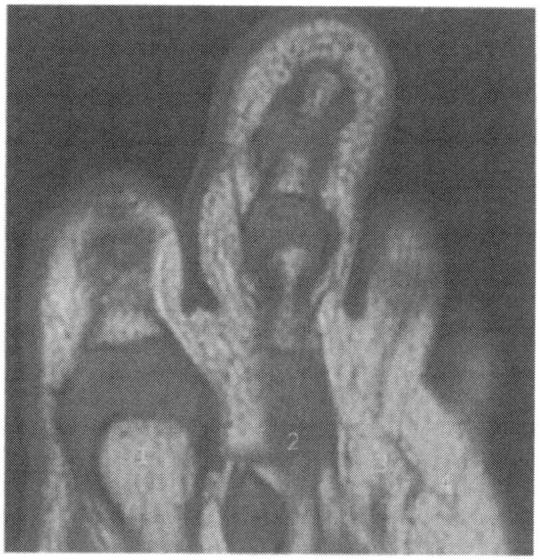

Figure 80C. T1 axial (footprint).

Figures 80A, 80B, 80C. There is diffuse subcutaneous fibrofatty and osseous overgrowth of the second digit. No soft tissue masses or calcifications are seen. Fluid-sensitive sequences (not shown) revealed no underlying soft tissue lesions or edema.

Diagnosis

Macrodystrophia lipomatosa of the second toe

Questions

1. T/F: Overgrowth of all mesodermal elements of the digit typically occurs with macrodystrophia lipomatosa.
2. T/F: Macrodystrophia lipomatosa may occur in the setting of fibrolipomatous hamartoma.

Discussion

Macrodystrophia lipomatosa is a nonhereditary overgrowth disorder that most commonly affects the second and third digits and is usually present at birth (1, 2). It occurs in the hands or feet but may also involve an entire limb (2). It is usually unilateral. There is overgrowth of both soft tissue and osseous components of the digit; however, there is usually disproportionate overgrowth of the subcutaneous fat (Answer to Question 1). Overgrowth tends to occur along the volar surface of the digit, which leads to dorsal angulation of the affected digit (2). The disease is benign and self-limited; digital overgrowth stops at puberty. Digital anomalies, such as polydactyly and syndactyly, may also be seen in the affected hand or foot.

Fibrolipomatous hamartoma is a rare condition of early adulthood that results in fatty proliferation of nerve bundles (3). The median and plantar nerves are most commonly affected. It may occur independently or in association with macrodystrophia lipomatosa (Answer to Question 2). When macrodystrophia lipomatosa is extensive, a determination of coexistent fibrolipomatous hamartoma by MRI is difficult (4).

On MRI, the dominant feature of macrodystrophia lipomatosa involving a digit or limb is subcutaneous fatty proliferation that is usually out of proportion to any hypertrophy of the adjacent muscle and bone. The subcutaneous fatty proliferation follows fat SI on both T1 and fat suppressed water sensitive sequences. PD FS or a 3D SPGR FS sequences are helpful to delineate the physes of the digits should epiphysiodesis be considered to limit future digital overgrowth. Multiplanar delineation of the physes is particularly helpful since overgrowth may be asymmetric and the surgeon may wish to correct angulation deformities.

Other causes of macrodactyly include tenosynovitis related to inflammatory arthritis, particularly psoriatic arthritis (sausage digit), vascular malformations (Figures 80D–80F), neurofibromatosis, and rare entities such as multiple subcutaneous lipomas associated with Bannayan-Rublacava-Riley syndrome. Macrodactyly in the setting of neurofibromatosis is due to the presence of plexiform or diffuse neurofibromas associated with venolymphatic malformations, soft tissue and osseous overgrowth (5). Fibrolipomatous hamartoma is distinguishable from a neurofibroma by the presence of diffuse fatty proliferation within the affected nerve.

This patient subsequently underwent second-digit debulking surgery. Surgical pathology showed mature fibroadipose tissue.

Orthopedic Perspective

Macrodactyly needs to be distinguished from the other causes of an enlarged digit as listed above. In most cases, surgical intervention is not needed, although on occasion an epiphysiodesis, osteotomy, debulking procedure, or amputation may be indicated. In such cases, defining the extent of the abnormality is useful.

What the Clinician Needs to Know

1. Distinguish macrodactyly from other causes of digit enlargement. Has a vascular malformation been excluded?
2. Define extent of hamartomatous proliferation.
3. Has a coexisting fibrolipomatous hamartoma of the nerve been excluded?
4. What is the main soft tissue composition of the macrodactyly?

Answers

1. True.
2. True.

Additional Example

Venolymphatic Malformation of the Second and Third Toe

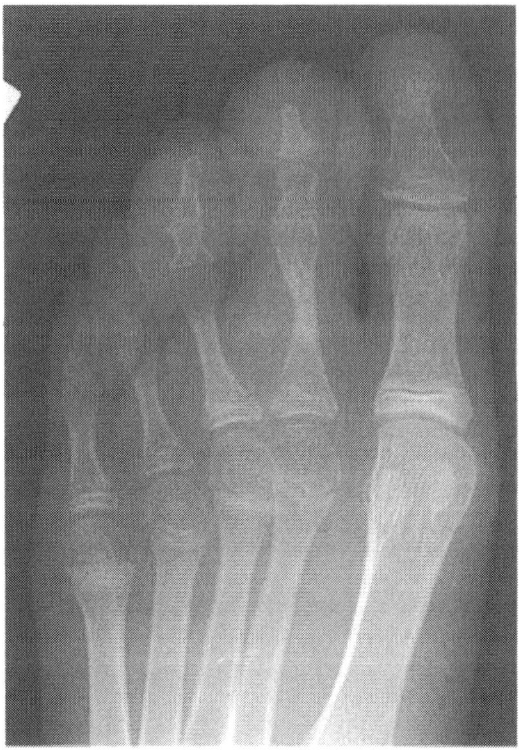

Figure 80D. AP radiograph.

Figure 80E. Axial (footprint) STIR.

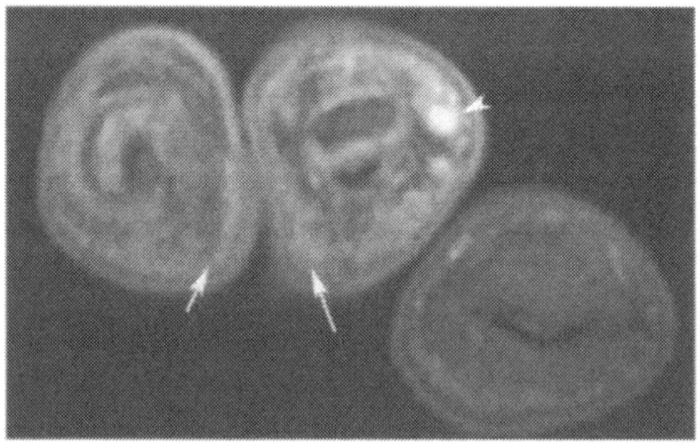

Figure 80F. Coronal T1 post-Gd FS.

Findings

This is a 12-year-old boy with macrodactyly of the 2nd and 3rd digits of his foot.

Figure 80D. AP radiograph demonstrates enlargement of the soft tissue and elongation of the proximal phalanges of the 2nd and 3rd toes.

Figure 80E. There are multiple tiny hyperintense foci consistent with cysts within the subcutaneous soft tissues of the 2nd and 3rd digits. The normal volume and distribution of hypointense fat within the digit excludes macrodystrophia lipomatosa.

Figure 80F. There is fine septal enhancement **(arrows)** as well as a large draining vein **(arrowhead)**. This was pathologically proven to be a venolymphatic malformation with a minor venous component.

Pitfalls and Pearls

1. Macrodystrophia lipomatosa most commonly affects the second and third digit of the hand or foot.
2. Fibrolipomatous hamartoma most commonly affects the median nerve of the wrist and hand.

References

1. Ly JQ, Beall DP. Quiz case: Macrodystrophia lipomatosa. *Eur J Radiol* 2003; 47:16–18.
2. Wang YC, Jeng CM, Marcantonio DR, Resnick D. Macrodystrophia lipomatosa: MR imaging in three patients. *Clin Imaging* 1997; 21:323–327.
3. De Maeseneer M, Jaovisidha S, Lenchik L, et al. Fibrolipomatous hamartoma: MR imaging findings. *Skeletal Radiol* 1997; 26:155–160.
4. Brodwater BK, Major NM, Goldner RD, Layfield LJ. Macrodystrophia lipomatosa with associated fibrolipomatous hamartoma of the median nerve. *Pediatr Surg Int* 2000; 16:216–218.
5. Herring JA, Tachdjian MO. Chapter 30. Orthopaedic-related syndromes. In: *Tachdjian's Pediatric Orthopaedics*. 3rd ed. Philadelphia: W.B. Saunders Company, 2002; 1585–1684.

Case 81

History

This is a 13-year-old boy with a 5 day history of left arm swelling and a left axillary mass. He was otherwise healthy. Laboratory work, including complete blood cell count, was normal.

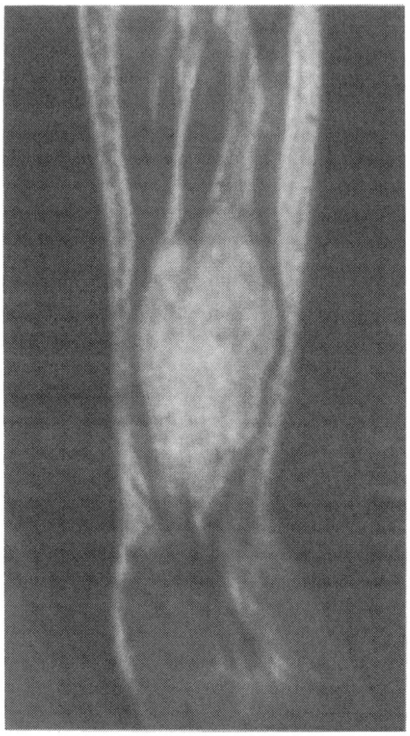

Figure 81A. Coronal T2 (no FS) of the left forearm.

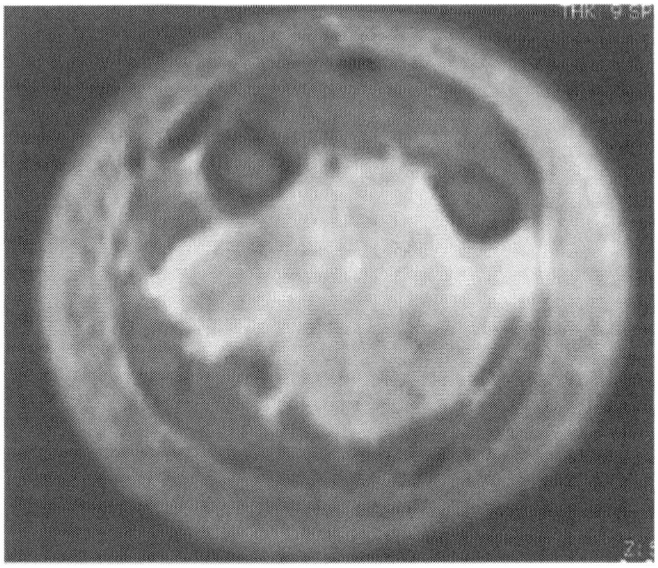

Figure 81B. Axial T1 post-Gd FS.

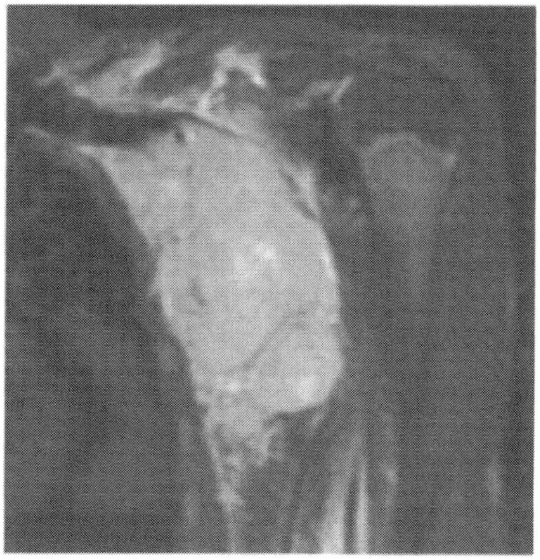

Figure 81C. Coronal STIR of the axilla.

Figures 81A, 81B. There is a multicompartmental mass within the volar aspect of the forearm. It is hyperintense on T2 and demonstrates heterogeneous enhancement. The mass abuts, but does not invade the ulna and radius. There is significant subcutaneous edema.

Figure 81C. There is also a large lobulated hyperintense mass within the axilla, abutting the chest wall.

Diagnosis

Rhabdomyosarcoma with axillary metastasis

Questions

1. T/F: Distal extremity rhabdomyosarcomas have a more favorable outcome compared with proximal extremity rhabdomyosarcomas.
2. T/F: Calcification is a common radiographic finding of rhabdomyosarcoma.

Discussion

Rhabdomyosarcoma is the fourth most common solid tumor in children and the most common soft tissue sarcoma in children under the age of 15 (1, 2). Two thirds of cases present under 10 years of age. The most common location is the head/neck region, followed by the genitourinary tract, extremities, and much less commonly the gastrointestinal tract, thorax, and retroperitoneum. Rhabdomyosarcoma has been associated with fetal alcohol syndrome, basal cell nevi syndrome, Li-Fraumeni syndrome, and neurofibromatosis. It may occur in areas of prior radiation therapy (3–5).

The three histologic subtypes of rhabdomyosarcoma are alveolar, embryonal, and pleomorphic. Embryonal histology usually occurs in patients from birth to 15 years of age and alveolar histology is found in patients from 10 to 25 years. Pleomorphic histology in patients 20 years and older (6). Head/neck and genitourinary (nonbladder/prostate) locations and alveolar histology are associated with a good prognosis. Rhabdomyosarcoma involving the extremities is associated with a poorer prognosis. Within the extremities, the most common histologic subtype is alveolar and approximately 50% of patients have regional lymphatic spread at the time of presentation (2). In a study of 189 cases of rhabdomyosarcoma, Andrassy et al. observed that proximal or distal location within the extremity did not affect prognosis (Answer to Question 1) (7). This contrasts with synovial cell sarcoma where distal extremity tumors are associated with a better prognosis than proximal extremity tumors (8).

The MRI features of rhabdomyosarcoma are nonspecific and overlap with other benign and malignant soft tissue masses. Rhabdomyosarcomas tend to be solid masses that are hypointense on T1, hyperintense on fluid-sensitive sequences, and show heterogeneous tumoral enhancement (9). The margins may be infiltrative or well-defined due to pseudocapsule formation. Tumoral necrosis, cysts with fluid-fluid levels (Figures 81D–81F), and high flow vessels may be seen. Rarely, calcification is seen within the tumor or related to adjacent bone destruction (Answer to Question 2) (10). Compared with synovial cell sarcoma, rhabdomyosarcoma is less likely to have internal calcifications, osseous destruction, and intratumoral cyst formation (11).

MRI documentation of tumor size, regional nodal spread, metastases, and osseous and neurovascular invasion is important to determine prognosis. Defining whether the tumor has extended beyond a single anatomic compartment is often helpful for surgical and radiotherapy planning. In one study, tumor size greater than 5 cm had a 91% incidence of local extension with extracompartmental spread and regional lymph node involvement (Figures 81D–81F) (12). A tumor is considered intracompartmental if the outermost margin is bounded by natural anatomic barriers such as fascia, tendon, ligaments, cortical bone, articular cartilage, or joint capsule (13). If the fat plane is preserved around a neurovascular structure, then it is not considered invaded.

The differential diagnosis is limited in this case because of metastatic disease in the axilla. Alternative considerations include other soft tissue sarcomas such as synovial cell sarcoma, malignant peripheral nerve sheath tumor, and lymphoma. Multiple infantile hemangiomas and hemangioendotheliomas would not be considered due to the patient's age and the sudden appearance of these masses.

This patient presented because of acute arm swelling, likely related to venolymphatic outflow obstruction by the large axillary metastasis. He underwent nodal biopsy, which was positive for alveolar rhabdomyosarcoma. He was treated with local radiation therapy and a combination of vincristine, actinomycin, and cyclophosphamide. A pancreatic metastasis was identified at 2-year follow-up. He expired 5 years after initial diagnosis from multiorgan failure related to intra-abdominal metastasis.

Orthopedic Perspective

Treatment of the rhabdomyosarcoma is surgical whenever possible. Imaging of the primary tumor is imperative to assess resectability. The relationship of the mass to the adjacent neurovascular structures and bones is vital. The number of compartments involved has a direct impact on the type of resection that may be employed. The goal is complete surgical resection with negative margins, but this is often difficult to achieve without substantial functional loss. Some centers prefer to treat with neoadjuvant chemotherapy before surgery, hoping for a response that will make the resection easier, but this approach is not universally employed. Radiotherapy is also frequently employed either as an adjunct to surgical excision or as definitive local control, as in this case. Assessing the radiation field and whether growth plates will be involved becomes an issue. If lymph node involvement is suspected by imaging, a biopsy is done for confirmation because the treatment protocols are different if there is distant disease present at diagnosis.

What the Clinician Needs to Know

1. The clinician needs to distinguish sarcomas like rhabdomyosarcoma from other soft tissue masses such as vascular malformations, myositis ossificans, and abscesses.
2. The extent of the mass locally as well as the presence of nodal and distant disease must be known to establish a prognosis and to determine the correct treatment protocol.
3. The extent of the primary tumor must be carefully defined with respect to neurovascular structures, bones, and growth plates to determine resectability and functional consequences of either resection or radiotherapy.

Answers

1. False.
2. False.

Additional Example

Rhabdomyosarcoma of the Calf

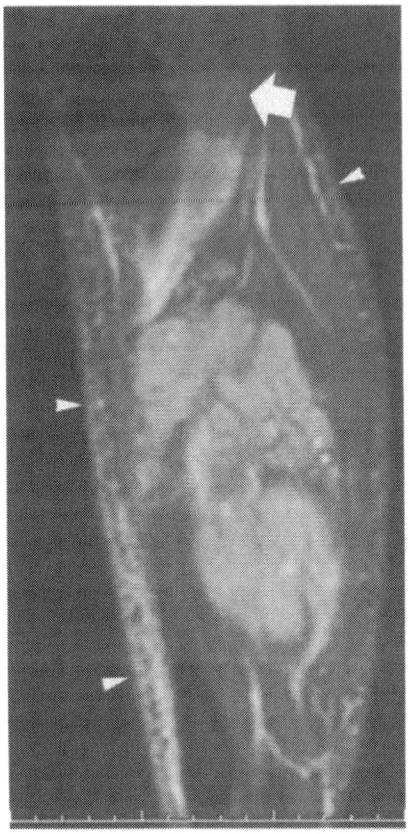

Figure 81D. Sagittal STIR of the left calf.

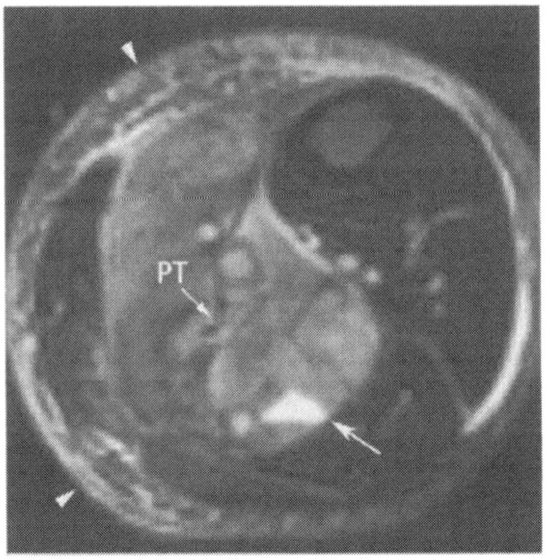

Figure 81E. Axial T2 FS.

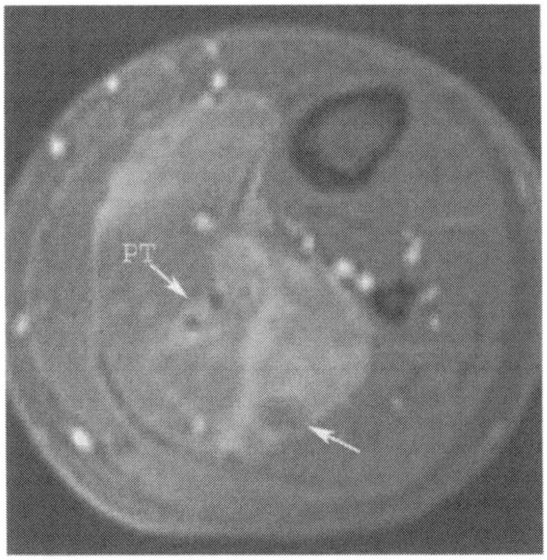

Figure 81F. Axial T1 post-Gd FS.

Findings

This is a 13-year-old girl with a six-week history of left calf and back pain.

Figures 81D, 81E, 81F. There is a sharply marginated multicompartmental mass arising from the soleus muscle. There is considerable subcutaneous edema **(arrowheads)**, but the SI of the surrounding calf musculature is relatively normal. Along the caudal margin of the mass, a fluid-fluid level is identified **(thin arrows)**. The mass encases the tibial nerve, posterior tibial artery, and posterior tibial vein complex (PT). A popliteal fossa pathologic node is also seen **(thick arrow)**. Open biopsy showed alveolar rhabdomyosarcoma.

Pitfalls and Pearls

Do not confuse high flow vessels often seen with soft tissue sarcomas with hemangiomas and high flow vascular malformations.

References

1. McHugh K, Boothroyd AE. The role of radiology in childhood rhabdomyosarcoma. *Clin Radiol* 1999; 54:2–10.
2. Arndt CA, Crist WM. Common musculoskeletal tumors of childhood and adolescence. *N Engl J Med* 1999; 341:342–352.
3. Tateishi U, Hasegawa T, Miyakawa K, Sumi M, Moriyama N. CT and MRI features of recurrent tumors and second primary neoplasms in pediatric patients with retinoblastoma. *AJR Am J Roentgenol* 2003; 181:879–884.
4. Kim EE, Valenzuela RF, Kumar AJ, Raney RB, Eftekari F. Imaging and clinical spectrum of rhabdomyosarcoma in children. *Clin Imaging* 2000; 24:257–262.
5. Cavalier ME, Davis MM, Croop JM. Germline p53 mutation presenting as synchronous tumors. *J Pediatr Hematol Oncol* 2005; 27:441–443.
6. Suzuki Y, Ehara S, Shiraishi H, Nishida J, Murooka G, Tamakawa Y. Embryonal rhabdomyosarcoma of foot with expansive growth between metatarsals. *Skeletal Radiol* 1997; 26:128–130.
7. Andrassy RJ, Corpron CA, Hays D, et al. Extremity sarcomas: An analysis of prognostic factors from the Intergroup Rhabdomyosarcoma Study III. *J Pediatr Surg* 1996; 31:191–196.
8. Tateishi U, Hasegawa T, Beppu Y, Satake M, Moriyama N. Synovial sarcoma of the soft tissues: Prognostic significance of imaging features. *J Comput Assist Tomogr* 2004; 28:140–148.
9. Laor T. MR imaging of soft tissue tumors and tumor-like lesions. *Pediatr Radiol* 2004; 34:24–37.
10. Brasch RC, Kim OH, Kushner JH, Rosenau W. Ossification in a soft tissue embryonal rhabdomyosarcoma. *Pediatr Radiol* 1981; 11:99–101.
11. McCarville MB, Spunt SL, Skapek SX, Pappo AS. Synovial sarcoma in pediatric patients. *AJR Am J Roentgenol* 2002; 179:797–801.
12. LaQuaglia MP, Ghavimi F, Penenberg D, et al. Factors predictive of mortality in pediatric extremity rhabdomyosarcoma. *J Pediatr Surg* 1990; 25:238–243; discussion 243–234.
13. De Schepper AM, De Beuckeleer L, Vandevenne J, Somville J. Magnetic resonance imaging of soft tissue tumors. *Eur Radiol* 2000; 10:213–223.

Case 82

History

This is a 14-year-old boy with chronic right knee pain for 3 months.

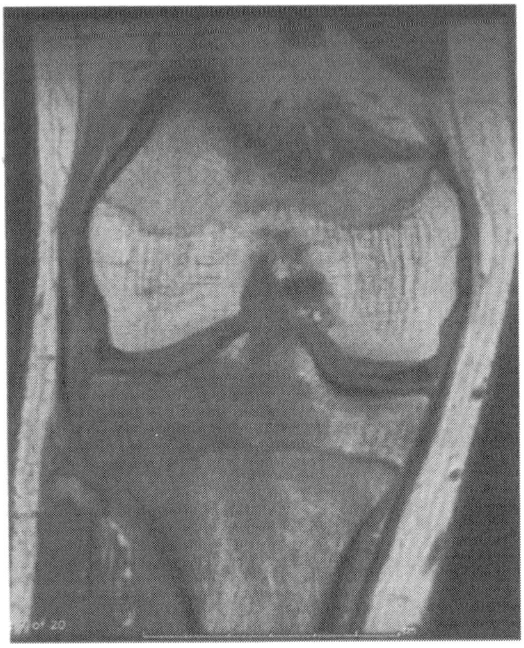

Figure 82A. Coronal T1 of the right knee.

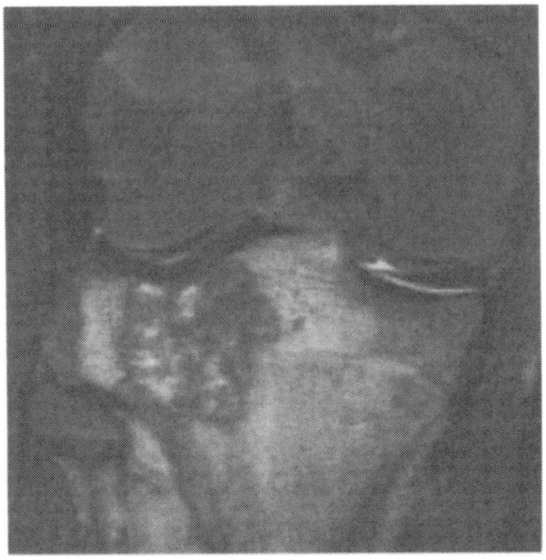

Figure 82B. Coronal STIR.

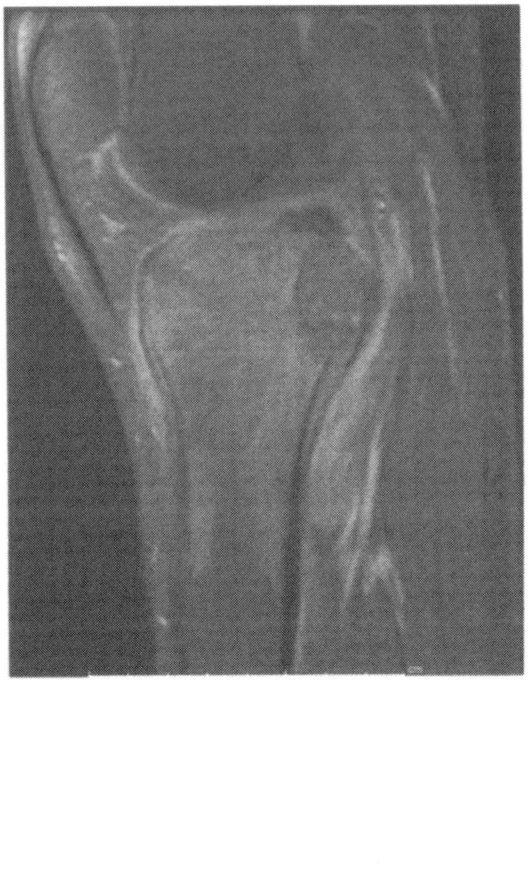

Figure 82C. Sagittal T1 post-Gd FS.

Figures 82A, 82B, 82C, (82C with annotations). There is a well-defined lesion centered in the posterolateral aspect of the proximal tibial epiphysis that extends across the physis into the epiphysis. It is associated with striking marrow and juxtacortical soft tissue edema. The lesion is isointense on T1, heterogeneously hyperintense on STIR, and demonstrates mild enhancement. A thin hypointense rim surrounds the lesion **(arrowhead)**. Periosteal new bone formation and enhancement are also evident **(arrow)**.

Figure 82D. There is a fairly well-defined osteolytic lesion arising from the posterior epiphysis of the tibia **(arrowheads)**. There is mature periosteal reaction **(arrow)**.

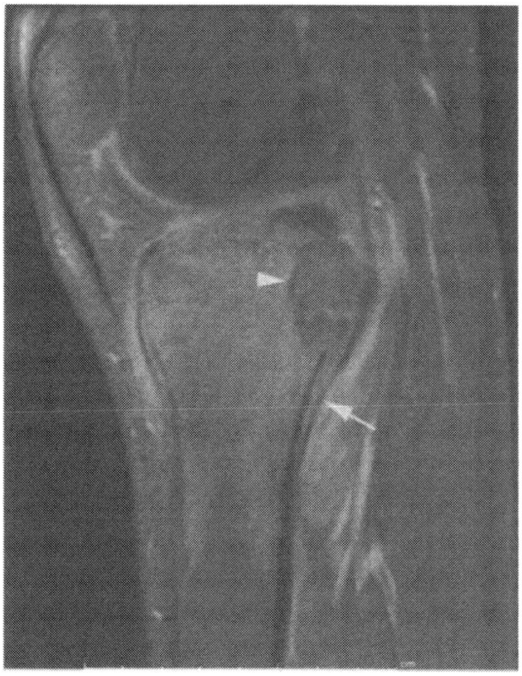

Figure 82C* Annotated.

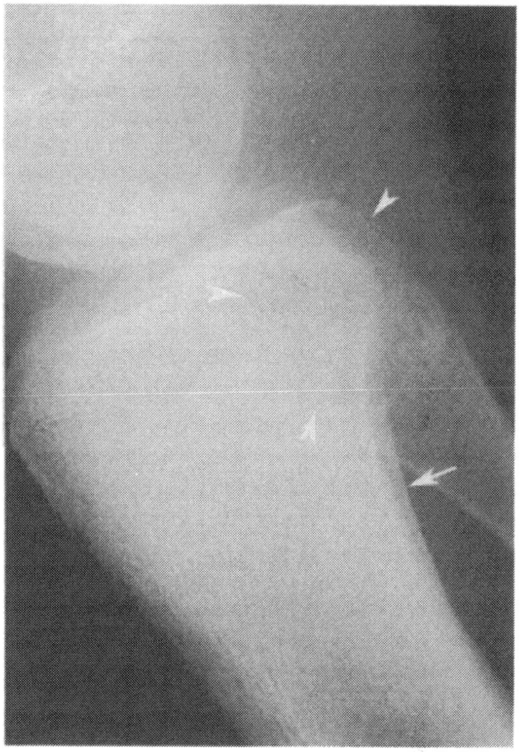

Figure 82D. Lateral radiograph.

Diagnosis

Chondroblastoma

Questions

1. T/F: Chondroblastomas rarely cross the physis.
2. On MRI, what distinguishes chondroblastoma from other cartilaginous tumors?

Discussion

Chondroblastomas are rare, benign cartilage tumors that occur most often in patients 10–26 years of age. The mean age of presentation in the long tubular bones is 16 years and 28 years in flat bones (1). With rare exceptions, chondroblastomas arise from epiphyses or epiphyseal equivalent regions. The most common location is the knee (divided equally between the femoral condyles and tibia plateaus), followed by the proximal femur and humerus (1).

On radiography, chondroblastomas are osteolytic, geographic lesions with sharp, sclerotic margins. They may be eccentric in location and cause endosteal erosion and cortical destruction. Periosteal reaction related to an epiphyseal chondroblastoma may be identified and is typically found in the neighboring metaphysis. Chondroid calcification may rarely be identified on plain radiography (2). In Bloem et al.'s series, 51% of epiphyseal chondroblastomas of the appendicular skeleton (n = 80) crossed the physis and extended into the metaphysis (Answer to Question 1); 5% were confined to the metaphysis (1).

On MRI, chondroblastoma is a lobular mass that is hypo- to isointense on T1, variably intense on fluid-sensitive sequences, and demonstrates variable enhancement (3). Chondroblastomas tend to be more heterogeneous than other chondroid tumors, and often contain areas of hypointensity on fluid sensitive sequences. This observation has been pathologically correlated with the presence of dense immature chondroid matrix, hemosiderin, and calcification within the tumor (Answer to Question 2) (4). A low SI rim surrounding the tumor may be identified on all imaging sequences (5). Fluid-fluid levels, septations, and secondary aneurysmal bone cyst (ABC) may develop within chondroblastomas (Figures 82E, 82F) (1). MRI nicely delineates chondroblastomas when they breach cortex and extend into the adjacent soft tissues and joint (Figures 82G–82I).

The key MR imaging feature of chondroblastomas is perilesional hyperintensity on fluid sensitive sequences and gadolinium enhancement, reflecting local inflammatory reaction in bone marrow, adjacent soft tissues, and joint (synovitis). Therefore, the presence of a subchondral epiphyseal tumor in the immature skeleton, associated with significant edema, is usually a chondroblastoma (Answer to Question 2).

The recurrence risk of chondroblastoma is relatively low. In one series, a 15% recurrence rate was noted after operative treatment (most commonly curettage) (6). The most common location for recurrence was the proximal femur (capital femoral epiphysis and greater trochanteric apophysis). Secondary ABC, open growth plate, and tumor size were not statistically significant risk factors for tumor recurrence in this study.

The differential diagnosis for an epiphyseal chondroblastoma includes epiphyseal osteomyelitis with secondary septic arthritis, epiphyseal ganglion cysts, Langerhans cell histiocytosis (LCH), and rarely epiphyseal osteosarcoma. A purely epiphyseal giant cell tumor may be considered in the skeletally mature. Osteomyelitis with an epiphyseal Brodie's abscess may be considered when there is significant marrow edema and synovitis. Contrast enhanced images are critical in this situation since a rim enhancing fluid collection will favor epiphyseal osteomyelitis with a Brodie's abscess, whereas a centrally enhancing mass suggests chondroblastoma. If matrix calcification and a sclerotic rim are present, the diagnosis of chondroblastoma is favored.

This lesion was curetted and pathologically confirmed to represent a chondroblastoma. His symptoms resolved by his second follow-up appointment.

Orthopedic Perspective

When faced with a painful radiolucency of the epiphysis in a child near adolescence, chondroblastoma is the most likely neoplasm. Because of its inflammatory nature, it must be distinguished from osteomyelitis and LCH. The presence of calcifications within the lesion (best seen by CT) can be helpful in making that assessment, and CT-directed needle biopsies are very useful in establishing the diagnosis. The presence of a secondary ABC can also be misleading. The relationship of the tumor to the growth plate and adjacent articular surface is important to assess by MRI. However, since chondroblastomas typically occur near skeletal maturity, growth issues are usually not a major concern. Destruction of the subchondral bone and collapse of the articular cartilage are major problems especially when large lesions occur in weightbearing bones. Treating chondroblastomas before this happens is optimal. In some sites, such as the proximal femur, this can be a major surgical challenge. Exposing the lesion without damaging the blood supply to the femoral head, and obtaining a complete curettage is essential. The role of radiofrequency heat ablation for chondroblastoma has not been defined, but it may be useful in small lesions where the integrity of the subchondral bone is not in question.

What the Clinician Needs to Know

1. Distinguish chondroblastoma from other lesions that may occur in the epiphysis of the immature skeleton.
2. What is the relationship of the lesion to the growth plate, especially in younger children?
3. What is the status of the subchondral cortex and the articular cartilage?
4. On follow-up studies, is there evidence of recurrent chondroblastoma?
5. For sites such as the proximal humerus and femur, axial imaging is helpful in planning the most direct surgical approach.

Answers

1. False.
2. Chondroblastomas are much more heterogeneous. In addition, they demonstrate significant perilesional marrow and juxtacortical soft tissue edema in the absence of a pathologic fracture.

Additional Examples

Chondroblastoma with Secondary ABC

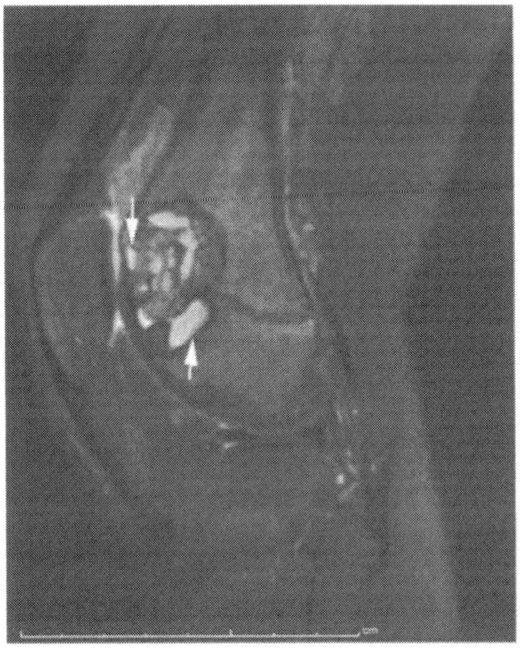

Figure 82E. Sagittal T2 FS of the right knee.

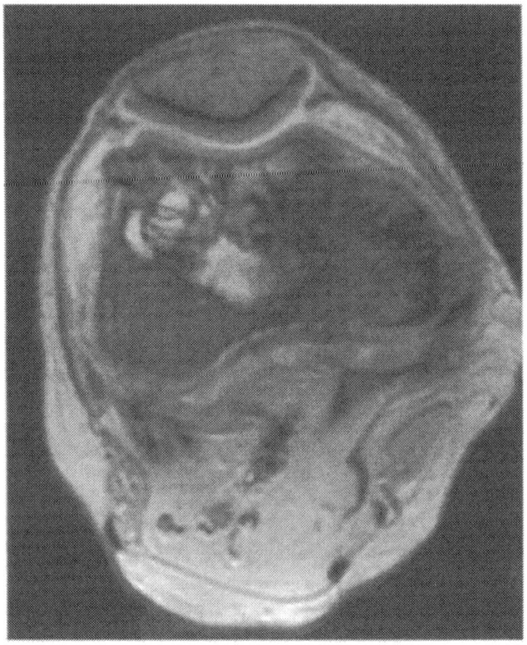

Figure 82F. Axial MPGR.

Findings

This is a 13-year-old girl who carried a provisional diagnosis of JRA.

Figures 82E, 82F. There is a complex epiphyseal/metaphyseal mass abutting the articular surface of the femoral condyle that has septations and fluid-fluid levels **(arrows)**. There is marrow and soft tissue edema as well as a reactive joint effusion. This was pathologically proven to be a chondroblastoma with secondary ABC.

Chondroblastoma with Intra-articular Extension

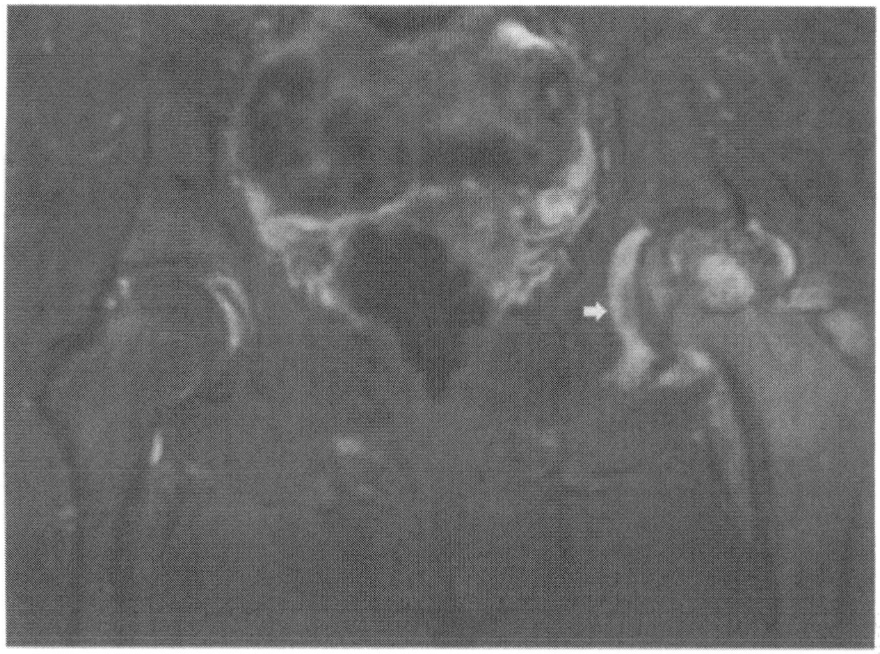

Figure 82G. Coronal STIR.

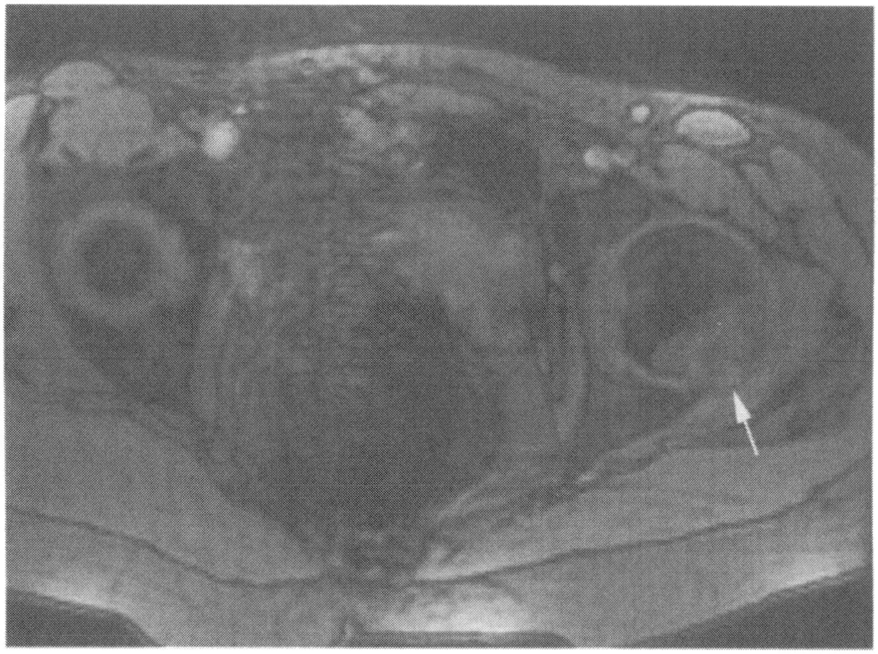

Figure 82H. Axial MPGR.

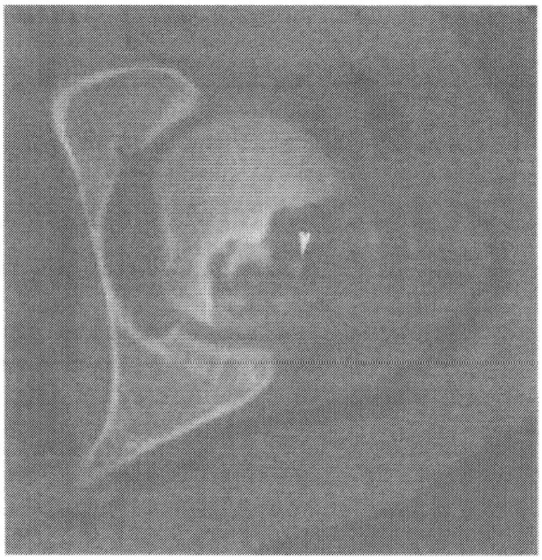

Figure 82I. CT.

Findings

This is a 13-year-old girl with a 2-year history of chronic left hip pain attributed to arthritis.

Figures 82G, 82H. An eccentric left capital femoral epiphyseal mass is present with components that follow the SI of cartilage **(thin arrow)**. Hypointensity is also evident within the lateral aspect of this lesion and may represent blood products, calcification, and/or dense cellularity. Significant marrow edema extending to the subtrochanteric femur and a reactive effusion **(thick arrow)** are present.

Figure 82I. CT demonstrates chondroid matrix calcification **(arrowhead)** and shows that the cortex is breached by this mass. This was pathologically proven to be a chondroblastoma.

Pitfalls and Pearls

In the older child, chondroblastoma is the most likely diagnosis for an epiphyseal osteolytic lesion with significant surrounding edema.

References

1. Bloem JL, Mulder JD. Chondroblastoma: A clinical and radiological study of 104 cases. *Skeletal Radiol* 1985; 14:1–9.
2. Kaim AH, Hugli R, Bonel HM, Jundt G. Chondroblastoma and clear cell chondrosarcoma: Radiological and MRI characteristics with histopathological correlation. *Skeletal Radiol* 2002; 31:88–95.
3. Weatherall PT, Maale GE, Mendelsohn DB, Sherry CS, Erdman WE, Pascoe HR. Chondroblastoma: Classic and confusing appearance at MR imaging. *Radiology* 1994; 190:467–474.
4. Jee WH, Park YK, McCauley TR, et al. Chondroblastoma: MR characteristics with pathologic correlation. *J Comput Assist Tomogr* 1999; 23:721–726.
5. Oxtoby JW, Davies AM. MRI characteristics of chondroblastoma. *Clin Radiol* 1996; 51:22–26.
6. Ramappa AJ, Lee FY, Tang P, Carlson JR, Gebhardt MC, Mankin HJ. Chondroblastoma of bone. *J Bone Joint Surg Am* 2000; 82-A:1140–1145.

Case 83

History

This is a 15-year-old girl who jumped 5 stairs and landed on her left heel. She has had constant throbbing in the back of her left foot since then. The radiographs were normal (not shown).

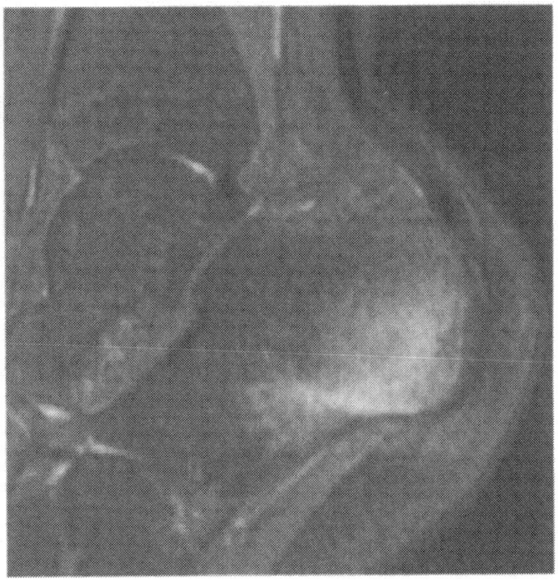

Figure 83A. Sagittal STIR of the calcaneus.

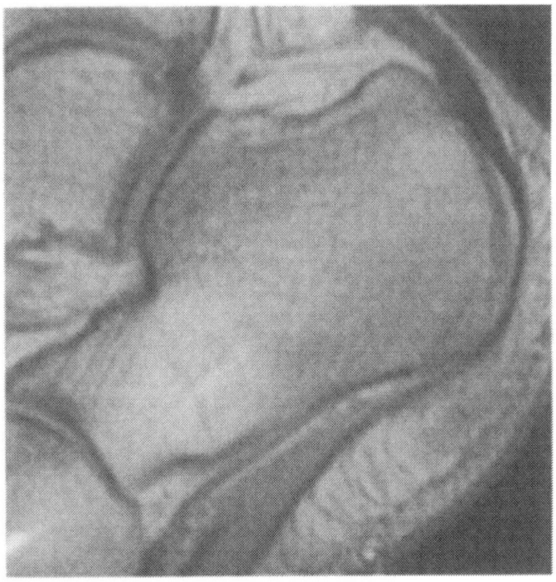

Figure 83B. Sagittal PD.

Figure 83A. There is focal increased SI present in the plantar aspect of the posterior calcaneal body.

Figures 83B, (83B with annotations). A discrete vertical oblique fracture line is identified that extends to the cortex **(arrow)**. There is a prominent but normal vertically oriented trabeculae located anteriorly **(arrowhead)**, without associated edema.

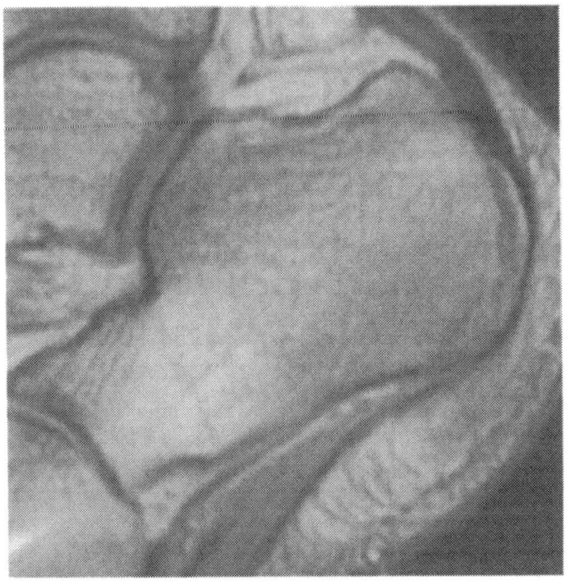

Figure 83B* Annotated.

Diagnosis

Nondisplaced extra-articular calcaneal fracture

Questions

1. T/F: In the setting of acute trauma, intra-articular calcaneal fractures are more common then extra-articular fractures.
2. T/F: Hematogenous osteomyelitis of the calcaneus is more common than direct inoculation.

Discussion

Calcaneal fractures may occur from direct impact on the posterior calcaneus, indirect injury related to the pull at muscular insertions, and vertical loading forces that drive the lateral process of the talus into the posterior subtalar joint (1, 2). Occult fractures of the calcaneus in the young child may be considered a variant of a toddler's fracture (3). Altered loading forces due to a leg length discrepancy or a pes cavus deformity may increase the risk for calcaneal injuries. Intra-articular fractures are more common than extra-articular fractures (Answer to Question 1) (1). The posterior subtalar joint is the most frequent site of intra-articular extension (Figures 83C, 83D).

In addition to nonspecific bone marrow edema, MRI features of fractures include the presence of a discrete fracture line with continuous trabecular disruption or fracture with displacement. A nondisplaced acute traumatic fracture and a stress fracture from chronic overuse are indistinguishable on MRI. Care should be taken not to mistake a normally prominent trabecular pattern (Figure 83B) or vascular channels for a fracture line. When marrow edema is present without a discrete fracture line, the term *stress reaction* should be applied. When noncontiguous disrupted trabeculae are present in the setting of marrow edema, the term *bone contusion* should be applied (4). Bone contusion actually represents multiple trabecular microfractures. There is imaging overlap between stress reaction, bone contusion, and a nondisplaced fracture since they represent a spectrum of increasing severity of bone injury. Stress reaction may progress to a stress fracture if the patient is not appropriately treated.

MRI differentiation between trauma and infection may be challenging since both can produce significant marrow and juxtaosseous soft tissue edema. Hematogenous osteomyelitis of the calcaneus usually arises from the posterior aspect of the body, since the largest metaphyseal equivalent region is located here (Figures 83E, 83F) (5, 6). Ossification of the calcaneal apophysis does not appear until 4 to 7 years in girls and 4 to 10 years in boys (7). Nonspecific findings of calcaneal osteomyelitis include: marrow edema, soft tissue edema, or nonenhancing fluid collections. Definitive MRI features of osteomyelitis include the presence of a cloacal tract (bone to periosteum), sinus tract (bone to skin), or sequestrum. Intraosseous and soft tissue fluid collections with rim enhancement may be seen with infection (abscess) as well as after trauma (post-traumatic seroma or hematoma). If blood products are evident as hyperintensity on T1W or as susceptibility artifact on GRE sequences, then a traumatic etiology is favored. Calcaneal osteomyelitis is most commonly the result of inoculation from direct penetration of the skin, rather than from hematogenous osteomyelitis (Answer to Question 2) (5). Therefore the soft tissues of the heel should be examined for an inoculation pathway.

The difficulty distinguishing calcaneal osteomyelitis from fracture on MRI grounds is compounded by two related reasons. Many children with calcaneal trauma are not immediately diagnosed. In Wiley et al.'s series of pediatric calcaneal fractures, 10 of 23 children under the age of 10 had a delay in diagnosis for various reasons (1). Therefore, once healing has begun, determining the etiology of marrow edema may be impossible. Additionally, there is often an antecedent history of trauma in many patients with hematogenous calcaneal osteomyelitis. In Rasool's calcaneal osteomyelitis series, 5 of 14 children had a history of trauma (5).

MRI was utilized in this case because there was still a high clinical suspicion for calcaneal injury despite negative plain radiographs. Once the diagnosis of a stress fracture was confirmed, this patient was treated with casting and did well.

Orthopedic Perspective

Occult calcaneal fractures may result from a large landing force onto the hindfoot. For nondisplaced fractures, radiographs may be negative and the diagnosis can be delayed. When there is high clinical concern for a radiographically occult fracture, patients may be referred to either CT or MRI. The treatment of nondisplaced fractures is casting. Displaced fractures, particularly displaced intra-articular fractures of the posterior facet, are often treated by open reduction or percutaneous pin fixation. The goal of surgical reduction of intra-articular fractures is to preserve functionality of the subtalar joint and prevent premature degenerative changes.

What the Clinician Needs to Know

Does the fracture extend to the articular surface, is it displaced, are there loose bodies within the subtalar joint, and are there other fractures?

Answers

1. True.
2. False.

Additional Examples

Stress Fracture with Extension to the Posterior Subtalar Joint

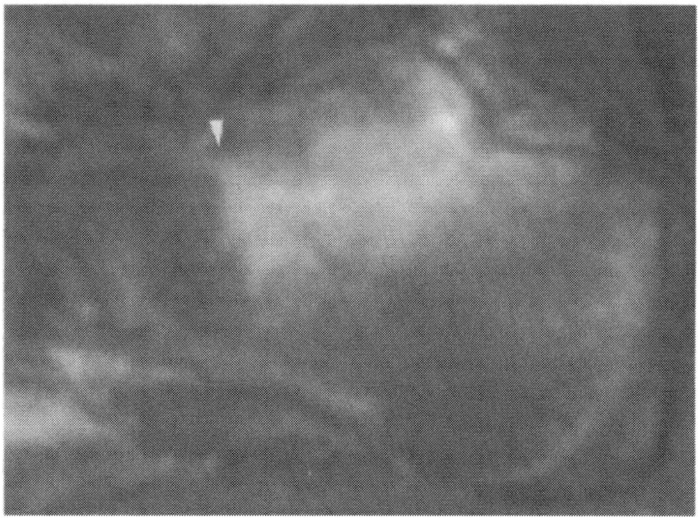

Figure 83C. Sagittal STIR of the right foot.

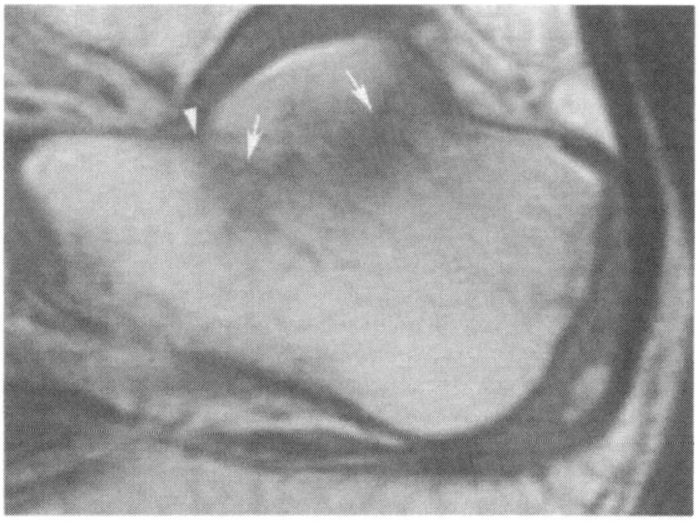

Figure 83D. Sagittal PD.

Findings

This 6-year-old boy had chronic foot pain.

Figures 83C, 83D. There is diffuse increased STIR SI seen along the superior aspect of the posterior body that extends to the anterior margin of the posterior subtalar joint **(arrowheads)**. There are discontinuous hypointense fracture lines identified best on PD **(arrows)**.

Hematogeneous Osteomyelitis of the Calcaneal Body

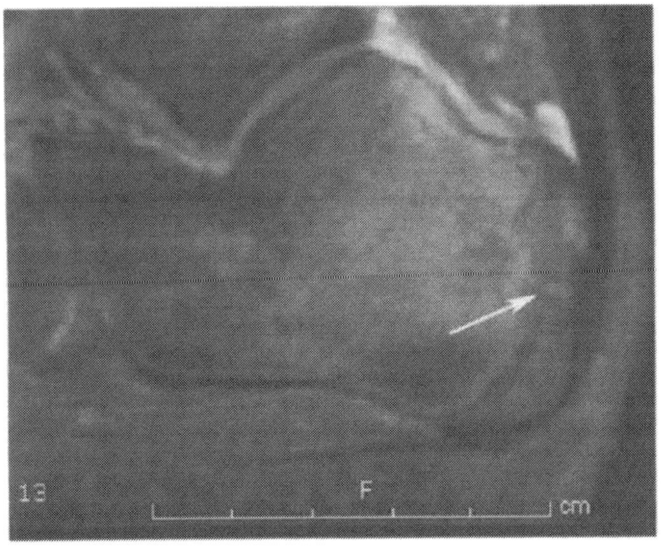

Figure 83E. Sagittal STIR of the right calcaneus.

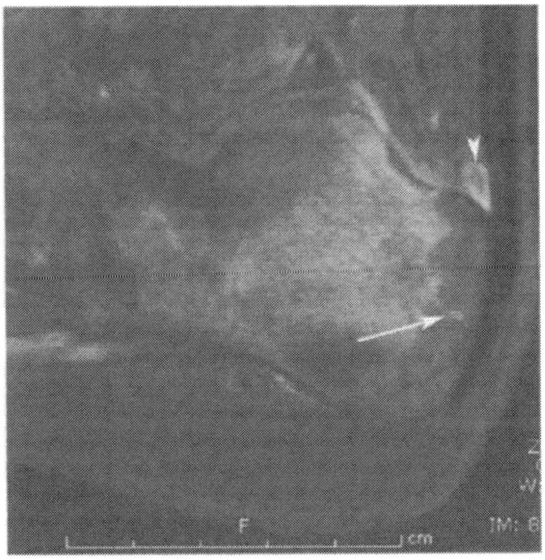

Figure 83F. Sagittal T1 post-Gd FS.

Findings

This 10-year-old boy had a 10-day history of right heel pain.

Figures 83E, 83F. There is diffuse, increased STIR SI and enhancement seen throughout the calcaneal body. There is a small amount of fluid present within the retrocalcaneal bursa with thin rim enhancement **(arrowhead)**. There is also transphyseal extension identified **(arrows)**. Blood cultures were positive for *Staphylococcus aureus*.

Pitfalls and Pearls

1. Calcaneal fractures often extend to the posterior subtalar joint. Make sure to describe whether the fracture is intra- or extra-articular.
2. Calcaneal stress fracture and osteomyelitis may be indistinguishable on MRI. Be sure to obtain T1 or PDW images to assess for a fracture line.

References

1. Wiley JJ, Profitt A. Fractures of the os calcis in children. *Clin Orthop* 1984:131–138.
2. Weber JM, Vidt LG, Gehl RS, Montgomery T. Calcaneal stress fractures. *Clin Podiatr Med Surg* 2005; 22:45–54.
3. Kim CW, Shea K, Chambers HG. Heel pain in children: Diagnosis and treatment. *J Am Podiatr Med Assoc* 1999; 89:67–74.
4. Beltran J, Shankman S. MR imaging of bone lesions of the ankle and foot. *Magn Reson Imaging Clin N Am* 2001; 9:553–566, xi.
5. Rasool MN. Hematogenous osteomyelitis of the calcaneus in children. *J Pediatr Orthop* 2001; 21:738–743.
6. Kleinman PK. A regional approach to osteomyelitis of the lower extremities in children. *Radiol Clin North Am* 2002; 40:1033–1059.
7. Heneghan MA, Wallace T. Heel pain due to retrocalcaneal bursitis: Radiographic diagnosis (with an historical footnote on Sever's disease). *Pediatr Radiol* 1985; 15:119–122.

Case 84

History

This is a 17-year-old girl dancer. She has left hip pain with occasional snapping and locking.

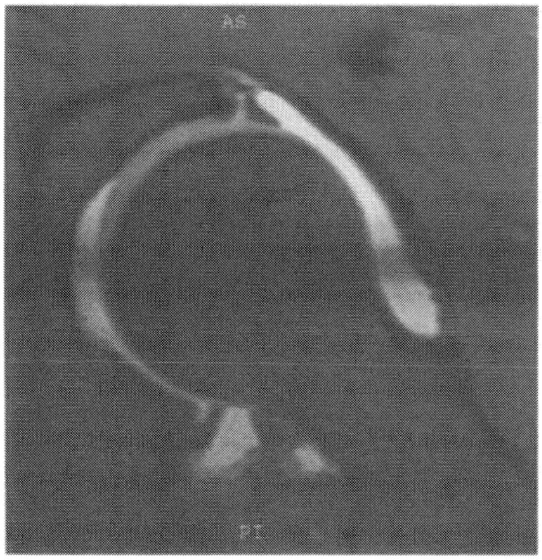

Figure 84A. Direct left hip gadolinium arthrogram. Radial T1 FS section extending from the anterosuperior (AS) to the posteroinferior (PI) quadrants of the acetabulum.

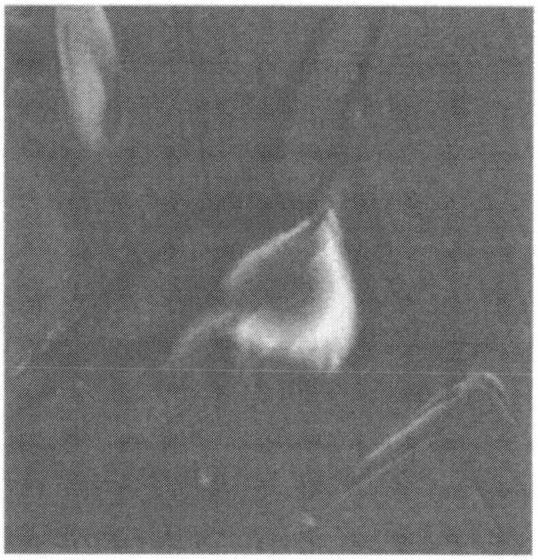

Figure 84B. Coronal T1 FS, anterior acetabular rim.

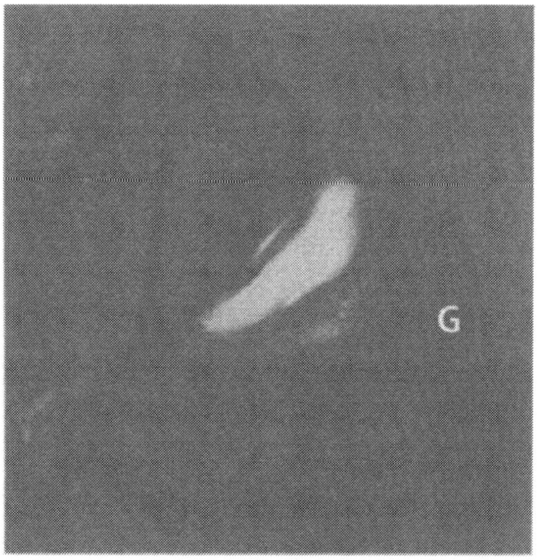

Figure 84C. Coronal T1 FS, posterior acetabular rim. Greater trochanter (G).

Figures 84A, 84B, 84C, (84A, 84B, 84C, with annotations). There is an undersurface defect that parallels the anterosuperior osteolabral junction **(arrows)**. The labrum maintains a triangular shape and is normal SI. There is also a linear contrast collection that parallels the posterior osteolabral junction **(arrowheads)**. The posterior labrum has normal shape and SI. G = Greater trochanter.

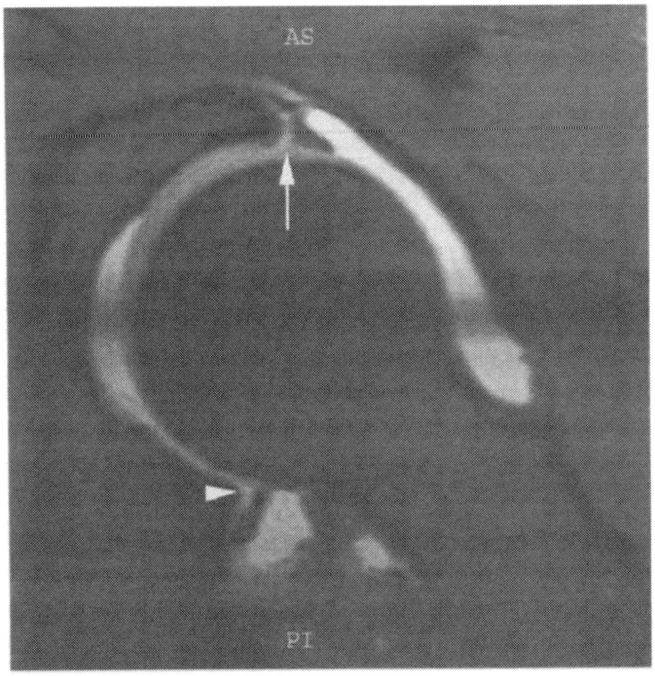

Figure 84A* Annotated.

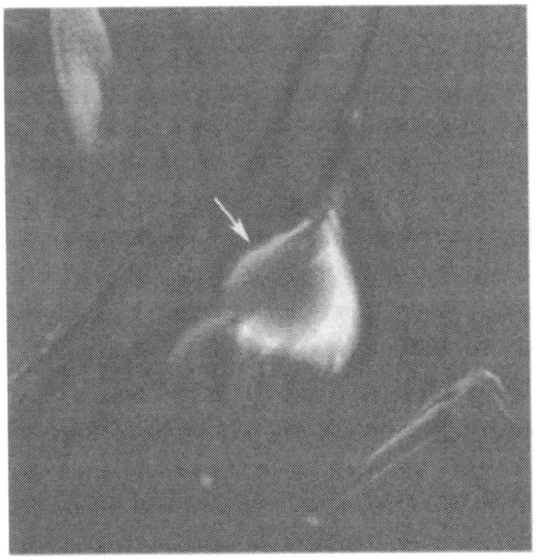

Figure 84B* Annotated.

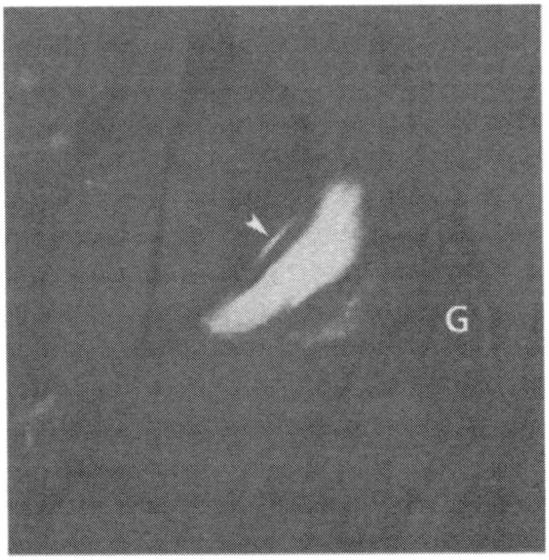

Figure 84C* Annotated.

Diagnosis

Anterosuperior acetabular osteolabral separation

Questions

1. What is the most common location for labral tears?
2. T/F: A normal acetabular sublabral sulcus is located both anteriorly and posteriorly.

Discussion

Acetabular labral degeneration, tears, and osteolabral separation are most commonly noted in patients with femoroacetabular impingement (see Case 71), hip dysplasia (see Case 37), and normal hips predisposed to overuse injury. Most labral pathology is located in the anterosuperior quadrant of the acetabulum (Answer to Question 1) (1, 2). The labrum is composed of fibrocartilage. It is hypointense on all imaging sequences and usually has a normal triangular shape. With early degeneration, labral hypertrophy, ganglia, and increased intrasubstance SI may be seen. Labral tears usually begin along the undersurface of the labrum and should be distinguished from an osteolabral separation. Labral tears usually occur in the setting of preexisting labral degeneration (Figure 84D).

A sublabral sulcus is a normal variant of the labrum that may collect contrast material and simulate an undersurface labral tear or osteolabral separation (Figures 84A, 84C) (3). In general, any anterior or anterosuperior undersurface labral defect that is defined with direct Gd arthrography should be considered abnormal (Answer to Question 2). This principle does not apply to the posterior labrum, a common site for a sublabral sulcus (3), but an unusual site for labral tears. The transverse acetabular ligament–labral recess should also not be misinterpreted as labral pathology (Figure 84E). Labral anatomy is better delineated by obtaining radial sequences (Figure 84F).

Labral degeneration and tears in the setting of nondysplastic hips are classically seen in professional ballet dancers (2). These patients are predisposed to hip degeneration because of excessive hip twisting, hyperextension, and hyperabduction. Chronic hyperextension and hyperabduction places increased loading forces along the anterosuperior acetabular labrum and superior femoral neck. These supraphysiologic forces may cause a femoroacetabular impingement effect (4). These symptoms may be exacerbated once secondary degenerative changes develop with deformities of the femoral head/neck region, such as ring osteophytes, leading to a true CAM-type femoroacetabular impingement (see Case 71).

Other hip region injuries that may occur in the professional ballet dancer include myotendinous injuries of the hamstrings and adductors, snapping hip syndrome (tensor fascia lata friction rub over the greater trochanter), greater trochanteric bursitis, femoral neck stress fractures, and iliopsoas tendinopathy, a condition that is uniquely exacerbated during the developpé maneuver (5).

This patient underwent hip arthroscopy and the anterosuperior (9:00 to 12:00 positions) labral tear was confirmed and subsequently debrided. The posterior labrum was normal.

Orthopedic Perspective

Labral tears are being identified as a cause of hip pain in the adolescent athlete with increased frequency due to greater awareness of this diagnosis, the advent of MR arthrography of the hip, and the use of hip arthroscopy. Labral tears typically present as groin pain with mechanical symptoms. Anterosuperior labral tears are painful with hip flexion, adduction, and internal rotation. Imaging features that affect conservative and surgical management of labral tears include: the presence of underlying hip dysplasia, femoroacetabular impingement, loose bodies, ligamentum teres tears, and presence of articular cartilage degeneration. Labral tears associated with CAM-type femoroacetabular impingement are usually treated by open management; labral tears associated with DDH are often referred for possible osteotomy and repair; and isolated labral tears are treated arthroscopically. Labral tears and osteolabral separation are usually treated similarly during arthroscopy. Labral degeneration is usually watched clinically unless it is very symptomatic and the clinician is left without option but to perform diagnostic arthroscopy.

What the Clinician Needs to Know

1. What is the exact location of the labral tear? What is the size of the tear?
2. Is the labral tear secondary to femoroacetabular impingement? If so, is the deformity related to CAM-type or pincer-type femoroacetabular impingement?
3. Is the labral tear due to DDH?
4. Are there associated internal derangements such as chondral injury, loose bodies, and ligamentum teres tears?

Answers

1. Anterosuperior.
2. False.

Additional Examples

Anterior Labral Tear with Degeneration

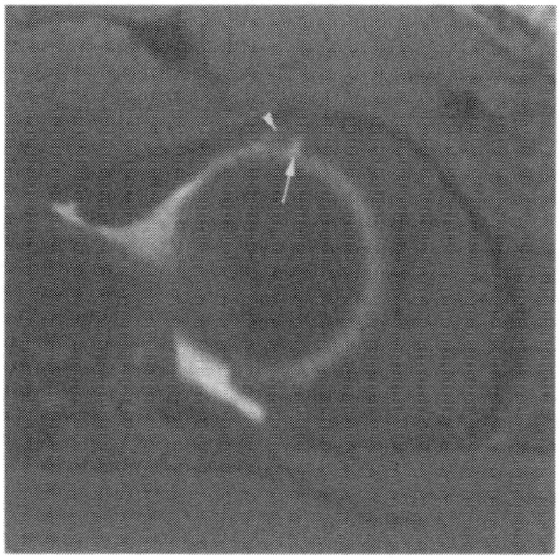

Figure 84D. Axial-oblique T1 FS of the right hip after direct Gd arthrography.

Findings

This is a 15-year-old female dancer who had right hip pain whose symptoms did not resolve despite physical therapy.

Figure 84D. The anterior labrum appears small, frayed, and there is increased intrasubstance SI present **(arrowhead)**. There is an intrasubstance anterior labral tear delineated by contrast **(arrow)** that splits the labrum. The labral tear was confirmed surgically.

Normal Transverse Acetabular Ligament

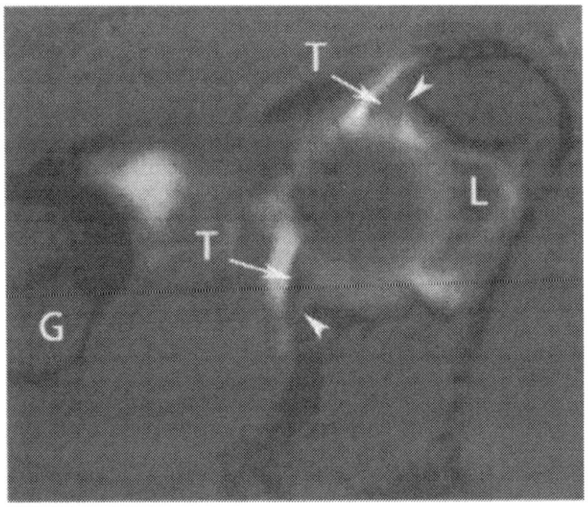

Figure 84E. Axial-oblique T1 FS through the most inferior portion of the right hip after direct Gd arthrography.

Findings

This is an 18-year-old female dancer with right pain.

Figure 84E. The transverse acetabular ligament/labral junction forms a normal recess **(arrowheads)** that is often delineated after direct Gd arthrography. An anterosuperior labral tear was noted on other sections (not shown). Ligamentum teres (L), transverse acetabular ligament (T), greater trochanter (G).

Radial Sequence Acquisition for Labral Assessment

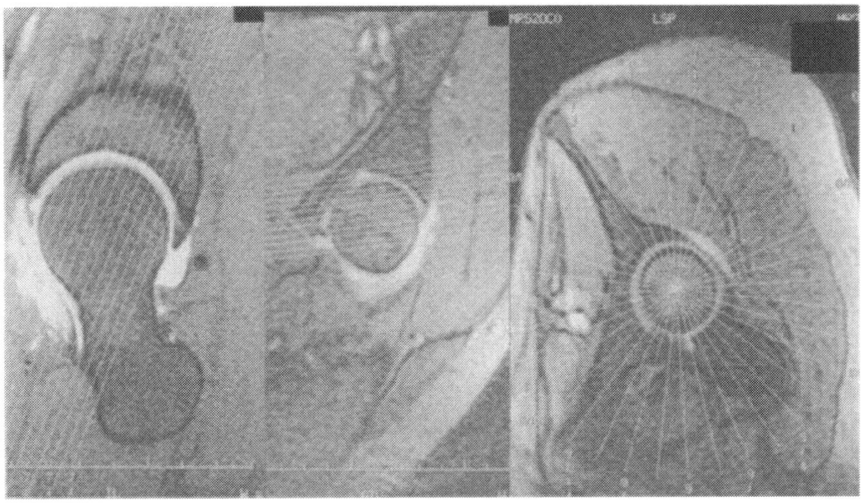

Figure 84F. Radial sequence localizers.

Findings

Figure 84F. By convention, the labrum is divided into a clock, and the findings are described based on the anatomic orientation at arthroscopy. The observer looks up at the clock from the perspective of the femoral neck. When viewing the left hip, the superior labrum is 12:00, anterior labrum is 9:00, posterior labrum is 3:00, and the ligamentum teres and transverse ligament are located at approximately the 6:00 position. If the right hip is viewed, the anterior labrum is 3:00 and the posterior labrum is 9:00, to coincide with the arthroscopist's perspective. Radial sequences that conform to the different positions of the clock are obtained using 4 localizer images in planes orthogonal to the acetabulum:

1. A straight coronal localizer is obtained (not shown).
2. Off the coronal localizer, an oblique axial localizer is obtained parallel to the femoral neck (left).
3. Based on the oblique axial localizer (left), an oblique coronal localizer is obtained (center).
4. Based on the oblique coronal localizer, an oblique sagittal localizer is obtained (right). From the oblique sagittal localizer, 16 radial sections are proscribed at 10 degree increments.

Pitfalls and Pearls

1. Do not confuse the transverse acetabular ligament/labral recess or the posterior sublabral sulcus for a labral tear.
2. Although indirect (intravenous) gadolinium enhanced arthrography has a role in the evaluation of articular pathology of the hip (see Case 37), optimal assessment of the acetabular labrum is achieved with direct gadolinium arthrography.

References

1. Leunig M, Podeszwa D, Beck M, Werlen S, Ganz R. Magnetic resonance arthrography of labral disorders in hips with dysplasia and impingement. *Clin Orthop* 2004:74–80.
2. McCarthy J, Noble P, Aluisio FV, Schuck M, Wright J, Lee JA. Anatomy, pathologic features, and treatment of acetabular labral tears. *Clin Orthop Relat Res* 2003:38–47.
3. Dinauer PA, Murphy KP, Carroll JF. Sublabral sulcus at the posteroinferior acetabulum: A potential pitfall in MR arthrography diagnosis of acetabular labral tears. *AJR Am J Roentgenol* 2004; 183:1745–1753.
4. Ganz R, Parvizi J, Beck M, Leunig M, Notzli H, Siebenrock KA. Femoroacetabular impingement: A cause for osteoarthritis of the hip. *Clin Orthop* 2003:112–120.
5. Sammarco GJ. The dancer's hip. *Clin Sports Med* 1983; 2:485–498.

Case 85

History

What do these three patients have in common?

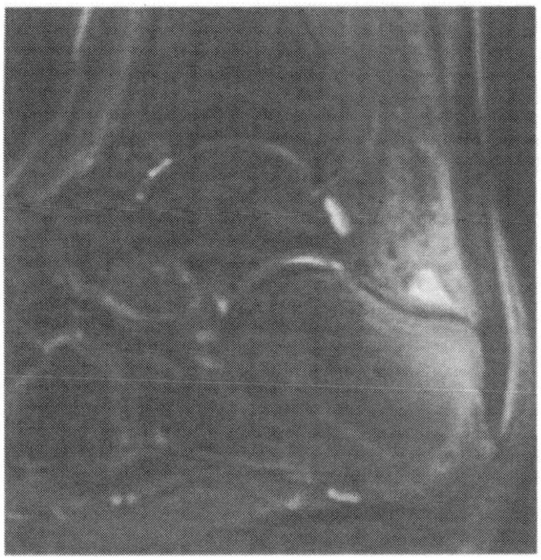

Figure 85A. Patient 1, 16-year-old girl. Sagittal T2 FS.

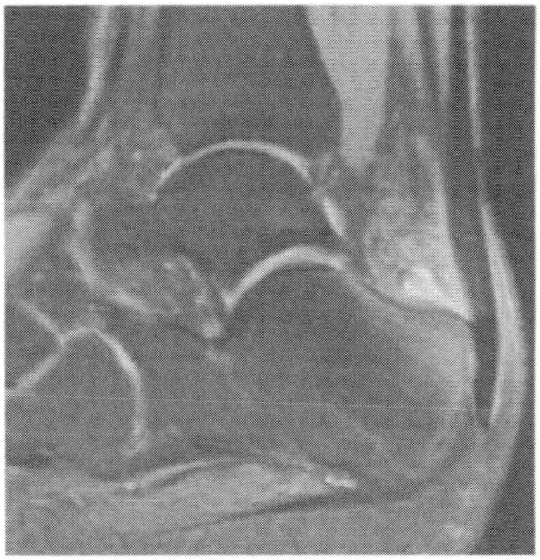

Figure 85B. Patient 1. Sagittal T1 post-Gd FS.

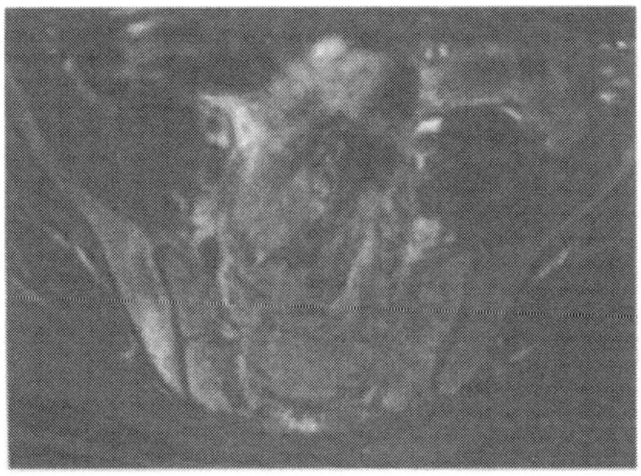

Figure 85C. Patient 2, 14-year-old boy. Axial T2 FS.

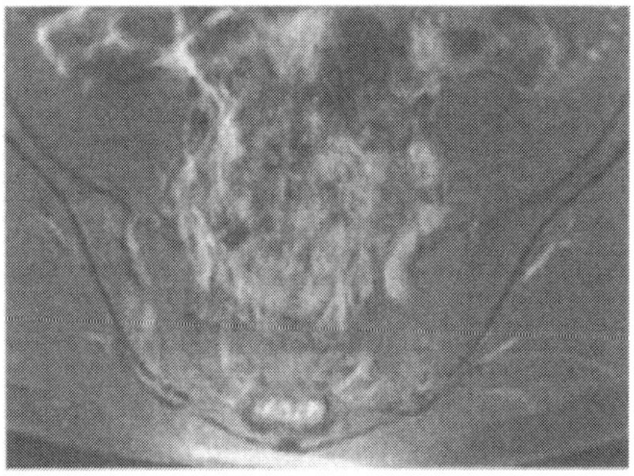

Figure 85D. Patient 2. Axial T1 post-Gd FS.

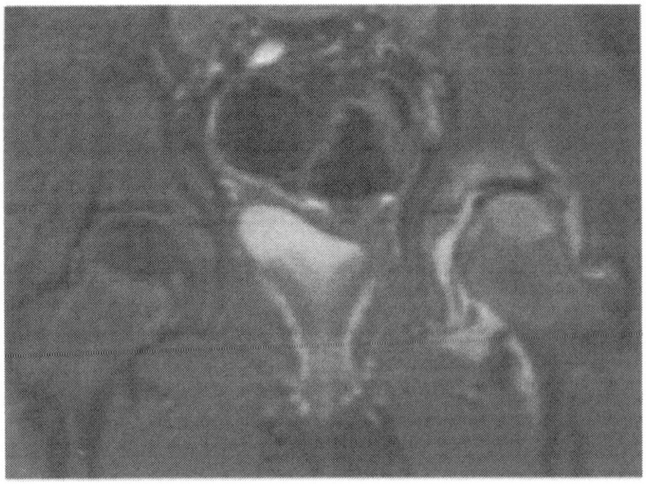

Figure 85E. Patient 3, 8-year-old girl. Coronal STIR.

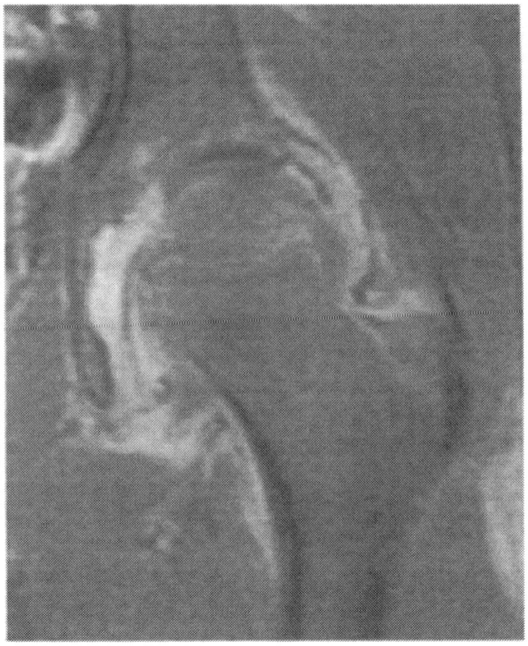

Figure 85F. Patient 3. Coronal T1 post-Gd FS of the left hip.

Patient 1

Figures 85A, 85B. The Achilles tendon is thickened with increased intrasubstance SI. There is hyperintensity on STIR and enhancement within the soft tissues posterior to the tendon, extending to its calcaneal insertion. There are diffuse inflammatory changes within Kager's fat pad and increased fluid SI and synovial enhancement of the retrocalcaneal bursa. There is marrow edema and enhancement of the posterior calcaneus, centered at the Achilles tendon insertion.

Patient 2

Figures 85C, 85D. There is increased T2 SI along the inferior iliac margin of the right SI joint. Following gadolinium, there is diffuse synovial enhancement of the right SI joint. The left SI joint is normal.

Patient 3

Figures 85E, 85F. There is significant left hip synovial enhancement, juxta-articular soft tissue edema, and marrow edema in the acetabulum and capital femoral epiphysis. The right hip is normal.

Diagnosis

Seronegative spondyloarthropathies

Questions

1. T/F: Osteopenia is equally common in juvenile rheumatoid arthritis (JRA) and the juvenile seronegative spondyloarthropathies.
2. What is the most specific extra-articular feature of the spondyloarthropathies?

Discussion

The juvenile seronegative spondyloarthropathies include: juvenile ankylosing spondylitis (AS), juvenile psoriatic arthritis, arthritis associated with inflammatory bowel disease (IBD), and Reiter's disease/reactive arthritis (Table 85A) (1). The term *spondyloarthropathy* may be misleading since peripheral arthritis is more common than axial involvement in these disorders in the pediatric population. For this reason, the International League Association for Rheumatology (ILAR) have adopted an alternative classification system that distinguishes psoriatic arthritis, and groups the other three arthritic categories into one group: enthesitis related arthritis (2).

The ILAR diagnostic criteria (2 of 3 positive) for psoriatic arthritis include: dactylitis, nail pitting or onycholysis, or a first-degree relative with psoriasis. The ILAR criteria (2 of 5 positive) for enthesitis related arthritis include: HLA-B27 positive, sacroiliac joint and/or inflammatory lumbrosacral pain, acute anterior uveitis, onset of arthritis in a male over 6 years of age, and spondyloarthropathy or acute anterior uveitis in a first-degree relative (2).

Patients with the juvenile seronegative spondyloarthropathies are usually HLA-B27 positive, under 16 years of age at presentation, and often present with extra-articular inflammatory changes such as enthesitis and tenosynovitis (1, 3, 4). Unilateral sacroiliac joint disease may be detectable at presentation, but bilateral, often asymmetric involvement develops in most cases. Except for psoriatic arthritis, these patients are usually male and appendicular large joint involvement is usually confined to the lower extremity (1). Juvenile psoriatic arthritis tends to occur in females and has a predilection for both the small joints of the hands and feet and the large joints such as the knee, ankle, and hip (Figures 85E, 85F).

On plain radiography, patients with juvenile spondyloarthropathies tend to develop bony proliferation rather than the osteopenia typically seen in JRA (Answer to Ques-

Table 85A. The juvenile spondyloarthropathies.

	Small joint	Large joint	Other
Psoriasis	Present (DIP, sausage digit)	Present	Arthritis may precede skin lesions (1).
Ankylosing spondylitis	Uncommon	Present	
IBD	Rare	Present	Two patterns: either appendicular large joint or axial (sacroiliac and spondylitis). Arthritis may precede GI symptoms.
Reiter's	Present (usually the feet)	Present	Triad of arthritis, urethritis, and conjunctivitis is rarely seen in the pediatric population.

tion 1). As with JRA, patients may develop periostitis, large peripheral joint involvement (particularly the knee), myositis may coexist, and osseous erosions are a late finding (1, 5, 6).

When large joint inflammatory changes are present on MRI, there are no features that confidently distinguish the spondyloarthropathies from JRA, pyogenic arthritis, and Lyme arthritis. All of these inflammatory processes may cause synovial enhancement, joint effusion, tenosynovitis, and juxta-articular marrow edema. The most specific extra-articular feature of the spondyloarthropathies is enthesitis (Answer to Question 2). The common sites for enthesitis include the Achilles tendon attachment to the calcaneus, plantar fascia attachment to the calcaneus, the patellar tendon attachment to the tibial tuberosity, and the quadriceps tendon insertion onto the patella (4). On MRI, tendinous thickening associated with increased SI within the substance of the tendon and peritendinous soft tissues are frequently present. Marrow edema and erosions at the tendinous insertional site may also be seen. A detailed discussion on tenosynovitis related to juvenile psoriatic arthritis is presented in Case 69.

Patient 1 had a history of inflammatory bowel disease. Patient 2 has ankylosing spondylitis. Patient 3 has psoriatic arthritis.

Orthopedic Perspective

The spondyloarthropathies often present with joint pain or enthesopathy. The diagnosis is made based on clinical and laboratory grounds by a rheumatologist. The juvenile spondyloarthropathies are, however, great mimickers of other entities, particularly in patients without the typical psoriatic skin rash or abdominal manifestations of inflammatory bowel disease. Patients may be referred for MRI based on concerns of osteomyelitis/septic arthritis. Although there is considerable overlap of these entities, the presence of multifocal involvement, tenosynovitis, and enthesitis on MRI should suggest a rheumatologic condition. The diagnosis ultimately rests on laboratory studies, exclusion of an infectious etiology, and clinical follow-up.

What the Clinician Needs to Know

1. Can the diagnosis of osteomyelitis/septic arthritis be completely excluded?
2. Provide baseline information for patients with known juvenile spondyloarthropathies to guide medical therapy.

Answers

1. False.
2. Enthesitis.

Pitfalls and Pearls

Enthesitis is a characteristic feature of seronegative spondyloarthropathies and is not usually seen in patients with JRA.

References

1. Buchmann RF, Jaramillo D. Imaging of articular disorders in children. *Radiol Clin North Am* 2004; 42:151–168, vii.

2. Petty RE, Southwood TR, Manners P, et al. International League of Associations for Rheumatology classification of juvenile idiopathic arthritis: Second revision, Edmonton, 2001. *J Rheumatol* 2004; 31:390–392.
3. Azouz EM, Duffy CM. Juvenile spondyloarthropathies: Clinical manifestations and medical imaging. *Skeletal Radiol* 1995; 24:399–408.
4. Azouz EM. Arthritis in children: Conventional and advanced imaging. *Semin Musculoskelet Radiol* 2003; 7:95–102.
5. Cohen PA, Job-Deslandre CH, Lalande G, Adamsbaum C. Overview of the radiology of juvenile idiopathic arthritis (JIA). *Eur J Radiol* 2000; 33:94–101.
6. Fleckenstein JL, Reimers CD. Inflammatory myopathies: Radiologic evaluation. *Radiol Clin North Am* 1996; 34:427–439, xii.

Case 86

History

This is a 5-year-old boy with chronic left leg swelling and deformity.

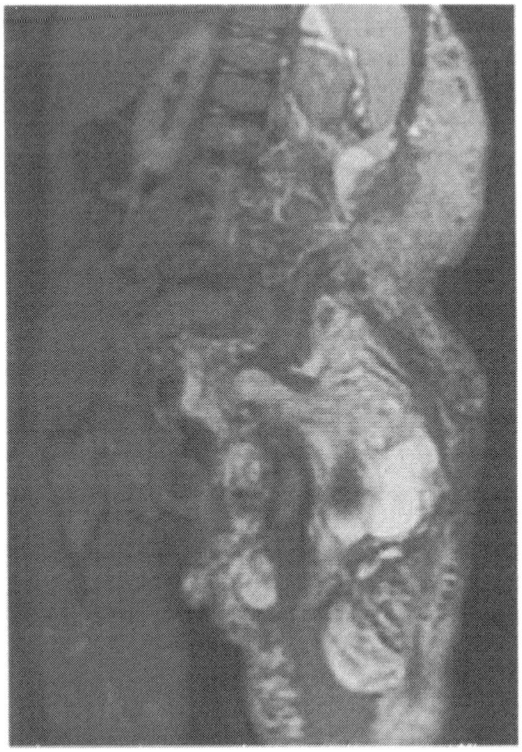

Figure 86A. Coronal STIR.

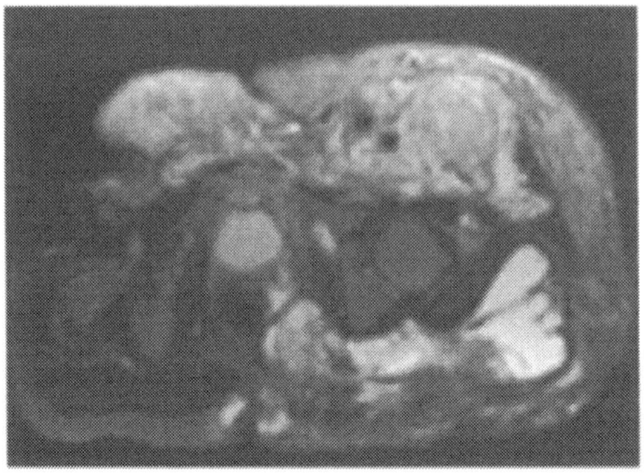

Figure 86B. Axial T2 FS.

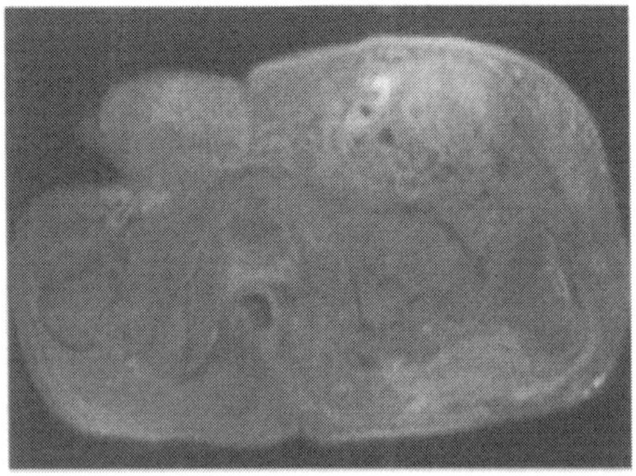

Figure 86C. Axial T1 post-Gd FS.

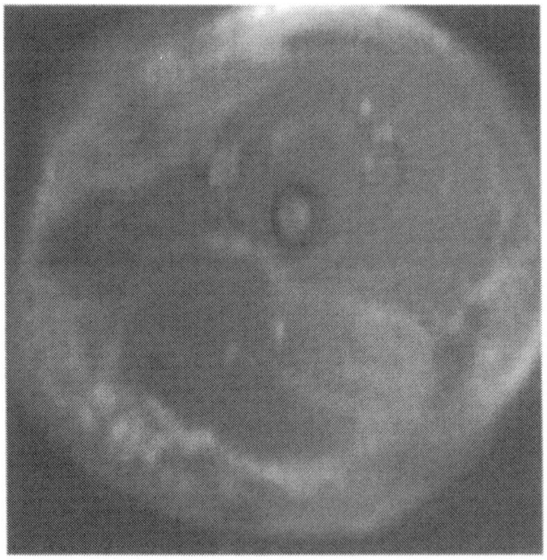

Figure 86D. Axial T1 post-Gd FS (mid-left thigh).

Figure 86A. Heterogeneous increased STIR SI is seen in the superficial and deep soft tissues of the left flank, pelvis, and thigh.

Figures 86B, 86C, (86C with annotations). Variably sized fluid filled structures with septations are identified **(thin arrows)** consistent with macrocystic lymphatic malformations (LM). A more solid appearing T2 hyperintense enhancing component occupies the superficial and deep soft tissues of the anterior pelvis **(thick arrow)**, compatible with microcystic LM.

Figures 86D, (86D with annotations). A venous malformation is present along the posteromedial thigh **(arrowheads)**. Mixed enhancing tissue is noted in the deep thigh consistent with microcystic LM (*).

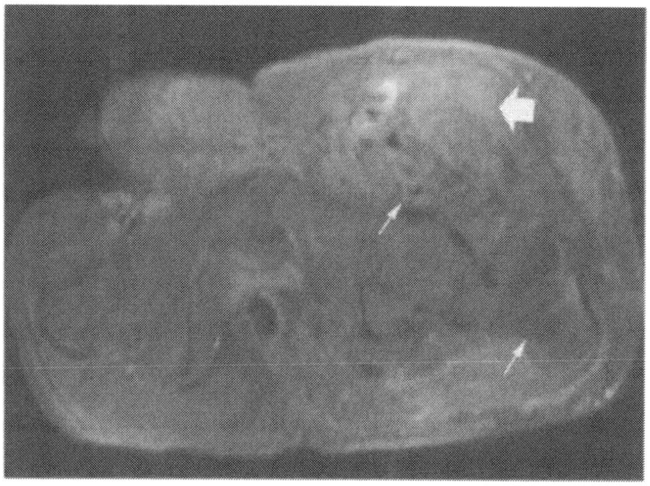

Figure 86C* Annotated.

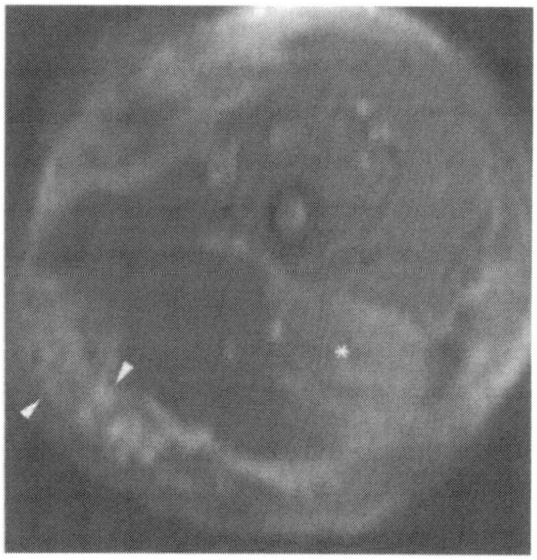

Figure 86D* Annotated.

Diagnosis

Klippel-Trenaunay syndrome (KTS)

Questions

1. What are the three types of vascular malformation that are present in patients with Klippel-Trenaunay syndrome?
2. What is the marginal vein of Servelle?

Discussion

Klippel-Trenaunay syndrome (KTS) is composed of slow flow vascular malformations, including capillary, venous, and lymphatic malformations (LM), as well as deep and superficial vein anomalies (Answer to Question 1) (1). A geographic cutaneous capillary stain is usually present over the affected extremity. Limb overgrowth with increased circumference is often present. Less commonly, limb undergrowth is seen. In one series, 13 of 49 individuals had more than one limb involved and 17 of 49 individuals demonstrated extremity length discrepancies (2). The increased limb circumference is related to increased subcutaneous fat deposition, bulk created by vascular malformations, dermal thickening, and lymphedema. Isolated venous malformations (VM) or lymphatic malformations should be distinguished from KTS.

Anomalous varicose veins due to interrupted or absent superficial and deep venous systems are typical of KTS. The marginal vein of Servelle (Figures 86E, 86F) is a characteristic anomalous superficial varicose vein seen in KTS, and is found along the lateral thigh and calf (Answer to Question 2) (3). Venous outflow tracts may be better defined using 2D time-of-flight (TOF) MRV or gadolinium enhanced dynamic MR angiography.

Microcystic and macrocystic LM may coexist (Figures 86B, 86C). Both may be hypointense on T1 and moderately to markedly hyperintense on fluid-sensitive sequences. The septal and rim enhancement seen with microcystic LM may be confluent and mistaken for a solid soft tissue neoplasm (Figure 86C) (3). Capillary malformations may also coexist with microcystic lymphatic malformations and may be indistinguishable by MRI and histology (4).

Venous malformations associated with Klippel-Trenaunay syndrome should be distinguished from LM. Venous malformations are usually well-defined, lobular structures that are isointense on T1, hyperintense on T2, and demonstrate tubular and lobular enhancement (see Case 67) as opposed to the septal and rim enhancement that may be seen with LM. Venolymphatic malformations are also seen and are frequently found at the interface between a large draining vein and a microcystic LM (4).

Other overgrowth syndromes associated with complex vascular malformations include Parkes-Weber, Proteus, and Bannayan-Riley-Ruvalcaba syndromes. Parkes-Weber syndrome, in contrast to Klippel-Trenaunay syndrome, is associated with high flow vascular malformations (arteriovenous malformations and arteriovenous fistulas) and hypertrophy of osseous structures and muscle. Additional finding includes a cutaneous port-wine stain over the affected extremity (cutaneous capillary malformations) (1). Proteus syndrome is an overgrowth syndrome comprised of complex vascular malformations including slow and high flow lesions, skeletal deformities, verrucous nevi, macrocephaly, and tumors of adipose, connective tissue, vascular, and neurogenic origin (1). Bannayan-Riley-Ruvalcaba syndrome (BRR) (Figures 86G–86I) shares similar features with Proteus syndrome, but patients may also develop gastrointestinal hamartomous polyps and soft tissue tumors, particularly lipomas (75%) (5).

This patient had a known history of Klippel-Trenaunay syndrome. This exam was ordered prior to a series of sclerotherapy and debulking surgeries.

What the Clinician Needs to Know

1. What is the specific draining venous anatomy? Is there evidence of deep venous thrombosis to explain any recent extremity enlargement?
2. What are the largest lymphatic and venous malformations? Which lesions are most amenable for sclerotherapy?
3. Is there an intra-abdominal component of the vascular malformation?

Answers

1. Capillary, venous, and lymphatic malformations.
2. Anomalous draining vein seen in patients with Klippel-Trenaunay syndrome. It is located along the lateral thigh and calf.

Additional Examples

Klippel-Trenaunay Syndrome with Bilateral Lower Extremity Involvement

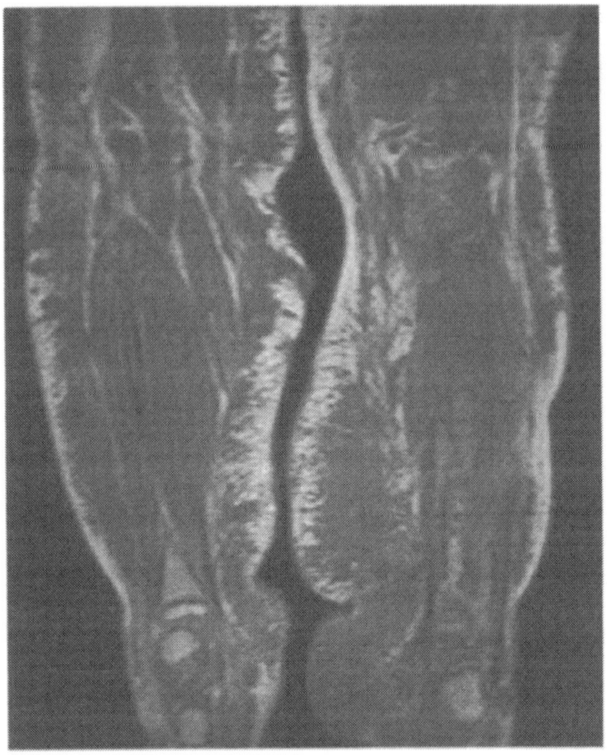

Figure 86E. Coronal T1.

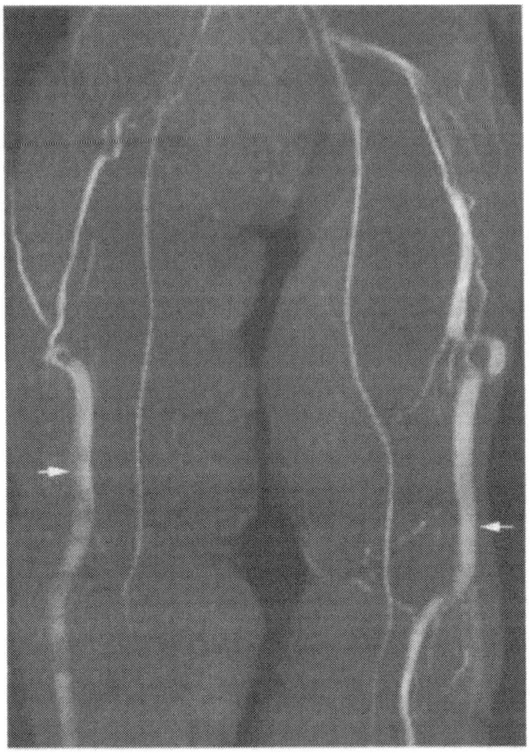

Figure 86F. 2D TOF MRV of both lower extremities.

Findings

This is a 3-year-old girl with known Klippel-Trenaunay syndrome with chronic, diffuse leg swelling bilaterally.

Figure 86E. There is diffuse infiltration of the subcutaneous fat in both lower extremities, compatible with microcystic lymphatic malformations and edema.

Figure 86F. On MRV, the marginal veins of Servelle **(arrows)** are identified bilaterally along the lateral thigh and calf.

Bannayan-Riley-Ruvalcaba Syndrome

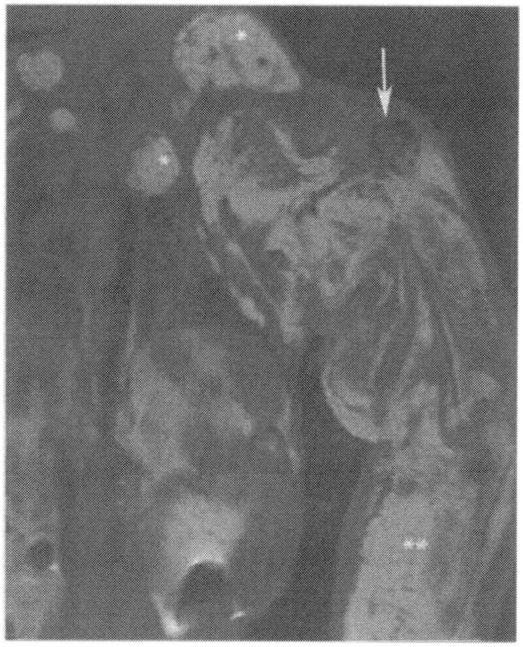

Figure 86G. Coronal T1 of the left flank and arm.

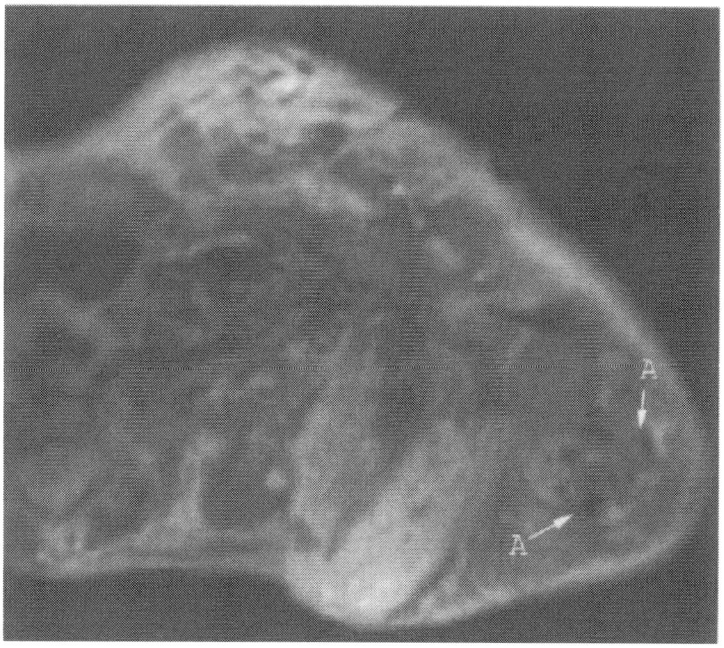

Figure 86H. Axial T2 FS.

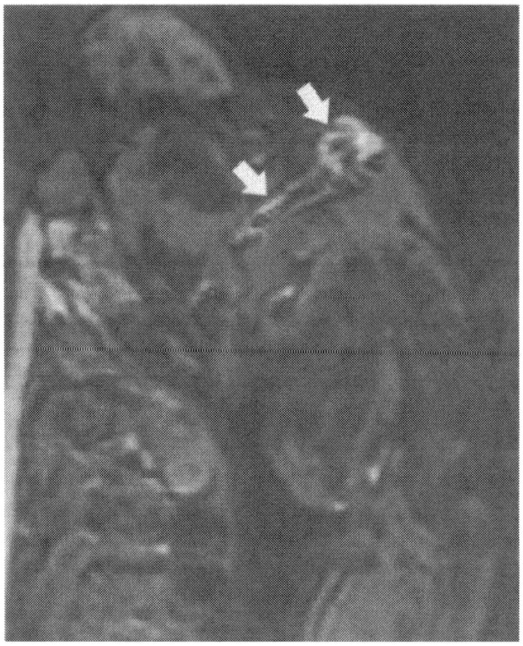

Figure 86I. Coronal 3D SPGR FS post-Gd dynamic MRA, early phase.

Findings

This is a 3-year-old girl who has a history of Bannayan-Riley-Ruvalcaba syndrome.

Figure 86G, 86H, 86I. Well-defined (*) and infiltrative (**) lipomas are identified in the neck, thoracic inlet, and arm. A complex vascular malformation with a high-flow arteriovenous component is identified **(arrow in Figure 86G)**, apparent as flow voids on T2W (A). Tubular enhancement is present on early phase dynamic MRA **(thick arrows)**.

Pitfalls and Pearls

The complications of Klippel-Trenaunay syndrome relate to poor lymphatic and venous outflow causing subcutaneous lymphedema, venous thrombosis, hemorrhage within vascular malformations, and infection (1).

References

1. Burrows PE, Laor T, Paltiel H, Robertson RL. Diagnostic imaging in the evaluation of vascular birthmarks. *Dermatol Clin* 1998; 16:455–488.
2. Berry SA, Peterson C, Mize W, et al. Klippel-Trenaunay syndrome. *Am J Med Genet* 1998; 79:319–326.
3. Konez O, Burrows PE. Magnetic resonance of vascular anomalies. *Magn Reson Imaging Clin N Am* 2002; 10:363–388, vii.
4. Burrows PE. Personal communication. 2005.
5. Gorlin RJ, Cohen MM, Jr., Condon LM, Burke BA. Bannayan-Riley-Ruvalcaba syndrome. *Am J Med Genet* 1992; 44:307–314.

Case 87

History

This is a 13-year-old girl with a 4- to 6-week history of left elbow pain. She is an avid basketball player (and is left handed), but there is no history of a specific traumatic event. She recently noticed that her left arm was enlarging. She is afebrile and otherwise healthy.

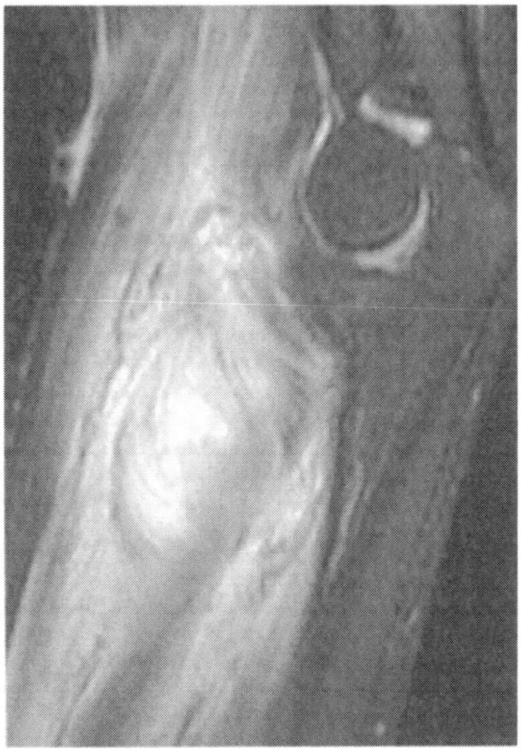

Figure 87A. Sagittal T2 FS of the left forearm.

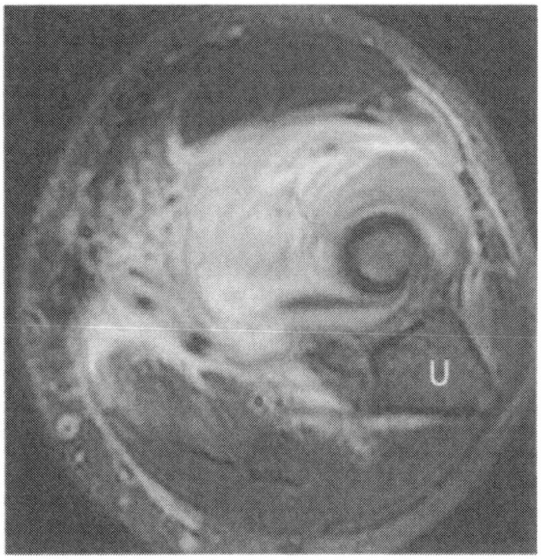

Figure 87B. Axial T2 FS.

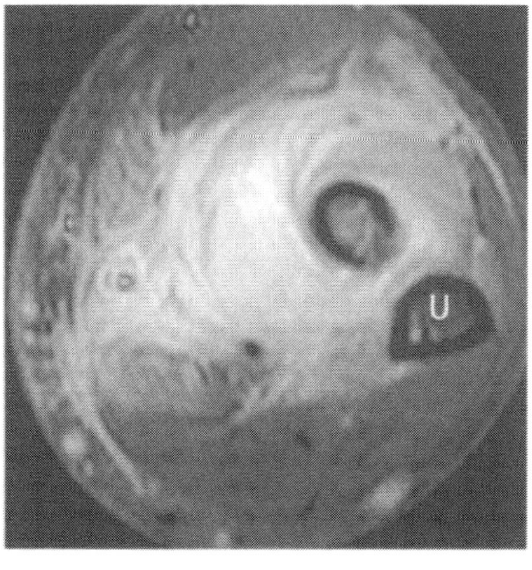

Figure 87C. Axial T1 post-Gd FS.

Figures 87A, 87B, 87C, (87C with annotations). There is infiltrative increased T2 SI and enhancement involving the biceps brachii and supinator muscles, deep and superficial fascia, and subcutaneous fat. There is also increased T2 SI within the bone marrow and loss of cortical definition of the radius **(arrow)**. Ulna (U).

Figure 87D. There is peripheral flocculent early calcification and a central lucency identified along the ventral aspect of the proximal forearm.

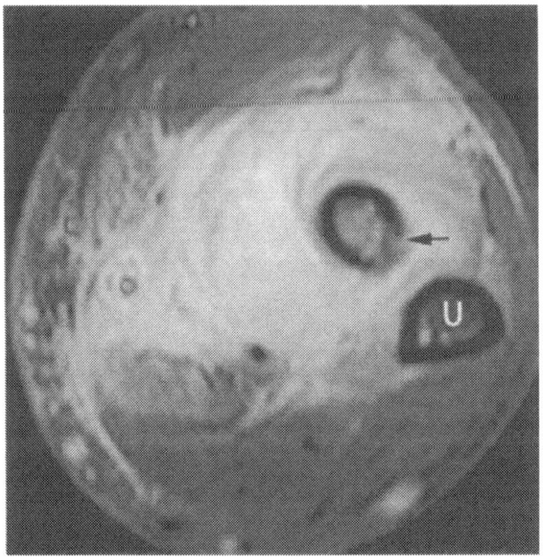

Figure 87C* Annotated.

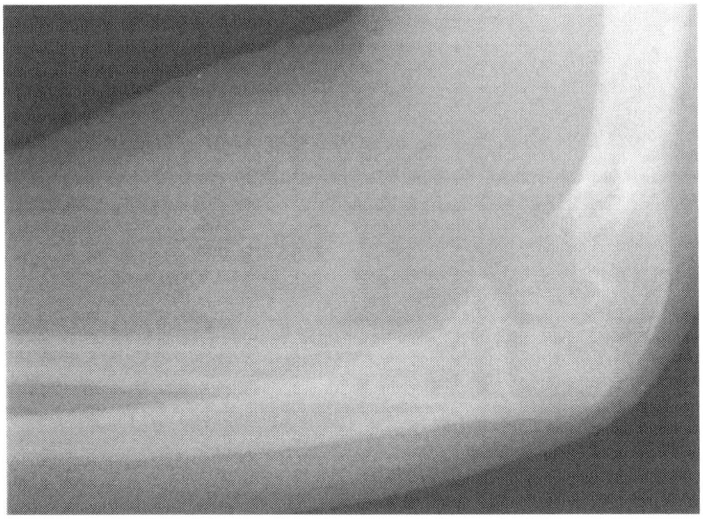

Figure 87D. Lateral radiograph obtained 18 days after MRI and 10 days following percutaneous biopsy.

Diagnosis

Myositis ossificans traumatica (MOT)

Questions

1. Where are the two most common locations for myositis ossificans?
2. How much time is required following a traumatic event for the appearance of peripheral calcifications with myositis ossificans?

Discussion

Myositis ossificans is a non-neoplastic inflammatory reaction characterized by fibroblastic proliferation and subsequent heterotopic bone formation within muscle (1). Myositis ossificans may be separated into three different types: myositis ossificans traumatica (MOT), nontraumatic myositis ossificans, and fibrodysplasia ossificans progressiva (FOP) (2). Nontraumatic myositis ossificans usually occurs in patients with systemic disorders such as paraplegia, and is often associated with generalized heterotopic ossification. FOP is an autosomal dominant disorder characterized by severe heterotopic ossification that involves muscle, but may also involve fascia, tendons, and ligaments (3). In patients with FOP, heterotopic ossification develops spontaneously or is related to minor trauma. FOP presents during the first decade of life with abnormal ossification first appearing over the back and neck, eventually developing throughout the muscles of the torso and extremities. FOP is associated with skeletal deformities, which include: great toe malformations (short first metatarsal and proximal phalanx) (Figure 87E) and similar but less frequent deformities of the thumb, and short broad femoral necks.

Patients without a history of trauma are usually challenging because the features of early MOT may overlap with aggressive sarcomas by imaging and histology. The most common location of MOT is in the thigh (quadriceps) and arm (brachialis) (Answer to Question 1) (4). MOT may also be subclassified based on location within a muscle compartment. It may be completely intramuscular, juxtacortical, or intermediate (at muscle insertion sites) (5).

Like soft tissue sarcomas, early MOT may be isointense on T1, hyperintense on T2, and will show diffuse or rim enhancement (4). Early MOT often shows ill-defined infiltrative margins with soft tissue edema and enhancement that extend into multiple soft tissue compartments (6). Bone marrow edema and periosteal reaction may be seen as well. Early MOT may mimic a soft tissue abscess or a necrotic tumor because of the presence of rim enhancement and central fluid-fluid levels (7). An early clue that suggests MOT and weighs against a sarcoma is the conspicuous presence of an infiltrative edema pattern and absence of a solid soft tissue mass. In general, most soft tissue sarcomas have well-defined margins, appear solid, and demonstrate minor adjacent edema, unless there has been recent trauma or biopsy.

The characteristic, peripheral rim of floccular calcification seen with subacute MOT may be identifiable by 4 to 6 weeks after recalled or occult injury (Answer to Question 2) (4). This contrasts with the typical central calcification seen in necrotic soft tissue tumors. On MRI, the peripheral rim will be hypointense on all imaging sequences (Figures 87F–87H). However, CT is the study of choice in the follow-up of suspected MOT because of superior sensitivity for and better characterization of early calcification. Without this characteristic peripheral rim of calcification, the diagnosis remains nonspecific and an aggressive sarcoma or infection cannot be excluded. Finally, the chronic changes of MOT reflect the imaging characteristics of lamellar bone, with a cortical and medullary osseous architecture. Chronic MOT shares a similar imaging appearance with bony cortex and marrow (7). Adjacent soft tissue edema is usually absent when MOT is chronic.

MOT should not be confused with heterotopic ossification occurring within non-muscular soft tissues. Deep/superficial fascia and subcutaneous soft tissue bone formation should be labeled heterotopic ossification rather than MOT. MOT should be distinguished from heterotopic ossification related to an avulsion injury (Figures 87I–87L). Other causes of extraosseous calcification and ossification include sarcomas (extraskeletal osteosarcomas, synovial cell sarcoma), tumoral calcinosis, and dermatomyositis. Sarcomas usually demonstrate calcification centrally. Tumoral calcinosis tends to occur in African Americans in a periarticular location, most often the shoulder. Calcification seen with dermatomyositis tends to be plate-like and involves the subcutaneous soft tissues.

This patient was referred for biopsy of a soft tissue mass. Since she reported no specific traumatic event, there was a concern for a soft tissue tumor or unusual infection. A CT-guided biopsy was nondiagnostic, but showed no evidence of tumor. A follow-up radiograph (Figure 87D) showed the characteristic pattern of peripheral calcification of MOT.

Orthopedic Perspective

Myositis ossificans traumatica (MOT) presents as a mass and the history of trauma is often forgotten by the patient, but must be sought by careful questioning. MOT is usually not treated surgically, so the major concern for the orthopedist is distinguishing MOT from a soft tissue sarcoma or an abscess. The forearm case shown was particularly troublesome since there was no clear trauma history and the lesion was very painful and enlarging. A needle biopsy was required to establish the diagnosis. As is often the case, the mineralization pattern characteristic of MOT was not evident until a CT (not shown) was obtained showing the characteristic mineralization pattern (more mature in the periphery). As noted above, CT is frequently more helpful and the MRI findings may mislead the inexperienced observer since the pattern of mineralization is not as well seen by MRI. These cases can be observed and with time the size of the mass reduces. If it persists and is symptomatic, the bony lump can be excised, but it is generally considered wise to wait at least six months to permit spontaneous regression and to reduce the chance of recurrent muscle ossification postoperatively.

What the Clinician Needs to Know

1. Distinguish MOT from sarcoma or abscess.
2. A biopsy at times is necessary, but often the characteristic mineralization pattern seen by radiographs or CT will be diagnostic.

Answers

1. Thigh and arm.
2. 4–6 weeks.

Additional Examples

Fibrodysplasia Ossificans Progressiva

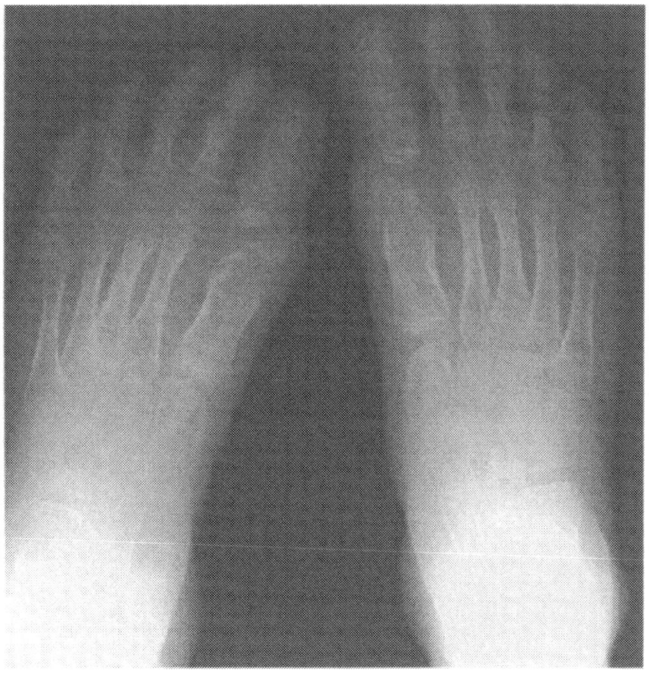

Figure 87E. AP radiograph of the foot.

Findings

This is a 4-year-old girl with known fibrodysplasia ossificans progressiva.

Figure 87E. Note characteristic shortened and dysplastic appearing first proximal phalanx deformities bilaterally.

Myositis Ossificans Traumatica

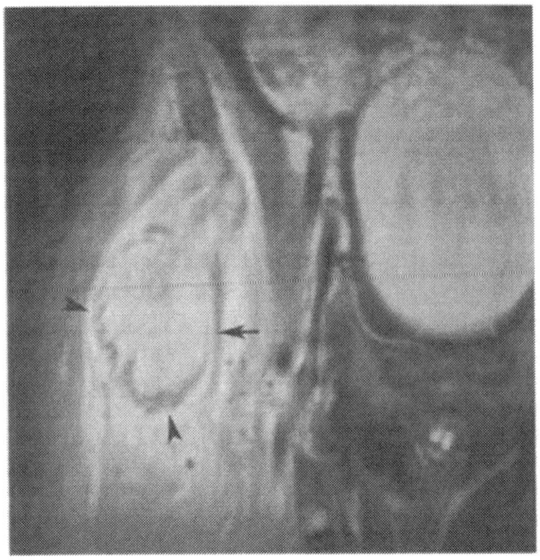

Figure 87F. Coronal STIR.

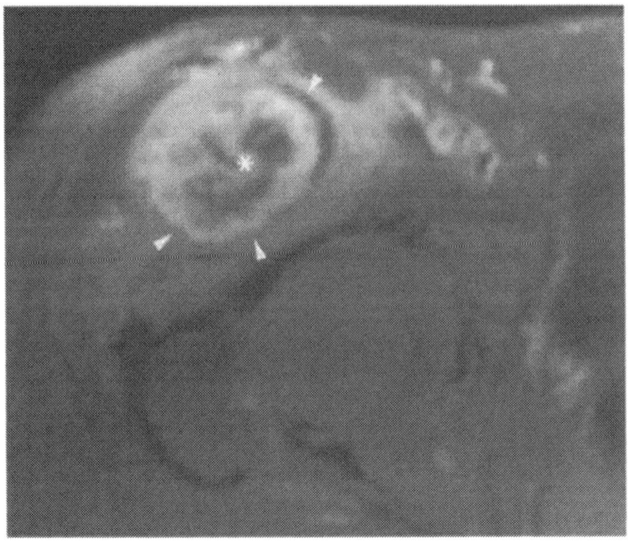

Figure 87G. Axial T1 post-Gd FS.

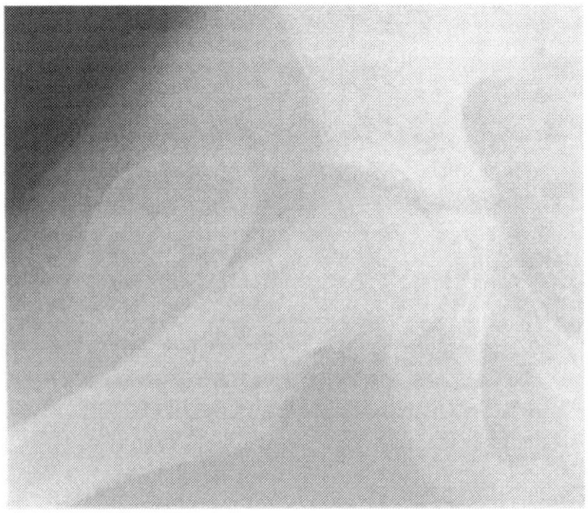

Figure 87H. Frogleg lateral radiograph obtained 2 months later.

Findings

This is a 9-year-old boy with gradual anterior thigh swelling over several months.

Figures 87F, 87G. There is a soft tissue mass centered within the rectus femoris muscle and fascia with surrounding infiltrative edema and enhancement. A hypointense peripheral rim is evident **(arrowheads)**. The myotendinous junction is deviated medially **(arrow)**. On the post-Gd sequence, the nonenhancing center (*) simulates a necrotic tumor or abscess.

Figure 87H. A follow-up radiograph 2 months later confirms peripheral calcification, characteristic of myositis ossificans.

Mature Heterotopic Ossification After Anterior Superior Iliac Spine (ASIS) Avulsion Injury

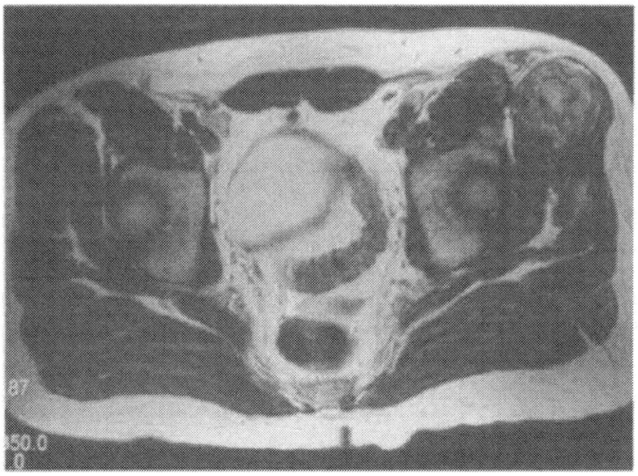

Figure 87I. Axial T1.

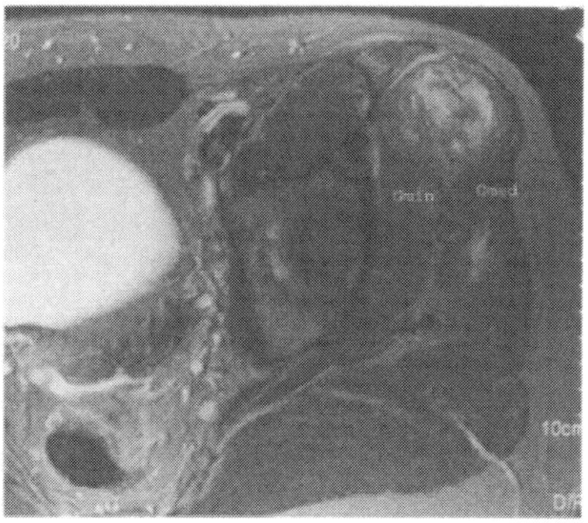

Figure 87J. Axial T2 FS.

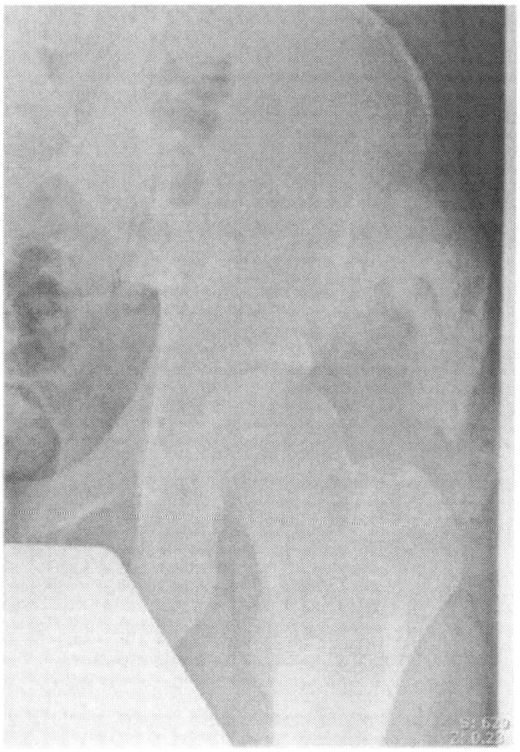

Figure 87K. AP radiograph.

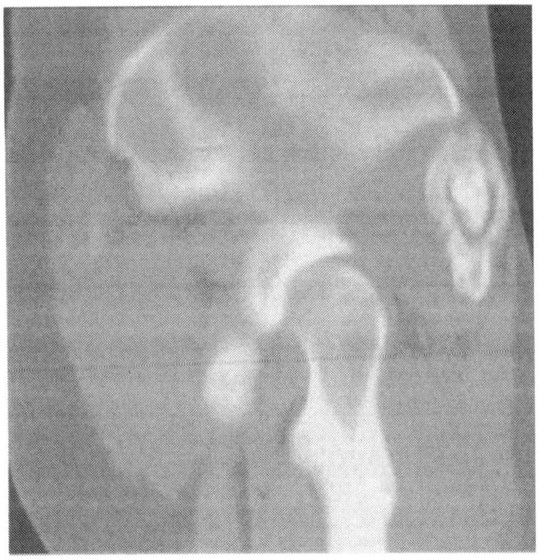

Figure 87L. CT with sagittal reformat of the left hip.

Findings

This is a 17-year-old high school football player with left hip immobility after an anterior superior iliac spine (ASIS) avulsion injury.

Figures 87I, 87J. There is maturing heterotopic ossification present between the deep fascia of the gluteus minimi (gmin) and gluteus medius (gmed), evidenced by residual mild increased T2 SI. Note that the mass follows fat SI, indicating central fatty marrow metaplasia.

Figures 87K, 87L. The radiograph and CT demonstrate mature ossification extending from the ASIS.

Pitfalls and Pearls

1. Soft tissue sarcomas have central calcification, whereas MOT has early peripheral calcification that may progress to central calcification.
2. MOT is diagnosed by correlating plain radiography and CT with a clinical history of antecedent trauma. The early and subacute MRI features in isolation are nonspecific and often misleading.
3. The term *myositis ossificans* should apply only to ossification within muscle. Use the term *heterotopic ossification* for benign extraskeletal ossification elsewhere.

References

1. Nuovo MA, Norman A, Chumas J, Ackerman LV. Myositis ossificans with atypical clinical, radiographic, or pathologic findings: A review of 23 cases. *Skeletal Radiol* 1992; 21:87–101.
2. Gindele A, Schwamborn D, Tsironis K, Benz-Bohm G. Myositis ossificans traumatica in young children: Report of three cases and review of the literature. *Pediatr Radiol* 2000; 30:451–459.
3. Mahboubi S, Glaser DL, Shore EM, Kaplan FS. Fibrodysplasia ossificans progressiva. *Pediatr Radiol* 2001; 31:307–314.
4. Parikh J, Hyare H, Saifuddin A. The imaging features of post-traumatic myositis ossificans, with emphasis on MRI. *Clin Radiol* 2002; 57:1058–1066.
5. Ehara S, Shiraishi H, Abe M, Mizutani H. Reactive heterotopic ossification: Its patterns on MRI. *Clin Imaging* 1998; 22:292–296.
6. Hanquinet S, Ngo L, Anooshiravani M, Garcia J, Bugmann P. Magnetic resonance imaging helps in the early diagnosis of myositis ossificans in children. *Pediatr Surg Int* 1999; 15:287–289.
7. De Smet AA, Norris MA, Fisher DR. Magnetic resonance imaging of myositis ossificans: Analysis of seven cases. *Skeletal Radiol* 1992; 21:503–507.

Case 88

History

This is a 15-year-old girl with right-sided hip pain.

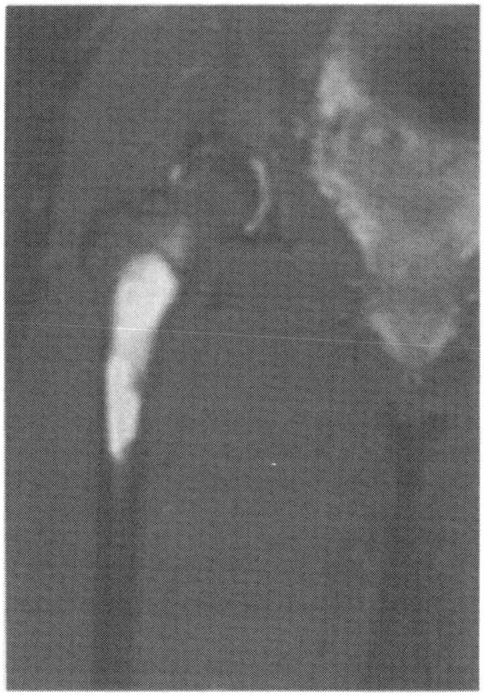

Figure 88A. Coronal STIR.

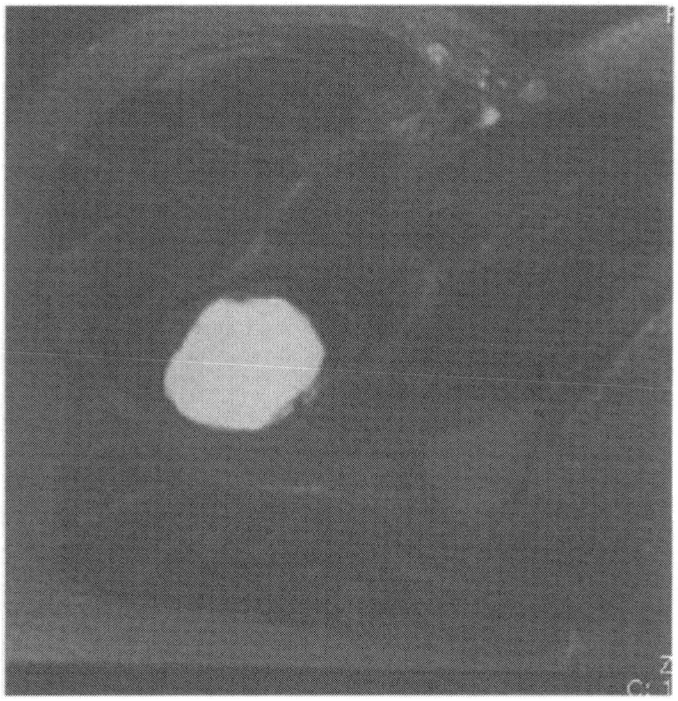

Figure 88B. Axial T2 FS.

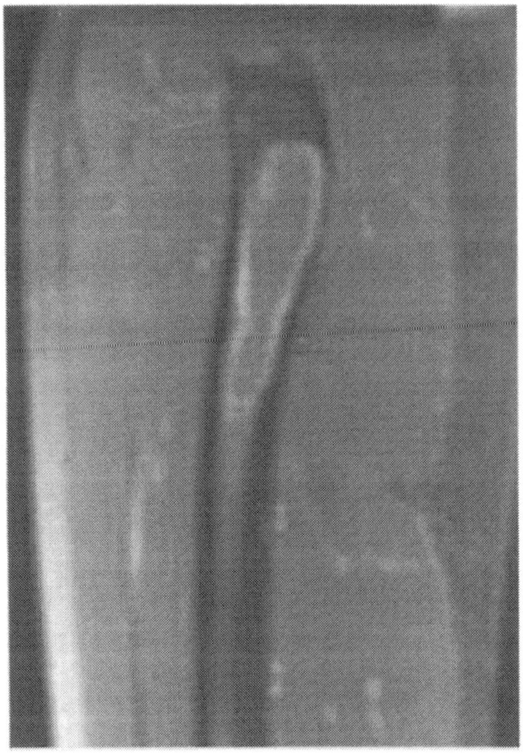

Figure 88C. Sagittal T1 post-Gd FS.

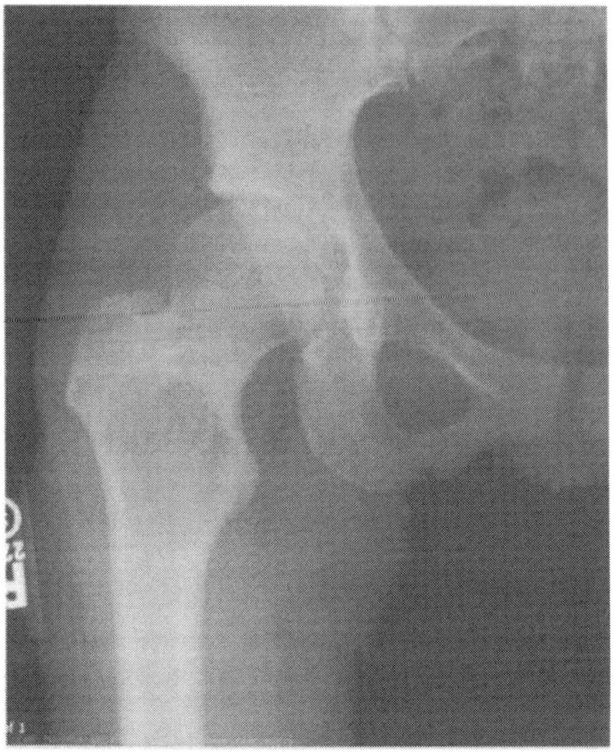

Figure 88D. AP radiograph of the right hip.

Figures 88A, 88B, 88C, 88D. On radiography and MRI, there is a sharply defined osteolytic lesion present in the proximal right femur. This lesion follows fluid SI on all imaging sequences. Mild thin peripheral enhancement is present. No cortical disruption, internal matrix, septations, or soft tissue mass is seen.

Diagnosis

Unicameral bone cyst (UBC) of the proximal femur

Questions

1. What are the two most common locations for a unicameral bone cyst?
2. T/F: MRI easily distinguishes a healing UBC from an aneurysmal bone cyst.

Discussion

A unicameral bone cyst (UBC) is a benign cystic lesion of bone that is most commonly seen in patients age 9 to 15 years. Two thirds are located in the proximal humerus, followed by the proximal femur and proximal tibia (Answer to Question 1) (1). These benign bone lesions are predisposed to pathologic fracture because they replace normal bony architecture.

On radiography, UBCs are usually located centrally within the medullary canal of the metaphysis and rarely demonstrate transphyseal extension into the epiphysis (2). With longitudinal growth of the affected bone, they may migrate into the diaphysis. UBCs are radiolucent with no internal matrix and have well-defined, usually sclerotic margins. An uncomplicated UBC shows no evidence of septation or periosteal reaction. When complicated by pathologic fracture, a fallen fragment sign may be seen (Figure 88E), representing a cortical fragment layering within the dependent portion of the cyst.

On MRI, uncomplicated unicameral bone cysts follow fluid SI on all imaging sequences and usually demonstrate thin peripheral enhancement, likely related to the cyst's thin fibrovascular membrane (3). An enhancing soft tissue component or cortical destruction is absent. With acute injury, the internal characteristics of a UBC will vary on T1 and T2W sequences due to the presence of blood products. Fluid-fluid levels may also be seen. During healing, intra- and extraosseous enhancing granulation tissue (Figures 88E–88G), internal septations, internal calcification, and bone expansion may be seen. The constellation of these findings may make it impossible to distinguish the lesion from an aneurysmal bone cyst (ABC) (Answer to Question 2) (3).

The plain radiographic features, location, and age are usually sufficient for diagnosis. MRI evaluation is usually reserved for cases where radiographic features are atypical, or the symptoms cannot be explained based on radiographs alone. Alternative considerations for a UBC-like lesion include Langerhans cell histiocytosis, fibrous dysplasia, ABC, and enchondroma. Rarely malignancies such as an osteosarcoma (Figures 88H–88M) and Ewing's sarcoma (see Case 53) will masquerade as a benign appearing cystic lesion of bone. In the context of a pathologic fracture, it may be impossible to exclude an aggressive neoplasm on MRI. Any enhancing soft tissue mass in an otherwise benign appearing cystic lesion of bone should be viewed with caution and close follow-up is appropriate.

In this patient, the MRI was requested to define the full extent of the cystic lesion. An intraoperative intraosseous cystogram was performed, confirming a UBC. Because the femoral UBC was large and located in a weightbearing bone, steroid injection was performed and the patient did well.

Orthopedic Perspective

The main concern for the orthopedist is to determine if the patient with a UBC is a fracture risk, and if so, whether treatment is necessary. Most UBCs eventually involute and fill in with bone over time, but until this occurs, the lesion poses a fracture risk for the child. The treatment options are to curette and bone graft the cyst, or to inject with steroid and demineralized bone matrix. It is important that the patient and parents know that a fracture through a cyst does not routinely lead to healing of the cyst (although it may). The callus formation initially seen on radiographs may erroneously suggest cyst healing. It is also important that the relationship of the cyst to the growth plate be assessed. UBCs occasionally involve the growth plate and lead to growth arrest. MRI is particularly good at assessing the relationship of the cyst to the growth plate and occult transgressions to the growth plate may only be apparent with this technique.

What the Clinician Needs to Know

1. Is this a bone cyst or other lesion? Is there an enhancing soft tissue component?
2. What is the fracture risk?
3. Does a pathologic fracture explain atypical radiographic features?

Answers

1. Proximal humerus followed by proximal femur.
2. False.

Additional Examples

UBC with Healing After Fracture

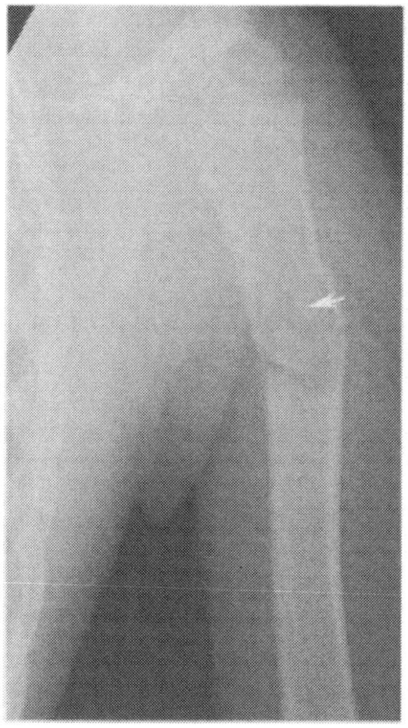

Figure 88E. AP radiograph of the left humerus.

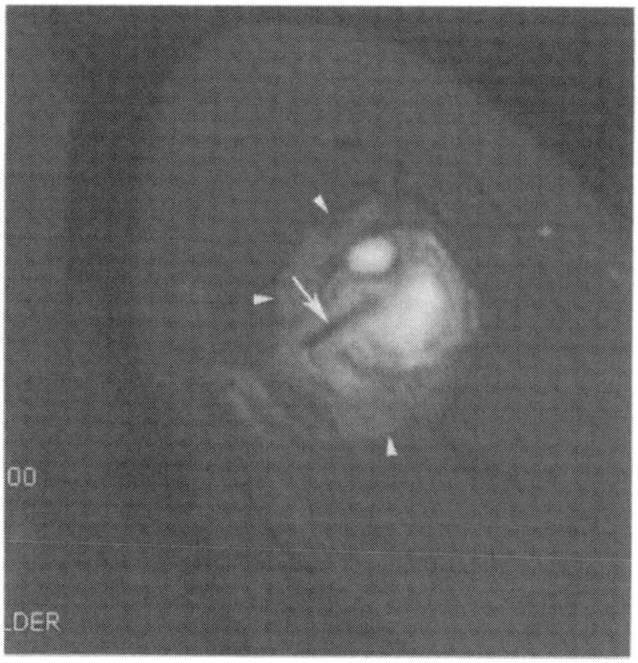

Figure 88F. Axial T2 FS through the lesion (one month later).

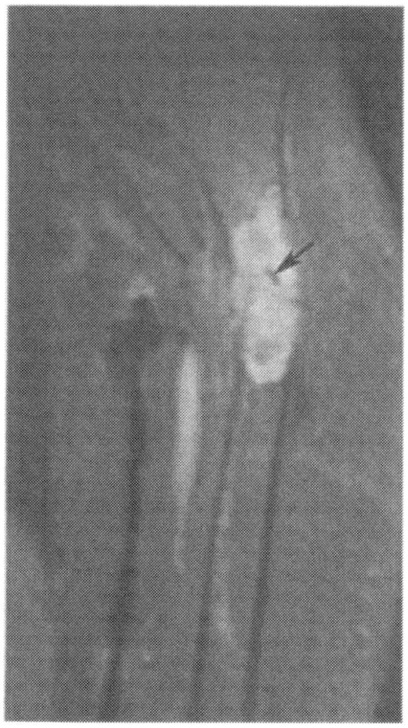

Figure 88G. Coronal T1 post-Gd FS.

Findings

This is a 5-year-old boy who fell on his arm while jumping on a bed.

Figure 88E. There is a pathologic fracture with a fallen fragment **(arrow)** through a well-defined osteolytic lesion located in the proximal diaphysis of the humerus.

Figures 88F, 88G. Heterogeneous lesional T2 hyperintensity and enhancement is evident compatible with granulation tissue. Juxtacortical soft tissue T2 hyperintensity is also evident **(arrowheads)**. The fallen fragment sign is evident **(arrows)**. This was a pathologically proven unicameral bone cyst. Without the initial radiograph or history, the MRI features would be nonspecific and indistinguishable from LCH, fibrous dysplasia, enchondroma or even a sarcoma.

Osteosarcoma Mimicking a UBC

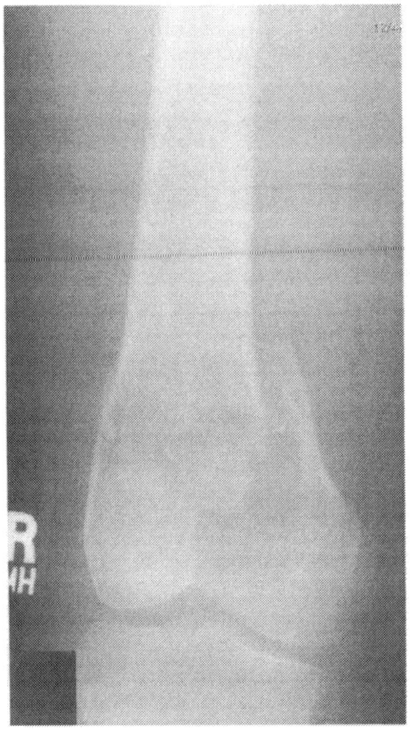

Figure 88H. AP radiograph of the right femur.

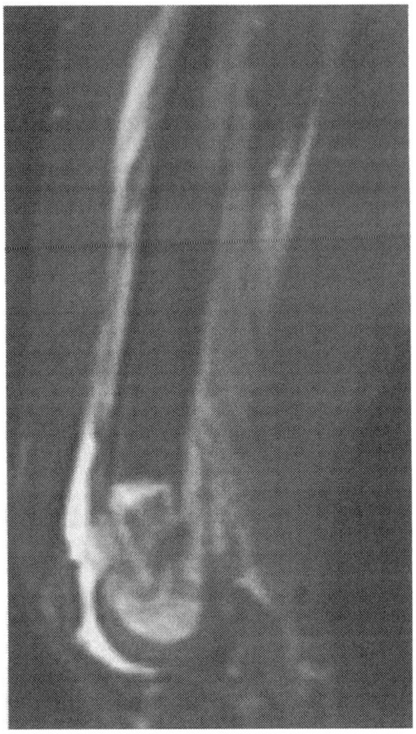

Figure 88I. Sagittal STIR.

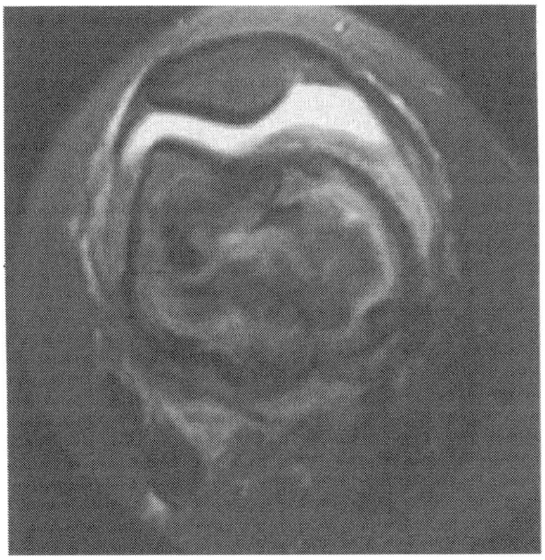

Figure 88J. Axial T2 FS through the distal femoral metaphysis.

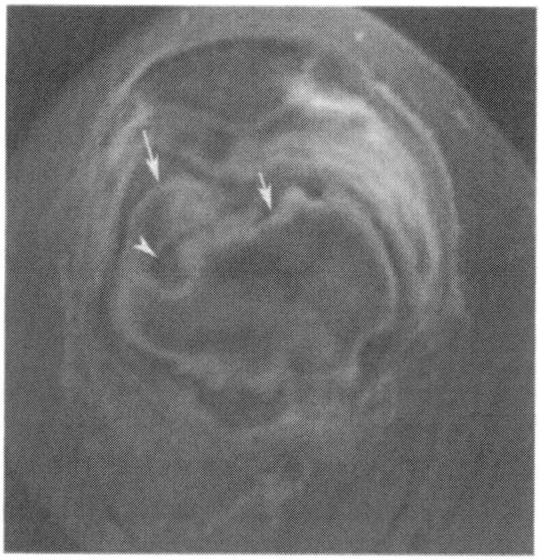

Figure 88K. Axial T1 post-Gd FS through the distal femur metaphysis.

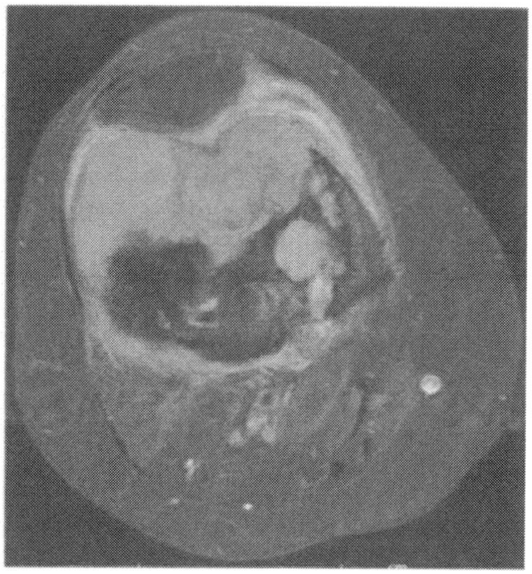

Figure 88L. Axial T1 post-Gd FS through the distal femur metaphysis (two years later).

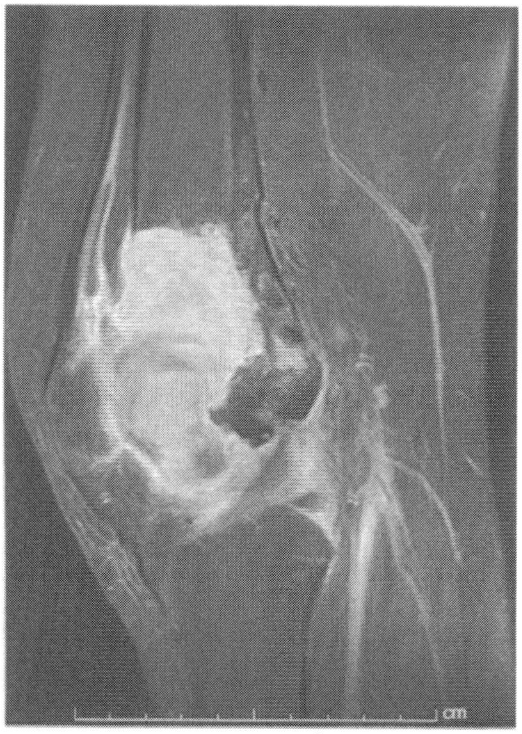

Figure 88M. Sagittal T1 post-Gd FS (two years later).

Findings

This is a 16-year-old girl who had minor right knee trauma.

Figure 88H. A comminuted, pathologic fracture through a lytic lesion of the distal right femoral metaphysis and epiphysis is evident.

Figures 88I, 88J. A joint effusion and diffuse edema is noted in the deep soft tissues. Internal heterogeneous SI is due to mixed blood products, but no fluid-fluid levels are evident.

Figure 88K. There is rim enhancement **(arrows)**, but no solid enhancing components are identified. A fallen fragment is identified on the MRI **(arrowhead)**, but is not appreciable on the radiograph. No discrete soft tissue mass is seen. Based on the imaging findings of a cystic metaphyseal lesion with substantial epiphyseal extension, an ABC was considered in addition to a UBC. Pathology of the curetted specimen was consistent with a UBC with pathologic fracture. A bridging external fixation device was applied, and subsequently the cyst was injected with demineralized bone and bone marrow.

Figures 88L, 88M. Because of persistent knee pain after apparent healing, a repeat MRI was performed. There is a large enhancing soft tissue mass present within the distal femur that extends into the knee joint. A repeat biopsy showed intermediate to high grade osteosarcoma.

Pitfalls and Pearls

1. A healing UBC may have aggressive characteristics on MRI. Get the original studies before drawing any conclusions.
2. On occasion, an aggressive lesion will simulate a UBC. In the absence of a fracture, be wary of periosteal reaction or enhancing soft tissue components.

References

1. Sullivan RJ, Meyer JS, Dormans JP, Davidson RS. Diagnosing aneurysmal and unicameral bone cysts with magnetic resonance imaging. *Clin Orthop Relat Res* 1999:186–190.
2. Meyer JS, Dormans JP. Differential diagnosis of pediatric musculoskeletal masses. *Magn Reson Imaging Clin N Am* 1998; 6:561–577.
3. Margau R, Babyn P, Cole W, Smith C, Lee F. MR imaging of simple bone cysts in children: Not so simple. *Pediatr Radiol* 2000; 30:551–557.

Case 89

History

This is a 4-year-old boy with left genu valgum. The right knee was normal (not shown).

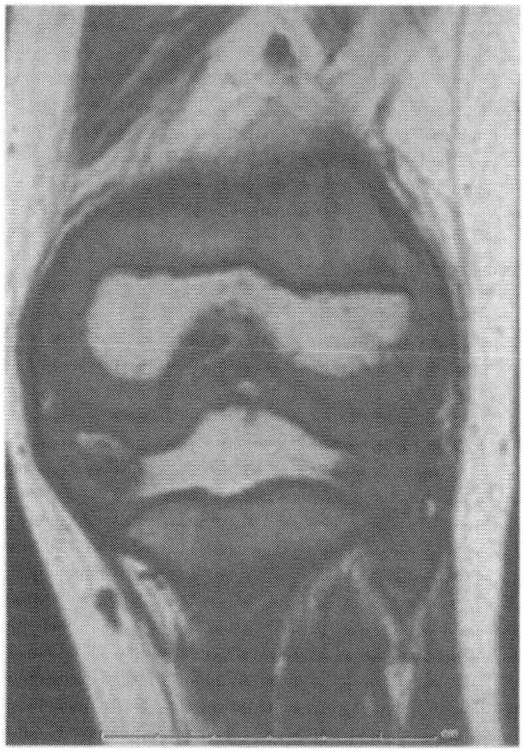

Figure 89A. Coronal T1 of the left knee.

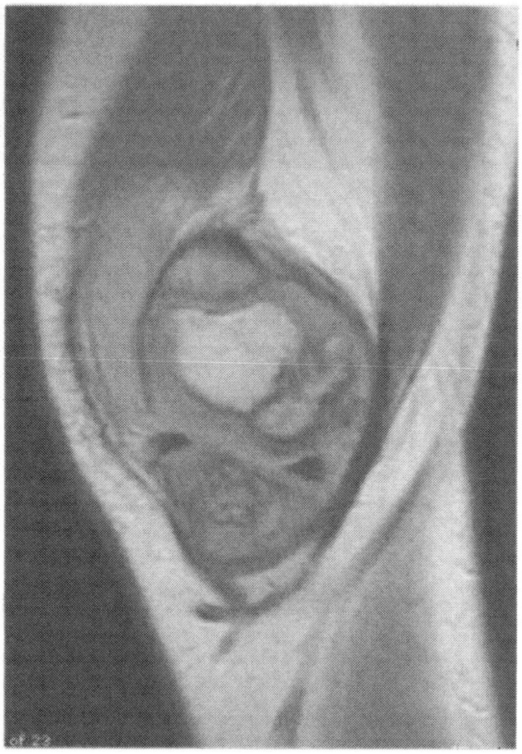

Figure 89B. Sagittal PD (medial compartment).

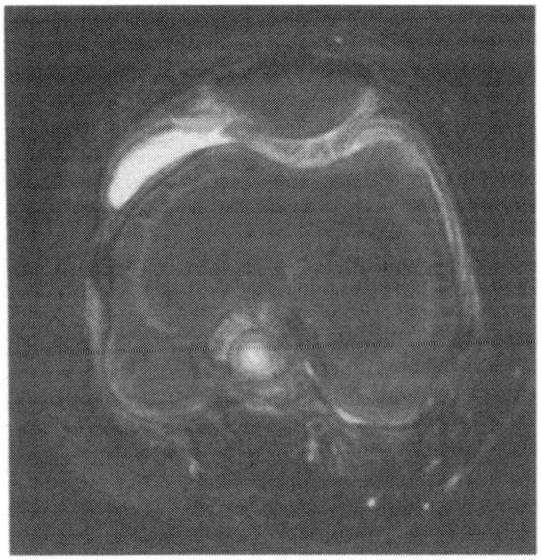

Figure 89C. Axial T2 FS.

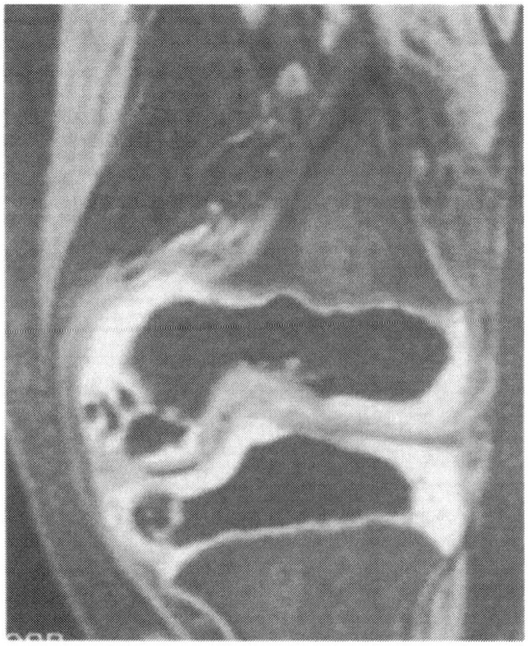

Figure 89D. Coronal-oblique 3D SPGR FS reformat.

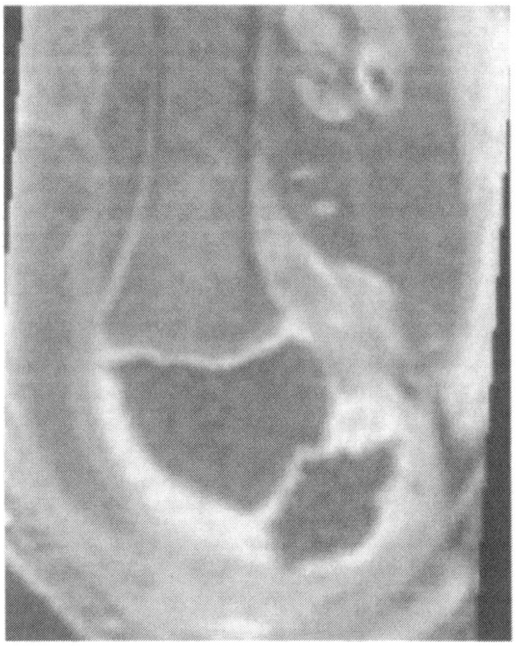

Figure 89E. Sagittal 3D SPGR FS reformat (medial compartment).

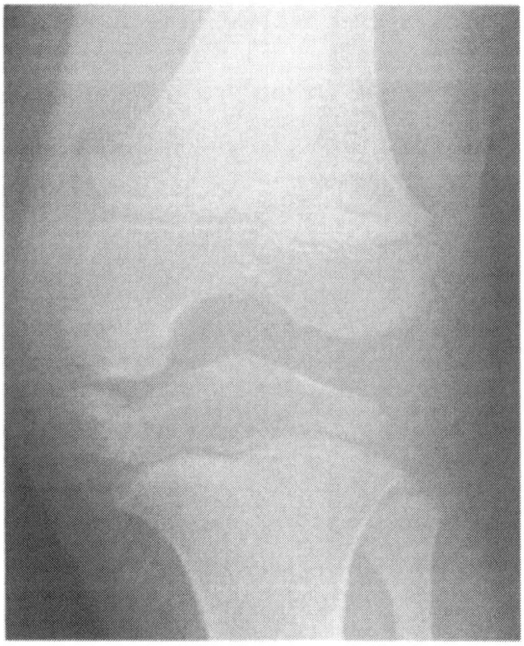

Figure 89F. AP radiograph.

Figures 89A, 89B, 89C, 89D, 89E, 89F. Two irregular bony excrescences are noted to arise from the medial femoral condyle and tibial plateau. The lesions have ossific centers containing fatty marrow and are surrounded by tissue that follows cartilage SI on all sequences. A moderate sized joint effusion is also identified.

Figure 89G. Less well developed epiphyseal dysplastic changes with irregular ossification are identified.

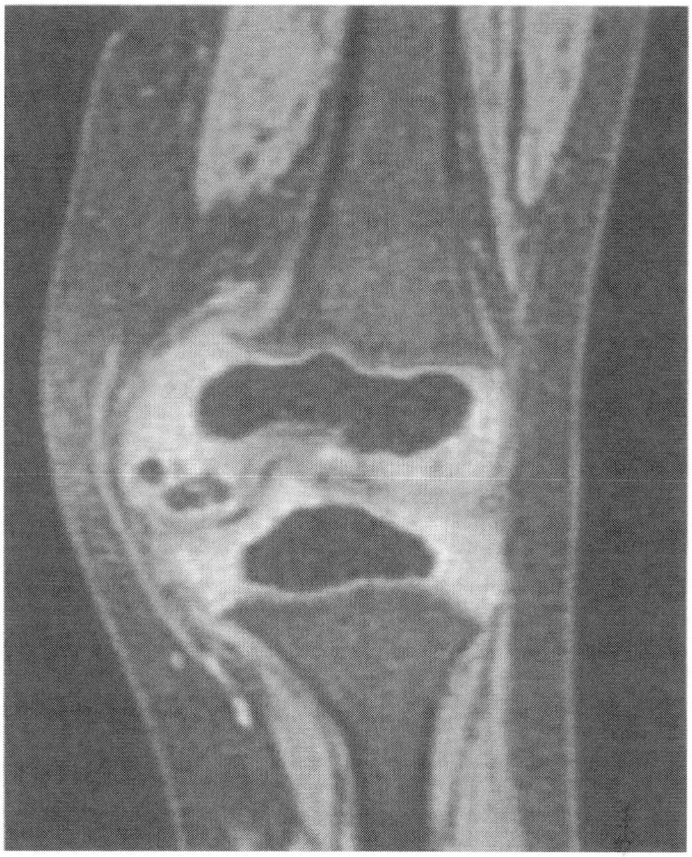

Figure 89G. Coronal-oblique 3D SPGR FS, 2 years earlier.

Diagnosis

Dysplasia epiphysealis hemimelica, classical form

Questions

1. What are the three forms of Trevor's disease and which is most common?
2. T/F: Trevor's disease preferentially involves the medial compartment of the lower extremity.

Discussion

Dysplasia epiphysealis hemimelica (AKA, Trevor's disease tarsal aclasis) is a rare condition characterized by single or multiple epiphyseal osteochondromas that most commonly affects the lower extremities. The typical age of presentation is from 2 to 14 years with a male : female ratio of 3 : 1 (1). It is a self-limited process, but can cause significant angular deformity and early degenerative changes if left untreated.

Trevor's disease may be separated into localized, classical (most common), and generalized forms (Answer to Question 1) (2). The localized form describes a single epiphyseal osteochondroma and most commonly occurs in the ankle or hindfoot (Figures 89H–89J). The classical form describes "kissing" or abutting epiphyseal osteochondromas involving a single joint, as present in this case. The generalized form is multiple epiphyseal osteochondromas involving several joints of a single extremity. In all forms of Trevor's disease, the epiphyseal osteochondromas are typically confined to either the medial or lateral compartment of the joint. The medial compartment is 2–3 times more frequently affected than the lateral compartment (Answer to Question 2) (1). Although Trevor's disease is most commonly found in the lower extremities, there are case reports of upper extremity involvement (3, 4).

On radiography Trevor's disease appears as an intra-articular mass arising from the epiphysis. Lobulated ossification may be identified within the lesion, representing aberrant secondary centers of ossification. The involved epiphysis usually demonstrates earlier maturation compared with the contralateral unaffected epiphysis (1). A varus or valgus joint deformity may result, depending on lesion location. With ossification, the lesion may mimic synovial osteochondromatosis or juxta-articular soft tissue calcifications related to myositis ossificans, tumoral calcinosis, or an extraskeletal osteosarcoma.

On MRI, the epiphyseal exostosis is continuous with the cartilaginous or ossified epiphysis with extension into the joint. On fluid sensitive sequences, the cartilaginous component of the exostosis may have slightly higher SI than the normal cartilage at the chondro-osseous junction due to the presence of a higher proteoglycan content, and thus higher water content (5, 6). The articulating surface of the exostosis may be lobulated and irregular. The secondary ossification center of the exostosis usually exhibits an SI similar to the adjacent epiphyseal ossification center. The secondary ossification center of the exostosis may be multiple and appear fragmented. When well formed, it may be continuous with the marrow of the epiphysis, or separated by a sharp or irregularly ossified cleavage plane.

The purpose of MRI evaluation of known Trevor's disease is to properly classify the lesion and to assess for complications. This may be helpful when the epiphyseal exostosis is predominantly cartilaginous and the radiographs underestimate the true size and extent of the lesion. Complications from Trevor's disease that are well delineated by MRI include early degenerative changes, osteocartilaginous loose bodies, physeal arrest, and marrow and soft tissue edema related to altered weightbearing (6, 7).

The unilateral genu valgus deformity and joint effusion in this patient were secondary to altered knee mechanics from the abutting osteochondromas. At surgery, the femoral epiphyseal osteochondroma was resected, as well as multiple small loose bodies, which were not seen on MRI.

Orthopedic Perspective

The diagnosis of Trevor's disease is usually apparent from radiographs. Small lesions are not problematic, but large lesions that involve the growth plate or articular surfaces are. Nonarticular portions of the exostosis can be resected if they are symptomatic or are causing an angular deformity, but if there is involvement of the articular surface with distortion as shown in the first case here, treatment is problematic. It is usually not possible to sculpt the exostosis from the secondary ossification center and maintain a smooth articular surface, and in many cases observation is the best option. The joint may adapt to the configuration of the exostosis and be superior to what the surgeon can create. Also, if resecting the exostosis involves violating the growth plate, angular deformities can arise from treatment. The 3D configuration of the lesion by CT or MRI reconstruction is also helpful when planning surgical excision.

What the Clinician Needs to Know

1. What is the extent of the exostosis with respect to the growth plate and articular surface, and are both sides of the joint involved?
2. Three dimensional imaging is useful in planning excision.

Answers

1. Localized, classical (most common), and generalized.
2. True.

Additional Example

Dysplasia Epiphysealis Hemimelica, Localized to the Talus

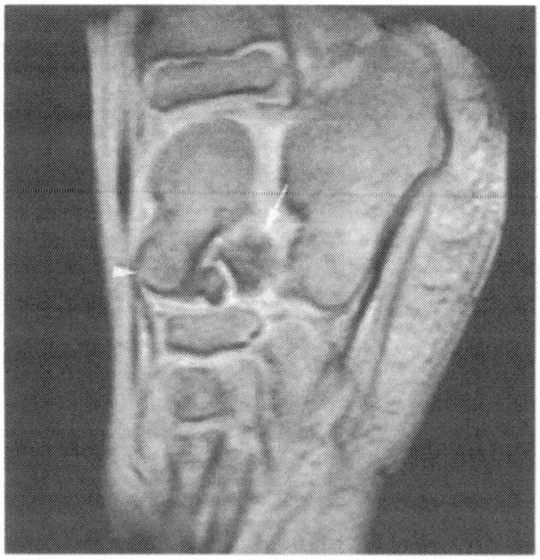

Figure 89H. Sagittal MPGR of the left foot.

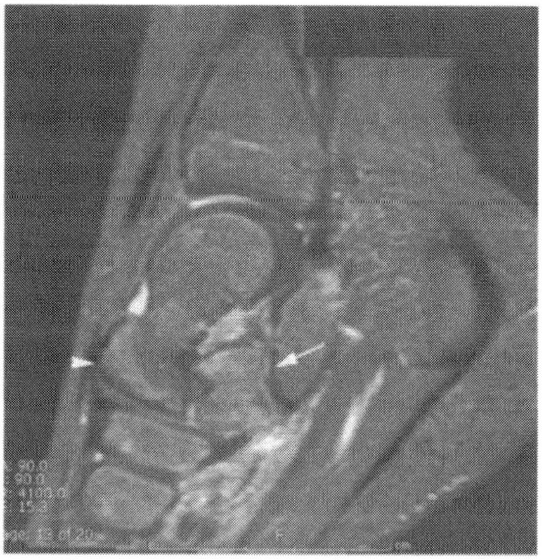

Figure 89I. Sagittal STIR.

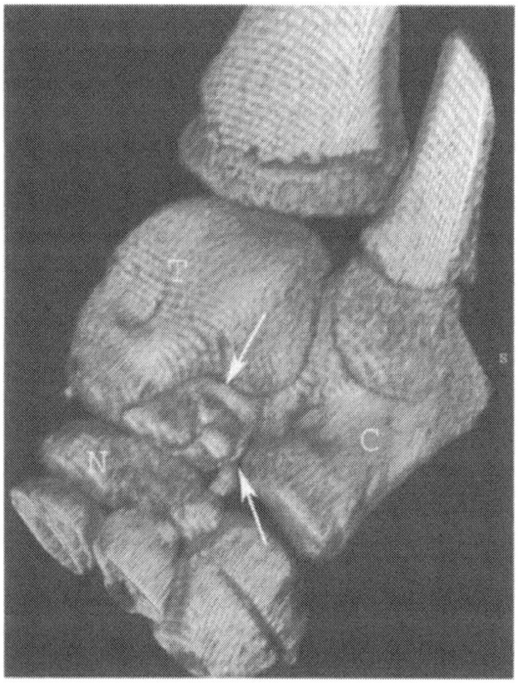

Figure 89J. CT with 3D reformat.

Findings

This is a 7-year-old boy with chronic left foot pain.

Figures 89H, 89I, 89J. There is a localized osteochondroma **(arrows)** arising from the talus that extends into the medial and anterior facets of the subtalar joint. There is talonavicular subluxation **(arrowheads)** due to elevation of the anterior talus by the mass. Edema within the accessory ossification center is identified on the STIR image. At surgery, a single talar osteochondroma was identified that contained multiple accessory ossification centers. The talar osteochondroma was resected and a talonavicular joint arthrotomy and chondroplasty of the talar head were performed. No loose bodies were identified. Talus (T), calcaneus (C), and navicular (N).

Pitfalls and Pearls

1. Unilateral genu valgum and medial joint space soft tissue density in the very young child may be the initial presentation of Trevor's disease.
2. Use MRI, in particular 3D SPGR FS sequences, to assess for early complications related to Trevor's disease, such as physeal arrest, early degenerative changes, and loose bodies.

References

1. Silverman FN. Dysplasia epiphysealis hemimelica. *Semin Roentgenol* 1989; 24:246–258.
2. Azouz EM, Slomic AM, Marton D, Rigault P, Finidori G. The variable manifestations of dysplasia epiphysealis hemimelica. *Pediatr Radiol* 1985; 15:44–49.
3. Rao SB, Roy DR. Dysplasia epiphysealis hemimelica: Upper limb involvement with associated osteochondroma. *Clin Orthop Relat Res* 1994:103–109.
4. Azouz EM, Slomic AM, Archambault H. Upper extremity involvement in Trevor disease. *J Can Assoc Radiol* 1984; 35:209–211.
5. Iwasawa T, Aida N, Kobayashi N, Nishimura G. MRI findings of dysplasia epiphysealis hemimelica. *Pediatr Radiol* 1996; 26:65–67.
6. Lang IM, Azouz EM. MRI appearances of dysplasia epiphysealis hemimelica of the knee. *Skeletal Radiol* 1997; 26:226–229.
7. Kuo RS, Bellemore MC, Monsell FP, Frawley K, Kozlowski K. Dysplasia epiphysealis hemimelica: Clinical features and management. *J Pediatr Orthop* 1998; 18:543–548.

Case 90

History

This is an 8-year-old boy with left knee pain for 1 month.

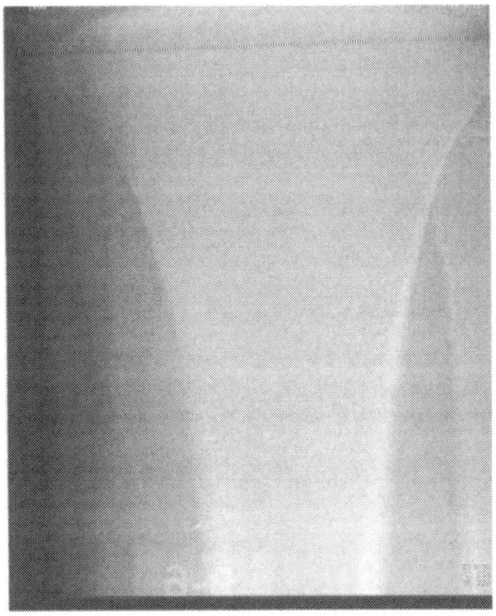

Figure 90A. AP radiograph of the left proximal tibia.

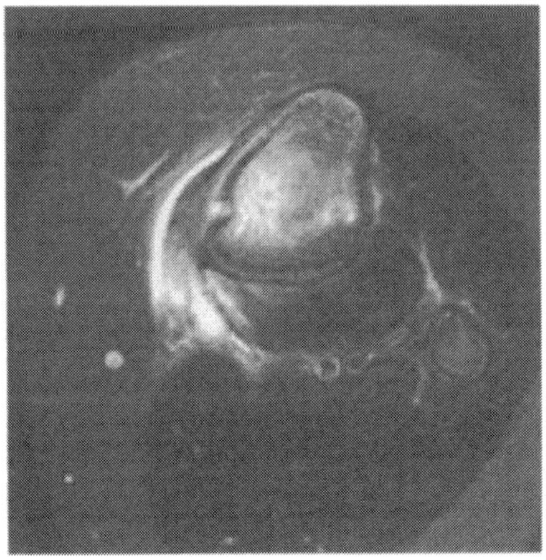

Figure 90B. Axial T2 FS.

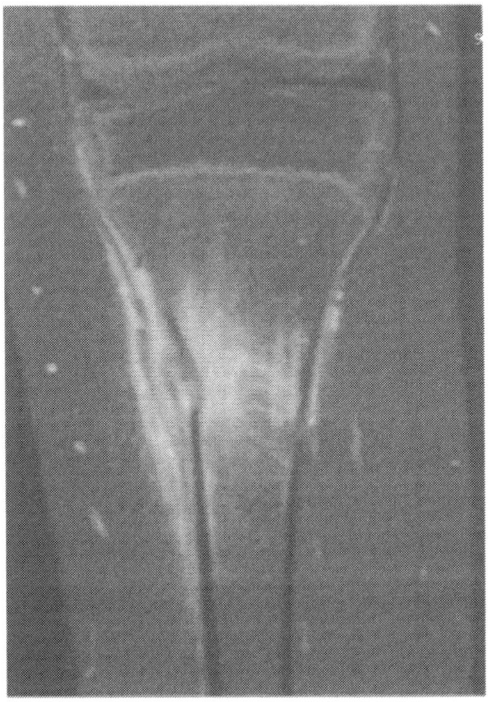

Figure 90C. Coronal T1 post-Gd FS.

Figure 90A. There is cortical indistinctness and faint periosteal reaction along the medial aspect of the proximal tibia metadiaphysis.

Figures 90B, 90C (90B, 90C with annotations). T2 hyperintensity and enhancement are seen in the medial metadiaphysis and adjacent soft tissues. There is periosteal elevation and enhancement **(arrowheads)** and reactive changes in the adjacent bone marrow. A nidus arising from the cortex is identified that is hyperintense on T2 and enhances **(arrows)**. Although there is juxtacortical edema and enhancement, no soft tissue mass is evident.

Figure 90D. CT demonstrates aggressive appearing periosteal reaction. A focal osteolytic lesion is identified within the cortex **(arrow)**, compatible with a nidus.

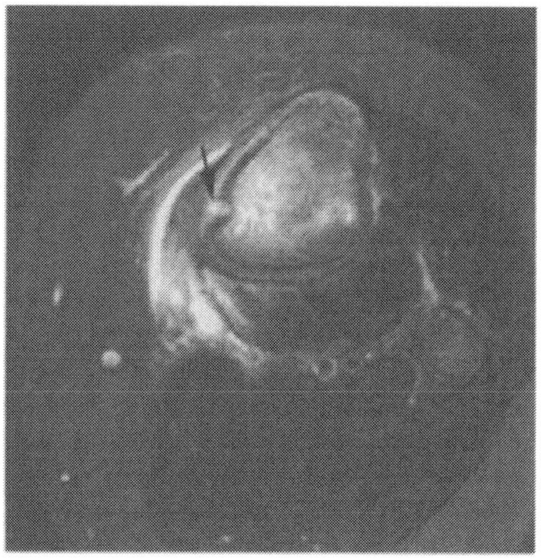

Figure 90B* Annotated.

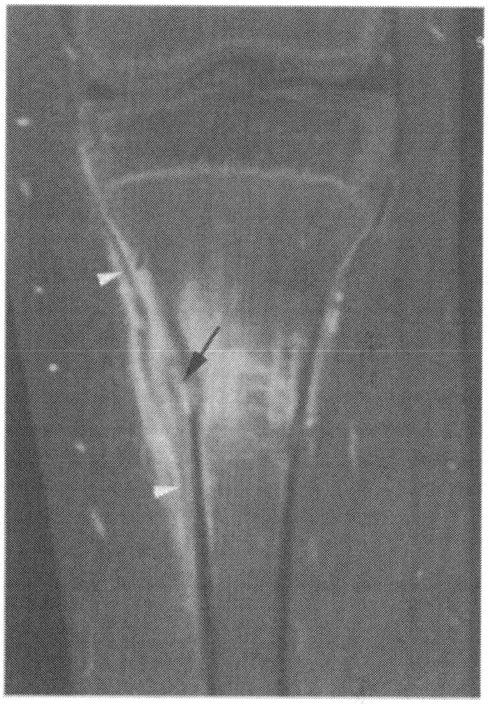

Figure 90C* Annotated.

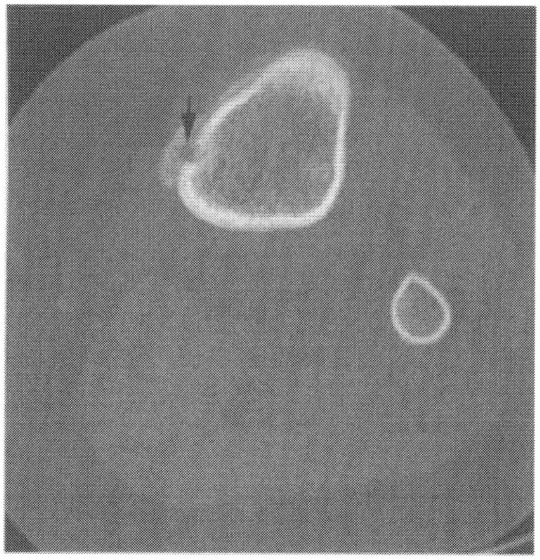

Figure 90D. CT.

Diagnosis

Osteoid osteoma

Questions

1. T/F: All osteoid osteomas arise from the bony cortex.
2. What are the most important differential considerations for an osteoid osteoma when a soft tissue mass is absent?

Discussion

The classic clinical presentation of osteoid osteomas is bone pain that is worse at night and is relieved with aspirin. The majority of patients are males and about half present between 10 to 20 years of age. Most osteoid osteomas arise from the diaphysis or meta-diaphysis. In a study of 225 cases of osteoid osteomas, the most common location was the femur, followed by the tibia (1). Osteoid osteomas may arise from the cortex (most common), medullary bone, and periosteum (least common) (Answer to Question 1). When greater than 1.5 cm in diameter, the lesions are generally considered osteoblastomas, which have an identical histology (Figures 90E–90H) (2). In contrast to osteoid osteomas, osteoblastomas have a predilection for the axial skeleton (osteoblastoma: 36%; osteoid osteoma: 10%) and have a higher incidence of recurrence after local excision (osteoblastoma: 10%; osteoid osteoma: 4.5%) (3).

On radiography, osteoid osteomas are characterized by solid periosteal reaction and bony proliferation surrounding a nidus. The nidus represents the osteoid osteoma, whereas the periosteal reaction and bony proliferation represent a secondary effect of the osteoid osteoma. The nidus may be radiolucent, partially or completely calcified, and is usually centered within the zone of thick periosteal reaction. Sometimes the nidus may be eccentrically positioned with respect to often exuberant endosteal reaction. Intra-articular osteoid osteomas tend to be the intramedullary subtype and often lack periosteal reaction, since perichondrium, rather than periosteum, covers the intra-articular cortex (1).

The typical MRI features of osteoid osteomas include marrow and extraosseous soft tissue edema, nidus, and periosteal reaction. The nidus may be hypointense on T1, hypo- or hyperintense on fluid sensitive sequences, and demonstrate enhancement (4). Soft tissue edema may be more pronounced with osteoid osteomas located close to the cortical surface and the nidus is usually found in the center of the edema (5). Although edema is characteristic of the lesion, Davies et al. showed that 6 out of 43 osteoid osteomas demonstrated minimal or no edema (6). In addition, they found an accuracy of only 65% (28/43 patients) for MRI in the identification of the osteoid osteoma nidus (6). On the other hand, Spouge et al. noted that MRI was able to identify the nidus in all 10 osteoid osteomas in their series, with 2 intra-articular lesions that were not initially seen by CT (4). Liu et al. identified the nidus in all 11 of their patients by using dynamic gadolinium enhanced MR imaging (7). In their study, the nidus showed rapid arterial enhancement and partial washout in 9 of 11 cases, and was distinguishable from marrow and extraosseous soft tissue edema, which demonstrated delayed enhancement.

MRI is limited as a primary investigative tool for the evaluation of osteoid osteomas. It may suggest other neoplastic or inflammatory process since the nidus may be inconspicuous and soft tissue and marrow edema are nonspecific. Therefore, radiographic and CT correlation is often essential to arrive at the correct diagnosis. The MRI diagnosis of an intra-articular osteoid osteoma may pose an even greater challenge since effusion and synovitis may predominate, and periosteal reaction and bony sclerosis may be absent.

The main differential considerations for osteoid osteoma in the absence of a discrete mass are stress injury and osteomyelitis (Answer to Question 2). A linear hypointense zone on T1 or PD images that is perpendicular to the cortex is a feature of a stress fracture that permits differentiation from an osteoid osteoma. Intra-articular osteoid osteomas may mimic chondroblastoma, epiphyseal osteomyelitis, and pyogenic and nonpyogenic arthritis.

In this patient, the diagnosis of an osteoid osteoma was made preoperatively based on the combination of CT and MRI features. The patient underwent curettage of the lesion and subsequent relief of symptoms.

Orthopedic Perspective

The main challenge for the orthopedist is to find the tumor. Osteoid osteoma can present with a variety of confusing symptoms. In addition to the classic pain syndrome worse at night and relieved by aspirin, they may present as a joint effusion, painful scoliosis, limb-length discrepancy, or local gigantism in a digit. Some patients have been considered to have psychiatric problems until the nidus was located, and occasionally the first presentation is a gastrointestinal bleed from aspirin consumption. The unsuspecting clinician may order an MRI that may show soft tissue edema surrounding the lesion and mimic a soft tissue sarcoma or even a bone sarcoma due to extensive periosteal reaction. If the diagnosis of osteoid osteoma is not considered, unnecessary biopsies may be performed. Imaging is also useful in planning surgical approaches and documenting complete resection or curettage, but currently osteoid osteomas are treated primarily by radiofrequency heat ablation, obviating the need for extensive operations. One useful technique is to employ intraoperative nuclear medicine localization for difficult lesions such as the spine. By imaging before and after excision, a complete eradication can be assured.

What the Clinician Needs to Know

1. Distinguish osteoid osteoma from other lesions, especially stress fracture, osteomyelitis, osteoblastoma, and aggressive neoplasm.
2. Avoid misinterpretation of MRI findings that may suggest bone or soft tissue sarcoma. CT is most helpful in this regard.
3. The best approach is surgical excision if necessary. Document complete resection intraoperatively with radionuclide imaging.

Answers

1. False.
2. Stress injury and osteomyelitis.

Additional Example

Osteoblastoma of the C5 Pedicle

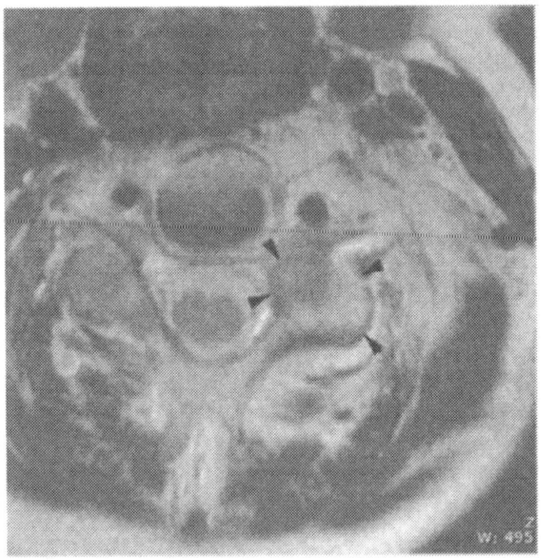

Figure 90E. Axial T2 at the level of C5.

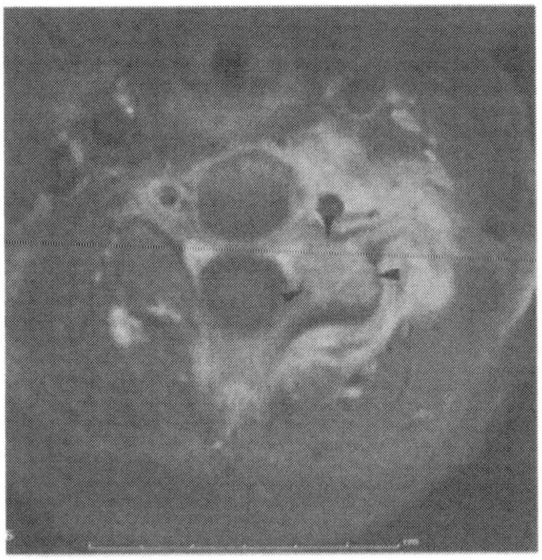

Figure 90F. Axial T1 post-Gd FS.

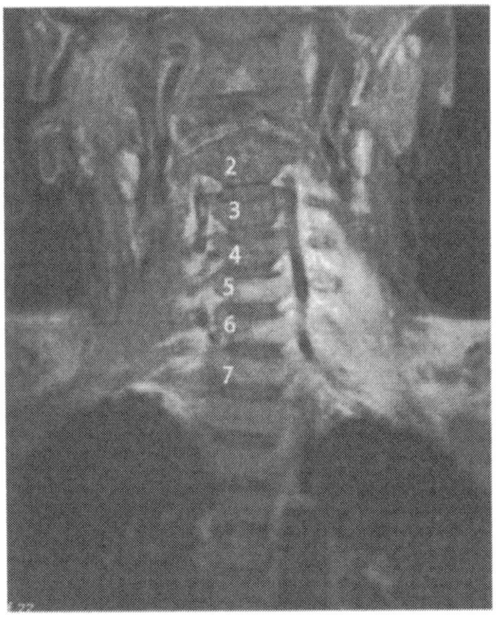

Figure 90G. Coronal T1 post-Gd FS.

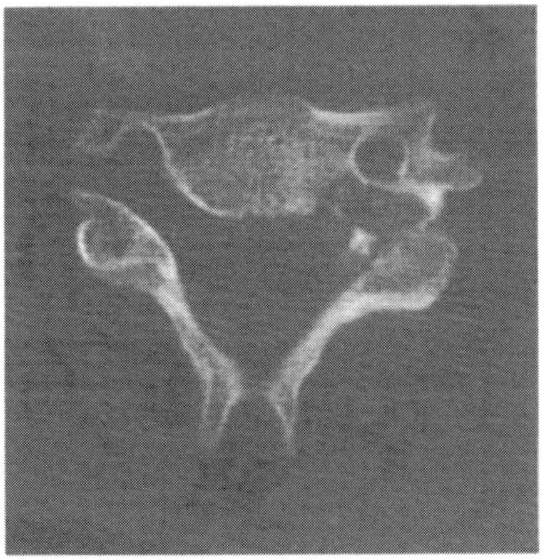

Figure 90H. CT.

Findings

This is a 12-year-old boy with a long standing history of neck stiffness.

Figures 90E, 90F, 90G. There is a mildly expansile mass in the left pedicle of C5 that is heterogeneously hyperintense on T2 and enhances **(arrowheads)**. There is significant soft tissue and marrow edema. On the coronal sequence, marrow enhancement extends to involve the C5 and C6 vertebral bodies.

Figure 90H. Matrix calcification is seen in an osteolytic lesion associated with cortical disruption. This lesion was pathologically proven to be an osteoblastoma.

Pitfalls and Pearls

1. Be sure to review radiographs at the time of the MRI.
2. For suspected osteoid osteoma, CT is the preferred method of evaluation.
3. The MRI features of osteoid osteoma may be potentially misleading when a nidus is not identified. The nonspecific soft tissue and intraosseous edema generated from osteoid osteomas may mimic infection, trauma, and even sarcoma.

References

1. Kransdorf MJ, Stull MA, Gilkey FW, Moser RP, Jr. Osteoid osteoma. *Radiographics* 1991; 11:671–696.
2. Woertler K. Benign bone tumors and tumor-like lesions: Value of cross-sectional imaging. *Eur Radiol* 2003; 13:1820–1835.
3. Jackson RP, Reckling FW, Mants FA. Osteoid osteoma and osteoblastoma: Similar histologic lesions with different natural histories. *Clin Orthop Relat Res* 1977:303–313.
4. Spouge AR, Thain LM. Osteoid osteoma: MR imaging revisited. *Clin Imaging* 2000; 24:19–27.
5. Nogues P, Marti-Bonmati L, Aparisi F, Saborido MC, Garci J, Dosda R. MR imaging assessment of juxta cortical edema in osteoid osteoma in 28 patients. *Eur Radiol* 1998; 8:236–238.
6. Davies M, Cassar-Pullicino VN, Davies AM, McCall IW, Tyrrell PN. The diagnostic accuracy of MR imaging in osteoid osteoma. *Skeletal Radiol* 2002; 31:559–569.
7. Liu PT, Chivers FS, Roberts CC, Schultz CJ, Beauchamp CP. Imaging of osteoid osteoma with dynamic gadolinium-enhanced MR imaging. *Radiology* 2003; 227:691–700.

Case 91

History

This is a 12-year-old boy with right thigh pain for 3 months.

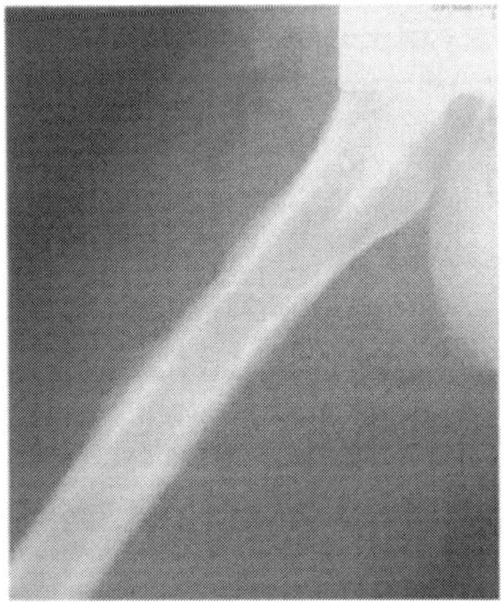

Figure 91A. Frog lateral radiograph of the right femur.

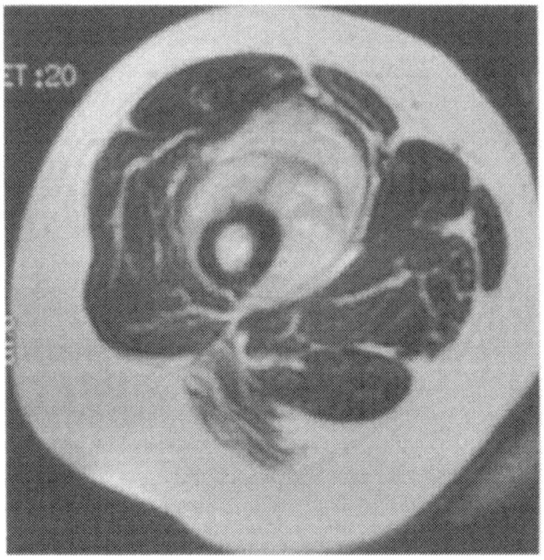

Figure 91C. Axial T2 of the proximal diaphysis.

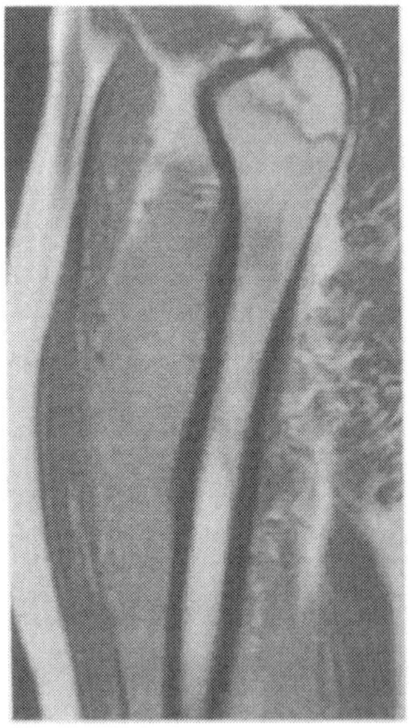

Figure 91B. Sagittal T1.

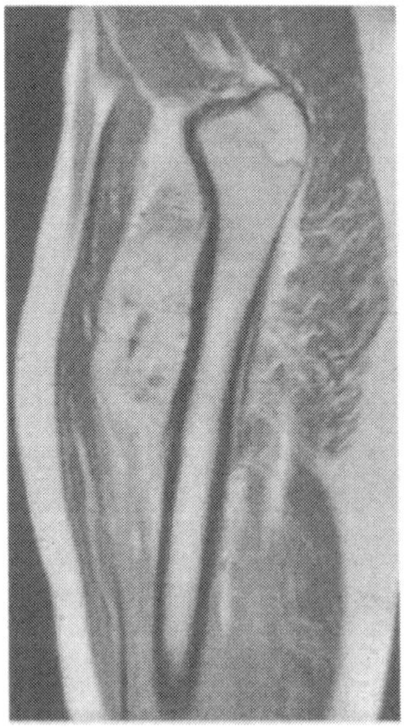

Figure 91D. Sagittal T1 post-Gd.

Figure 91A. Interrupted periosteal new bone formation is noted along the anterior surface of the femoral diaphysis and fine periosteal new bone is present posteriorly. No gross cortical or medullary destruction is evident.

Figures 91B, 91C, 91D (91C with annotations). A large juxtacortical soft tissue mass is identified that is slightly hyperintense on T1, moderately hyperintense on T2, and demonstrates significant tumoral enhancement. The mass surrounds approximately 60% of the femur. There is mild hypointensity on T1 within the marrow in the proximal diaphysis. On the T2W image, there are faint hypointense striations within the mass that run perpendicular to the cortex **(arrowheads)**. Cortical erosion **(arrow)** is evident, but the mass and marrow appear separate.

Figure 91E. Within a slightly hypodense extraosseous soft tissue mass **(arrows)**, there is perpendicular (hair-on-end) periosteal reaction **(arrowhead)**, corresponding to the striated hypointensities noted in Figure 91C.

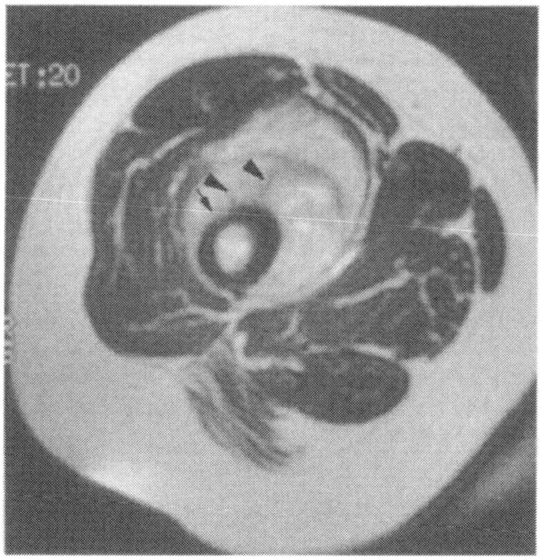

Figure 91C* Annotated.

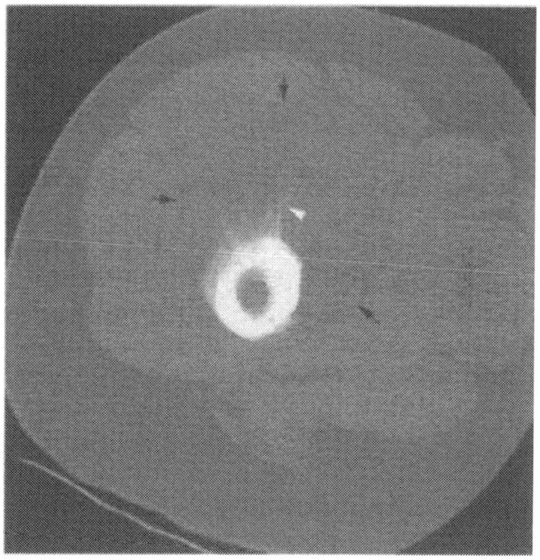

Figure 91E. Axial CT through the proximal diaphysis of the right femur.

Diagnosis

Periosteal osteosarcoma

Questions

1. What are the four subtypes of surface osteosarcomas?
2. T/F: Medullary extension of periosteal osteosarcomas is common.

Discussion

Periosteal osteosarcoma accounts for approximately 25% of all juxtacortical osteosarcomas. The other three subtypes of surface osteosarcomas are intracortical (rarest), parosteal (most common, 65%), and high grade surface osteosarcomas (10%) (Answer to Question 1) (1). Periosteal osteosarcoma, like conventional intramedullary osteosarcoma, most often occurs in patients 20 to 30 years of age and most commonly involves the femur and tibia. (2). Periosteal osteosarcoma has a better prognosis than conventional osteosarcoma. The tumor usually occurs in the diaphysis or metadiaphysis, whereas conventional osteosarcoma tends to involve the metaphysis.

Parosteal osteosarcoma is more common than periosteal osteosarcoma and tends to occur in older patients (30 to 50 years of age) with a female predilection. It has a better prognosis compared with periosteal and conventional osteosarcoma, and is characteristically found in the posterior aspect of the distal femoral metaphysis (3).

On radiography, periosteal osteosarcomas may demonstrate both benign appearing (thick and regular) and aggressive (hair-on-end) periosteal reaction (2). Cloud-like osteoid calcification may be identified associated with a soft tissue mass. In Murphey et al.'s series of periosteal osteosarcomas, cortical thickening with cortical scalloping was present in 68% of cases (Figure 91A). Hair-on-end periosteal reaction often extended into an extraosseous soft tissue mass (2). The medullary cavity is grossly preserved since periosteal osteosarcomas are usually confined to the surface of bone and rarely demonstrate bony invasion.

On MRI, the classic appearance of a periosteal osteosarcoma is a well-defined surface based extraosseous soft tissue mass without medullary canal invasion (Figures 91F–91H). A median of 50% to 55% circumferential encasement of cortical bone was observed in Murphey et al.'s series (2). The rare instance of intramedullary invasion should be distinguished from reactive marrow edema (Answer to Question 2). In Murphey et al.'s series, marrow SI alternations in the absence of direct tumor invasion were evident in 67% (8 of 12) cases (2). Reactive marrow edema is favored if the cortex is intact, marrow SI alterations are feathery without evidence of a discrete mass, and the surface tumor is not continuous with the marrow signal abnormality. The soft tissue component of periosteal osteosarcomas is usually heterogeneously isointense on T1 and heterogeneously hyperintense on fluid sensitive sequences due to the high water content. Periosteal osteosarcomas often have a large cartilaginous component that may lead to an erroneous pathologic diagnosis of chondrosarcoma, but these lesions are more properly characterized as chondroblastic osteosarcomas.

The alternative considerations for this lesion include diaphyseal conventional osteosarcoma, Ewing's sarcoma, periosteal chondrosarcomas, and trauma. Osteochondroma, bizarre parosteal osteochondromatous proliferation (BPOP), or a periosteal chondroma would not be considered here since aggressive periosteal reaction was evident on radiography, and the presence of an intact underlying cortex excludes the possibility of a sessile osteochondroma. Ewing's sarcoma would be a distinct possibility in this case, but the lack of significant medullary involvement and permeative bone destruction would be atypical. The aggressive, interrupted periosteal reaction and enhancing soft tissue mass would exclude a traumatic etiology. Although conventional osteosarcomas are more often metaphyseal, an osteosarcoma arising from the diaphysis is more likely to represent a conventional osteosarcoma (Figures 91I–91K) than a periosteal osteosarcoma. Periosteal osteosarcoma may be excluded if cross-sectional imaging points to a medullary origin of the tumor.

In this patient, the tumor was pathologically proven to be periosteal osteosarcoma (chondroblastic predominant). Metastatic work-up was negative. He received 4 months of chemotherapy and then underwent radical resection with preservation of the proximal femur and reconstruction with an intercalary allograft.

Orthopedic Perspective

The evaluation, diagnosis, and biopsy of patients with surface osteosarcomas are similar to the other sarcomas discussed in this text. The main concern for the orthopedist is establishing the diagnosis with histological grade of the lesion, and determining whether the medullary cavity is involved with tumor. It is usually possible to distinguish a low grade parosteal osteosarcoma because it presents as a lobulated bony mass on the surface of the bone, usually the distal femur. With time these may penetrate the cortex and this must be evaluated prior to treatment that is surgical alone. At times a hemicortical resection can be performed, but the extension of the lesion on the surface of the bone and into the adjacent joint must be carefully assessed. In periosteal osteosarcoma, response to chemotherapy can be assessed by imaging and, as in this case, a good response can facilitate surgical excision. As in parosteal osteosarcoma, the precise extension of the tumor around the bone and any medullary extension must be precisely defined.

What the Clinician Needs to Know

1. The extent of the tumor on the surface of the bone.
2. The presence, if any, of extension into the medullary cavity and into the adjacent joint.

Answers

1. Periosteal, intracortical, parosteal, and high grade surface.
2. False.

Additional Examples

Periosteal Osteosarcoma of the Tibia

(Courtesy of Jae Suk Oh, BS and Daniel Rosenthal, MD)

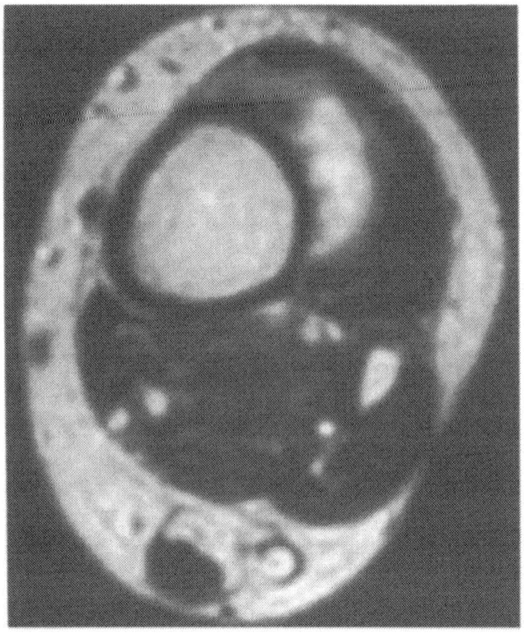

Figure 91F. Axial T2 of the left proximal tibia diaphysis.

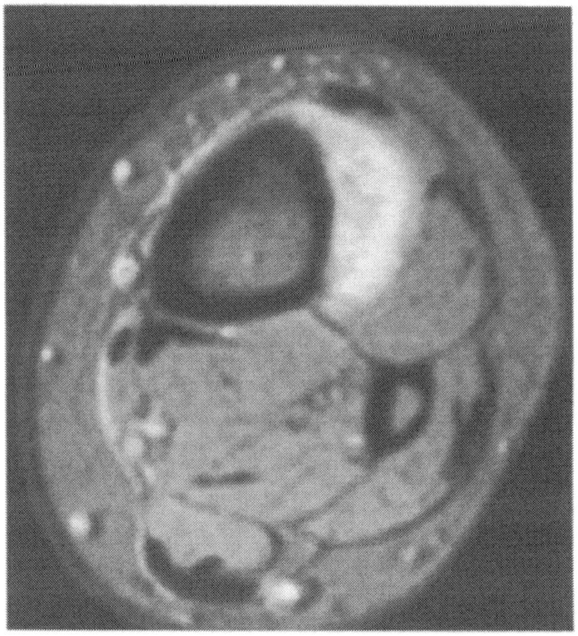

Figure 91G. Axial T1 post-Gd FS.

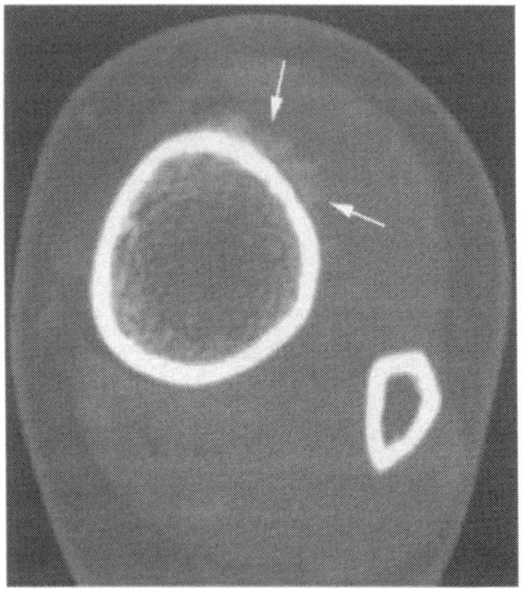

Figure 91H. CT.

Findings

This is a 14-year-old girl with a growing anterior tibia mass.

Figures 91F, 91G. There is a surface based mass that is hyperintense on T2 and demonstrates diffuse enhancement. The mass partially encircles the anterolateral cortex of the tibia. The underlying cortex and marrow are normal.

Figure 91H. Hair-on-end periosteal reaction is evident on CT **(arrows)**. The underlying cortex is intact. This was a pathologically proven periosteal osteosarcoma.

Conventional Osteosarcoma, Diaphyseal Involvement

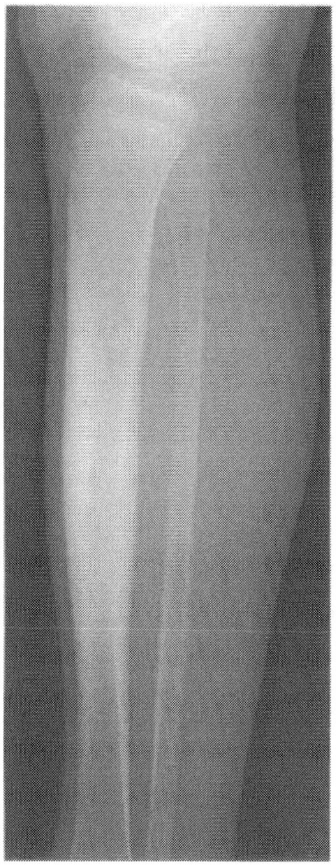

Figure 91I. Lateral radiograph of the right tibia.

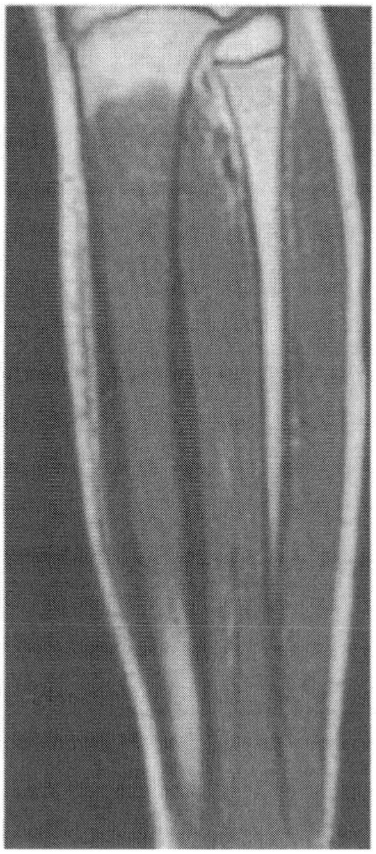

Figure 91J. Sagittal T1.

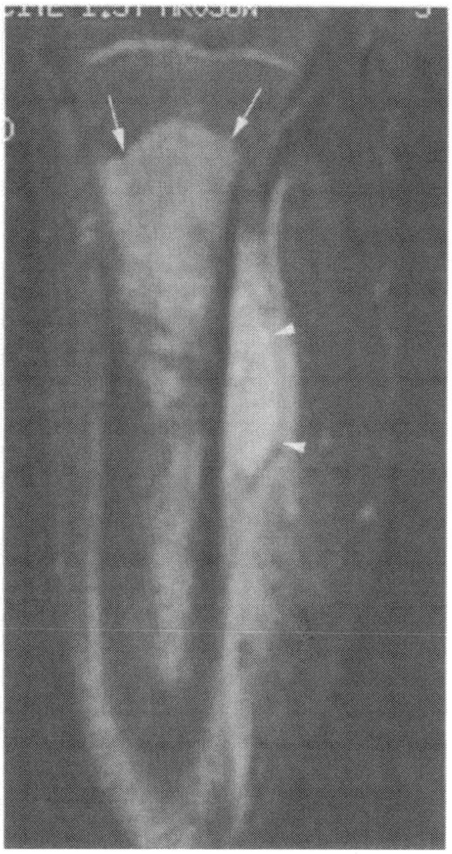

Figure 91K. Coronal T2 FS.

Findings

This is an 11-year-old boy who complained of right calf pain for 3 weeks.

Figure 91I. There is diffuse cortical thickening and medullary sclerosis seen in the mid- and proximal diaphysis of the tibia. The regular organized cortical thickening suggests an indolent process.

Figures 91J, 91K. Juxtacortical soft tissue extension is present **(arrowheads)**. Diffuse tumoral replacement of the marrow with sharply defined margins **(arrows)** is seen. This was a pathologically proven diaphyseal conventional osteosarcoma.

Pitfalls and Pearls

1. In children, the high chondroid matrix composition of periosteal osteosarcomas should not be confused with periosteal chondrosarcomas on imaging or histology. Periosteal chondrosarcomas is a disease seen in older patients (40s–50s).
2. Don't exclude the diagnosis of osteosarcoma just because an aggressive lesion is situated in the diaphysis of a long bone.

References

1. Murphey MD, Robbin MR, McRae GA, Flemming DJ, Temple HT, Kransdorf MJ. The many faces of osteosarcoma. *Radiographics* 1997; 17:1205–1231.
2. Murphey MD, Jelinek JS, Temple HT, Flemming DJ, Gannon FH. Imaging of periosteal osteosarcoma: Radiologic-pathologic comparison. *Radiology* 2004; 233:129–138.
3. Raymond AK. Surface osteosarcoma. *Clin Orthop Relat Res* 1991:140–148.

Case 92

History

This is an 11-year-old girl with a 6-month history of left knee pain with activity.

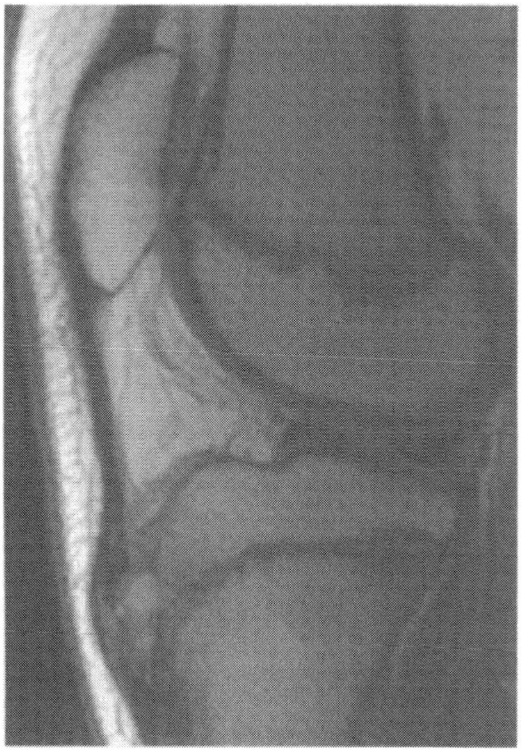

Figure 92A. Sagittal PD of the left knee.

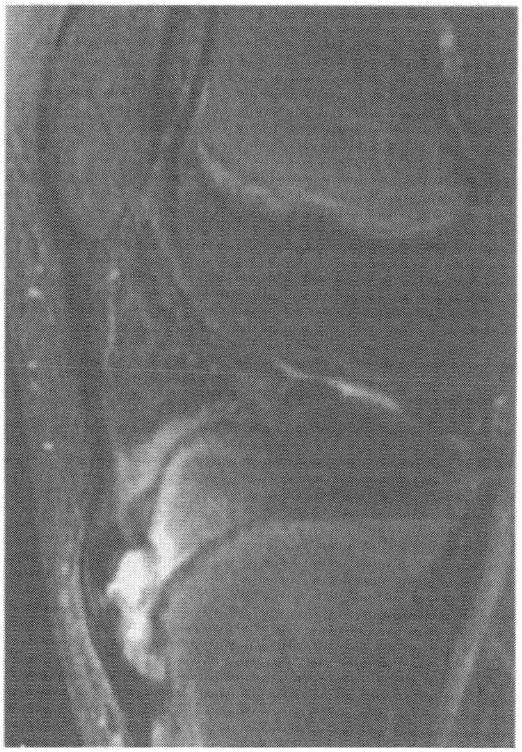

Figure 92B. Sagittal T2 FS.

Figure 92A. There is irregularity of the tibial tubercle with thickening and heterogeneous SI in the distal patellar tendon.

Figures 92B (92B with annotations). There is diffuse hyperintensity in the tibial tubercle apophysis, extending to the anterior aspect of the tibial epiphysis and Hoffa's fat pad **(arrow)**.

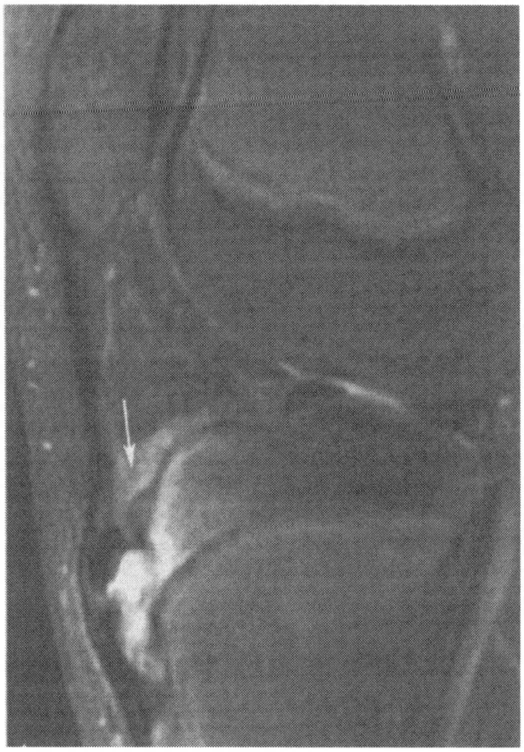

Figure 92B* Annotated.

Diagnosis

Osgood-Schlatter disease

Questions

1. What is the weakest component of the patella tendon extensor mechanism in children?
2. T/F: Patella alta is associated with Osgood-Schlatter disease.

Discussion

The MRI spectrum of overuse syndromes of the patella tendon includes two osteochondroses and patella tendinopathy. The two osteochondroses are Osgood-Schlatter disease (tibial tubercle) and Sinding-Larsen-Johansson disease (inferior patella pole) (Figure 92C, 92D). *Jumper's knee* (patella tendinopathy) is a separate term that may be used when there is patella tendinopathy without inferior patella pole or tibial tubercle edema (Figure 92E).

Osgood-Schlatter disease affects children 10 to 15 years of age (1). Up to 50% of cases may be bilateral (2). The etiology of Osgood-Schlatter disease remains somewhat controversial, but the generally accepted etiology is a chronic avulsive injury at the tibial tubercle insertion of the patella tendon. The chondro-osseous junction of the tibial tubercle apophysis is considered the weakest component of the patella tendon extensor mechanism in the immature skeleton (Answer to Question 1). Therefore, injury to the chondro-osseous junction is more likely to occur before patella tendon injury.

The plain radiographic findings of Osgood-Schlatter disease include: tibial tubercle fragmentation, overlying soft tissue density, obscuration of the patella tendon, increased density in Hoffa's fat pad, and patella alta (Answer to Question 2). Acute avulsion fractures of the tibial tubercle in children with known Osgood-Schlatter disease have also been reported (3). The tibial tubercle apophysis begins to ossify at 8–12 years in girls and 9–14 years in boys, and ossification may appear fragmented (4). Thus the radiographic diagnosis of Osgood-Schlatter disease is not possible in the absence of patellar tendon thickening or a clearly displaced secondary ossification center of the apophysis.

Although Osgood-Schlatter disease is generally a clinical diagnosis, MRI may be used in problematic cases. The MRI criteria for diagnosing Osgood-Schlatter disease include inflammatory changes of the tibial tubercle apophysis and the distal patella tendon insertion (5). MRI can also delineate edema in the anterior tibia epiphysis, Hoffa's fat pad, and the infrapatellar bursae. The anterior tibial epiphysis may be affected since the tibial tubercle apophysis is an extension of the tibial epiphysis and reactive marrow edema may extend posteriorly to this region (Figure 92F).

This patient was given a brace and asked to limit her activities. Her symptoms from Osgood-Schlatter disease resolved with conservative therapy.

Orthopedic Perspective

Extensor mechanism dysfunction is a common cause of knee pain in the pediatric athlete. Pain can occur at multiple locations along the extensor mechanism including quadriceps tendonitis, patellofemoral dysfunction, Sinding-Larsen-Johansson syndrome, patellar tendonitis (jumper's knee), and Osgood-Schlatter disease.

Osgood-Schlatter disease is usually diagnosed clinically and MRI is reserved for those cases with atypical presentation or who are unresponsive to nonoperative management. Nonoperative treatment is the mainstay of treatment with activity restriction, physical therapy, and bracing/straps as needed. Upon closure of the tibial tubercle apophysis, symptoms usually resolve. Residual pain can persist in cases with a large bony prominence or ossicle(s). Occasionally, a more acute presentation of tibial tubercle pain can be superimposed on chronic Osgood-Schlatter disease symptoms. These cases are treated as acute tibial tubercle apophyseal fractures with cast immobilization.

What the Clinician Needs to Know

1. Distinguish between an acute avulsion fracture of the tibial tubercle and Osgood-Schlatter disease.
2. Distinguish between an acute inferior patella sleeve fracture, Sinding-Larsen-Johansson syndrome, and normal developmental variant.
3. Assess the entire knee extension mechanism for other causes for anterior knee pain.

Answers

1. Chondro-osseous junction of the tibial tubercle.
2. True.

Additional Examples

Sinding-Larsen-Johansson Disease

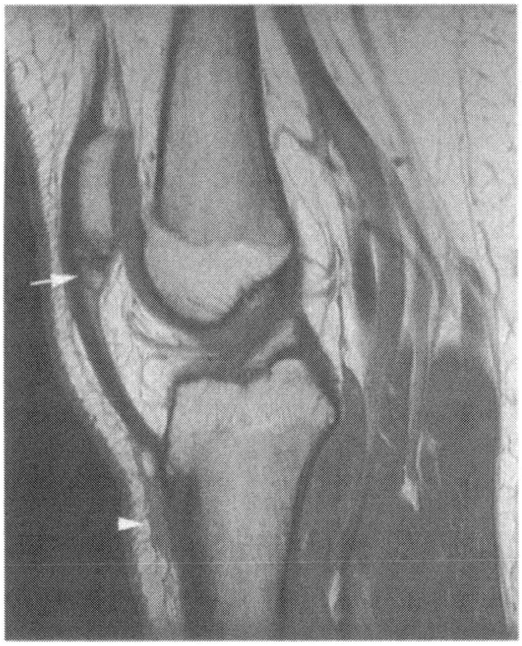

Figure 92C. Sagittal PD.

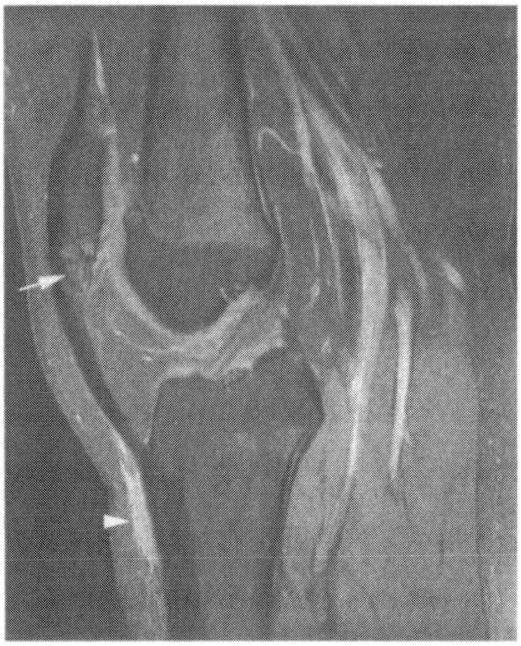

Figure 92D. Sagittal STIR.

Findings

This is a 15-year-old girl with chronic anterior knee pain.

Figures 92C, 92D. There is fragmentation and edema of the inferior patella **(arrows)**, classic features of Sinding-Larsen-Johansson disease. There is also prepatella subcutaneous edema anterior to the tibial tubercle **(arrowheads)**.

Jumper's Knee

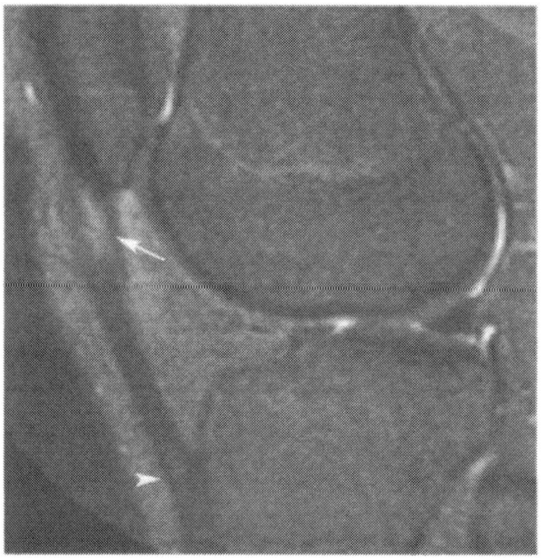

Figure 92E. Sagittal STIR of the right knee.

Findings

This is a 15-year-old girl with infrapatellar pain following a knee injury.

Figure 92E. There is increased SI in the proximal patella tendon **(arrow)** and the adjacent Hoffa's fat pad near the patella tendon insertion. The patellar secondary ossification center and cartilage are normal. Therefore, the diagnosis is a jumper's knee rather than Sinding-Larsen-Johansson disease. Incidental note of magic angle artifact in a normal thickness inferior patella tendon, near the tibial tubercle **(arrowhead)**.

Osgood-Schlatter Disease

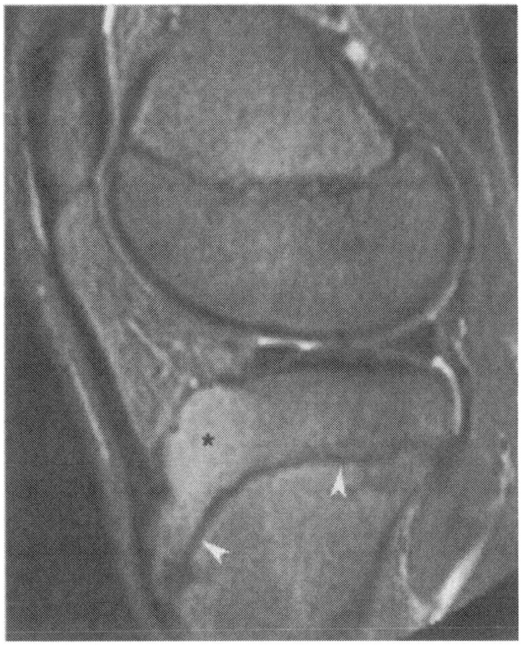

Figure 92F. Sagittal STIR of the right knee.

Findings

This 15-year-old boy had chronic anterior knee pain.

Figure 92F. There is increased fluid SI present in the tibial tubercle that extends into the anterior tibial epiphysis (*). Note that the proximal tibial physis extends anteriorly and undercuts the tibial tubercle **(arrowheads)**. Since the findings in this case are restricted to the secondary ossification center, the term *tibial tubercle apophysitis* is applicable.

Pitfalls and Pearls

1. MRI and radiographic findings of isolated tibial tubercle fragmentation, without soft tissue or marrow edema, is most likely a normal developmental variation; however, old inactive Osgood-Schlatter disease can produce a similar appearance.
2. Sinding-Larsen-Johansson and Osgood-Schlatter disease are both considered osteochondroses. MRI diagnosis requires fragmentation and signal abnormality of the inferior patella pole or tibial tubercle. If the signal abnormality is confined to the patella tendon, the term jumper's knee is more appropriate.

References

1. Harcke HT, Mandell GA, Maxfield BA. Trauma to the growing skeleton. In: *Caffey's Pediatric Diagnostic Imaging,* 10th ed., Kuhn JP, Slovis TL, Haller JO, eds., Section 9, Part 8. Philadelphia: Mosby, 2003; 2269–2303.

2. Stevens MA, El-Khoury GY, Kathol MH, Brandser EA, Chow S. Imaging features of avulsion injuries. *Radiographics* 1999; 19:655–672.
3. Baltaci G, Ozer H, Tunay VB. Rehabilitation of avulsion fracture of the tibial tuberosity following Osgood-Schlatter disease. *Knee Surgery, Sports Traumatology, Arthroscopy* 2004 Mar; 12(2):115–118.
4. Herring JA. Lower extremity injuries. In: *Tachdjian's Pediatric Orthopaedics*, 3rd ed., Chapter 42. Philadelphia: W.B. Saunders Company, 2002; 2251–2438.
5. Rosenberg ZS, Kawelblum M, Cheung YY, Beltran J, Lehman WB, Grant AD. Osgood-Schlatter lesion: Fracture or tendinitis? Scintigraphic, CT, and MR imaging features. *Radiology* 1992; 185:853–858.

Case 93

History

This is a 14-year-old boy who fractured his right humerus several months ago.

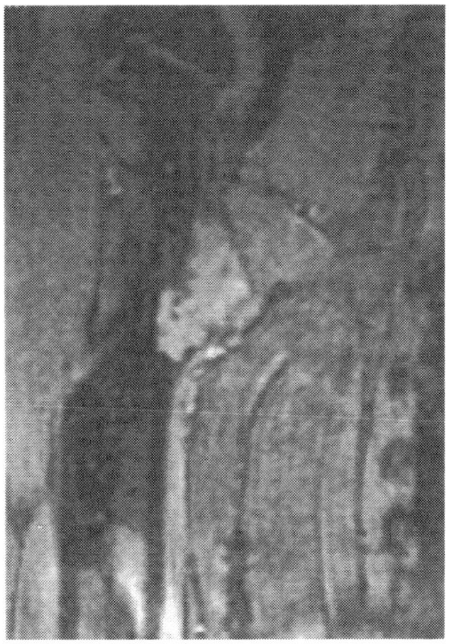

Figure 93A. Coronal MPGR of the right humerus.

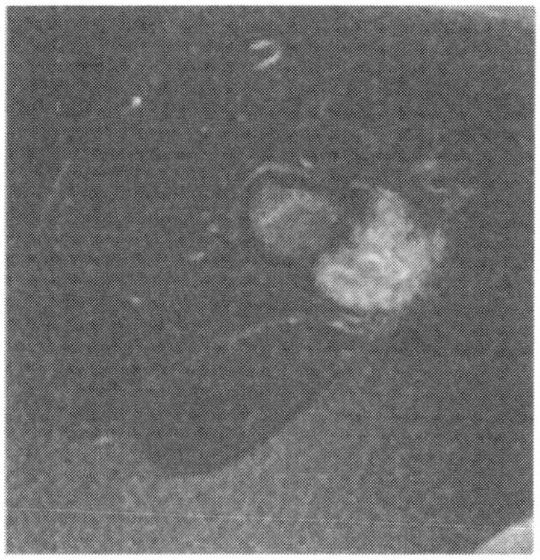

Figure 93B. Axial T2 FS.

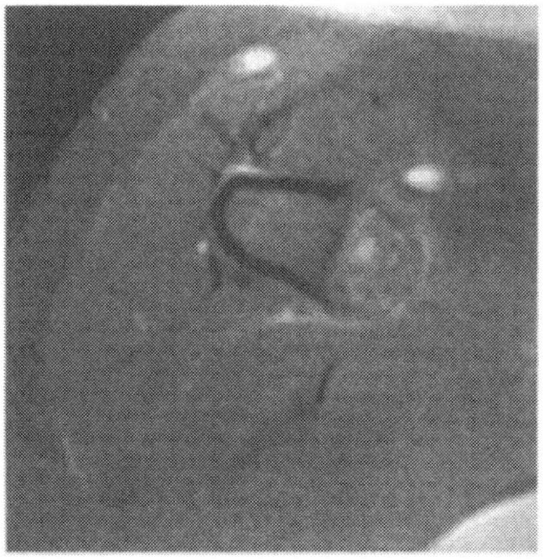

Figure 93C. Axial T1 post-Gd FS.

Figure 93D. Coronal T1 post-Gd FS.

Figures 93A (93A with annotations). There is a well-defined hyperintense multilobulated mass involving the surface of the proximal humeral diaphysis. Although cortically based, the lesion does encroach on the medullary cavity **(arrow)**. A thin hypointense line is identified overlying the mass **(arrowheads)** that is continuous with the periosteum.

Figures 93B, 93C, 93D. This mass is heterogeneously hyperintense on the T2W sequence, following the SI of cartilage, and demonstrates mild heterogeneous enhancement. There is minor juxtacortical soft tissue edema and enhancement.

Figure 93E. Although this exophytic mass is cortically based, the sharply sclerotic margins extend to the medullary cavity **(thin white arrow)**. A thin cortical shell surrounds this lesion **(thick white arrow)**. Note scattered foci of mineralization **(thin black arrow)** within the otherwise radiolucent exophytic component of the mass. A healing mid-diaphyseal fracture is also seen.

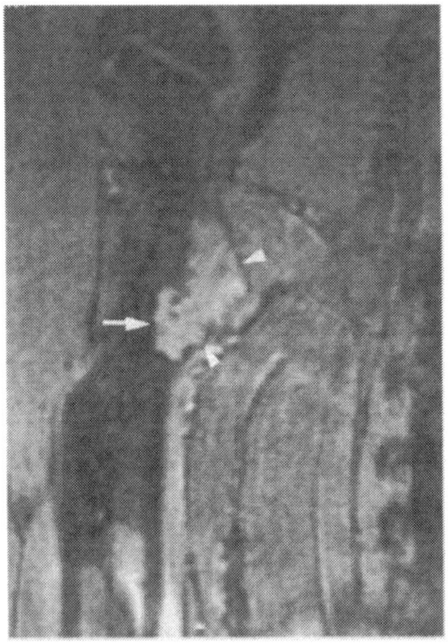

Figure 93A* Annotated.

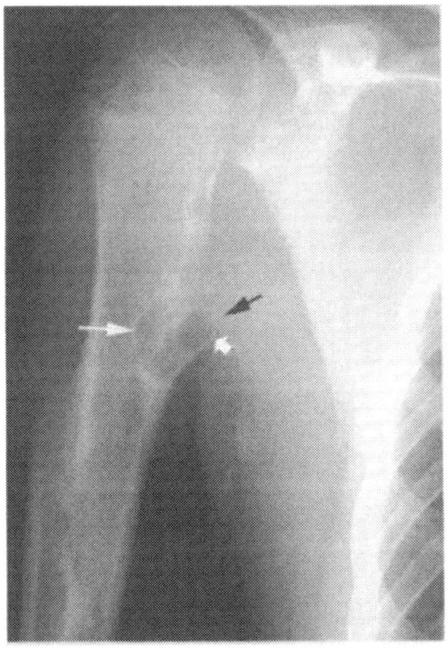

Figure 93E. AP radiograph of the right humerus.

Diagnosis

Periosteal chondroma

Questions

1. What would be the diagnosis if this lesion originated in the medullary cavity?
2. Which malignant surface based tumor may mimic a periosteal chondroma?

Discussion

Periosteal chondroma (AKA juxtacortical chondroma) is a benign cartilaginous tumor that arises along the surface of the metaphysis or metadiaphysis of tubular bones. Histologically, it is identical to an enchondroma, but located in the juxtacortical region (Answer to Question 1). Most patients are males and present before the age of 30, usually during the second decade of life (1). The humerus is the most commonly affected bone, followed by the femur, tibia, and the short tubular bones of the hands (Figures 93F–93H) and feet. Rarely, these tumors coexist with enchondromas or osteochondromas (2, 3).

On radiography, a shell of calcification may be seen delineating the outer margin of a periosteal chondroma (1). Benign appearing, well-defined cortical scalloping may be present along parent bone. Soft tissue density and chondroid matrix calcification may be seen within the tumor matrix (Figure 93F).

On MRI, periosteal chondromas appear as a lobulated mass arising from the surface of the cortex associated with cortical scalloping. The tumors are hypointense to isointense on T1, hyperintense on fluid sensitive sequences, hyperintense on MPGR, and demonstrate heterogeneous enhancement. Areas that are hypointense on all imaging sequences may represent matrix calcification or fibrous septa. Infrequently, tumor extension and edema in the medullary canal may be seen (4). A thin hypointense outer periosteal layer may be identified. A peripheral enhancement pattern may be seen, and this pathologically correlates with fibrovascular bundles that surround cartilaginous lobules (5).

The differential diagnosis for a periosteal chondroma includes avulsive cortical irregularity (distal femur) and cortical aneurysmal bone cyst. Developing (preossified) and sessile osteochondromas also enter into the diagnosis if the underlying cortex is not well seen on radiographs (see Case 65). Additional considerations, especially when there is juxtacortical mineralization associated with the mass, include: Nora's lesion (AKA bizarre parosteal osteochondromatous proliferation [BPOP]), periosteal osteosarcoma (chondroblastic), synovial cell sarcoma, trauma, and tumoral calcinosis (Answer to Question 2). Distinguishing between a periosteal osteosarcoma and a periosteal chondroma may be difficult. Periosteal osteosarcomas share several radiographic and MRI features with periosteal chondromas, including enhancement (a feature shared by other cartilaginous tumors), matrix calcification, intramedullary reactive changes, and rarely true invasion (4, 6). Cortical scalloping may be present with both entities; however, cortical scalloping seen with periosteal osteosarcomas tends to be shallower and occurs in areas of thickened cortex. Periosteal osteosarcomas usually demonstrate aggressive periosteal reaction, a feature not seen with periosteal chondromas, unless there has been a pathologic fracture with healing.

In this patient, the lesion was found incidentally after he fractured the mid-shaft of his humerus. MRI evaluation was performed after the humerus fracture had partially healed. The lesion was surgically resected and pathology confirmed a periosteal chondroma.

Orthopedic Perspective

The main concern is in distinguishing periosteal chondroma from other lesions, especially periosteal osteosarcoma (intermediate grade chondroblastic juxtacortical osteosarcoma). Often a biopsy is necessary to establish the diagnosis and usually the chondroma can be resected at the same time. However, if the imaging is diagnostic and the patient is asymptomatic, periosteal chondromas can be observed, as can enchondromas. At times it may be difficult to distinguish periosteal chondroma from a sessile osteochondroma, but CT or MR imaging demonstrating a thick lobulated cartilaginous mass in the former and a contiguous cortex and medullary cavity with the lesion in the latter usually settles the issue. The cartilage cap of an osteochondroma is usually much thinner than the lobulated cartilage mass of a periosteal chondroma. Both lesions can be managed by observation or excision. At times, periosteal chondromas located near a joint can interfere with motion and excision will be necessary. In these cases, MRI is helpful in defining the adjacent neurovascular and joint structures and in planning the optimal surgical approach.

What the Clinician Needs to Know

1. Benign surface lesions, like periosteal chondroma, need to be distinguished from malignant ones, especially periosteal osteosarcoma.
2. Defining the extent of the periosteal chondroma is helpful in planning surgical excision.
3. For those lesions that are observed, MRI is useful to monitor changes in size of the lesion and the relationship to surrounding neurovascular structures.

Answers

1. Enchondroma.
2. Periosteal osteosarcoma (chondroblastic).

Additional Example

Periosteal Chondroma of the Hand

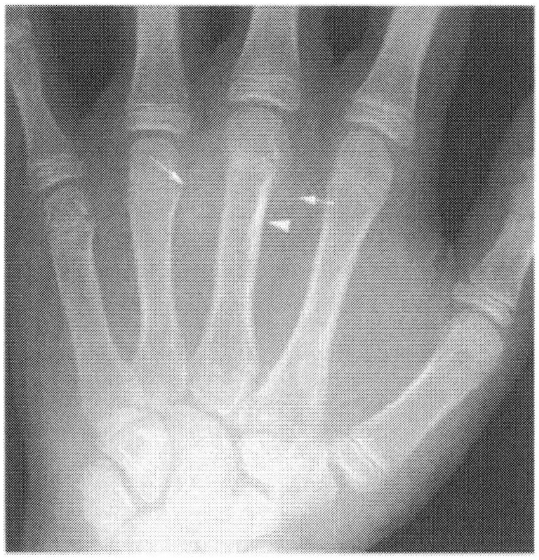

Figure 93F. PA radiograph of the hand.

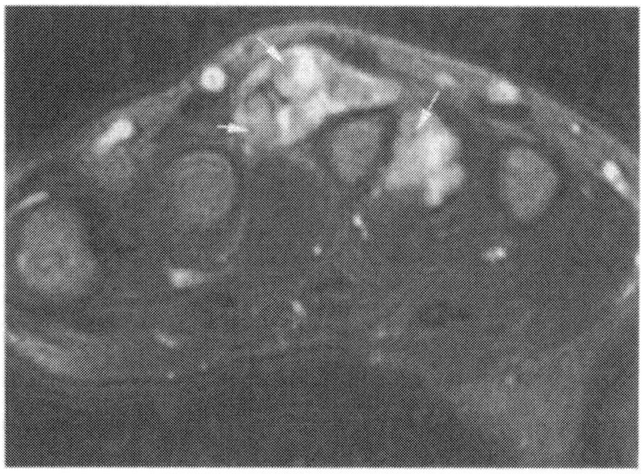

Figure 93G. Axial T2 FS.

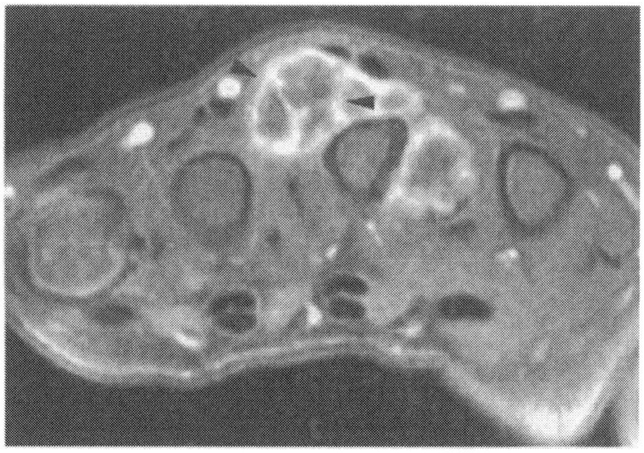

Figure 93H. Axial T1 post-Gd FS.

Findings

This is a 12-year-old boy with an enlarging dorsal hand mass.

Figure 93F. Multiple ring-and-arc chondroid calcifications are present in the soft tissues **(arrows)**. There is mild cortical scalloping of the 3rd metacarpal **(white arrowhead)**.

Figures 93G, 93H. The mass is heterogeneously hyperintense on the T2W sequence with areas of low SI **(arrows)**, correlating with the chondroid calcifications. Peripheral and septal enhancement is seen **(black arrowheads)**. This lesion was a pathologically proven periosteal chondroma.

Pitfalls and Pearls

The differential considerations for periosteal chondroma range from the benign to highly malignant, so exercise great care with these lesions, and don't hesitate to recommend a biopsy.

References

1. Robbin MR, Murphey MD. Benign chondroid neoplasms of bone. *Semin Musculoskelet Radiol* 2000; 4:45–58.
2. Ishida T, Iijima T, Goto T, Kawano H, Machinami R. Concurrent enchondroma and periosteal chondroma of the humerus mimicking chondrosarcoma. *Skeletal Radiol* 1998; 27:337–340.
3. Kahn S, Taljanovic MS, Speer DP, Graham AR, Dennis PD. Kissing periosteal chondroma and osteochondroma. *Skeletal Radiol* 2002; 31:235–239.
4. Robinson P, White LM, Sundaram M, et al. Periosteal chondroid tumors: Radiologic evaluation with pathologic correlation. *AJR Am J Roentgenol* 2001; 177:1183–1188.
5. Woertler K, Blasius S, Brinkschmidt C, Hillmann A, Link TM, Heindel W. Periosteal chondroma: MR characteristics. *J Comput Assist Tomogr* 2001; 25:425–430.
6. Murphey MD, Jelinek JS, Temple HT, Flemming DJ, Gannon FH. Imaging of periosteal osteosarcoma: Radiologic-pathologic comparison. *Radiology* 2004; 233:129–138.

Case 94

History

This is a 9-year-old girl with chronic left knee pain.

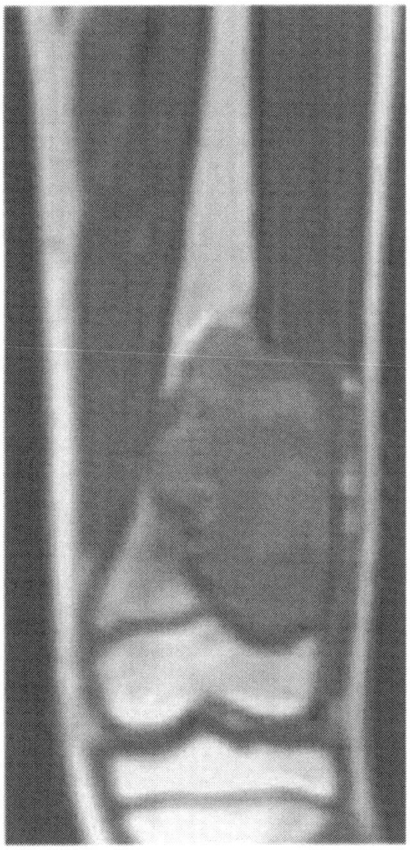

Figure 94A. Coronal T1 of the distal left femur.

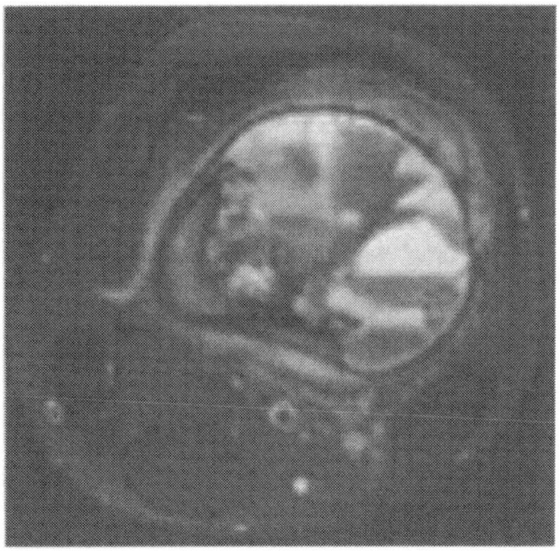

Figure 94B. Axial T2 FS.

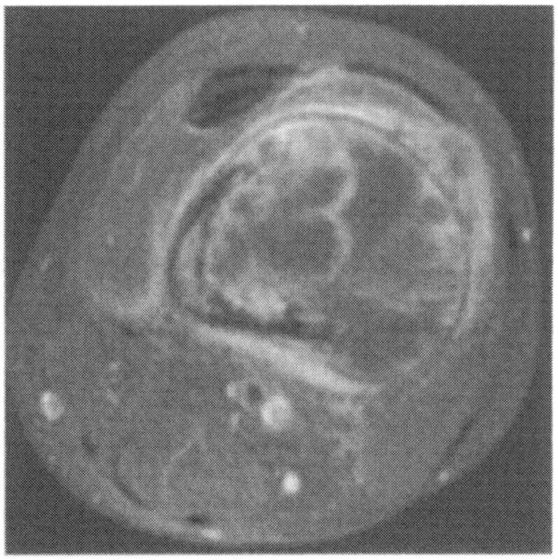

Figure 94C. Axial T1 post-Gd FS.

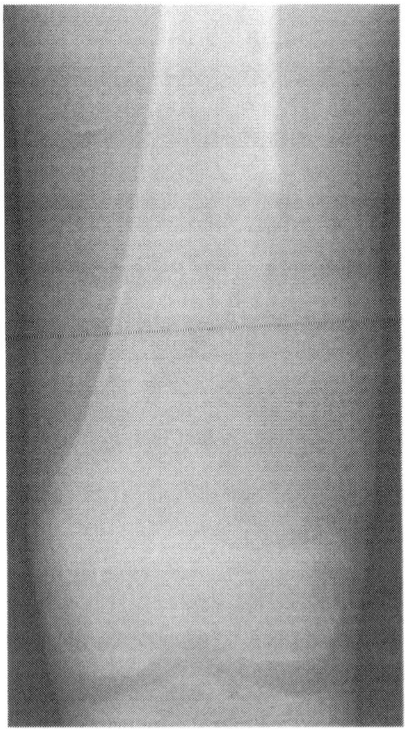

Figure 94D. AP radiograph.

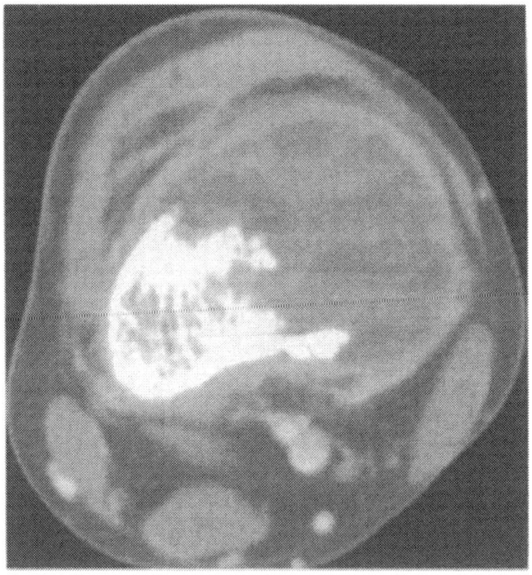

Figure 94E. CT.

Figures 94A, 94B, 94C (94C with annotations). A heterogeneous metaphyseal mass is identified that extends into the epiphysis. The diaphyseal and epiphyseal margins of the tumor are sharply marginated by the intramedullary fat. The mass distends the medullary cavity, destroys the cortex **(white arrowheads)**, and drapes and transgresses the periosteal envelope **(black arrowheads)**. Fluid-fluid levels are present. Septal nodularity as well as intra- and extraosseous soft tissue enhancement is seen **(white arrows)**.

Figures 94D, 94E (94D with annotations). There is a destructive, osteolytic lesion in the distal left femur with a nonsclerotic wide zone of transition. Aggressive, interrupted periosteal reaction is seen with a Codman's triangle **(black arrowhead)**. On CT, a large mass with fluid-fluid levels is identified extending into the extraosseous soft tissues. Osteoid matrix calcification is absent.

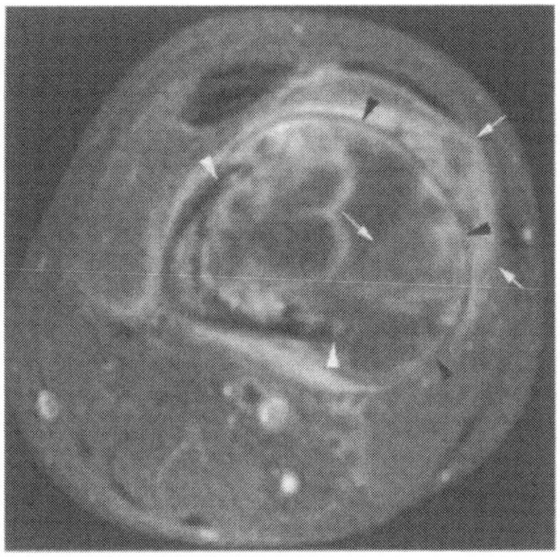

Figure 94C* Annotated.

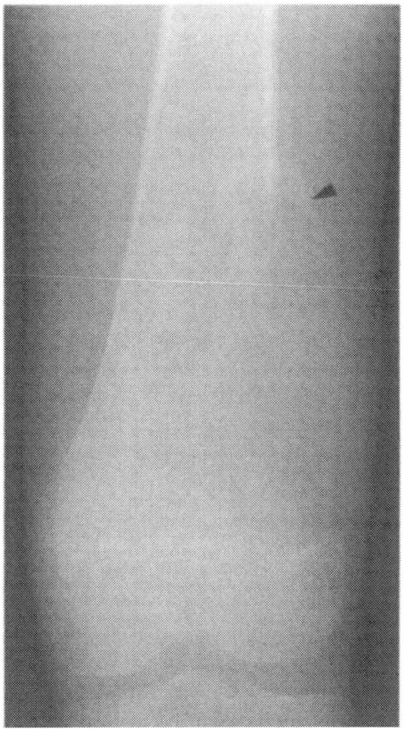

Figure 94D* Annotated.

Diagnosis

Telangiectatic osteosarcoma

Questions

1. T/F: By definition, matrix calcification is not present with telangiectatic osteosarcomas.
2. T/F: Aneurysmal bone cyst may simulate a telangiectatic osteosarcoma.

Discussion

Telangiectatic osteosarcoma is a rare, osteolytic variant of osteosarcoma, constituting approximately 11% of all osteosarcomas (1). Like conventional osteosarcoma, telangiectatic osteosarcoma most commonly occurs during the second and third decade of life and the knee region is most often affected (distal femur followed by the proximal tibia) (2). Approximately 8% of telangiectatic osteosarcomas arise from the diaphysis compared with 5% of conventional osteosarcomas (1). Greater than 90% of the tumoral mass must contain hemorrhagic, cystic, or necrotic components prior to treatment to be considered a telangiectatic osteosarcoma.

On radiography, telangiectatic osteosarcomas may demonstrate geographic, moth-eaten or permeative bone destruction with a wide zone of transition, soft tissue mass, and aggressive periosteal reaction (3). Although the dominant feature of these tumors is osteolytic bone destruction, one study demonstrated peripheral osteoid matrix calcification in 58% (21 of 36 cases) of biopsy proven lesions (Answer to Question 1) (4). In this same study, matrix calcification was evident on CT in 85% of cases.

On MRI, telangiectatic osteosarcomas are iso- to hyperintense on T1 and hyperintense on fluid sensitive sequences. These tumors are very heterogeneous on both T1 and fluid sensitive sequences due to the presence of blood products and mineralization. Characteristic fluid-fluid levels are better delineated on the fluid sensitive sequences. In Murphey et al.'s series, fluid-fluid levels were present in 89% of telangiectatic osteosarcomas (4). Post contrast images often demonstrate an enhancing soft tissue mass and/or thick, nodular septal enhancement.

Differential considerations for telangiectatic osteosarcomas include primary aneurysmal bone cyst (ABC), osteomyelitis, and other neoplasms such as Langerhans cell histiocytosis, Ewing's sarcoma, osteoblastoma, and lymphoma. ABCs (Figures 94F–94H) may demonstrate aggressive features on radiography and MRI that resemble those seen with telangiectatic osteosarcoma (Answer to Question 2). However, the presence of extensive cortical destruction and osteoid matrix calcification favors telangiectatic osteosarcoma over a primary ABC. Up to 12% of aneurysmal bone cysts demonstrate central enhancing soft tissue in addition to cysts on MRI (5), simulating a telangiectatic osteosarcoma. It is advisable to biopsy any solid soft tissue component within an ABC-like bone lesion to exclude telangiectatic osteosarcomas or other primary bone tumors that may undergo secondary ABC.

The lesion was a pathologically proven telangiectatic osteosarcoma. She underwent 5 months of chemotherapy and subsequently underwent extra-articular, radical resection of the distal femur and rotation plasty reconstruction.

Orthopedic Perspective

This case points out the difficulty of distinguishing osteosarcoma from ABC. The initial radiographic diagnosis was ABC and a needle biopsy was consistent with that diagnosis, despite the relatively aggressive radiographic margins. An open biopsy was done and on frozen section, ABC was still considered to be the most likely diagnosis. The final diagnosis was osteosarcoma. The issues regarding treatment and preoperative assessment are the same as for all osteosarcomas.

What the Clinician Needs to Know

Distinguishing telangiectatic osteosarcoma from ABC is critical to avoid mistreatment of these bone lesions, and at times this distinction can be very difficult.

Answers

1. False.
2. True.

Additional Example

Cortical Aneurysmal Bone Cyst with Aggressive Characteristics

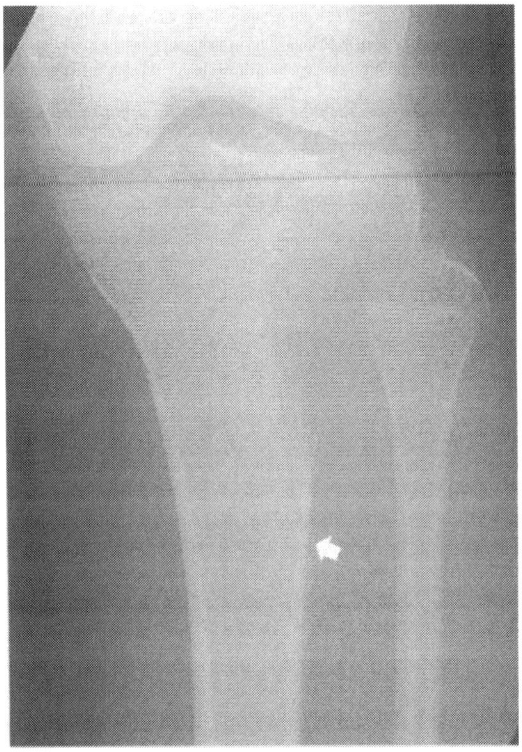

Figure 94F. AP radiograph of the left proximal tibia.

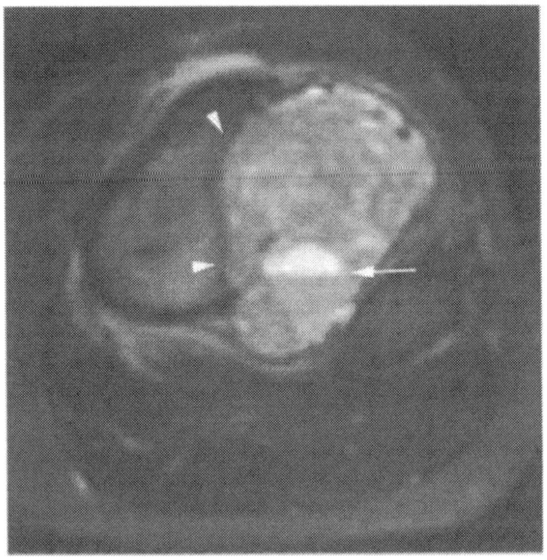

Figure 94G. Axial T2 FS.

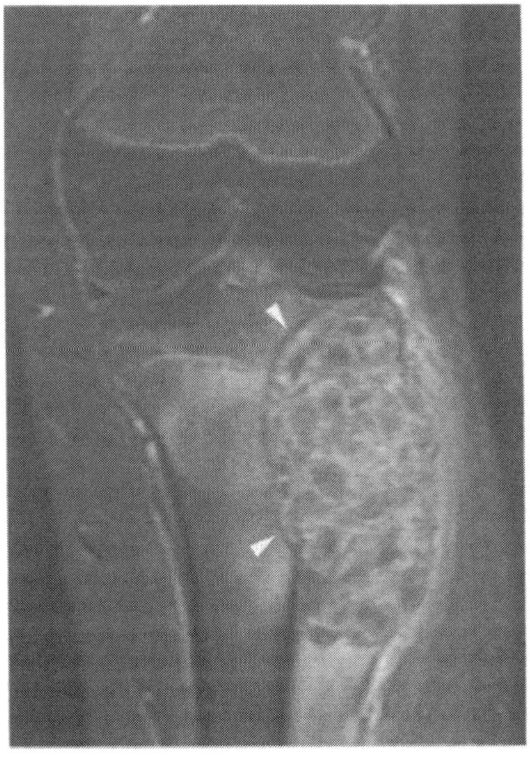

Figure 94H. Coronal T1 post-Gd FS.

Findings

This is a 14-year-old girl who fell 3 months ago. Since then, she has been having persistent pain and swelling over the posterolateral aspect of her left knee.

Figure 94F. There is an osteolytic mass originating from the metaphyseal cortex that extends into the lateral tibial plateau. There is an indistinct slightly sclerotic zone of transition and laminated periosteal reaction is present **(thick arrow)**.

Figures 94G, 94H. The zone of transition is well defined by a hypointense rim **(arrowheads)**. Fluid-fluid levels **(arrow)**, septal and rim enhancement, as well as extensive edema in the marrow and extraosseous soft tissues are evident. No discrete enhancing soft tissue mass is identified. The aggressive radiographic features of the lesion suggest a telangiectatic osteosarcoma, but the MRI findings are more consistent with an ABC. The pathologic diagnosis was a cortical ABC.

Pitfalls and Pearls

A telangiectatic osteosarcoma should be considered in any ABC-like tumor with aggressive cortical destruction, soft tissue mass, or matrix calcification.

References

1. Huvos AG, Rosen G, Bretsky SS, Butler A. Telangiectatic osteogenic sarcoma: A clinicopathologic study of 124 patients. *Cancer* 1982; 49:1679–1689.
2. Spina V, Montanari N, Romagnoli R. Malignant tumors of the osteogenic matrix. *Eur J Radiol* 1998; 27 Suppl 1:S98–109.
3. Vanel D, Tcheng S, Contesso G, et al. The radiological appearances of telangiectatic osteosarcoma: A study of 14 cases. *Skeletal Radiol* 1987; 16:196–200.
4. Murphey MD, wan Jaovisidha S, Temple HT, Gannon FH, Jelinek JS, Malawer MM. Telangiectatic osteosarcoma: Radiologic-pathologic comparison. *Radiology* 2003; 229:545–553.
5. Mahnken AH, Nolte-Ernsting CC, Wildberger JE, et al. Aneurysmal bone cyst: Value of MR imaging and conventional radiography. *Eur Radiol* 2003; 13:1118–1124.

Case 95

History

This is a 3-day-old boy born at term with a large mass arising from his left leg.

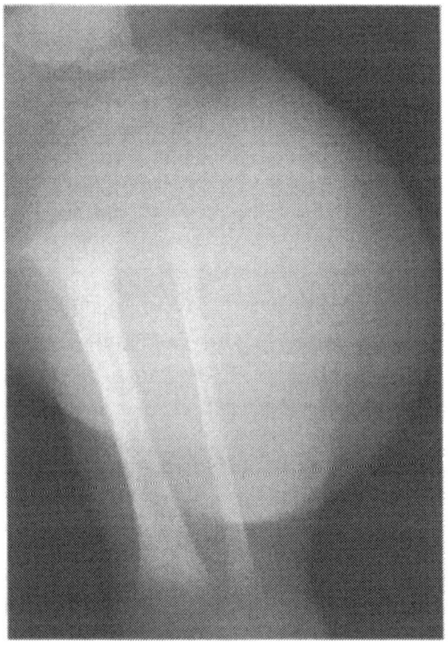

Figure 95A. AP radiograph of the left leg.

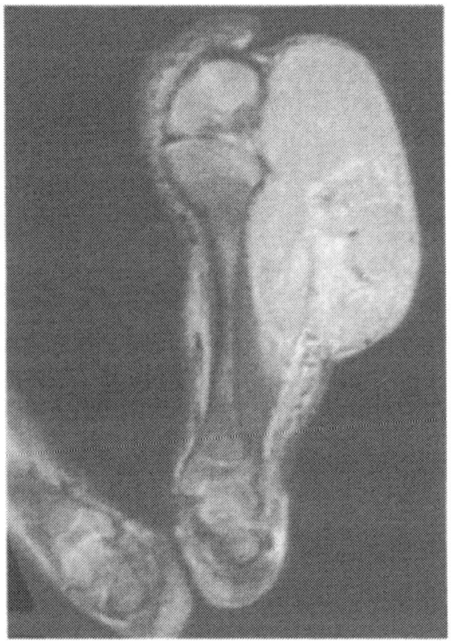

Figure 95B. Coronal STIR.

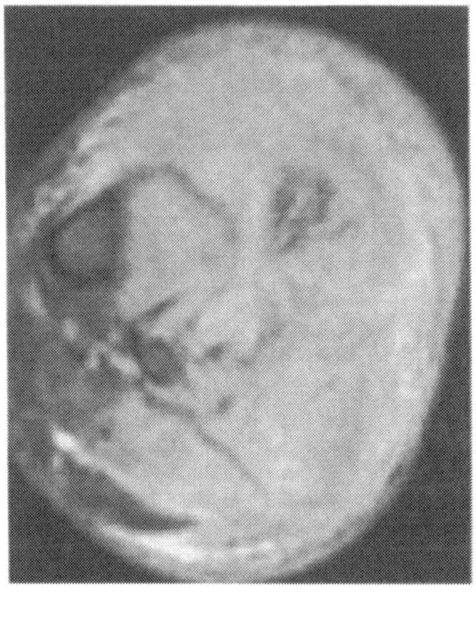

Figure 95C. Axial T2 FS.

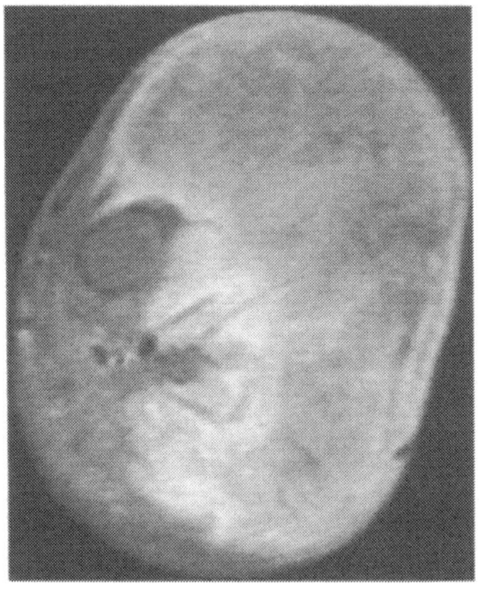

Figure 95D. Axial T1 post-Gd FS.

Figure 95A. There is a large, noncalcified soft tissue mass in the knee and lower leg. The proximal fibula is demineralized and the cortex is indistinct.

Figures 95B, 95C, 95D, (95D with annotations). There is a large heterogeneous SI mass extending through multiple compartments of the leg. The extensor mechanism and the peroneal and soleus muscles are affected. Diffuse fibular cortical destruction **(thin black arrow)** and focal tibial cortical destruction **(white arrowhead)** are seen. Note preservation of the medial tibial cortex **(thick white arrow)**. The anterior tibial artery/vein/deep peroneal nerve complex is not identified and probably encased by the mass. The peroneal artery and vein (Per) are displaced posteriorly by the mass. Multiple prominent flow voids **(black arrowheads)** extend from the peroneal vessels to the tumor. Posterior tibial artery/vein/nerve (PT). (95B Reprinted from Laor T, MR Imaging of soft tumors and tumor-like lesions. Pediatr Radiol 2004. 34:24–37 with kind permission of Springer Science and Business Media.)

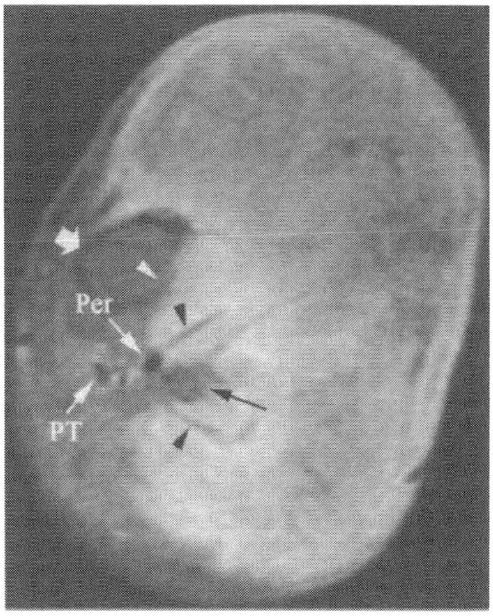

Figure 95D* Annotated.

Diagnosis

Infantile fibrosarcoma

Questions

1. T/F: Infantile fibrosarcoma may contain high-flow vessels and be mistaken for a hemangioma.
2. How is congenital hemangioma differentiated from infantile fibrosarcoma on clinical grounds?

Discussion

Infantile fibrosarcoma (AKA congenital fibrosarcoma) usually presents at or shortly after birth. It most commonly occurs in the extremities. Unlike adult fibrosarcoma, infantile fibrosarcoma has a good prognosis, with a greater than 90% 5-year disease-free survival when disease is localized (1). The incidence of distant metastasis for primary extremity and axial infantile fibrosarcomas is reported at 8% and 26% respectively and local recurrence after tumor resection is high for both locations (2).

On MRI, infantile fibrosarcomas are inhomogeneous and generally hypointense to iso-intense on T1, hyperintense on fluid sensitive sequences, and enhance following gadolinium. Infantile fibrosarcomas may be well-defined by a pseudocapsule, or be infiltrative with multicompartmental spread. They may contain hemorrhagic, necrotic, fibrous, and cystic components and demonstrate high flow vasculature (3–5). Tumors may be mistaken for a congenital hemangioma or hemangioendothelioma because of the presence of high flow vessels (Answer to Question 1). Areas of low SI within the lesion on all sequences may represent tumoral fibrosis (5).

The differential diagnosis for infantile fibrosarcoma includes congenital hemangioma, hemangioendothelioma, microcystic lymphatic and other vascular malformations, infantile myofibromatosis, and rarely infantile leukemia and rhabdomyosarcoma. Congenital hemangiomas and hemangioendotheliomas may share features with infantile fibrosarcoma, including multicompartmental spread, central tumoral necrosis, and abundant neovascularity (6). However, congenital hemangiomas are largest at birth and rapidly decrease in size, unlike infantile fibrosarcomas (Answer to Question 2). Congenital hemangioma should be distinguished from a proliferating infantile hemangioma, which presents after birth, usually within the first year of life. They are "light-bulb" bright on fluid sensitive sequences, show homogeneous intense lobular enhancement, have well-defined margins and usually do not show multicompartmental extension or tumoral necrosis. Solitary infantile myofibromatosis may be indistinguishable from infantile fibrosarcoma. However, multicentric infantile myofibromatosis often has characteristic metaphyseal osseous lesions that help distinguish this entity from infantile fibrosarcoma (see Case 43) (7). Infantile fibrosarcoma may be indistinguishable from infantile rhabdomyosarcoma, a rare lesion with a very poor prognosis. Approximately 50% of these highly malignant lesions demonstrate metastases at presentation (3).

This was a pathologically proven infantile fibrosarcoma. The patient was initially treated with chemotherapy (vincristine, actinomycin, cyclophosphamide, and mercaptoethanesulfonate [mesna]). At 4 months of age, the mass was resected and a follow-up MRI at 3 years of age showed no evidence of local recurrence.

Orthopedic Perspective

When a neonate presents with a soft tissue mass, the orthopedist must distinguish between myofibromatosis, fibrosarcomas, rhabdomyosarcomas, and vascular lesions.

Frequently a biopsy will be necessary to establish the diagnosis and the imaging can be useful to choose areas for sampling either by open or needle techniques. The soft tissue lesions in infancy are different from those seen in later childhood and adults and an experienced pathologist is essential. Infantile fibrosarcomas are treated surgically if they can be excised with minimal morbidity. Spontaneous regression has been described, but is not the rule. As in this case, imaging frequently shows that bones, nerves, and major vessels are involved, or are at least in close proximity, and multiple anatomic compartments may be involved. In these instances, an amputation would be the only way to secure a complete resection. Fortunately, infantile fibrosarcoma responds to chemotherapy in most instances and, as in this case, the response allows a more limited resection. This child had a resection following chemotherapy with close, but negative margins, and had no significant functional defect or growth deformity. He has remained disease free for more than 5 years.

What the Clinician Needs to Know

1. The clinician needs to distinguish infantile fibrosarcoma from other neoplasms and vascular lesions that occur in this age group.
2. An open or needle biopsy is often necessary and the imaging can direct needle placement.
3. The anatomic extent of the tumor needs to be defined to determine if treatment will be primarily surgical, or whether chemotherapy should be employed as well.

Answers

1. True.
2. Congenital hemangiomas are largest at birth and will then decrease in size. Infantile fibrosarcomas may be large at birth but will usually grow thereafter.

Pitfalls and Pearls

The differential diagnosis for a large soft tissue mass in the newborn includes: congenital hemangioma, kaposiform hemangioendothelioma, infantile fibrosarcoma, infantile myofibromatosis, and vascular tumors and malformations. Although not diagnostic, MRI can often point to specific diagnoses, providing important guidance to the surgeon and the pathologist.

References

1. Neifeld JP, Berg JW, Godwin D, Saizberg AM. A retrospective epidemiologic study of pediatric fibrosarcomas. *J Pediatr Surg* 1978; 13:735–739.
2. Blocker S, Koenig J, Ternberg J. Congenital fibrosarcoma. *J Pediatr Surg* 1987; 22:665–670.
3. McCarville MB, Kaste SC, Pappo AS. Soft-tissue malignancies in infancy. *AJR Am J Roentgenol* 1999; 173:973–977.
4. Fink AM, Stringer DA, Cairns RA, Nadel HR, Magee JF. Pediatric case of the day: Congenital fibrosarcoma (CFS). *Radiographics* 1995; 15:243–246.
5. Lee MJ, Cairns RA, Munk PL, Poon PY. Congenital-infantile fibrosarcoma: Magnetic resonance imaging findings. *Can Assoc Radiol J* 1996; 47:121–125.
6. Konez O, Burrows PE, Mulliken JB, Fishman SJ, Kozakewich HP. Angiographic features of rapidly involuting congenital hemangioma (RICH). *Pediatr Radiol* 2003; 33:15–19.
7. Johnson GL, Baisden BL, Fishman EK. Infantile myofibromatosis. *Skeletal Radiol* 1997; 26:611–614.

Case 96

History

This is a 4-month-old boy with a history of a hip problem.

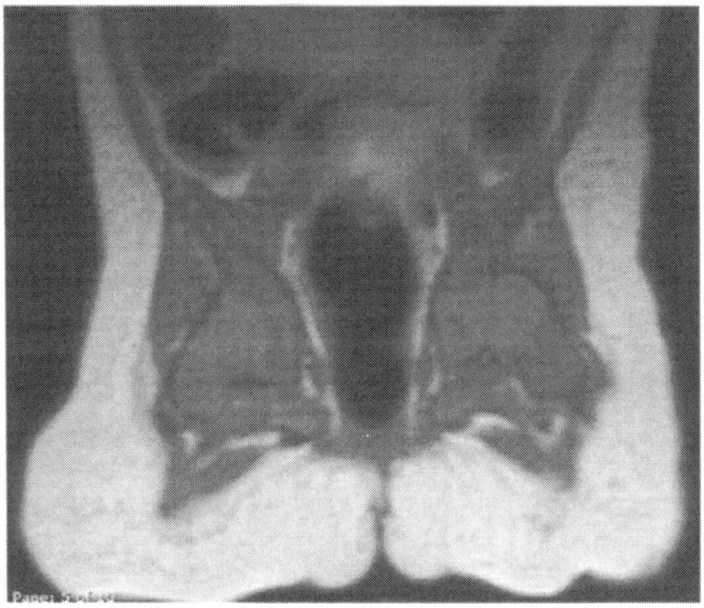

Figure 96A. Coronal PD.

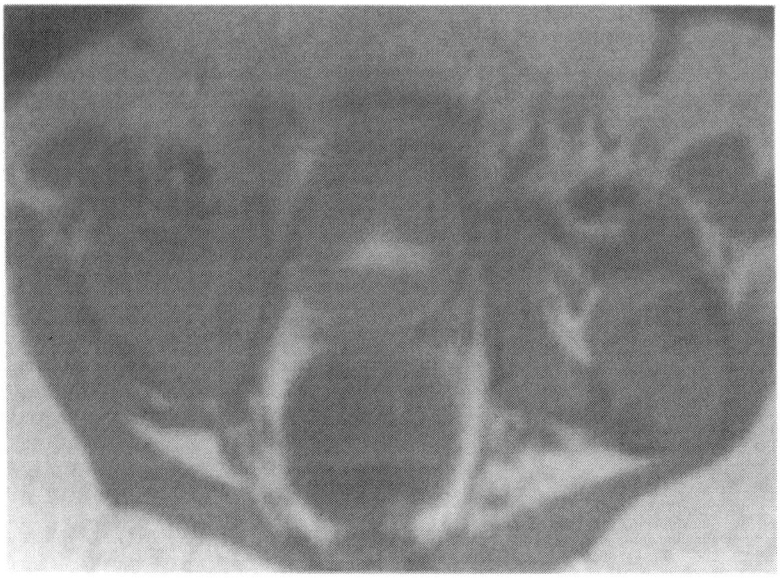

Figure 96B. Axial PD.

Figures 96A, 96B, (96A, 96B with annotations). The left hip is posterolaterally subluxed. The acetabulum is shallow, the ligamentum teres is thickened **(white arrowhead)**, and the fatty pulvinar is prominent **(black arrowhead)**. The superior labrum is hypertrophied **(arrow)**, but maintains a triangular shape with a sharp pointed tip. The right hip is normal.

Figures 96C, 96D. One month later, following left hip open reduction and capsulorrhaphy, psoas tendon release, and spica cast application. Changes from iliopsoas tendon release **(thick black arrow)** are seen. The left hip is now concentrically reduced. The left acetabular roof remains shallow. Acetabular hyaline cartilage **(thin white arrow)** covers much of the weightbearing surface. The right hip is normal.

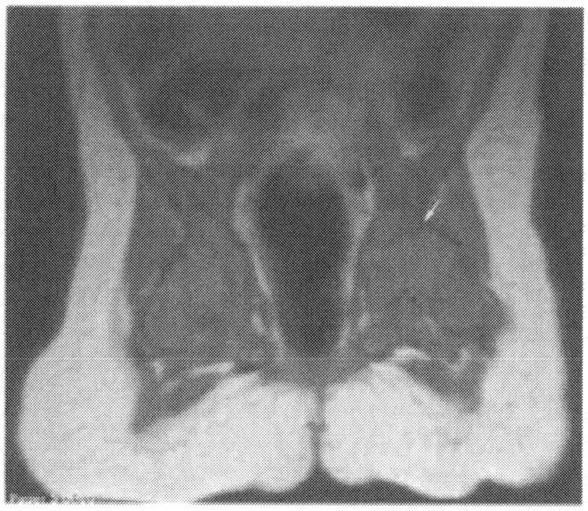

Figure 96A* Annotated.

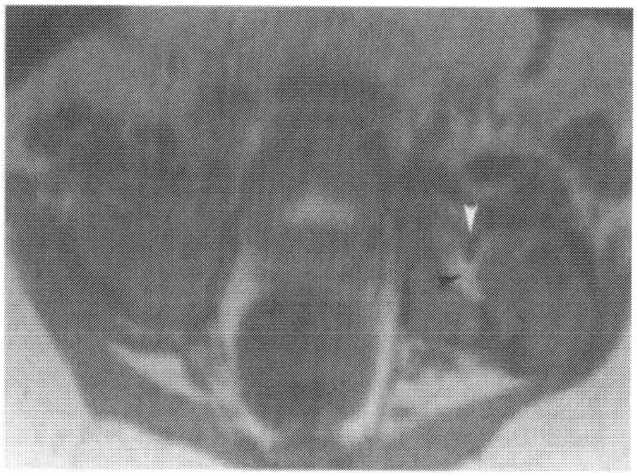

Figure 96B* Annotated.

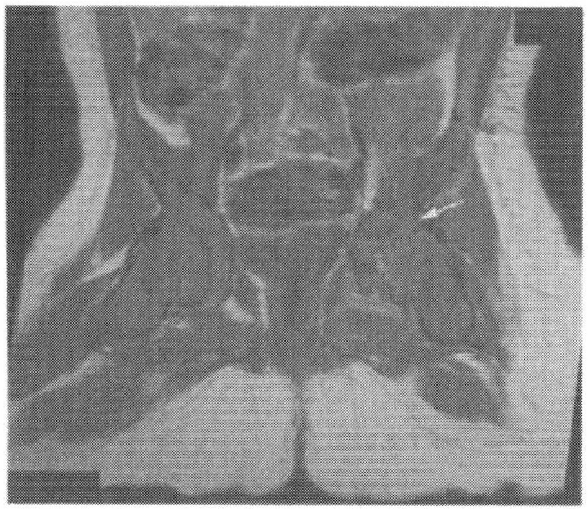

Figure 96C. Coronal PD one month later.

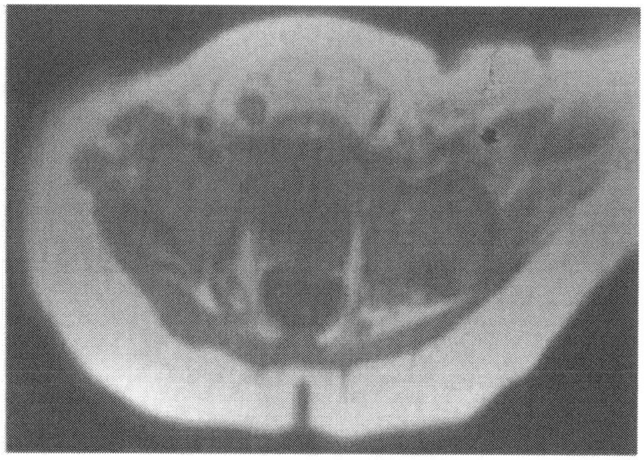

Figure 96D. Axial PD one month later.

Diagnosis

Developmental dysplasia of the left hip

Questions

1. What are some teratologic causes of developmental dysplasia of the hip?
2. What intra-articular structures may block normal hip reduction in DDH?

Discussion

Developmental dysplasia of the hip (DDH) represents a spectrum of acetabular dysplasia, femoral head dysplasia, and varying degrees of femoral head displacement. The estimated incidence of hip dislocation is 1 in 1000 and hip subluxation is 10 in 1000 (1). DDH is most commonly related to fetal positioning (e.g., breech presentation) and ligamentous laxity related to high estrogen levels. DDH is also associated with torticollis, clubfeet, and Native American ethnicity. The association with torticollis and clubfeet may be due to abnormal in utero fetal positioning. Teratologic associations with DDH include: arthrogryposis, proximal focal femoral deficiency, Larsen's syndrome (hereditary congenital multiple joint dislocations associated with facial abnormalities), neuromuscular disorders, mucopolysaccharidosis, diastrophic dwarfism and other rare disorders (Answer to Question 1) (2).

The early diagnosis of DDH in infants 1 to 6 months of age is made on clinical grounds and is generally confirmed by US. With the exception of grossly dislocated hips, ultrasound evaluation is not advocated during the first month of life because of common and normal ligamentous laxity in neonates. Beyond 6 months, the diagnosis is generally made with plain radiography.

MRI is not used for initial diagnosis of DDH, but may be used to evaluate failure of proper reduction and complications after reduction. Additionally, MRI can easily confirm proper reduction following spica casting without radiation or sedation. In the study of Laor et al. that utilized an axial and coronal PD sequence, the mean scan time was 3 minutes (range: 2:24–6:34) (3). Westhoff et al. found that 3 of 21 casted hips were not concentrically reduced when MRI was performed immediately after casting (4).

After open or closed reduction, the hip may be concentrically reduced, subluxed, or dislocated. A hip is subluxed if the femoral head partially extends beyond the confines of the acetabulum. The hip is dislocated when the femoral head is outside of the original acetabulum. If the hip is not concentrically reduced, it is important to assess for both internal and external obstacles to complete hip reduction. Internal obstacles include a thickened and interposed ligamentum teres or transverse acetabular ligament (Figures 96E–96G), hypertrophied pulvinar fat, and an inverted acetabular labrum (Answer to Question 2) (5, 6). External obstacles to complete reduction include an invaginated iliopsoas muscle, shortening of the short external rotators and adductor muscles, and adhesion of the capsule to the ilium. Additional obstacles to proper reduction include dysplastic shape of the acetabulum or femoral head. A hypertrophied pulvinar may be diagnosed by finding too much fat within the inferomedial hip joint. When characterizing the labrum, it is important to discuss its size, position, and morphology (hypertrophied, inverted, everted, sharp, blunted) and the relationship with respect to the dislocated hip. It is generally helpful to image the unaffected hip for comparison.

Proper treatment of DDH requires hip reduction with hip flexion and hyperabduction (7). This may cause kinking of the superior branches of the medial circumflex artery between the femur and superior labrum leading to femoral head ischemia (8). Jaramillo et al. showed that hips that had undergone a greater degree of abduction (>50 degrees) after DDH reduction were more likely to show decreased femoral head enhancement on gadolinium enhanced MRI (Figures 96H, 96I) (9). Some patients with ischemia may not develop ischemic necrosis, and long term studies have not been performed to determine the risk of ischemic necrosis in the absence of enhancement on MRI. However, the presence of gadolinium enhancement appears to mitigate against the possibility of later ischemic necrosis.

This patient with DDH was initially treated with a Pavlik harness, but serial ultrasound examinations showed persistent subluxation. A closed reduction in the operating room was unsuccessful. Following the open reduction and spica casting, radiographs showed concentric location of both femoral heads with near normal appearing acetabula (not shown). Clinical examination demonstrated normal full range of motion of the hips at one year follow-up.

Orthopedic Perspective

The treatment of DDH depends on the extent of involvement and the age of diagnosis. Infants less than 6 months old are typically treated in a Pavlik harness. Children who are 6 months to 18 months old are generally treated with closed reduction and spica casting. Children over 18 months are typically treated with open reduction and spica casting. Older children often require femoral and/or acetabular osteotomies. After closed or open reduction and spica casting, fluoroscopic images are obtained in the operating room. However, these planar images can be misleading with respect to the three-dimensional structure of the hip. MRI has replaced CT as the imaging modality of choice in these cases.

What the Clinician Needs to Know

1. Adequacy of closed or open reduction.
2. Intrinsic and extrinsic barriers to a concentric reduction.
3. Any signs of ischemic necrosis of the femoral head after hyperabduction hip reduction?

Answers

1. Arthrogryposis, proximal focal femoral deficiency, Larsen's syndrome, neuromuscular disorders, mucopolysaccharidosis, and diastrophic dwarfism.
2. Thickened and interposed ligamentum teres or transverse acetabular ligament, hypertrophied pulvinar fat, and an inverted acetabular labrum.

Additional Examples

DDH S/P Open Reduction

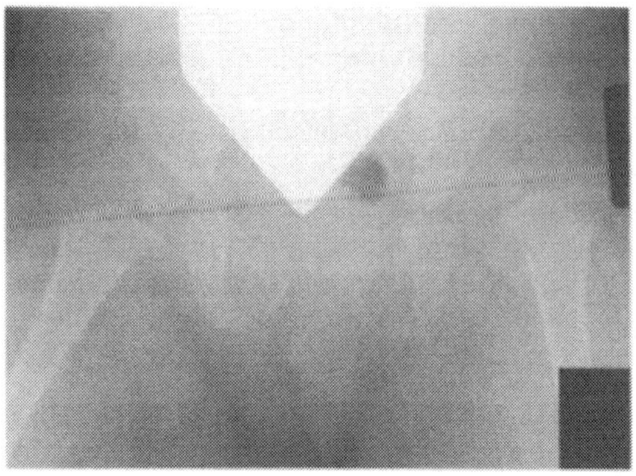

Figure 96E. AP radiograph.

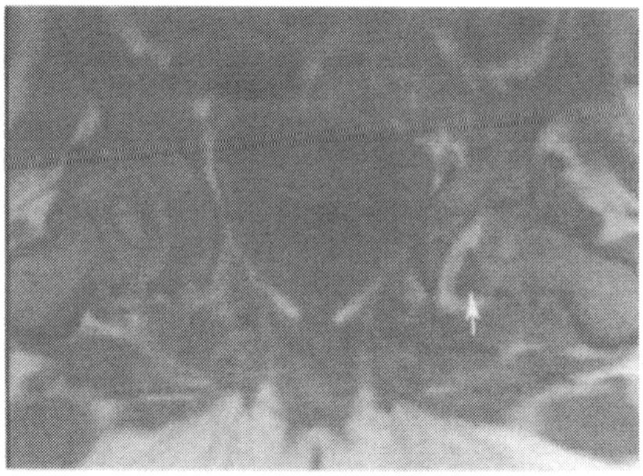

Figure 96F. Coronal PD after reduction.

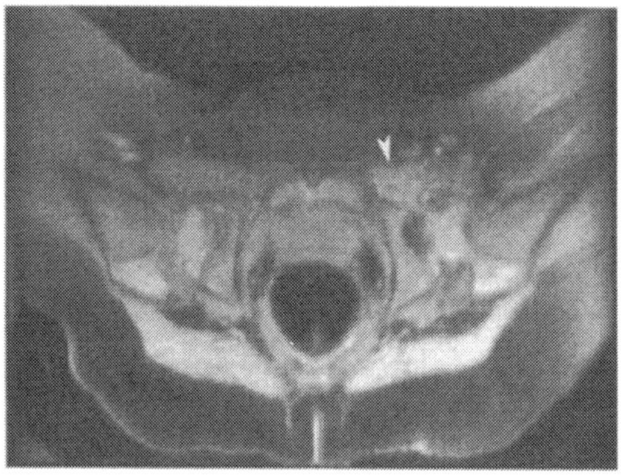

Figure 96G. Axial T1 post-Gd FS after reduction.

Findings

This is a 15-month-old girl with severe, bilateral DDH, left greater than right.

Figure 96E. Notice that the dislocated left capital femoral epiphysis is smaller than the right, and the acetabular roofs are both shallow and vertically oriented. She subsequently underwent open reduction with left adductor tenotomy.

Figure 96F. The postoperative MRI shows residual left posterolateral subluxation. Note thickened ligamentum teres **(arrow)** and hypertrophied pulvinar.

Figure 96G. Dynamic contrast study shows symmetric enhancement of the capital femoral epiphyses. The left adductor muscle is thickened and hyperintense related to the adductor tenotomy **(arrowhead)**.

Right DDH and Subsequent Ischemic Necrosis

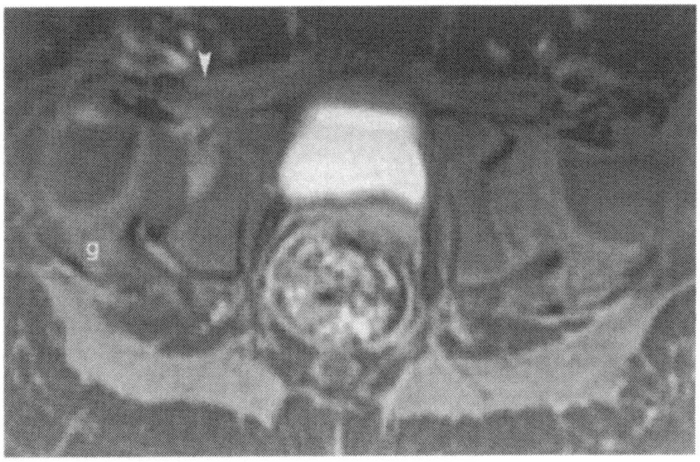

Figure 96H. Axial T1 Post-Gd FS.

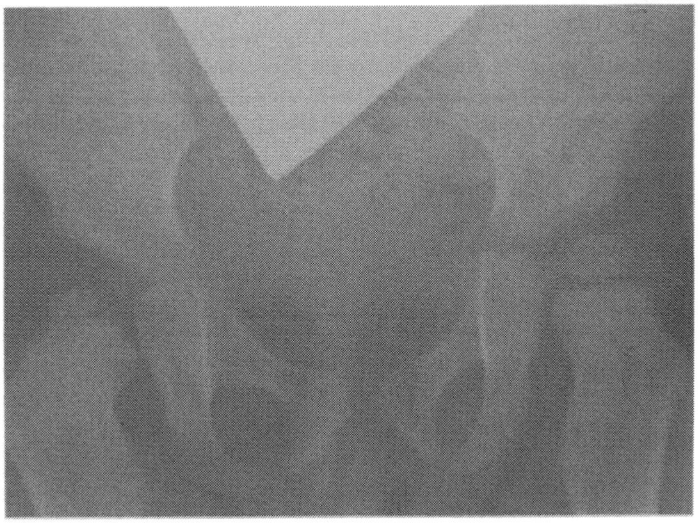

Figure 96I. AP radiograph 1 year later.

Findings

This 8-month-old girl had open reduction of right DDH.

Figure 96H. The dynamic enhanced MRI was performed immediately after reduction and right adductor tenotomy **(arrowhead).** The right femoral head is minimally posteriorly subluxed. There is global hypoperfusion to the cartilaginous right femoral head. Contrast this appearance with normal enhancement of the greater trochanter (g) and the left femoral head.

Figure 96I. A follow-up radiograph one year later showed persistent DDH and right femoral head ischemic necrosis.

Pitfalls and Pearls

Obstacles to concentric reduction may be significant in patients with hip dislocation when compared with patients with hip subluxation. Patients with hip dislocation are more likely to have ligamentous and capsular redundancy and iliopsoas tendon interposition because of the increased space between the acetabular fossa and femoral head.

References

1. Murray KA, Crim JR. Radiographic imaging for treatment and follow-up of developmental dysplasia of the hip. *Semin Ultrasound CT MR* 2001; 22:306–340.
2. Gruel CR, Birch JG, Roach JW, Herring JA. Teratologic dislocation of the hip. *J Pediatr Orthop* 1986; 6:693–702.
3. Laor T, Roy DR, Mehlman CT. Limited magnetic resonance imaging examination after surgical reduction of developmental dysplasia of the hip. *J Pediatr Orthop* 2000; 20:572–574.
4. Westhoff B, Wild A, Seller K, Krauspe R. Magnetic resonance imaging after reduction for congenital dislocation of the hip. *Arch Orthop Trauma Surg* 2003; 123:289–292.
5. Aoki K, Mitani S, Asaumi K, Akazawa H, Inoue H. Utility of MRI in detecting obstacles to reduction in developmental dysplasia of the hip: Comparison with two-directional arthrography and correlation with intraoperative findings. *J Orthop Sci* 1999; 4:255–263.
6. Johnson ND, Wood BP, Jackman KV. Complex infantile and congenital hip dislocation: Assessment with MR imaging. *Radiology* 1988; 168:151–156.
7. Ramsey PL, Lasser S, MacEwen GD. Congenital dislocation of the hip: Use of the Pavlik harness in the child during the first six months of life. *J Bone Joint Surg Am* 1976; 58:1000–1004.
8. Ogden JA. Changing patterns of proximal femoral vascularity. *J Bone Joint Surg Am* 1974; 56:941–950.
9. Jaramillo D, Villegas-Medina O, Laor T, Shapiro F, Millis MB. Gadolinium-enhanced MR imaging of pediatric patients after reduction of dysplastic hips: Assessment of femoral head position, factors impeding reduction, and femoral head ischemia. *AJR Am J Roentgenol* 1998; 170:1633–1637.

Case 97

History

This is a 15-year-old girl with left knee pain of 3 weeks duration.

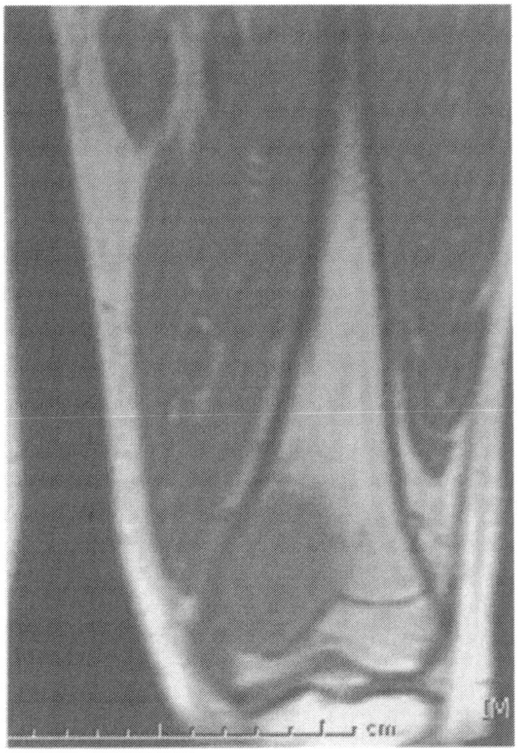

Figure 97A. Coronal T1 of the left distal femur.

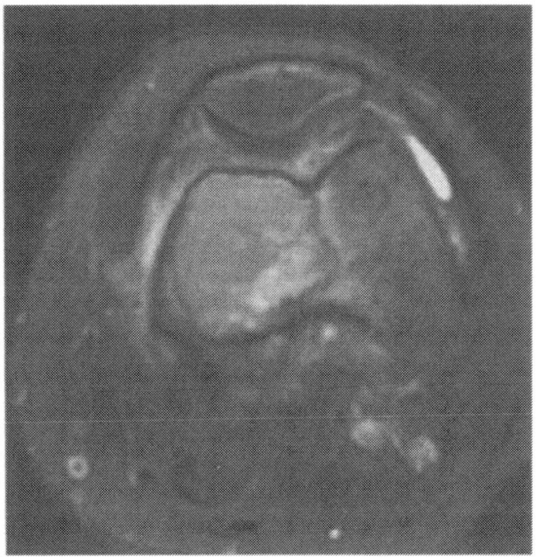

Figure 97B. Axial T2 FS.

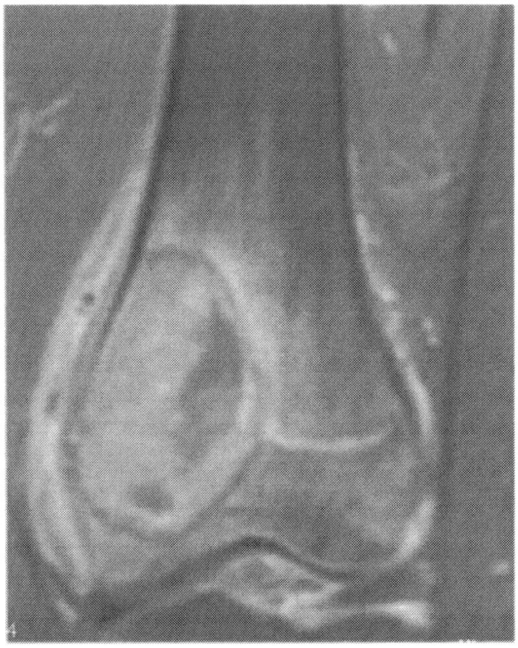

Figure 97C. Coronal T1 post-Gd FS.

Figure 97A. There is a distal femoral metaphyseal lesion, extending across the maturing physis, to within 1 cm of the articular surface. On T1, the lesion is mainly isointense with loss of the fat plane between the overlying cortex and adjacent muscle.

Figures 97B, 97C, (97C with annotations). The tumor is predominantly isointense on T2 and demonstrates significant enhancement. Within the medial aspect of the lesion, there is a nonenhancing focus **(white arrow)** that corresponds to a zone of hyperintensity on the T2W image. The lesion is circumscribed by a discrete, hypointense zone **(white arrowhead)** with adjacent marrow edema and enhancement. The medial metaphyseal cortex is thinned **(black arrow)** with adjacent periosteal and soft tissue enhancement, but no gross extraosseous mass is evident.

Figure 97D. This distal femoral lesion is osteolytic with poorly defined margins. It crosses a physis that is nearly fused. Note indistinctness of the adjacent metaphyseal cortex and faint periosteal reaction **(black arrow)**.

Figures 97E, 97F. Recurrent tumor is evident **(white arrowheads)** surrounding a fibular bone graft (F).

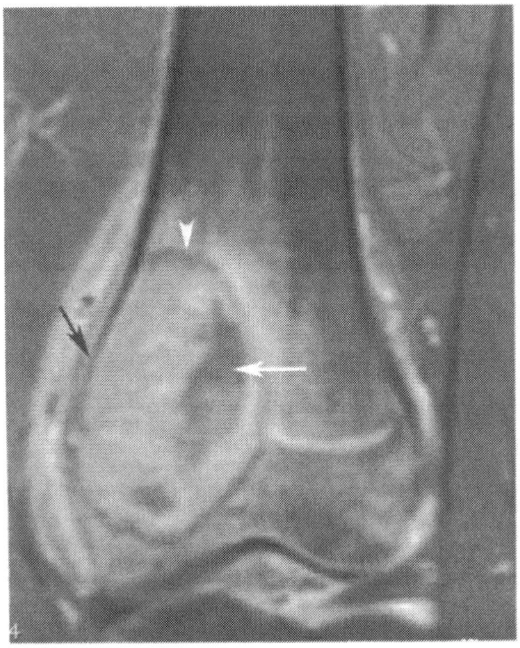

Figure 97C* Annotated.

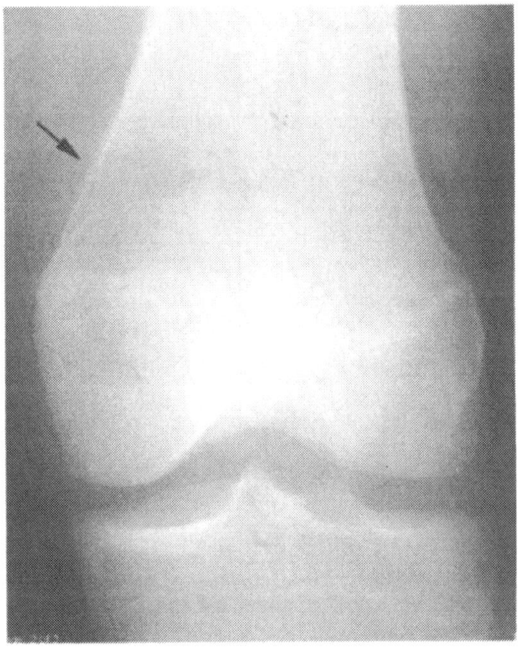

Figure 97D. AP radiograph of the left knee.

Figure 97E. Coronal T1 post-Gd FS 14 months after surgical treatment.

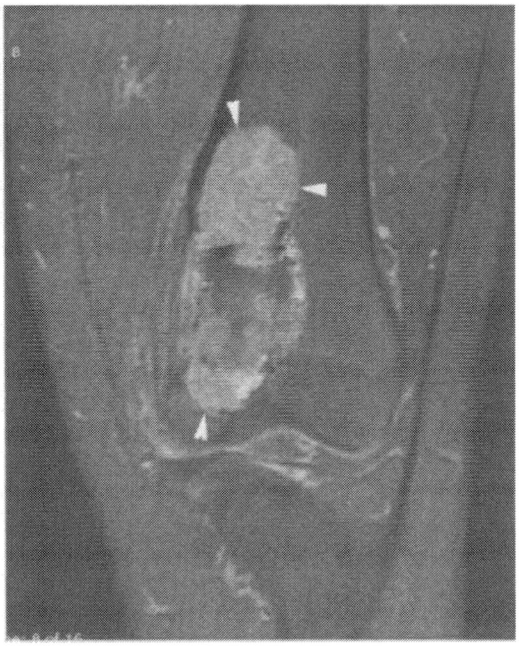

Figure 97F. Coronal T1 post-Gd FS, anterior to Figure 97E.

Diagnosis

Giant cell tumor of bone

Questions

1. T/F: Giant cell tumors do not metastasize.
2. T/F: Most giant cell tumors of bone in children are located in the epiphysis.

Discussion

Giant cell tumor (GCT) in the skeletally immature patient is rare, representing approximately 5.7% of the 876 cases of giant cell tumor referred to the Armed Forces Institute of Pathology (1). Based on this series, the tibia is most commonly affected. GCT is a histologically benign tumor characterized by multinucleated giant cells with osteoclastic activity. There is a relatively high local recurrence rate with contemporary surgical management, ranging from 7% after wide resection to 25% after curettage and bone grafting/methylmethacrylate plugging (Figures 97E, 97F) (2). Distant metastases may rarely occur, despite benign tumor histology (Answer to Question 1). The lung is the most common location for metastatic disease.

In contrast to GCT in adults, the tumor in children is frequently localized to the metaphysis rather than the epiphysis (Answer to Question 2). In Kransdorf et al.'s series, 48 of 50 GCTs were centered in the metaphysis (1). Once physeal maturation has occurred, epiphyseal extension to within 1 cm of the articular surface is often seen (Figures 97G–97I). On plain radiography, giant cell tumors are osteolytic, geographic lesions with a narrow zone of transition, and the margins are usually nonsclerotic. They may be central or eccentric in location, expansile, and demonstrate endosteal scalloping, cortical disruption, and periosteal reaction.

On MRI, both cystic and solid elements may be identified. The solid, highly cellular components of GCT are usually low to intermediate SI on T1 and fluid-sensitive sequences and demonstrate intense enhancement. Blood products may also be identified within giant cell tumors, and are hypointense on all imaging sequences as well as demonstrating blooming artifact on GRE sequences. The cystic components of these tumors are hypointense on T1, hyperintense on fluid sensitive sequences, may contain fluid-fluid levels, and demonstrate rim and/or septal enhancement. The margins of giant cell tumors are usually well defined by MRI with a thin marginal rim of hypointensity seen on all imaging sequences surrounding the entire tumor, presumably representing osseous sclerosis or pseudocapsule (2). With cortical disruption, extraosseous soft tissue and intra-articular extension may be seen. Approximately 14% of lesions demonstrate secondary aneurysmal bone cyst (ABC) formation (bony expansion, fluid-fluid levels, and multiple septations) (2). The solid enhancing component of giant cell tumors is characteristic and not a typical feature of primary ABC. On occasion it may be impossible to distinguish GCT from primary ABC, even on histologic grounds.

The differential diagnosis for giant cell tumors includes primary ABC, telangiectatic osteosarcoma, and subacute or chronic osteomyelitis with intraosseous abscess. All of these lesions may occur in the metaphysis, extend to the epiphysis, and demonstrate fluid-fluid levels. A soft tissue mass is typically absent with primary ABC, although approximately 12% will demonstrate central diffuse enhancement (3). A telangiectatic osteosarcoma will often demonstrate aggressive periosteal reaction, and show osteoid matrix calcification and soft tissue mass (4). Osteomyelitis is characterized by extensive edema that is usually out of proportion to the amount of enhancing solid soft tissue mass (granulation tissue).

Giant cell reparative granuloma is a reactive, tumor-like condition that biologically and histologically behaves differently than giant cell tumors. Giant cell reparative granuloma is usually indistinguishable from giant cell tumor of bone by MRI, but it tends to show fewer cystic components (2). It may occur in long tubular bones, but is most frequent in the jaw and the short tubular bones of the hands and feet (5).

The diagnosis of a giant cell tumor was pathologically proven in this patient. Curettage, bone packing with bone chips, and a fibular allograft were performed, but there was a local recurrence 14 months later (Figures 97E, 97F). A repeat curettage was performed.

Orthopedic Perspective

GCT is typically found in young adults and presents as an epiphyseal/metaphyseal radiolucency. Since GCT is rare in childhood, it is seldom considered in the differential diagnosis in this age group. In children it usually starts as a metaphyseal lesion eventually extending into the epiphysis. The orthopedist is primarily concerned with making the correct diagnosis and distinguishing GCT from ABC, giant cell reparative granuloma ("solid ABC"), Langerhans cell histiocytosis, infection, and most importantly osteosarcoma. Once the diagnosis is confirmed, the extent of the lesion in the bone and the presence of soft tissue extension are important to assess for planning treatment. Although locally very aggressive, treatment usually consists of a thorough curettage with an attempt to extend well beyond the area of radiographic abnormality to limit the likelihood of recurrence, which occurs about 25% of the time. The resulting cavity is filled with bone graft or polymethylmethacrylate cement. Giant cell tumors limited to an expendable bone, like the example of the fibular GCT, are treated with resection, which has a very low recurrence rate. Since GCT has a small metastatic potential and is occasionally multicentric, a chest CT and bone scan are essential components of the initial evaluation.

What the Clinician Needs to Know

1. Distinguish GCT from other bone tumors and infections, especially osteosarcoma.
2. The precise extent of the lesion for surgical planning.
3. Is there multicentric involvement or pulmonary metastasis?

Answers

1. False.
2. False.

Additional Example

Epiphyseal Giant Cell Tumor of the Fibula

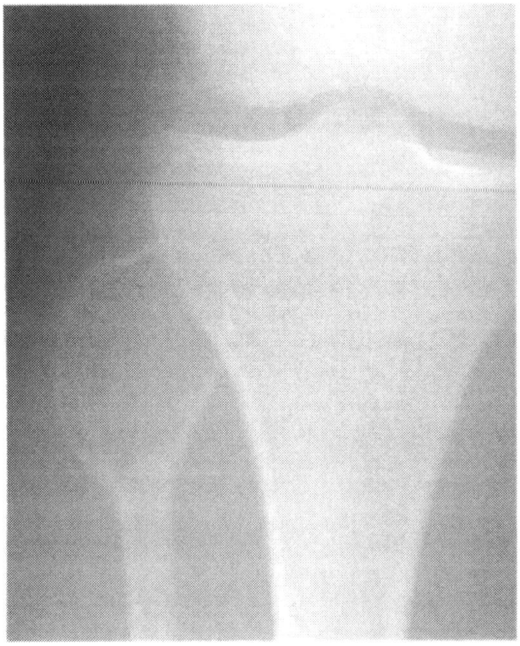

Figure 97G. AP radiograph of the right knee.

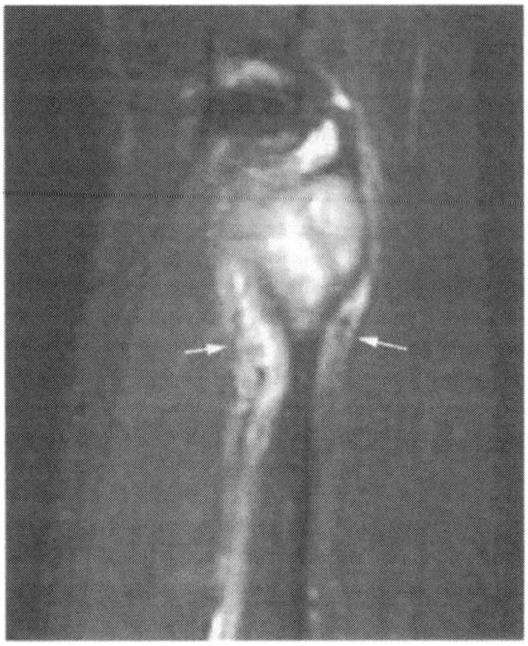

Figure 97H. Sagittal STIR through the right fibula.

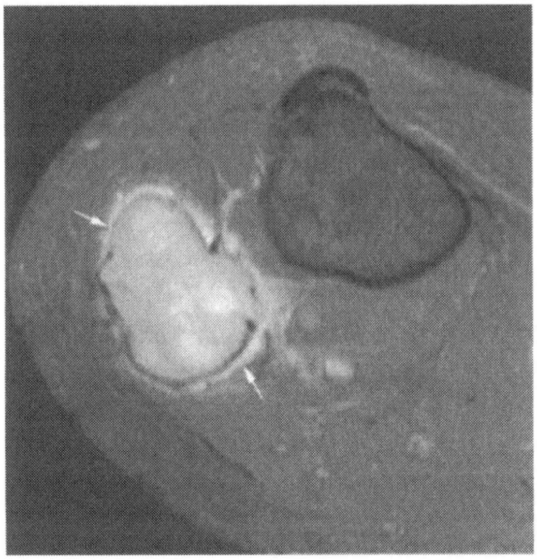

Figure 97I. Axial T1 post-Gd FS.

Findings

This is a 15-year-old girl with persistent right knee pain.

Figure 97G. There is an expansile epiphyseal lesion of the proximal fibula that extends to the articular surface. The proximal fibular physis is not identified and the adjacent tibial physis is fused.

Figures 97H, 97I. There is diffuse increased STIR SI and enhancement of the mass with transcortical extension to the extraosseous soft tissues **(arrows)**. This was surgically confirmed to represent a giant cell tumor.

Pitfalls and Pearls

1. The solid component of giant cell tumors often demonstrates hypointense to intermediate SI on all imaging sequences, indicating its highly cellular composition.
2. GCT confined to the metaphysis presents a diagnostic challenge since more common primary bone tumors, both benign and malignant, occur in this region in childhood.

References

1. Kransdorf MJ, Sweet DE, Buetow PC, Giudici MA, Moser RP, Jr. Giant cell tumor in skeletally immature patients. *Radiology* 1992; 184:233–237.
2. Murphey MD, Nomikos GC, Flemming DJ, Gannon FH, Temple HT, Kransdorf MJ. From the archives of AFIP. Imaging of giant cell tumor and giant cell reparative granuloma of bone: Radiologic-pathologic correlation. *Radiographics* 2001; 21:1283–1309.
3. Mahnken AH, Nolte-Ernsting CC, Wildberger JE, et al. Aneurysmal bone cyst: Value of MR imaging and conventional radiography. *Eur Radiol* 2003; 13:1118–1124.
4. Murphey MD, wan Jaovisidha S, Temple HT, Gannon FH, Jelinek JS, Malawer MM. Telangiectatic osteosarcoma: Radiologic-pathologic comparison. *Radiology* 2003; 229:545–553.
5. Ilaslan H, Sundaram M, Unni KK. Solid variant of aneurysmal bone cysts in long tubular bones: Giant cell reparative granuloma. *AJR Am J Roentgenol* 2003; 180:1681–1687.

Case 98

History

This 13-year-old boy injured his right knee after falling off a motorbike. The initial plain radiographs showed a small effusion but no fracture (not shown).

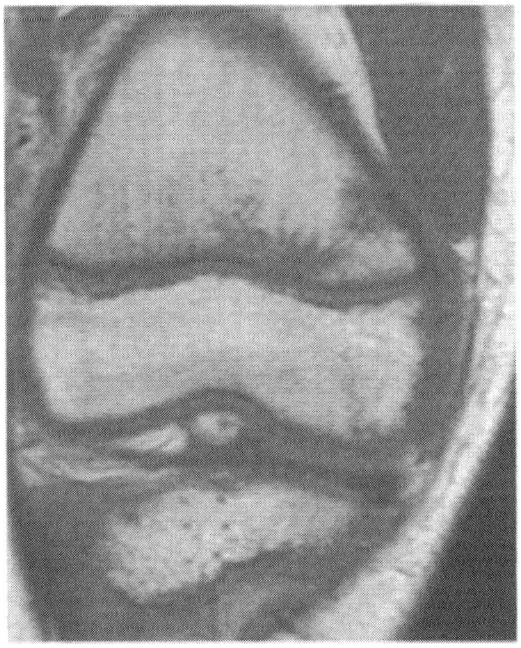

Figure 98A. Coronal T1 of the right knee.

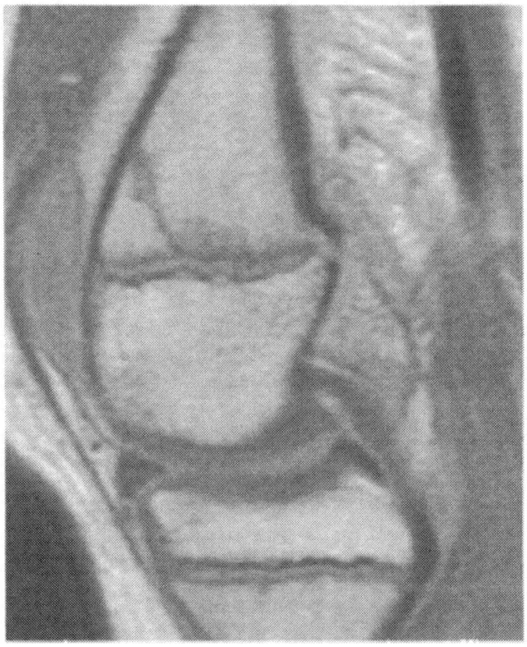

Figures 98B. Sagittal PD.

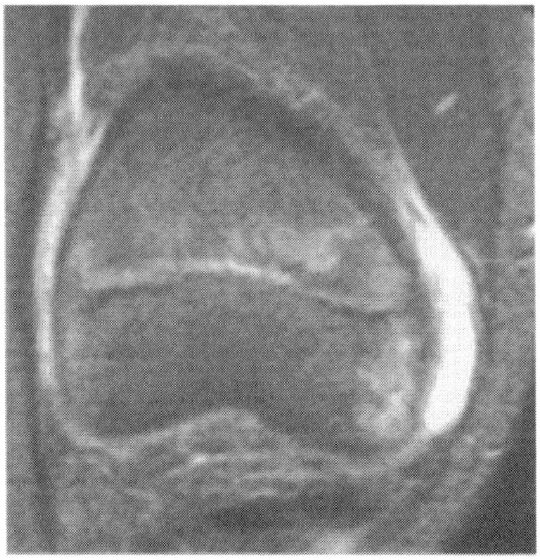

Figure 98C. Coronal STIR.

Figures 98A, 98B, (98B with annotations). There is a nondisplaced, oblique fracture in the anteromedial metaphysis of the femur that extends to the physis. The fracture line is hypointense on T1W and PD sequences **(arrow)**.

Figures 98C, (98C with annotations). The physis is of normal width, but there is central physeal STIR hyperintensity **(arrowhead)** that extends from the fracture entry site along the metaphysis **(arrow)**. There is also mild STIR hyperintensity within the medial femoral condyle **(thick arrow)** and a moderate joint effusion is present (*).

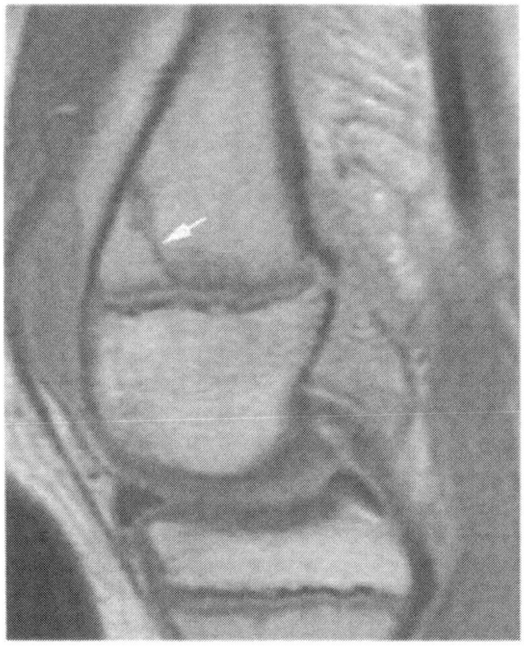

Figure 98B* Annotated.

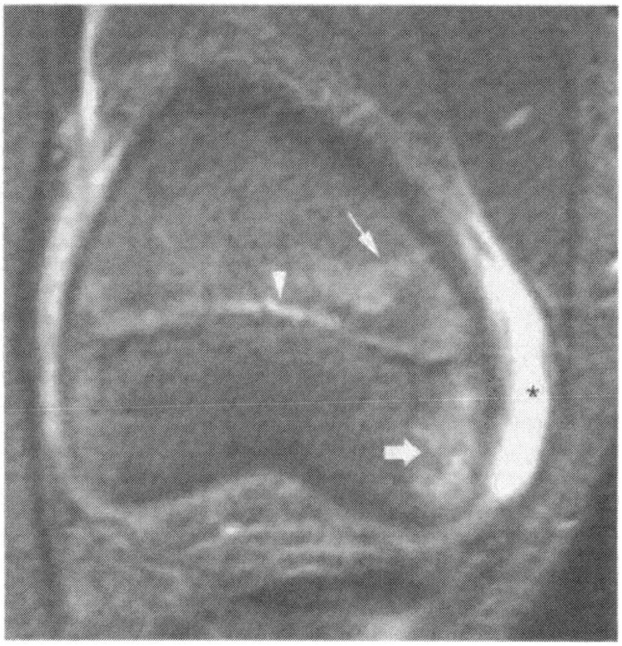

Figure 98C* Annotated.

Diagnosis

Salter-Harris type 2 fracture of the distal femur

Questions

1. What are the typical MRI features of a fracture line?
2. T/F: MRI is helpful in the evaluation of osteomyelitis complicating closed fractures.

Discussion

MRI evaluation of acute skeletal trauma includes: assessment of radiographically occult fractures, bone contusions, bone stress reaction, soft tissue injuries including ligamentous and tendinous pathology, better defining the extent of known fractures, and identifying chondral/osteochondral fractures. Close et al. reviewed 315 consecutive knee MRIs and found 8 physeal fractures (1). Management in 7 of the 8 children changed because the extent of the fracture was better defined compared with radiographs. For two of these patients, the fracture was found only after MRI.

Indications to proceed to MRI with negative radiographs include: persistent pain, refusal to bear weight on the affected extremity, or a high clinical suspicion for a fracture. If a hemarthrosis is present without radiographic abnormality, MRI may be used to help explain the hemarthrosis (Figure 98D). Wessel et al. found an etiology for the majority of 29 children who had a hemarthrosis but no evidence of fracture on plain radiographs (2). The most common osseous abnormality was bone contusions. MRI missed 3 chondral lesions and 2 osteochondral fractures that were subsequently found during arthroscopy or arthrotomy.

The MRI appearance of a nondisplaced fracture is linear decreased SI on T1 and variable hypointensity or hyperintensity on fluid sensitive sequences (Answer to Question 1). Significant marrow edema is often evident surrounding the fracture line. The fracture plane should be described as it relates to the joint, physis, and articular cartilage (Figure 98E, 98F). With subacute fractures, there may be corresponding contrast enhancement along the fracture line margin, reflecting granulation tissue. Cartilage sensitive sequences, including PD FS and 3D SPGR FS, or equivalent may be used to characterize physeal and osteochondral injury.

Patients with closed fractures are occasionally referred to MRI for the evaluation of possible osteomyelitis shortly after their injury (Figures 98G, 98H). Unlike open fractures, osteomyelitis complicating closed fractures is extremely rare. In a meta-analysis of 895 closed tibia fractures in adults, the incidence of osteomyelitis was 0% for closed treatment, 0.4% with plate fixation, 1% with reamed nailing, and 1.5% with unreamed nailing (3). No large study evaluating osteomyelitis in children after closed fractures has been performed, although case reports exist (4). In contrast, the reported incidence of osteomyelitis following open fractures is 2% to 16% (5). Differentiating a healing fracture from one associated with osteomyelitis is difficult by MRI because subperiosteal hematomas, reactive granulation tissue, and marrow edema related to a healing fracture are usually indistinguishable from the MRI changes of osteomyelitis (Answer to Question 2). Subperiosteal hematomas related to a fracture may show rim enhancement and may be mistaken for a subperiosteal abscess. Marrow edema is nonspecific since it may be seen with trauma or infection. Therefore, in patients with closed fractures and the suspicion of osteomyelitis, the MRI must be carefully correlated with the clinical findings.

The indication for MRI in this patient was to assess for internal derangement related to his motorbike accident since the plain radiographs were normal. The Salter-Harris

(SH) type 2 fracture was isolated without associated internal derangement. This patient was treated with casting and did well on follow-up.

Orthopedic Perspective

As seen in this case, the management of acute traumatic effusion in the setting of negative plain radiographs may be significantly altered based on the results of MRI. Hemarthrosis may result from a SH fracture requiring cast immobilization or secondary to internal derangements, such as ligamentous, meniscal, and chondral injuries, which often requires arthroscopic surgery. MRI may also give prognostic information in regard to growth disturbance and early degenerative changes and may help guide how closely to follow a patient after a fracture. For instance, SH fractures may lead to secondary physeal growth disturbance. SH fractures that extend to the articular surface (Figure 98E) may also lead to secondary degenerative changes.

What the Clinician Needs to Know

1. Correctly identify a cause of an acute traumatic effusion. Is the effusion secondary to a radiographically occult fracture and/or internal derangement?
2. Does the SH fracture extend to the articular surface?

Answers

1. Hypointense on T1, variable T2 SI, and enhancement, usually during the subacute phase.
2. False.

Additional Examples

SH 3 Fracture of the Distal Femur with Hemarthrosis

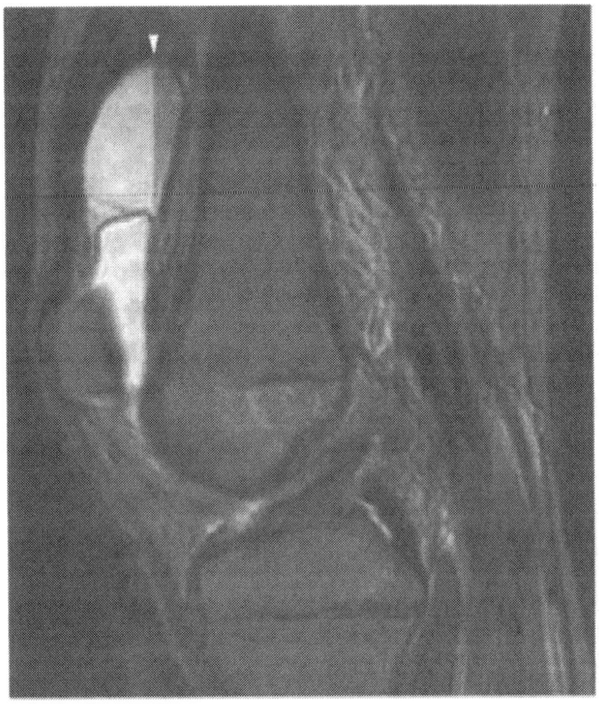

Figure 98D. Sagittal T2 FS of the right knee.

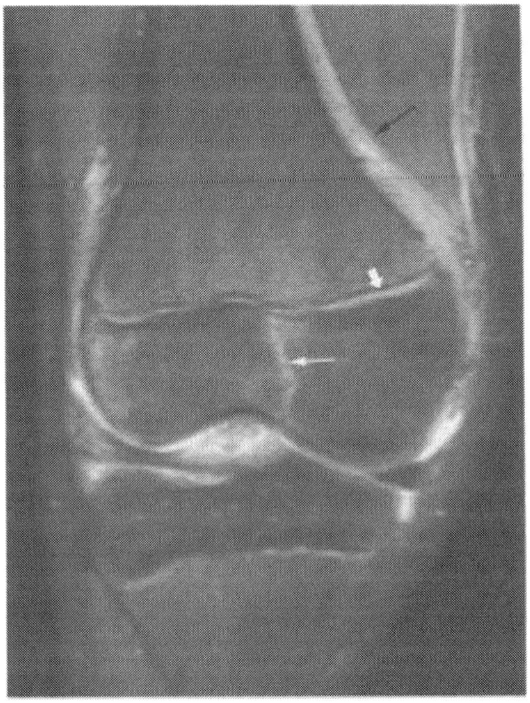

Figure 98E. Coronal PD FS.

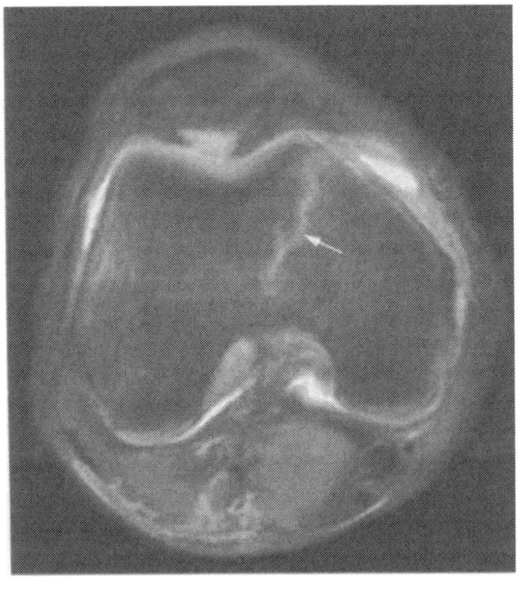

Figure 98F. Axial PD FS.

Findings

This 13-year-old boy injured his knee while playing football.

Figure 98D. There is a large joint effusion with a fluid-fluid level **(white arrowhead)**, indicating a hemarthrosis.

Figures 98E, 98F. There is a SH 3 fracture extending from the physis to the articular cartilage of the medial femoral condyle **(white thin arrows)**. The medial physis is slightly widened **(thick arrow)**. The medial femoral condylar articular cartilage at the fracture site appears intact. There is also juxtacortical edema present **(black thin arrow)**.

Subacute SH 2 Fracture of the Distal Femur with Subperiosteal Hematoma and Bone Infarction

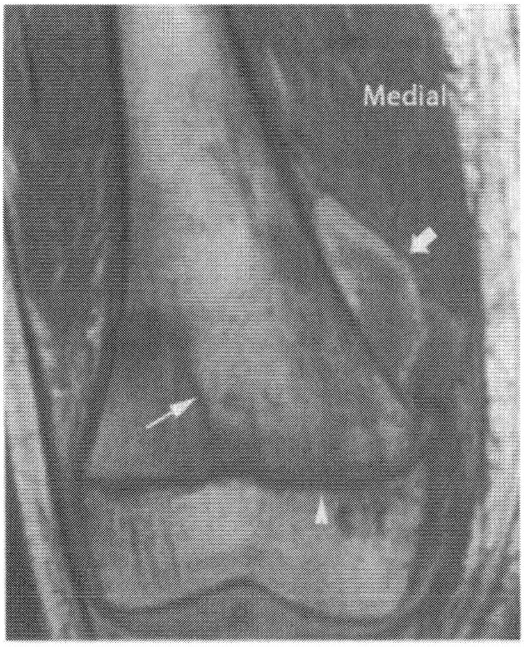

Figure 98G. Coronal T1 of the right distal femur.

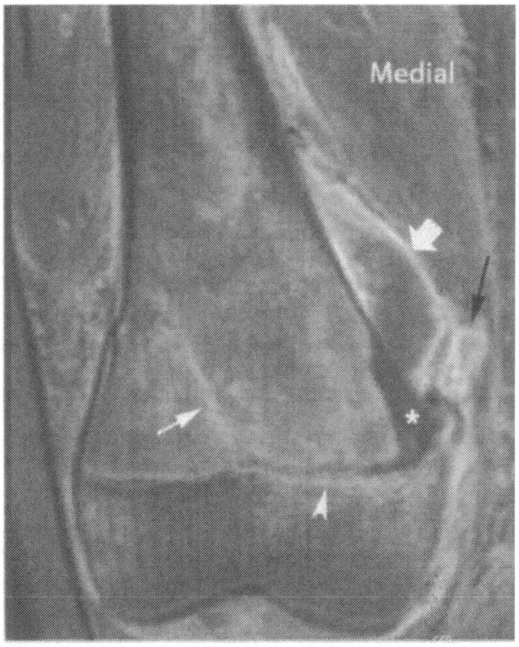

Figure 98H. Coronal T1 post-Gd FS.

Findings

This 14-year-old boy had a closed SH 2 fracture of his distal femur 10 days ago while skiing. He had persistent pain, fever, and elevated sedimentation rate and the study was requested to evaluate for osteomyelitis.

Figures 98G, 98H. There is a fracture of the lateral aspect of the distal femoral metaphysis **(white thin arrow)** that extends into the physis **(arrowheads)**, constituting an SH 2 injury. The medial physis is abnormally widened **(arrowheads)**. The fracture margins show enhancement, which is consistent with granulation tissue. There is also a medial subperiosteal fluid collection **(thick arrows)** and a smaller inferior component that is multilocular **(black thin arrow)**. Note a triangular nonenhancing focus (*) along the medial metaphysis that is hyperintense on pre-Gd image (hypointense on T2 FS sequences—not shown), consistent with devascularized bone. Foci of metaphyseal bone infarction are not unusual with SH fractures of the distal femur. At surgery, the subperiosteal fluid collection contained blood, and all cultures, including bone and joint fluid, were negative.

Pitfalls and Pearls

1. MRI may be falsely positive for loose bodies in the setting of trauma since both osteochondral fragments and blood products may produce filling defects in a joint effusion.
2. Do not mistake normal trabecular variation for a fracture on T1 and PD sequences. Look for other clues that are seen with a true fracture, such as marrow edema, chondral injury, and subperiosteal fluid collections.

References

1. Close BJ, Strouse PJ. MR of physeal fractures of the adolescent knee. *Pediatr Radiol* 2000; 30:756–762.
2. Wessel LM, Scholz S, Rusch M, et al. Hemarthrosis after trauma to the pediatric knee joint: What is the value of magnetic resonance imaging in the diagnostic algorithm? *J Pediatr Orthop* 2001; 21:338–342.
3. Coles CP, Gross M. Closed tibial shaft fractures: management and treatment complications: A review of the prospective literature. *Can J Surg* 2000; 43:256–262.
4. Veranis N, Laliotis N, Vlachos E. Acute osteomyelitis complicating a closed radial fracture in a child: A case report. *Acta Orthop Scand* 1992; 63:341–342.
5. Ehara S. Complications of skeletal trauma. *Radiol Clin North Am* 1997; 35:767–781.

Case 99

History

This is a 9-year-old girl with a 2-month history of left heel pain. There is no history of trauma.

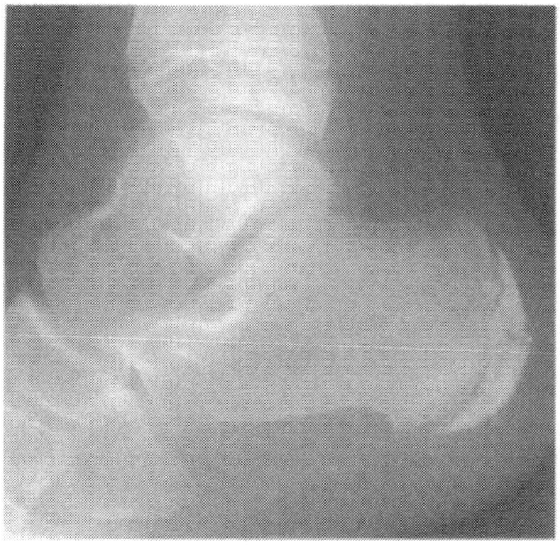

Figure 99A. Lateral radiograph of the LEFT foot.

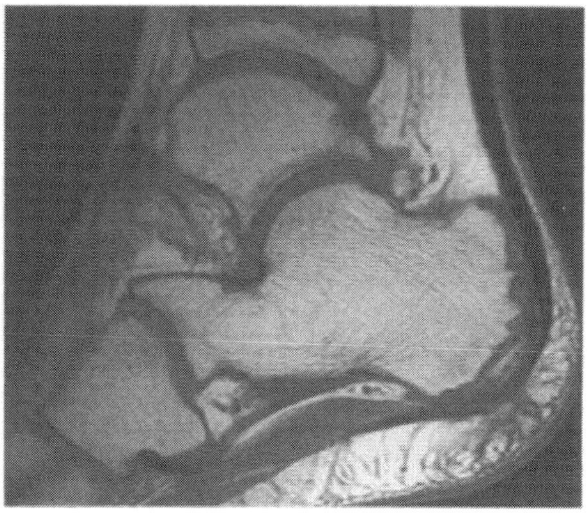

Figure 99B. Sagittal T1.

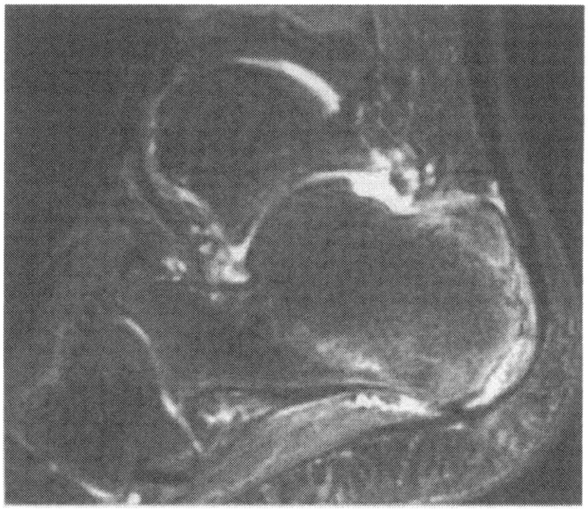

Figure 99C. Sagittal T2 FS.

Figure 99A. The calcaneal apophysis is dense and bipartite; this is a normal finding. Kager's triangle and the Achilles tendon are identified and are also normal.

Figures 99B, 99C, (99C with annotations). There is decreased T1 SI and moderate T2 hyperintensity in the calcaneal apophysis **(arrowhead)**. A small amount of fluid is present in the retrocalcaneal bursa **(thin arrow)**. There is also mild increased T2 SI within the body of the calcaneus **(thick arrows)** compatible with stress reaction.

Figure 99D. For comparison the calcaneus and calcaneal apophysis are normal in the right foot.

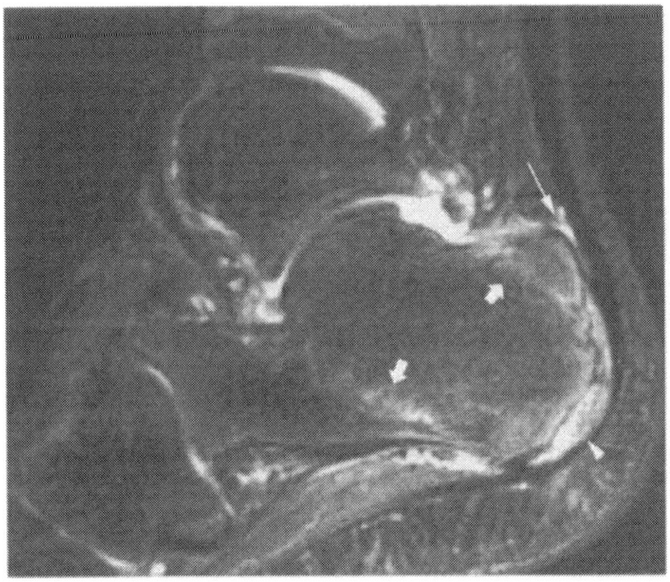

Figure 99C* Annotated.

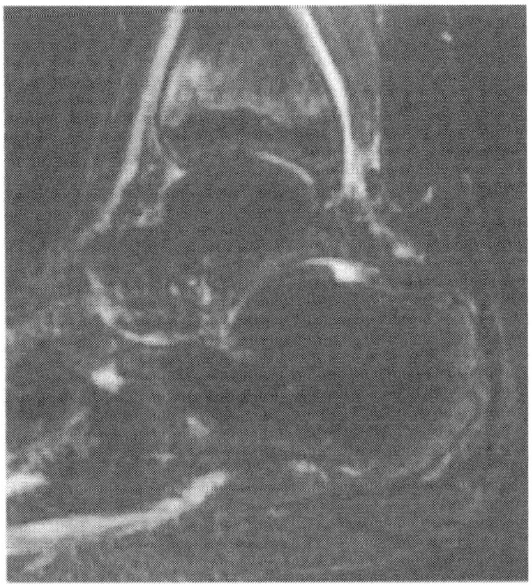

Figure 99D. Sagittal T2 with FS of the *right* foot.

Diagnosis

Sever's disease (calcaneal apophysitis)

Question

1. What are three osteochondroses of the foot?

Discussion

Three osteochondroses of the foot are Sever's disease (calcaneal apophysis), Kohler's disease (navicular) (Figure 99E), and Freiberg's infraction (metatarsal heads) (Figures 99F–99H) (Answer to Question 1). Kohler's disease tends to occur in boys from 3 to 10 years of age, Freiberg's infraction in athletic adolescent girls, and Sever's disease in children 7 to 10 years of age (1, 2).

Osteochondrosis is a nonspecific term that has traditionally been used to describe a symptomatic physeal growth disturbance (e.g., Blount's disease) or irregular fragmentation and/or sclerosis of an epiphysis or apophysis (Table 99A). Symptoms are related to nonspecific "epiphysitis" or "apophysitis." The etiologies of most osteochondroses are poorly understood, and thus many theories have been proposed depending on the site of involvement. Theories include chronic microtrauma, ischemic necrosis, and chronic avulsion injury from tendinous attachments, pyogenic infection, and nonpyogenic inflammatory disorders (3–6).

This contrasts with osteochondritis dissecans (OCD), a disorder generally attributed to chronic microtrauma (7). It is characterized by focal subchondral fragmentation and tends to occur in particular areas, such as the capitellum, talus, and femoral condyles. The majority of osteochondroses resolve spontaneously independent of treatment. OCD, on the other hand, may lead to loose bodies and secondary degenerative changes. There is likely overlap between the osteochondroses and OCD, depending on patient age, anatomic site, and the underlying pathologic process.

The radiologic description of most osteochondroses occurred before the advent of MRI. In some instances, the early reports of osteochondroses were probably descriptions of the normal irregular patterns of secondary ossification (8). Differentiation of these normal variants from pathologic processes was impossible on imaging grounds and tenuous on clinical criteria.

MRI provides important anatomic and physiologic data that assist in distinguishing normal patterns of ossification from true "epiphysitis" or "apophysitis." MRI better

Table 99A. Selected osteochondroses.

NAME	LOCATION	AGE (years)
Kohler's disease	Navicular	3–10
Freiberg's infraction	Metatarsal head	13–18
Sever's disease	Calcaneal apophysis	7–10
Osgood-Schlatter disease	Tibial tuberosity apophysis	11–15
Sinding-Larsen-Johansson disease	Inferior patella	10–14
Blount's disease	Medial proximal tibia physis	Infantile: 1–3 Adolescence: 8–15
Scheuermann's disease	Vertebral body apophysis	13–17
Panner's disease	Capitellum	5–11

delineates the symptomatic secondary ossification center. Symptomatic sites may be hypointense on T1W sequences and show corresponding hyperintensity on fluid-sensitive sequences. T2 hypointensity may be seen as well, particularly if the dominant feature is epiphyseal/apophyseal sclerosis or ischemic necrosis. Since T2 SI may be variable within the marrow and cartilage adjacent to maturing secondary ossification centers, imaging the asymptomatic extremity is often helpful for comparison. When an osteochondrosis affects the physis (e.g., Blount's disease), physeal narrowing and fibrous/bony bar formation may be seen.

The clinical outcome of the osteochondroses is dependent on the location and the severity of the pathologic process. Sever's and Kohler's disease usually resolve spontaneously, independent of treatment. Freiberg's infraction, which has been generally accepted to represent trauma induced ischemic necrosis, may result in flattening of the metatarsal heads and secondary degenerative changes. (For a discussion of other forms of osteochondrosis, see Case 20, Blount's; Case 32, Scheuermann's; Case 70, Panner's; and Case 92, Osgood-Schlatter.)

This patient was given crutches, part-time casting, and physical therapy. Her symptoms of left heel pain resolved after 3 months.

Orthopedic Perspective

The diagnosis of Sever's disease is usually made clinically based on pain and tenderness to palpation of the calcaneal tuberosity. Treatment is with a heel cup and physical therapy to stretch the heel cord. The clinical diagnosis of Kohler's disease is made based on pain with weightbearing and tenderness over the mid-tarsal region. Although radiographs are often used to confirm the clinical diagnosis, distinguishing Koehler's disease from common normal variations of navicular ossification is often challenging. Treatment of Kohler's disease is with a short-leg cast. The clinical diagnosis of Freiberg's infraction is forefoot pain that is aggravated by metatarsal head palpation and with motion of the affected MTP joint. Radiographs are used to confirm the clinical diagnosis. Therapeutic options range from medical therapy to surgical options ranging from debridement to osteotomies of the metatarsal head.

MRI evaluation may be indicated for Sever's disease, Kohler's disease, and Freiberg's infraction when symptoms are atypical, radiographs are normal, or alternative etiologies for foot pain are being considered, such as an occult stress fracture, osteomyelitis, or neoplasm.

What the Clinician Needs to Know

1. Distinguish normal patterns of secondary ossification center development from a true epiphysitis or apophysitis.
2. The degree of epiphyseal/tarsal collapse and secondary degenerative changes when present.

Answer

1. Sever's disease, Kohler's disease, and Freiberg's infraction.

Additional Examples

Kohler's Disease of the Right Foot

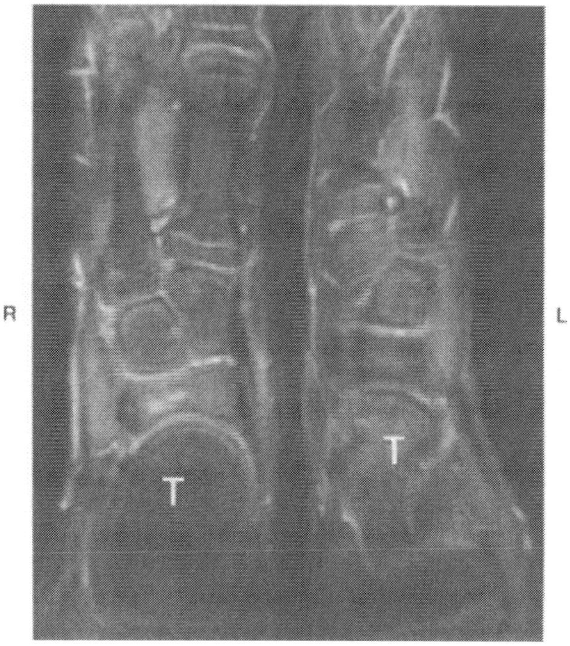

Figure 99E. Axial (footprint) STIR of both feet.

Findings

This 5-year-old boy had right foot pain for months.

Figure 99E. MRI demonstrates right navicular flattening and increased STIR SI. Note the normal SI of the left navicular. Talus (T).

Freiberg's Infraction

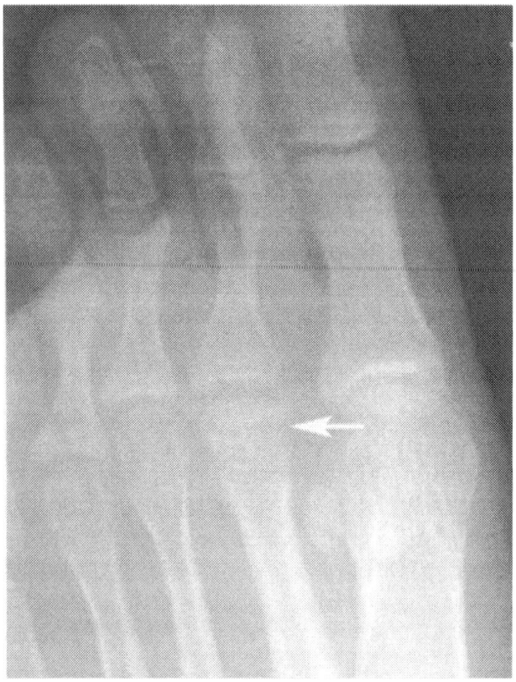

Figure 99F. AP radiograph of the foot.

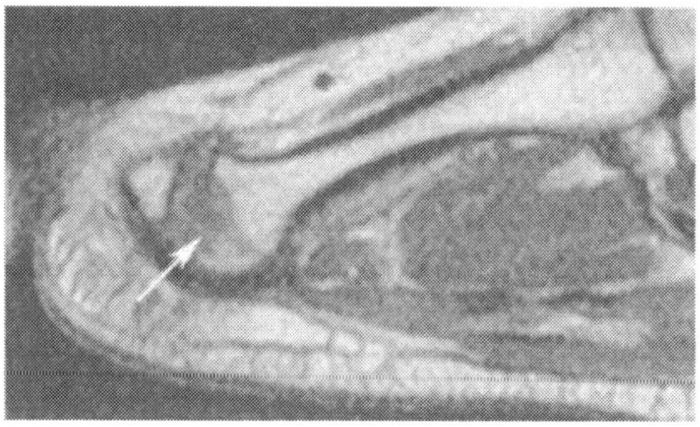

Figure 99G. Sagittal T1 (through second metatarsal head).

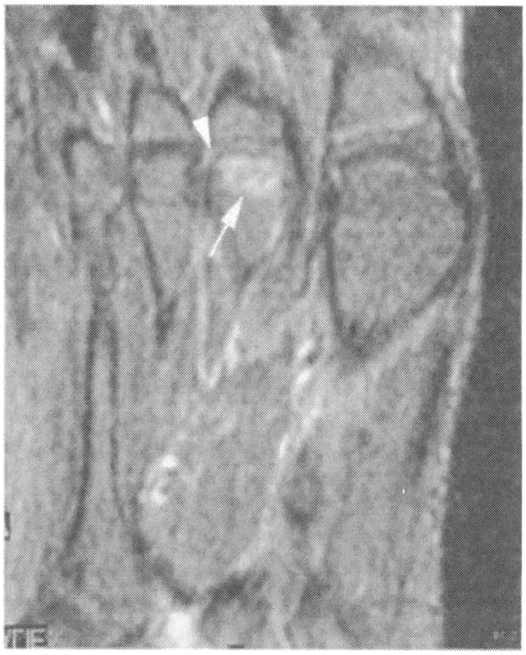

Figure 99H. Axial (footprint) STIR.

Findings

This is a 15-year-old girl with chronic foot pain.

Figure 99F. The second metatarsal head is flattened with subchondral lucency and sclerosis **(arrow)**. The second MTP joint is wide.

Figures 99G, 99H. There is decreased T1 SI and corresponding STIR hyperintensity within the second metatarsal head **(arrows)**. There is also a tiny joint effusion within the second MTP joint **(arrowhead)**.

Pitfalls and Pearls

1. Koehler's disease should be distinguished from Mueller-Weiss syndrome, a spontaneous osteonecrosis of the navicular affecting adult women (9).
2. The irregular, sclerotic, and fragmented appearance of the developing epiphysis, apophysis, or epiphyseal-like centers is often confused with pathology. Use MRI as a problem-solving tool to differentiate these normal variations in ossification from an osteochondrosis.

References

1. Harty MP. Imaging of pediatric foot disorders. *Radiol Clin North Am* 2001; 39:733–748.
2. Adirim TA, Cheng TL. Overview of injuries in the young athlete. *Sports Med* 2003; 33:75–81.
3. Ogden JA, Ganey TM, Hill JD, Jaakkola JI. Sever's injury: A stress fracture of the immature calcaneal metaphysis. *J Pediatr Orthop* 2004; 24:488–492.
4. Medlar RC, Lyne ED. Sinding-Larsen-Johansson disease: Its etiology and natural history. *J Bone Joint Surg Am* 1978; 60:1113–1116.
5. Rosenberg ZS, Kawelblum M, Cheung YY, Beltran J, Lehman WB, Grant AD. Osgood-Schlatter lesion: Fracture or tendinitis? Scintigraphic, CT, and MR imaging features. *Radiology* 1992; 185:853–858.
6. Klein EW. Osteochondrosis of the capitellum (Panner's disease): Report of a case. *Am J Roentgenol Radium Ther Nucl Med* 1962; 88:466–469.
7. Jaramillo D, Shapiro F. Musculoskeletal trauma in children. *Magn Reson Imaging Clin N Am* 1998; 6:521–536.
8. Resnick D. Osteochondroses. In: *Diagnosis of Bone and Joint Disorders*, 4th ed., Chapter 74. WB Saunders. 2002; 3686–3741.
9. Haller J, Sartoris DJ, Resnick D, et al. Spontaneous osteonecrosis of the tarsal navicular in adults: Imaging findings. *AJR Am J Roentgenol* 1988; 151:355–358.

Case 100

History

This is a 1-year-old girl who refuses to bear weight on her left foot.

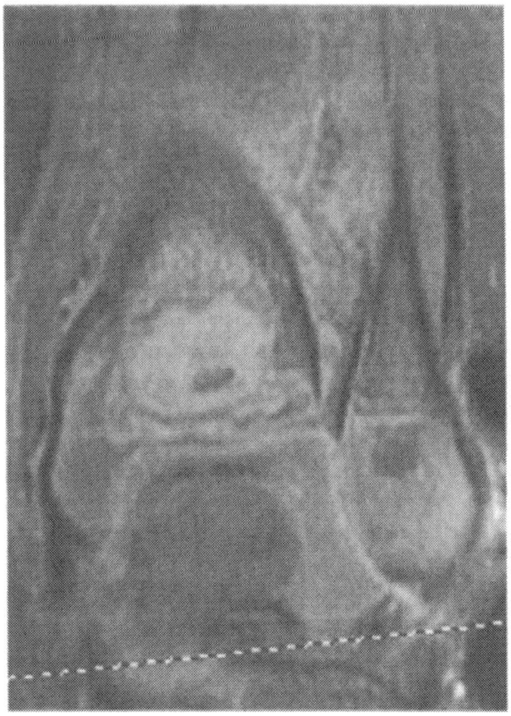

Figure 100A. Coronal PD FS, left ankle.

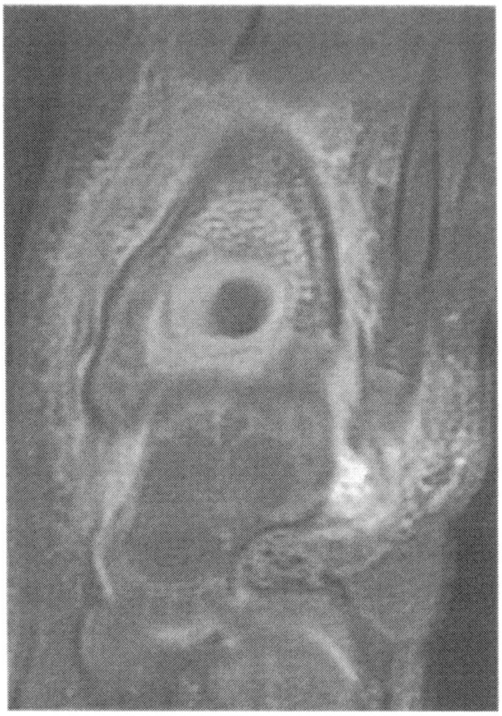

Figures 100B. Coronal T1 post-Gd FS.

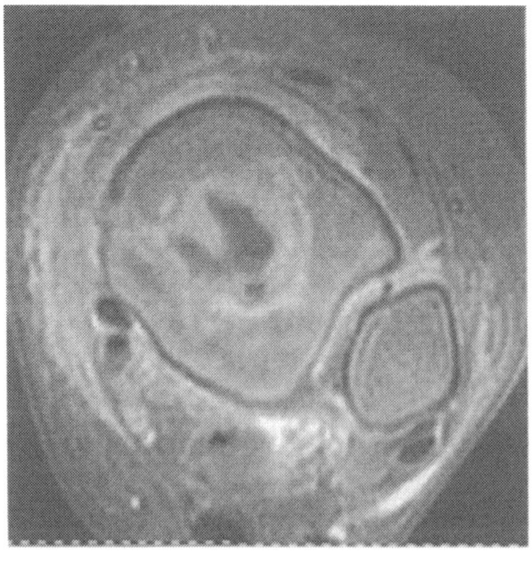

Figure 100C. Axial T1 post-Gd FS.

Figures 100A, (100A with annotations). There is a round, multilayered lesion centered at the distal tibia physis. There is a hyperintense focus with a sharply circumscribed hypointense (sclerotic) rim **(arrowheads)**. There is a zone of surrounding bone marrow edema (E). The well-circumscribed hypointense center represents the remnant of the ossified distal tibial epiphysis **(thick arrow)**, correlating with the radiograph (Figures 100D).

Figures 100B, 100C, (100B, 100C with annotations). There is a conspicuous nonenhancing central abscess (A) surrounded by enhancing granulation tissue (*). Marrow enhancement (E) and a sclerotic margin **(arrowheads)** correlate with the pre-contrast image. The epiphyseal remnant noted on the PD FS sequence is sequestered, and obscured by the adjacent abscess. A cortical break **(thin arrow in Figure 100C)** is present along the medial margin of the tibial metaphysis. This fistulous tract, or cloaca, communicates with the thickened and enhancing juxta-cortical soft tissues.

Figure 100D. Initial radiograph obtained 1 month earlier is normal.

Figure 100E. The radiograph obtained at the time of MRI shows diffuse smooth subperiosteal new bone formation along the tibial shaft, and an osteolytic metaphyseal lesion with a sclerotic rim **(arrowhead)**. A small, sclerotic epiphyseal remnant is seen **(thick arrow)**.

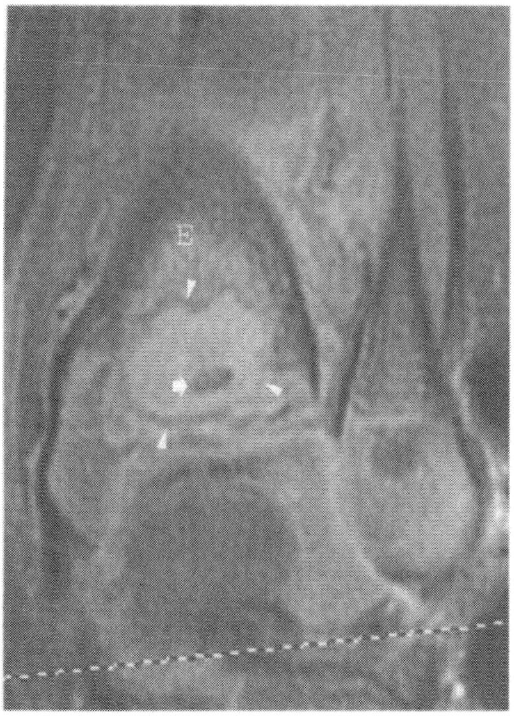

Figure 100A* Annotated.

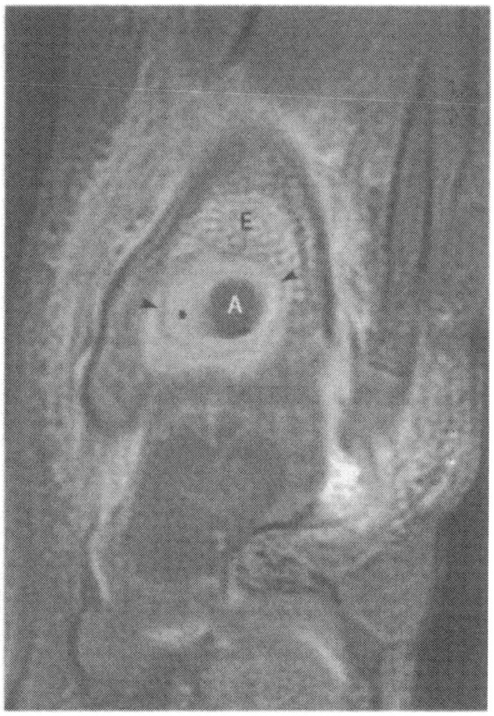

Figure 100B* Annotated.

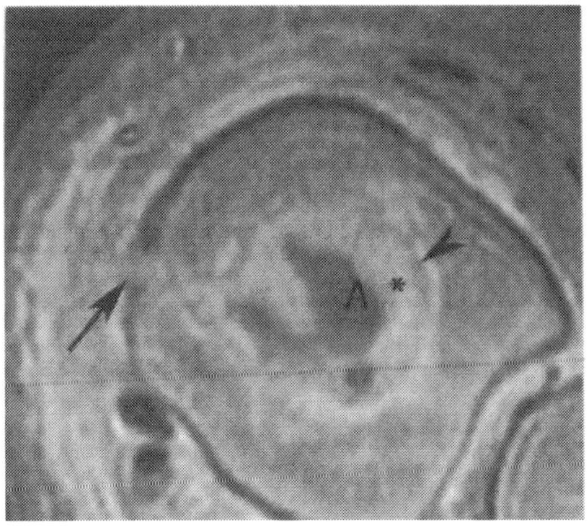

Figure 100C* Annotated detail.

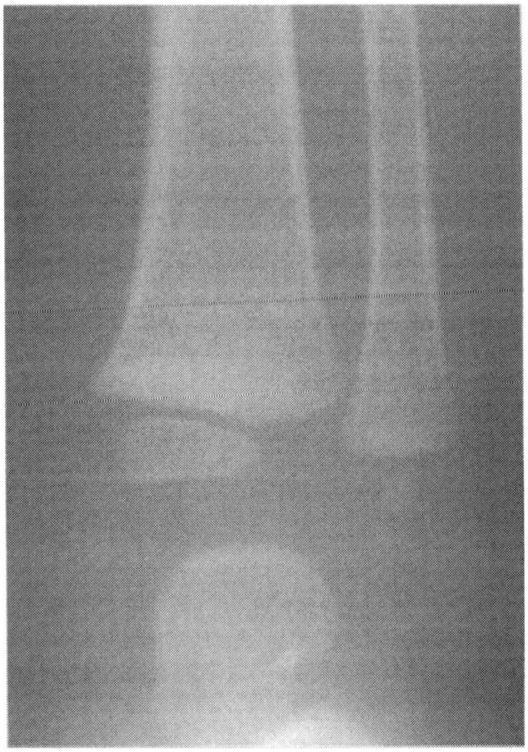

Figure 100D. AP radiograph 1 month earlier.

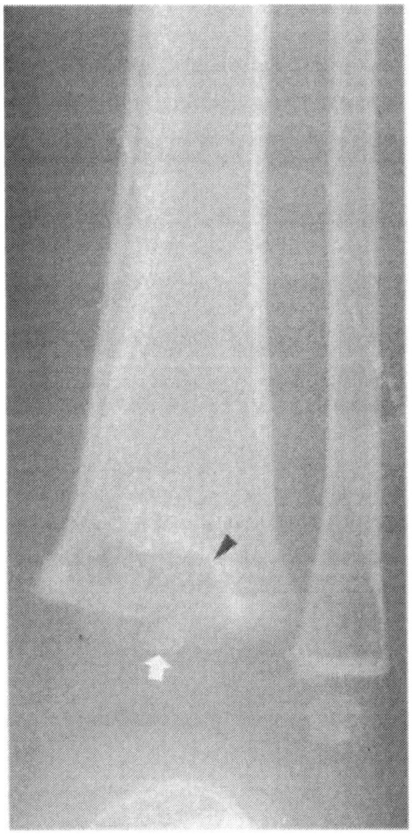

Figure 100E. AP radiograph at the time of MRI.

Diagnosis

Subacute osteomyelitis with a Brodie's abscess

Questions

1. What MRI sign characterizes a Brodie's abscess?
2. What imaging findings differentiate an abscess cavity from a bony sequestrum?

Discussion

This case demonstrates a Brodie's abscess and a variety of other features of subacute osteomyelitis:

- *Brodie's abscess:* Brodie's abscess has a characteristic target sign (Answer to Question 1) (1). Brodie's abscesses are usually located in the metaphysis of the long bones; however, in this case the abscess is centered at the physis. A characteristic multilayered appearance is present:
  1. The center is composed of pus. Therefore, it is hyperintense on fluid sensitive sequences and does not enhance. In this case, a small hypointense focus is present within the pus, which represents the sequestered epiphyseal remnant within the Brodie's abscess cavity.
  2. Vascular rich granulation tissue surrounds the pus. It is slightly hyperintense on T1W, intermediate to hyperintense on T2W sequences (but hypointense compared with the pus filled center), and strongly enhances. This appearance has been referred to as the penumbra sign (2).
  3. The outer edge represents sclerotic bone reaction (1). It is of variable thickness and is hypointense on all imaging sequences. Reactive edema and enhancement are present beyond the outer edge of the Brodie's abscess.
- *Cloaca:* A cloaca is a fistulous tract between a medullary bone abscess cavity and the extraosseous soft tissues. Abscess fluid under pressure escapes through the eroded and weakened cortex to the extraosseous soft tissues.
- *Sequestrum:* Classically, the term sequestrum applies to a necrotic bone fragment within the medullary cavity of a long bone. It represents a devascularized bone fragment surrounded by granulation tissue and normal bone. Therefore, it follows cortical bone SI on all imaging sequences: it is hypointense on fluid-sensitive sequences and does not enhance. However, the margin of the sequestrum may show modest enhancement, representing granulation tissue. This contrasts with the appearance of abscess fluid, which is hyperintense on fluid sensitive sequences and does not enhance (Answer to Question 2). The sequestrum may not be visible on radiography during the early stages of formation (Figures 100F, 100G). MRI may delineate a sequestrum before it becomes evident by radiographs.
- *Epiphyseolysis:* This term can be applied to loss of epiphyseal mineralization due to an associated bacterial infection. Historically, it has been referred to as the vanishing epiphysis due to hyperemia associated with septic arthritis (3). In this case, the loss of epiphyseal mineralization appears to be related, at least in part, to direct bacterial destruction of the secondary ossification center.
- *Involucrum:* An involucrum represents a sheath of reactive bone formation that surrounds dead bone (sequestrum). With healing, an involucrum may incorporate with the sequestrum and appear as cortical thickening.

This patient was initially thought to have an occult fracture, based on the initial normal radiographs. When her symptoms did not improve, the diagnosis of osteomyelitis was made after follow-up radiographs and MRI. The abscess cavity was surgically debrided and grew *Staphylococcus aureus*.

Orthopedic Perspective

Subacute osteomyelitis with a Brodie's abscess, cloaca, and sequestrum are treated with surgical drainage and antibiotics. Antibiotics alone will not penetrate the abscess or effectively treat the dead bone of the sequestrum. MRI guides surgical treatment by defining the size and location of the Brodie's abscess, cloaca, and sequestrum.

What the Clinician Needs to Know

1. The degree of involvement of the physis and epiphysis. This may help predict future growth arrest and deformity.
2. The size of the abscess and sequestrum, as well as the extent of soft tissue inflammatory changes. What is the best surgical approach for drainage?

Answers

1. Target sign.
2. An abscess cavity is hyperintense on T2W sequences. A sequestrum is hypointense on all sequences.

Additional Example

Staphylococcus aureus Osteomyelitis with Early Intramedullary Sequestrum

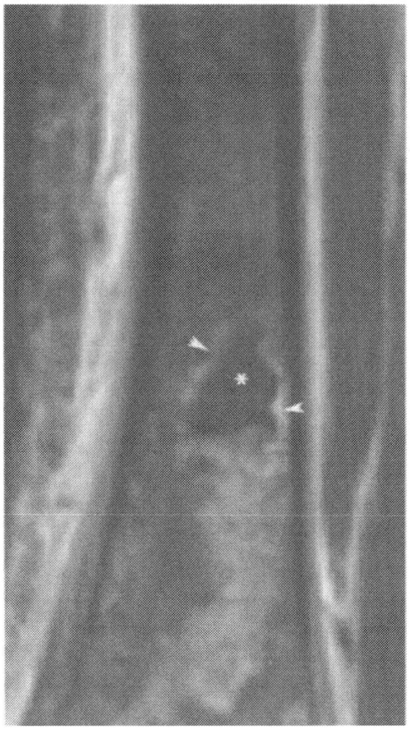

Figure 100F. Sagittal STIR of the left ankle.

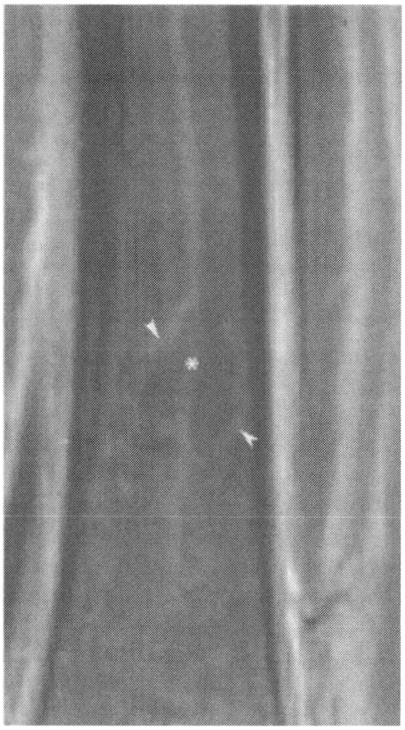

Figure 100G. Sagittal T1 post-Gd FS.

Findings

This is a 13-year-old boy who presented with ankle swelling. Plain radiographs were normal (not shown).

Figures 100F, 100G. STIR and post-Gd images demonstrate a low SI geographic area (*) within the metadiaphysis consistent with a sequestrum. There is a thin rim of increased STIR SI and enhancement reflecting granulation tissue along the outer margin of the sequestrum **(arrowheads)**.

Pitfalls and Pearls

The differential diagnosis for an osteolytic epiphyseal lesion includes neoplasms, such as Langerhans cell histiocytosis, leukemia, metastatic neuroblastoma, and chondroblastoma.

References

1. Gylys-Morin VM. MR imaging of pediatric musculoskeletal inflammatory and infectious disorders. *Magnetic Resonance Imaging Clinics of North America* 1998; 6:537–559.
2. Davies AM, Grimer R. The penumbra sign in subacute osteomyelitis. *Eur Radiol* 2005; 15:1268–1270.
3. Wood BP. The vanishing epiphyseal ossification center: A sequel to septic arthritis of childhood. *Radiology* 1980; 134:387–389.

Case 101

History

This 15-year-old girl had bilateral anterior shin pain for the last 3 weeks. She is a track runner in high school. She has anterior point tenderness 6 cm distal to her knees bilaterally.

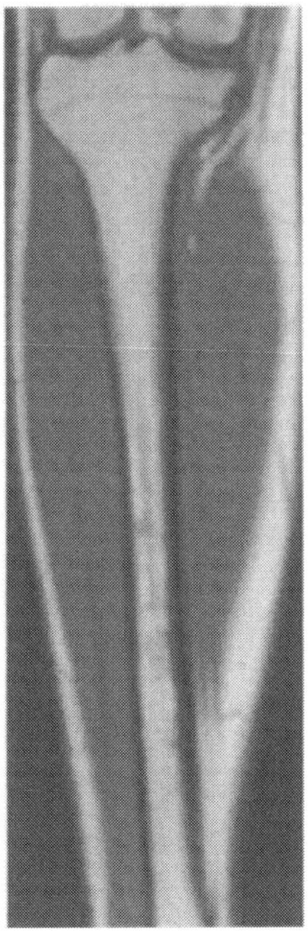

Figure 101A. Coronal T1 of the right tibia.

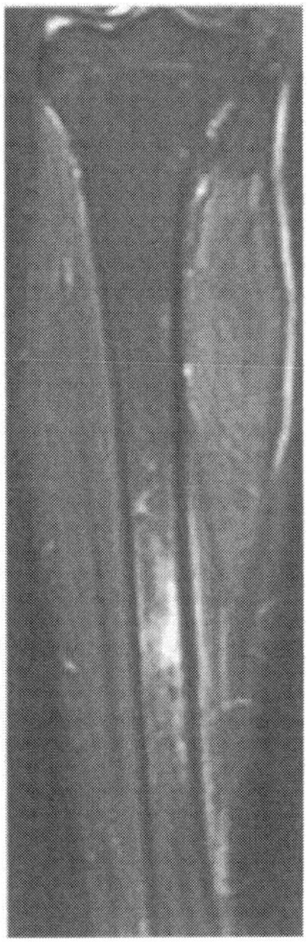

Figure 101B. Coronal STIR.

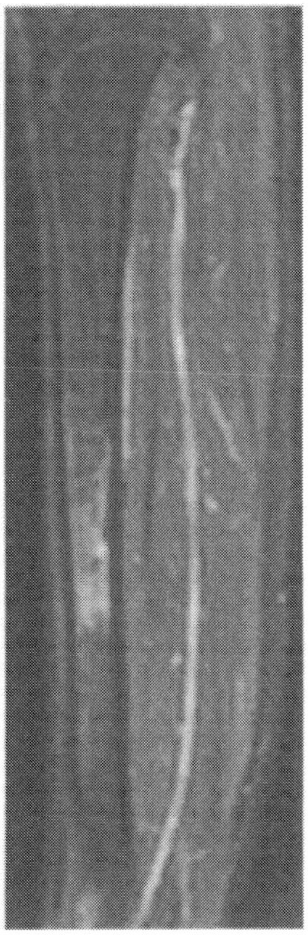

Figure 101C. Sagittal STIR.

Figures 101A, 101B, 101C, (101B with annotations). There is mild T1 hypointensity and mild STIR hyperintensity within the medullary cavity **(arrowhead)** and endosteal surface of the mid-tibia. Focal periosteal hyperintensity is evident **(arrow)**. A discrete fracture line is not seen. Similar signal abnormality was also present in the left tibia (not shown).

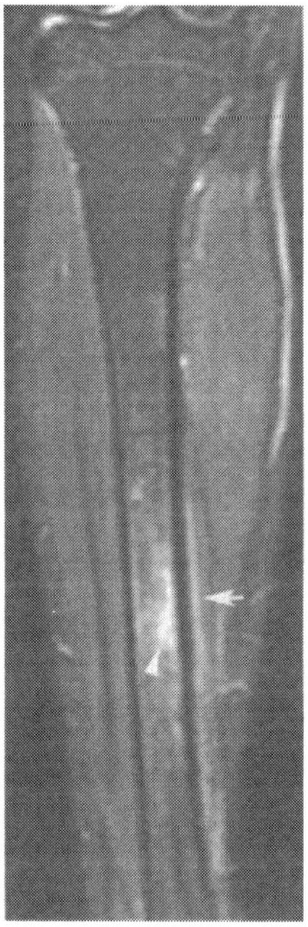

Figure 101B* Annotated.

Diagnosis

Grade 3 tibial stress reaction

Questions

1. T/F: Shin splints are easily distinguishable from stress reaction on MRI.
2. What is the most common location for a tibial stress fracture in children? Which athletes get anterior tibial stress fractures?

Discussion

The term "shin splints" is a clinical and scintigraphic diagnosis that is difficult to distinguish from tibial stress reaction on MRI (Answer to Question 1). Shin splints, tibial stress reaction, and tibial stress fracture are related entities with imaging features that depend on the severity of stress injury (1). Tibial stress injury appears to result from a combination of periosteal stripping at muscular insertions and cortical weakening related to tensile forces (2). This leads to periosteal, endosteal, and marrow edema. When marrow edema is evident without a discrete fracture line through marrow and cortex, the term stress reaction should be used. A fracture line is usually perpendicular to the shaft and is hypointense on T1 and may be hypo- to hyperintense on fluid sensitive sequences.

Fredericson et al. has proposed an MRI grading system for tibial stress reaction (3):

MRI grading system for tibial stress reaction.

Grade	Periosteal Edema T2	Marrow Edema T2	Marrow Edema T1	Other
1	+	none	none	
2	+	+	none	
3	+	+	+	
4	+	+	+	Fracture line present through marrow and cortex

nl = normal, + = hypointense T1 SI or hyperintense T2 SI.

As is apparent from this scheme, differentiation of stress reaction from fracture may be subtle, and a stress reaction may progress to fracture. Therefore, it may be prudent to use the term *stress fracture/reaction* to direct appropriate therapy.

The pathophysiology of periosteal stripping at muscular insertions with shin splints is evident on bone scan by modest linear radiotracer uptake located anteromedially that parallels the cortex (4). Uptake is present only on delayed imaging, whereas stress fractures often show increased uptake on all phases of triple-phase bone scintigraphy. Stress fractures, unlike shin splints, show relatively intense focal uptake along the posteromedial tibia.

In children, tibial stress fractures tend to occur in the posterior aspect of the proximal tibial diaphysis (Figure 101D–101F). In adults, tibial stress fractures tend to occur in the posterior mid- to distal tibial (5). The exception is in athletes who jump. These patients tend to develop anterior tibial diaphyseal stress fractures because of excessive tensile forces along the anterior convexity of the tibia (Answer to Question 2) (6).

Tibial stress reaction without a discrete fracture may follow a similar distribution depending on age or sport activity.

Stress reaction in the tibia may mimic osteomyelitis, sarcomas, and residual red marrow. If a discrete fracture line with a typical edema pattern is identified, and there is no intraosseous or juxtacortical mass effect, the diagnosis of stress reaction/fracture is straightforward. If findings are equivocal, follow-up imaging including post-gadolinium sequences may be useful. Residual red marrow tends to be flame shaped and occurs in the metadiaphyses of the tubular bones. Residual red marrow is isointense on T1 and mildly hyperintense on fluid sensitive sequences and should not be confused with stress reaction (7).

This patient was treated conservatively for one month, but continued to have bilateral shin pain with activity. Despite orthotics and rest, she still had bilateral shin pain at 2-month follow-up. She was subsequently lost to follow-up.

Orthopedic Perspective

Leg pain in a runner can occur due to stress reaction, stress fracture, or exertional compartment syndrome. In the female athlete, irregular periods and disordered eating may result in osteoporosis, predisposing to stress fracture. Stress reaction and stress fracture of the tibia are usually tender to palpation near the junction of the middle and distal thirds. Since stress reaction likely represents an earlier stage on the continuum to stress fracture, both are treated with activity restriction until asymptomatic. A gradual return to activity is then allowed after addressing muscle-tendon imbalance and alignment issues, such as pes planus and pronation.

What the Clinician Needs to Know

1. Location, severity, and degree of stress reaction.
2. Distinguish between stress fracture, stress reaction, and exertional compartment syndrome.

Answers

1. False.
2. The most common location for stress fracture is the posterior aspect of the proximal tibia in children. Anterior tibial stress fractures occur in athletes who jump.

Additional Example

Stress Fracture of the Proximal Tibia

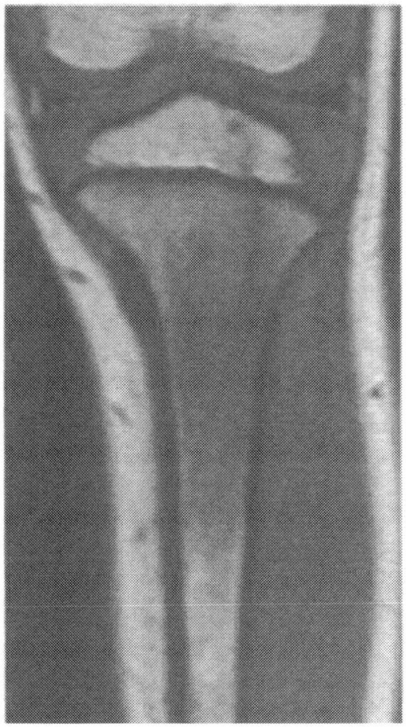

Figure 101D. Coronal T1 of the left tibia.

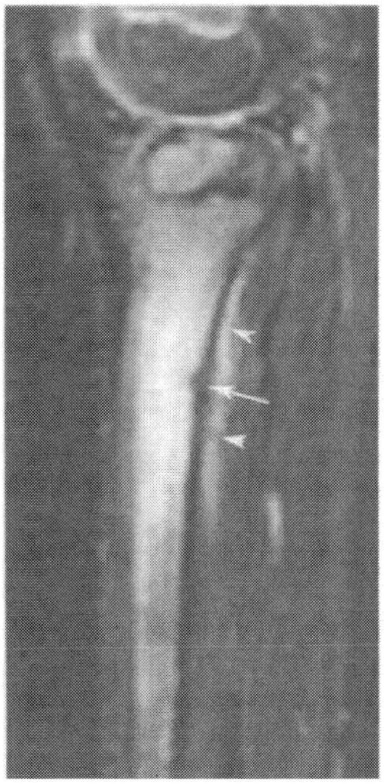

Figure 101E. Sagittal STIR.

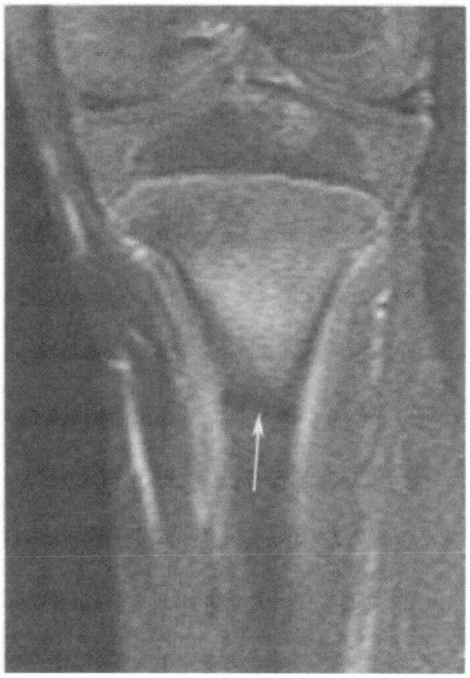

Figure 101F. Coronal T1 post-Gd FS.

Findings

This 6-year-old boy complained of left calf pain.

Figures 101D, 101E, 101F. There is diffuse hypointensity on T1, hyperintensity on STIR, and enhancement in the proximal left tibial marrow. There is posterior periosteal reaction **(arrowheads)** and a hypointense transverse stress fracture line is present **(arrows)**.

Pitfalls and Pearls

1. Strictly speaking, the term *stress fracture* should apply only when a discrete fracture line is evident. However, if there are impressive stress reaction changes, it may be advisable to use the term *stress reaction/fracture* to direct appropriate management.
2. The term *shin splints* is often used based on clinical grounds and bone scintigraphy. With MRI, however, the term *stress reaction* is preferable.
3. Do not mistake normal red marrow for stress reaction. Coronal or sagittal images may be helpful to define the characteristic flame shaped appearance of residual red marrow.

References

1. Spitz DJ, Newberg AH. Imaging of stress fractures in the athlete. *Radiol Clin North Am* 2002; 40:313–331.
2. Anderson MW, Ugalde V, Batt M, Gacayan J. Shin splints: MR appearance in a preliminary study. *Radiology* 1997; 204:177–180.
3. Fredericson M, Bergman AG, Hoffman KL, Dillingham MS. Tibial stress reaction in runners: Correlation of clinical symptoms and scintigraphy with a new magnetic resonance imaging grading system. *Am J Sports Med* 1995; 23:472–481.
4. Zwas ST, Elkanovitch R, Frank G. Interpretation and classification of bone scintigraphic findings in stress fractures. *J Nucl Med* 1987; 28:452–457.
5. Daffner RH. Stress fractures: Current concepts. *Skeletal Radiol* 1978; 2:221–229.
6. Beals RK, Cook RD. Stress fractures of the anterior tibial diaphysis. *Orthopedics* 1991; 14:869–875.
7. States LJ. Imaging of metabolic bone disease and marrow disorders in children. *Radiol Clin North Am* 2001; 39:749–772.

Case 102

History

This is a 5-year-old boy who fell off a trampoline and injured his left elbow. Radiographs, including stress views, were negative for fracture or effusion (not shown).

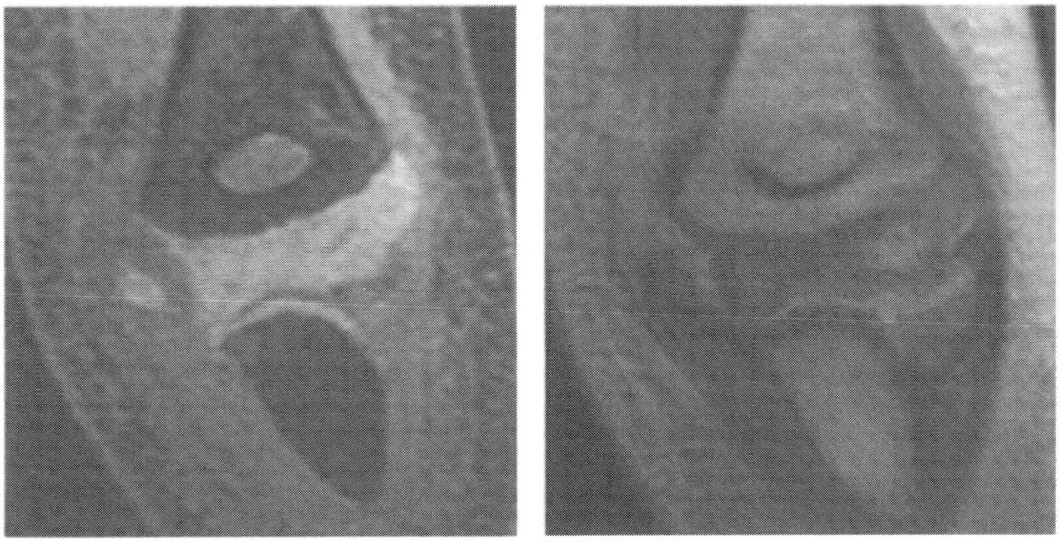

Figure 102A. Coronal 3D SPGR FS of the left elbow.

Figure 102B. Coronal PD.

Figure 102A. The cartilaginous medial epicondyle is avulsed. The fracture line extends to the medial condyle (trochlea), but does not involve the articular surface.

Figures 102B, (102B with annotations). The common flexor-pronator tendon origin on the avulsed medial epicondyle is identified **(arrow)**.

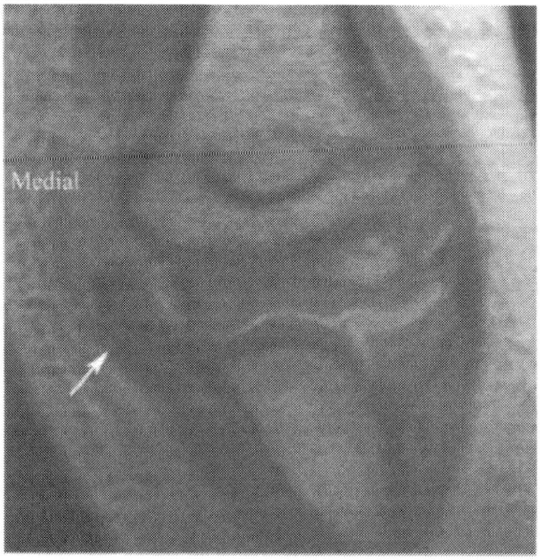

Figure 102B* Annotated.

Diagnosis

Medial epicondylar avulsion fracture

Question

1. T/F: Pure cartilaginous elbow fractures always result in periosteal new bone formation on follow-up radiographs.

Discussion

Valgus injuries to the medial epicondyle vary based on patient age. In the child and adolescent, medial epicondylitis and physeal injury may be seen (1). This is because the physis and epicondyle are the weakest component of the flexor-pronator tendon/medial collateral ligament complex that attaches to the medial epicondyle. With physeal injury, the medial epicondylar avulsion fragment may displace into the elbow joint, a finding that is easily overlooked (2). The intra-articular location of the avulsed medial epicondyle may mimic a trochlear ossification center. This mistake may be avoided by knowing that the medial epicondyle ossification center usually appears before the trochlea ossification center (Table 102A) (3). Once the physis is fused, small avulsion flake fractures at the flexor-pronator tendon/medial collateral ligament complex attachment, as well as tendinous/ligamentous injuries, are more frequently encountered. Occasionally, physeal injuries and flake fractures may coexist.

In the infant and young child, the medial epicondyle and distal humerus epiphysis is a single cartilaginous unit. Medial epicondylar avulsion fractures are quite uncommon prior to the appearance of the secondary ossification center. These fractures may go undiagnosed because the cartilaginous epiphysis is invisible radiographically. The fracture course within the unossified medial epicondyle is unpredictable. The fracture line may involve only the epicondyle, or may extend to the articular surface of the medial condyle. A more predictable avulsion fracture at the medial epicondylar physis is usually seen once the epicondylar ossification center has appeared.

Radiography may also be misleading when a medial epicondylar avulsion fracture is purely extra-articular and a joint effusion is absent. Since epiphyseal cartilage is relatively avascular, a hemarthrosis may also be absent, even if the cartilaginous fracture extends to the joint (4). Follow-up radiography to assess for periosteal new bone formation may also be misleading. Cartilaginous fractures do not always incite periosteal new bone since secondary ossification centers are covered by perichondrium, rather than periosteum (Answer to Question 1) (5). Periosteal new bone formation will be evident on follow-up if the physeal injury is associated with periosteal stripping.

MRI evaluation of the pediatric elbow should focus on delineating cartilaginous anatomy, which is best seen with GRE or PD sequences with FS, and a 3D SPGR FS acquisition or similar sequence is particularly useful for complex injuries that may

Table 102A. Appearance of elbow ossification centers: CRITOE (8).

Capitellum: 1–2 years
Radial head: 3–6 years
Medial (internal) epicondyle: 4 years
Trochlea: 8 years
Olecranon: 3–6 years
Lateral (external) epicondyle: 10 years

benefit from multiplanar reformations. PD sequences are valuable to delineate ligamentous and tendinous anatomy. A determination of the extent of fragment displacement, articular surface involvement, the presence of loose bodies, as well as other radiographically occult fractures, ligamentous sprains and tendinous injury help guide treatment (Figures 102C–102E).

The term "little leaguer's elbow" describes a spectrum of elbow injuries found in young throwing athletes. This term generally refers to medial epicondylitis (Figure 102F, 102G) but it also encompasses osteochondritis dissecans of the capitellum and radial head seen in the same athletic group (6, 7). With chronic valgus forces related to throwing, tensile force is applied to the medial compartment resulting in medial epicondylitis and tendinopathy. Compressive forces applied to the lateral compartment result in osteochondritis dissecans in the capitellum and radial head.

This patient was referred for MRI when radiographs with stress views were normal (not shown). With the results of the MRI, the patient was treated conservatively with casting. The patient was well and out of cast at two month follow-up. There was no evidence of elbow deformity. His range of motion of the left elbow showed full extension but lacked 10 to 15 degrees of flexion.

Orthopedic Perspective

Medial epicondyle fractures occur due to a valgus loading force on the elbow. Fractures may occur in isolation, or in association with elbow dislocation with occasional intra-articular entrapment of the medial epicondyle. Treatment of medial epicondyle fractures is based on associated injuries and the amount of displacement. Isolated fractures with displacement less than 5 mm are treated nonoperatively with cast immobilization. Fractures associated with elbow dislocations or those with displacement greater than 5 mm are usually treated surgically with open reduction and internal fixation.

What the Clinician Needs to Know

1. Measurement of medial epicondyle displacement.
2. Are there associated injuries of the capitellum and radial head?
3. Is there also a tear of the supporting medial collateral ligaments or evidence of flexor/pronator tendinopathy?

Answer

1. False.

Additional Examples

Medial Epicondylar Avulsion Fracture

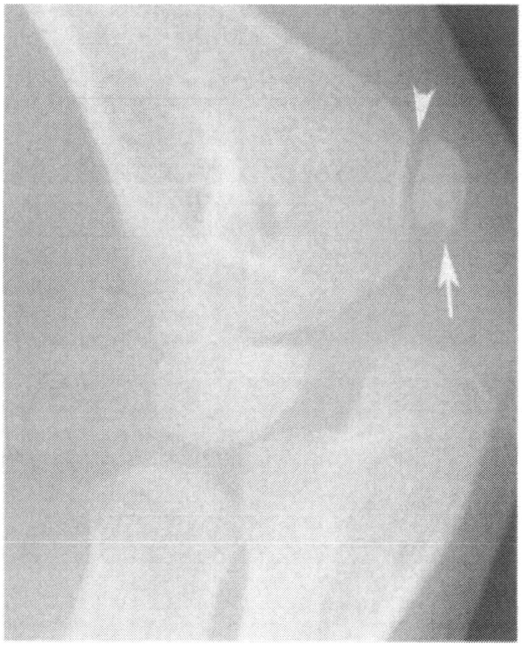

Figure 102C. Oblique radiograph of the left elbow.

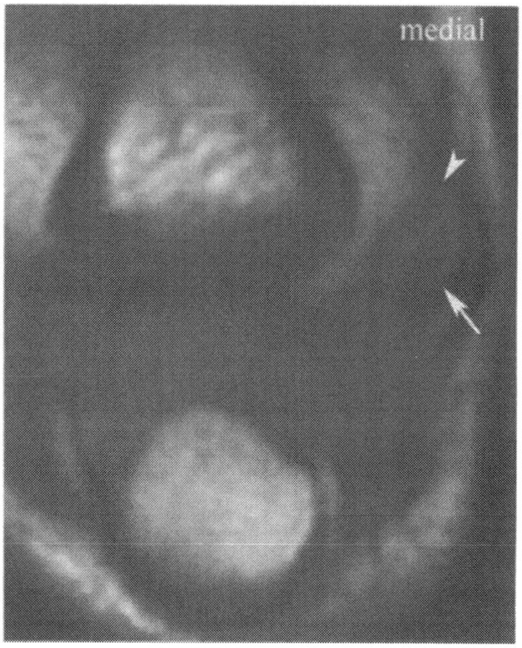

Figure 102D. Coronal T1.

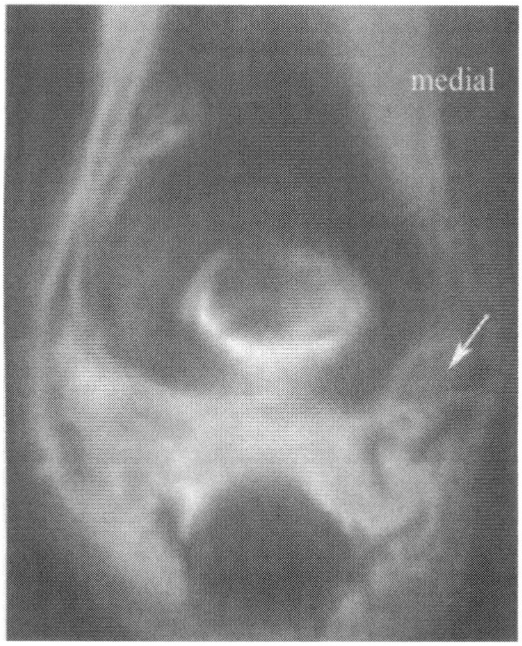

Figure 102E. Coronal PD FS.

Findings

This 6-year-old girl injured her left elbow when she fell off monkey bars.

Figures 102C, 102D, 102E. There is decreased T1 SI and increased PD SI within the medial epicondyle **(arrows)** and the associated physis **(arrowheads)**. There is subtle cortical irregularity of the medial epicondyle. The medial epicondylar physis is slightly widened, consistent with a physeal injury with osseous avulsion of the medial epicondyle.

Medial Epicondylitis (AKA Little Leaguer's Elbow)

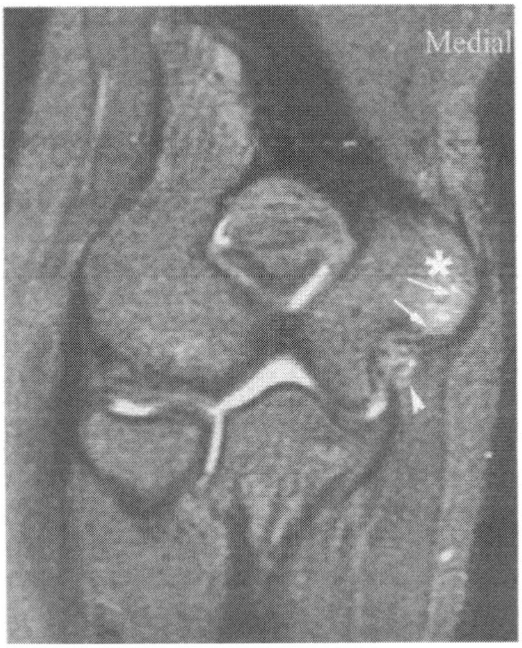

Figure 102F. Coronal STIR of the right elbow.

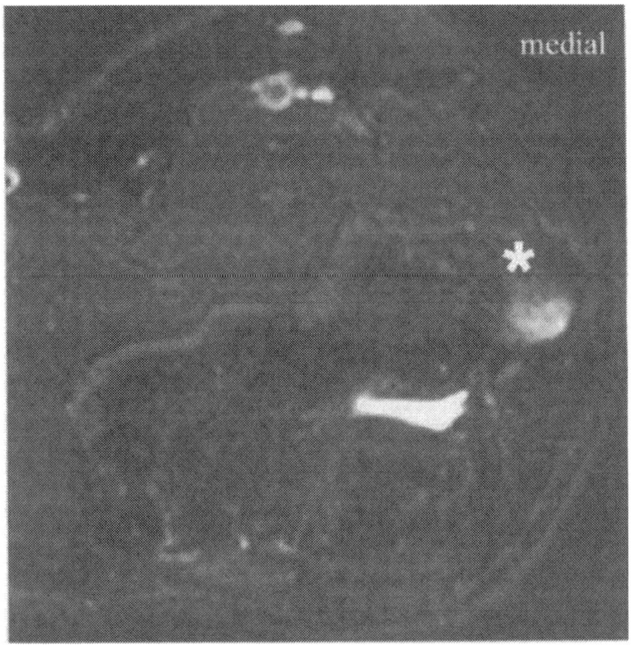

Figure 102G. Axial T2 FS.

Findings

This 14-year-old boy had chronic medial elbow pain when pitching.

Figures 102F, 102G. There is mild increased STIR and T2 SI in the medial epicondyle (*) as well as the anterior band of the ulnar collateral ligament **(arrowhead)**. The closing physis is visible **(arrows)**.

Pitfalls and Pearls

1. Significant elbow injuries may occur in the absence of a joint effusion in children.
2. Physeal injuries of the distal humerus in infants (<1 year) are often associated with child abuse.
3. Although the ulnar collateral ligament may be sprained with valgus injuries, it may normally show increased SI on fluid sensitive sequences in children due to increased elastin content in the anterior fibers compared with adults (9).

References

1. DaSilva MF, Williams JS, Fadale PD, Hulstyn MJ, Ehrlich MG. Pediatric throwing injuries about the elbow. *Am J Orthop* 1998; 27:90–96.
2. Kocher MS, Waters PM, Micheli LJ. Upper extremity injuries in the paediatric athlete. *Sports Med* 2000; 30:117–135.
3. Jaramillo D, Waters PM. MR imaging of the normal developmental anatomy of the elbow. *Magn Reson Imaging Clin N Am* 1997; 5:501–513.
4. Hall FM. Traumatic elbow effusions. *AJR Am J Roentgenol* 1999; 172:550–551.

5. Donnelly LF, Klostermeier TT, Klosterman LA. Traumatic elbow effusions in pediatric patients: Are occult fractures the rule? *AJR Am J Roentgenol* 1998; 171:243–245.
6. Gomez JE. Upper extremity injuries in youth sports. *Pediatr Clin North Am* 2002; 49:593–626, vi–vii.
7. Klingele KE, Kocher MS. Little league elbow: Valgus overload injury in the paediatric athlete. *Sports Med* 2002; 32:1005–1015.
8. Laor T, Jaramillo D, Oestreich AE. Musculoskeletal system. In: *Practical Pediatric Imaging. Diagnostic Radiology of Infants and Children,* 3rd ed., Chapter 5, Kirks DR, ed. Philadelphia: Lippincott-Raven, 1998; 327–510.
9. Kaplan LJ, Potter HG. MR imaging of ligament injuries to the elbow. *Magn Reson Imaging Clin N Am* 2004; 12:221–232, v–vi.

Appendix: List of Case Diagnoses

Case	Diagnoses
1	Lateral patellar dislocation with displaced osteochondral fracture
2	Chronic recurrent multifocal osteomyelitis
3	Pigmented villonodular synovitis
4	Ewing's sarcoma
5	Aneurysmal bone cyst
6	Synovial sarcoma
7	Juvenile dermatomyositis
8	Type 2 accessory navicular with stress change
9	Gymnast's wrist
10	Multiple epiphyseal dysplasia with valgus tibiotalar slant
11	Sternocostoclavicular hyperostosis
12	Rickets
13	Osteochondritis dissecans
14	Salter-Harris type 1 fracture
15	Legg-Calve-Perthes disease
16	Enchondromatosis
17	Conventional osteosarcoma
18	Septic arthritis
19	Talocalcaneal coalition
20	Infantile Blount's disease
21	Insufficiency fracture of the proximal femur
22	Ruptured Baker's cyst
23	Osteonecrosis
24	Primary bone lymphoma
25	Synovial venous malformation
26	Bony physeal bar
27	Septic arthritis after fracture and internal fixation
28	Neonatal brachial plexopathy with glenohumeral dysplasia
29	Proximal focal femoral deficiency
30	Epiphyseal osteomyelitis and abscess
31	Acute bone marrow infarction
32	L5-S1 spondylodiscitis
33	Avulsion injury of the lesser trochanter
34	Discoid lateral meniscus (type 2) with a radial tear
35	Toddler's fracture of the fibula
36	Normal developmental irregularity of the femoral condyles
37	Treated developmental dysplasia of the hip complicated by ischemic necrosis and physeal bar
38	Achondroplasia

39 Juvenile rheumatoid arthritis
40 Infantile hemangioma
41 Subcutaneous granuloma annulare
42 Necrotizing fasciitis
43 Infantile myofibromatosis, multicentric
44 Ganglion cyst
45 Acute tibial osteomyelitis with transphyseal extension
46 ACL complex injury with tibial eminence fracture
47 Lyme arthritis
48 Nonossifying fibroma
49 Langerhans cell histiocytosis
50 Granulocyte colony-stimulating factor effect on marrow
51 Posterior rib fractures due to child abuse
52 Anterior labroligamentous periosteal sleeve avulsion and Hills-Sachs deformity
53 Osteomyelitis mimicking Ewing's sarcoma
54 Bucket-handle medial meniscal tear and ACL tear
55 Iliac osteomyelitis and sacroiliitis
56 Lateral condylar fracture
57 Plexiform neurofibroma
58 Slipped capital femoral epiphysis
59 Macrocystic lymphatic malformation
60 Bilateral spondylolysis of L5
61 Ischiopubic synchondrosis osteomyelitis
62 Acute lymphoblastic leukemia
63 Lipoblastoma
64 Polyostotic fibrous dysplasia
65 Sessile osteochondroma of the proximal fibula
66 Iliac crest osteomyelitis
67 Venous malformation of the great toe
68 Hemophilic arthropathy
69 Psoriatic dactylitis
70 Panner's disease and normal trochlear pre-ossification center
71 CAM-type femoroacetabular impingement
72 Lipoma
73 Osteofibrous dysplasia with pathologic fracture
74 Neuroblastoma with bone metastases
75 Fibular osteomyelitis and secondary septic arthritis
76 Neonatal osteomyelitis and septic arthritis
77 Avulsive cortical irregularity
78 Juvenile rheumatoid arthritis with rice bodies
79 Desmoid fibromatosis
80 Macrodystrophia lipomatosa of the toe
81 Rhabdomyosarcoma with axillary metastasis
82 Chondroblastoma
83 Nondisplaced extra-articular calcaneal fracture
84 Anterosuperior acetabular osteolabral separation
85 Seronegative spondyloarthropathies
86 Klippel-Trenaunay syndrome
87 Myositis ossificans traumatica
88 Unicameral bone cyst of the proximal femur
89 Dysplasia epiphysealis hemimelica, classical form
90 Osteoid osteoma
91 Periosteal osteosarcoma
92 Osgood-Schlatter disease
93 Periosteal chondroma

94 Telangiectatic osteosarcoma
95 Infantile fibrosarcoma
96 Developmental dysplasia of the hip
97 Giant cell tumor of bone
98 Salter-Harris type 2 fracture of the distal femur
99 Sever's disease
100 Subacute osteomyelitis with a Brodie's abscess
101 Grade 3 tibial stress reaction
102 Medial epicondylar avulsion fracture

Index

R

Made in the USA
Coppell, TX
26 February 2021

50836164R00422